Computational Modeling of Masonry Structures Using the Discrete Element Method

Vasilis Sarhosis
Newcastle University, UK

Katalin Bagi
Budapest University of Technology and Economics, Hungary

José V. Lemos
National Laboratory for Civil Engineering, Portugal

Gabriele Milani
Technical University in Milan, Italy

A volume in the Advances in Civil and Industrial Engineering (ACIE) Book Series

ENGINEERING SCIENCE REFERENCE
An Imprint of IGI Global

Published in the United States of America by
Engineering Science Reference (an imprint of IGI Global)
701 E. Chocolate Avenue
Hershey PA, USA 17033
Tel: 717-533-8845
Fax: 717-533-8661
E-mail: cust@igi-global.com
Web site: http://www.igi-global.com

Library of Congress Cataloging-in-Publication Data

Names: Sarhosis, Vasilis, 1981- editor.
Title: Computational modeling of masonry structures using the discrete element method / Vasilis Sarhosis, Katalin Bagi, J.V. Lemos, and Gabriele Milani, editors.
Description: Hershey PA : Engineering Science Reference, [2016] | Includes bibliographical references and index.
Identifiers: LCCN 2016003307| ISBN 9781522502319 (hardcover) | ISBN 9781522502326 (ebook)
Subjects: LCSH: Masonry--Data processing. | Brickwork--Mathematics. | Stone buildings--Mathematical models. | Discrete element method.
Classification: LCC TH1199 .C678 2016 | DDC 624.1/830113--dc23 LC record available at https://lccn.loc.gov/2016003307

This book is published in the IGI Global book series Advances in Civil and Industrial Engineering (ACIE) (ISSN: 2326-6139; eISSN: 2326-6155)

British Cataloguing in Publication Data
A Cataloguing in Publication record for this book is available from the British Library.

Advances in Civil and Industrial Engineering (ACIE) Book Series

ISSN: 2326-6139
EISSN: 2326-6155

Mission

Private and public sector infrastructures begin to age, or require change in the face of developing technologies, the fields of civil and industrial engineering have become increasingly important as a method to mitigate and manage these changes. As governments and the public at large begin to grapple with climate change and growing populations, civil engineering has become more interdisciplinary and the need for publications that discuss the rapid changes and advancements in the field have become more in-demand. Additionally, private corporations and companies are facing similar changes and challenges, with the pressure for new and innovative methods being placed on those involved in industrial engineering.

The **Advances in Civil and Industrial Engineering (ACIE) Book Series** aims to present research and methodology that will provide solutions and discussions to meet such needs. The latest methodologies, applications, tools, and analysis will be published through the books included in **ACIE** in order to keep the available research in civil and industrial engineering as current and timely as possible.

Coverage

- Engineering Economics
- Structural Engineering
- Urban Engineering
- Production Planning and Control
- Coastal Engineering
- Productivity
- Earthquake engineering
- Quality Engineering
- Hydraulic Engineering
- Construction Engineering

IGI Global is currently accepting manuscripts for publication within this series. To submit a proposal for a volume in this series, please contact our Acquisition Editors at Acquisitions@igi-global.com or visit: http://www.igi-global.com/publish/.

The Advances in Civil and Industrial Engineering (ACIE) Book Series (ISSN 2326-6139) is published by IGI Global, 701 E. Chocolate Avenue, Hershey, PA 17033-1240, USA, www.igi-global.com. This series is composed of titles available for purchase individually; each title is edited to be contextually exclusive from any other title within the series. For pricing and ordering information please visit http://www.igi-global.com/book-series/advances-civil-industrial-engineering/73673. Postmaster: Send all address changes to above address.

Titles in this Series

For a list of additional titles in this series, please visit: www.igi-global.com

Advanced Research on Nanotechnology for Civil Engineering Applications
Anwar Khitab (Mirpur University of Science and Technology, Pakistan) and Waqas Anwar (Mirpur University of Science and Technology, Pakistan)
Engineering Science Reference • copyright 2016 • 339pp • H/C (ISBN: 9781522503446) • US $195.00 (our price)

Emerging Challenges and Opportunities of High Speed Rail Development on Business and Society
Raj Selladurai (Indiana University Northwest, USA) Peggy Daniels Lee (Kelley School of Business, Indiana University, USA) and George VandeWerken (Providence Bank, USA)
Engineering Science Reference • copyright 2016 • 289pp • H/C (ISBN: 9781522501022) • US $210.00 (our price)

Advanced Manufacturing Techniques Using Laser Material Processing
Esther Titilayo Akinlabi (Univeristy of Johannesburg, South Africa) Rasheedat Modupe Mahamood (University of Johannesburg, South Africa & University of Ilorin, Nigeria) and Stephen Akinwale Akinlabi (University of Johannesburg, South Africa)
Engineering Science Reference • copyright 2016 • 288pp • H/C (ISBN: 9781522503293) • US $175.00 (our price)

Handbook of Research on Applied E-Learning in Engineering and Architecture Education
David Fonseca (La Salle Campus Barcelona, Universitat Ramon Llull, Spain) and Ernest Redondo (Universitat Politècnica de Catalunya, BarcelonaTech, Spain)
Engineering Science Reference • copyright 2016 • 569pp • H/C (ISBN: 9781466688032) • US $310.00 (our price)

Emerging Design Solutions in Structural Health Monitoring Systems
Diego Alexander Tibaduiza Burgos (Universidad Santo Tomás, Colombia) Luis Eduardo Mujica (Universitat Politecnica de Catalunya, Spain) and Jose Rodellar (Universitat Politecnica de Catalunya, Spain)
Engineering Science Reference • copyright 2015 • 337pp • H/C (ISBN: 9781466684904) • US $235.00 (our price)

Robotics, Automation, and Control in Industrial and Service Settings
Zongwei Luo (South University of Science and Technology of China, China)
Engineering Science Reference • copyright 2015 • 337pp • H/C (ISBN: 9781466686939) • US $215.00 (our price)

Using Decision Support Systems for Transportation Planning Efficiency
Ebru V. Ocalir-Akunal (Gazi University, Turkey)
Engineering Science Reference • copyright 2016 • 475pp • H/C (ISBN: 9781466686489) • US $215.00 (our price)

IGI GLOBAL
DISSEMINATOR OF KNOWLEDGE
www.igi-global.com

701 E. Chocolate Ave., Hershey, PA 17033
Order online at www.igi-global.com or call 717-533-8845 x100
To place a standing order for titles released in this series, contact: cust@igi-global.com
Mon-Fri 8:00 am - 5:00 pm (est) or fax 24 hours a day 717-533-8661

Table of Contents

Detailed Table of Contents

In this chapter, a review on the mechanical behaviour of masonry is presented. The aim is to establish a base of knowledge and understanding of masonry that will underpin its mechanical characteristics and will inform the decisions towards the selection of the computational tool used which are going to be described in the following chapters. Initially, a brief description of the factors that influence the mechanical response of masonry and the variation of the material properties are discussed. The review then considers the possible causes of cracking in masonry and the different failure modes that may occur during loading. Principal findings from the review are summarised at the end of the chapter.

In this chapter, a review of the available methods and their challenges to simulate the mechanical behavior of masonry structures are presented. Different micro-modeling computational options are considered and compared with regard to their ability to define the initial state of the structure, realism in simulation, computer efficiency and data availability for their application to model low bond strength masonry structures. It is highlighted that different computational approaches should lead to different results and these will depend on the adequacy of the approach used and the information available. From the results analysis it is also highlighted that a realistic analysis and assessment of existing masonry structures using numerical methods of analysis is not a straight forward task even under full knowledge of current conditions and materials.

The "distinct element method" was proposed by Peter Cundall in 1971 for the analysis of rock slopes by means of rigid block or circular particle models. This method led to the UDEC and 3DEC codes, presently in wide use in rock engineering. Their application to masonry structures started in the 90's, as researchers found that they were also excellent tools to approach the highly nonlinear behavior of masonry, in particular the collapse processes of stone block structures under static or seismic loads. This chapter reviews the essential assumptions of UDEC and 3DEC, relating them to other methods and codes, and stressing the features that make them suitable for masonry analysis. Rigid and deformable blocks, contact mechanics, contact detection, and solution algorithms are examined. Key issues in the modelling of masonry are addressed, including: irregular block models; determination of collapse loads; large displacement analysis; computational efficiency issues in dynamic analysis. Practical examples taken from the published literature illustrate these issues.

"DDA" stands for "Discontinuous Deformation Analysis", suggesting that the displacement field of the analyzed domain shows abrupt changes on the element boundaries in the model. This chapter introduces the theoretical fundaments of DDA: mechanical characteristics of the elements together with the basic degrees of freedom, contact behavior, the equations of motion and their numerical integration with the help of Newmark's beta-method taking into account contact creation, loss and sliding with the help of an open-close iteration technique. Finally, a short overview on practical and scientific applications for masonry structures is given.

The Contact Dynamics method, developed still in the 1980s, was originally applied for granular assemblies because of its efficiency in simulating rapid granular flows or vibration problems of discrete systems. In the oldest models the elements were spherical and perfectly rigid, but later the application of polyhedral and deformable elements also became widespread, allowing for the reliable simulation of problems related to masonry structures. The basic unit of the analysis in Contact Dynamics is the pair of two randomly chosen elements. The essence of the method is to find the contact force vector between the two elements in such a way that during the analysed time step the elements should not overlap each other. At the considered time instant an iterative process sweeps along randomly chosen pairs over and over again, and gradually adjusts the estimated contact forces to get an improving approximation of a state that satisfies the dynamic equations of the system. The method is particularly advantageous for earthquake analysis of masonry structures.

This chapter presents a general overview of the combined Finite-Discrete Element Method (FEM/DEM) which is considered as a state-of-the-art technique for the mechanical analysis of masonry structures. In a FEM/DEM simulation each discrete element representing a stone block is discretized into finite elements in order to describe the deformability of the blocks. This chapter deals with the main steps of the FEM/DEM including contact detection, contact interaction, fracture and fragmentation algorithms, calculation of deformations and the time integration of the equation of motion. The FEM/DEM is advantageously used to simulate transition from continua to discontinua processes which may lead to the collapse of the structure. Some examples for practical applications found in the literature are mentioned.

Circular Particle Models (PM) are a class of discrete elements which has been increasingly used for detailed analysis in rock and concrete structures. There have been few applications to masonry, but the potential of these techniques appears significant, due to their proven ability to simulate fracture processes through random particle assemblies representing quasi-brittle materials at the grain scale. The present chapter presents the fundamentals of this approach and reviews some previous applications of PM models to masonry. The model capabilities are first exemplified by simple models involving a few irregular blocks formed by particles. Irregular stone masonry wall specimens under compression and under in-plane shear loading are then presented. In these models both the units and the mortar are represented by circular particles, and failure processes through the joints or through joints and stones are analyzed. The main issues regarding the use of these models are finally discussed.

This work concerns the numerical modelling of masonry dams using the Discrete Element Method. It begins with a review of the history of masonry dams and their behaviour. A numerical tool based on the Discrete Element Method developed specifically for the structural assessment of masonry dams is then presented. The mechanical calculations performed by the tool are discussed in detail, together with the approach used for the modelling of passive anchors and the modules for seismic analysis and hydromechanical analysis. Structural and hydraulic analyses of a diverse set of existing masonry dams conducted using the tool are then presented. The Discrete Element Method is shown to be capable of reproducing the structural behaviour of masonry dams and identifying their likely failure mechanisms as required for structural safety evaluations.

interacting with infill. A preliminary validation of the procedure is addressed for the prediction of the masonry arches limit state behavior where the stones are supposed infinite resistant and plastic hinges can occur exclusively on mortar joints, modeled as cohesive frictional interfaces. The sensitivity of the infill role varying mechanical properties of the infill is extensively discussed.

Masonry structures are comprised of a finite number of distinct interacting rock blocks that have a length scale relatively comparable to the structure. Therefore, they are ideal candidates for modeling as discrete systems. This chapter covers the Discrete Finite Element Method (DFEM) developed by the author to model discontinuous media consisting of blocks of arbitrary shapes. The DFEM is based on the finite element method incorporating contact elements. The DFEM considers blocks as sub-domains and represents them as solid elements. Contact elements are used to model block interactions such as sliding or separation. In this chapter, through some illustrative examples, the applicability of the DFEM to static and dynamic analysis of masonry structures, including arch bridges, walls, slopes, and underground openings, is discussed. The DFEM provides an efficient tool for researchers and practical engineers in designing, analyzing, and studying the behavior of masonry structures under static and dynamic loadings.

This chapter addresses the numerical modeling of freestanding rigid blocks by means of a semi-discrete approach. The pure rocking motion of single rigid bodies can be easily studied with the differential equation of motion, which can be solved by numerical integration or by linearization. However, when we deal with sliding and jumping motion of rigid bodies, the mathematical formulation becomes quite complex. In order to overcome this complexity, a Semi-Discrete Model (SMD) is proposed for the study of rocking motion of rigid bodies, in which the rigid body is considered as a mass element supported by springs and dashpots, in the spirit of deformable contacts between rigid blocks. The SMD can detect separation and sliding of the body; however, initial base contacts do not change, keeping a relative continuity between the body and its base. Extensive numerical simulations have been carried out in order to validate the proposed approach.

In this chapter, through some illustrative examples, the applicability of the Discrete Finite Element Method (DFEM) to analysis of unreinforced masonry structures such as rock pillars, open rock slopes, underground openings, tunnels, fault propagations, and fault-structure interactions is examined and discussed. In the numerical study, the behavior of contacts and blocks is assumed to be elasto-plastic or elastic. The Mohr-Coulomb yield criterion, representing material behavior of contacts, is implemented in the developed codes for DFEM used in the analysis. The secant stiffness method with the updated Lagrangian scheme is employed to deal with non-linear behavior. The constant strain triangular element with two degrees of freedoms at each node, formed by properly joining the corners and contact nodes of an individual block, is adopted for finite element meshing of the blocks. The DFEM provides an efficient and promising tool for designing, analyzing, and studying the behavior of unreinforced masonry structures.

Preface

Masonry is a combination of units such as stones, bricks or blocks usually laid in a cementitious or lime mortar. It is probably the oldest material used in construction and has proven to be both simple to build and durable. Over the years, existing masonry constructions have inevitably suffered damage with time. Earthquakes, soil settlements, material degradation and lack of maintenance are the main reasons for that. Careful and periodic assessment of such structures is necessary in order to evaluate their structural capacity and safety levels. However, performing the structural analysis of masonry construction is not an easy task.

Masonry is a material which is characterised by high non-linearity and discontinuity; it is the dry or mortar joints in a masonry structure which act as planes of weakness. When subjected to very low levels of stress, masonry behaves approximately in a linear elastic manner. This becomes increasingly non-linear after the formation of cracks and the subsequent redistribution of stress through the uncracked material as the structure approaches collapse. The behaviour of masonry is complicated further by the inherent variations in the constituent materials, variations in workmanship, the effects of deterioration caused by weathering processes and the development of other defects during the life of the masonry structure.

Accurate structural analysis is needed is needed to understand the behaviour of these many different forms of masonry constructions and avoid erroneous or defective conclusions. In particular it is important to understand the pre- and post-cracking behaviour to inform decisions concerning the maintenance needs, management of safety risks, assessment of levels of safety and the need for repair or strengthening. As experimental research is prohibitively expensive, it is fundamentally important to have available a computational model that can be used to predict the in-service and near-collapse behaviour with sufficient reliability.

The selection of the most appropriate method to use depends on, among other factors, the structure under analysis; the level of accuracy and simplicity desired; the knowledge of the input properties in the model and the experimental data available; the amount of financial resources; time requirements and the experience of the modeller. It should also be expected that different methods should lead to different results depending on the adequacy of the approach and the information available. Preferably, the approach selected to model masonry should provide the desired information in a reliable manner within an acceptable degree of accuracy and with least cost.

Engineering modelling of masonry structures is often based on continuum representations, using appropriate constitutive models, which provide an adequate solution for many practical cases. However, such models does not represent accurately the mechanical behaviour of masonry components. Over the last two decades, discontinuous models are applied with increasing frequency; since they intended at simulating the mechanical behaviour of masonry structures more accurately in a simplified micro modeling manner.

The present book focuses on the discrete element modelling of masonry structures, a designation that covers a variety of representations of a structure as a system of blocks (rigid or deformable) or particles. The possibility of frequent changes in the connectivity and the type of contact as well as marked non-linearity induced by the inability of the masonry joints to withstand tension makes the Discrete Element Method (DEM) a suitable approach for solving problems involving discontinuities as is the case with low bond strength masonry. Models based on the discrete element approach solve the equations of motion using a time stepping scheme and contact forces are obtained by means of the concept of joint stiffness, an approach related to the penalty methods. The resultant out of balance forces and motion parameters are calculated at each integration time step. Material deformability, complex contact interaction laws as well as failure and fracture criterial at the interfaces can be assigned. More recent implementations allow for an increased block deformability and block fracturing, comprising a combined discrete/finite element approach.

A review is presented of the main models based on the discrete element method and the available related numerical techniques that have been proposed for the analysis of masonry. The essential assumptions adopted by these models and numerical implementation issues are discussed. Differences between available models are illustrated by applications to various masonry problems including static and dynamic analysis of masonry arch bridges, walls, vaults, domes and ancient colonnades.

This book is composed of 17 chapters authored/co-authored by 25 outstanding researchers from 11 countries (Canada, France, Greece, Hungary, Iran,Italy, Mexico, Norway, Portugal, UK, USA), which were reviewed by 60 referees.

In Chapter 1, a review on the mechanical behaviour of masonry is presented. The aim is to establish a base of knowledge and understanding of masonry that will underpin its mechanical characteristics and will inform the decisions towards the selection of the computational tool used which are going to be described in the following chapters. Initially, a brief description of the factors that influence the mechanical response of masonry and the variation of the material properties are discussed. The review then considers the possible causes of cracking in masonry and the different failure modes that may occur during loading. Principal findings from the review are summarised at the end of the chapter.

In Chapter 2, a review of the available methods and their challenges to simulate the mechanical behavior of masonry structures are presented. Different micro-modeling computational options are considered and compared with regard to their ability to define the initial state of the structure, realism in simulation, computer efficiency and data availability for their application to model low bond strength masonry structures. It is highlighted that different computational approaches should lead to different results and these will depend on the adequacy of the approach used and the information available. From the results analysis it is also highlighted that a realistic analysis and assessment of existing masonry structures using numerical methods of analysis is not a straight forward task even under full knowledge of current conditions and materials.

In Chapter 3, the "distinct element method" is introduced. The distinct element method was proposed by Peter Cundall in 1971 for the analysis of rock slopes by means of rigid block or circular particle models. This method led to the UDEC and 3DEC codes, presently in wide use in rock engineering. Their application to masonry structures started in the 90's, as researchers found that they were also excellent tools to approach the highly nonlinear behavior of masonry, in particular the collapse processes of stone block structures under static or seismic loads. This chapter reviews the essential assumptions of UDEC and 3DEC, relating them to other methods and codes, and stressing the features that make them suitable for masonry analysis. Rigid and deformable blocks, contact mechanics, contact detection, and solution algorithms are examined. Key issues in the modelling of masonry are addressed, including: irregular block models; determination of collapse loads; large displacement analysis; computational efficiency issues in dynamic analysis. Practical examples taken from the published literature illustrate these issues.

In Chapter 4, the discontinuous deformation analysis (DDA) with polyhedral elements is presented. The basic degrees of freedom and the element behaviour are introduced. Also, the contact behaviour, the quantities in the equations of motion, and the time integration scheme are presented. A short comparison is given between DDA and 3DEC. Finally, applications of DDA to predict the mechanical behaviour masonry are shown.

In Chapter 5, the basic concepts of the Non-Smooth Contact Dynamics method are presented. For the sake of simplicity, detailed theoretical description is given only for the case of rigid elements; fundaments of modelling with deformable elements are then shortly summarized. Questionable features of NSCD such as non-uniqueness of the solution are emphasized. Finally, different applications of the Contact Dynamics method in the analysis of masonry structures are introduced.

In Chapter 6, the main processes of the combined finite-discrete element method are discussed including contact detection, contact interaction, fracture and fragmentation algorithms, calculation of deformations, parallelization and the time integration of the equations of motion. Different modelling approaches and examples connected to the structural analysis of masonry structures are mentioned which were found in the literature. Afterwards a brief comparison is made between FEM/DEM and other techniques belonging to the DEM.

In Chapter 7, the application of circular particle models to masonry structures is addressed. The potential of these techniques appears significant, due to their proven ability to simulate fracture processes through random particle assemblies representing quasi-brittle materials at the grain scale. The chapter presents the fundamentals of the numerical approach and reviews some previous applications to masonry. The capabilities of the model are exemplified by simple tests involving a few irregular blocks formed by particles. Examples of an irregular stone masonry wall specimen under compression and in-plane shear are also presented, considering failure processes through the joints and the stones.

In Chapter 8, the numerical modelling of masonry dams using the Discrete Element Method is presented. The chapter begins with a review of the history of masonry dams and their behaviour. A numerical tool based on the Discrete Element Method developed specifically for the structural assessment of masonry dams is then presented. The mechanical calculations performed by the tool are discussed in detail, together with the approach used for the modelling of passive anchors and the modules for seismic analysis and hydro-mechanical analysis. Structural and hydraulic analyses of a diverse set of existing masonry dams conducted using the tool are then presented. The Discrete Element Method is shown to be capable of reproducing the structural behaviour of masonry dams and identifying their likely failure mechanisms as required for structural safety evaluations.

In Chapter 9, the different aspects of simplified micro-modeling strategies including the Discrete Element Method (DEM) for modeling masonry-infilled frames are investigated. Masonry infill panels are represented by individual bricks and blocks separated by zero thickness interfaces representing mortar joints. A thorough overview of the different DEM studies performed on concrete and brick masonry-infilled frames are presented. The essential assumptions adopted by these models and numerical implementation issues are discussed outlined. Also, the advantages and disadvantages of modeling masonry infilled frames using the discrete element method are discussed. This 'discontinuum' approach, an alternative to modeling masonry as a homogenized continuum, is particularly suited for studying the mechanical behavior and interaction between of the individual brick/blocks and their interaction with the framed structure.

In Chapter 10, the effect of past drum dislocations on the vulnerability of classical columns and presents a performance-based framework for their seismic risk assessment. The vulnerability is numerically calculated through response estimations using detailed three-dimensional models based on the Discrete Element Method. Conditional limit-state probabilities are calculated and appropriate performance criteria are suggested. The proposed methodology is able to pinpoint cases where past damage affects the vulnerability of such structures

In Chapter 11, the most important structural components of historic unreinforced masonry buildings are considered. Gothic buttresses, arches of different shapes, and multidrum columns are analysed with the help of UDEC. Extensive parameter studies for the various types of structures are conducted, considering different geometric shapes and mechanical properties. The output of numerical simulations is compared with available analytical solutions. Modeling of dynamic problems is also addressed. The influence of stiffness, friction and damping parameters is assessed for the different types of masonry structures.

In Chapter 12, an overview of the major contributions for the analysis of limit stability of masonry arches during the last three decades is presented, together with the later developments of limit analysis. The solution of arch problems with DEM is addressed, and the numerical models are validated against analytical solutions of collapse loads for arches under gravity and lateral loading. An application of DEM to the safety assessment of an ancient tunnel structure is presented.

In Chapter 13, two engineering applications of the Distinct Element Method to the analysis of historic masonries are presented. In particular, the software UDEC, which implements DEM in a variety of engineering problems, is here used to analyse the Águas Livres aqueduct in Lisbon (Portugal) and multi-leaf masonry arch-tympana carrying systems of a basilica in Como (Italy). The second example analyses the arch-tympana carrying systems of a church in Italy. They present an unusual building technology, relying into a multiple-leaf arch, and a tympanum, made by a mixture of bad quality mortar and small stones. Again the structure is discretized into distinct elements and the load carrying capacity under dynamic excitation is evaluated, discussing the role played by the infill.

In Chapter 14, the feasibility of the utilization of a combined Finite Element/Discrete Element (FEM/DEM) approach to investigate the behaviour of masonry arch bridges is assessed. In particular, the Chapter proposes and discusses a possible approach to FEM/DEM modelling of two existing masonry arch bridges. Attention is paid to the assessment of the load carrying capacity of the structures by means of a suitable coupled FEM/DEM 2D approach.

In Chapter 15, the analysis of unreinforced masonry structures which are comprised of a finite number of distinct, interacting blocks that have length scales relatively comparable with the structure of interest, is concerned using the Discrete Finite Element Method (DFEM) developed by the author. DFEM is based on the finite element method incorporating contact elements: blocks are considered as solid sub-domains and contact elements are used to model block interactions such as sliding or separation.

In Chapter 16, a Semi-Discrete Model (SDM) is proposed for the study of freestanding blocks, in which the block is considered as a mass element supported by springs and dashpots, in the spirit of deformable contacts between rigid blocks. The advantages of the proposed model with respect to a full discrete model are discussed in further detail. The proposed model is able to reproduce the six motion states of the rigid bodies: rest, slide, rotation, slide-rotation, translation-jump and rotation-jump.

In Chapter 17, through some illustrative examples, the applicability of the Discrete Finite Element Method (DFEM) to the analysis of unreinforced masonry structures such as rock pillars, open rock slopes, underground openings, tunnels, fault propagations, and fault-structure interactions is examined and discussed. In the numerical study, the behaviour of contacts and blocks is assumed to be elasto-plastic or elastic. The Mohr-Coulomb yield criterion, representing material behaviour of contacts, is implemented in the developed codes for DFEM. The secant stiffness method together with the updated Lagrangian scheme is employed to deal with non-linear behaviour. The constant strain triangular element with two degrees of freedoms at each node, formed by properly joining the corners and contact nodes of an individual block, is adopted for finite element meshing of the blocks.

It is hoped that this book tool will be used by civil, architectural and mechanical engineering students, lecturers, researchers, asset managers, consultants and engineers to understand the different numerical approaches based on the discrete element method to be used for the structural assessment of masonry structures in their care.

Vasilis Sarhosis
Newcastle University, UK

Katalin Bagi
Budapest University of Technology and Economics, Hungary

José V. Lemos
Laboratório Nacional de Engenharia Civil, Portugal

Gabriele Milani
Politecnico di Milano, Italy
December 15, 2015

Chapter 1
On the Mechanical Behavior of Masonry

Vasilis Sarhosis
Newcastle University, UK

D. V. Oliveira
University of Minho, Portugal

P. B. Lourenco
University of Minho, Portugal

ABSTRACT

In this chapter, a review on the mechanical behaviour of masonry is presented. The aim is to establish a base of knowledge and understanding of masonry that will underpin its mechanical characteristics and will inform the decisions towards the selection of the computational tool used which are going to be described in the following chapters. Initially, a brief description of the factors that influence the mechanical response of masonry and the variation of the material properties are discussed. The review then considers the possible causes of cracking in masonry and the different failure modes that may occur during loading. Principal findings from the review are summarised at the end of the chapter.

INTRODUCTION

Masonry is a very common and traditional form of construction that has been used for centuries and around the world. Some of the most important cultural and historical monuments (like Parthenon, Pyramids, Colosseum; Segovia aqueduct etc.) have been constructed using masonry. Masonry constructions also represent the vast majority of the traditional buildings like church domes (Hagia Sophia, Istanbul; Maria del Fiore in Florence, Italy; St Peter in Rome, Italy) and gothic structures (Amiens Cathedral; Beauvais Cathedral). Most of these historic and heritage structures are old and are deteriorating over time. Research in the area of masonry constructions is therefore essential to understand their structural capacity, how they behave with the application of external load, to assess their design levels, their design potential and retrofitting measures. In spite the urgent need to understand the mechanical behavior of masonry, only recently researchers have shown interest in studying the behavior of structural masonry

DOI: 10.4018/978-1-5225-0231-9.ch001

in detail. This is mainly due to: a) the high complexity of masonry behavior when compared to other construction materials such as concrete and steel; and b) the absence of solid and comprehensive experimental and numerical research.

Masonry is a heterogeneous brittle material that consists of units and mortar joints. Masonry units usually consist of fired clay, concrete or calcium silicate bricks; concrete or fired clay blocks; adobe or various types, sizes and shapes of naturally occurring stones. The composition of the mortar joints is usually expressed in terms of the volume or weight ratio of the binder and the sand (or fine aggregate). The most commonly used binder in modern construction is Ordinary Portland Cement (OPC). This is sometimes supplemented by a small amount of hydrated lime which aids workability and cohesiveness. Water is added, not only to react with the OPC to produce the hydration products responsible for strength development, but also provide workability of the mortar in the fresh state. It is usually up to the mason or bricklayer to add the required quantity of water to obtain the desired workability. The strength of the mortar is classified as per the composition of its constituents e.g. parts of OPC to parts of sand to parts of water (OPC: sand: water).

From the different combinations of masonry units, mortars and unit bonding patterns, a large number of geometric arrangements and strength characteristics can be obtained. Generally, the main mechanical features of masonry can be characterised by the rigid nature of the masonry units which have a high resistance to compression; the deformability of the mortar joints with a low resistance to tension and the frictional properties of the unit/mortar joint interface. However, the characteristics and the mechanical properties of masonry may vary significantly even within the same structure. Extensive studies have been carried out in the past to investigate the factors influencing the mechanical behaviour and strength of masonry under loading, understand their specifications and how they behave and how to assess their structural capability and to design potential retrofitting and repair methods (Hendry, 1998; Rots, 1997; Van der Pluijm, 1999). The most important factors influencing the mechanical response of masonry are: Unit characteristics; mortar joint characteristics; brick/mortar bond characteristics; curing processes and workmanship. These are considered in more detail below.

FACTORS INFLUENCING THE MECHANICAL RESPONSE OF MASONRY

Unit Characteristics

Masonry is composed of individual units laid in and bound together by mortar. The common materials of units are brick, stone and concrete blocks. Masonry is generally a highly durable form of construction. The properties of the bricks in any typical structure will vary. Such variations may have an effect on the mechanical response to applied load or environmental changes (e.g. humidity and temperature). Some of the factors that may be responsible for variations in the bricks include: a) the brick making process; b) the natural variation in the composition and quality of the raw materials used in the brick making process; and c) deterioration due to ageing effects. The shape of the bricks will also vary as a result of differences in contraction during and after firing and moisture expansion.

According to McKibbins et al. (2006), bricks that are fired at the centre of a clamp in the kiln are subjected to burning and baking at higher temperatures than the rest and are of better quality and durability. Moreover, as the brick raw materials (i.e. clay or shale) are variable in moisture content, each batch of bricks will require a different amount of water to be added to the mix to ensure the right consistency at the end of the mixing process. Stone inclusions that may be found in the raw materials can also vary the mechanical properties of clay bricks, in some cases increasing the compressive strength. Beyond their original variability, the process of ageing and deterioration of bricks in old structures is another factor which can influence the current condition and physical characteristics of masonry.

Stone masonry construction is known since ancient times and can be seen all over the world. The type of stones used for the construction is usually variable and highly dependent on the local availability. Stone structures built without mortar rely on the skill of the craftsmen and the forces of gravity and frictional resistance. The primary function of masonry elements is to sustain a vertical gravity load. However, structural masonry elements are required to withstand combined shear, flexure and compressive stresses under earthquake or wind load combinations consisting of lateral as well as vertical loads. Stones of different shapes and types have been used in the past for the construction of walls, arches, flat roofs and domes. Usually historical buildings were built with:

- Sawn drystack or dry-stone masonry without bonding mortar;
- Irregular stone masonry with bonding mortar;
- Rubble masonry with irregular bonding mortar thickness; and
- A combination of the three techniques.

When bonding mortar was used, it was usually of low strength. Masonry with mortar joints can experience a significant loss of mortar due to combined chemical, physical and mechanical degradation and the behaviour of these constructions can then become similar to those made of dry joint masonry. The type of stone, the shape and size and the exposure to extreme environmental conditions will influence significantly the strength properties of the masonry construction.

Masonry units could also be made of concrete. Concrete blocks may be produced with hollow canters to reduce weight or improve insulation. Concrete blocks are usually come in many sizes and specifications to allow special construction features. U-shaped blocks or knockout blocks with notches to allow the construction of bond beams or lintel assemblies, using horizontal reinforcing grouted into place in the cavity. Blocks with a channel at the end, known as "jamb blocks", allow doors to be secured to wall assemblies. Blocks with grooved ends permit the construction of control joints, allowing a filler material to be anchored between the un-mortared block ends. Other features, such as corners known as "bullnoses" may be incorporated. Concrete masonry units may be formulated with special aggregates and thus their strength properties and characteristics will vary.

Mortar Joint Characteristics

The quality of mortar can also vary significantly even within the same structure. Some factors that may cause such variations are: a) the composition and quality of the materials used during the mortar making process; b) the interaction of the mortar with the adjacent bricks in the structure; c) the orientation of the joints in the structure and d) deterioration due to ageing effects.

Most mortars are a mix of cement, lime, sand and water. As a result the composition of the mortar may not be completely uniform throughout the structure, due to a lack of consistency in the batching of the mix constituents. Also, the properties of the mortar are influenced by the units surrounding them. According to Brocken and Pel (1995), mortar joints between masonry units with a high moisture content will cure in a different way to those between units of a lower moisture content.

Also, the characteristics of the mortar joints at different orientations could vary. According to Dialer (1990), the strength of the head or perpend joints is usually lower than the strength of the bed joints. This is a result of the greater degree of mortar shrinkage in the perpend joints and because these joints are often not filled fully with mortar. According to Mann and Muller (1982), such a reduction in strength and stiffness of the perpend joints can generate a discontinuous or non-uniform stress distribution in masonry subjected to in-plane loading. This is illustrated in Figure 1, in which σ_x is the normal stress, $\Delta\sigma_x$ is the change in normal stress and τ_{xy} is the complementary shear stresses. In addition, the processes of ageing and deterioration of mortar especially in old structures is another factor which can influence the current condition and physical characteristics of masonry. The reader should also be aware that construction of masonry wall systems is possible without the use of mortar. It occurs sometimes in stone masonry and it can be found in places where the lime was not locally available (or it was too expensive). In this case, the mechanical behaviour and strength of masonry construction will be completely different to that containing mortar joints.

Figure 1. Stress variations in a masonry wall subjected to in-plane loading as a result of differences in the bed and perpend joints
(Schlegel & Rautenstrauch, 2004)

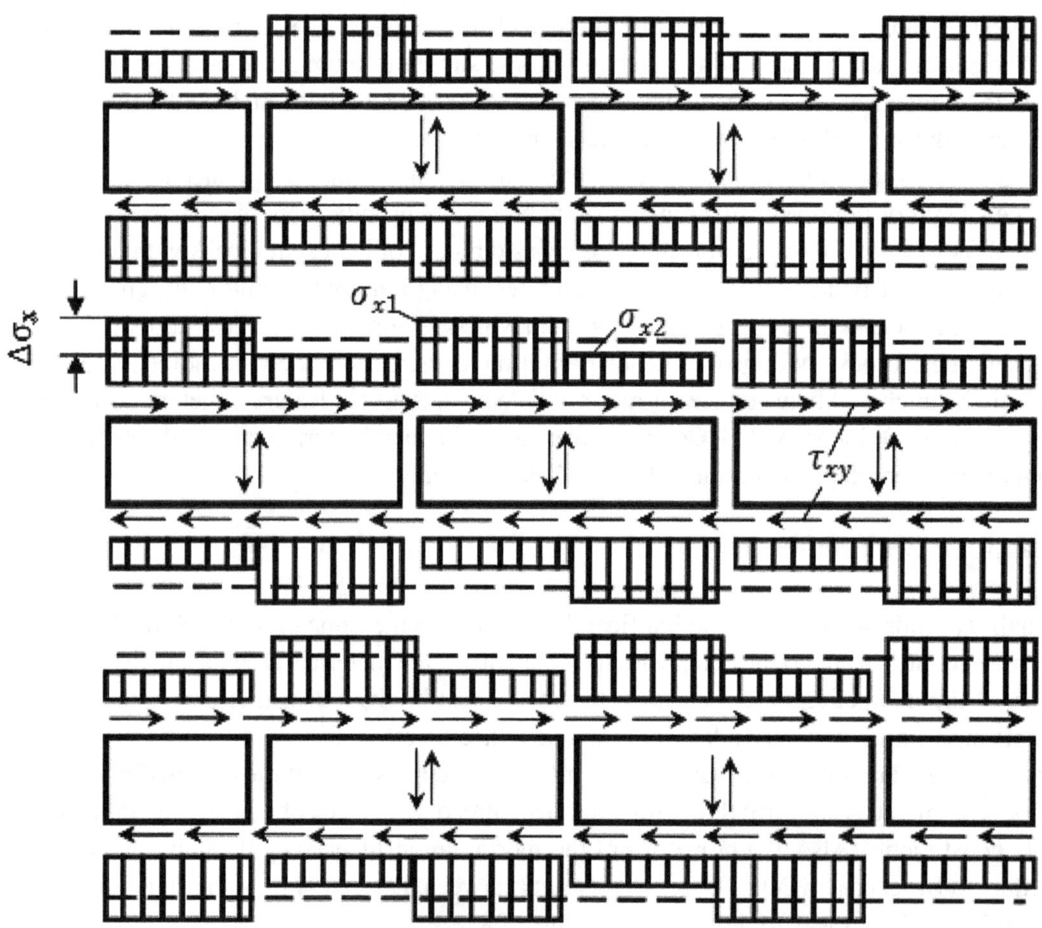

Brick/Mortar Bond Characteristics

The interaction between the bricks and the mortar joints is important in the mechanical behaviour of masonry as it has a considerable effect on load transfer and cracking. The value of the bond strength at the unit/mortar interface is very variable and has a large influence on brickwork's tensile strength and therefore its resistance to cracking. According to Grandet et al. (1972) and Lawrence et al. (1987) bonding of the brick and mortar is initially caused by the mechanical interlocking of the Calcium-Silicate Hydrates (C-S-H) and/or $Ca(OH)_2$ crystals which form in the surface micro-pores and cracks in the brick. The analysis of the bond interfaces using X-Ray and scanning electron microscopy techniques revealed no evidence of chemical reactions between the mortar and the brick materials. Furthermore, investigations by many researchers (Lawrence et al., 2008; Reda and Shrive, 2000; Sugo et al., 2001) have demonstrated that the formation of bond between the bricks and mortar is mechanical involving the transport of mortar fluids and fines to the brick/mortar interface followed by the hydration of the cementing materials.

Several researchers (Anderson & Held, 1986; Goodwin & West, 1982; Jung, 1988; Kjaer, 2010; Lawrence & Cao, 1988; Ostergard, 2010; Vermeltfoort et al., 2007) have investigated the factors that influence bond strength. Based on their studies, the most important factors are:

- The surface texture and the suction behaviour of units;
- The mortar composition;
- The grain size distribution of the aggregate in mortar; and
- The type of binders and the use of admixtures and additions for the preparation of the mortar.

The water suction of the masonry units is perhaps the most important intrinsic factor affecting the fresh mortar and, consequently, the properties of the hardened mortar and the properties of the brick/mortar interface. According to Hendry et al. (2004), bricks with a high suction rate will tend to remove water from the mortar leaving insufficient water for efficient hydration and the formation of C-S-H and $Ca(OH)_2$ required for high bond strength. If the suction rate is too low, the hydration products do not penetrate deep enough into the pores and the bond strength is reduced. Tensile bond strength is highly variable with moisture content of the bricks at the time of laying, as indicated in Figure 2.

Vermeltfoort et al. (2007) found that for a mortar of given consistency, the maximum bond strength was achieved when the brick contained an optimum amount of water. This occurred when the amount of "free" water in the mortar matched the initial rate of suction of the brick, which in turn is dependent on the amount of water in the brick at the time of laying.

In addition, the use of lime and plasticiser or air entraining agent increases the workability of mortar. Studies of the microstructure at the interface between the bricks and mortar and brick/mortar bond by Lawrence and Cao (1988) showed that the effect of lime on the interface microstructure is to facilitate the formation of the initial calcium rich film and to increase the amount of calcium hydroxide at the interface. The addition of lime enhances bonding but also increases the water demand of the mortar. However, due to this increase in water content, the strength of lime mortar tends to be low. The interface between bricks and mortar containing an air entraining agent has been found to contain a significant number of voids/air bubbles which can have a significant effect on the resulting bond. According to (Anderson & Held, 1976; Jung, 1988; Schubert, 1988) the general trend is that an increase in air content of the mortar will lead to a decrease in unit-mortar bond strength. However, recent studies have revealed

Figure 2. Variation of the brick/mortar tensile bond strength with moisture content of the bricks at the time of laying
(Hendry et al., 2004)

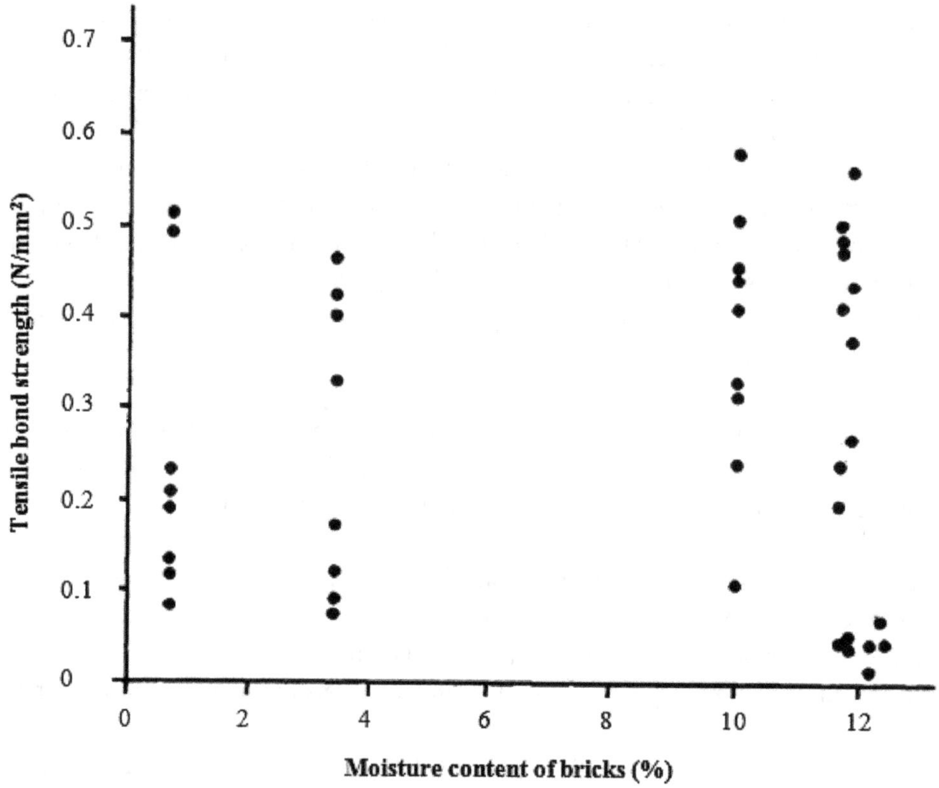

that it is the structure of the air content which determines and influence the bond strength rather than the amount of air (Kjaer, 2010).

Studies carried out by Ostergaard (2010) have demonstrated that the grain size distribution of aggregates for mortar has a significant effect on the bond strength of the unit/mortar interface. It was concluded that there is a decrease in bond strength with an increase in the grain size distribution of the mortar joints and this effect is noticeable for mortar with high cement content.

Finally, sand faced bricks are often used in construction for aesthetic reasons and/or for de-moulding the bricks from the steel plates during the brick making process. According to Vermeltfoort et al. (2007) bond strength is lower for sand faced bricks. It was found that the sand from the faces of the brick was well bonded to the mortar but readily separated from the brick thereby causing low bond strength. It was also found that the sand obstructs the transport of the binder (cement, lime) to the solid brick mass which further contributes to the low bond strength.

Curing Process

Moist curing of masonry after construction helps to ensure the maximum hydration of the cement in the mortar, thereby improving the bond, particularly to high suction bricks. The effects of curing conditions on the masonry strength have been investigated by many researchers (Anderson & Held, 1976; James, 1973; Marquis & Borchelt, 1986). They concluded that air cured masonry specimens have lower flexural bond strength compared with those wrapped in polyethylene sheeting or sprayed with a concrete curing compound. The difference in flexural bond strength between cured and not cured masonry increases for bricks with a high initial rate of suction (James, 1973). Also, as expected, masonry built on site, or in similar conditions, has lower flexural bond strength than masonry built in the laboratory (Marquis & Borchelt, 1986). However, even when masonry is constructed in environmentally controlled conditions in the laboratory, variations in curing and the properties of the masonry can still occur.

Presence of Moisture Content

The mechanical behaviour of masonry is highly variable. The presence of moisture in masonry, mainly conveyed by rising damp, is extremely frequent and plays a key role in the deterioration state of old masonry structures, owing to salt crystallisation, frost damage etc. In addition, the presence of moisture in the brick and mortar material pores may also directly influence the mechanical characteristics (compressive and tensile strength, elastic modulus) due to the interactions with the pore surface, enhancement of crack propagation velocity and other mechanisms. The contribution of moisture presence to the collapse of masonry built with ferruginous stone ashlars has been studied in Verstrynge (2014). From such studies, the presence of water in masonry found to lower both the compressive strength and stiffness of the stone. In the case of sandstone, which is a highly porous material, similar trends obtained. Blocks made of sandstone found to swell due to the water and hence decrease the internal friction of the stone Erguler 2009).

A decrease of the elastic modulus in highly moisture conditions was also found in other literature papers for different kinds of clay bearing rocks and buildings stones (see, e.g. Fjær 2009; Jimenez-Gonzalez & Scherer, 2004; Martìnez-Martìnez et al. 2011; Sassoni et al. 2013; Vásárhelyi & Ván 2006; Vasarhelyi, 2003). Among these studies, several mechanisms were observed to play a role in decreasing the mechanical performance of saturated clay bearing stones. These include: fracture energy reduction, capillary tension decrease, pore pressure increase, frictional reduction and chemical and corrosive deterioration (Erguler, 2009). It is worth mentioning that, in some cases, decrease in compressive strength and elastic modulus of up to 90% were registered for saturated clay beraring rocks in comparison with the oven dried ones (Gentilini et al. 2012; Franzoni et al. 2014). In relation to the shear behaviour of masonry in wet and dry conditions, studies suggest that saturation causes deterioration in strength and stiffness of up to one half.

Quality of Masonry Work

The strength of masonry is also highly affected by the quality of construction, i.e. site workmanship. The most obvious workmanship factors that may affect the mechanical characteristics of masonry include:

- Incorrect proportioning and mixing of the mortar;
- Incorrect or incomplete filling of the joints between the bricks with mortar;
- Incorrect area of adhesion between bricks and mortar e.g. frogged bricks should always be installed with the frog uppermost so that they are fully filled with mortar;

- The quality of the craftsmanship of the bricklayer or mason, e.g. ability to build walls to "line and level";
- Unfavourable curing conditions; and
- Disturbance of the masonry units or joints after laying.

In particular cases, these defects will present in varying degrees and the overall behaviour of the brickwork will reflect their combined effect. Assessing the overall effect of workmanship on the behaviour and strength of brickwork is not a straightforward issue. West et al. (1986) investigated the effect of bad workmanship. According to the study, the greatest effects of bad workmanship were found in masonry construction using mortar mixes weaker than 1:0.25:3 (OPC:lime:sand ration) parts of mortar mix properties. It is assumed that variations due to workmanship factors will be more heightened when using very weak mortars.

Variations in masonry properties due to bad workmanship effects are unavoidable even under closely controlled laboratory conditions. An extensive investigation on the degree of variation due to workmanship effects in masonry can be found elsewhere (Hendry, 1998).

FAILURES IN MASONRY

Movements in masonry may arise from the application of external load, foundation settlement, temperature changes, creep, and chemical reactions in the materials such as chemical attack or corrosion of any carbon steel components embedded in the mortar such as ties or reinforcement (Hendry, 1998). Restraining the movement of a brittle material such as masonry can result in cracking. Cracking and crushing in masonry structures may occur in:

- The units;
- The mortar;
- The brick/mortar interface; and/or in
- All of the above.

Cracks in masonry may not open uniformly but may close and open according to the type of stresses applied to them over a period of time. Usually cracks of 0.2 mm and wider are assumed to be significant because they are visible to the naked eye. If such cracks open up and propagate through the structure they may reduce its load carrying capacity and could lead, eventually, to collapse. The five basic failure modes of masonry as observed by Lourenço and Rots (1997) are shown in Figure 3. They identified that the occurrence of one of the modes of failure depends on the magnitude and direction of the shear and normal stress on the masonry. From Figure 3 it can be seen that (a) and (b) are joint mechanisms, (d) is a unit mechanism and (c), (e) are combined mechanisms involving the units and joints. However, variation in the stress-state within masonry can lead to combined failure modes in the structure (Garrity, 1995, 2004; Melbourne & Tomor, 2005; see also Figures 4 & 5). Experimental evidence (Abdou et al., 2006; Adami & Vintzileou, 2008; Garrity et al., 2010) has shown that at low values of normal stress, the principal failure mode of masonry with low strength mortar is either in the brick/mortar interface or in the mortar itself, resulting to joint opening due to cracking (Figure 3a) and/or frictional sliding along a bed or perpend joint (Figure 3b). The behaviour of masonry under tensile and combined compression and shear are considered further below.

Figure 3. Masonry failure modes
(Lourenço & Rots, 1997)

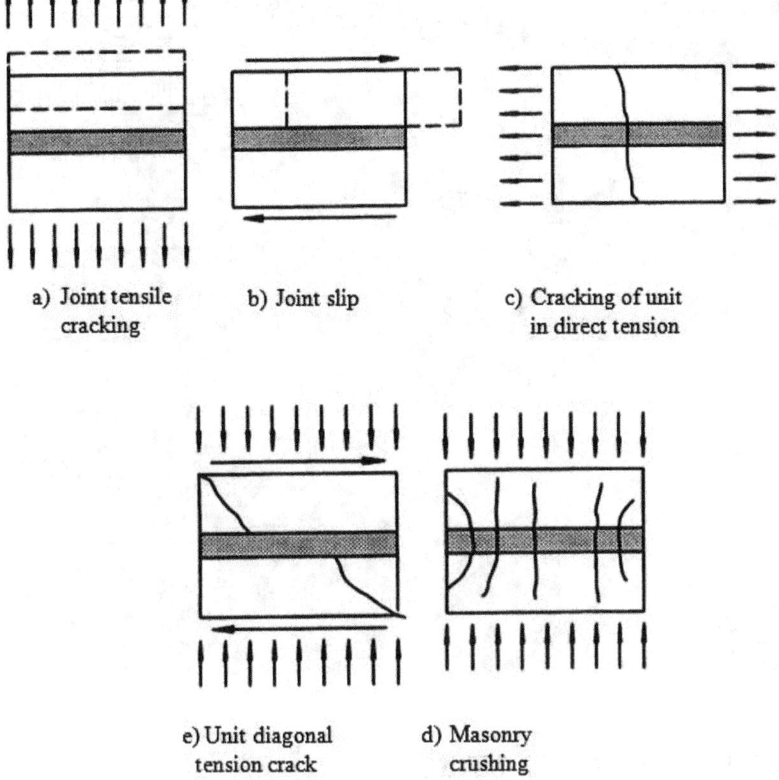

Figure 4. Crack patterns and failure modes for an arch constructed with brickwork
(Melbourne & Tomor, 2005)

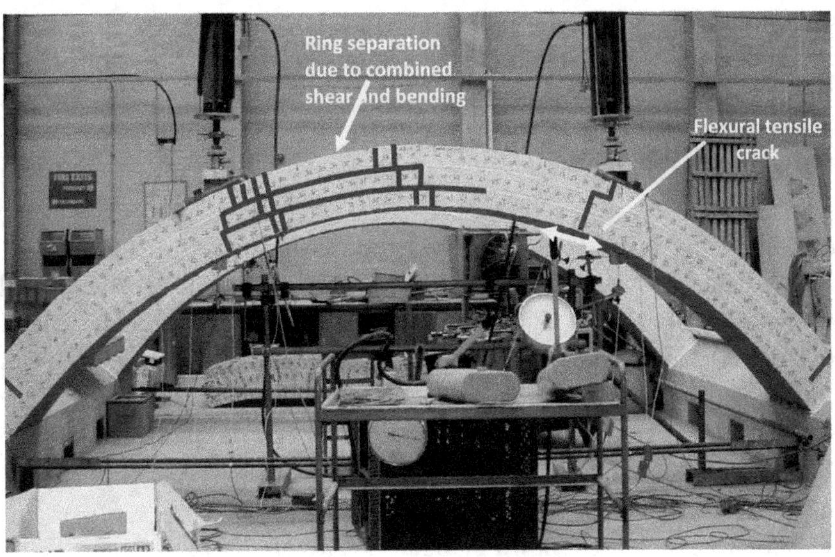

Figure 5. Cracking in a masonry wall panel due to flexural tension and combined shear and flexure (Garrity, 2004)

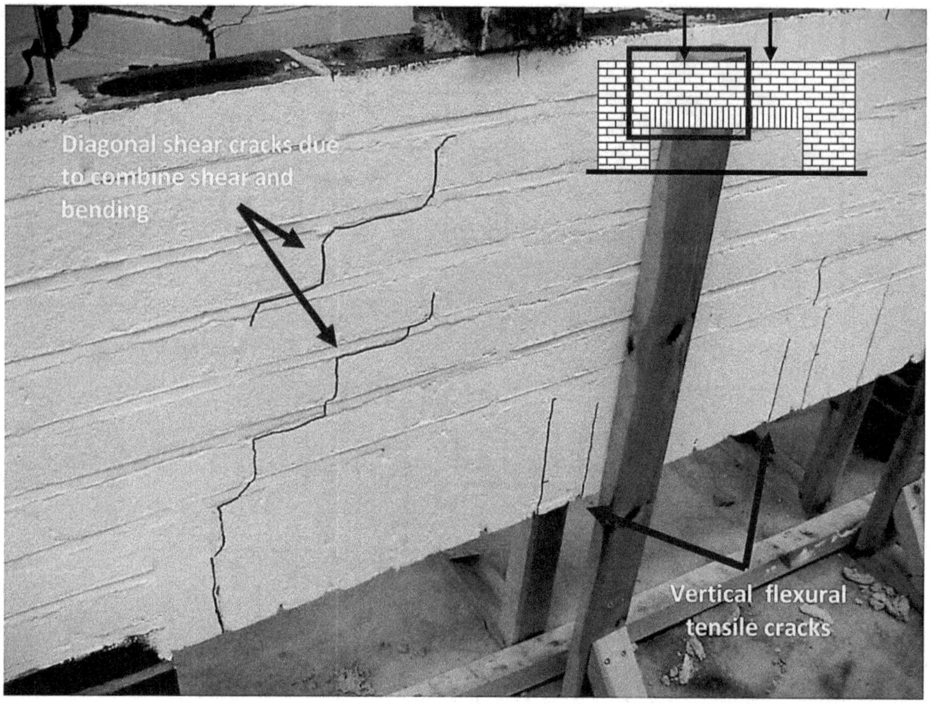

COMPRESSIVE STRENGTH OF MASONRY

Under uniaxial compressive loading, mortar tends to expand laterally more than the brick, because it has weaker mechanical properties. Due to the continuity between the bricks and mortar, ensured by cohesion and friction, mortar is confined laterally by the bricks. Shear stresses will then develop at the brick-mortar interface which produce a triaxial compressive stress state in the mortar and a bilateral horizontal tension coupled with vertical compression in the brick (see Figure 6). In this way, failure usually occurs by the development of cracks in the bricks, parallel to the loading direction.

TENSILE STRENGTH OF MASONRY

Masonry has a relatively high resistance to compressive stress but has little resistance to tensile stresses. From the literature (Adami & Vintzileou, 2008; Lawrence et al., 2008), the tensile strength (i.e. the resistance of masonry to tensile stresses) is highly dependent on many factors including the type of masonry unit, the composition of the mortar, the admixtures that may be included in the mortar and workmanship. In order to evaluate the behaviour of masonry under tension, different types of tests have been set up. According to Jukes and Riddington (1997), tensile strength tests can be divided into two main categories:

1. Direct tensile strength tests (as shown in Figure 7a & d) and
2. Flexural tensile strength tests as shown in Figure 7 (b & c).

Figure 6. Masonry prism under compressive loading vertical to bed joints and stress states of brick and mortar elements (C: compression; T: tension)
(Oliveira, 2003)

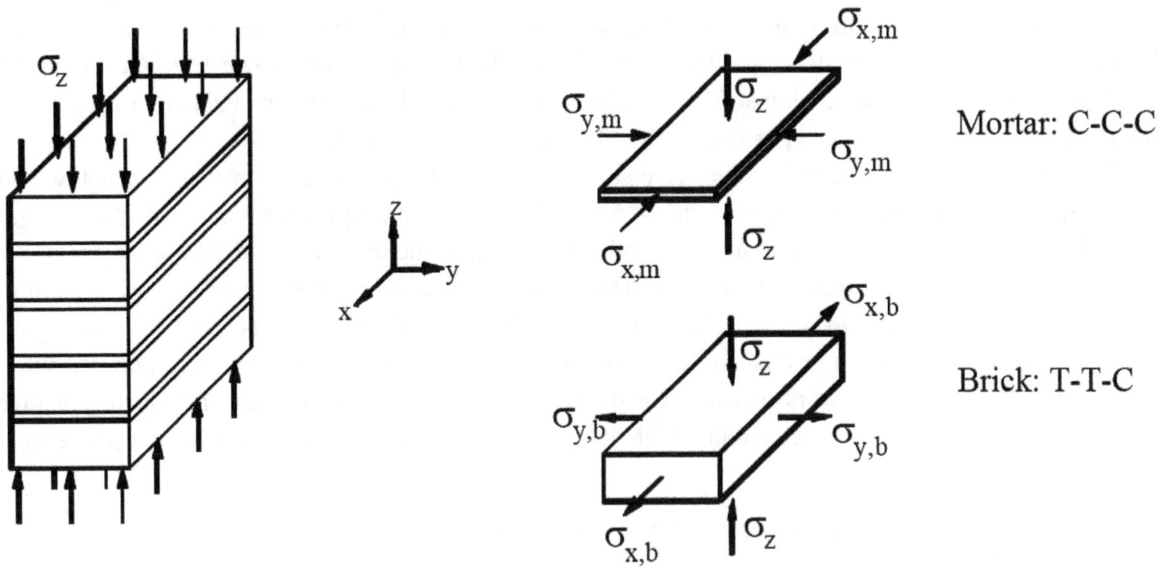

Figure 7. Different tensile bond strength tests: (a) direct tensile bond strength test; (b) flexural bond strength test; (c) bond wrench test; (d) splitting test

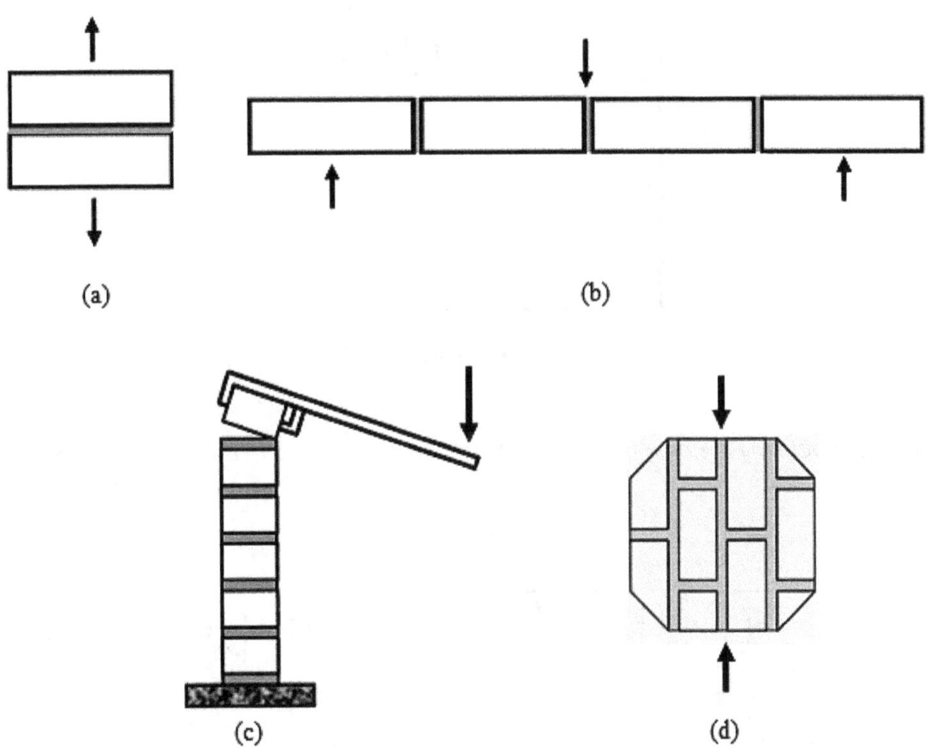

Tensile Strength

Tensile stresses can occur in masonry as a result of in-plane loading effects caused by eccentric gravity loads, thermal or moisture movements, foundation movement etc. For a tensile test, the relationship between the applied tensile stress and the elongation of the test specimen shown in Figure 8 can be obtained. This shows some tension softening behaviour after tensile failure. The behaviour prior to attaining the peak load is a close approximation to linear elastic behaviour.

Also, numerous experimental studies have shown that the tensile resistance of masonry is low and variable. As discussed previously, the tensile strength of brick masonry is governed mainly by the strength of the bond between bricks and mortar joints and the strength of the mortar. Van der Pluijm (1999) carried out tensile tests on masonry specimens made from solid clay units. A visual inspection of the failure surfaces showed that the bond area at failure was smaller than the full possible contact area. From Figure 9, the net bond area seems to be concentrated in the inner part of the specimen which can be attributed to a combined effect of poor workmanship. The difference in the net bond areas can significantly scatter in pre- and post-peak masonry behaviour which demonstrates the vast variability of masonry properties.

Figure 8. Schematic diagram of a deformation controlled tension test
(Van der Pluijm, 1999)

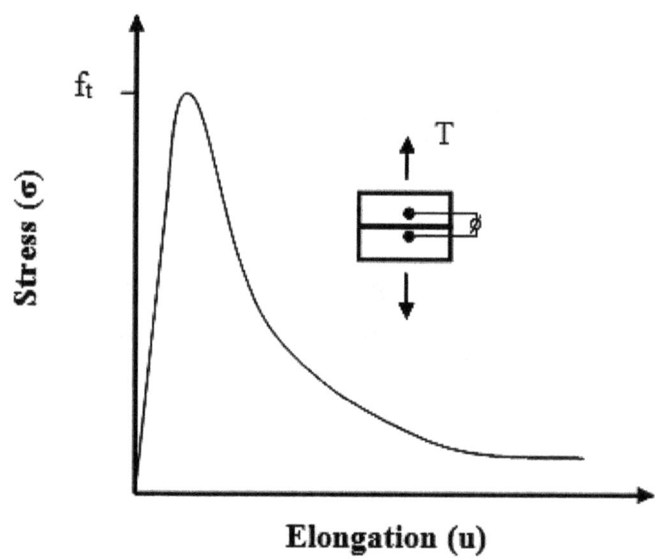

Figure 9. Tensile bond surface of soft mud clay brick with 1:2:9 (OPC:lime:sand) mortar
(Van der Pluijm, 1999)

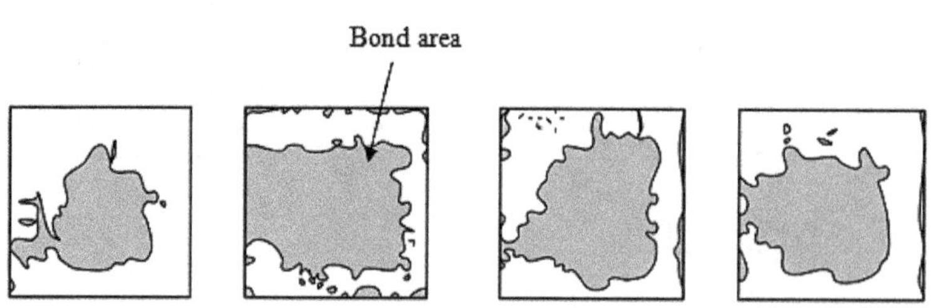

Tensile failure can occur either at the interface (Figure 10a & b) or in the mortar itself (Figure 10c). Failure modes of typical masonry units failed during tensile testing are shown in Figure 11.

The significance of the result obtained from tensile testing depends on the mode of failure. If failure occurs at the interface, the result obtained from the test represents the tensile bond strength, if, however, the mortar joint cracks then the test result gives the tensile strength of the mortar. In general, it is difficult to predefine the mode of failure in a tension test and there is no clear or reliable relation that exists between the tensile stress and the mode of failure.

From the literature (Hendry, 1998; Van der Pluijm, 1999), the tensile strength of brickwork typically ranges from 0 to 1N/mm^2. However, the variability of this property even in the same structure has to be kept in mind.

Flexural Bond Tensile Strength

The flexural bond strength of masonry is needed particularly for the design of masonry walls subjected to horizontal forces applied normal to the face of the wall, such as wind forces. As in the case of tension, the strength developed is dependent on the absorption characteristics of the bricks and also on the

Figure 10. Typical tensile test failure modes: (a) bond tensile failure of the top interface; (b) bond tensile failure of the bottom interface; (c) tensile cracking of the mortar

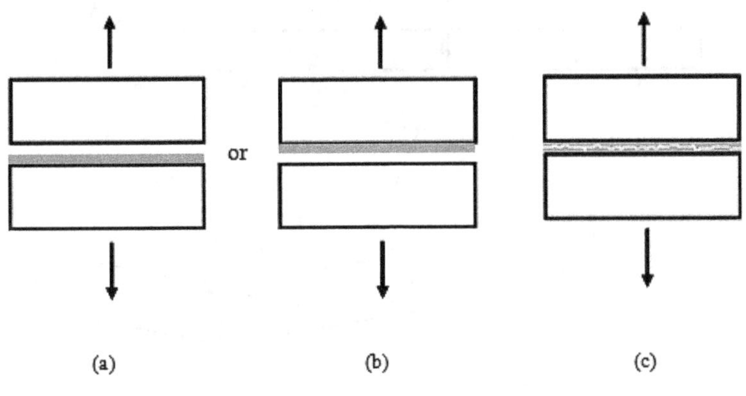

Figure 11. Tensile failure modes: (a) bond tensile failure of bottom interface; (b) tensile cracking of mortar (Adami & Vintzileou, 2008)

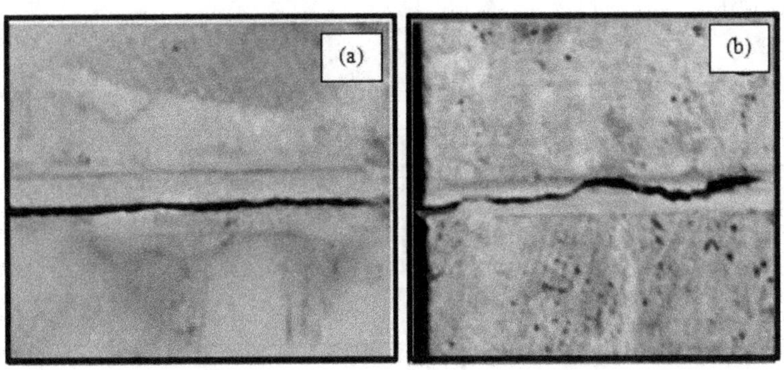

type of mortar used. Flexural tensile strength tests can also be used to obtain tensile bond strength data. However, as identified by Oliveira (2003), flexural tests measure the bond strength at the edge of the mortar joint, but the strength at that point may not be representative of the strength of the full depth joint. Flexural tensile failure modes can either be in the mortar or at the brick/mortar interface, see Figure 12. According to Hendry et al. (2004), the flexural tensile strength of clay brickwork ranges from about 0.8 to 2.0 N/mm^2 depending on whether the plane of failure is parallel or perpendicular to the bed joint.

COMBINED COMPRESSION AND SHEAR STRENGTH OF MASONRY

The strength of masonry in combined shear and compression is of importance in relation to the resistance of masonry to lateral loading. Several tests on masonry panels and specimens subjected to this type of loading have been carried out to understand its behaviour and to identify limiting stresses to be used in design, see Figure 13. A discussion regarding the adequacy of the different test methods will not be given here; the reader is referred to Atkinson et al. (1989) and Jukes and Riddington (1997) for further information.

Figure 12. Flexural tensile stresses in a masonry panel (plane of failure perpendicular to the bed joints)

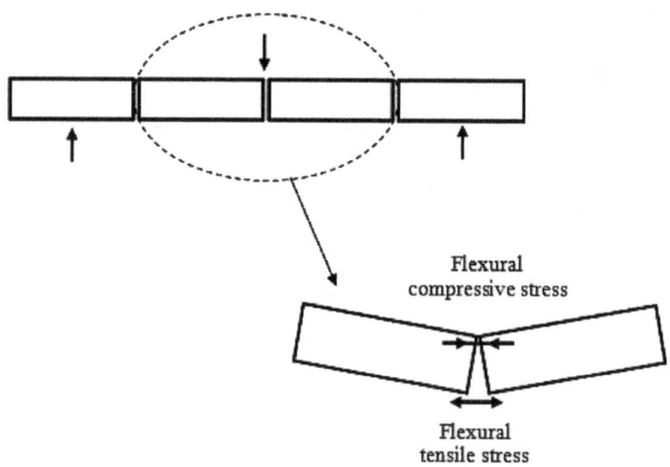

Figure 13. Different types of shear tests a) direct shear test, b) Van der Pluijim test (Van der Pluijim, 1999) and c) triplet test

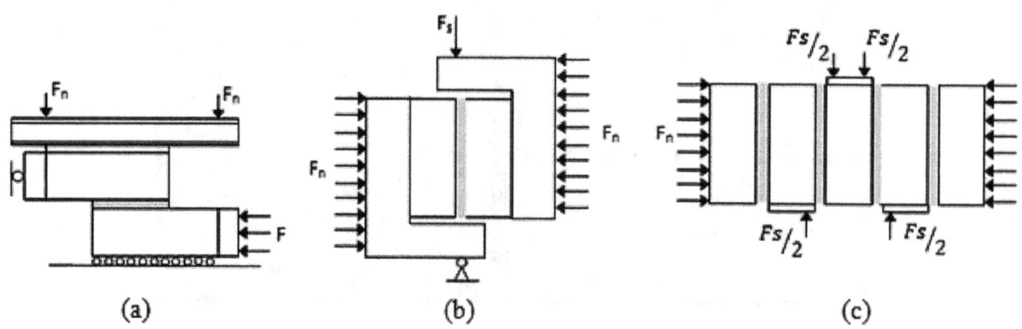

From the tests, it has been found that the shear strength depends on the mortar strength and results from the combination of two mechanisms: a) bond strength and b) frictional resistance. Typical results of the shear stress plotted against normal pre-compression for a shear test are shown in Figure 14. From Figure 14, the failure behaviour of masonry joints under shear accompanied with vertical pre-compression levels can be represented by the Coulomb friction law. The relationship between the shear stress (τ) and the normal stress (σ) is linear and is given by:

$$\tau = c + \sigma \tan \varphi$$

where c represents the cohesion or shear bond strength and is the shear strength at zero pre-compression and $tan\varphi$ is the tangent of the friction angle of the interface between unit and mortar joint and is not necessary equal to the coefficient of dry friction (Van der Pluijm, 1992). However, for high normal compressive stresses, the validity of the Coulomb failure is lost and cracking/crushing of the units occurs. According to Hendry (1998), for clay bricks, this limit is about 2N/mm^2. For a shear test with normal pre-compression, the relationship between the shear stress and the shear displacement is of the form shown in Figure 15. The descending part of the graph between ultimate shear strength of a test τ_u and τ_{fr} is described by Van der Pluijm (1999) as softening of the cohesion.

The values of cohesion and friction angle that define the brick/mortar interface may vary considerably according to the different unit/mortar combinations (Sarhosis et al. 2015). BS5628 (2005) gives design

Figure 14. Typical relationship between the shear strength of brickwork and the vertical pre-compression obtained from a shear test
(Hendry et al., 2004)

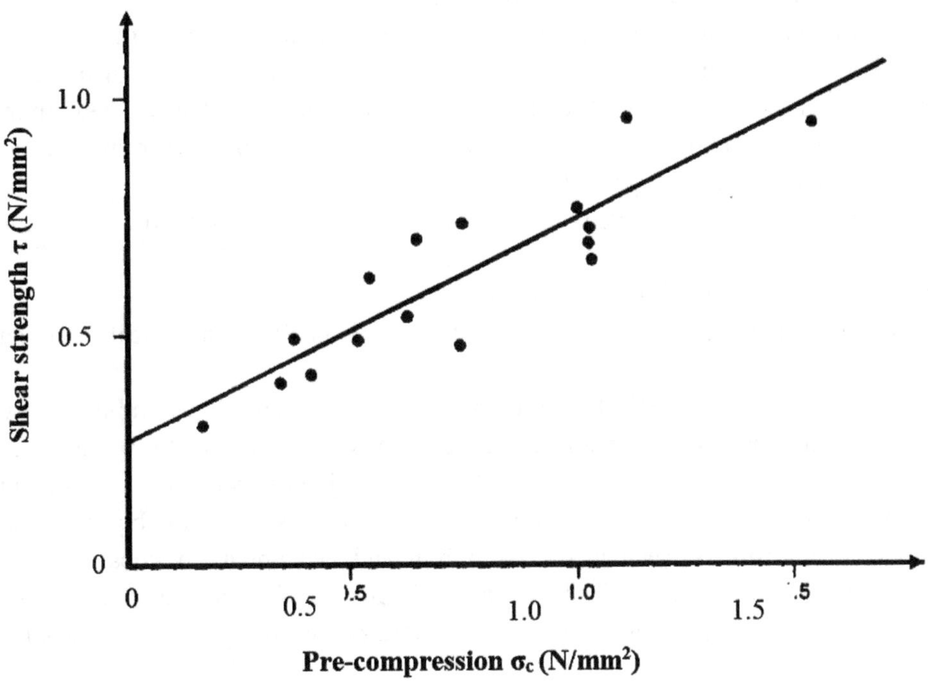

Figure 15. Schematic diagram of a deformation controlled shear test under constant normal compression (Van der Pluijm, 1993)

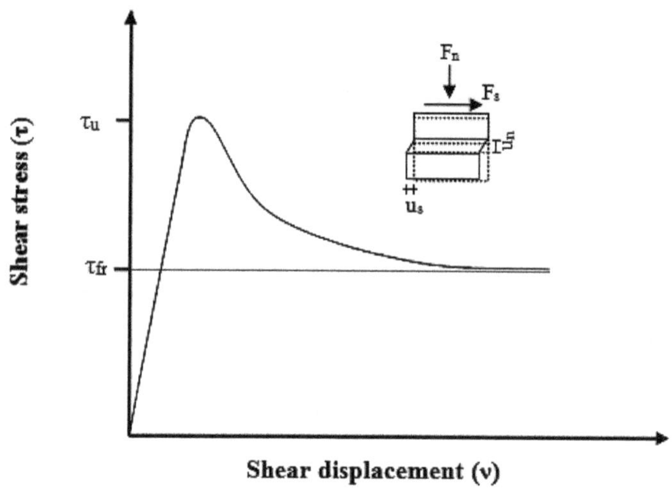

values for cohesion ranging from 0.35N/mm^2 to 1.75 N/mm^2 and tanφ equal to 0.6 for mortar designation (i), (ii) and (iii). In EC6, a table is provided which gives values for cohesion for different unit/mortar combinations and a constant value of tanφ equal to 0.4. In the literature, the value of tanφ usually varies between 0.7 and 1.2 (35^0 to 50^0) depending on the materials used (Lourenço et al. 2004). The published values of the cohesion parameter are reported to vary between 0.1 and 1.8 N/mm^2 (Lourenço, 1998b; Hendry, 1998; Jukes & Riddington, 1994; Van der Pluijm, 1999). According to experimental studies on masonry wall panels having a height to length ratio of 1.0 or more and units with a compressive strength of between 20 and 50 N/mm^2 laid in 1:1:6 (OPC:lime:sand) mortar, the value of cohesion has been estimated to be equal to 0.2 N/mm^2 (Hendry et al., 2004).

Another relevant feature of masonry joints is the so-called dilatancy angle (ψ). This is a measure of the volume change upon shearing. From Figure 15, the ratio between the normal displacement (u_n) and the shear displacement (u_s) gives tanψ. The dilatancy angle is positive but tends to zero upon increasing normal confining stress (Van der Pluijm, 1999). A typical relationship between the normal stress and the dilation angle is illustrated in Figure 16.

The failure modes generally recorded during shear tests for masonry with low strength mortar are:

Mode a: Shear failure of the top/bottom interface.
Mode b: A fracture plane at each brick-mortar interface accompanied by a vertical crack in the mortar.

These failure modes are schematically represented in Figure 17 and shown in Figures 18 & 19.

The mode of failure that may occur during shear testing is determined by the shear bond strength at the unit/mortar interface and the shear strength of the mortar joint in relation to the amount of shearing under vertical pre-compression. However, as far as the author is aware, there is no evidence of any research into the relationship between the mode of failure and the amount of vertical compressive stress for very low values of compression. When evaluating the results of shear tests, the shear strength will be the lower value of the mortar strength or the brick/mortar bond strength (i.e. whichever fails first).

Figure 16. Effect of normal stress on the dilation angle
(Van der Pluijm, 1993)

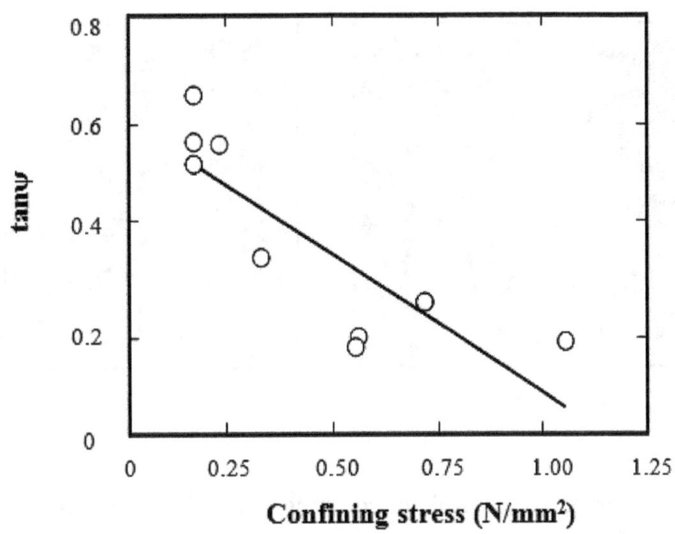

Figure 17. Typical shear failure modes: (a) shear failure of the top interface; (b) shear failure of mortar. Arrows indicate forces in the brick specimen.

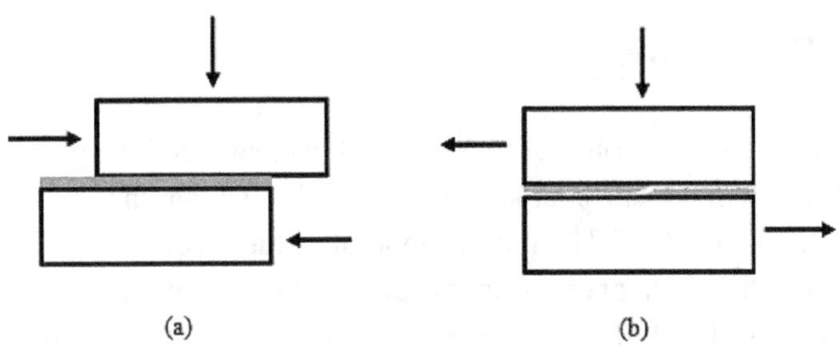

Figure 18. Failure mode during shear test
(Abdou et al., 2006)

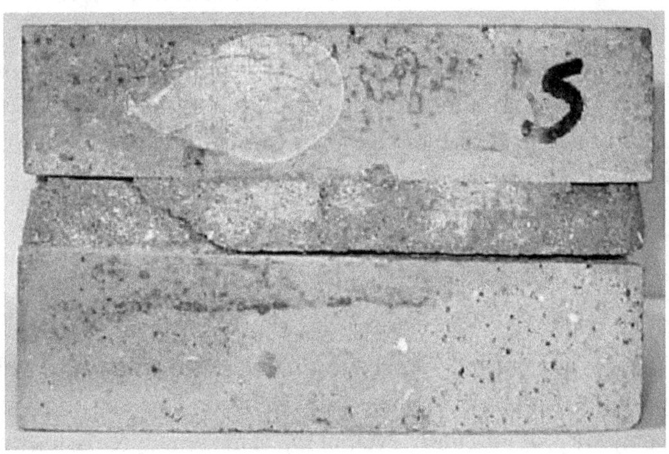

Figure 19. Failure modes (a) and (b) from the triplet shear test
(Beattie, 2003)

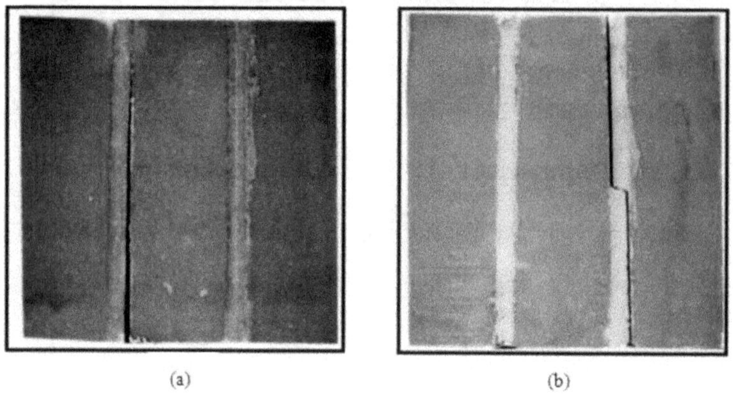

Overall, for masonry subjected to in-plane combined axial loading and shear where the levels of compressive stress are sufficiently low to avoid a crushing failure, it seems that the behaviour of the masonry can be best described by a constitutive law which is based on or encompasses the Mohr-Coulomb failure criterion.

OUT OF PLANE BEHAVIOUR

Out of plane loading refers to the response of masonry to horizontal actions perpendicular to their plane. Such loading can be due to wind or earthquake. Experience has shown that masonry structures are most vulnerable to the out of plane loading. The major mode of failure is due to overturning (Casarin, 2006). Most of the times, such failure is happening in structural elements of the building which may lead in total collapse. Typical out of plain failure modes are presented in Figure 20. Figure 20a shows the out of plane failure of the façade of a building due to poor connection between the diaphragms and the external walls. An out of plain failure with the overturning of some of the façade walls of the top of building due to Christchurch earthquake in 2011, is shown in Figure 20b. The lack of connection between external and internal walls of the buildings did not allow the box behaviour to develop and led to catastrophic collapse. As we can see, the walls have completely detached. Figure 20c shows a partial overturning of façade at the weakest component (openings). Last, Figure 20d illustrates overturning of a corner wedge due to ineffective connection between external walls, insufficient anchoring of the floors to the perimeter walls and the presence of openings near the edges. As can be seen from Figure 20d, the crack pattern follows the opening of the façade.

CONCLUSION

Masonry is a heterogeneous, anisotropic composite material. The non-linear behaviour of masonry is controlled by the mortar joints which act as planes of weakness. The characteristics and the mechanical properties of masonry may vary significantly even in the same structure. Brick properties may vary due

Figure 20. Failure modes due to out of plane loading during earthquakes: (a) Overturning of façade; (b) Overturning of portions of the façade; (c) Partial overturning of façade (effect of openings) (Binda et al. 2006); (d) Overturning of corner wedge
(Lagomarsino 1998; Beolchini et al. 2002).

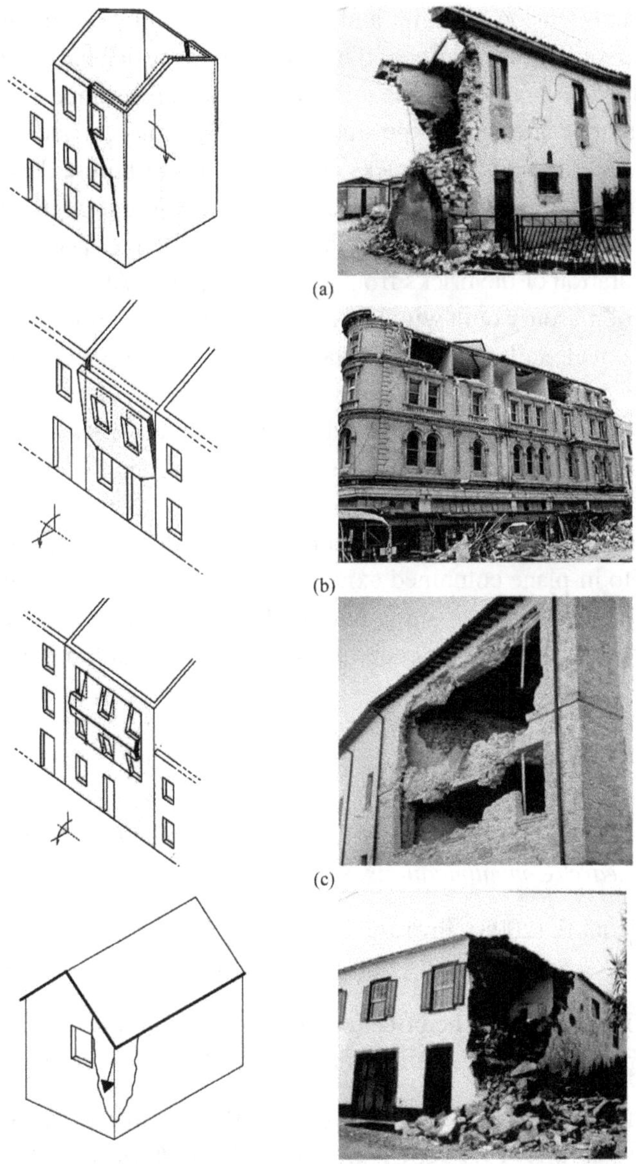

to the brick making process; variation in the raw material or as a result of deterioration. Mortar joint properties may also vary considerably due to the variability of the mixed constituents and their position (horizontal or vertical) in the structure. There may be non-uniform distributions of stress due to applied loading caused by differences in the stiffness of the perpend and bed joints. The most important factors that influence the mechanical response of brick masonry are: a) the brick characteristics; b) the mortar joint characteristics; c) the brick/mortar bond characteristics; d) the curing process; and d) the quality and workmanship. Restraining the movement of a brittle material such as masonry can result in cracking. Cracks in masonry can occur in the masonry units, in the mortar, at the unit/mortar interface or in all of these. Cracks in masonry may not open uniformly but may close and open according to the type of applied stress. Possible failure modes that can occur in masonry are: joint tensile cracking, joint slip cracking, masonry unit tension cracking and masonry crushing. The main mechanical features of masonry can be characterised by the rigidity of the masonry units which have a comparatively high resistance to compression, the deformability of the joints with a weak resistance to tension and the frictional properties of the brick/mortar interface. Experimental investigations carried out by several researchers have demonstrated that the occurrence of cracks and the modes of failure depend on the direction and magnitude of shear and normal stress applied to the masonry. For low strength masonry (such as historic masonry structures), cracking tends to be along the brick/mortar interfaces and failure usually results from de-bonding or separation of the bricks from the mortar joints. At low values of normal stress, the principal failure modes of masonry built with low strength mortar are either: a). De-bonding at the brick mortar interface or b). Tensile failure of the mortar resulting in an opening of the joint (i.e. cracking) and frictional sliding. Several small scale tests have been carried out in the past by many researchers to investigate the strength of masonry in tension and in combined compression and shear. However, depending on the type of test and the materials tested a large variation in the tensile and shear strength has been obtained. Values of tensile strength can range from 0.0 to 1.0 N/mm^2. Values of cohesive strength are reported to vary from 0.1 to 1.8 N/mm^2 depending on the materials tested and the test method used. For masonry subjected to in-plane combined axial loading and shear, where the levels of compressive strength are sufficiently low to avoid crushing failure, the behaviour of masonry can best be described by a constitutive law which is based on or encompasses the Mohr-Coulomb failure criterion.

REFERENCES

Abdou, L., Ami, S. R., Meftah, F., & Mebarki, A. (2006). Experimental investigations of the joint-mortar behaviour. *Mechanics Research Communications*, *33*(3), 370–384. doi:10.1016/j.mechrescom.2005.02.026

Adami, C. E., & Vintzileou, E. (2008). Investigations of the bond mechanism between stones or bricks and grouts. *Materials and Structures*, *41*(2), 255–267. doi:10.1617/s11527-007-9235-z

Anderson, C., & Held, L. C. (1986). The effect of sand grading on mortar properties and tensile bond strength of brickwork specimens. *Proceedings of the Brick Masonry Society*, *1*, 1–6.

Atkinson, R. H., Amadei, B. P., Saeb, S., & Sture, S. (1989). Response of masonry bed joint in direct shear. *Journal of Structural Engineering*, *115*(9), 2276–2296. doi:10.1061/(ASCE)0733-9445(1989)115:9(2276)

Beattie, G. (2003). *Joint fracture in Reinforced and Unreinforced Masonry under Quasi-static and Dynamic Loading*. (Ph.D thesis). University of Liverpool, Liverpool, UK.

Beolchini, G.C., Cifani, G., Pucci, G.G., et al. (2002) *Repertorio dei meccanismi di danno, delle tecniche di intervento e dei relativi costi negli edifici in muratura*. Academic Press.

Binda, L., Cardani, G., Saisi, A., & Valluzzi, M.R. (2006). *Vulnerability analysis of the historical buildings in seismic area by a multilevel approach*. Academic Press.

Brocken, H. J. P., & Pel, L. (1995). Moisture transport over the brick/mortar interface. In: National Research Council Canada (NRCC). *International Symposium on Moisture Problems in Building Walls.*

BS 5628. (2005). *Code of practice for the use of masonry. Materials and components, design and workmanship.*

Dialer, C. (1990). *Fracture and deformation behaviour of stress, biaxial tests on scale model masonry.* (MSc dissertation). Technical University of Munich, Germany.

Erguler, Z. A., & Ulusay, R. (2009). Water-induced variations in mechanical properties of clay-bearing rocks. *Int J Rock Mech Min, 46*(2), 355–370. doi:10.1016/j.ijrmms.2008.07.002

Fjær, E. (2009). Static and dynamic moduli of a weak sandstone. *Geophysics, 74*(2), WA102–WA112. doi:10.1190/1.3052113

Franzoni, E., Gentilini, C., Graziani, G., & Bandini, S. (2014). Towards the assessment of the shear behaviour of masonry in on-site conditions: A study on dry and salt/ water conditioned brick masonry triplets. *Construction & Building Materials, 65*, 405–416. doi:10.1016/j.conbuildmat.2014.05.002

Garrity, S. W. (1995). Testing of small scale masonry arch bridges with surface reinforcement. In *Proceedings of the 6th International Conference on Structural Faults and Repair.* Engineering Technics Press.

Garrity, S. W. (2004). *Independent testing of the Bersche-Rolt system for strengthening a brickwork lintel containing a soldier course. Technical Report.* Bradford, UK: University of Bradford, School of Engineering, Design and Technology.

Garrity, S. W., Ashour, A. F., & Chen, Y. (2010). An experimental investigation of retro-reinforced clay brick arches. In *Proceedings of the 8th International Masonry Conference.*

Gentilini, C., Franzoni, E., Bandini, S., & Nobile, L. (2012). Effect of salt crystallisation on the shear behaviour of masonry walls: An experimental study. *Construction & Building Materials, 37*, 181–189. doi:10.1016/j.conbuildmat.2012.07.086

Goodwin, J.F., & West, H.W.H. (1982). *A review of the literature on brick/mortar bond.* The British Ceramic Research Association. Technical Note No. 308.

Grandet, J., Javelza, R., Perrin, B., & Thenoz, B. (1972). Role of ettrignite in the binding of the mechanical type between clay and Portal cement paste. *Revue Terre Cuite, 48*, 21–28.

Hendry, A. W. (1998). *Structural masonry* (2nd ed.). London, UK: Palgrave Macmillan. doi:10.1007/978-1-349-14827-1

Hendry, A. W., Sinha, B. P., & Davis, S. R. (2004). *Design of masonry structures* (3rd ed.). London, UK: Taylor and Francis.

James, J. A. (1973). Investigation of the effect of workmanship and curing conditions on the strength of brickwork. In L. Fortig, & K. Gobel (eds.), *Proceedings of the 3rd International Brick Masonry Conference.*

Jimenez-Gonzalez, I., & Scherer, G. W. (2004). Effect of swelling inhibitors on the swelling and stress relaxation of clay bearing stones. *Environ Geol, 46*(3-4), 364–377. doi:10.1007/s00254-004-1038-8

Jukes, P., & Riddington, J. R. (1997). A review of masonry joint shear strength test methods. *Masonry International, 11*(2), 37–43.

Jung, E. (1988). The binding between mortar and brick. In J.W. Courcy (Ed.), *Proceedings of the 8th International Brick and Block Masonry Conference (IBMAC)*. Trinity University.

Kjaer, E. (2010). Bond strength: What is influencing the bond strength? In what way the bond strength influencing the masonry properties. In *Proceedings of the 8th International Masonry Conference*.

Lagomarsino, S. (1998). Seismic Damage Survey of the Churches in Umbria.*Proc. of the Workshop on Seismic Performance of Monuments*.

Lawrence, S. J., & Cao, H. T. (1987). An experimental study of the interface between brick and mortar. In *Proceedings of the 4th North American Masonry Conference*. University of California.

Lawrence, S. J., & Cao, H. T. (1988). Microstructure of the interface between brick and mortar. In: J.W. Courcy (Ed.), *Proceedings of the 8th International Brick and Block Masonry Conference (IBMAC)*. Trinity University.

Lawrence, S. J., Sugo, H. O., & Page, A. W. (2008). Masonry bond strength and the effects of supplementary cementitious materials. *Australian Journal of Structural Engineering, 8*(2), 101–115.

Lourenço, P. B. (1998a). Experimental and numerical issues in the modelling of the mechanical behaviour of masonry. In P. Roca, J. L. Gonzalez, E. Onate, & P. B. Lourenço (Eds.), *Structural Analysis of Historical Constructions II* (pp. 57–92). Barcelona, Spain: CIMNE.

Lourenço, P. B. (1998b). Sensitivity analysis of masonry structures. In *Proceedings of the 8th Canadian Masonry Symposium*.

Lourenço, P. B., Barros, J. O., & Oliveira, J. T. (2004). Shear testing of stack bonded masonry. *Construction & Building Materials, 18*(2), 125–132. doi:10.1016/j.conbuildmat.2003.08.018

Lourenço, P. B., & Rots, J. G. (1997). A multi-surface interface model for the analysis of masonry structures. *Journal of Engineering Mechanics, 123*(7), 660–668. doi:10.1061/(ASCE)0733-9399(1997)123:7(660)

Mann, W., & Muller, H. (1982). Failure of shear-stressed masonry – An enlarged theory, tests and application to shear walls. In H.W.H West (Ed.), *Proceedings of the British Ceramic Society.*

Marquis, E. L., & Borchelt, G. (1986). Bond strength comparison of laboratory and field tests. In *Proceedings of the 4th Canadian Masonry Symposium*.

Martìnez-Martìnez, J., Benavente, D., & Garcìa-del-Cura, M. A. (2011). Comparison of the static and dynamic elastic modulus in carbonate rocks. *Bulletin of Engineering Geology and the Environment*, 1–6.

McKibbins, L., Melbourne, C., Sawar, N., & Galliard, C. S. (2006). *Masonry arch bridges, condition appraisal and remedial treatment*. Report C656. CIRIA, London, UK.

Melbourne, C., & Tomor, A. K. (2005). *Effect of weak/deteriorated masonry on the performance of arch bridges (Arch tests R-S). Test Report*. Salford, UK: University of Salford.

Oliveira, D. V. (2003). *Experimental and numerical analysis of blocky masonry structures under cyclic loading*. (PhD thesis). University of Minho, Portugal.

Ostergaard, J. (2010). *The influence of the grain size distribution on the mortar and masonry properties. Test Report*. Denmark: Danish Technological Institute.

Reda, T. M. M., & Shrive, N. G. (2001). The use of pozzolans to improve bond strength. In *Proceedings of the 9th Canadian Symposium*. University of Rots.

Sarhosis, V., Garrity, S. W., & Sheng, Y. (2015). Influence of the brick-mortar interface on the mechanical response of low bond strength masonry lintels. *Engineering Structures*, *88*, 1–11. doi:10.1016/j.engstruct.2014.12.014

Sarhosis, V., & Sheng, Y. (2014). Identification of material parameters for low bond strength masonry. *Engineering Structures*, *60*, 100–110. doi:10.1016/j.engstruct.2013.12.013

Sassoni, E., Franzoni, E., Pigino, B., Scherer, G. W., & Naidu, S. (2013). Consolidation of calcareous and siliceous sandstones by hydroxyapatite: Comparison with a TEOS-based consolidan. *Journal of Cultural Heritage*, *14*(1), 103–108. doi:10.1016/j.culher.2012.11.029

Schlegel, R., & Rautenstrauch, K. (2004a). Failure analyses of masonry shear walls. In H. Konietzky (Ed.), *Numerical modelling of discrete materials in geotechnical engineering, civil engineering and earth sciences* (pp. 15–20). UK: Taylor and Francis Group London.

Schubert, I. P. (1988). The influence of mortar on the strength of masonry. In J.W. Courcy (Ed.), *Proceedings of the 8th International Brick and Block Conference*. Elsevier.

Sugo, H. O., Page, A. W., & Lawewnce, S. J. (2001). The development of mortar/unit bond. In *Proceedings of the* 9th*Canadian masonry symposium*. University of New Brunswick.

Van der Pluijm, R. (1992). Material properties of masonry and its components under tension and shear. In *Proceedings of the 6th Canadian Masonry Symposium*. University of Saskatchewan.

Van der Pluijm, R. (1993). Shear behaviour of bed joints. In: A.A. Hamid, & H.G. Harris (Eds.), *Proceedings of the 6th North American Masonry Conference*. Drexel University.

Van der Pluijm, R. (1999). *Out-of-plane bending of masonry behaviour and strength.* (Ph.D thesis). Eindhoven University of Technology, The Netherlands.

Vasarhelyi, B. (2003). Some observations regarding the strength and deformability of sandstones in dry and saturated conditions. *Bulletin of Engineering Geology and the Environment*, *62*(3), 245–249. doi:10.1007/s10064-002-0186-x

Vásárhelyi, B., & Ván, P. (2006). Influence of water content on the strength of rock. *Engineering Geology*, *84*(1), 70–74. doi:10.1016/j.enggeo.2005.11.011

Vermeltfoort, A. T., Martens, D. E. W., & van Zijl, G. P. A. G. (2007). Brick mortar interface effects on masonry under compression. *Canadian Journal of Civil Engineering*, *34*(1), 1475–1485. doi:10.1139/L07-067

Verstrynge, E., Adriaens, R., Elsen, J., & Van Balen, K. (2014). Multi-scale analysis on the influence of moisture on the mechanical behavior of ferruginous sandstone. *Construction & Building Materials*, *54*, 78–90. doi:10.1016/j.conbuildmat.2013.12.024

West, H. W. H., Hodgkinson, H. R., Haseltine, B. A., & De Vekey, R. C. (1986). Research results on brickwork and aggregate blockwork since 1977. *The Structural Engineer*, *64*(11), 255–267.

ADDITIONAL READING

Al-Chaar, G., Issa, M., & Sweeney, S. (2002). Behavior of masonry-infilled nonductile reinforced concrete frames. *Journal of Structural Engineering*, *128*(8), 1055–1063. doi:10.1061/(ASCE)0733-9445(2002)128:8(1055)

Al-Chaar, G. L., & Mehrabi, A. (2008). *Constitutive models for nonlinear finite element analysis of masonry prisms and infill walls (No. ERDC/CERL TR- 08- 19)*. US Army Corps of Engineering.

Almeida, J. C., Lourenco, P. B., & Barros, J. A. (2002). Characterization of brick and brick–mortar interface under uniaxial tension, in: Proceedings of 7th Int. Seminar on Structural Masonry, Brazil: CEFET-MG. 67–76.

Amato, G., Cavaleri, L., & Fossetti, M. (2008). Infilled frames: Influence of vertical load on the equivalent diagonal strut model. Presented at the 14 WCEE, Beijing, China.

Amde M, Martin J V, and Colville J. 2007. "The Effects of Moisture on Compressive Strength and Modulus of Brick Masonry." TMS Journal (September): 31–40.

Braga, F., Gigliotti, R., Laterza, M., D'Amato, M., & Kunnath, S. (2012). Modified steel bar model incorporating bond-slip for seismic assessment of concrete structures. *Journal of Structural Engineering*, *138*(11), 1342–1350. doi:10.1061/(ASCE)ST.1943-541X.0000587

Buonopane, S., & White, R. (1999). Pseudodynamic testing of masonry infilled reinforced concrete frame. *Journal of Structural Engineering*, *125*(6), 578–589. doi:10.1061/(ASCE)0733-9445(1999)125:6(578)

Calvi, G. M., & Bolognini, D. (2001). Seismic response of reinforced concrete frames infilled with weakly reinforced masonry panels. *Journal of Earthquake Engineering*, *5*(2), 153–185. doi:10.1080/13632460109350390

Calvi, G.M., Bolognini, D., Penna, A., (2004). Seismic performance of masonry-infilled RC frames: benefits of slight reinforcement. SÍSMICA 2004 - 6 o Congreso Nacional de Sismología e Engeñaría Sísmica.

Costa, A. A., Arêde, A., Costa, A., & Oliveira, C. S. (2011). In situ cyclic tests on existing stone masonry walls and strengthening solutions. *Earthquake Engineering & Structural Dynamics*, *40*(4), 449–471. doi:10.1002/eqe.1046

Croci, G. (1998). *The Conservation and Structural Restoration of Architectural Heritage, International Series on Advances in Architecture*. Southampton, UK: Computational Mechanics Publications.

Fiorato, A. E., Sozen, M. A., & Gamble, W. L. (1970). *An investigation of the interaction of reinforced concrete frames with masonry filler walls (No. UILU-ENG-70-100)*. Urbana- Champaign, IL: Dept. of Civil Engineering, University of Illinois.

Flanagan, R. D., & Bennett, R. M. (1999). Arching of masonry infilled frames: Comparison of analytical methods. *Practice Periodical on Structural Design and Construction*, *4*(3), 105–110. doi:10.1061/(ASCE)1084-0680(1999)4:3(105)

Foti, D. On the numerical and experimental strengthening assessment of tufa masonry with FRP, Mechanics of Advanced Materials and Structures, 2013, vol. 20 (02): 163-175, ISSN: 1537-6532, doi: 10.1080/15376494.2012.743634

Foti, D. Shape Optimization of Rectified Brick Blocks for the Improvement of the out-of-Plane Behavior of Masonry, International Journal of Mechanics, vol. 7, p. 417-424, 2013. ISSN: 1998-4448.

Franzoni, E., Gentilini, C., Graziani, G., & Bandini, S. (2015). Compressive Behaviour of Brick Masonry Triplets in Wet and Dry Conditions. *Construction & Building Materials*, *82*, 45–52. doi:10.1016/j.conbuildmat.2015.02.052

Gabrielsen, B. L., & Kaplan, K. (1997). Arching in masonry walls subjected to out-of-plane forces. *Earthquake Resistant Masonry Construction. NBS Building Science Series*, *106*, 283–313.

Gumaste, K. S., Nanjunda Rao, K. S., Venkatarama Reddy, B. V., & Jagadish, K. S. (2006). Strength and elasticity of brick masonry prisms and wallets under compression. *Materials and Structures*, *40*(2), 241–253. doi:10.1617/s11527-006-9141-9

Hamburger, R. O., & Meyer, J. D. (2006). The performance of steel-frame buildings with infill masonry walls in the 1906 San Francisco earthquake. *Earthquake Spectra*, *22*(1), 43–67. doi:10.1193/1.2185656

Haselton, C.B., Goulet, C.A., Mitrani-Reiser, J., Beck, J.L., Deierlein, G.G., Porter, K.A., (2007). An assessment to benchmark the seismic performance of a code-conforming reinforced concrete moment-frame building (No. PEER 2007/12).

Haselton, C. B., Liel, A. B., Lange, S. T., & Deierlein, G. G. (2008). Beam-column element model calibrated for predicting flexural response leading to global collapse of RC frame buildings (No. PEER 2007/03). CA, Berkeley.

Hashemi, A., & Mosalam, K. (2007). *Seismic evaluation of reinforced concrete buildings including effects of masonry infill walls, PEER Technical Report*. Pacific Earthquake Engineering Research Center.

Lepage, A. (1997) A Method for Drift-Control in Earthquake-Resistant Design of RC Building Structures. PhD Thesis, University of Illinois, Urbana, 251.

Lourenço, P. B., Oliveira, D. V., Roca, P., & Orduña, A. (2005). Dry joint stone masonry walls subjected to in-plane combined loading. *Journal of Structural Engineering*, *131*(11), 1665–1673. doi:10.1061/(ASCE)0733-9445(2005)131:11(1665)

Magenes, G., Penna, A., Galasco, A., & Rota, M. (2010) Experimental characterisation of stone masonry mechanical properties, 8th International Masonry Conference 2010, Dresden, pp. 247-256.

Matjaz, D., & Fajfar, P. (2005). Simplified Non-linear Seismic Analysis of Infilled Reinforced Concrete Frames. *Earthquake Engineering & Structural Dynamics*, *34*(1), 49–66. doi:10.1002/eqe.411

Nanni, A., & Tumialan, J. G. (2001). *In-plane and out-of-plane behavior of masonry walls strengthened with FRP system, CIES 01-24*. Dept. of Civil Eng., University of Missouri-Rolla.

Nanni, A., Tumialan, J. G., & Myers, J. J. (1999). *Field evaluation of masonry walls strengthened with FRP composites at theMalcom Bliss Hospital, CIES 99-12*. Dept. of Civil Eng., University of Missouri-Rolla.

Negro, P., & Verzeletti, G. (1996). Effect of Infills on the Global Behaviour of Frames: Energy Considerations from Pseudo-dynamic Tests. *Earthquake Engineering & Structural Dynamics*, *25*(8), 753–773. doi:10.1002/(SICI)1096-9845(199608)25:8<753::AID-EQE578>3.0.CO;2-Q

Rodriguez, M. E. (2008) Selecting Earthquake Records for Nonlinear Dynamic Analysis of Structures, *Proc Seismic Engineering International Conference Commemorating the 1908 Messina and Reggio Calabria Earthquake*. doi:10.1063/1.2963824

Saneinejad, A., & Hobbs, B. (1995). Inelastic Design of Infilled Frames. *Journal of Structural Engineering*, *121*(4), 634–650. doi:10.1061/(ASCE)0733-9445(1995)121:4(634)

Uranjek, M., & Bokan-Bosiljkov, V. (2015). Influence of Freeze–thaw Cycles on Mechanical Properties of Historical Brick Masonry. *Construction & Building Materials*, *84*, 416–428. doi:10.1016/j.conbuildmat.2015.03.077

Varela-Rivera, L., Navarrete-Macias, D., Fernandez-Baqueiro, L. E., & Moreno, E. I. (2011). Out-of-plane behaviour of confined masonry walls. *Engineering Structures*, *33*(5), 1734–1741. doi:10.1016/j.engstruct.2011.02.012

Vasarhelyi, B. (2003). Some Observations Regarding the Strength and Deformability of Sandstones in Dry and Saturated Conditions. *Bulletin of Engineering Geology and the Environment*, *62*(3), 245–249. doi:10.1007/s10064-002-0186-x

Vásárhelyi, B. (2005). Statistical Analysis of the Influence of Water Content on the Strength of the Miocene Limestone. *Rock Mechanics and Rock Engineering*, *38*(1), 69–76. doi:10.1007/s00603-004-0034-3

Watt, D., & Colston, B. (2000).. . *Investigating the Effects of Humidity and Salt Crystallisation on Medieval Masonry*, *35*, 737–749.

Wood, S. (1991). Performance of Reinforced Concrete Buildings Duringthe 1985 Chile Earthquake: Implications for the Design of Structural Walls. *Earthquake Spectra*, *7*(4), 607–639. doi:10.1193/1.1585645

Zarnic, R., Gosti, S., Crewe, A. J., & Taylor, C. A. (2001). Shaking Table Tests of 1:4 Reduced-Scale Models of Masonry Infilled Reinforced Concrete Frame Buildings. *Earthquake Engineering & Structural Dynamics*, *30*(6), 819–834. doi:10.1002/eqe.39

KEY TERMS AND DEFINITIONS

Bond Strength: The strength that keeps building material and mortar together.

Brickwork: Masonry produced by laying one brick over the other separated with or without mortar.

Compressive Strength: Max compressive force resisted per unit area of net cross sectional area of masonry.

Dilation Angle: The angle which measures the uplift of one unit (i.e. brick or block) over the other.

Friction Angle: Force resisting the relative motion of two surfaces of solid blocks.

Load: External forces applied to the structure.

Tensile Strength: Max stress that masonry can withstand while being stretched or pulled and before fail.

Chapter 2
Micro-Modeling Options for Masonry

Vasilis Sarhosis
Newcastle University, UK

ABSTRACT

In this chapter, a review of the available methods and their challenges to simulate the mechanical behavior of masonry structures are presented. Different micro-modeling computational options are considered and compared with regard to their ability to define the initial state of the structure, realism in simulation, computer efficiency and data availability for their application to model low bond strength masonry structures. It is highlighted that different computational approaches should lead to different results and these will depend on the adequacy of the approach used and the information available. From the results analysis it is also highlighted that a realistic analysis and assessment of existing masonry structures using numerical methods of analysis is not a straight forward task even under full knowledge of current conditions and materials.

The following abbreviations are used in the manuscript:

- **BEM:** Boundary element method.
- **DDA:** Discontinuous deformation analysis.
- **DEM:** Discrete/distinct element method.
- **FEM:** Finite element method.
- **LAM:** Limit analysis method.
- **PFC:** Particle flow code.

INTRODUCTION

Masonry is one of the oldest building materials. It is composed of masonry units (i.e. brick, blocks) bonded with or without mortar. Despite its simplicity of construction, the mechanical behavior of masonry remains a challenge. Masonry is a heterogeneous anisotropic material where mortar joints act as a plane

DOI: 10.4018/978-1-5225-0231-9.ch002

of weakness. The failure mechanism of the material includes tensile failure of units and joints, shear failure of joints and compressive failure of the composite. At low levels of stress, masonry behaves as a linear elastic material. Its behavior becomes increasingly non-linear as the load applied on it increases and cracks develop and propagate. Cracking in masonry structures may be induced by deformation in bending/shear or volumetric changes of the component bricks, blocks, or mortar arising from natural expansion or shrinkage, temperature change, corrosion, or associated reactions (Cook & Pegam 1993; Sarhosis et al. 2015). Experience demonstrates that many masonry constructions have collapsed in the past. In Europe, there are several examples demonstrating this, like the ones related to the earthquake of Lisbon in 1755, where several monuments and hundreds of constructions were heritage and collapsed. Fatigue and strength degradation, accumulated damage due to traffic, wind and temperature loads, soil settlements and the lack of structural understanding of the original constructors are some of the factors that contribute to the deterioration as well as the continuous degradation of masonry structures. The loss is even more dominant when damage occurs at historic and cultural structures where damage is most of the time non-reversible. Recent examples of losses of the cultural and architectural damage can be found, for example, in Italy: Campanile of St. Marcus in Venice (total collapse in 1902 after being repeatedly struck by lightning), Civic Tower of Pavia (total collapse in 1989 with hardly any warning) and Cathedral of Noto (collapse of the dome in 1996). Famous examples of historical constructions in risk due to soil settlements are the Cathedral of Mexico City and the Tower of Pisa, and constructions in risk due to a deficient structural conception are the Cathedral of Pavia and the Cathedral of Florence (Lourenço, 2002).

Research is needed to be able to understand the behaviour of masonry construction, exhibiting highly non-linear characteristics. In particular it is important to understand the pre- and post-cracking behaviour to inform decisions concerning the maintenance needs, management of safety risks, assessment of levels of safety and the need for repair or strengthening. As experimental research is prohibitively expensive, it is fundamentally important to have available a computational model that can be used to predict the in-service and near-collapse behaviour with sufficient reliability. Once such a model has been established, it can be used to investigate a range of complex problems and scenarios that would not, otherwise, be possible.

A review of the current strategies for modelling structural masonry will be given with an emphasis on those considered to be appropriate for modelling low bond strength or dry joint masonry. The different available approaches are considered and compared with regard to their ability to define the initial state of the structure, realism in simulation, computer efficiency and data availability for their application to model masonry structures.

CHALLENGES FOR MODELLING MASONRY

The need to predict the in-service behaviour and load carrying capacity of masonry structures led researchers to develop several numerical methods and computational tools which are characterized by their different levels of complexity. For a numerical model to adequately represent the behaviour of a real structure, both the constitutive model and the input material properties must be selected carefully by the modeller. However, it should be considered that analysis and assessment of masonry structures may be achieved with little site-specific data and awareness that deformability and strength properties of both brick and mortar may vary considerably. For example, it is practically impossible to obtain complete data of boundary stresses for each individual brick of a masonry structure and it is impossible to measure accurately the strength of each mortar joint in every direction. The possibility of performing non-destructive and minor destructive tests is very interesting since it allows gathering valuable data which can be fed to the computational models. However, it may often happens that even such methods tend not to provide sufficient characterisation of the structural materials needed for advanced modelling. With historical forms of construction, in-situ destructive testing or the testing of samples of masonry removed from the structure that are large enough to be representative, is not usually possible. Even with masonry structures that are not classified as being of historical importance, it can be difficult to interpret the results of non-destructive or minor destructive tests and to relate them to the material parameters required for use in a computational model. Additionally, it is usually difficult to obtain specimens of masonry that are small and manageable enough for laboratory and representative of the structure. Another difficulty in the analysis of masonry (and especially historic) structures is the generation of the current loading level and distribution (Lemos, 2004). A direct application of self-weight and other expected loads may not reflect correctly the load paths and distribution that has resulted from a building's long history. According to Roca et al. (2010), load transfer and distribution may activate specific resisting phenomena (contact problems, friction, eccentric loading). In essence, this is a result of the difficulty to capture accurately and realistically the geometry of the structure under investigation. These factors often demand parametric studies, in addition to the usual uncertainties about material and interface properties. In addition, with the increase of complexity and sophistication of the numerical developments, models may not be fully understood by the engineers and are often classified as "black box" exercises which can play little or no role in scientific explanation. Therefore, it should be recommended that a few simple (in geometry) numerical models should be developed primarily by the modeler/engineer in order to understand the dominant mechanisms affecting the behavior of the masonry composite system under different material properties and loading configurations. It would be appropriate if these models are validated against controlled laboratory tests or in-situ tests carried out (such as flat jacks) and static or dynamic monitoring. Once the behavior of the system is understood and the results could be validated with those of the laboratory, then it is possible to develop simple idealizations for the predictive capability and response of more complicated structures such as old and deteriorated masonry structures. In addition, the use of sophisticated methods of analysis and computational models may require material parameters that could be difficult to be obtained from experimental testing; some of these material parameters may have only a mathematical significance. The engineering judgment and the modeler's experience it is necessary in such cases.

DESIRABLE FEATURES FOR THE SELECTION OF THE APPROPRIATE MODELLING APPROACH

A broad range of numerical methods is available today, ranging from the classical plastic solution methods (Heyman, 1998) to the most advanced non-linear computational formulations (e.g. finite element and discrete element methods of analysis). The selection of the most appropriate method to use depends on, among other factors, the structure under analysis; the level of accuracy and simplicity desired; the knowledge of the input properties in the model and the experimental data available; the amount of financial resources; time requirements and the experience of the modeller (Lourenço, 2002). It should also be expected that different methods should lead to different results depending on the adequacy of the approach and the information available. Preferably, the approach selected to model masonry should provide the desired information in a reliable manner within an acceptable degree of accuracy and with least cost. Considering the mechanical behaviour and characteristics of masonry (Hendry 1998; Hendry et al. 2004), the criteria that could be used to select the most appropriate model to best simulate the behaviour of masonry include:

- To allow for the opening, closing and shear sliding of the joints taking into account any resulting changes in strength and stiffness;
- To be able to capture tensile failure; debonding and shear slip at the brick/mortar interfaces;
- To be able to simulate failure of the brick/mortar interface and failure in the mortar joint;
- The mechanical actions in the masonry should include the components of bonding, frictional and cohesive resistance and the dilatational characteristics of the interface;
- To allow the user to assign different material properties at different locations of the structure. This can be important when different mechanical characteristics of the perpend and bed joints need to be assigned or when repair methods such as re-pointing of mortar needs to be simulated;
- To be able to capture realistically the onset of the formation of the first significant cracking and the subsequent propagation of cracking throughout the structure;
- To be able to provide a realistic estimate of the surface crack width;
- To be able to simulate the post-cracking and near-collapse behaviour of masonry;
- To be able to provide a load vs. displacement relationship for any part of the structure; and
- To be capable of incorporating and simulating, with a satisfactory degree of accuracy, various strengthening methods including the addition of steel reinforcement.

STRATEGIES FOR MODELLING MASONRY

At present, the approaches that have been developed for the computational representation of masonry tend to focus on the individual masonry units (i.e. the bricks, blocks etc) and the mortar, or on a macro-modelling approach in which masonry is considered as composite material. Depending on the level of accuracy and simplicity required the following modelling strategies can be used (Lourenço, 1996); these are summarised in Figure 1:

- **Detailed Micro-Modelling:** Both the masonry units and the mortar are discretised and modelled with continuum elements while the unit/mortar interface is represented by discontinuous elements accounting for potential crack or slip planes, as shown in Figure 1(b);
- **Simplified Micro-Modelling:** Expanded units are modelled as continuous elements while the behaviour of the mortar joints and unit-mortar interface is lumped in discontinuous elements, see Figure 1(c);
- **Macro-Modelling:** The units, mortar joints and unit-mortar interfaces are smeared out in a homogeneous anisotropic continuum; see Figure 1(d).

Detailed micro-modelling is probably the most accurate tool available today to simulate the real behaviour of masonry as the elastic and inelastic properties of both the units and the mortar can be realistically taken into account. With this method, a suitable constitutive law is introduced in order to

Figure 1. Modelling strategies for masonry: (a) typical masonry specimen; (b) detailed micro-modelling; (c) simplified micro-modelling; (d) macro-modelling
(Lourenço, 1996)

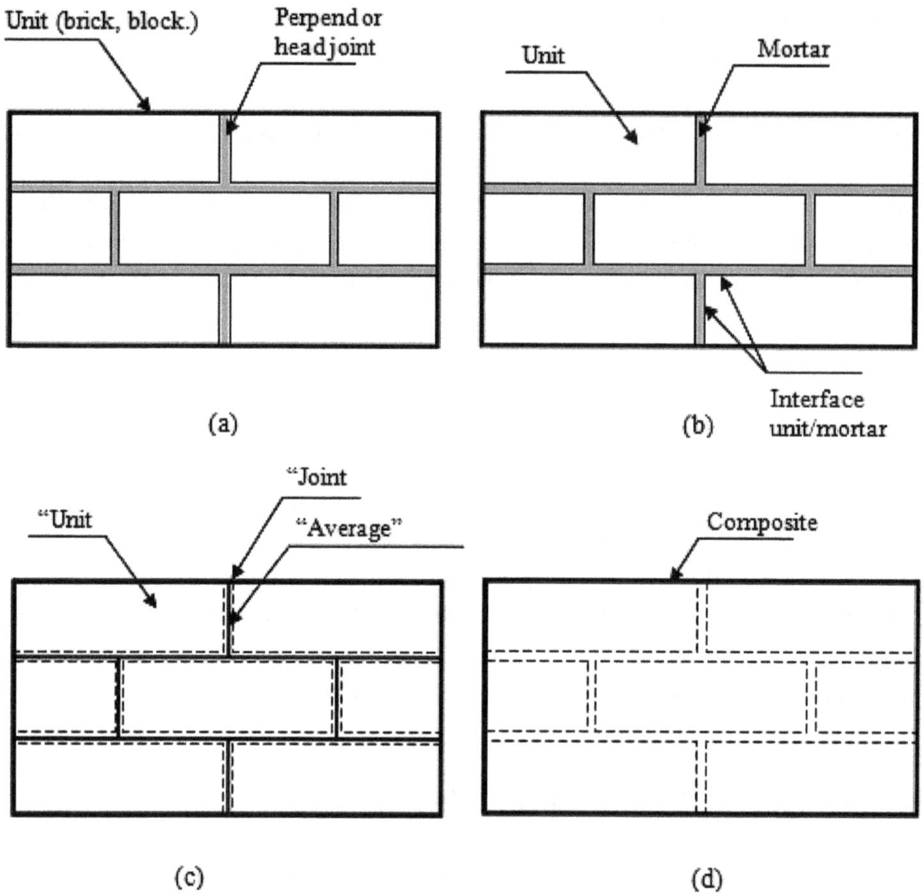

reproduce not only the behaviour of the masonry units and mortar, but also their interaction. However, any analysis with this level of refinement requires large computational effort to analyse. Thus, this method is mainly used to simulate tests on small specimens in order to determine accurately the stress distribution in the masonry materials (Lourenço & Pina-Henriques, 2006; Papa, 2001; Rots, 1991; Rots 1997; Zucchini & Lourenço, 2006).

The drawback of the large computational effort required by detailed micro-modelling is partially overcome by the simplified micro-modelling strategy. In this case, each joint, consisting of mortar and two unit-mortar interfaces, is lumped into an "average" interface while the units are slightly expanded in size in order to keep the geometry unchanged. With this approach it is possible to consider masonry as a set of elastic blocks bonded together by potential fracture slip lines at the joints. Cracking in the masonry units can also be simulated by assigning potential vertical zero thickness interfaces (unit to unit) at the unit's centre lines as shown in Figure 2.

The behaviour of the masonry is based on the knowledge of the properties of each constituent and the interface. Material properties are normally obtained from experimental tests on the masonry units,

Figure 2. Simplified micro-modelling strategy for masonry allowing for the failure of the masonry units

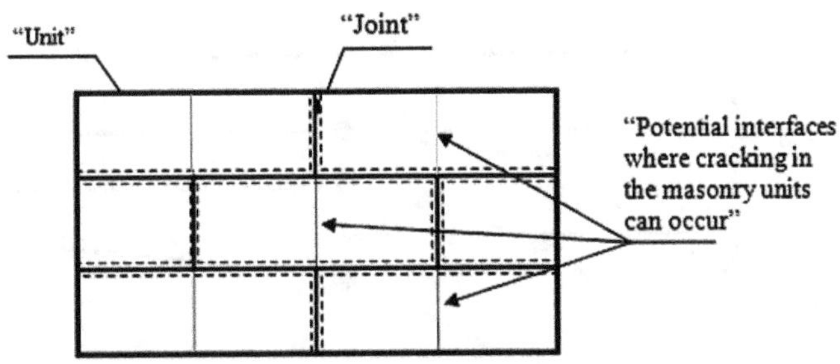

the mortar and on masonry samples. However, according to Lofti and Shing (1994), Gambarotta and Lagomarsino (1997a) and Lourenço and Rots (1997) accuracy is lost since Poisson's effect on the mortar cannot be included and, as a result, the brick-mortar interaction can only be partially described. None of the aforementioned researchers quantify the loss of accuracy resulting from the loss of Poisson's effect on the mortar joints. In order to overcome the exclusion of the Poisson's ratio for the mortar, Gambarotta and Lagomarsino (1997) suggested that the brick model should incorporate the compressive failure of masonry, which actually involves both bricks and mortar. Also, according to Schlegel and Rautenstrauch (2004a), the fact that the dimensions of the bricks have to be expanded slightly has an effect on the stiffness and rotation of the masonry units and thus the accuracy of the model is reduced. The reduction in accuracy is demonstrated in Figure 3. For a 100 mm high stone with a length to height (L_{st}/h_{st}) ratio of 3, an increase in the stone unit dimensions by 10 mm in each direction, to cater for the simplified micro-modelling strategy using average interfaces, would give an L_{st}/h_{st} value of 2.54 (Sarhosis 2012). According to Figure 3, this would reduce the ultimate load from 400kN (L_{st}/h_{st} = 3.0) to approximately 380 kN (L_{st}/h_{st} = 2.54), a reduction of 20 kN or 5%. If a similar relationship is assumed to be valid for clay brickwork, and the value of L_{st}/h_{st} for a clay brick is 215/65 or 3.3, allowing for a 10mm increase in the length and height values, to cater for 10mm thick mortar joints, gives a value of L_{st}/h_{st} of 225/75 = 3.0. Figure 3 indicates that the reduction in the ultimate load is also likely to be small, although it is acknowledged that the relationship presented by Schlegel and Rautenstrauch (2004a) was proposed for stones of at least a 100 mm thickness rather than the much smaller clay bricks. Finally, another issue that should be taken into consideration when modelling masonry, using the simplified micro-modelling strategy, is that the numerical problem quickly grows in complexity as the structure size increases.

With the macro-modelling approach there is no distinction between individual masonry units and the mortar joints. Masonry is considered as a homogeneous anisotropic composite such that the joints and any cracks are smeared out in the continuum. The homogeneous behaviour of masonry is described in terms of average stress and strains assuming different elastic and inelastic properties along the material axes (Lourenço, 1998a). Although this approach is very attractive for large-scale masonry structures, because of the reduced time and memory requirements as well as user friendly mesh generation, it is not adequate for detailed studies and for capturing failure mechanisms (Lourenço, 1996).

Along with advanced computational strategies and methods, artificial intelligence techniques such as artificial neural networks and cellular automata have become increasingly common in the field of structural engineering. Such techniques have been used by several researchers to model the behaviour of masonry structures (Liu et al., 2006; Rafiq et al., 1998; Zhou, 2002; Zhou & Rafiq, 2006). However,

Figure 3. Ultimate loads for different variations of stone height
(Schlegel & Rautenstrauch, 2004a)

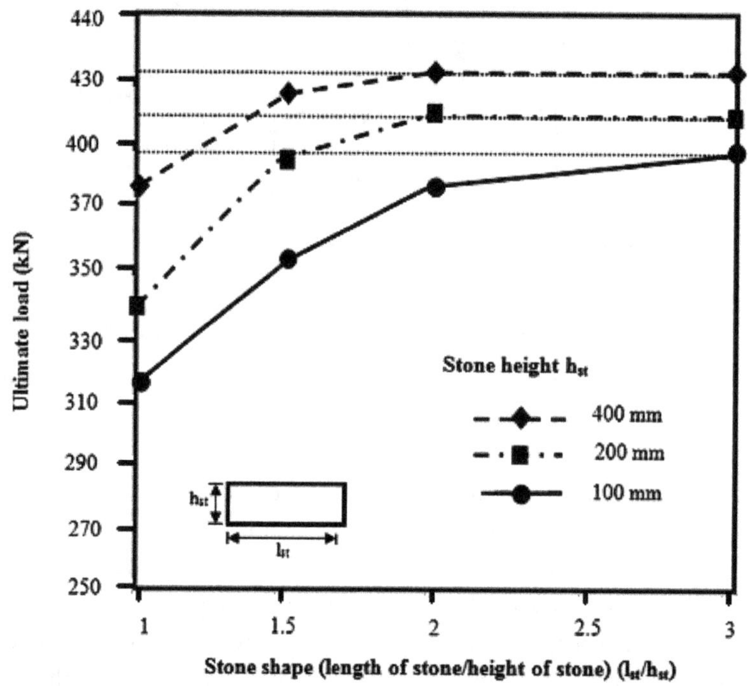

the following issues need to be highlighted when considering the use of artificial intelligence techniques to model masonry:

- The applications of artificial intelligence in the structural analysis of masonry have not yet been able to produce more accurate analysis than the conventional analytical techniques (Zhou & Rafiq, 2006);
- The current experimental data and records were designed for the conventional methodologies instead of those relating to artificial intelligence;
- Due to the variability of masonry properties even within the same structure, a large volume of experimental data is required to build an effective artificial intelligence model. However, this can be very computationally expensive (Zhang et al., 2010).

Following the descriptions above, it is clear that each modelling strategy has different fields of application. The simplified micro-modelling can give a better understanding of the local behaviour of low bond strength or historic masonry. The micro-modelling approaches are suitable for structural elements with strongly heterogeneous states of stress and strain. Their primary aim is to closely represent the masonry based on the knowledge of the properties of each constituent and the interface. Within simplified micro-modelling, initiation of cracks can be localized and crack propagation up to failure can be investigated. Also, the most observed failure mechanisms of masonry can be realistically simulated with reduced time and computer memory requirements (Lourenço, 2001; Lourenço, 2002; Roca et al., 2010; Rots, 1997).

ANALYSIS OF MASONRY STRUCTURES WITH INTERFACE ELEMENTS

The main methods available for modelling masonry structures using the simplified micro-modelling approach include: a) the discontinuous finite element method; b) the discrete element method; and c) the limit analysis method. Such methods are reviewed below with regard to their capability to represent masonry structures, computer efficiency, material input parameters availability and representation of real masonry structures. Consideration is also given to their suitability for carrying out the analysis of low bond strength masonry structures. In spite of the modern developments in engineering and computing, the understanding and simulation of masonry is still facing significant difficulties associated with the computational effort, the understanding of the behaviour of masonry, data acquisition, the accuracy of the constitutive laws and input material parameters and the limited realism of the available methods of analysis.

Discontinuum Finite Element Models

When modelling masonry with a discontinuum finite element model the mortar joints are represented as discontinuities where a potential crack, slip or crushing failure can occur. Discontinuities are generally introduced using interface elements, for which the constitutive model is in direct relation with the stress vector and the relative displacement vector along the interface (Oliveira, 2003). Thus, for an accurate simulation of masonry behaviour, it is essential to obtain a constitutive model for the interface elements which is able to capture realistically the behaviour of masonry and be able to simulate all the failure mechanisms. Page (1978) was probably the first researcher to use interface elements for modelling brick masonry. He assumed elastic behaviour of the bricks and a non-linear response for the joints. After Page, several constitutive models were developed based on the theory of plasticity (Chakrabarty, 2006) and damage mechanics (Voyiadjis & Kattan, 2005).

Lourenço (1996) developed a composite constitutive model for the static analysis of interface elements based on the theory of plasticity. The composite interface model includes a tension cut-off for tensile failure (mode I), a Coulomb friction envelope for shear failure (mode II) and a cap mode for compressive failure, see Figure 4.

Figure 4. Interface model for masonry
(Lourenço & Rots, 1997)

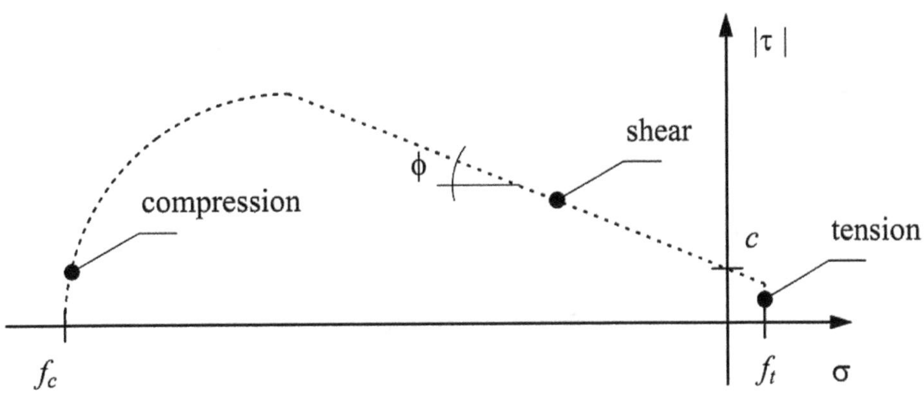

The structural element used by Lourenço in his study was a shear wall with an opening; see Figures 5(a) and 5(b). The numerical model was checked both qualitatively and quantitatively against experimental data and a high degree of correlation was found as shown in Figure 5(c). A complete discussion of the numerical results has been reported elsewhere (Lourenço et al., 1997; Lourenço, 1998a).

Also, Gambarotta and Lagomarsino (1997a) developed a constitutive model for mortar joints based on the principles of damage mechanics. Such model considers both mortar damage and debonding of the

Figure 5. Result of the deformed mesh analysis at a displacement of: (a) 1mm (b) 25mm; (c) load – displacement diagram
(Lourenço, 1996)

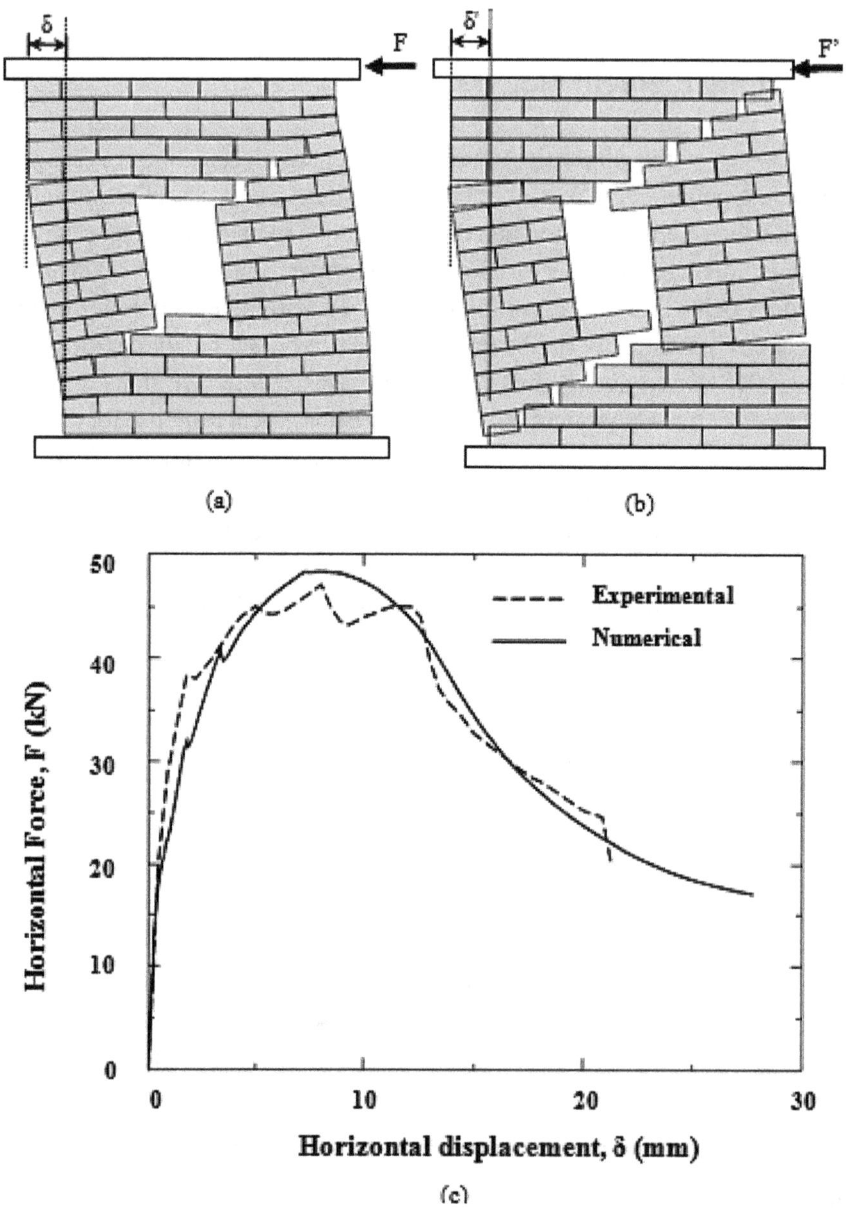

brick- mortar interface which takes place when opening and frictional sliding are activated. The inelastic strain components (i.e. sliding and opening of the mortar joints) are assumed to be linearly dependent on the mean stress and a damage variable related to the damage mechanics approach. Sliding of the units is limited by the presence of friction at the brick-mortar interface. This model has been applied to the analysis of brick masonry walls under constant vertical loads and horizontal in-plane cyclic forces. Although this approach has been found to simulate the inelastic behaviour of masonry (i.e. opening and sliding of joints), it was too computationally expensive to analyse full scale masonry wall panels.

Pegon et al. (2001) carried out an analysis of the historic masonry structure shown in Figure 6. They used interface elements ruled by an improved elasto-plastic Coulomb friction law with material softening and an allowance for dilatancy. The analysis was performed both in 2D and 3D using the computer code CASTEM 2000 (Millard, 2003). The 2D model was found to predict deformation and damage patterns quite accurately and, as a result, it was considered to be useful to assess the stability and damage of other structures under combined vertical and lateral loads. The 3D analysis provided an improved description of the behaviour of the column of the structure and was used to calibrate the material parameters for the joints found from the 2D analysis. In spite of the apparent success, Pegon et al. (2001) considered that the 3D modelling proved to be too time consuming especially when it was used to carry out a series of parametric studies. Difficulties were also experienced when attempting to mesh the stone blocks and joints.

Discrete Element Models

The discrete element method (DEM) is characterized by modelling the materials as an assemblage of distinct blocks or particles interacting along their boundaries. According to its developers, Cundall (1971), the name "discrete" only applies to a computer approach if: a) it allows finite displacements and rotations

Figure 6. Monastery San Vicente de Fora, Lisbon (left) and deformation pattern using interface elements with CASTEM 2000 (right)
(Pegon et al., 2001)

of discrete bodies and b) new contacts between the blocks or particles are automatically recognized and updated as the calculation progresses. The formulation of the method was proposed initially by Cundall (1971) for the study of jointed rock, modelled as an assemblage of rigid blocks. Later this approach was extended to other fields of engineering requiring a detailed study of the contact between blocks or particles such as soil and other granular materials (Ghaboussi & Barbosa, 1990). More recently the approach was applied successfully to model historic masonry structures by Alexandris (2004); Lemos (1995) and Sincraian and Lemos (1999) in which the collapse modes were typically governed by mechanisms in which the deformability of the blocks plays little or no role. Also, the possibility of frequent changes in the connectivity and the type of contact as well as marked non-linearity induced by the inability of the masonry joints to withstand tension makes DEM a suitable method for solving problems involving discontinuities as is the case with low bond strength masonry. The Distinct Element Method, the Discrete-Finite Element Method, Discontinuous Deformation Analysis (DDA) and the particle flow models are all different formulations of the DEM with important applications to masonry structures. These are considered in more detail below.

Distinct Element Method

The distinct element method is an explicit method based on finite difference principles, derived from Cundall's original work. It is presented in the UDEC (Universal Distinct Element Code) and 3DEC software, developed for commercial use by Itasca Ltd for either the static or dynamic analysis of 2 and 3-dimensional structures (Itasca 2004a, 2004b). In the distinct element method masonry bricks or blocks are represented as an assembly of rigid or deformable blocks which may take any arbitrary geometry. Rigid blocks do not change their geometry as a result of any applied loading. Deformable blocks are internally discretised into finite difference triangular zones. These zones are continuum elements as they occur in the finite element method (FEM). However, unlike FEM, in the distinct element method a compatible finite element mesh between the blocks and the joints is not required. Mortar joints are represented as zero thickness interfaces between the blocks. Representation of the contact between blocks is not based on joint elements, as occurs in the discontinuum finite element models. Instead the contact is represented by a set of point contacts with no attempt to obtain a continuous stress distribution through the contact surface. The assignment of contacts allows the interface constitutive relations to be formulated in terms of the stresses and relative displacements across the joint. As with FEM, the unknowns are the nodal displacements and rotations of the blocks. However, unlike FEM, the unknowns in the distinct element method are solved explicitly by differential equations from the known displacement while Newton's second law of motion gives the motion of the blocks resulting from known forces acting on them. So, large displacements and rotations of the blocks are allowed with the sequential contact detection and update of tasks automatically. This differs from FEM where the method is not readily capable of updating the contact size or creating new contacts. This method is also applicable for quasi-static problems using artificial viscous damping controlled by an adaptive algorithm. Typical examples of distinct element analysis carried out for masonry structures using UDEC and 3DEC include:

- **Masonry Wall Panels:** Dialer (1992); Rots (1997); Dialer (2002); Zhuge (2002); Schlegel and Rautenstrauch (2004a); Churilov and Dumova-Jovanoska (2008); Sarhsosis (2014a); Sarhosis (2015).
- **Dry Stone Retaining Walls:** Walker et al. (2006);

- **Foundation Movements in Houses:** Zhuge et al. (2003);
- **Narrow Pier:** Rots (1997);
- **Masonry-Infilled Steel Frames with Opening:** Mohebkhah et al. (2008);
- **Stone Masonry Arches and Aqueducts:** Lemos (1998), see Figure 7; Sincraian and Lemos (1999); Roberti (2001); Roberti and Calvetti (1998); Lemos (2004); Tóth et al. (2009); Tran et al. (2014); Sarhosis (2014b); and
- **Column-Architrave Structures Under Seismic Action:** Papastamatiou et al. (1993); Papantonopoulos et al. (2002); Psycharis et al (2003); Stefanou (2011).

The drawbacks of the method are the unavailability of sophisticated constitutive models for the internal elements when deformable blocks are used and the difficulty in obtaining representative values for the material properties from small-scale experiments.

Discontinuous Deformation Analysis (DDA)

Discontinuous deformation analysis (DDA) is a discrete element method developed by Shi and Goodman (1988) for solving stress-displacement problems for a jointed rock mass. Later, the method was applied successfully to model masonry arches by Ma et al. (1996); Bićanić et al. (2001) and Bićanić et al. (2003), see Figure 8. With DDA, the blocks are considered to be deformable but with a uniform stress and strain distribution. Contacts are considered rigid (the "hard contact approach", Cundall & Hart, 1989) and no interpenetration of blocks is permitted. This is opposed to the distinct element method where a "soft contact" approach is adopted and blocks can overlap when they are in compression. The hard contact approach is limited as shear displacement occurs only if the blocks slide. With the soft contact approach the contact forces are a function of the relative shear forces and sliding which is more realistic when modelling masonry (Lemos, 2007). Also, DDA uses an implicit algorithm based on a global stiffness matrix (unlike UDEC), for simultaneous solutions of the equations of equilibrium by minimizing

Figure 7. The deformed shape of a masonry arch bridge modelled using 3DEC
(Lemos, 1998)

the potential energy of the blocky rock mass system. Although DDA and the distinct element method (DEM) are similar in: a) the sense of representation of contacts; b) both can simulate the behaviour of interacting discrete bodies; and c) they can recognize new contacts between bodies during calculations, they are quite different. Their fundamental differences are represented in Table 1, (Bićanić et al., 2001; Bobet et al., 2009).

Particle Flow Models

Particle-flow models are another type of discrete element method that has been receiving increased attention in various fields of engineering. Particle-flow models simulate the mechanical behaviour of a system comprised of a collection of arbitrary shaped particles. They were initially proposed to model the micro-mechanical behaviour of granular material such as soils (Cundall & Strack, 1979). Particle-flow models have also been used to model masonry structures (Thavalingam et al., 2001; Tóth, 2004). The particle flow code (PFC) is the commercial software developed by Itasca Ltd for the analysis of 2D and 3D cases (Itasca PFC 2004). PFC models the movement and interaction of particles using the distinct element method, described earlier. Discs or spherical shaped particles are used for the 2D and 3D

Figure 8. Deformed shape of an arch bridge using discontinuous deformation analysis with simplified deformable blocks
(Bićanić et al., 2003)

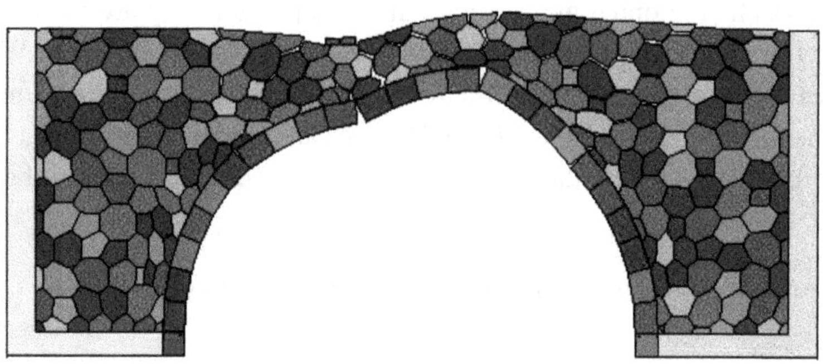

Table 1. Differences between DEM and DDA

Distinct Element Method (DEM)	Discontinuous Deformation Analysis (DDA)
• Each block is discretised into the FE mesh; • Each block is treated separately during the analysis; • The unknowns are displacements, usually obtained from the velocities (of rigid block centroids, or deformable block grid-points). • The strains are calculated from displacements, and the stresses from strains; • Contacts are resolved by defining the contact displacements and forces in terms of block overlap; • Uses an explicit procedure to solve the equilibrium equation; • Unbalanced forces drive the solution process and damping is used to dissipate energy.	• Each block has a uniform stress state; • The displacements are the unknowns; • Interpenetration of blocks is prevented by adding springs to the contacts; • Uses implicit method to achieve equilibrium; • As a technique of discontinuous analysis it resembles and follows the procedures developed for FEM; • Total potential energy of the system is minimized to find the solution.

cases, respectively, and the particles can be displaced independently from one another and interact only at contacts or interfaces between the adjacent particles. The particle models are computationally very efficient because detection of any contact between the discs or spheres requires only the calculation of the distance between the particle centres. However, as the size of the structure to be analysed increases, these calculation tasks can be very time consuming. Parallel processing techniques are required in order to make this model more attractive in the future. Typical analysis of masonry arch bridges carried out with PFC^{2D} and PFC^{3D} are illustrated at Figures 9 and 10 respectively. The masonry units are represented by larger particles with smaller particles representing the mortar. Unit/mortar interface effects are simulated by assigning different bond strengths between the two types of particles. Further background information on the theory and applicability of the PFC software can be found elsewhere (Itasca PFC, 2004).

Figure 9. Semi-circular masonry arch as derived from analysis with PFC^{2D}
(Tóth, 2004)

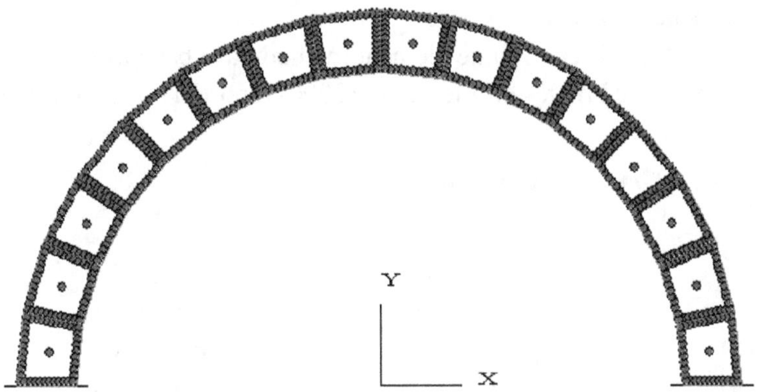

Figure 10. Model of a backfilled semi-circular masonry arch with PFC^{3D}
(Thavalingam et al., 2001)

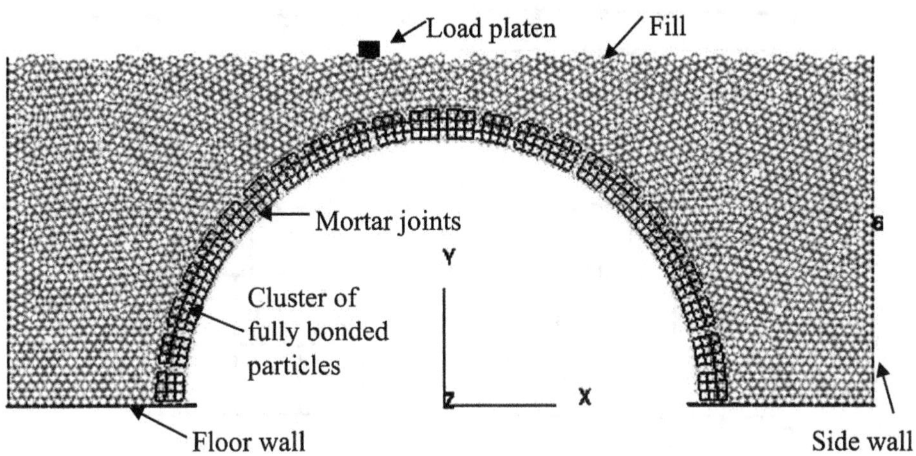

Discrete-Finite Element Models

As the name implies, discrete-finite element models are those that combine features of the DE and FE methods. They have been developed to reduce the computational cost. The method considers deformable blocks represented by a mesh of triangular elements which may split and separate during the analysis. Like the finite element technique, non-linear material models are used to define the characteristics of the masonry units and the mortar. Munjiza et al. (1995) developed a method based on fracture mechanics criteria for simulating fracturing problems. Later, Mamaghani (1999) used a standard FE formulation for the block representation. Having assembled the global stiffness matrix, iterations are performed at each load increment to account for non-linear behaviour. The method has been applied to the stability analysis of different masonry structures such as masonry arches and pyramids, see Figure 11.

Also, Owen et al. (1998) used finite and discrete elements to assess the ultimate load carrying capacity of a damaged two span masonry arch bridge and to examine the efficiency of alternative repair strategies. Masonry blocks were represented by deformable discrete elements while the fill was represented by spherical discrete elements. The structure was subjected to abnormal vehicular loading in order to determine the load distribution through the backfill and into the arch rings. The numerical load–displacement prediction of a three ring arch bridge with an anchor stitching system was also presented, and compared with experimental results.

Finally, applications of the discrete finite element method have been used to design the strengthening measures for masonry arch bridges. Mullett et al. (2006) reported on the development of a model to

Figure 11. Failure mode of a masonry pyramid
(Mamaghani et al., 1999)

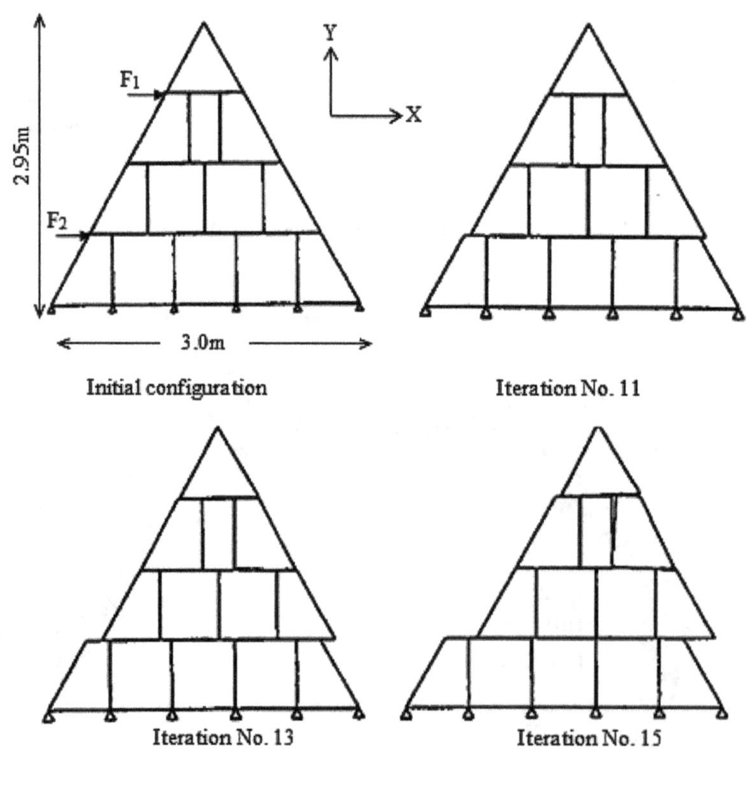

assess the load-carrying capacity of a masonry arch bridge and to identify the suitability of the Archtec strengthening system. The analysis was carried out using the explicit mode of the ELFEN (Rockfield engineering) package which was developed by Owen in 1986; see Figure 12.

Limit Analysis Models

Limit analysis models for masonry structures consider the static or kinematic equilibrium of rigid blocks connected through joints. According to Orduña and Lourenço (2003), the following hypotheses are normally adopted: a) limit load occurs at small overall displacements, which is true for most cases and b) masonry has zero tensile strength, which can be justified with the relatively low tensile strength of mortar joints in historic construction. In reality, however, zero tensile strength only occurs for dry joints. Experimental works carried out by several researchers (Garrity, 2010, Van der Pluijm, 1999) have clearly demonstrated that; a) shear failure at the joint is perfectly plastic, an assumption fully supported by experimental results (Van der Pluijm, 1999); and b) hinging failure mode at a joint occurs for a compressive load independent from the rotation. This hypothesis is debatable as crushing behaviour has minor importance in the response of masonry structures except for very shallow arches, pillars, towers and massive vertical structures (Roca et al, 2010). Based on the pioneering work of Heyman (1969), Livesley (1978) first developed the limit analysis method for simulating discretised behaviour in masonry structures. According to Livesley's method, rigid blocks, typically larger than the physical blocks to account for the zero thickness interface, are analysed using either the equilibrium or kinematic (work) form and are then solved using linear programming techniques (Ahmed & Gilbert, 2003; Vanderbei, 2001). Within non-standard limit analysis, block interfaces can be regarded as elements and the blocks as nodes. The equilibrium and kinematic compatibility equations are formulated in terms of finite sets of quantities (i.e. nodal displacements and forces or element stresses and strains). So, if the equilibrium form is formulated, the displacement and hinge rotations will become available. Many other researchers have applied limit analysis for the assessment of masonry structures. Gilbert and Melbourne (1994) applied the kinematic theorem in order to analyse a masonry arch bridge. Begg and Fishwick (1995) have released the restriction of the associated flow rule or the normality condition for masonry arches, including shear failure. Later, Baggio and Trovalusci (1998) noted that extending the formulation to non-

Figure 12. Finite/discrete element method applied to a masonry arch bridge
(Mullett et al., 2006)

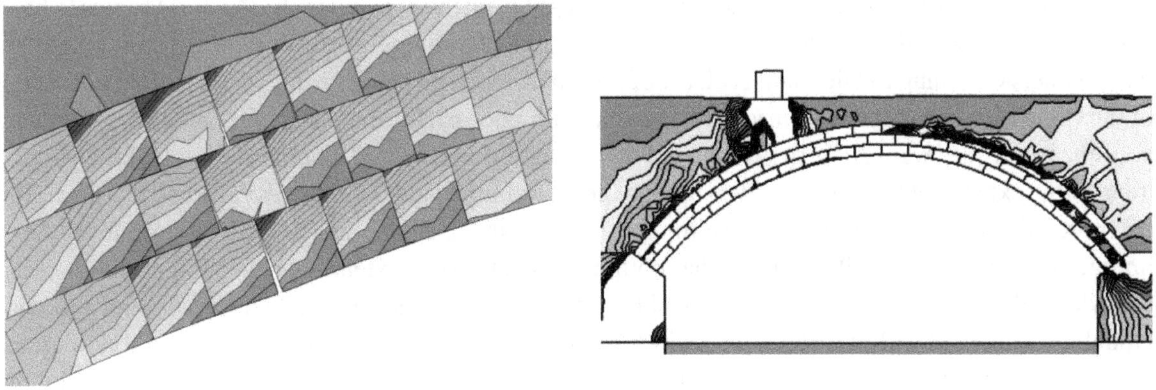

associate flow results in a non-linear mathematical problem of significantly larger size in comparison with the lower size linear problem resulting from classical theory. Ferris and Tin-Loi (2002) analysed a shear wall panel with an opening. The model involved rigid block assemblages, characterized by frictional (non-associative) and tensionless interfaces. The method was validated against results obtained from shear wall tests on panels with an opening. Also, Orduna and Lourenço (2003) proposed a cap model which takes into account the non-associated flow rules and the limited compressive strength of masonry. Later, Gilbert et al. (2006) carried out a limit analysis procedure for rigid block assemblages using a non associative frictional joint model to account for sliding between blocks. Contacts were governed by a modified Mohr-Coulomb failure surface. Failure surfaces were updated at each iteration until a converged solution was obtained. The model provided reasonable estimates of the ultimate load carrying capacity for a wide range of problems and has proved to be particularly efficient for solving large masonry structures such as façades and arch bridges. An extensive review of limit analysis applied to masonry arch bridges is presented elsewhere (Gilbert, 2007).

COMPARATIVE STUDIES

As described previously, there are several models available to model low bond strength masonry that are based on the simplified micro-modelling approach. Separate comparative studies to identify the capabilities and limitations of each computational model have been carried out by several researchers. Such studies are mainly focused on comparing the results of large scale experiments against those obtained from the different computational models under review.

Thavalingam et al. (2001) analyzed and compared the load-carrying capacity of an experimental backfilled masonry arch bridge model using a non-linear finite element technique with joint/interface elements (FEM based DIANA) and two models derived from the discrete element method (DDA and PFC). The load displacement curves obtained from the numerical simulations are shown in Figure 13. Of the three models, PFC seems to give better predictions of the collapse load obtained from the experimental study. The curves also show the ability of the discrete models to simulate the post peak structural response of a masonry arch bridge.

Furthermore, Giordano et al. (2002) investigated the applicability of the non-linear FEM (based on CASTEM 2000) and the Distinct Element method (based on UDEC) by comparing the computed results with those obtained from experimental tests on a full-scale masonry specimen forming part of the cloisters façades of the Sao Vicente de Fora monastery in Lisbon. Figure 14 compares the experimental and the numerical results. The load-displacement curve obtained from the analysis with CASTEM seems to be in good agreement with the experimental envelope, although it is slightly stiffer compared with the experimental one. According to Giordano et al. (2002), such differences could be due to the simple modelling of the stone blocks and the masonry infill, meshed with poor-performing triangular elements, in which non-linearity is allowed. Also, the DE model developed by Lemos (2001) reproduces well the general trend of the experimental envelope, even though the maximum load is slightly overestimated. The reason for this could be due to the simple modelling of the masonry in-fill by linear elastic blocks. In spite the limitations of each model, it is evident that both methods, to some extent, managed to grasp the global behaviour of the experimental test and thus both of them can be used to model masonry. For the joint model implemented with the FEM, the main difficulty is the need to provide an easy methodology for re-meshing contacts when large displacements are allowed. Such a limitation is easily overcome by the

Figure 13. Comparison of experimental against numerical results
(Thavalingam et al., 2001)

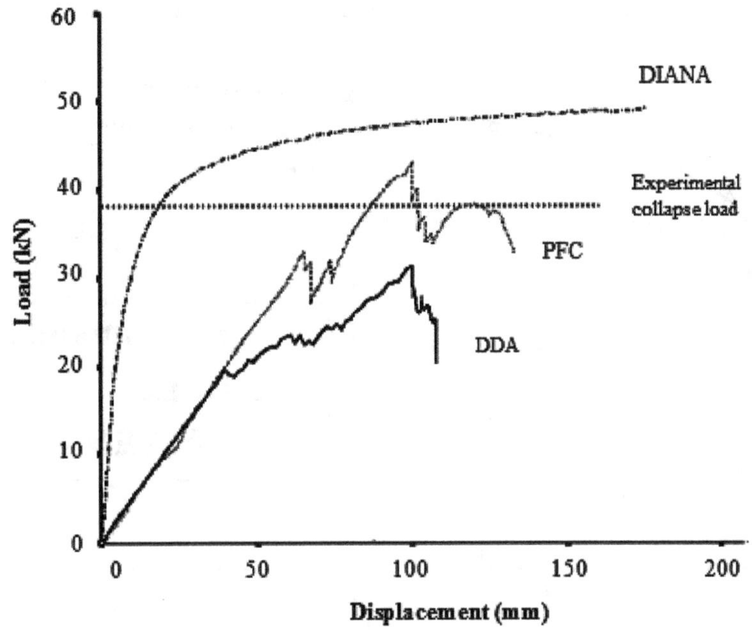

DE method which is able to accommodate compatible meshes and large displacements. The main drawback identified for the DE model is the poor constitutive law for the internal elements when deformable blocks are taken into account. However, this is not the case when modelling low bond strength masonry where deformability of the units plays little or no role at all in the behaviour and the predominant mode of failure occurs in the mortar joints or at the unit/mortar interfaces.

In order to assess the suitability of the distinct element method for masonry, Zhuge (2002), carried out numerical studies with UDEC on shear wall panels with an opening, similar to those carried out by Lourenço (1996) using the discontinuum finite element method, see Figure 5. The results obtained compare well with those obtained experimentally and when using the finite element method. UDEC managed to capture the tensile and shear failure of the joints as well as the final collapse mode, although it failed to predict the possible compressive failure at the bottom of the pier. Suggestions were made to incorporate an orthotropic material model and a failure criteria for the blocks.

Schlegel and Rautenstrauch (2004b) carried out comparative studies of two different computational models. A continuum model (ANSYS) was compared with a discontinuum model (3DEC) to simulate the behaviour of a masonry arch bridge with spandrel walls under vertical in-plane load. It was found that the ultimate load capacity of the bridge predicted using the continuum model was 5% less than that predicted using 3DEC. According to Schlegel and Rautenstrauch (2004b), the difference in the results arises from the discontinuum blocking effects. Also, it was found that as the stone size in the arch was increased, so the vertical load that the masonry arch could carry for a given vertical displacement was increased. The results of the study are presented in Figure 15.

Comparative computational studies on a stone masonry wall with dry joints unable to carry tension and resistant to sliding by friction have been reported by Baggio and Trovalusci (1995). The non-linear finite element analysis (ANSYS), the distinct element method (UDEC) and a non-standard limit analy-

Figure 14. Comparison of experimental against numerical results
(Giordano et al., 2002)

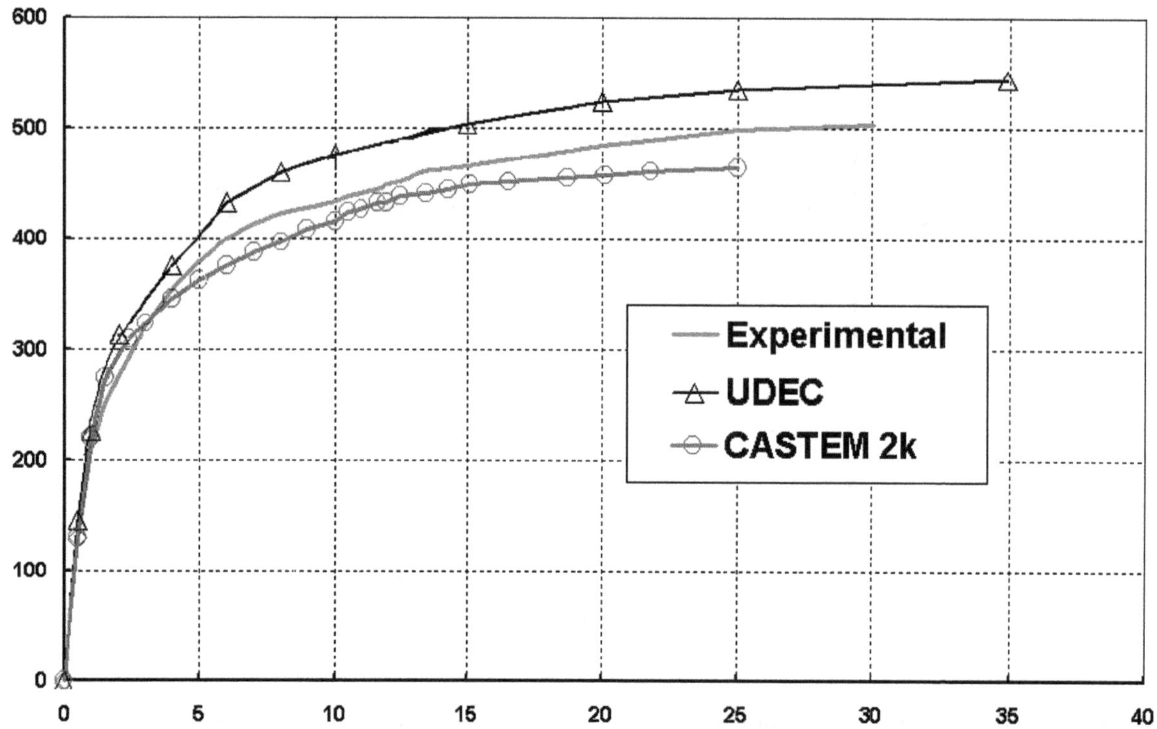

Figure 15. Load against displacement relationship
(Schlegel & Rautenstrauch, 2004b)

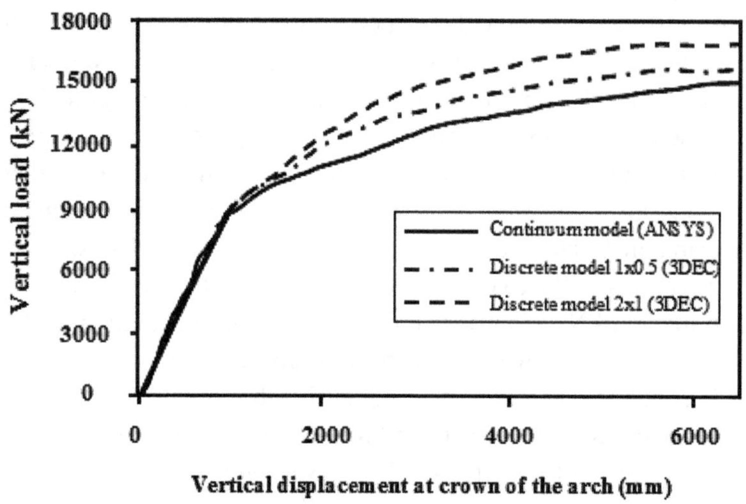

sis (LA), restoring the analogy of the non-associated flow rules (Boothby, 1994; Livesley, 1992) were compared. The results were compared qualitatively against experiments and good agreement was found for all the methods, see Figure 16.

A study on a small masonry wall panel using a non-linear finite element analysis and a non-standard limit analysis, Baggio and Trovalusci (1993), concluded that the FE approach better succeeds in the evaluation of the strength of the walls because the imperfections of the actual structure (previously defined as "fictitious elastic" effects) are better modelled by deformable joint elements. In contrast, the limit analysis method better describes the collapse mechanism and gives an evaluation of the upper limit strength value of an ideal system.

DISCUSSION OF THE APPROPRIATE MODEL TO USE FOR SIMULATING THE MECHANICAL BEHAVIOUR OF MASONRY

From the above review of the literature and the comparative studies carried out by several researchers, the advantages and limitations of each method of computational modelling have been identified. These are summarised in Tables 2 to 4 below.

Figure 18 compares four of the most common methods used to model masonry with respect to the ability to: a) define the initial state; b) availability of input data; c) applicability to simulate crack propagation and damaged masonry; d) ease of model implementation, and d) the computing costs. The differences are shown qualitatively using the size of the marker. The larger markers represent beneficial

Figure 16. Wall panel shear test results from: a) Experiment; b) Non-linear FE analysis; c) Limit analysis; d) Distinct element method
(Baggio & Trovalusci, 1995)

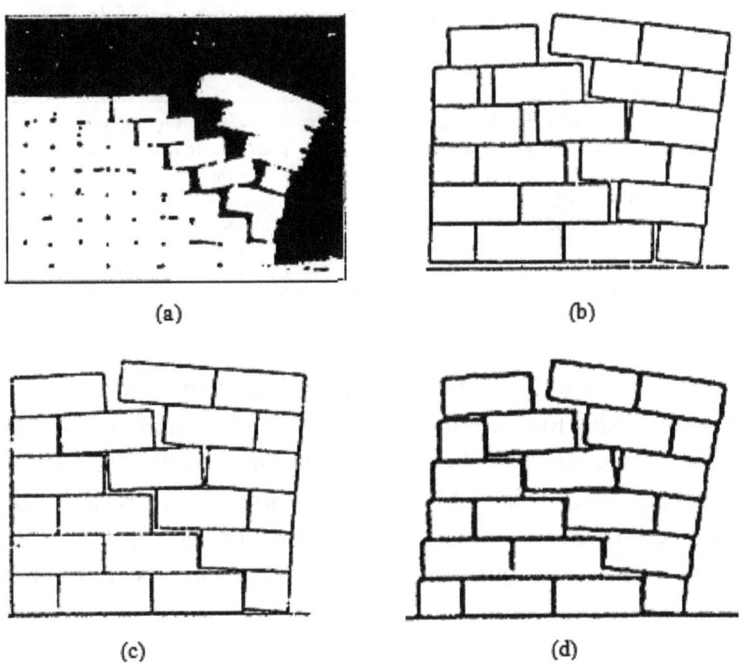

Table 2. Comparison of micro-modelling approaches: discontinuum finite element models

	Capabilities	Limitations
Discont. finite element models	• Can give good predictions of structural behaviour; • Suitable for small deformations of the unit/joint interfaces.	• Computationally expensive when dealing with large structures; • Difficulty to mesh blocks and joints when complex block arrangements are modelled or 3D cases; • The increase of sophistication of the models has largely increased the number of parameters necessary for the material models but dilation effects are not included; • When large displacements occur in the interfaces it is not possible to provide easy re-meshing, update existing contacts and/or create new ones; • Local stress singularities at the corners of the masonry units often occur creating inaccuracies; • No warning of masonry unit overlap; • The user needs to incorporate the interface constitutive law into the FE model; • Pre-defined crack patterns are required.

Table 3. Comparison of micro-modelling approaches: discrete element models

	Capabilities	Limitations
Discrete element models	• Large displacements and rotations of the units are allowed; • Ability to mesh the blocks independently, without the need to match nodal points (as in FEM), see Figure 17; • Suitable for parallel processing; • Same algorithm for static and dynamic analysis; • Capable of modelling post peak behaviour; • Overlap of units is restricted and dilation effects are included.	• Difficult to code; • Poor performance in terms of stress distribution accuracy (Giordano et al., 2002); • Computationally expensive when dealing with large structures.

Table 4. Comparison of micro-modelling approaches: limit analysis models

	Capabilities	Limitations
Limit Analysis models	• Suitable for the casual users and practising engineers; • Able to simulate fairly good the structural behaviour of masonry; • Reduced number of input material properties; • Easy to discuss and understand alternative solutions; • Quick method.	• As the complexity of the structure increases so does the time to obtain results; • Suitable only for dry joint masonry due to the assumption of zero tensile strength; dilation effects are not included; • The lack of customized packages increases the time to prepare the model.

characteristics while the small ones less beneficial ones. Also, from Figure 18, it is observed that standard FEM approaches are lacking to model accurately low bond strength or historic masonry where the predominant mode of failure occurs at the brick-mortar interface, while DEM is extremely well suited, at least on static and pseudo-static loading. Also, the DDA method is nearly as good as the DEM, except in the case of pre-deformed structures or when large deformations are expected. The DEM also has the advantage of being independent upon mesh size. The LAM method has a wider range of limitations.

Figure 17. Contact representation between blocks for the discontinuum finite element model and distinct element models
(Lemos 2007)

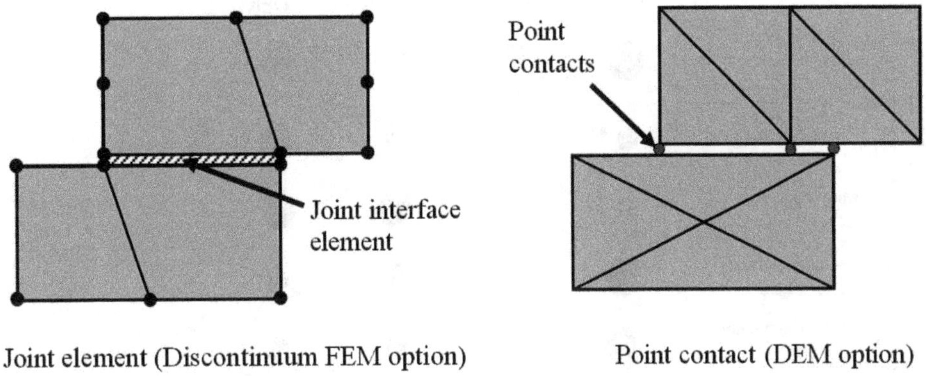

The complexity of DEM and DDA model generation, computational effort, and high input data requirements are all sources for concern. When properly applied, it does however perform well, especially for establishing a masonry structure's initial state of stress.

CONCLUSION

A review of the literature on the available approaches that can simulate the mechanical behaviour of low bond strength masonry under static load has been carried out. Many different computational methods are available to model masonry. It should be expected that the use of different methods will lead to different results depending on the adequacy of the approach and the information available. For a computational model to accurately represent the behaviour of masonry, both the constitutive model and the input material properties must be chosen carefully by the modeller in order to take into account the variation of masonry properties and the range of stress state types that exists in masonry structures. At present, there are two main computational approaches used to model brick masonry namely, micro-modelling (detailed or simplified) of the brick units and mortar, or macro-modelling of masonry as a composite. Detailed micro-modelling is probably the most accurate tool available today to simulate masonry behaviour. However, the method is only suitable to model small specimens of masonry as it requires a large computational effort to complete the analysis. On the other hand, the macro-modelling approach is suitable for large scale masonry structures, owing to the reduced time and memory requirements as well as user friendly mesh generation; it is not adequate for detailed studies and for capturing failure mechanisms. A simplified micro-modelling approach can give a good representation of the local behaviour of low bond strength masonry. With the simplified micro-modelling approach expanded units are modelled as continuous elements while the behaviour of the mortar joints and unit-mortar interface is lumped in discontinuous elements. The initiation of cracks can be localized and crack propagation up to failure can be investigated. Also, the most common failure mechanisms of masonry can be realistically simulated with reduced time and computer memory requirements (Lourenço & Rots 1997). The main available

Figure 18. Comparison of the main methods of analysis for masonry
(after Morabadi & Laefer, 2014).

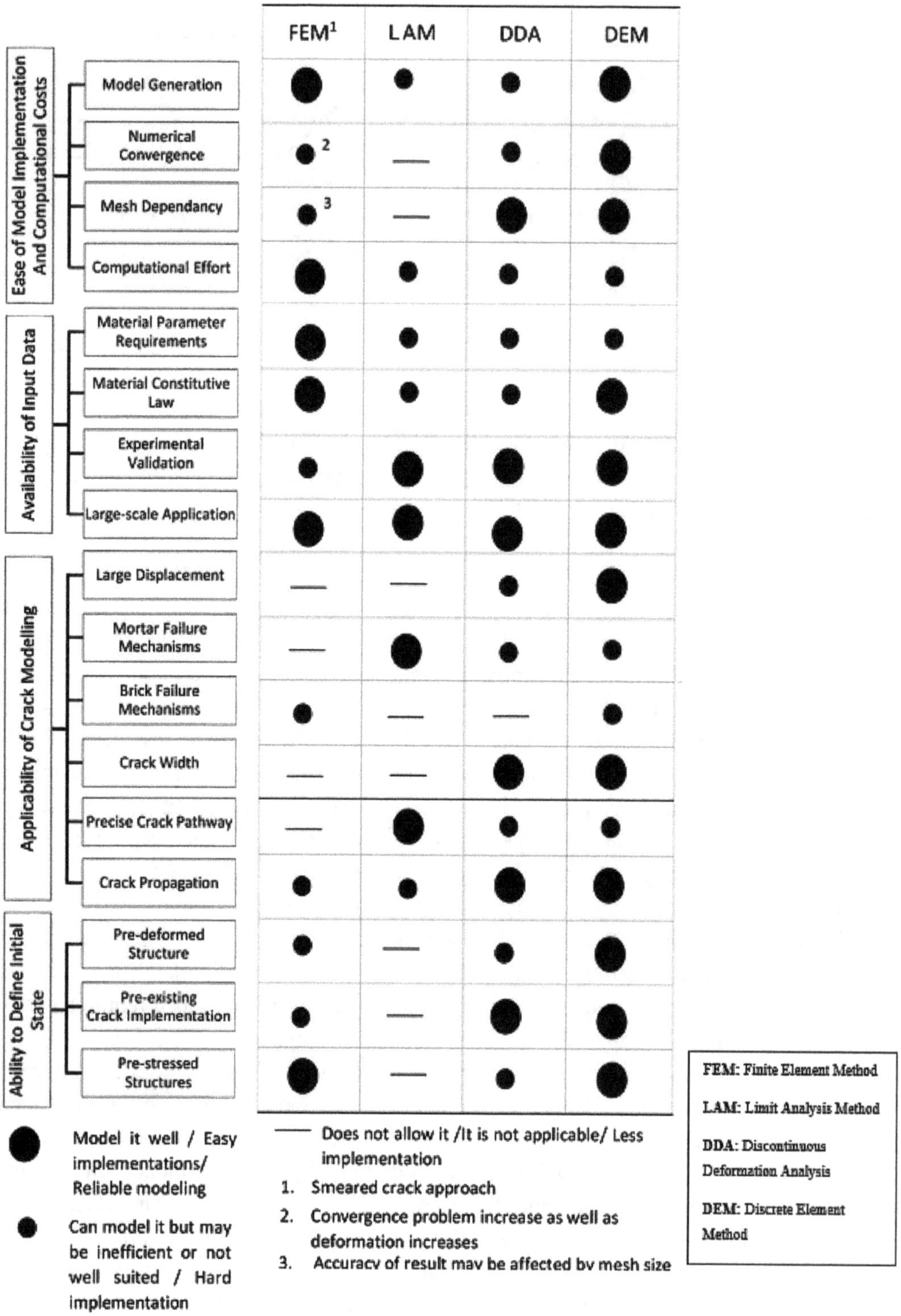

methods for modelling masonry structures with the simplified micro-modelling approach include: a) the discontinuous finite element method; b) the discrete element method; and c) the limit analysis method. By comparing each method separately, the Distinct Element Method (DEM) found to be the preferred method to use where block deformability plays little or no part in the behaviour and large displacements and rotations of the blocks is required. This represents the common situation with low bond strength masonry and historic masonry. The available models derived from the Discrete Element method are: a) the distinct element; b) the discontinuous deformation analysis; c) the particle flow codes; and d) the combined FE/DE method. Although DEM seems to be appropriate in most of the cases, especially for avoiding convergence problems under large displacements, it is less effective when computational effort and material parameter requirements are considered.

REFERENCES

Ahmed, H. M., & Gilbert, M. 2003. The computational efficiency of two rigid block analysis formulations for application to masonry structures. In *9th International Conference on Civil and Structural Engineering Computing*. Egmond-aan-Zee, Holland: Civil-Comp Press.

Alexandris, A., Protopapa, E., & Psycharis, I. 2004. Collapse mechanisms of masonry buildings derived by the distinct element method. In *13th World Conference on Earthquake Engineering*.

Baggio, C., & Trovalusci, P. 1993. Discrete models for jointed block masonry walls. In *The Sixth North American Masonry Conference*. The Masonry Society TMS.

Baggio, C., & Trovalusci, P. (1995). Stone assemblies under in-plane actions comparison between non-linear discrete approaches. In J. Middleton & G.N. Pande (Eds.), Computer Methods in Structural Masonry – 3. Books and Journals International Ltd.

Baggio, C., & Trovalusci, P. (1998). Limit analysis for no-tension and frictional three-dimensional discrete systems. *Mechanics of Structures and Machines*, *26*(3), 287–304. doi:10.1080/08905459708945496

Begg, D. W., & Fishwick, R. J. (1995). Numerical analysis of rigid block structures including sliding. In: J. Middleton & G.N. Pande (eds.), *Proceedings of the 3rd International Symposium in Computational Methods Structural Masonry*. Books and Journals International.

Bićanić, N., Ponniah, D., & Robinson, J. (2001). Computational modelling of masonry, brickwork and blockwork structures. In J. W. Bull (Ed.), *Discontinuous Deformation Analysis of Masonry Bridges* (pp. 177–196). Edinburgh, UK: Saxe-Coburg Publications.

Bićanić, N., Stirling, C., & Pearce, C. J. (2003). Discontinuous modelling of masonry bridges. *Computational Mechanics*, *31*(12), 60–68. doi:10.1007/s00466-002-0393-0

Bobet, A., Fakhimi, A., Johnson, S., Morris, J., Tonon, F., & Yeung, M. R. (2009). Numerical models in discontinuous media, review of advances for rock mechanics applications. *Journal of Geotechnical and Geoenvironmental Engineering*, *135*(11), 1547–1561. doi:10.1061/(ASCE)GT.1943-5606.0000133

Boothby, T. E. (1994). Stability of masonry piers and arches including sliding. *Journal of Engineering Mechanics*, *120*(2), 304–319. doi:10.1061/(ASCE)0733-9399(1994)120:2(304)

Chakrabarty, J. (2006). *Theory of plasticity*. London, UK: Butterworth Heinemann.

Cook, D. A., & Pegam, G. (1993). Analysis of crack damage in three Georgian facades in the city of Bath, England, with a simple approach to the repair problem. In Structural repair and Maintanence of historical building III. Computational Mechanics Publications.

Cundall, P. A. (1971). A computer model for simulating progressive large scale movements in blocky rock systems. I: *Proceedings of the Symposium of the International Society of Rock Mechanics*.

Cundall, P. A., & Hart, R. D. (1989). Numerical modelling of discontinua. In *Proceedings of the 1st US Conference on Discrete Element Methods*. CSM Press.

Cundall, P. A., & Strack, O. D. L. (1979). A discrete numerical model for granular assemblies. *Geotechnique*, *29*(1), 47–65. doi:10.1680/geot.1979.29.1.47

Dialer, C. (1992). A distinct element approach for the deformation behaviour of shear stressed masonry panels. In *6th Canadian Masonry Symposium*. University of Saskatchewan.

Dialer, C. (2002). Typical masonry failures and repairs, a German Engineer's view. *Progress in Structural Engineering and Materials*, *4*(3), 332–339. doi:10.1002/pse.125

Ferris, M. C., & Tin-Loi, F. (2002). Limit analysis of frictional block assemblies as a mathematical program with complementary constrains. *International Journal of Mechanical Sciences*, *43*(1), 209–224. doi:10.1016/S0020-7403(99)00111-3

Gambarotta, L., & Lagomarsino, S. (1997a). Damage models for the seismic response of brick masonry shear walls. Part I: The mortar joint model and its applications. *Journal of Earthquake Engineering and Structural Dynamics*, *26*(4), 423–439. doi:10.1002/(SICI)1096-9845(199704)26:4<423::AID-EQE650>3.0.CO;2-#

Garrity, S. W., Ashour, A. F., & Chen, Y. (2010). An experimental investigation of retro-reinforced clay brick arches. In *Proceedings of the 8th International Masonry Conference*.

Ghaboussi, J., & Barbosa, R. (1990). Three-dimensional discrete element method for granular materials. *International Journal for Numerical and Analytical Methods in Geomechanics*, *14*(7), 451–472. doi:10.1002/nag.1610140702

Gilbert, M. (2007). Limit analysis applied to masonry arch bridges, state of the art and recent developments. In *ARCH'07 - 5th International Conference on Arch Bridges*. University of Minho.

Gilbert, M., Casapulla, C., & Ahmed, H. M. (2006). Limit analysis of masonry block structures with non-associative frictional joint using linear programming. *Computers & Structures*, *84*(1), 873–887. doi:10.1016/j.compstruc.2006.02.005

Gilbert, M., & Melbourne, C. (1994). Rigid block analysis of masonry structures. *The Structural Engineer*, *72*(21), 356–361.

Giordano, A., Mele, E., & Luca, A. (2002). Modelling of historical masonry structures. Comparison of different approaches through a case study. *Engineering Structures*, *24*(8), 1057–1069. doi:10.1016/S0141-0296(02)00033-0

Hendry, A. W. (1998). *Structural masonry* (2nd ed.). London, UK: Palgrave Macmillan. doi:10.1007/978-1-349-14827-1

Hendry, A. W., Sinha, B. P., & Davis, S. R. (2004). *Design of masonry structures* (3rd ed.). London, UK: Taylor and Francis.

Heyman, J. (1969). The safety of masonry arches. *International Journal of Mechanical Sciences*, *11*(4), 363–385. doi:10.1016/0020-7403(69)90070-8

Heyman, J. 1998. Structural Analysis: A historical approach. University Press. doi:10.1017/CBO9780511529580

Itasca. (2004a). *UDEC - Universal Distinct Element Code Manual. Theory and Background.* Itasca Consulting Group.

Itasca. (2004b). *UDEC -Universal Distinct Element Code Manual. User's manual.* Itasca Consulting Group.

Lemos, J. V. (1995). Assessment of the ultimate load of a masonry arch using discrete elements. In J. Middleton & G. N. Pande (Eds.), *Computer Methods in Structural Masonry – 3* (pp. 294–302). Swansea, UK: Books and Journals International.

Lemos, J. V. (1998). Discrete element modelling of the seismic behaviour of stone masonry arches. In G. N. Pande, J. Middleton, & B. Kralj (Eds.), *Computer Methods in Structural Masonry — 4* (pp. 220–227). London, UK: E & FN Spon.

Lemos, J. V. (2001). Modelling the behaviour of a stone masonry arch structure under cyclic loads. In T. G. Hughes & G. N. Pande (Eds.), *Computer Methods in Structural Masonry - 5* (pp. 101–108). Swansea, UK: Computers & Geotechnics Ltd.

Lemos, J. V. (2004). Modelling stone masonry dynamics with 3DEC. In H. Konietzky (Ed.), *Numerical modelling of discrete materials in geotechnical engineering, civil engineering and earth sciences* (pp. 7–13). London, UK: Taylor and Francis Group.

Lemos, J. V. (2007). Discrete element modelling of masonry structures. *International Journal of Architectural Heritage*, *1*(2), 190–213. doi:10.1080/15583050601176868

Liu, L. P., Tang, D. X., & Zhai, X. M. (2006). Failure criteria for grouted concrete block masonry under bi-axial compression. *Advances in Structural Engineering*, *9*(2), 229–239. doi:10.1260/136943306776987001

Livesley, R. K. (1978). Limit analysis of structures formed from rigid blocks. *International Journal for Numerical Methods in Engineering*, *12*(12), 1853–1871. doi:10.1002/nme.1620121207

Lofti, H. R., & Shing, P. B. (1994). Interface model applied to fracture of masonry structures. *Journal of Structural Engineering*, *120*(1), 63–80. doi:10.1061/(ASCE)0733-9445(1994)120:1(63)

Lourenço, P. B. (1996). *Computational strategies for masonry structures.* (Ph.D thesis). Delft University of Technology, Delft, The Netherlands.

Lourenço, P. B. (1998a). Experimental and numerical issues in the modelling of the mechanical behaviour of masonry. In P. Roca, J. L. Gonzalez, E. Onate, & P. B. Lourenço (Eds.), *Structural Analysis of Historical Constructions II* (pp. 57–92). Barcelona, Spain: CIMNE.

Lourenço, P. B. (2001). *Analysis of historical constructions: From thrust-lines to advanced simulations*. Guimaraes, Portugal: Historical Constructions.

Lourenço, P. B. (2002). Computations on historic masonry structures. *Progress in Structural Engineering and Materials*, *4*(3), 301–319. doi:10.1002/pse.120

Lourenço, P. B., & Pina-Henriques, J. (2006). Validation of numerical and continuum numerical methods for estimating the compressive strength of masonry. *Computers & Structures*, *84*(29), 1977–1989. doi:10.1016/j.compstruc.2006.08.009

Lourenço, P. B., & Rots, J. G. (1997). A multi-surface interface model for the analysis of masonry structures. *Journal of Engineering Mechanics*, *123*(7), 660–668. doi:10.1061/(ASCE)0733-9399(1997)123:7(660)

Ma, M. A., Pan, A. D. E., Luan, M., & Gebara, J. M. (1996). Seismic analysis of stone arch bridges using discontinuous deformation analysis. In *Proceedings of the 11th World Conference on Earthquake Engineering*.

Mamaghani, I. H. P. (1999). Analysis of masonry structures under static and dynamic loading by discrete finite element method. *Structural Engineering / Earthquake Engineering*, *16*(2), 75–86.

Millard, A. (2003). CASTEM 2000, User Guide, Report CEA 93/007. Saclay, France: Academic Press..

Moradabadi, E., & Laefer, D. F. (2014). Numerical modelling options for cracked masonry buildings. *9th International Masonry Conference*, Guimarães, Portugal.

Mullett, P. J., Briggs, M., & Minton, K. (2006). Archtec-Strengthening and preventing masonry arch bridges in Cumbria. In *Proceedings of the 11th International Conference on Structural Faults and Repair*. Edinburgh Press.

Munjiza, A., Owen, D. R. J., & Bićanić, N. (1995). A combined finite/discrete element method in transient dynamics of fracturing solids. *Engineering Computations*, *12*(2), 145–174. doi:10.1108/02644409510799532

Oliveira, D. V. C. (2003). *Experimental and numerical analysis of blocky masonry structures under cyclic loading*. (Ph.D thesis). University of Minho, Minho, Portugal.

Orduña, A., & Lourenço, P. B. (2003). Cap model for limit analysis and strengthening of masonry structures. *Journal of Structural Engineering*, *129*(10), 1367–1375. doi:10.1061/(ASCE)0733-9445(2003)129:10(1367)

Owen, D. R. J., Peric, D., & Petrinic, N. (1998). Finite/discrete element models for assessment and repair of masonry structures. In A. Sinopoli (Ed.), *Proceedings of the 2nd International Conference on Arch Bridges: History, Analysis, Assessment, Maintenance and Repair*.

Page, A. W. (1978). Finite element model for masonry. *Journal of the Structural Division*, *104*(8), 1267–1285.

Papa, E. (2001). Damage and failure modes. In J. W. Bull (Ed.), *Computational modelling of masonry, brickwork and blockwork structures* (pp. 1–26). Stirling, UK: Saxe-Coburg Publications. doi:10.4203/csets.6.1

Papantonopoulos, C., Psycharis, I. N., Papastamatiou, D. Y., Lemos, J. V., & Mouzakis, H. P. (2002). Numerical prediction of the earthquake response of classical columns using the distinct element method. *Earthquake Engineering & Structural Dynamics*, *31*(9), 1699–1717. doi:10.1002/eqe.185

Papastamatiou, D., & Psycharis, I. (1993). Seismic response of classical monuments – A numerical perspective developed at the Temple of Apollo Bassae, Greece. *Terra Nova*, *5*(1), 591–601. doi:10.1111/j.1365-3121.1993.tb00309.x

Pegon, P., Pinto, A. V., & Geradin, M. (2001). Numerical modelling of stone-block monumental structures. *Computers & Structures*, *79*(22), 2165–2181. doi:10.1016/S0045-7949(01)00070-0

Psycharis, I. N., Lemos, J. V., Papanastasiou, D. Y., Zambas, C., & Papantonopoulos, C. (2003). Numerical study of the seismic behaviour of a part of the Parthenon Pronaos. *Earthquake Engineering & Structural Dynamics*, *32*(13), 2063–2084. doi:10.1002/eqe.315

Rafiq, M. Y., Bugmann, G., & Easterbrook, D. J. (1998). Artificial neural networks for modelling some of the activities of the computational stage of the design process. In K.C.P. Wang, T. Adams, M.L. Maher, & A. Songer (Eds.), *Proceedings of the International Computing Congress.*

Roca, P., Cervera, M., Gariup, G., & Pela, L. (2010). Structural analysis of masonry historical constructions. Classical and advanced approaches. *Archives of Computational Methods in Engineering*, *17*(3), 299–325. doi:10.1007/s11831-010-9046-1

Rots, J. G. (1991). Numerical simulation of cracking in masonry. *HERON*, *36*(2), 49–63.

Rots, J. G. (1997). *Structural masonry: An experimental/numerical basis for practical design rules.* A.A. Balkema Publishers.

Sarhosis, V. (2012). *Computational modelling of low bond strength masonry.* (PhD thesis). University of Leeds, UK.

Sarhosis, V., Garrity, S. E., & Sheng, Y. (2015). Influence of the brick-mortar interface on the mechanical response of low bond strength masonry lintels. *Engineering Structures*, *88*, 1–11. doi:10.1016/j.engstruct.2014.12.014

Sarhosis, V., Oliveira, D. V., Lemos, J. V., & Lourenço, P. B. (2014b). The effect of the angle of skew on the mechanical behaviour of arches. *Journal of Mechanics Research Communications*, *61*, 53–49. doi:10.1016/j.mechrescom.2014.07.008

Sarhosis, V., Oliveira, D. V., & Lourenço, P. B. (2015). On the mechanical behaviour of masonry structures. In V. Sarhosis, J. V. Lemos, K. Bagi, & G. Milani (Eds.), *Computational Modeling of Masonry Structures Using the Discrete Element Method.* IGI Global.

Sarhosis, V., & Sheng, Y. (2014a). Identification of material parameters for low bond strength masonry. *Engineering Structures*, *60*, 100–110. doi:10.1016/j.engstruct.2013.12.013

Schlegel, R., & Rautenstrauch, K. (2004a). Failure analyses of masonry shear walls. In H. Konietzky (Ed.), *Numerical modelling of discrete materials in geotechnical engineering, civil engineering and earth sciences* (pp. 15–20). Taylor and Francis Group London.

Schlegel, R., & Rautenstrauch, K. (2004b). Comparative computations of masonry arch bridges using continuum and discontinuum mechanics. In H. Konietzky (Ed.), *Numerical modelling of discrete materials Numerical modelling of discrete materials in geotechnical engineering, civil engineering and earth sciences* (pp. 3–6). London, UK: Taylor and Francis Group.

Shi, G. H., & Goodman, R. E. (1988). Discontinuous deformation analysis – a new method for computing stress, strain and sliding of block systems. In: P.A. Cundall, R. Sterling, & A. Starfield (eds.), *Proceedings of the 29th U.S. Symposium: Key Questions in Rock Mechanics*. Balkema.

Sincraian, G. E., & Lemos, J. V. (1999). Seismic analysis of a stone masonry aqueduct using Discrete Elements. In *Proceedings of the 8th Canadian Conference on Earthquake Engineering*.

Stefanou, I., Psycharis, I., & Georgopoulos, I. O. (2011). Dynamic response of reinforced masonry columns in classical monuments. *Journal of Construction and Building Materials*, *25*(1), 4325–4337. doi:10.1016/j.conbuildmat.2010.12.042

Thavalingam, A., Bicanic, N., Robinson, J. I., & Ponniah, D. A. (2001). Computational framework for discontinuous modelling of masonry arch bridges. *Computers & Structures*, *79*(19), 1821–1830. doi:10.1016/S0045-7949(01)00102-X

Toth, A. R., Orban, Z., & Bagi, K. (2009). Discrete element analysis of a masonry arch. *Mechanics Research Communications*, *36*(4), 469–480. doi:10.1016/j.mechrescom.2009.01.001

Van der Pluijm, R. (1999). *Out-of-plane bending of masonry behaviour and strength.* (Ph.D thesis). Eindhoven University of Technology, The Netherlands.

Vanderbei, T. J. (2001). *Linear Programming: Foundations and extensions* (2nd ed.). London, UK: Springer. doi:10.1007/978-1-4757-5662-3

Voyiadjis, G. Z., & Kattan, P. I. (2005). *Damage mechanics*. London, UK: Taylor and Francis. doi:10.1201/9781420027839

Walker, P., McCombie, P., & Claxton, M. (2006). Plane strain numerical model for drystone retaining walls. *Proceedings of the ICE - Geotechnical Engineering*, *160*(2), 97–103. doi:10.1680/geng.2007.160.2.97

Zhang, Y., Zhou, G. C., Xiong, Y., & Rafiq, M. Y. (2010). Techniques for predicting the cracking pattern of masonry wallet using artificial neural networks and cellular automata. *Journal of Computing in Civil Engineering*, *24*(2), 161–172. doi:10.1061/(ASCE)CP.1943-5487.0000021

Zhou, G. C. (2002). *Application of stiffness/strength corrector and cellular automata in predicting response of laterally loaded masonry panel.* (Ph.D thesis). University of Plymouth, Plymouth, UK.

Zhou, G. C., & Rafiq, M. Y. (2006). A CA & ANN techniques of predicting failure load and failure pattern of laterally loaded masonry panel. In *Proceedings of the 11th Joint International Conference on Computing and Design Making in Civil and Building Engineering*.

Zhuge, Y. (2002). Micro-modelling of masonry shear panels with distinct element approach. In L. Chowdhury & Fragomeni (Eds.), Advances in Mechanics of Structures and Materials. Swets & Zeitinger.

Zhuge, Y., Jin, F., & Hunt, S. (2003). The prediction of damage to masonry houses caused by foundation movements. *Advances in Structural Engineering*, *7*(1), 81–93. doi:10.1260/136943304322985783

Zucchini, A., & Lourenço, P. B. (2006). Mechanics of masonry in compression: Results from a homogenisation approach. *Computers & Structures*, *85*(3), 193–204.

ADDITIONAL READING

Abdou, L., Ami, S. R., Meftah, F., & Mebarki, A. (2006). Experimental investigations of the joint-mortar behaviour. *Mechanics Research Communications*, *33*(3), 370–384. doi:10.1016/j.mechrescom.2005.02.026

Adami, C. E., & Vintzileou, E. (2008). Investigations of the bond mechanism between stones or bricks and grouts. *Materials and Structures*, *41*(2), 255–267. doi:10.1617/s11527-007-9235-z

Ahmed, H. M., & Gilbert, M. 2003. The computational efficiency of two rigid block analysis formulations for application to masonry structures. In: Barry Topping (ed.), *9th International Conference on Civil and Structural Engineering Computing*. Egmond-aan-Zee, Holland: Civil-Comp Press, pp.21-33.

Alexandris, A., Protopapa, E., & Psycharis, I. 2004. Collapse mechanisms of masonry buildings derived by the distinct element method. In: *13th World Conference on Earthquake Engineering*. 1 - 6 August, Vancouver, Canada, Paper No.548.

Atkinson, R. H., Amadei, B. P., Saeb, S., & Sture, S. (1989). Response of masonry bed joint in direct shear. *Journal of Structural Engineering*, *115*(9), 2276–2296. doi:10.1061/(ASCE)0733-9445(1989)115:9(2276)

Barrentine, B. L. (2003). *An introduction to design of experiments: A simplified approach experiments with three factors* (pp. 27–35). USA: The American Society for Quality.

Begg, D. W., & Fishwick, R. J. 1995. Numerical analysis of rigid block structures including sliding. In: J. Middleton & G.N. Pande (eds.), *Proceedings of the 3rd International Symposium in Computational Methods Structural Masonry*. Books and Journals International, Portugal, pp.177-183.

Bićanić, N., Stirling, C., & Pearce, C. J. (2003). Discontinuous modelling of masonry bridges. *Computational Mechanics*, *31*(12), 60–68. doi:10.1007/s00466-002-0393-0

Brocken, H. J. P., & Pel, L. 1995. Moisture transport over the brick/mortar interface. In: National Research Council Canada (NRCC), *International Symposium on Moisture Problems in Building Walls*. Porto, Portugal, pp.397-401.

Chakrabarty, J. (2006). *Theory of plasticity*. London, UK: Butterworth Heinemann.

Churilov, S., & Dumova-Jovanoska, E. 2008. Calibration of a numerical model for masonry with applications to experimental results. Journal of Architecture, Civil Engineering and Environment, The Silesian University of Technology, 3, pp.41-48.

Dialer, C. (2002). Typical masonry failures and repairs, a German Engineer's view. *Progress in Structural Engineering and Materials*, *4*(3), 332–339. doi:10.1002/pse.125

Draper, N. R., & Smith, H. (1998). *Applied regression analysis*. Wiley-Interscience. doi:10.1002/9781118625590

Gambarotta, L., & Lagomarsino, S. (1997a). Damage models for the seismic response of brick masonry shear walls. Part I: The mortar joint model and its applications. *Journal of Earthquake Engineering and Structural Dynamics*, *26*(4), 423–439. doi:10.1002/(SICI)1096-9845(199704)26:4<423::AID-EQE650>3.0.CO;2-#

Garrity, S. W. 1995. Testing of small scale masonry arch bridges with surface reinforcement. In: *Proceedings of the 6th International Conference on Structural Faults and Repair*. Engineering Technics Press, Edinburgh, Scotland, UK, pp.409-418.

Garrity, S. W. (2004). *Independent testing of the Bersche-Rolt system for strengthening a brickwork lintel containing a soldier course. Technical Report*. Bradford, UK: University of Bradford, School of Engineering, Design and Technology.

Garrity, S. W., Ashour, A. F., & Chen, Y. 2010. An experimental investigation of retro-reinforced clay brick arches. In: *Proceedings of the 8th International Masonry Conference*. Dresden, July 2010, Germany, pp.733-742. [CD-ROM proceedings].

Gideon, P. A., & Zijl Van, G. (2004). Modelling masonry shear-compression: The role of dilatancy highlighted. *Journal of Engineering Mechanics*, *30*(11), 1289–1297.

Gilbert, M., Casapulla, C., & Ahmed, H. M. (2006). Limit analysis of masonry block structures with non-associative frictional joint using linear programming. *Computers & Structures*, *84*(1), 873–887. doi:10.1016/j.compstruc.2006.02.005

Grandet, J., Javelza, R., Perrin, B., & Thenoz, B. (1972). Role of ettrignite in the binding of the mechanical type between clay and Portal cement paste.[in French]. *Revue Terre Cuite*, *48*, 21–28.

Haseltine, B. A., Kirtschig, K., & Macchi, G. (1998). *Eurocode No. 6: Common unified rules for structural masonry structures. Luxemburg*. Commission of the European Communities.

Hendry, A. W. (1998). *Structural masonry* (2nd ed.). London, UK: Palgrave Macmillan. doi:10.1007/978-1-349-14827-1

Heyman, J. (1998). *Structural Analysis: A historical approach*. Cambridge, UK: University Press. doi:10.1017/CBO9780511529580

James, J. A. 1973. Investigation of the effect of workmanship and curing conditions on the strength of brickwork, In: L. Fortig, & K. Gobel (eds.), *Proceedings of the 3rd International Brick Masonry Conference*. Essen, Germany, pp.192-201.

Jukes, P., & Riddington, J. R. (1997). A review of masonry joint shear strength test methods. *Masonry International*, *11*(2), 37–43.

Khalaf, M. F. (2005). New tests for the determination of masonry tensile bond strength. *Journal of Materials in Civil Engineering*, *17*(6), 715–732. doi:10.1061/(ASCE)0899-1561(2005)17:6(725)

Lawrence, S. J., & Cao, H. T. 1987. An experimental study of the interface between brick and mortar. In: *Proceedings of the 4th North American Masonry Conference*. 5-10 August, University of California, Los Angeles, USA, pp.48.1 - 48.14.

Liu, L. P., Tang, D. X., & Zhai, X. M. (2006). Failure criteria for grouted concrete block masonry under biaxial compression. *Advances in Structural Engineering*, *9*(2), 229–239. doi:10.1260/136943306776987001

Livesley, R. K. (1978). Limit analysis of structures formed from rigid blocks. *International Journal for Numerical Methods in Engineering*, *12*(12), 1853–1871. doi:10.1002/nme.1620121207

Lofti, H. R., & Shing, P. B. (1994). Interface model applied to fracture of masonry structures. *Journal of Structural Engineering*, *120*(1), 63–80. doi:10.1061/(ASCE)0733-9445(1994)120:1(63)

Ma, M. A., Pan, A. D. E., Luan, M., & Gebara, J. M. 1996. Seismic analysis of stone arch bridges using discontinuous deformation analysis. In: *Proceedings of the 11th World Conference on Earthquake Engineering*, Amsterdam, The Netherlands, Paper No.155. [CD-ROM proceedings].

Mann, W., & Muller, H. 1982. Failure of shear-stressed masonry – An enlarged theory, tests and application to shear walls. In: H.W.H West (ed.), *Proceedings of the British Ceramic Society*, 30, pp.223-235

Marquis, E. L., & Borchelt, G. 1986. Bond strength comparison of laboratory and field tests. In: *Proceedings of the 4th Canadian Masonry Symposium*. June, Federation New Brunswick, pp.94-204.

McKibbins, L., Melbourne, C., Sawar, N., & Galliard, C. S. 2006. *Masonry arch bridges, condition appraisal and remedial treatment*. Report C656. CIRIA, London, UK.

Orduña, A., & Lourenço, P. B. (2003). Cap model for limit analysis and strengthening of masonry structures. *Journal of Structural Engineering*, *129*(10), 1367–1375. doi:10.1061/(ASCE)0733-9445(2003)129:10(1367)

Papa, E. (2001). Damage and failure modes. In J. W. Bull (Ed.), *Computational modelling of masonry, brickwork and blockwork structures* (pp. 1–26). Stirling, Scotland, UK: Saxe-Coburg Publications. doi:10.4203/csets.6.1

Papantonopoulos, C., Psycharis, I. N., Papastamatiou, D. Y., Lemos, J. V., & Mouzakis, H. P. (2002). Numerical prediction of the earthquake response of classical columns using the distinct element method. *Earthquake Engineering & Structural Dynamics*, *31*(9), 1699–1717. doi:10.1002/eqe.185

Papastamatiou, D., & Psycharis, I. (1993). Seismic response of classical monuments – A numerical perspective developed at the Temple of Apollo Bassae, Greece. *Terra Nova*, *5*(1), 591–601. doi:10.1111/j.1365-3121.1993.tb00309.x

Rafiq, M. Y., Bugmann, G., & Easterbrook, D. J. 1998. Artificial neural networks for modelling some of the activities of the computational stage of the design process. In: K.C.P. Wang, T. Adams, M.L. Maher, A. Songer (eds.), *Proceedings of the International Computing Congress*. 18-21 October, Boston, Massachusetts, ASCE Reston, USA, pp.631-643.

Roberti, M. G., & Calvetti, F. 1998. Distinct element analysis of stone arches. In: Sinopolo (ed.), *Proceedings of the Second International Conference on Arch Bridges*. Italy, Venice, pp.181-186.

Roberts, M. J., & Russo, R. (1999). *A student's guide to Analysis of Variance*. London: Routledge.

Schubert, I. P. 1988. The influence of mortar on the strength of masonry. In: J.W. Courcy (ed.), *Proceedings of the 8th International Brick and Block Conference*. Elseiver, London, pp.162-174.

Shi, G. H., & Goodman, R. E. 1988. Discontinuous deformation analysis – a new method for computing stress, strain and sliding of block systems. In: P.A. Cundall, Sowden, A.M., 1990. The maintenance of brick and stone structures. Taylor and Francis, London, UK.

Stefanou, I., Psycharis, I., & Georgopoulos, I. O. (2011). Dynamic response of reinforced masonry columns in classical monuments. *Journal of Construction and Building Materials*, *25*(1), 4325–4337. doi:10.1016/j.conbuildmat.2010.12.042

Thavalingam, A., Bicanic, N., Robinson, J. I., & Ponniah, D. A. (2001). Computational framework for discontinuous modelling of masonry arch bridges. *Computers & Structures*, *79*(19), 1821–1830. doi:10.1016/S0045-7949(01)00102-X

Tran, V. H., Vincens, E., Morel, J. C., Dedecker, F., & Le, H. H. (2014). 2D-DEM modelling of the formwork removal of a rubble stone masonry bridge. *Engineering Structures*, *75*, 448–456. doi:10.1016/j.engstruct.2014.05.048

KEY TERMS AND DEFINITIONS

Computer Simulation: A simulation run on a computer to reproduce behavior of a system.

Constitutive Law: A series of equations describing the mechanical characteristics of masonry.

Distinct Element Method: An approach for computing the motion and effect of a large number of small particles.

Finite Element Method (FEM): A numerical technique where it uses subdivision of a whole problem domain into simpler parts and finds approximate solutions to boundary value problems for partial differential equations.

Low Bond Strength Masonry: Masonry where the bond strength is relatively low and determines the mechanical behaviour.

Material Parameters: A series of values used to describe the strength and stiffness of masonry

Numerical Models: Any of numerical methods for computing the motion and effect of a large number of small particles.

Unreinforced Masonry: Masonry that is not strengthened/braced with reinforcement beams.

Chapter 3
The Basis for Masonry Analysis with UDEC and 3DEC

José V. Lemos
National Laboratory for Civil Engineering Portugal

ABSTRACT

The "distinct element method" was proposed by Peter Cundall in 1971 for the analysis of rock slopes by means of rigid block or circular particle models. This method led to the UDEC and 3DEC codes, presently in wide use in rock engineering. Their application to masonry structures started in the 90's, as researchers found that they were also excellent tools to approach the highly nonlinear behavior of masonry, in particular the collapse processes of stone block structures under static or seismic loads. This chapter reviews the essential assumptions of UDEC and 3DEC, relating them to other methods and codes, and stressing the features that make them suitable for masonry analysis. Rigid and deformable blocks, contact mechanics, contact detection, and solution algorithms are examined. Key issues in the modelling of masonry are addressed, including: irregular block models; determination of collapse loads; large displacement analysis; computational efficiency issues in dynamic analysis. Practical examples taken from the published literature illustrate these issues.

INTRODUCTION

The development of the finite element method and its wide acceptance by the engineering community during the 1960's supplied a powerful tool for structural analysis. Joint elements were proposed by Goodman et al. (1968) to represent joints or other thin discontinuities, which play a key role in the behavior of rock or masonry. Cundall (1971) proposed a different approach to the study of discontinuous systems, by viewing them as an assembly of blocks in mechanical interaction through sets of point contacts. He employed an explicit solution algorithm based on dynamic relaxation instead of the common matrix methods. With these tools, he was able to simulate the full evolution of collapse processes of blocky media taking into account the changes in geometry and contact due to large displacements, including total block separation. The initial aim was the analysis of rock slopes, but this approach was soon recognized to be of interest in many other engineering fields dealing with complex, discontinuous structures with

DOI: 10.4018/978-1-5225-0231-9.ch003

strongly nonlinear behavior. Papastamatiou and Psycharis (1993) and Psycharis et al. (1993) showed the potential of this method for the study of the dynamics of block masonry structures, in their analysis of the drum columns of the Apollo temple at Bassae. Extensive application to other types of masonry constructions would follow.

Cundall named his approach the "distinct element method", but the designation "discrete element method (DEM)" is nowadays more common, encompassing not only the codes derived from Cundall' work, such as UDEC (Itasca, 2014) and 3DEC (Itasca, 2013), but also many other formulations and codes currently used in masonry analysis (Lemos, 2007a). It is interesting to note that Cundall's 1971 paper already included an example of a circular particle model, a class of DEM models which became quite widespread after the work of Cundall and Strack (1979).

In this chapter, the essential features of UDEC and 3DEC are reviewed, with an emphasis on clarifying differences and similarities with respect to the alternative approaches within the family of discrete elements. The key ingredients required to model the discontinuous nature of masonry will be stressed. A discussion of the main types of problems for which these codes are suitable follows, illustrated by examples taken from the literature. While the present chapter refers mainly to UDEC and 3DEC, the discussion of their application to masonry is based on general concepts so that it is also relevant for users of other DE formulations and codes. The practical issues examined, such as the choice of a rigid or deformable block representation, the generation of complex geometries, the simulation of reinforcement elements, or the specific requirements for an efficient seismic analysis, are encountered by all those using DE models in this field.

THE EVOLUTION OF UDEC AND 3DEC

Early Rigid Block Codes

The original DEM codes developed by Cundall (1971) allowed the analysis of systems of rigid bodies in mechanical interaction, either polygonal blocks or circular particles. The intended application was the analysis of the stability of slopes in jointed rock, taking into account the possible separation between the rock blocks during the failure process. The key requirements were the proper simulation of the nonlinear behavior of the joints, assuming no tensile strength and frictional sliding. The main assumptions were:

- Polygons behave as rigid bodies.
- Mechanical interaction represented by point contacts.
- Contact behavior characterized by: normal stiffness, shear stiffness, and friction angle.
- Solution based on the integration of the equations of motion by means of an explicit algorithm.
- Static solutions obtained by dynamic relaxation using artificial damping.
- Large displacement analysis, including automatic contact detection and update.

Many of the features in this list would be retained in the future codes, as they proved to be very effective in the study of many other types of discrete systems.

The 2D Code UDEC: Block Deformability

From the early codes, evolved the 2D code UDEC developed by Cundall in 1980. It shared many of the features in the previous list, but there were significant enhancements, the main one regarding the incorporation of block deformability. Two options were tried. The simplest formulation assumed a uniform state of stress in each block. This feature was also adopted by the DDA method of Shi & Goodman (1988). To simulate more general deformation patterns, UDEC also introduced blocks, then designated as "fully-deformable", with an internal mesh of triangular elements. Although formulated from the standpoint of the finite-difference method (which uses the designation of zones), their performance is exactly the same as uniform strain finite elements, governed by linear interpolation shape functions. Therefore, these deformable blocks in UDEC are equivalent to those used in discrete-finite element methods (e.g., Munjiza, 2004). Subsequent versions of UDEC incorporated many new options, from more elaborate joint and block constitutive models to structural elements, or adaptive damping schemes, which will be discussed below.

The 3DEC Code

The 3D counterpart of UDEC, designated 3DEC, was developed in 1988. The handling of the contact problem in 3D, using the common-plane concept (Cundall, 1988), was the main innovation. Particular care was devoted to implement efficient contact detection and update techniques, which are critical for large 3D systems. Either rigid or deformable blocks, with internal meshes of tetrahedral elements (or zones), as shown in Figure 1, were available (Hart et al., 1988). A procedure to create non-convex blocks by joining convex components was advanced. Many other features resembled the 2D code, namely the explicit solution algorithm for static and dynamic problems, joint and block constitutive models, or structural elements.

Figure 1. Rigid and deformable polyhedral blocks in 3DEC

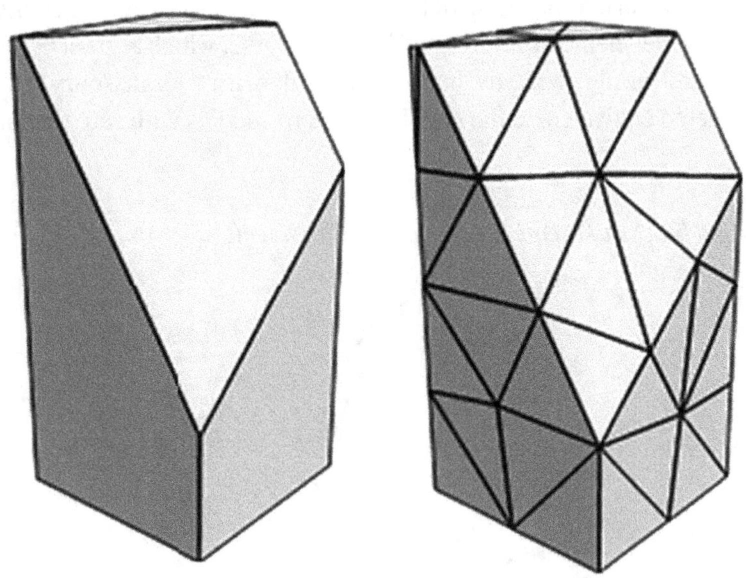

The FISH Language

The current versions of the codes have an internal programming language, FISH, common to other Cundall's codes. It is an extremely powerful tool, as it gives the user the ability to access the entire internal data structure and to change selected variables during the simulation. It may be applied during model generation, for example, allowing the parametrization of the geometry, so that other models of the same type but different dimensions can be created without effort. It also permits monitoring the results at any location in the model during the time stepping procedure. In this way the user may create his own behavior indicators, or tailor the built-in ones to better serve his needs. Afterwards, histories of these variables can be created to follow their evolution during the run. At any stage, spatial contours of user-defined variables may be plotted, to reveal the specific features of the response of a given structural system.

FUNDAMENTAL ASSUMPTIONS AND MODELING FEATURES

Block Representation

In UDEC, blocks are represented by polygons, either convex or concave. 3DEC requires polyhedral blocks to be convex, mainly to simplify the contact operations. Non-convex blocks may be formed by "joining" (i.e., "gluing together") the faces of convex sub-blocks, which are then constrained to move together, without relative displacements. For example, in the simplified model of a pointed arch of Figure 2, each cross-section is assumed to behave as a single block, which is formed by 2 convex blocks rigidly attached.

While, for masonry applications, non-convex blocks may have limited interest, this restriction influences the way a model is generated, particularly when blocks are created by successive cuts, because each cutting plane is required to split the entire block in two. The joining procedure will still allow the composite blocks to behave as intended, but a profusion of model construction joined cuts tends to deteriorate the efficiency.

Deformable blocks have an internal mesh of triangular zones (or elements) in UDEC, and a tetrahedral mesh in 3DEC. Higher-order elements are also available in 3DEC, which allow a better approximation of shells in bending or elasto-plastic materials, but they are seldom used for masonry. The deformable block models may be considered equivalent to the simplified micro-models with zero thickness joint elements,

Figure 2. Joining rigid blocks to create a non-convex cross-section

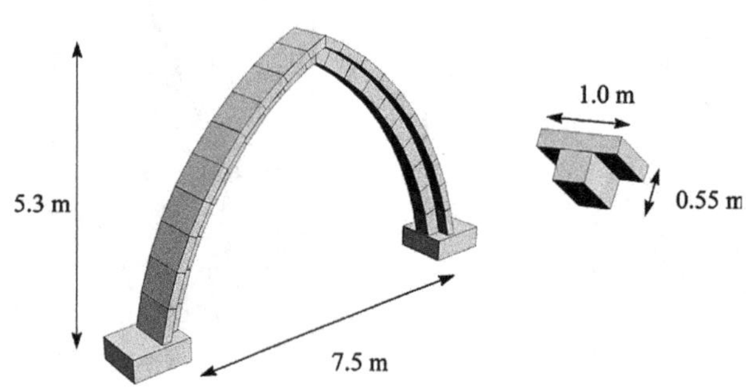

common in FE analysis. In principle, detailed micro-models may also be created, by incorporating thin deformable blocks to simulate the mortar.

Constitutive Models for Block Materials

Deformable blocks, with internal meshes, may be assumed elastic or display nonlinear behavior. The Mohr-Coulomb elasto-plastic model is perhaps the most widely used, allowing the simulation of tensile and shear failure in the block material. Currently, both codes provide facilities for user-defined constitutive models. These are written in C++ and can be linked to the executable at run time. The procedures to program these models can be found in the manuals, which also provide examples of the C++ code for the built-in models, a useful starting point to develop more elaborate functions. This feature increases the power and generality of the codes, particularly as tested constitutive models for different types of materials are becoming available on the web.

Contact Representation

Point Contacts

A distinctive feature of DE codes is the numerical representation of contact between the model components. In UDEC and 3DEC, as in most DE codes, the point contact assumption is employed (Lemos, 2007a). It essentially means that the mechanical interaction forces between 2 blocks are calculated and applied at a set of discrete points. These forces are a function of the relative displacement between the 2 blocks at that point. There is no integration of stresses on the contact surface as in FE joint elements. However, each contact point is assigned an area, and all the areas add up to the total contact surface (for face-to-face contacts). Therefore, joint stresses can be evaluated at each point contact, and the standard joint constitutive models, relating normal and shear stresses with joints displacements, can be employed. In general, the stress distribution on the joint plane obtained with point contacts is less smooth and accurate than when joint elements are used. The approximation can be improved by inserting more point contacts, which can be achieved by splitting the block edges or faces (even for the case of rigid blocks).

This type of contact formulation has the advantage of handling arbitrary types of interaction, e.g., true point contacts, such as vertex-to-face or edge-to-edge, in addition to face-to-face configurations. The latter will involve several point contacts. This approach is also very convenient to address large displacement problems, with changes in contact types and connectivity, as point contacts locations are continuously updated as blocks move, and new contacts are automatically detected.

There is a substantial difference in the geometric representation of contacts in the 2D and 3D codes, as explained in the following.

UDEC Contact Representation

In UDEC, Cundall implemented a contact representation based on the assumption of "rounded corners", in which the polygon vertices are approximated by a circular arch (Figure 3). This approach allows vertex-to-vertex contacts to be handled very conveniently, as the contact normal is always defined by the line connecting the centers of the 2 arcs, and varies smoothly as the blocks displace and rotate. It also permits a continuous transition from a vertex-to-vertex to a vertex-to-face condition (i.e., vertex-to-edge in 2D), as illustrated in Figure 4. For rigid blocks, a face-to-face condition (i.e., edge-to-edge in 2D) will be represented by 2 vertex-to-face contacts. For deformable blocks, with internal triangular element meshes, there may be additional nodal points (grid-points) along the edge, so more vertex-to-face contacts will exist.

The rounded corner model provides a very effective and robust simulation of the contact physical problem. The only drawback is that the user need to specify the rounding length, defined as the distance between the start of the arc segment and the theoretical sharp vertex (Figure 3). Typically, this parameter has no noticeable effect on the results, and values between 1% and 10% of the average edge length are common. Note that the minimum allowable edge length, measured between the theoretical sharps corners, will be twice the rounding length. This provides a criterion to select the rounding length, as a function of the level of geometric detail intended for the model.

3DEC Contact Representation

The rounded corner representation performs very well in 2D, but its generalization to 3D is somewhat cumbersome. Cundall (1988) devised an alternative approach for 3DEC, based on a method to define uniquely a "common-plane" between 2 convex polyhedral, defined as the plane that maximizes the gap, or minimizes the overlap, between the blocks. Figure 5 illustrates the concept in 2D, for clarity. The

Figure 3. UDEC rounded corner concept. The sharp angle is approximated by a circular arch

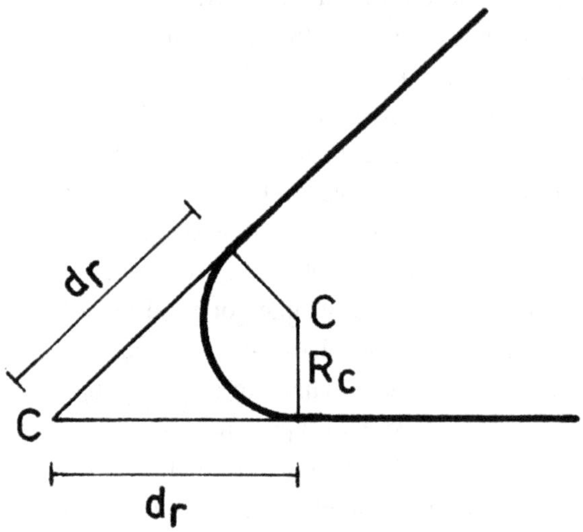

Figure 4. Smooth transition from a vertex-to-vertex to a face-to-face interaction in UDEC

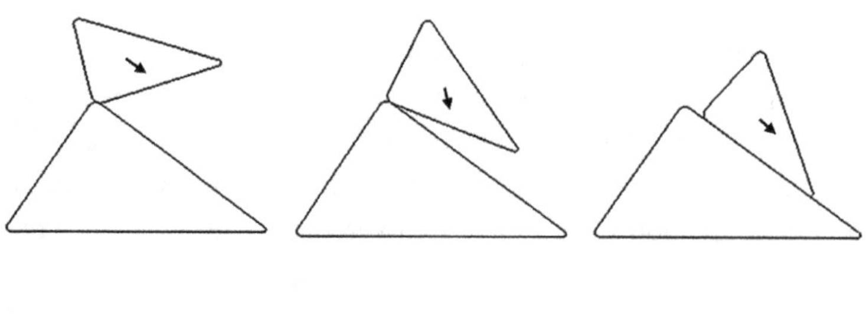

common-plane defines the contact normal. In 3D, 2 types of elementary point contacts (also termed sub-contacts) may exist, vertex-to-face and edge-to-edge (Figure 6). These may represent true pointwise interactions (in which the vertex-to-face includes the particular cases of vertex-to-edge and vertex-to-vertex), or be associated to simulate a face-to-face interaction. The sum of the areas of the point contacts equals the total contact area. For true point interactions, a minimum contact area is defined. This minimum area multiplied by the joint stiffness gives the stiffness of the point contact, and should not be too low, in order to avoid excessive overlap between blocks. It is not really physically meaningful to have a very small minimum area, since the interaction between the blunt vertices of a masonry unit and a joint surface has a measurable stiffness.

Joint Constitutive Models

The joint constitutive model provides the joints normal and shear stresses given the relative displacements between the blocks. The normal and shear stiffness parameters define the elastic deformability, and the strength parameters govern the nonlinear behavior. Coulomb friction models are the most widely used for dry masonry joints. A full Mohr-Coulomb model, with tensile and shear strength, is also available. This built-in model assumes brittle failure, with a drop from peak to residual strength envelopes, as soon as either normal or shear failure takes place. However, more elaborate assumptions can be implemented. As for block constitutive models, UDEC and 3DEC allow user-defined joint models. In particular, soft-

Figure 5. Illustration of the 'common plane' concept in 3DEC (shown in 2D, for clarity)

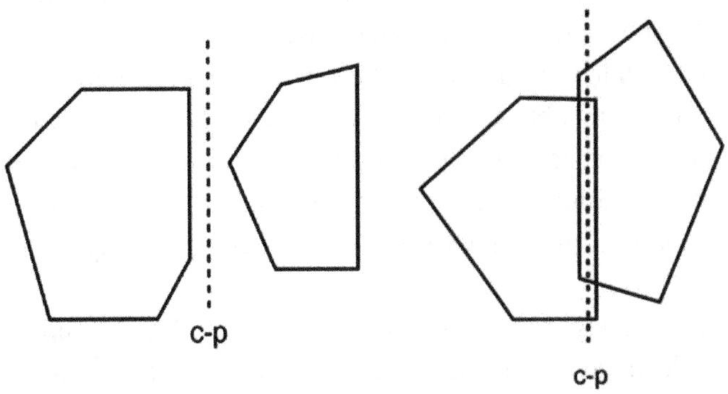

Figure 6. Face-to-face and edge-to-edge block interactions represented by elementary point contacts of 2 types: vertex-to-face (VF) and edge-to-edge (EE)

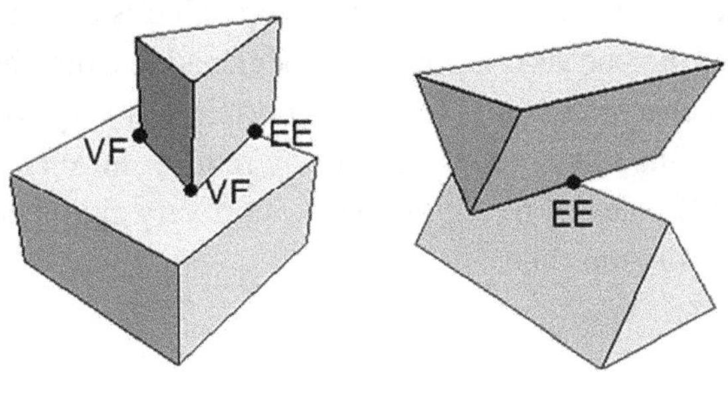

ening laws adequate for masonry joints may be introduced by modifying the coding examples provided for built-in models.

Solution Algorithm

The time stepping algorithm employed by UDEC and 3DEC is one of the features commonly associated with the codes, even if it is just a solution technique, not intrinsically bound to the formulation. Dynamic relaxation techniques, using explicit algorithms, were a means to solve static problems quite common in the finite-difference community in the '50s and '60s. The finite element codes in the '60s took advantage of the developments in computers then taking place to resort to matrix solution methods, and implicit dynamic algorithms. However, explicit algorithms continued to be competitive, given their robustness for strongly nonlinear problems, and efficiency for very large systems. Cundall's option for dynamic relaxation in the original rigid block code was largely based on its ability to handle a system of blocks undergoing a collapse mechanism involving large displacements. A stiffness matrix approach would be less appropriate or efficient to handle unconnected blocks or rapidly changing contact conditions.

This solution algorithm is based on the integration of the equations of motion in the time domain by an explicit method, the central difference method. For rigid blocks, the variables are the degrees of freedom of a rigid body: in 2D, 2 displacements and 1 rotation; in 3D, 3 displacements and 3 rotations. For deformable blocks, the variables are the nodal degrees of freedom, 2 in 2D and 3 in 3D.

In dynamic analysis with rigid blocks, the inertia tensor is assembled, and Euler's equations are solved, considering the coupling terms for non-isotropic inertia tensors. If a block is formed by several joined sub-blocks, the geometric parameters of the composite block are determined. For deformable blocks, the mass assigned to a given node is the sum of the contributions from the elements containing the node.

The algorithm is really a dynamic one, but may also be used for static or quasi-static problems. For static analysis, the time scale is fictitious, and a large amount of damping has to be introduced to eliminate vibratory movement and obtain convergence to a state of equilibrium or to a failure mode. A discussion of damping options for static analysis of masonry is given below in the section on determination of collapse loads. In static solutions, inertial masses (but not weights) may be scaled to accelerate convergence. In dynamic analysis, the real masses have to be used, and Rayleigh damping is the standard, as discussed below.

The disadvantage of explicit algorithm is that they require small time steps to be numerically stable. The codes calculate the allowable time step, as a function of the masses and stiffnesses present in the system. As the evaluation of the theoretical time step would be too time consuming, an approximation is employed, which is usually on the safe side, but it may be reduced by the user. This time step is reduced drastically by very stiff materials or joints, and also by small or narrow internal elements (zones). It is instructive for the new user to see this by trying different values of the block Young's modulus, the joint stiffness or the mesh size. For static analysis, the time step value is arbitrary since masses are scaled, but high contrasts in stiffness or zone sizes across the model lead to a noticeable increase in the number of steps required to achieve convergence. It is really the presence of very stiff and soft elements (due to material properties or size) in the same model that mostly reduces efficiency. In some cases, it may be possible to replace a very stiff material by a rigid block, or a very rigid contact spring by a rigid connection, with substantial gains in run time.

Large Displacements: Contact Updates

Large displacement analysis is required in some masonry studies, for example, problems involving block rocking in which the response is strongly influenced by the motion amplitude, and the block overturning depends on the location of the mass center. However, for many collapse analyses, usually performed under quasi-static conditions, it is possible to identify the failure mechanism while still in the small displacement range. Continuing the analysis into the large displacement regime may be instructive, as it helps to visualize and understand the structural behavior, but it is often not essential for the determination of the collapse load.

Large displacement analysis is the default option in UDEC and 3DEC. It may be switched off, so that the model geometry is not updated, which is helpful in particular situations. It also saves some checking for new contact and geometrical calculations, but this is often not very significant, since contact updates are triggered by effective movement of the blocks. Details of the actual criteria, and how to modify them, can be found in the code manuals. The general rule is that new contacts need to be detected before the blocks actually touch. Therefore, a virtual contact is created when the approaching blocks are within a short distance, so that afterwards detailed checking of the actual interaction is performed by the code at every step. The amount of movement required to trigger contact detection and update depends on the block dimensions. The default values are a fraction of the smallest edge length allowed in the model, i.e., the geometric resolution intended by the user. In UDEC, this minimum length is twice the rounding length, and in 3DEC is given by the parameter "atol", which can be specified at the start of the run. Therefore, setting these reference lengths to very small values tends to reduce the code efficiency, as it increases the local geometric detail, and requires triggering contact detection and updates more often.

Structural Elements

While the structural element available in UDEC/3DEC were originally intended for rock mechanics applications, they are also useful for masonry analysis. The most elementary type is the local reinforcement or "axial" element which is simply a spring attached to 2 points across a joint. At present, in spite of its name, it may be composed of an axial spring and a shear spring (Figure 7). Both springs have non-elastic behavior, defined in terms of yield strength and rupture strain, which breaks the element.

The "cable" elements are reinforcement elements inserted through the blocks and grouted, mainly for geotechnical applications. The "beam" elements, however, can be useful in masonry models. They can be assembled into simple frame structures, and are connected to the block faces by means of breakable contacts. Local reinforcement elements only introduce stiffness components in the model. Cable and beam elements, however, involve adding structural nodes at the element end-points, for which the equations of motion have to be solved during the time stepping algorithm. The nodes of beam elements interact with block faces through structural point contacts, with a Mohr-Coulomb type of nonlinear behavior, so that they can break off the block or slide.

It is also possible to create other types of reinforcement resorting to FISH functions. For example, ties connecting 2 arbitrary points in blocks at some distance, without links to intermediate blocks, can be implemented quite easily. At every step, the function needs to retrieve the displacements on the 2 points, calculate the force based on the relative movement, possibly checking for element failure, and apply it at 2 anchor points. This FISH procedure was used by Lemos et al. (2000), while standard local

Figure 7. Local reinforcement structural element, composed by a normal and a shear spring, both with nonlinear behavior
(Itasca, 2014)

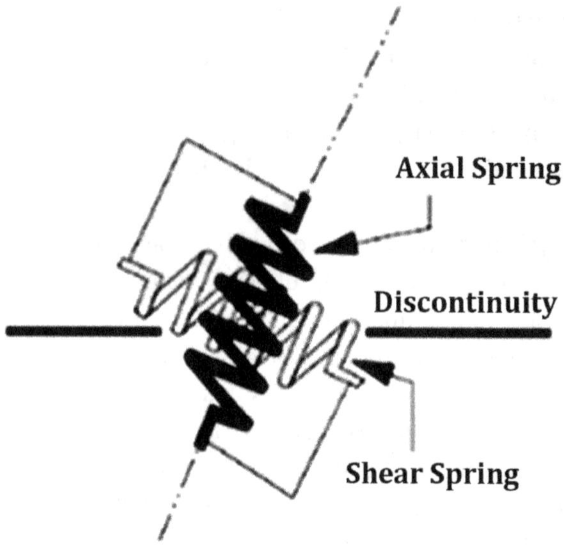

reinforcement elements were used by Psycharis et al. (2003), to simulate clamps and dowels connecting columns and architraves, and also by Stefanou et al. (2011), and Catki et al. (2014).

MASONRY APPLICATION ISSUES

Model Generation

Model generation in rock mechanics problems often starts by creating a large block, encompassing the entire domain. Then, successive operations of cutting and deleting blocks allow the definition of the ground and excavation surfaces, and the insertion of rock fractures. This procedure can sometimes be applied to masonry structures, but in most cases it is more convenient to create the blocks one by one. In UDEC, any polygonal configuration can be created by "block" commands. In 3DEC, only convex polyhedra are permitted. Simple shapes can be built with "poly brick" or "poly prism" commands; and shapes with arbitrary convex polygonal faces are specified by the "poly face" commands. FISH functions can be employed to program the intended block shapes, or they can be built externally with CAD software, and then text data files built according to the code's input format. For example, Lemos et al. (2011) created a Voronoi polygon pattern to simulate block irregularity and study its effect on the out-of-plane strength of walls (Figure 8).

Creating non-convex blocks in 3DEC can be achieved with the procedure of "joining" convex block components. In rigid block models, this scheme leads to the creation of a single composite rigid body, for which the motion equations are performed, using the global mass and inertia tensor. The contacts with adjacent blocks, however, are still detected and processed for each component block, so the common-plane logic can be applied. In deformable block models, the "joining" procedure really applies to the

Figure 8. Modeling out-of-plane failure of wall with an irregular block pattern
(Lemos et al., 2011)

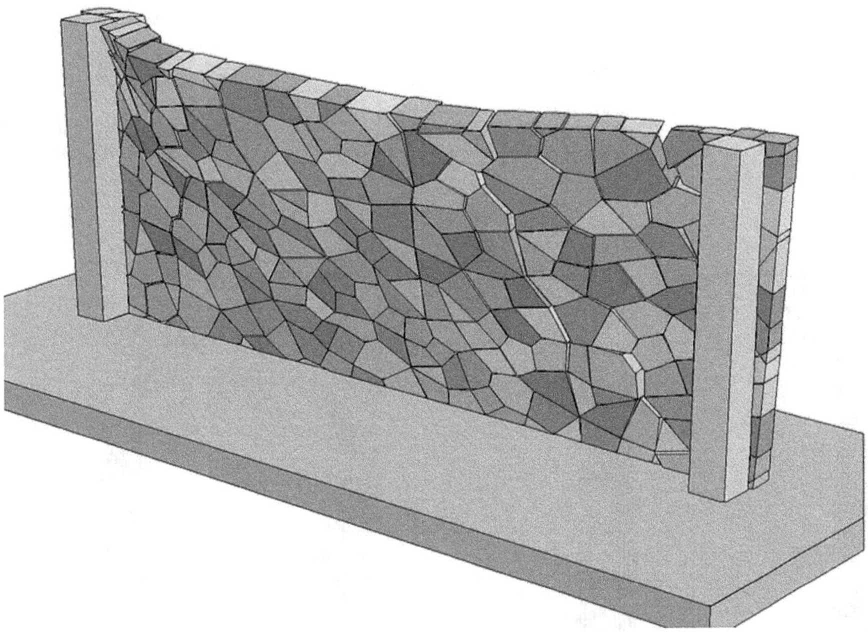

joints, leading to the establishment of rigid connection between the coincident grid-points of the various sub-blocks, which are thus constrained to move together. Therefore, no relative movement between the component blocks takes place on the common faces, and the composite block behaves as a single FE mesh.

Rigid vs. Deformable Blocks

In a rigid block model all the system deformation is concentrated at the joints between the blocks. In the elastic range, the contact normal and shear stiffnesses govern the overall deformability. There are many masonry structures which fail by global mechanisms, for example voussoir arches or walls under out-of-plane loading, and may therefore be well represented by a rigid block system. In these cases, the DE model may be viewed as just a generalization of classical limit equilibrium techniques. For static analysis of simple systems, the joint stiffness often do not influence the failure loads. If realistic deformation estimation is required, the contact stiffnesses need to provide the whole deformability, taking into account the joints and units. For complex geometries, structures with different materials, or whenever weak units are likely to fail, it may be difficult to achieve a good representation with rigid blocks. Deformable blocks become a preferable option in these cases, but they require more input data, as both the joints and unit materials must be characterized. Fine meshes have been sometimes used to examine the stress paths in blocky structures, which are strongly influenced by the joint locations, as in study of a colonnade under in-plane seismic loads by Sarhosis et al. (2015) (Figure 9).

The internal mesh in deformable blocks is automatically generated given a user-defined average element size. Fine meshes are obviously more demanding in terms the run times, so the aim of the model should determine how fine the mesh needs to be. If a detailed stress analysis inside the block is not the objective, then a coarse FE mesh may be suitable. In computational terms, a static solution with a

Figure 9. Stress analysis in deformable blocks with fine meshes
(Sarhosis et al. 2015)

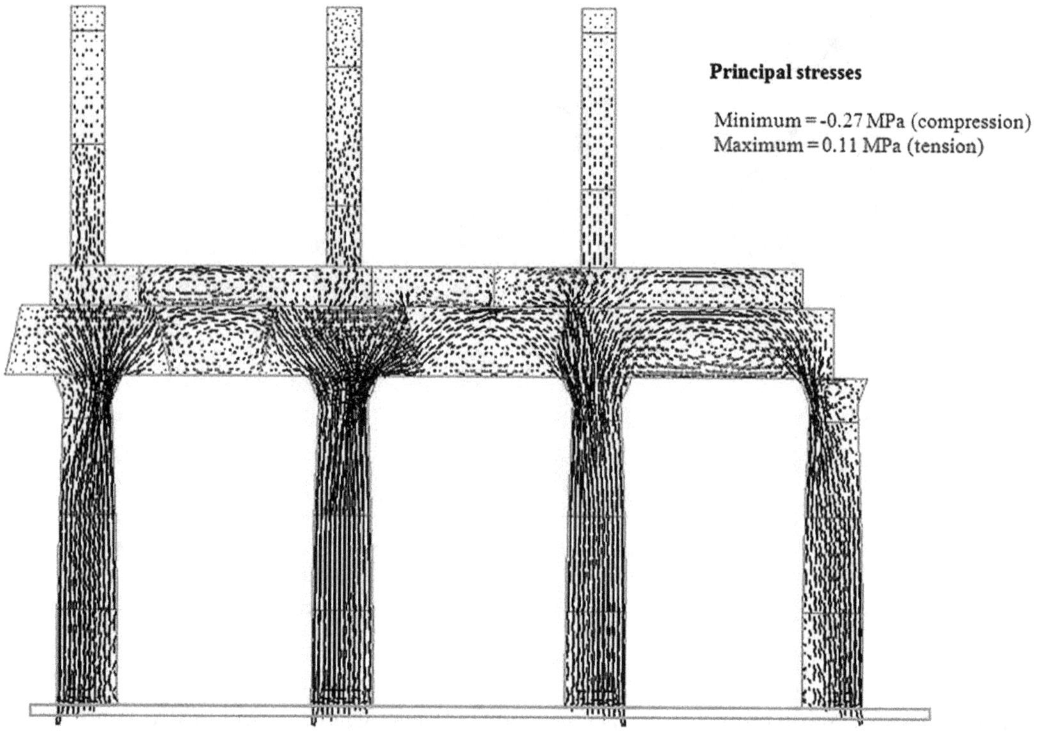

coarse-mesh deformable block model is not significantly greater than that of a rigid block model. The dynamic case is very different, as discussed below. Therefore, for static analysis, deformable blocks with coarse meshes may be used for most problems, particularly if the assignment of joint properties is more straightforward. An exception to this rule is the case of models with blocks of very different Young's modulus or size, as already discussed. For these cases, making the stiffer blocks rigid, and the other blocks deformable would be a possible solution. There are other cases in which the combination of both types of blocks is useful, for example, to model the soil foundation with a large deformable block, and the masonry with rigid blocks.

Dynamic analysis is the field where rigid block models continue to be essential. In static analysis, the mass scaling techniques make deformable blocks competitive. However, the dynamic response needs to be based on true masses, and the presence of very small or thin elements in the mesh will demand very small time steps to maintain numerical stability of the explicit algorithm. Lemos (2007b) compared the time steps for different representations of a simple structure, showing that, for some choices of material properties, time steps with deformable blocks may be an order of magnitude smaller. The Rayleigh damping parameters also affect the time step as explained in the section about dynamics.

Determination of Collapse Loads

The analysis of collapse mechanisms and the corresponding loads is one of the main applications of DE codes in masonry studies. In many cases, this is performed under static (or quasi-static) conditions.

Either load-control or displacement-control may be used. The former is much simpler and computationally efficient, but it is does not give information on the process after the peak load is reached. Displacement-control is able to provide a post-peak structural softening, but it usually requires longer run times, as low prescribed velocities are required to produce a smooth response. Even if the time scale is not real in quasi-static analysis, the user must still prescribe a velocity to the loaded boundary surfaces (or to the boundary rigid blocks). It is important to verify whether the deformation and failure process is smooth, or vibratory motion arises, which would indicate that the applied velocity is too high and should be reduced. Monitoring the motion (or stresses) at critical points in the model, and the total load being applied, is important to verify and understand the model results. The total force being applied at the prescribed velocity boundaries is usually obtained with a FISH function that, at every step, adds the reactions at those particular boundary grid-points or rigid blocks.

Whenever possible, load-control should be selected as it is the simpler and faster option. It is advisable to apply the loads in increments, particularly from the point at which the nonlinear behavior becomes important, to better simulate the physical processes taking place as the structure moves and deforms. The user may choose between 2 types of damping in a quasi-static simulation: viscous damping or Cundall's "local damping". Viscous damping, selected by the "damp auto" option, involves inserting mass-proportional damping in the equations of motion, controlled by an adaptive scheme, designed to insure a faster convergence to the solution. As the damping constant varies during the cycling process, the response of the model may display cyclic variations. In addition, a damping peak could slow down a failure process, giving the wrong suggestion of equilibrium in a model that is actually ailing slowly. So, it is useful to step a little beyond the beginning of a plateau in the displacement history to ensure that the model is actually stable and the key displacement histories are unchanged. The alternative "local damping" formulation is a type of non-viscous energy dissipation, appropriate for quasi-static analysis. It applies a damping force, to blocks and grid-points, opposite to the direction of velocity. However, unlike viscous damping, its magnitude is not proportional to the velocity, but to the magnitude of the unbalanced force. Local damping does not have the drawbacks associated with "damp auto", and it is generally preferable for the evaluation of collapse loads. There are a few systems, however, for which viscous damping provides smoother or faster solutions. So, if a large number of runs needs to be done, it may be advisable to experiment with both procedures in a trial test.

Dynamic Analysis with Rigid Blocks

Seismic analysis with rigid block models has been an important application of DE models, as the literature survey in the next section demonstrates. Drum columns, arches, walls and other structures frequently meet the conditions that justify a rigid block representation. The computational advantage of this option for explicit time stepping codes was already discussed.

In dynamic problems, a realistic representation of the structural stiffness is crucial as it governs the dominant vibration modes of the response. Since joint stiffnesses completely determine the global elastic deformability, they have to be calibrated to obtain a good match of the overall structural behavior. Therefore, the stiffness parameters should not be based on the actual joint surface properties, as would be the case in a deformable block model. Measurements of the natural frequencies by in situ tests based on ambience noise usually provide very useful information for this purpose. Of course, these values correspond to very small levels of deformation, not representative of movements under strong motion records, but they may be a good starting point, in the absence of other data.

Damping in Dynamic Analysis

The choice of damping for dynamic simulations always poses some difficulties, as there is often not enough experimental data to support this decision for various types of masonry structures. For nonlinear analysis with explicit time stepping codes, Rayleigh damping is the common option, since it is not possible to define damping ratios per eigenmode. Rayleigh damping involves 2 components: the mass proportional component, which provides energy dissipation inversely proportional to frequency; and the stiffness-proportional component, which increases linearly with frequency. Adding the 2 components leads to a U-shaped curve, with a relatively uniform plateau in the vicinity of the minimum. A common choice is to consider the minimum of the curve to be located at the fundamental frequency, and assign to it the desired fraction of critical damping. The damping for higher frequencies will be derived from the curve, and are outside user control.

For rocking problems, Pena et al. (2007) obtained a good match with experimental tests, using only stiffness-proportional damping, which, for rigid blocks, corresponds to adding a dashpot at the joints, in parallel with the joint stiffness spring. They also related this spring-dashpot contact to the classical restitution coefficients. Papantonopoulos et al. (2002) modelled shaking table tests of a drum column with 3DEC. They recommend using stiffness damping, but they also had good results for the response peaks without any viscous damping, i.e., relying only on joint friction as energy dissipation mechanism. Of course, the tail of the response was underdamped. In seismic assessment of simple structures made of strong rock blocks this could be a reasonable conservative assumption. For larger models, stiffness-proportional damping may be difficult to apply because it requires a reduction of the time step for numerical stability. The time step is automatically calculated by the code, and a rough run time estimate can tell if the simulation is feasible. The alternative is to apply only the mass-proportional component of Rayleigh damping, which does not demand a time step reduction. This is often the only workable solution. The desired fraction of critical damping is assigned to the fundamental frequency. All the higher frequencies will have a lower damping ratio. As a consequence, the dynamic response will contain considerable higher frequency noise, which could be filtered if necessary.

Eigenvalue Calculation

The current 3DEC code has a simple built-in procedure for calculation of eigenvalues for rigid block models, which is very useful in structural dynamics problems. If all the contacts are assumed to be elastic, then it is possible to build a stiffness matrix of a rigid block system, which is a function of the joint normal and shear stiffness. Under very low amplitude vibration, as is the case of in situ ambient noise characterization tests, the assumption of elastic response is acceptable. Therefore, these tests give helpful information for the choice of joint stiffnesses for dynamic analysis, notwithstanding the obvious approximations implied in a rigid block DE model.

As an alternative to the eigenvalue calculation, it is usually also feasible to obtain a rough estimation of the fundamental frequency by setting the model to run under free vibration without damping, by applying an initial perturbation intended to excite this mode. For example, for a tall structure, we can apply a horizontal gravity force for a couple of steps, then turn it off and plot the time response, where it may be possible to identify a low frequency harmonic component.

A rigid block system obviously seems a poor choice to represent an elastic structure. However, if the number of blocks is large, and the joints are elastic, it will be capable of simulating the main traits of behavior of the elastic system, although requiring more degrees-of-freedom than a deformable block formulation to achieve the same level of accuracy. Lemos (2007b) compared eigenvalue solutions of 3DEC rigid block models with the analytical solutions for beams and plates, concluding that the approximations obtained are acceptable for practical purposes. One of the examples concerned a cantilever wall, with three sections of different thickness, for which a numerical solution of the vibration modes, based on the Mindlin plate theory, is available (Liu & Buchanan, 2004). Table 1 compares the solution first 4 frequencies with a FEM continuum model (using 20-node bricks) and a 3DEC rigid block model. The mode shapes of the latter are shown in Figure 10.

Bonded Block Models

The interest for micro- and meso-scale modelling of geomaterials has been expanding, as numerical tools have proved an important tool in the understanding of their fundamental behavior. Bonded-particle models, using circular or spherical particles, have been used for some time in rock mechanics, and their extension to masonry appears promising (Azevedo et al., 2015). Bonded-block models are more complex and time consuming, but polygonal particles provide a closer approximating to the underlying material grain arrangements than circular particles.

Several authors have used this type of model in the past as an approximation to the material failure of masonry units or rock specimens. Sincraian and Lemos (1999) analyzed the fracture of a stone masonry pillar with UDEC, using bonded rigid blocks, where the bonded joints defined a few potential paths for crack development. Munjiza (2004) presented elaborate models of fracturing continua, in

Table 1. Natural frequencies (Hz) of the stepped cantilever wall problem

Mode	Mindlin Plate Solution	FEM	Rigid Blocks
1	1.40	1.40	1.36
2	2.52	2.50	2.99
3	5.46	5.37	5.48
4	6.18	6.10	6.68

Figure 10. Stepped cantilever wall problem. Mode shapes obtained with a rigid block model.
(Lemos, 2007b)

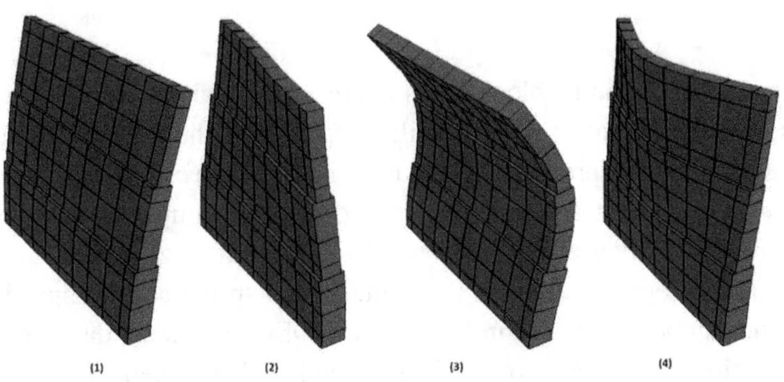

which elements of the mesh could progressively break off and become discrete blocks. Pina-Henriques and Lourenço (2006) applied FE micro-models with very fine meshes to examine in detail the fracture of bricks and brick-mortar specimens. Larger model sizes are now feasible, even in 3D. Garza-Cruz & Pierce (2014) used a 3DEC model with about 50,000 bonded tetrahedral blocks to study the spalling of rock near underground excavations. Each block has a single FE, so it is essentially a mesh in which the faces of every element can potential fail in tension or shear and allow a fracture pattern to develop.

These models are particularly useful to examine the behavior of complex materials and constructions, such as traditional masonry with irregular units, or failure processes under complex stress states and general load paths. They also constitute an invaluable aid in the interpretation of the results of laboratory experiments. The application of constitutive models that include fracture mechanics criteria is important to ensure consistency of numerical results.

Model Verification and Error Detection

In 2D simulations, it is possible to visualize the entire domain, so the detection of errors in the input data tends to be evident. In complex 3D problems, model building becomes more difficult and time consuming. Verification of data requires good visualization tools, which allow rapid 3D perspectives, as well as cross-sections through the model. Typical errors involve the wrong assignment of material properties and, particularly, of joint properties. The properties actually assigned to the various joints and model regions must be thoroughly verified, preferably using the graphical interface. Boundary conditions and applied loads may also be a source of errors. In UDEC/3DEC it is often useful to step a few cycles and stop to visualized the velocity field pattern that is starting to develop. It is an easy way to detect a wrong location of fixed points or loads. The results of elastic solutions are usually more intuitive and easier to foresee. It is a good practice to perform a trial elastic run before the nonlinear analyses, by assigning large values to the strength of joints and blocks. If the output does not meet the expectations, then it is useful to understand the differences before proceeding to more complex analyses. The liberal use of histories of displacement and stresses at various locations is also recommended, in order to build a complete picture of the system performance under static or dynamic loads. Contours of key variables on the block surfaces or cross-sections of the model are also essential to verify and interpret the global model response, since relying only on curves showing a single global indicator may be misleading.

APPLICATION EXAMPLES

Block Rocking

The rocking motion of a free-standing block under cyclic horizontal actions plays a major role in many masonry problems, namely the out-of-plane stability of walls, or the failure of columns and other tall structures. The classical approach proposed by Housner involves the concept of a restitution coefficient, which controls the energy dissipation in the sequence of alternate impacts of the block corners on the rigid base. This approach is very useful for analytical solutions of single blocks under various types of excitation. The alternative representation of the contact by a spring and dashpot is more suitable for numerical models and can be easily used in UDEC and 3DEC, where the dashpot simply corresponds to the stiffness-proportional component of Rayleigh damping. This scheme has been shown by several authors to provide responses similar to Housner's solution, e.g., Pena et al. (2007), which also compared the UDEC model with results of shaking table tests of single rock blocks of different dimensions under harmonic and earthquake motions.

Columns

The study of the seismic behavior of drum columns in classical temples was among the first masonry problems addressed with UDEC (Papastamatiou & Psycharis, 1993). 3DEC models were compared with shaking table tests of a marble column performed at NTU Athens, showing the capability of the numerical representation to address the complexity of the dynamic phenomenon, characterized by a marked variability and sensitivity to input parameters, discussed by Papantonopoulos et al. (2002). The ability to represent the existing damage in the drums, and the role of reinforcement elements, was exemplified by Psycharis et al. (2003), in the 3DEC modelling of the restoration options of the column-architrave structure of the Partheneon Pronaos (Figure 11). The effect of cracked drums on structural safety and the implementation of reinforcement techniques were addressed by Stefanou et al. (2013), based on data collected in laboratory tests.

Many other studies of stone columns have been performed by several researchers. Ambraseys & Psycharis (2011) review the methods to assess stability of columns and statues mounted on columns, including a 3DEC analysis of a case study. Dimitri (2011) examined the effect of harmonic impulses of different frequencies and durations, as well as the role damping and material properties. Baggio et al. (2002) modelled the colonnades of the Coliseum hypogeum with UDEC. A 3DEC model of the roman temple in Évora, Portugal, was developed by Nayeri (2012), who represented accurately the present shapes of drum columns and architraves measured by laser scanning. The joint stiffness parameters were calibrated with the results of in situ dynamic characterization tests, in order to approximate the measured frequencies and the lower mode shapes. The issues of variability of the rocking behavior, evident in laboratory tests, as well as the uncertainties about material properties or the seismic excitation, require

Figure 11. Models of columns with damaged drum under seismic loading
(Psycharis et al., 2003)

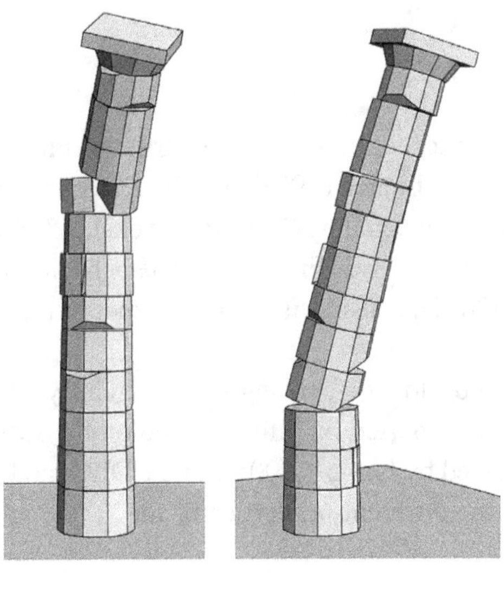

more elaborate methods of assessing the output of numerical calculations. A methodology for seismic risk assessment of classical columns was proposed by Psycharis et al. (2013), in which 3DEC simulations are used to assist the development of fragility curves.

The seismic stability of drum columns and column-architrave structures is usually well addressed with rigid bock models. However, detailed analysis with deformable blocks can provide an insight into the internal flow of stresses and the interplay of the various structural components. Giamundo et al. (2014) used this approach to study a colonnade in Pompeii (Figure 9). For this type of model, seismic stability assessment may be achieved by means of static pushover analysis.

Towers and Other Tall Structures

The stability of different types of tall structures, always a major concern in seismic regions, has been addressed by several authors with 3DEC. Early 3D models for dynamic analysis were usually highly simplified, in order to reduce run times, namely resorting to block sizes much larger than the real ones, e.g., Lemos et al. (2000). Oliveira et al. (2002) employed 3DEC models to reproduce the observed large displacements and rotations displayed by several simple tall structures (such as a tower, a chimney and a statue), during the 1998 Azores earthquake. More elaborate models are possible nowadays, for example, the minaret models presented by Catki et al. (2014, 2015), which compared the damage and failure modes produced by various types of recorded and synthetic input records (Figure 12). Lemos et al. (2015) analyzed the observed rotations of a stone obelisk displaced during the Lorca earthquake with a rigid block model, stressing the importance of measuring the natural frequencies in situ in order to calibrate the joint stiffness parameters. The dynamics of church spires were studied by DeJong & Vibert (2012), who compared the numerical results with a physical model experiment.

Aqueducts and other arched structures were also analyzed with rigid block models (e.g., Sincraian et al., 1998; Drei & Oliveira, 2001). Sincraian and Lemos (1999) employed a 2D UDEC model to analyze the out of plane behavior of a pillar of the Lisbon aqueduct under seismic loads. In this model, the possible cracking of the stone blocks was investigated by inserting joints of various orientations in the critical blocks, which were assigned the tensile and cohesive strength corresponding to the intact stone properties. These joints represented predefined potential failure paths, as is presently done in the "bonded block models".

Arches

The dynamics of free-standing arches, often present in partially ruined structures, were analyzed with simple 3DEC rigid block models by Lemos (1998). In-plane mechanisms displayed by circular arches under impulsive loads were studied in detail by De Lorenzis et al. (2007), applying both analytical solutions and numerical simulations with UDEC (Figure 13). The behavior of arches supported by buttresses was simulated by Dimitri et al. (2011), considering various types of block representations of the support walls and buttresses.

The previous works used rigid blocks, focusing on the arch dynamic response. Static analysis of an arcade composed by 3 arches supported by pillars, based on the structure of the S. Vicente de Fora monastery in Lisbon, was presented by Lemos (2001). The UDEC model employed deformable blocks, allowing the representation of the different properties of the stones and fill materials. The numerical

Figure 12. Model of a minaret and view of internal structure
(Catki et al., 2014)

Figure 13. Deformation of arch subject to a base dynamic impulse
(De Lorenzis et al., 2007)

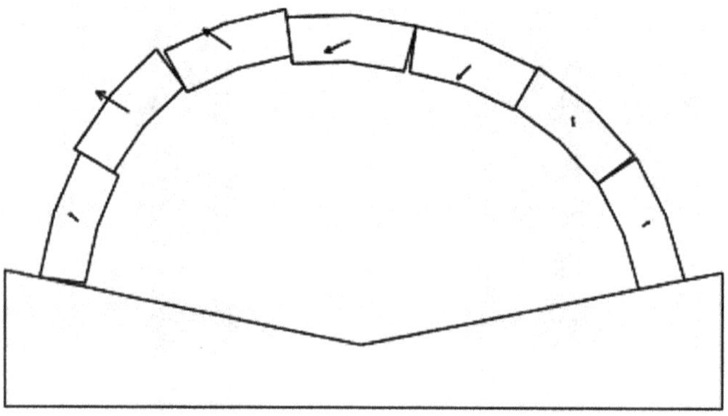

model was able to simulate quite well the cyclic response observed in quasi-static tests of a physical model conducted at the ISPRA laboratory.

Arch Bridges

The carrying capacity of masonry arch bridges is traditionally evaluated by analytical solutions, based on the plasticity limit theorems, well described in the classical works by Heyman. Discrete element models provide an alternative approach, very useful for cases in which in geometry does not meet the assumptions of the classical techniques, and in general for 3D analysis. Lemos (1995) presented a 3DEC model of a masonry bridge including the arch and the spandrel walls, and studied the effects of the load distribution across the width. Sarhosis et al. (2014) addressed the case of skew bridges, examining the effect of the arch shape and the angle of skew on the bridge capacity. Rigid block models are often sufficient for these models which neglect the fill contribution to the arch strength. However, this can also be taken into account, as done by Toth et al. (2009), with a UDEC model in which the fill material was simulated by deformable blocks with elasto-plastic behavior. Tran et al. (2014) employed also deformable blocks, with very refined internal meshes, to calculate the displacements during arch loading. The considerable overlap that presently exists between the numerical methods is highlighted by the model proposed by Milani and Lourenço (2012), which even if formulated within a FE framework, represents the arch bridge and the fill by rigid polyhedra with deformable nonlinear interfaces.

Vaults and Domes

The failure modes of groin vaults were analyzed by McInerney & DeJong (2014), who compared 3DEC numerical models with simplified analytical solutions. In particular, the sensitivity of the structure to support displacements was examined. Simon & Bagi (2014) applied 3DEC models to the determination of the minimum thickness of oval domes, which allowed a critical discussion of the classical empirical and analytical approaches to the design of these structures (Figure 14). Lengyel and Bagi (2015) investigated the role played by ribs in groin vaults, comparing 3DEC rigid block models and finite element models using shell elements. The results show the ability of discontinuum models to represent realistically the contact between rib and vault.

Figure 14. Collapse mechanism for spherical dome (colors indicate displacement magnitude
(Simon & Bagi, 2014)

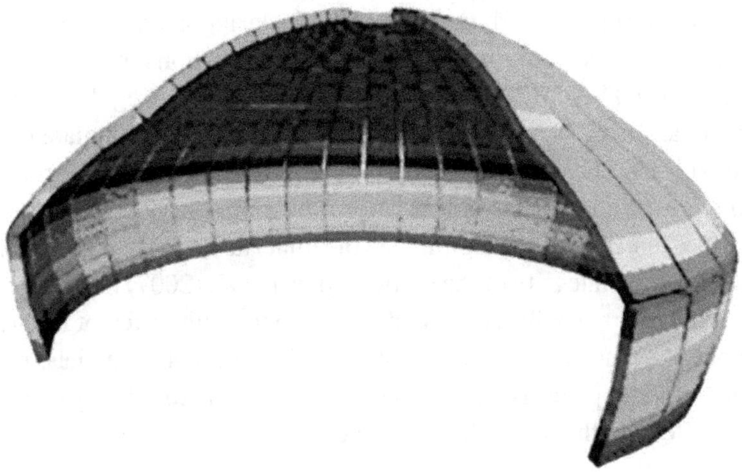

Walls and Panels

A substantial literature exists on FE micro-modelling of brick wall panels under in-plane loads. UDEC models with deformable blocks have also been used for this type of problem (e.g., Sincraian & Azevedo, 1998; Schlegel & Rautenstrauch, 2004; Churilov & Dumova, 2008). De Felice and Giannini (2001) studied the out-of-plane behavior of masonry walls, considering the interlocking with cross-walls, using regular block patterns. Asteris et al. (2013) review the state of the art of modelling infilled frames, considering FE and DE models. Sarhosis et al. (2014) applied UDEC to the study of steel frames with masonry infill, examining the effect of window openings, and comparing the cases of point and distributed horizontal loads.

The key issue of the calibration of numerical models with experimental data was addressed by Sarhosis & Sheng (2014). Using a UDEC model of a brick panel with an opening, they developed a methodology to identify the numerical model parameters from laboratory tests, which was then applied successfully to predict the results of other experiments (Figure 15).

Figure 15. Failure mode of panel under vertical load
(Sarhosis & Sheng, 2014)

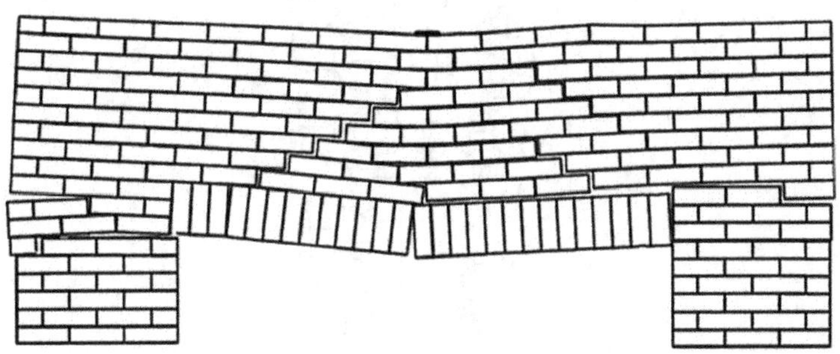

Irregular Stone Walls

The ability of DE models to represent arbitrarily shaped blocks has been employed by various researchers to address the representation of traditional masonry constructions, displaying irregular units and complex patterns. Roberti and Spina (2001) used a UDEC model with irregular polygonal blocks in their analysis of the ancient Sardinian 'Nuraghe' structures. De Felice (2011) simulated a 3-leaf cross section of a traditional wall with UDEC, simulating closely the observed irregular blocks geometries, namely in the very detailed representation of the inner leaf of rubble masonry (Figure 16). Static and dynamic analyses were performed, and the results highlighted the importance of the connection between the outer wall leaves in maintaining the integrity of the section. Meyer et al. (2007) also addressed the 3-leaf wall problem, focusing on the internal collapse under dynamic input with different frequencies. Lemos et al. (2011) studied the out-of-plane behavior of masonry walls with 3DEC models, using static pushover analysis. Regular and irregular patterns were considered, namely Voronoi polygons and irregular course geometries, and their effect on the failure loads were examined.

Figure 16. Analysis of failure modes of 3-leaf wall sections
(De Felice, 2011)

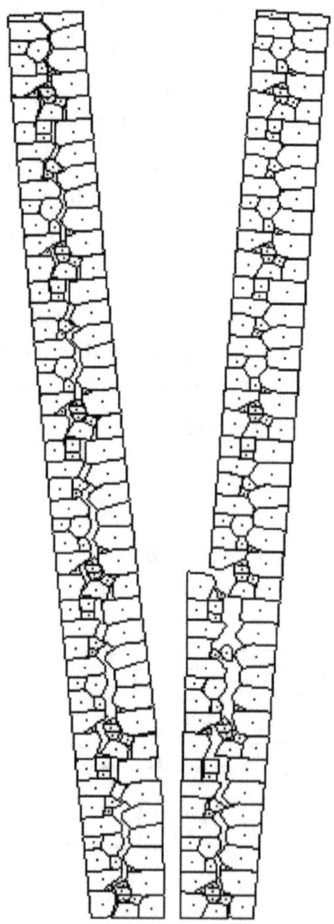

Buildings and Other Structures

Large and complex building always pose a challenge to analysts. First, the generation of the model is time-consuming. Then, run times may be forbidding, particularly in dynamic analysis. Simplification is always required, since it is impossible to represent all the blocks. In fact, many details of the construction are not visible and have to be inferred. A simplified model that represents well the main components and their interactions is often capable of providing realistic results, even if the block sizes are larger than the real ones, i.e., if a numerical block is in effect a macro-block assembling various real blocks. The model must represent the possible collapse mechanisms. Even if the blocks are larger than the reality, if there are enough of them, and joints to provide the failure paths, then the collapse modes and loads will still be well approximated.

Constructions with simpler shapes create fewer difficulties. The collapse mechanisms of traditional 1- and 2-storey houses under earthquake loading were analyzed by Alexandris et al. (2004), in 2D and 3D. The models were used to evaluate alternative intervention options. Psycharis et al. (2011) studied the behavior of the Parthenon walls with a rigid block model, examining, in particular, the effects of different seismic motions (Figure 17). Zanotti (2015) created a complex 3D model of a church, considering different walls sections made of irregular rubble masonry and regular adobe blocks, and performed static pushover analysis to evaluate seismic performance.

Models can also be useful to investigate the strength of historical hypotheses, in the context of discussions about the events that led to the destruction of monuments or buildings. Alexandris et al. (2014) performed analyses of single column models to discuss the destruction of the temple of Zeus at Olympia. Young et al. (2015) developed a 3DEC model of the ancient temple of Nemea to investigate if an earthquake was a plausible cause of collapse.

Figure 17. Model for dynamic analysis of the Parthenon walls
(Psycharis et al., 2011)

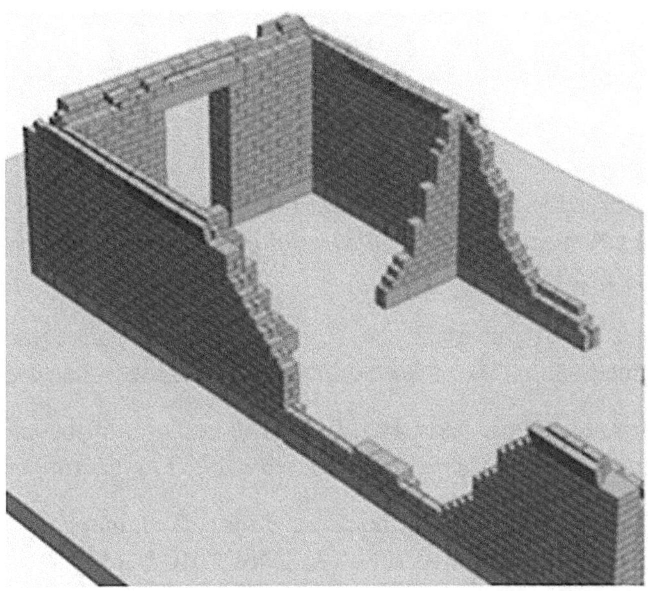

Studies of soil-structure interaction with DE codes have also been published. Powrie et al. (2002) provide an application to drystone retaining walls, with deformable blocks for the back-fill soil. Alexakis & Makris (2013) analyzed the tunnel entrance to the ancient stadium at Nemea, using deformable blocks to assess the stresses in the roof arch.

CONCLUSION

The expanding group of discrete element methods exhibits a significant variety of fundamental assumptions and numerical implementation features. The choice of the most appropriate tool for the analysis of a masonry structure, and the ability to use it judiciously, requires a good understanding of those essential concepts. In this chapter, the UDEC and 3DEC codes, representative of a central line of DE formulations, were examined, with the purpose of providing some insight and guidance for their application to masonry structures.

The analysis of simple block structures and structural components, under static or dynamic loads, is handled in a very straightforward manner by modern DE codes. Difficulties in the choice of input parameters reflect mostly the shortage of experimental data, of a proper characterization of the materials and the construction, and are thus common to any modeling attempt. On the other hand, large and complex structures still pose a challenge to discontinuous idealizations, as it is mandatory to simplify and this requires experience and a good insight into the expected structural behavior. The problem is more serious in dynamic problems, where partial models are often not feasible due to the inability to establish correct boundary conditions. Some amount of experimentation with different representations is normally necessary until a good solution is found. Therefore, developments in model generation procedures applicable to complicated structures are essential to improve the analysis effectiveness.

At the other end of the analysis scale, the meso- and micro-scale, we can expect an increased use of DE block models to investigate the fundamental behavior of materials, often tied to traditional construction techniques, and explore new rehabilitation methodologies. In this field, close interaction between numerical techniques and laboratory experiments, is the key to advance the state of knowledge and our predictive capabilities.

REFERENCES

Alexakis, H., & Makris, N. (2013). Structural stability and bearing capacity analysis of the tunnel-entrance to the stadium of ancient Nemea. *International Journal of Architectural Heritage*, *7*(6), 673–692. doi:10.1080/15583058.2012.662262

Alexandris, A., Protopapa, E., & Psycharis, I. (2004). Collapse mechanisms of masonry buildings derived by the distinct element method. In *Proc. 13th World Conf. Earthquake Engineering*.

Alexandris, A., Psycharis, I., & Protopapa, E. (2014). The collapse of the ancient temple of Zeus at Olympia revisited. In *2nd European Conference on Earthquake Engineering and Seismology*.

Ambraseys, N., & Psycharis, I. N. (2011). Earthquake stability of columns and statues. *Journal of Earthquake Engineering*, *15*(5), 685–710. doi:10.1080/13632469.2010.541549

Asteris, P. G., Cotsovos, D. M., Chrysostomou, C. Z., Mohebkhah, A., & Al-Chaar, G. K. (2013). Mathematical micro-modeling of infilled frames: State of the art. *Engineering Structures*, *56*, 1905–1921. doi:10.1016/j.engstruct.2013.08.010

Azevedo, N. M., Lemos, J. V., & Almeida, J. R. (2015). Discrete Element Particle Modelling of Stone Masonry. In Sarhosis et al. (Eds.), *Computational Modeling of Masonry Structures Using the Discrete Element Method*. IGI Global.

Baggio, C., De Felice, G., Sguerri, L., & Callipo, C. (2002). The resistance to earthquakes of structures made by dry stone blocks: The case of the Coliseum hypogeum. In *Proc. 12th European Conference on Earthquake Engineering*. Elsevier.

Cakti, E., Oliveira, C. S., Lemos, J. V., Saygili, O., Görk, S., & Zengin, E. (2015). Ongoing Research on Earthquake Behavior of Historical Minarets in Istanbul. In I. N. Psycharis et al. (Eds.), *Seismic Assessment, Behavior and Retrofit of Heritage Buildings and Monuments, Computational Methods in Applied Sciences, 37*. Springer.

Cakti, E., Saygili, O., Lemos, J. V., & Oliveira, C. S. (2014). A parametric study on earthquake behavior of masonry minarets. *Tenth U.S. National Conference on Earthquake Engineering, Frontiers of Earthquake Engineering*.

Churilov, S., & Dumova-Jovanoska, E. (2008). Calibration of a numerical model for masonry with applications to experimental results. Journal of Architecture, Civil Engineering and Environment, 3, 41-48.

Cundall, P. A. (1971). A computer model for simulating progressive large scale movements in blocky rock systems. In *Proc. Symp. Rock Fracture (ISRM)*.

Cundall, P. A. (1988). Formulation of a three-dimensional distinct element model - Part I: A scheme to detect and represent contacts in a system composed of many polyhedral blocks. *International Journal of Rock Mechanics and Mining Sciences*, *25*(3), 107–116. doi:10.1016/0148-9062(88)92293-0

Cundall, P. A., & Strack, O. D. L. (1979). A discrete numerical model for granular assemblies. *Geotechnique*, *29*(1), 47–65. doi:10.1680/geot.1979.29.1.47

De Felice, G. (2011). Out-of-plane seismic capacity of masonry depending on wall section morphology. *International Journal of Architectural Heritage*, *5*(4-5), 466–482. doi:10.1080/15583058.2010.530339

De Felice, G., & Giannini, R. (2001). Out-of-plane seismic resistance of masonry walls. *Journal of Earthquake Engineering*, *5*(2), 253–271. doi:10.1080/13632460109350394

De Lorenzis, L., DeJong, M., & Ochsendorf, J. (2007). Failure of masonry arches under impulse base motion. *Earthquake Engineering & Structural Dynamics*, *36*(14), 2119–2236. doi:10.1002/eqe.719

DeJong, M. J., & Vibert, C. (2012). Seismic response of stone masonry spires: Computational and experimental modeling. *Engineering Structures*, *40*, 566–574. doi:10.1016/j.engstruct.2012.03.001

Dimitri, R., De Lorenzis, L., & Zavarise, G. (2011). Numerical study on the dynamic behavior of masonry columns and arches on buttresses with the discrete element method. *Engineering Structures*, *33*(12), 3172–3188. doi:10.1016/j.engstruct.2011.08.018

Drei, A., & Oliveira, C. S. (2001). The seismic behaviour of the 'Aqueduto da Amoreira' in Elvas using distinct element modeling. In P. B. Lourenço & P. Roca (Eds.), *Historical Constructions* (pp. 903–911). Guimarães.

Garza-Cruz, T. V., & Pierce, M. (2014). A 3DEC Model for Heavily Veined Massive Rock Masses. In *Proceedings of the 48th US Rock Mechanics/Geomechanics Symposium*. American Rock Mechanics Associated.

Giamundo, V., Sarhosis, V., Lignola, G. P., & Cosenza, E. (2014). Discrete element modelling of the archaeological colonnade in Pompeii. In *9th International Masonry Conference*.

Goodman, R. E., Taylor, R. L., & Brekke, T. L. (1968). A model for the mechanics of jointed rock. *Journal of the Soil Mechanics and Foundations Division*, *94*(3), 637–659.

Hart, R. D., Cundall, P. A., & Lemos, J. V. (1988). Formulation of a three-dimensional distinct element model - Part II: Mechanical calculations. *International Journal of Rock Mechanics and Mining Sciences*, *25*(3), 117–125. doi:10.1016/0148-9062(88)92294-2

Itasca. (2013). *3DEC - Three-dimensional Distinct Element Code*. Itasca Consulting Group.

Itasca. (2014). *UDEC - Universal Distinct Element Code*. Itasca Consulting Group.

Lemos, J. V. (1995). Assessment of the ultimate load of a masonry arch using discrete elements. In J. Middleton & G.N. Pande (Eds.), Computer Methods in Structural Masonry. Books & Journals International.

Lemos, J. V. (1998). Discrete element modelling of the seismic behaviour of stone masonry arches. In J. Middleton, G. N. Pande, & B. Kralj (Eds.), *Computer Methods in Structural Masonry - 4* (pp. 220–227). London: E&FN Spon.

Lemos, J. V. (2001). Modelling the behaviour of a stone masonry arch structure under cyclic loads. In T.G. Hughes & G.N. Pande (Eds.), Computer Methods in Structural Masonry - 5. Computers & Geotechnics Ltd.

Lemos, J. V. (2007a). Discrete element modeling of masonry structures. *International Journal of Architectural Heritage*, *1*(2), 190–213. doi:10.1080/15583050601176868

Lemos, J. V. (2007b). Numerical issues in the representation of masonry structural dynamics with discrete elements. In M. Papadrakakis, D.C. Charmpis, N.D. Lagaros, Y. Tsompanakis (Eds.), Compdyn2007. Rethymno.

Lemos, J. V., Costa, A. C., & Bretas, E. M. (2011). Assessment of the seismic capacity of stone masonry walls with block models. In M. Papadrakakis, M. Fragiadakis & N.D. Lagaros (Eds.), Computational Methods in Earthquake Engineering. Springer. doi:10.1007/978-94-007-0053-6_10

Lemos, J. V., Oliveira, C. S., & Navarro, M. (2015). 3D nonlinear behavior of an obelisk subjected to the Lorca May 11, 2011 strong motion record. *Engineering Failure Analysis*, *58*, 212–228. doi:10.1016/j.engfailanal.2015.09.001

Lemos, J. V., Sincraian, G. E., & Oliveira, C. S. (2000). Modelling the seismic behaviour and structural restoration of a damaged masonry tower. In *Proc. 5th Int. Congress on Restoration of Architectural Heritage*. CICOP.

Lengyel, G., & Bagi, K. (2015). Numerical analysis of the mechanical role of the ribs in groin vaults. *Computers & Structures*, *158*, 42–60. doi:10.1016/j.compstruc.2015.05.032

Liu, Y. J., & Buchanan, R. B. (2004). Free vibration of stepped cantilever Mindlin plates. *Journal of Sound and Vibration*, *271*(3-5), 1083–1092. doi:10.1016/S0022-460X(03)00777-6

McInerney, J., & DeJong, M. J. (2014). Discrete element modelling of groin vault displacement capacity. *International Journal of Architectural Heritage*. doi:10.1080/15583058.2014.923953

Meyer, P., Ochsendorf, J., Germaine, J., & Kausel, E. (2007). The impact of high-frequency/low-energy seismic waves on unreinforced masonry. *Earthquake Spectra*, *23*(1), 77–94. doi:10.1193/1.2431211

Milani, G., & Lourenço, P. B. (2012). 3D non-linear behavior of masonry arch bridges. *Computers & Structures*, *110-111*, 133–150. doi:10.1016/j.compstruc.2012.07.008

Munjiza, A. (2004). *The combined finite-discrete element method*. John Wiley. doi:10.1002/0470020180

Nayeri, S. A. (2012). *Seismic Assessment of the Roman Temple in Évora, Portugal*. (MS thesis). University of Minho, Guimarães, Portugal.

Oliveira, C. S., Lemos, J. V., & Sincraian, G. E. (2002). Modelling large displacements of structures damaged by earthquake motions. *European Earthquake Engineering*, *3*, 56–71.

Papantonopoulos, C., Psycharis, I. N., Papastamatiou, D. Y., Lemos, J. V., & Mouzakis, H. P. (2002). Numerical prediction of the earthquake response of classical columns using the distinct element method. *Earthquake Engineering & Structural Dynamics*, *31*(9), 1699–1717. doi:10.1002/eqe.185

Papastamatiou, D., & Psycharis, I. (1993). Seismic response of classical monuments – A numerical perspective developed at the Temple of Apollo Bassae, Greece. *Terra Nova*, *5*(6), 591–601. doi:10.1111/j.1365-3121.1993.tb00309.x

Peña, F., Prieto, F., Lourenço, P. B., Costa, A. C., & Lemos, J. V. (2007). On the dynamics of rocking motion of single rigid-block structures. *Earthquake Engineering & Structural Dynamics*, *36*(15), 2383–2399. doi:10.1002/eqe.739

Pina-Henriques, J., & Lourenço, P. B. (2006). Masonry compression: A numerical investigation at the meso-level. *Engineering Computations*, *23*(4), 382–407. doi:10.1108/02644400610661163

Powrie, W., Harkness, R. M., Zhang, X., & Bush, D. I. (2002). Deformation and failure of drystone retaning walls. *Geotechnique*, *52*(6), 435–446. doi:10.1680/geot.2002.52.6.435

Psycharis, I., Drougas, A., & Daisou, M. (2011). Seismic Behaviour of the Walls of the Parthenon: A Numerical Study. In Computational Methods in Earthquake Engineering. Computational Methods in Applied Sciences. doi:10.1007/978-94-007-0053-6_12

Psycharis, I. N., Fragiadakis, M., & Stefanou, I. (2013). Seismic reliability assessment of classical columns subjected to near-fault ground motions. *Earthquake Engineering & Structural Dynamics*, *42*, 2061–2079.

Psycharis, I. N., Lemos, J. V., Papastamatiou, D. Y., Zambas, C., & Papantonopoulos, C. (2003). Numerical study of the seismic behaviour of a part of the Parthenon Pronaos. *Earthquake Engineering & Structural Dynamics*, *32*(13), 2063–2084. doi:10.1002/eqe.315

Psycharis, I. N., Papastamatiou, D. Y., Cundall, P. A., & Lorig, L. (1993) Numerical and analytical modeling of Greek classical columns.*Proc. 3rd Pan American Congress of Applied Mechanics - PACAM III.*

Roberti, G. M., & Spina, O. (2001). Numerical analysis of the Sardinian 'Nuraghe. In T. G. Hughes & G. N. Pande (Eds.), *Computer Methods in Structural Masonry - 5* (pp. 190–197). Swansea, UK: Computers & Geotechnics Ltd.

Sarhosis, V., Lignola, G. P., & Asteris, P. (2015). Seismic Vulnerability of Ancient Colonnade: Numerical analysis of the two storey colonnade of the Forum in Pompeii. In P. G. Asteris & V. Plevris (Eds.), Seismic Assessment and Rehabilitation of Historic Structures. IGI Global.

Sarhosis, V., Oliveira, D. V., Lemos, J. V., & Lourenço, P. B. (2014). The effect of skew angle on the mechanical behaviour of masonry arches. *Mechanics Research Communications*, *61*, 53–59. doi:10.1016/j.mechrescom.2014.07.008

Sarhosis, V., & Sheng, Y. (2014). Identification of material parameters for low bond strength masonry. *Engineering Structures*, *60*, 100–110. doi:10.1016/j.engstruct.2013.12.013

Sarhosis, V., Tsavdaridis, K., & Giannopoulos, G. (2014). Discrete Element Modelling of masonry infilled steel frames with multiple window openings subjected to lateral load variations. *Open Construction and Building Technology Journal*, *8*(1), 93–103. doi:10.2174/1874836801408010093

Schlegel, R., & Rautenstrauch, K. (2004). Failure analyses of masonry shear walls. In H. Konietzky (Ed.), *Numerical modelling of discrete materials in geotechnical engineering, civil engineering and earth sciences* (pp. 15–20). London: Taylor & Francis.

Shi, G.-H., & Goodman, R. E. (1988). Discontinuous deformation analysis – A new method for computing stress, strain and sliding of block systems. In Cundall, Sterling & Starfield (Eds.), Key Questions in Rock Mechanics. Balkema.

Simon, J., & Bagi, K. (2014). Discrete element analysis of the minimum thickness of oval masonry domes. *International Journal of Architectural Heritage*. doi:10.1080/15583058.2014.996921

Sincraian, G. E., & Azevedo, J. J. (1998). Numerical simulation of the seismic behaviour of stone masonry structures using the discrete element method. In Bisch, Labbé & Pecker (Eds.), *Proc. 11th European Conf. on Earthquake Eng*. Balkema.

Sincraian, G. E., & Lemos, J. V. (1999). Seismic analysis of a stone masonry aqueduct using discrete elements. In *Proc. 8th Canadian Conference on Earthquake Engineering*.

Sincraian, G. E., Oliveira, C. S., & Lemos, J. V. (1998). Assessment of the seismic behaviour of a stone masonry aqueduct using the discrete element method. In Bisch, Labbé & Pecker (Eds.), *Proc. 11th European Conf. on Earthquake Eng*. Balkema.

Stefanou, I., Psycharis, I., & Georgopoulos, I. O. (2011). Dynamic response of reinforced masonry columns in classical monuments. *Journal of Construction and Building Materials*, *25*(1), 4325–4337. doi:10.1016/j.conbuildmat.2010.12.042

Toth, A. R., Orban, Z., & Bagi, K. (2009). Discrete element modelling of a stone masonry arch. *Mechanics Research Communications*, *36*(4), 469–480. doi:10.1016/j.mechrescom.2009.01.001

Tran, V. H., Vincens, E., Morel, J. C., Dedecker, F., & Le, H. H. (2014). 2D-DEM modelling of the formwork removal of a rubble stone masonry bridge. *Engineering Structures*, *75*, 448–456. doi:10.1016/j.engstruct.2014.05.048

Young, M. P., Schultz, A. E., & Lemos, J. V. (2015). Seismic analysis of the Panhellenic Sanctuary of Nemea, Greece. In *Proc. 12th North American Masonry Conference.*

Zanotti, S. (2015). *Seismic Analysis of the Church of Kuño Tambo (Peru).* (MS thesis). University of Minho, Guimarães, Portugal.

Chapter 4
The DDA Method

Katalin Bagi
Budapest University of Technology and Economics, Hungary

ABSTRACT

"DDA" stands for "Discontinuous Deformation Analysis", suggesting that the displacement field of the analyzed domain shows abrupt changes on the element boundaries in the model. This chapter introduces the theoretical fundaments of DDA: mechanical characteristics of the elements together with the basic degrees of freedom, contact behavior, the equations of motion and their numerical integration with the help of Newmark's beta-method taking into account contact creation, loss and sliding with the help of an open-close iteration technique. Finally, a short overview on practical and scientific applications for masonry structures is given.

INTRODUCTION

DDA is a numerical technique to simulate the mechanical behavior of a collection of polyhedral blocks. "DDA" stands for "Discontinuous Deformation Analysis", suggesting that the displacement field shows abrupt changes on the element boundaries in the model. (This is a fundamental difference from a finite element model where continuity conditions have to be satisfied at the coinciding nodes of neighboring elements.)

The first DDA model was published by Gen-Hua Shi in his PhD dissertation in 1988 and in a journal paper Shi, 1992. That code was a two-dimensional model, with uniform-strain polygonal elements, having undeformable contacts that transmitted normal and tangential forces between the elements. Shi originally suggested his model for the analysis of fractured rocks. In the next decades several versions of DDA were published: three-dimensional models were born (Shi, 2001), different element shapes were applied (e.g. spheres, Zhao et al, 2000), higher-order displacement fields inside the elements were used (MacLaughlin, 1997) etc. The method has been widely applied for the analysis of masonry structures like arches, columns or stone bridges (Thavalingam et al, 2001; Scheldt et al, 2002; Kamai et al, 2005; Kamai & Hatzor, 2008; Rizzi et al, 2014 etc).

DOI: 10.4018/978-1-5225-0231-9.ch004

This chapter focuses on DDA with polyhedral elements. The basic degrees of freedom and the element behavior are introduced in Section 2. Section 3 focuses on the contact behavior. Section 4 summarizes the quantities in the equations of motion, and Section 5 introduces how the time integration is done: the main idea of the calculation of a single time step is explained. Then in Section 6 applications of DDA to masonry problems are shown.

THE ELEMENTS

The elements in Shi's model have arbitrary polyhedral shapes. Each element has a reference point, which is the centre of gravity of the element. Denote the position of the reference point of element p by (x^p, y^p, z^p).

The element can translate and rotate as a rigid body (6 degrees of freedom); and in addition to that, a uniform deformation field (characterized by the usual small strain tensor) belongs to the element (another 6 degrees of freedom). The element has altogether 12 degrees of freedom, and the 12 unknown kinematic characteristics are summarized into the generalized displacement vector of the element:

$$\mathbf{u}^p = \begin{bmatrix} u_x^p \\ u_y^p \\ u_z^p \\ \varphi_x^p \\ \varphi_y^p \\ \varphi_z^p \\ \varepsilon_x^p \\ \varepsilon_y^p \\ \varepsilon_z^p \\ \gamma_{yz}^p \\ \gamma_{zx}^p \\ \gamma_{xy}^p \end{bmatrix}$$

Assuming small displacements, the translations of an arbitrary (x, y, z) point on the element can uniquely be determined with the help of the $\mathbf{u}^p$ generalized displacement vector of the element:

$$
\begin{bmatrix} u_x(x,y,z) \\ u_y(x,y,z) \\ u_z(x,y,z) \end{bmatrix} =
$$

$$
= \begin{bmatrix}
1 & 0 & 0 & 0 & (z-z^p) & -(y-y^p) & (x-x^p) & 0 & 0 & 0 & \frac{(z-z^p)}{2} & \frac{(y-y^p)}{2} \\
0 & 1 & 0 & -(z-z^p) & 0 & (x-x^p) & 0 & (y-y^p) & 0 & \frac{(z-z^p)}{2} & 0 & \frac{(x-x^p)}{2} \\
0 & 0 & 1 & (y-y^p) & -(x-x^p) & 0 & 0 & 0 & (z-z^p) & \frac{(y-y^p)}{2} & \frac{(x-x^p)}{2} & 0
\end{bmatrix}
\begin{bmatrix} u_x^p \\ u_y^p \\ u_z^p \\ \varphi_x^p \\ \varphi_y^p \\ \varphi_z^p \\ \varepsilon_x^p \\ \varepsilon_y^p \\ \varepsilon_z^p \\ \gamma_{yz}^p \\ \gamma_{zx}^p \\ \gamma_{xy}^p \end{bmatrix}
$$

or shortly:

$$
\mathbf{u}(x,y,z) = \mathbf{T}^p(x,y,z)\,\mathbf{u}^p .
$$

Note that these relations are valid for small incremental rotations only (the issue of large rotations will be discussed at the end of Section 2).

This translation vector is particularly important for those points on the element boundary where a contact is formed with a neighbouring element. The difference between the translations of the two material points belonging to the two elements that form the contact is considered the *contact deformation*. This relative translation determines the forces transmitted through the contact, according to the contact constitutive relations.

Note that the role of matrix **T** is similar to that of the transition matrix in FEM; the difference is that in DDA the deformation of the element is also taken into account.

The forces acting on the elements are either external (like gravity, or velocity-dependent drag forces) or contact forces expressed by the neighbouring elements. Consider element p, and consider all the forces acting on it, either external or contact forces. Reducing them to the reference point of the element 6 scalar componnts are received. In addition, corresponding to the uniform strain, a uniform stress field is assigned to element p which has 6 more scalar characteristics. So, altogether 12 components form the $\mathbf{f}^p$ generalized reduced force vector of element p:

$$\mathbf{f}^p = \begin{bmatrix} f_x^p \\ f_y^p \\ f_z^p \\ m_x^p \\ m_y^p \\ m_z^p \\ V^p\sigma_x^p \\ V^p\sigma_y^p \\ V^p\sigma_z^p \\ V^p\tau_{yz}^p \\ V^p\tau_{zx}^p \\ V^p\tau_{xy}^p \end{bmatrix}$$

It is important to emphasize that the uniform stress field in the element does not hold a direct relationship with the contact forces: the usual static boundary condition is not valid for the stress field at the boundary of a DDA element and the contact forces acting on it.

Note that if an element performing an incremental generalized displacement $d\mathbf{u}^p$ is acted upon by a generalized reduced force $\mathbf{f}^p$, the scalar product $(d\mathbf{u}^p)^{\mathrm{T}}\mathbf{f}^p$ means work increment. (This is the reason to multiply the lower 6 components of $\mathbf{f}^p$ with the V^p volume of the element.)

Consider now the force $\mathbf{F}^c$ acting on the analysed element in point c. This is reduced to the reference point with the help of the transpose of the matrix $\mathbf{T}$, so the term $(\mathbf{T}^T\mathbf{F}^c)$ is added to other forces when compiling the reduced force $\mathbf{f}^p$:

$$\mathbf{T}^{pT}\mathbf{F}^c = \begin{bmatrix} 1 & & \\ & 1 & \\ & & 1 \\ 0 & -(z-z^p) & (y-y^p) \\ (z-z^p) & 0 & -(x-x^p) \\ -(y-y^p) & (x-x^p) & 0 \\ (x-x^p) & 0 & 0 \\ 0 & (y-y^p) & 0 \\ 0 & 0 & (z-z^p) \\ 0 & \dfrac{(z-z^p)}{2} & \dfrac{(y-y^p)}{2} \\ \dfrac{(z-z^p)}{2} & 0 & \dfrac{(x-x^p)}{2} \\ \dfrac{(y-y^p)}{2} & \dfrac{(x-x^p)}{2} & 0 \end{bmatrix} \begin{bmatrix} F_x^c \\ F_y^c \\ F_z^c \end{bmatrix}$$

Independently of whether a contact force $\mathbf{F}^c$ is considered or any other concentrated external effect, the model of Shi gives the same simplified relation between forces and stresses as seen in the last six rows of the above relation. This approximately means as if calculating the average uniform stress caused by all the forces which act on the element.

In the practical analysis of fractured rock masses the initial self-stress must be taken into consideration for the reliable modelling. This can be done with the lower six components of $\mathbf{f}^p$. (The details of how to compile $\mathbf{f}^p$ from the different mechanical effects can be found in Shi, 2001.)

In the original DDA model the constitutive relations represent a linearly elastic, isotropic material. However, in principle any other kind of constitutive relations can be applied in the model, assuming that a potential energy can be assigned to the deformations of the elements.

DDA models with higher order strain fields (e.g. MacLaughlin, 1997) work in the same manner, consequently higher order stress fields appear, which means more kinematic and static variables belonging to the elements.

Another possibility to have a non-uniform strain field in the elements is to subdivide them into sub-blocks. This technique was applied by Pérez-Aparicio et al (2013). They successfully simulated the collapse of covered arches with different geometries and loads. The subdivision into sub-blocks allows for a more reliable prediction of collapses due to material failure.

Jiang and Zheng (2015) analysed an issue observed in the original version of DDA when the blocks undero large rotations. For a small rotation φ, the first-order approximations $\cos \varphi \approx 1$ and $\sin \varphi \approx \varphi$ (applied above) are acceptable, but at large rotations they lead to cumulative errors and as a result, a false volume expansion occurs that was observed by several authors (see Jiang and Zheng, 2015 for an overview). To avoid this error, Jiang and Zheng (2015) suggested to fix a local coordinate frame to each block, which moves and rotates together with the block. After the open-close iteration procedure determines the incremental motions of the block done during the analysed time step, the strain increments are transformed to the local coordinate frame and added to the accumulated strain of the block. Using this improvement, the authors succeeded to avoid the occurrence of false volume expansion regardless of the magnitude of block rotation.

THE CONTACTS

When two elements move into a position where they overlap each other, a contact is formed. This is represented in the mechanical calculations by the two material points forming the first entrance. Overlapping is resisted by penalty functions in the original DDA method: a compressional contact force occurs whose magnitude depends linearly on the depth of the overlap (it is practically equivalent to an elastic, no-tension contact model), and tangential forces are treated in the same way. Coulomb-friction gives a limit to the tangential force magnitude. More specifically, assume that the contact is formed by the material points pc and qc on the boundary of elements p and q. In terms of the increment of the generalized displacement vectors, $d\mathbf{u}^p$ and $d\mathbf{u}^q$, the increment of the relative translation vector at the contact can be expressed as $\mathbf{T}^q(x^{qc}, y^{qc}, z^{qc})\, d\mathbf{u}^q - \mathbf{T}^p(x^{pc}, y^{pc}, z^{pc})\, d\mathbf{u}^p$. The mechanical model of the contact behavior specifies how the contact force components are calculated from the relative translation components. In the simplest case constant normal and tangential stiffnesses (corresponding to constant penalty parameters) are applied, and the contact starts to slide when the tangential force reaches the Coulomb-limit.

Several DDA simulations have been performed on the sliding motion of a single block on an inclined surface (e.g. Doolin and Sitar, 2002; Tsesarsky et al, 2005; see an overview in MacLaughlin and Doolin, 2006). Tsesarsky et al (2005) recommended improving DDA by incorporating displacement-dependent friction, and Bakun-Mazor et al (2012) found that velocity-dependent friction should also be included.

In reality a tensional resistance can also occur between masonry blocks, and this may also be important to incorporate into the model. Seemingly, cohesion can easily be included in the contact model: if the separation or the relative tangential displacement of the two material points forming the contact exceeds a threshold, the contact is deleted; but until then, a tensional and a shear force occur proportionally to the gap size and the relative tangential displacement respectively. However, the performance of this simple model is questionable. Wang et al (2013) introduced a displacement-dependent shear strength model for the contacts, and illustrated the advantages on landslide simulations. In an example of a block sliding on a plane they pointed out that while the original DDA differed up to a 100% from the analytical solution, the displacement-dependent contact model gave an excellent coincidence with the theoretical results.

THE EQUATIONS OF MOTION

The theoretical basis of the derivation of the equations of motion in DDA is the potential energy function containing terms for element deformability, contact deformability (or penalty functions), D'Alembert forces, concentrated and body loads, initial stress and prescribed displacement history. The stationary theorem of the potential energy states that the first derivatives of the potential energy (according to every component of the generalized displacement vectors of the elements) have to be zero. This condition yields the system of dynamic equations, which can be arranged to the following form ("generalized equations of motion"):

$$\mathbf{Ma}(t) + \mathbf{Cv}(t) + \mathbf{K}(t)\mathbf{u}(t) = \mathbf{f}(t)$$

where the generalized displacement vector and reduced force vector for the total system containing N elements are: $\mathbf{u}(t) = \begin{bmatrix} \mathbf{u}^1(t) \\ \mathbf{u}^2(t) \\ \vdots \\ \mathbf{u}^N(t) \end{bmatrix}$; $\mathbf{f}(t) = \begin{bmatrix} \mathbf{f}^1(t, u(t), v(t)) \\ \mathbf{f}^2(t, u(t), v(t)) \\ \vdots \\ \mathbf{f}^N(t, u(t), v(t)) \end{bmatrix}$.

The generalized velocities and accelerations are

$$\mathbf{v}(t) = \frac{d\mathbf{u}(t)}{dt}, \quad \mathbf{a}(t) = \frac{d^2\mathbf{u}(t)}{dt^2}$$

The exact derivations can be found in Shi, 2001, detailed at a computer-programming level.

Corresponding to every element, there are 12 scalar equations in the generalized equations of motion. The first 3 equations set the link between translational accelerations of the element and the first 3 components in the reduced force vector. In the second 3 equations the rotational accelerations and the three moment components are related. The last 6 equations express the relations between stresses and the external loads acting on the element.

The matrix of inertia, **M**, has a block-diagonal structure:

$$\mathbf{M} = \begin{bmatrix} \mathbf{M}^1 & & & \\ & \mathbf{M}^2 & & \\ & & \ddots & \\ & & & \mathbf{M}^N \end{bmatrix},$$

where the $\mathbf{M}^p$ block belonging to the p-th element can be expressed from the $\mu(x, y, z)$ density function of the element, with the help of matrix **T** which was seen above:

$$\mathbf{M}^p = \int_{V^p} \mathbf{T}^T(x,y,z) \cdot \mathbf{T}(x,y,z) \cdot \mu(x,y,z)\, dV$$

(again, the details of the derivations can be found in Shi, 2001).

TIME INTEGRATION

The time integration is performed with the help of Newmark's β–method. The motion of the elements along time is simulated in a series of small finite time intervals.

Consider the time interval (t_i, t_{i+1}). At its beginning, t_i, the state of the system is known: the vectors $\mathbf{u}_i$, $\mathbf{v}_i$, $\mathbf{f}(t_i, \mathbf{u}_i, \mathbf{v}_i)$ are known, and they satisfy the generalized equations of motion. The aim of the calculations is to ensure that the equations of motion

$$\mathbf{0} = \mathbf{f}(t, \mathbf{u}(t), \mathbf{v}(t)) - \mathbf{M} \cdot \mathbf{a}(t) - \mathbf{C}\mathbf{v}(t) - \mathbf{K}(t)\mathbf{u}(t)$$

would be satisfied also at t_{i+1}:

$$\mathbf{0} = \mathbf{f}(t_{i+1}, \mathbf{u}_{i+1}, \mathbf{v}_{i+1}) - \mathbf{M}_{i+1} \cdot \mathbf{a}_{i+1} - \mathbf{C}_{i+1}\mathbf{v}_{i+1} - \mathbf{K}_{i+1}\Delta\mathbf{u}_{i+1}$$

where the aim is to find that $\Delta\mathbf{u}_{i+1} = \mathbf{u}_{i+1} - \mathbf{u}_i$ generalized displacement increment which takes the system to the end of the timestep into a state where the equations of motion are satisfied.

In addition to $\Delta\mathbf{u}_{i+1}$ the unknown $\mathbf{v}_{i+1}$ and $\mathbf{a}_{i+1}$ are also contained in the generalized equations of motion. According to the Newmark β-method with parameters $\beta = 1/2$ and $\gamma = 1$, approximate these unknowns in terms of $\Delta\mathbf{u}_{i+1}$ in the following way:

$$\mathbf{a}_{i+1} = \frac{1}{\Delta t^2 / 2}\left(\Delta\mathbf{u}_{i+1} - \Delta t \cdot \mathbf{v}_i\right)$$
$$\mathbf{v}_{i+1} = \mathbf{v}_i + \Delta t \cdot \mathbf{a}_{i+1}$$

The mechanical meaning of these approximations is that the acceleration $\mathbf{a}_{i+1}$ is valid in the whole (t_i, t_{i+1}) interval, independently of the accelerations at the end of the previous time interval, i.e. at t_i. (Note that for $\beta = 1/2$ and $\gamma = 1$ the Newmark-method is unconditionally stable.)

Inserting the above expressions for $\mathbf{a}_{i+1}$ and $\mathbf{v}_{i+1}$ into the equations of motion, $\Delta\mathbf{u}_{i+1}$ remains the only unknown. A new notation is introduced:

$$\mathbf{f}(t_{i+1}, \mathbf{u}_{i+1}, \mathbf{v}_{i+1}) - \mathbf{M}_{i+1} \cdot \mathbf{a}_{i+1} - \mathbf{C}_{i+1}\mathbf{v}_{i+1} - \mathbf{K}_{i+1}\Delta\mathbf{u}_{i+1} := \mathbf{r}(t_{i+1}, \Delta\mathbf{u}_{i+1})$$

which means that the aim is to find that$\Delta\mathbf{u}_{i+1}$ which makes $\mathbf{r}$ zero at t_{i+1}. The vector $\mathbf{r}(t, \Delta\mathbf{u})$ will be called "residual", and its magnitude expresses the error according to which the equations of motion are violated.

In DDA the unknown $\Delta\mathbf{u}_{i+1}$ is calculated with the help of the Newton-Raphson iteration method where the Jacobian matrix of $\mathbf{r}(t, \Delta\mathbf{u})$ by $\Delta\mathbf{u}$ is needed:

$$\mathcal{K}(t, \Delta\mathbf{u}) = \frac{d\mathbf{r}(t, \Delta\mathbf{u})}{d\Delta\mathbf{u}}$$

This matrix is the sum of the usual stiffness matrix based on the contact stiffnesses, and of other terms due to the inertia, the stiffnesses of the material of the elements, and additional terms taking into account effects like self stress etc. In his publications Shi gave these derivations in detail (e.g. Shi, 2001), so the calculation of the Jacobian is straightforward. Hence, the residual for a given t and $\Delta\mathbf{u}$ can uniquely be calculated – the interested reader should consult the literature for the exact details.

Focus now on how to determine the unknown $\Delta\mathbf{u}_{i+1}$. Let its first approximation be a zero vector, and for the beginning of the calculations apply the already known position, velocity and internal forces of the system valid at t_i:

$$\Delta\mathbf{u}_{i+1}^{(0)} := \mathbf{0}.$$

If $\mathbf{r}(t_{i+1}, \Delta\mathbf{u}_{i+1}^{(0)}) = \mathbf{0}$ were valid, the calculation would be ready: the equations of motion would be satisfied. However, typically this is not the case, so a better approximation can be produced as

$$\Delta\mathbf{u}_{i+1}^{(1)} := \Delta\mathbf{u}_{i+1}^{(0)} - \mathcal{K}(t_{i+1}, \Delta\mathbf{u}_{i+1}^{(0)})^{-1} \cdot \mathbf{r}(t_{i+1}, \Delta\mathbf{u}_{i+1}^{(0)}).$$

Check now whether $\mathbf{r}(t_{i+1}, \Delta\mathbf{u}_{i+1}^{(1)})$ is sufficiently small. The iteration can be terminated if its norm is below a pre-defined threshold value; otherwise, a next approximation has to be prepared, according to the general formula:

$$\Delta\mathbf{u}_{i+1}^{(k+1)} := \Delta\mathbf{u}_{i+1}^{(k)} - \mathcal{K}(t_{i+1}, \Delta\mathbf{u}_{i+1}^{(k)})^{-1} \cdot \mathbf{r}(t_{i+1}, \Delta\mathbf{u}_{i+1}^{(k)}).$$

The iteration is continued until $\left|\mathbf{r}(t_{i+1}, \Delta\mathbf{u}_{i+1}^{(k+1)})\right|$ decreases below the threshold set by the user. (In some methods the norm $\left|\Delta\mathbf{u}_{i+1}^{(k+1)} - \Delta\mathbf{u}_{i+1}^{(k)}\right|$ is also checked, which is particularly useful in the case of nearly-singular $\mathcal{K}$.)

After finding $\Delta\mathbf{u}_{i+1}$ with sufficient accuracy, the next task is to check whether the topology of the system remains unchanged, or the displacements led to contacts being lost, created or slipped. If an overlap is found between two elements not in contact at the beginning of the timestep, DDA returns to the beginning of the timestep, puts a compressional spring between the two elements, and the calculation of the timestep is repeated. Similarly, if a contact is lost because of the calculated displacements, the contact is removed and the calculation of the timestep is repeated without that contact. If the friction limit is exceeded by a tangential force in a contact, then the timestep is repeated by keeping the shear force at the friction limit, and setting the shear stiffness to zero.

Obviously, during the re-calculation of a timestep with a modified contact, other contacts may be modified or created. In this case the calculation is repeated again and again ("open-close iterations"), until no change is experienced any more. If convergence is not attained within a specified number of iterations (typically six to eight), the time step is reduced (usually to its one third), and the analysis for that time step is repeated using the reduced time step. This is a recursive process, meaning if convergence is not attained with the reduced time step, the time step is further reduced (usually to one third again, i.e., 1/9 of the original time step). This recursive process increases the number of matrix iterations required for each time step and decreases the time step length used over the course of an analysis. The increase in computational time involved in the implicit method is not merely related to the cost of a single matrix inversion per time step. Multiple matrix inversions may be required during each time step, especially in systems where a large number of contacts are breaking and forming.

After finding the new state which satisfies the equations of motion as well as all conditions related to the contact behaviour, the next timestep can be analysed.

The most important parameters affecting the numerical behaviour of DDA are the contact normal stiffness, the time step length and the applied damping (if any). Yagoda-Biran and Hatzor (2016) discussed the existing suggestions previously published in the literature, and added their own experiences. They found, for instance, that depending on the actual physical features of the analysed problem, the value of the contact stiffness should be about 2-4 orders of magnitude lower than the product of the Young modulus of the intact blocks and the average block size. The interested reader can find the details in Yagoda-Biran and Hatzor (2016).

Regarding the behaviour of the open-close iteration technique, Khan (2010) and Khan et al (2010) compared the DDA solution of a 2D slope sliding problem with a solution given by UDEC (where an explicit time integration is used). It was found that DDA may require several times or orders of magnitude longer computational time than UDEC. This is partly because of the open-close iterations decreasing the time step length. Other reasons (e.g. type of the applied equation solver) are also extensively discussed by Khan (2010).

It has to be emphasized that the convergence of the DDA method has not been proven yet. With constant topology the applied parameters of the Newmark method ensure numerical stability, but if open-close iterations have to be performed, convergence cannot be ensured any more.

Finally, a short comparison is given between DDA and the commercial code 3DEC which is probably the most popular discrete element software in practical masonry analysis (see Chapter 3 of the present book):

- In 3DEC the *elements* are either perfectly rigid, or subdivided into tetrahedral finite elements; in DDA the elements are uniform-strain blocks (or higher order in some versions) so a strongly varying stress field within a block cannot reliably be described in DDA.
- The *basic unknowns* in 3DEC are the three translations and three rotation components for rigid blocks, or the nodal translations for tetrahedral subdivisions; in DDA the uniform-strain blocks have 12 degrees of freedom (translations, rotations and strain components).
- 3DEC applies an *explicit* time integration technique (the method of central differences), while DDA is based on an *implicit* method (Newmark's β–method).
- From numerical point of view, 3DEC is only conditionally stable hence the *time step length* is very limited; in DDA the implicit nature of the integration allows for larger time steps, though it should be chosen with care (Yagoda-Biran & Hatzor, 2016). Application of *damping* is indispensable in 3DEC not only when the aim is to find the equilibrium state of a complex system, but also to ensure numerical stability; in DDA the decision to use or not to use some kind of a damping depends on the judgement of the user.
- The memory requirements of 3DEC are lower and the calculations are, in general, faster than in DDA, particularly if the topology of the modelled system significantly varies during the process (e.g. several contacts open up or many pairs of blocks slide along each other to form new contacts with other blocks); DDA is more advantageous than 3DEC in situations when the topology does not change much and the simulated phenomenon is quasi-static or, at least, the system remains close to equilibrium.

APPLICATIONS

DDA has been applied in the analysis of scientific issues as well as in practical engineering problems. A few characteristic examples are given below.

Bicanic and Stirling (2001) applied DDA to approximately check the minimal wall thickness of a semicircular arch predicted by Heyman. They built an arch model with a 8% lower thickness and another arch with 6% higher thickness than the analytical prediction, and found that with high frictional resistance and cohesion in the joints and zero tensile strength, the first arch collapsed and the second arch was stable. Then they analysed how the arch behaviour changed when applying tensional resistance, zero friction or zero cohesion. They also analysed the effect of applying a crushing criterion for large compressive stresses at the corners of the blocks.

Rizzi et al (2014) analysed a similar problem. Correcting a few errors in Heyman's analysis, different improved solutions were derived or re-derived, and then DDA-simulated experiments on discretized arches were performed.

A damaged arch of a 4th-century church in Mamshit was modelled by Kamai and Hatzor (2005). They simulated different dynamic effects which had possibly led to the damage that could be seen on the arch, and also investigated the necessary damping to be applied in the numerical model. They could find an acceptable approximative peak ground acceleration that could cause the damages. A similar back-analysis was done on the model of a Roman tower in Avdat. Though both structures could possibly experience more than one earthquake during their history and the separation of these different earthquakes was far beyond the scope of the analysis, the structural damages of the two structures were successfully reproduced by DDA. Similar back analysis was reported by Kamai and Hatzor (2008) on an asymmetric arch in the Nimrod fortress.

Yagoda-Biran and Hatzor (2010) dealt with on the collapsed columns of the Byzantine cathedral in Susita. In order to verify the contact normal stiffness in the model, rocking motion of a free-standing column was repeatedly simulated using different values for the stiffness, until good agreement was reached between simulation and the analytical solution. Then the verified model was applied for finding the required the peak ground acceleration for overturning a typical Susita column.

Jiang et al (2014) analysed a centuries-old masonry seawall at Qiantang River. (Besides its historical and cultural value, this seawall plays a basic role in protecting a prosperous area in southeast China against floods and tides.) The authors applied a displacement-dependent cohesional shear strength model in the contacts, and the soil domain of the model was discretized in such a way that the joints would allow for sliding along the potential failure surfaces predicted by the most important soil mechanics theories. In addition to gravity, effects like mortar or pile corrosion, soil strength degradation etc. were simulated and failure modes with corresponding safety factors were determined.

These examples demonstrate the high potential of DDA not only as a scientific research tool but also as a technique that can assist the solution of practical structural mechanical problems related to masonry structures. Though DDA simulations of realistic problems can still be rather time-consuming today, the development of hardware, intelligent parallelization, and powerful equation solvers may enable the application of DDA in the everyday engineering practice.

ACKNOWLEDGMENT

The above study was supported by the Hungarian National Research Fund under grant no. OTKA 100770.

REFERENCES

Bakun-Mazor, D., Hatzor, Y. H., & Glaser, D. (2012). Dynamic sliding of tetrahedral wedge: The role of interface friction. *International Journal for Numerical and Analytical Methods in Geomechanics, 36*(3), 327–343. doi:10.1002/nag.1009

Bicanic, N., & Stirling, C. (2001). DDA analysis of the Couplet/Heyman minimum thickness arch problem. In N. Bicanic (Ed.), Procs. ICADD-4.

Doolin, D. M., & Sitar, N. (2002). Displacement accuracy of discontinuous deformation analysis method applied to sliding block. *Journal of Engineering Mechanics, 128*(11), 1158–1168. doi:10.1061/(ASCE)0733-9399(2002)128:11(1158)

Hatzor, Y. H., Arzi, A. A., & Tsesarsky, M. (2002). Realistic dynamic analysis of jointed rock slopes using DDA. In Y. Hatzor (Ed.), *Procs. ICADD-5* (pp. 47–56). Abingdon, UK: Balkema.

Jiang, H., Wang, L., Li, L., & Guo, Z. (2014). Safety evolution of an ancient maonry seawall structure with modified DDA method. *Computers and Geotechnics*, *55*, 277–289. doi:10.1016/j.compgeo.2013.09.012

Jiang, W., & Zheng, H. (2015). An efficient remedy for the false volume expansion of DDA when simulating large rotation. *Computers and Geotechnics*, *70*, 18–23. doi:10.1016/j.compgeo.2015.07.008

Kamai, R., & Hatzor, Y. H. (2005). Dynamic back analysis of structural failures in archeological sites to obtain paleo-seismic parameters using DDA. In M. MacLaughlin, N. Sitar (Eds.), Procs. ICADD-7.

Kamai, R., & Hatzor, Y. H. (2008). Numerical analysis of block stone displacements in ancient masonry structures: A new method to estimate historic ground motions. *International Journal for Numerical and Analytical Methods in Geomechanics*, *32*(11), 1321–1340. doi:10.1002/nag.671

Khan, M. S. (2010). *Investigation of Discontinuous Deformation Analysis for Application in Jointed Rock Masses.* (PhD Thesis). Department of Civil Engineering, University of Toronto.

Khan, M. S., Riahi, A., & Curran, J. H. (2010). Effects of time-step size on the efficiency of Discontinuous Deformation Analysis. In *Procs.6th Asian Rock Mechanics Symposium.*

Law, H. K., & Lam, I. P. (2003). Evaluation of seismic performance for tunnel retrofit project. *Journal for Geotechnical and Geoenviromental Engineering*, *129*(7), 575–589. doi:10.1061/(ASCE)1090-0241(2003)129:7(575)

MacLaughlin, M. M. (1997). *Discontinuous deformation analysis of the kinematics of landslides.* (PhD Dissertation). Dept. of Civil and Envr. Eng., University of California Berkeley.

MacLaughlin, M. M., & Doolin, D. M. (2006). Review of validation of the discontinuous deformation analysis (DDA) method. *International Journal for Numerical and Analytical Methods in Geomechanics*, *30*(4), 271–305. doi:10.1002/nag.427

Mazor, D. B. (2011). *Modelling dynamic rock mass deformattion with the numerical DDA method.* (PhD Dissertation). Ben-Gurion University of the Negev.

Pérez-Aparicio, J. L., Bravo, R., & Ortiz, P. (2013). Refined element discontinuous numerical analysis of dry-contact masonry arches. *Engineering Structures*, *48*, 578–587. doi:10.1016/j.engstruct.2012.09.027

Rizzi, E., Rusconi, F., & Cocchetti, G. (2014). Analytical and numerical DDA analysis on the collapse mode of circular masonry arches. *Engineering Structures*, *60*, 241–257. doi:10.1016/j.engstruct.2013.12.023

Scheldt, T., Lu, M., & Myrvang, A. (2002). Numerical analysis of Gjovick cavern. In Y. Hatzor (Ed.), *Procs. ICADD-5* (pp. 125–132). Abingdon, UK: Balkema.

Shi, G.-H. (1988). *Discontinuous deformation analysis – A new model for the statics and dynamics of block systems.* (PhD thesis). University of California Berkeley.

Shi, G.-H. (1992). Discontinuous deformation analysis: A new numerical model for the statics and dynamics of deformable block structures. *Engineering Computations*, *9*(4), 157–168. doi:10.1108/eb023855

Shi, G.-H. (2001). Three dimensional discontinuous deformation analysis. In N. Bicanic (Ed.), Procs. ICADD-4.

Thavalingam, A., Bicanic, N., Robinson, J. I., & Ponniah, D. A. (2001). Computational framework for discontinuous modelling of masonry arch bridges. *Computers & Structures*, *79*(19), 1821–1830. doi:10.1016/S0045-7949(01)00102-X

Tsesarsky, M., Hatzor, Y. H., & Sitar, N. (2005). Dynamic displacement of a block on an inclined plane: Analytical, experimental and DDA results. *Rock Mechanics and Rock Engineering*, *38*(2), 153–167. doi:10.1007/s00603-004-0043-2

Wang, L., Jiang, H., Yang, Z., Xu, Y., & Zhu, X. (2013). Development of discontinuous deformation analysis with displacement-dependent interface shear strength. *Computers and Geotechnics*, *47*, 91–101. doi:10.1016/j.compgeo.2012.06.006

Yagoda-Biran, G., & Hatzor, Y. H. (2010). Constraining paleo PGA values by numerical analysis of overturned columns. *Earthquake Engineering & Structural Dynamics*, *39*, 463–472.

Yagoda-Biran, G., & Hatzor, Y. H. (2016). Benchmarking the numerical Discontinuous Deformation Analysis mehod. *Computers and Geotechnics*, *71*, 30–36. doi:10.1016/j.compgeo.2015.08.003

Zhao, Sh., Salami, M. R., & Rahman, M. Sh. (2000). Three dimensional spherical DDA model for granular media. In J.L. Tassoulas (Ed.), *Procs.14th Engineering Mechanics Conf.*

Zhu, W., Zhang, Q., & Jing, L. (1999). Stability analysis of the ship-lock slopes of the Three Gorges project by three-dimensional FEM and DEM techniques. In B. Amadei (Ed.), Procs. ICADD-3. American Rock Mechanics Association.

KEY TERMS AND DEFINITIONS

Generalized Displacement Vector in DDA: A hypervector that consists of as many blocks as the number of the elements. The p-th block contains the translation vector of the reference point of element p, the rotation vector about the reference point, and the components of the strain tensor of p.

Implicit Method: A time integration method is implicit if the approximated numerical solution of the differential equation of the analyzed initial value problem is calculated at the end of a time interval from the numerical solution at the beginning of the interval in such a way that the differential equation is satisfied at the end of the time interval according to any required exactness.

Open-Close Iterations: The contacts in a DDA model may be in three different states: locked, sliding or open. After calculating the characteristics of the system at the end of the analyzed time interval, the assumed contact states may turn out to be inconsistent with the contact forces or element positions. In this case DDA modifies the contact states, and repeats the calculation of the time interval with the new contact states. This procedure is called open-close iterations.

Time Integration: A procedure to quantitatively approximate the solution of an initial value problem described by a differential equation. The time integration provides quantitative approximations of the state variables of the modelled system at a series of separate time instants.

Uniform Strain Element: An element type (originally applied in DDA model) in which the strain tensor is the same in every point.

Chapter 5
The Contact Dynamics Method

Katalin Bagi
Budapest University of Technology and Economics, Hungary

ABSTRACT

The Contact Dynamics method, developed still in the 1980s, was originally applied for granular assemblies because of its efficiency in simulating rapid granular flows or vibration problems of discrete systems. In the oldest models the elements were spherical and perfectly rigid, but later the application of polyhedral and deformable elements also became widespread, allowing for the reliable simulation of problems related to masonry structures. The basic unit of the analysis in Contact Dynamics is the pair of two randomly chosen elements. The essence of the method is to find the contact force vector between the two elements in such a way that during the analysed time step the elements should not overlap each other. At the considered time instant an iterative process sweeps along randomly chosen pairs over and over again, and gradually adjusts the estimated contact forces to get an improving approximation of a state that satisfies the dynamic equations of the system. The method is particularly advantageous for earthquake analysis of masonry structures.

INTRODUCTION

The NSCD (Non-Smooth Contact Dynamics) method was presented to the public at the end of the 1980ies by M. Jean and J.J. Moreau (Moreau, 1988; Jean & Moreau, 1992). Its first main field of application was granular mechanics: in comparison to previous discrete element techniques, the NSCD method turned out to be particularly fast and efficient when simulating granular flows, rapid avalanches, segregation, vibration problems of granular materials etc. Since the individual deformations of the grains are usually negligible in these problems, models consisting of perfectly rigid elements were mostly applied at that time. In the first versions the elements were mostly spherical, but later the application of polyhedral elements also became widespread. Contact Dynamics models brought significant scientific achievements in the field of the dynamics of granular materials. (While the original papers on NSCD were rather abstract and not very helpful in providing practice-oriented explanations how the method really worked, the paper of Unger and Kertész (2003) brought a leap forward: it gave a clear, detailed, code-writer-oriented introduction to the line of thought of the method, providing valuable help for those

DOI: 10.4018/978-1-5225-0231-9.ch005

who wanted to write their own code and also for those who just wanted to understand the main concept lying behind the software which they were applying in their researches.)

For masonry structures polygonal or polyhedral elements are obviously more suitable than spheres, and the deformability of the elements is often also important to take into consideration. Jean and Moreau (1992) and Jean (1999) introduced the basics of modelling masonry walls with deformable rectangular blocks in NSCD, and Jean developed the software called LMGC that gave realistic results for the quasi-static selfweight problem of a planar wall (Jean, 1999). Dubois extended the method, developed the open code LMGC90 (Dubois & Jean, 2006), and offered it to the research community not only for using it but also for further developments. LMGC90 can model rigid or deformable, 2D or 3D bodies of spherical or polyhedral shape. Since its release in the early 2000ies several scientists and engineers have applied it for different quasi-static or dynamic problems related to masonry mechanics.

Another available Contact Dynamics software is SOLFEC (Koziara & Bićanić, 2008). SOLFEC aimed at providing a user friendly platform for testing formulations and solution methods. The code implements different (rigid, uniform-strain and finite element) block models, contact detection algorithms, and time integration techniques. SOLFEC is particularly powerful in modelling element deformability with the help of FEM subdivision. In order to have reasonable computation times for real problems, parallelization is also applied in SOLFEC.

The approach of the Contact Dynamics method is very different from other discrete element techniques often applied for masonry analysis, 3DEC (Cundall, 1988) or DDA (Shi, 1992) for instance. In NSCD the basic unit of the analysis is the pair of two randomly chosen elements (contacting or non-contacting). The essence of the method is to find the contact forces transmitted between the two elements of the pair in such a way that during the analyzed time step the two elements should not overlap each other. The contact force is set to zero if the elements would not touch each other without this contact force even at the end of the timestep, and a non-zero vector is chosen (satisfying conditions corresponding to the mechanical behavior of the contact) if the two elements have to be slowed down in order to avoid overlap. So the motion of the system is numerically simulated in time through finite time steps, but in such a way that at the considered time instant an iterative process sweeps along randomly chosen pairs of the system over and over again, and gradually adjusts the estimated contact forces to get an improving approximation of a state that satisfies the dynamic equations of the system.

Contact recognition and the determination of its geometrical data (i.e. point of action of the contact force, and the normal direction) for polyhedral shapes require more sophisticated techniques than the treatment of spherical elements. The "common plane concept", a very efficient solution of the problem, is an advantageous and widely applied possibility, and it will be introduced in a forthcoming section.

Most discrete element methods represent the deformability of the elements either by using an internal FEM mesh in the elements (e.g. 3DEC), or by concentrating the deformations into the contacts, like in the case of PFC (Cundall & Strack, 1979). The calculation of the contact forces between the elements are based on the stiffness characteristics of the contacts in those techniques. The philosophy of NSCD is different. According to the oldest NSCD models, the elements are perfectly rigid, and the contact forces are not related to any stiffness data: the contact forces are calculated to ensure the dynamic equations of the elements, and in addition, they must not violate requirements like the Coulomb limit for friction or the no-tension requirement in cohesionless contacts, but their calculation does not apply any constitutive relations. For statically highly indeterminate systems like e.g. a masonry wall, there exist several alternative force systems that satisfy the equations of motion; NSCD produces randomly one of them, while several equally valid solutions can be received if the problem is calculated repeatedly with the method starting from the same initial state, but considering the pairs of elements in different random orders. This non-uniqueness partly explains the doubts why NSCD is not widespread in the analysis of quasi-static problems. On the other hand, since the method is computationally very efficient for the simulation of dynamic problems, NSCD is more popular in earthquake simulations or vibration analysis. It has to be emphasized that there is a general lack of validation studies about the simulation of quasi-static as well as dynamic problems. The attempts of Ceh et al (2015a, 2015b) are very promising in this respect; similar verification examinations (static as well as dynamic) could potentially resolve the existing skepticism of the engineering community and give a fair general evaluation of the strengths and limitations of the technique. So this is a basically important task for future researches.

The aims of this chapter are (1) to provide an insight into how the method works; (2) to call the attention on the possible problems and questionable issues the user should be aware of; and (3) to present a collection of characteristic applications.

This Chapter is built up as follows. The basic concepts are presented in Section 2: for the sake of simplicity, in section 2 rigid elements are considered only. Section 3 focuses on deformable elements and calls the attention on the most important differences from the rigid-element model. Finally, Section 4 introduces different applications of the Contact Dynamics method in the analysis of masonry structures.

MODELLING WITH RIGID ELEMENTS

Geometrical Characteristics of the Pair of Two Elements

The Closest Points

Rigid elements in 3D have six degrees of freedom, 3 translational and 3 rotational. A reference point is defined on each element, usually coinciding with its centre of gravity (see Figure 1 for an illustration in 2D). Let us consider an arbitrarily selected pair of two elements not necessarily in contact with each other. State variables and the geometrical characteristics of this pair will be collected now.

Figure 1. Distance g^{pq} between elements p and q and the definition of vectors $\mathbf{r}^{pc}$ and $\mathbf{r}^{qc}$

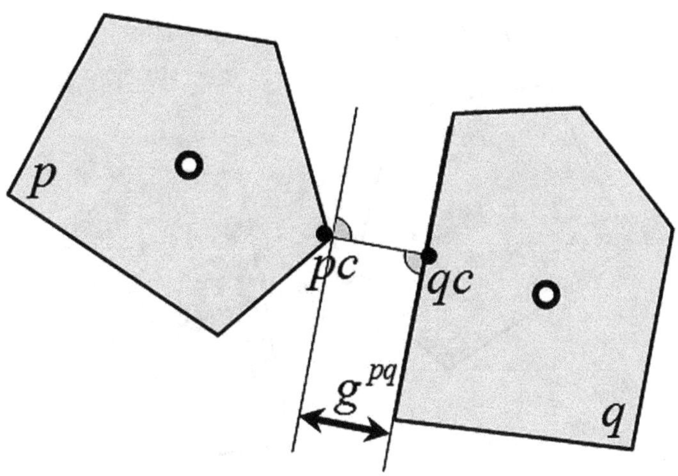

Figure 1 illustrates two elements, p and q. (Element p is the first and q is the second entity of the pair denoted by c.) In order to decide whether p and q are in contact at an actual time instant, the two nearest points on the boundary of p and q are determined (material points pc and qc respectively). In general they are node-to-face neighbours, as shown in Figure 1: in other words, assume for simplicity that the problem to find the two closest points has a unique solution. (The non-unique cases, i.e. edge-to-face neighbours when an edge lies in parallel to a face, or face-to-face neighbours when two planar faces lie parallel to each other, can be modelled as the simultaneous neighbouring pairs of a small finite number of node-to-face pairs).

Several possible algorithms are available to find the two closest points. The "common plane concept" is one of the most efficient techniques so this is applied in LMGC90 (Dubois & Mozul, 2013). This technique is also used in other discrete element codes, e.g. in 3DEC. The common plane can be imagined as a plate that is held loosely between the two polyhedral blocks. If the blocks are brought closer slowly, the plate will be moved by the blocks, and finally will become trapped at some particular position when the blocks come into contact. Whatever the shape and orientation of the blocks (provided they are convex), the plate will take up a position that defines the sliding plane for the two blocks. To carry the analogy a bit further, imagine that the plate is now repelled by the blocks even when they do not touch. As the blocks are brought together, the plate will take up a position midway between them, at a maximum distance from both. The algorithm for locating and moving the common plane can be formulated as an optimization problem: "Maximize the gap between the common plane and the closest vertex". The optimization algorithm applies translations and rotations to the common plane in order to maximize the gap.

The normal vector of the common plane will serve as the contact normal, and the points having the shortest distances from the common plane are pc and qc. After finding the two closest points, sum up their distances from the common plane, and let g^{pq} denote this sum which is equal to the gap width between the two blocks.

Assign an $(\mathbf{n}, \mathbf{t}, \mathbf{w})$ local coordinate frame to the pair in the way shown in Figure 2. The unit base vector $\mathbf{n}$ is directed towards the nearest point of p (first element of the pair). Base vectors $\mathbf{t}$ and $\mathbf{w}$ are an arbitrary pair of orthogonal unit vectors parallel to the contact plane. The g^{pq} distance is the $\mathbf{n}$-directional projection of the vector connecting the two closest points, oriented from q to p.

Figure 2. The local coordinate system of pair c

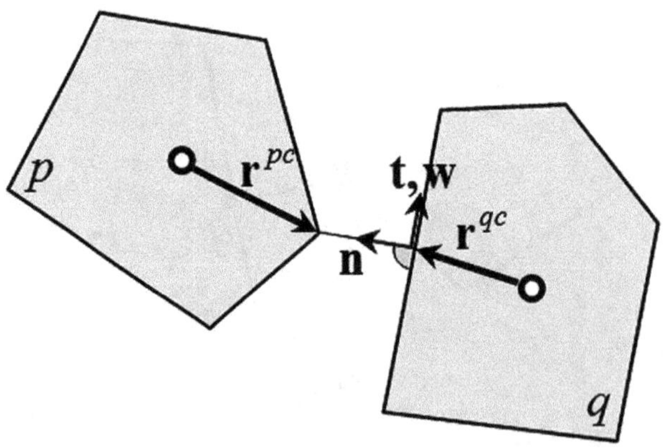

If this distance equals zero, the two elements form a point-like contact, and a concentrated contact force can be transmitted between the elements. Since the contact and the elements do not deform, the two elements must move in such a way that there would be no overlap between them: the gap width g^{pq} cannot become negative. In addition, the two elements can slide along each other (i.e. the two material points forming the contact point can translate relatively to each other in a direction perpendicular to **n**) only if the frictional limit is reached in the contact.

The Reduced Contact Forces

When a contact exists, contact forces can be transmitted between material points *pc* and *qc* (they are at the same position in this case). These forces are the basic unknowns of the method; the essence of the NSCD technique is to determine them with an iterative solver (see Sections 2.3 and 2.4) at discrete time instants. The force exerted on *p* by *q* is $\mathbf{f}^{pc}$, and the force acting on *q* by *p* is $\mathbf{f}^{qc}$ (note that $\mathbf{f}^{pc} = -\mathbf{f}^{qc}$):

$$\mathbf{f}^{pc} = \begin{bmatrix} F_x^{pc}(t) \\ F_y^{pc}(t) \\ F_z^{pc}(t) \end{bmatrix} \text{ and } \mathbf{f}^{qc} = \begin{bmatrix} -F_x^{pc}(t) \\ -F_y^{pc}(t) \\ -F_z^{pc}(t) \end{bmatrix},$$

They can be reduced to the reference points of *p* and *q* with the help of the $\mathbf{B}^{pc}(t)$ and $\mathbf{B}^{pc}(t)$ *transition matrices*, respectively:

$$\mathbf{B}^{pc}(t) = \begin{bmatrix} 1 & 0 & 0 \\ 0 & 1 & 0 \\ 0 & 0 & 1 \\ 0 & -r_z^{pc} & r_y^{pc} \\ r_z^{pc} & 0 & -r_x^{pc} \\ -r_y^{pc} & r_x^{pc} & 0 \end{bmatrix} \quad ; \quad \mathbf{B}^{qc}(t) = \begin{bmatrix} 1 & 0 & 0 \\ 0 & 1 & 0 \\ 0 & 0 & 1 \\ 0 & -r_z^{qc} & r_y^{qc} \\ r_z^{qc} & 0 & -r_x^{qc} \\ -r_y^{qc} & r_x^{qc} & 0 \end{bmatrix}.$$

The vectors $\mathbf{r}^{pc}$ and $\mathbf{r}^{qc}$ point from the reference points to the contact. The reduced forces are then:

$$\mathbf{f}_{red}^{pc}(t) = \mathbf{B}^{pc}(t) \cdot \mathbf{f}^{pc}(t) \quad ; \quad \mathbf{f}_{red}^{qc}(t) = \mathbf{B}^{qc}(t) \cdot \mathbf{f}^{qc}(t).$$

Summarize these two forces into the 12-scalar *reduced contact force vector* of the pair:

$$\mathbf{f}^{pq}(t) = \begin{bmatrix} \mathbf{f}_{red}^{pc}(t) \\ \mathbf{f}_{red}^{qc}(t) \end{bmatrix}.$$

This vector will take part in the equations of motion of the pair.

The Relative Velocity

The vector $\mathbf{v}^{pq}$ denotes the velocity vector of the (p, q) pair:

$$\mathbf{v}^{pq}(t) = \begin{bmatrix} \mathbf{v}^{p}(t) \\ \mathbf{v}^{q}(t) \end{bmatrix},$$

which can be expressed as the time derivatives of the displacements:

$$\mathbf{v}^{p}(t) = \begin{bmatrix} \frac{du_x^p(t)}{dt} \\ \frac{du_y^p(t)}{dt} \\ \frac{du_z^p(t)}{dt} \\ \frac{d\text{Æ}_x^p(t)}{dt} \\ \frac{d\text{Æ}_y^p(t)}{dt} \\ \frac{d\text{Æ}_z^p(t)}{dt} \end{bmatrix} \quad ; \quad \mathbf{v}^{q}(t) = \begin{bmatrix} \frac{du_x^q(t)}{dt} \\ \frac{du_y^q(t)}{dt} \\ \frac{du_z^q(t)}{dt} \\ \frac{d\text{Æ}_x^q(t)}{dt} \\ \frac{d\text{Æ}_y^q(t)}{dt} \\ \frac{d\text{Æ}_z^q(t)}{dt} \end{bmatrix}.$$

Consider now those two material points, *pc* and *qc*, which form the contact. Their velocities, $\mathbf{v}^{pc}$ és $\mathbf{v}^{qc}$, can be expressed again with the help of the $\mathbf{B}^{pc}(t)$ and $\mathbf{B}^{pc}(t)$ transition matrices:

$$\begin{aligned} \mathbf{v}^{pc}(t) &= \mathbf{B}^{pcT}(t) \cdot \mathbf{v}^{p}(t) \\ \mathbf{v}^{qc}(t) &= \mathbf{B}^{qcT}(t) \cdot \mathbf{v}^{q}(t) \end{aligned}.$$

The relative velocity of the contact is the difference of the velocities of the two material points *pc* and *qc*:

$$\text{¼}^{pq}(t) = \begin{bmatrix} \text{¼}_x^{pq}(t) \\ \text{¼}_y^{pq}(t) \\ \text{¼}_z^{pq}(t) \end{bmatrix} := \mathbf{v}^{pc}(t) - \mathbf{v}^{qc}(t)$$

which yields

$$\text{¼}^{pq}(t) = \mathbf{v}^{pc}(t) - \mathbf{v}^{qc}(t) = \left(\mathbf{B}^{pcT}(t) \cdot \mathbf{v}^{p}(t)\right) - \left(\mathbf{B}^{qcT}(t) \cdot \mathbf{v}^{q}(t)\right).$$

This vector shows the relative translational velocity of point *pc* with respect to point *qc*, so if the two points are already in contact, the **n**-component of this vector must not be negative in order to avoid overlapping. The tangential component has to be zero if the contact is not sliding.

Mechanical Conditions

In Coulomb-frictional, no-tension contact models the following requirements are to be satisfied in every pair:

1. If the two elements touch each other, i.e. if $g^{pq} = 0$, then a contact force can be transmitted. If $g^{pq} > 0$, then there is no contact, and no contact force exist in the pair. The case $g^{pq} < 0$ is not possible.
2. The normal component of the contact force, N^{pq}, can only be negative, i.e. compressional, but otherwise its magnitude ($N^{pq} = \mathbf{f}^{pcT} \cdot (-\mathbf{n})$ or $N^{pq} = \mathbf{f}^{qcT} \cdot \mathbf{n}$) is arbitrary. The (N^{pq}, g^{pq}) relation is shown in Figure 3.

The magnitude of the tangential force $\mathbf{T}^{pq}$ (having a *t*- and a *w*-directional component) is limited by the Coulomb friction law:

$$\left|\mathbf{T}^{pq}\right| \leq -\nu \cdot N^{pq}$$

where ν is the friction coefficient. Figure 4 illustrates this limitation: the vector of the tangential force must point from the origin either to inside the cone (non-sliding contact) or just to the surface of the cone (sliding contact).

As long as $\mathbf{T}^{pq}$ is below the friction limit, the tangential component of the relative translation must be zero (elastic contact deformation is not possible). When reaching the friction limit, the contact starts to slide: the $v_t^{pq}(t)$ relative velocity has to be just opposite to the direction of the $\mathbf{T}^{pq}$ force (the tangential relative translation must not have a component perpendicularly to the frictional force reaching the limit), and the magnitude of $\mathbf{T}^{pq}$ is equal to $-\nu \cdot \mathbf{N}^{pq}$. The magnitude of the relative translation is not limited in the contact model, this can be determined from kinematical considerations.

Figure 3. Normal force vs. gap width in Contact Dynamics

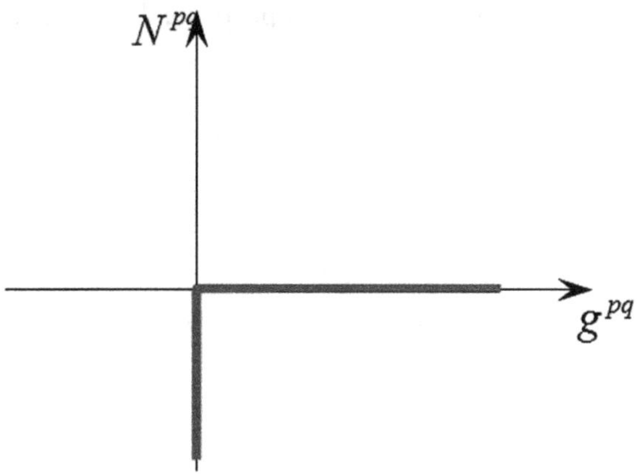

Figure 4. Coulomb limit for the tangential force

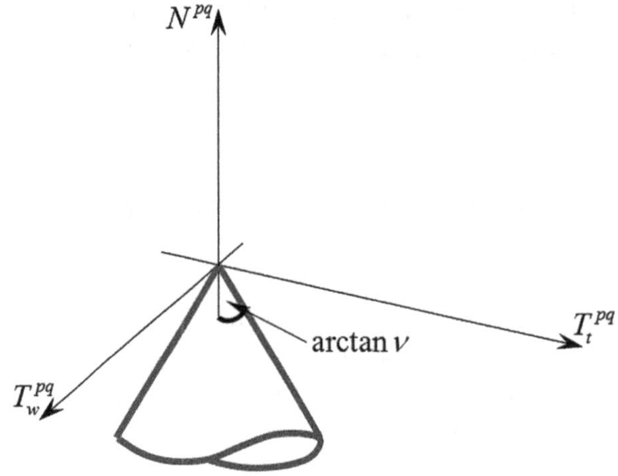

Without presenting the details, it is important to mention that Acary and Jean (2000) introduced more sophisticated and realistic representations of the real joint behaviour. They gave suggestions how to model finite tensional or shear resistance, elastic contact behaviour, brittle cohesion or progressive damage behaviour. These extensions of the joint modelling allow the user to give a realistic prediction for structures with mortared contacts. A wide range of such possibilities is available in LMGC90.

The Equations of Motion

Contact Dynamics is a time-stepping method: its fundamental unknowns are the time-dependent positions and velocities of the elements. They are searched for pair by pair.

Assume that at t_i the state of the system is known: according to the exactness numerically prescribed, the positions and velocities of the elements are given: $\mathbf{u}^p(t_i) \cong \mathbf{u}_i^p$; $\mathbf{v}^p(t_i) \cong \mathbf{v}_i^p$, and the external forces acting on the elements ($\mathbf{f}_i^{p,ext}$) and contact forces for all c ($\mathbf{f}_i^{pc}$) are also known. The external forces are reduced to the reference points; the contact forces act in the point-like contacts. The time-dependence of the external forces is also known (e.g. that the gravitational force is constant), so the external forces are given also in $t_{i+1} = t_i + \Delta t$ ($\mathbf{f}_{i+1}^{p,ext}$). From these data the state of the system at t_{i+1} (contact forces, the positions and the velocities of the elements) is searched for. Contact Dynamics applies the *implicit* version of the Euler method for this purpose. The basic step for the p and q pair can be written as:

$$\begin{bmatrix} \mathbf{v}_{i+1}^p \\ \mathbf{v}_{i+1}^q \end{bmatrix} := \begin{bmatrix} \mathbf{v}_i^p \\ \mathbf{v}_i^q \end{bmatrix} + \Delta t \cdot \begin{bmatrix} (\mathbf{M}^p)^{-1} & \\ & (\mathbf{M}^q)^{-1} \end{bmatrix} \cdot \begin{bmatrix} \mathbf{f}_{i+1}^p \\ \mathbf{f}_{i+1}^q \end{bmatrix};$$

$$\begin{bmatrix} \mathbf{u}_{i+1}^p \\ \mathbf{u}_{i+1}^q \end{bmatrix} := \begin{bmatrix} \mathbf{u}_i^p \\ \mathbf{u}_i^q \end{bmatrix} + \Delta t \cdot \begin{bmatrix} \mathbf{v}_{i+1}^p \\ \mathbf{v}_{i+1}^q \end{bmatrix}.$$

Here $\mathbf{f}^{p}_{i+1}$ and $\mathbf{f}^{q}_{i+1}$ denote the resultants of the external and all contact forces acting on *p* and *q* respectively, being reduced to the reference points. Note that according to the applied implicit scheme, the velocities and accelerations belonging to the end of the time interval are considered to be valid along the whole interval.

The reduced forces are

$$\begin{bmatrix} \mathbf{f}^{p}_{i+1} \\ \mathbf{f}^{q}_{i+1} \end{bmatrix} := \begin{bmatrix} \mathbf{f}^{p,ext}_{i+1} + \sum_{(pk)} \mathbf{B}^{pk} \cdot \mathbf{f}^{pk}_{i+1} \\ \mathbf{f}^{q,ext}_{i+1} + \sum_{(qk)} \mathbf{B}^{qk} \cdot \mathbf{f}^{qk}_{i+1} \end{bmatrix}.$$

Summation over index *pk* runs along all contacts of element *p*, including the just analysed contact *c* as well. Similarly, index *qk* runs along all contacts of *q*.

The transition matrices are assumed to be constant during the (t_i, t_{i+1}) time interval, and equal to their values at t_i. Indeed, if the displacement increments are small during the timestep, the modification of the vectors pointing from the reference points to the contacts is negligible.

Collect the mass and rotational inertia of the elements into the matrices $\mathbf{M}^p$ and $\mathbf{M}^q$, which have the following form in the case of non-spherical elements:

$$\mathbf{M}^p = \begin{bmatrix} m^p & & & & & \\ & m^p & & & & \\ & & m^p & & & \\ & & & I^p_{xx} & I^p_{xy} & I^p_{xz} \\ & & & I^p_{yx} & I^p_{yy} & I^p_{yz} \\ & & & I^p_{zx} & I^p_{zy} & I^p_{zz} \end{bmatrix}.$$

The lower right block depends on the actual orientation of the element so it varies with time.

In order to determine the velocity of the pair ($\mathbf{v}^{pq}_{i+1}$), the resultants $\mathbf{f}^{p}_{i+1}$ and $\mathbf{f}^{q}_{i+1}$ should be known. So, in addition to the external forces acting at t_{i+1}, the contact forces should also be known at the end of the timestep. The Contact Dynamics models search for these contact forces with the help of an *iterative solver* (which has to be performed over and over again at every timestep, as the contact forces change with time).

The Iterative Solver

The analysis of a time step consists of several iteration steps. In each iteration step the solver sweeps along all pairs of neighbouring or nearly contacting elements. When considering a given pair, an approximation is given (based on the equations of motion of the pair), so that the conditions assumed on the mechanical behaviour would be satisfied: no overlap; Coulomb-friction etc. Then a next pair is considered. After all pairs were swept over, a next iteration step starts: the solver sweeps along the pairs from the beginning again. This iteration step is repeated over and over again, until the next approximations are already sufficiently close to the previous ones. When such a state is reached, the contact forces belonging to t_{i+1} have been found, and the next time step can follow.

The length of the timestep can be relatively large in comparison to models using explicit time integration (e.g. PFC or 3DEC). As explained by Radjai and Richefeu (2009), the limit on the timestep length is given by the occurrence of cumulative numerical errors leading to undesired excess overlaps between the particles. They suggest that a typical value for time step length is 10^{-4} sec for a system that consists of 10^4 rigid elements.

The approximation of the contact force in the pair (p, q) is based on the equations of motion of that pair. Before turning onto the details, a few notations have to be introduced:

For element p, reduce to the reference point all those forces (external and contact forces) acting at t_{i+1}, except from the force expressed by element q through contact c:

$$\mathbf{f}_{red,i+1}^{p,no\_c} := \mathbf{f}_{i+1}^{p,ext} + \sum_{pk \neq pc} \mathbf{f}_{red,i+1}^{pk} .$$

The $\mathbf{f}_{i+1}^{pk}$ contact force is only an actual approximation of the force indeed acting in contact pk at t_{i+1}; it receives new and new values during the iterations. (At the beginning of the analysis of the time step the contact forces are approximated to be the same as their final, just determined values at the end of the previous time step, which is the same as the beginning of the just analysed timestep.)

Similarly, reduce all the forces acting at t_{i+1} on q – except from that force acting in qc – to the reference point of q:

$$\mathbf{f}_{red,i+1}^{q,no\_c} := \mathbf{f}_{i+1}^{q,ext} + \sum_{qk \neq qc} \mathbf{f}_{red,i+1}^{qk} ,$$

and collect the two vectors into a hypervector:

$$\mathbf{f}_{red,i+1}^{pq,no\_c} := \begin{bmatrix} \mathbf{f}_{red,i+1}^{p,no\_c} \\ \mathbf{f}_{red,i+1}^{q,no\_c} \end{bmatrix} .$$

Summarize the two transition matrices belonging to c into a hypermatrix:

$$\mathbf{B}^{pq} := \begin{bmatrix} \mathbf{B}^{pc} \\ -\mathbf{B}^{qc} \end{bmatrix}$$

and the matrices of inertia of p and q into a block-diagonal matrix, whose inverse is:

$$\left(\mathbf{M}^{pq}\right)^{-1} := \begin{bmatrix} \left(\mathbf{M}^{p}\right)^{-1} & \mathbf{0} \\ \mathbf{0} & \left(\mathbf{M}^{q}\right)^{-1} \end{bmatrix} .$$

Later the following two matrices will also be necessary:

$$\widetilde{\left(\mathbf{M}^{pq}\right)^{-1}} := \mathbf{B}^{pqT}\left(\mathbf{M}^{pq}\right)^{-1}\mathbf{B}^{pq},$$

and

$$\widetilde{\mathbf{M}^{pq}} := \left(\mathbf{B}^{pqT}\left(\mathbf{M}^{pq}\right)^{-1}\mathbf{B}^{pq}\right)^{-1}.$$

And now the equations of motion of the pair (p, q) can be compiled. First, the equations belonging to the end of the time interval can separately be written as:

$$\frac{1}{\Delta t}\begin{bmatrix}\mathbf{v}_{i+1}^{p} - \mathbf{v}_{i}^{p}\\ \mathbf{v}_{i+1}^{q} - \mathbf{v}_{i}^{q}\end{bmatrix} = \left(\mathbf{M}^{pq}\right)^{-1}\cdot\begin{bmatrix}\mathbf{f}_{red,i+1}^{p,no\_c} + \mathbf{f}_{red,i+1}^{pc}\\ \mathbf{f}_{red,i+1}^{q,no\_c} + \mathbf{f}_{red,i+1}^{qc}\end{bmatrix}.$$

Multiply both sides by $\mathbf{B}^{pqT}$ from the left:

$$\frac{1}{\Delta t}\left[\mu_{i+1}^{pq} - \mu_{i}^{pq}\right] = \mathbf{B}^{pqT}\left(\mathbf{M}^{pq}\right)^{-1}\cdot\mathbf{f}_{red,i+1}^{pq,no\_c} + $$
$$+\mathbf{B}^{pqT}\left(\mathbf{M}^{pq}\right)^{-1}\mathbf{B}^{pq}\cdot\mathbf{f}_{i+1}^{pc}$$

(it was taken into consideration that $\mathbf{f}_{i+1}^{qc} = -\mathbf{f}_{i+1}^{pc}$). After some rearrangements:

$$\mu_{i+1}^{pq} - \left(\mu_{i}^{pq} + \Delta t\cdot\mathbf{B}^{pqT}\left(\mathbf{M}^{pq}\right)^{-1}\cdot\mathbf{f}_{red,i+1}^{pq,no\_c}\right) = \Delta t\cdot\widetilde{\left(\mathbf{M}^{pq}\right)^{-1}}\mathbf{f}_{i+1}^{pc}.$$

It is easy to notice that on the left side the vector in the parentheses means the relative velocity which would occur in the contact at t_{i+1} if $\mathbf{f}_{i+1}^{pc}$ is zero, i.e. if there is no force in the contact. This vector will have a special importance in the forthcoming derivation, so a special notation is given to it:

$$\mu_{i+1}^{pq,no\_c} := \left(\mu_{i}^{pq} + \Delta t\cdot\mathbf{B}^{pqT}\left(\mathbf{M}^{pq}\right)^{-1}\cdot\mathbf{f}_{red,i+1}^{pq,no\_c}\right).$$

The equations of motion can now be written as:

$$\mu_{i+1}^{pq} = \mu_{i+1}^{pq,no\_c} + \Delta t\cdot\widetilde{\left(\mathbf{M}^{pq}\right)^{-1}}\mathbf{f}_{i+1}^{pc}$$

where μ_{i+1}^{pq} and $\mathbf{f}_{i+1}^{pc}$ are the unknowns. So the equations of motion give the relation between the unknown contact force and the unknown relative velocity belonging to the contact. This will be the starting point of the forthcoming calculations.

Finally the normal and tangential components of the relative velocity vector of the contact will be needed:

$$¼_n^{pq} = \mathbf{n}^T \cdot \mu^{pq} \; ; \quad \mu_{tw}^{pq} = \mu^{pq} - ¼_n^{pq} \cdot \mathbf{n}$$

(Remember that μ^{pq} denoted the velocity of the material point *pc* relative to the material point *qc*. Hence a positive $¼_n^{pq}$ means increasing gap between the two material points.) Since the vector μ_{i+1}^{pq,no_c} belonging to the time instant t_{i+1} can directly be calculated from the already existing approximations of all other contact forces except from *c*, the components $¼_{n,i+1}^{pq,no\_c}$ and $\mu_{tw,i+1}^{pq,no\_c}$ can also be determined, while the components of the vector μ_{i+1}^{pq} are unknowns.

The unknown μ_{i+1}^{pq} and $\mathbf{f}_{i+1}^{pc}$ vectors are determined in three steps:

First decide whether the two elements will be in contact at t_{i+1}: calculate how large will the gap be between them, assuming zero contact force:

$$g_{i+1}^{pq,no\_c} = g_i^{pq} + ¼_{n,i+1}^{pq,no\_c} \cdot \Delta t \, .$$

A positive result means that there will be no contact at t_{i+1}, and the analysis of another pair can immediately follow. A negative result, on the other hand, means that without a contact force the elements *p* and *q* would overlap, so an $\mathbf{f}^{pc}$ contact force is needed to avoid the overlap. In this case Step 2. follows.

Step 2: The contact force should modify the velocities of the two elements in such a way that instead of overlapping, they would exactly touch each other at the end of the time step. In Step 2. the aim is to determine $\mathbf{f}_{i+1}^{pc}$ that satisfies the following two conditions:

a. At t_{i+1} the gapwidth between *p* and *q* is exactly zero: $g_i^{pq} + ¼_{n,i+1}^{pq} \cdot \Delta t = 0$
b. The contact does not slide, so the tangential component of the relative translation is zero: $\left| \mu_{tw,i+1}^{pq} \right| = 0$

To satisfy these two conditions, the relative velocity of the contact should be:

$$\mu_{i+1}^{pq} = -\frac{1}{\Delta t} g_i^{pq} \cdot \mathbf{n}$$

(the negative sign means that if the gapwidth was larger than zero, then *p* should get closer to *q* to touch it).

The $\mathbf{f}_{i+1}^{pc}$ has to be such a force that if continuously acting between *p* and *q* during (t_i, t_{i+1}), at t_{i+1} the relative velocity would be just equal to $¼_{i+1}^{pq}$. From the equations of motion, this force turns out to be equal to:

$$\mathbf{f}_{i+1}^{pc} = \frac{1}{\Delta t}\widetilde{\mathbf{M}^{pq}} \cdot \left(-\frac{1}{\Delta t} g_i^{pq}\mathbf{n} - \mu_{i+1}^{pq,no\_c} \right).$$

Now the question is whether this force violates the constitutive conditions. There were two conditions on the components of the contact forces. The first one required the normal force a compression. This is automatically satisfied because of Step 1. The second one was the Coulomb-condition:

$$\left|\mathbf{T}_{i+1}^{pc}\right| \leq -\nu \cdot N_{i+1}^{pc}$$

If this holds for the calculated $\mathbf{f}_{i+1}^{pc}$, then the analysis of the (*p*, *q*) pair is ready, and a next pair can follow. However, if the tangential component exceeds the friction limit, then the calculated contact force cannot be transmitted in the contact: the contact slides, which means that the tangential component of μ_{i+1}^{pq} is not zero, and the calculation based on zero tangential component should be corrected. This correction is done in Step 3.

Step 3. In a sliding contact the tangential force component has to satisfy the following to conditions, and – as the third condition – the equations of motion:

a. The contact is sliding, so the magnitude of the tangential force component is equal to the Coulomb-limit: $\left|\mathbf{T}_{i+1}^{pc}\right| = -\nu \cdot N_{i+1}^{pc}$
b. The direction of the tangential relative velocity is just opposite to the direction of the tangential force component: $\frac{\mathbf{T}_{i+1}^{pc}}{\left|\mathbf{T}_{i+1}^{pc}\right|} = -\frac{\mu_{tw,i+1}^{pq}}{\left|\mu_{tw,i+1}^{pq}\right|}$

The equations of motion:

$$\mathbf{f}_{i+1}^{pc} = -\frac{1}{\Delta t}\widetilde{\mathbf{M}^{pq}} \cdot \left(\frac{1}{\Delta t} g_i^{pq}\mathbf{n} + \mu_{i+1}^{pq,no\_c} - \mu_{tw,i+1}^{pq} \right)$$

From these conditions the unknowns $\mathbf{f}_{i+1}^{pc}$ és μ_{i+1}^{pq} can be calculated, and the analysis of the (*p*, *q*) pair is ready. The next pair can follow.

These calculations introduced above give an approximation for the contact force in a pair, assuming that all other contact forces are unchanged and keep their values last approximated. When turning to the next pair, the latest approximations in other pairs are applied. Proceeding from pair to pair this way, an approximation is received for the whole system of contact forces. By sweeping through the complete set of contacts and nearly-contacting pairs over and over again, the results get closer and closer to what should exist at t_{i+1}. (Note that convergence is still an open issue: a precise proof does not exist in the literature.) The modifications caused by the consecutive iteration cycles cause smaller and smaller modifications in the contact forces; and the iteration can be terminated as the modifications decrease under a prescribed threshold. Now the state belonging to t_{i+1} has been found, and a new time step can be analysed.

The order according to which the pairs are considered within an iteration step is ***random***; the only requirement is that every pair should be considered once within a step. In the next iteration step the starting pair and the ordering are different, prescribed also by a random number generator.

If the same problem is analysed twice, by starting the random number generator from two different initiations, the two resulting contact force systems will be different. This non-uniqueness of the solution has been emphasized by e.g. Jean (1999) or Moreau (2006). Indeed, for a statically highly indeterminate system several equilibrated force systems can be found, and without the flexibility data the "correct" one cannot be selected. In addition, the possibility of frictional sliding makes the solution history dependent. Moreau (2006) gave a very interesting discussion on the non-uniqueness of the solution. However, experiences on granular assemblies, e.g. Radjai and Richefeu (2009), show that though the order of the pairs greatly affect the individual contact forces and even the topology of the system, the overall, "macro" characteristics like average stress tensor or frequency diagram of contact force magnitudes remain the same, apart from slight statistical deviations. This conclusion might be valid for masonry systems too, but the existing investigations up to the present are still insufficient to draw reliable conclusions.

MODELLING WITH DEFORMABLE ELEMENTS

Elements and Contacts

Blocks in NSCD can be made deformable either by applying a uniform strain field in the whole block (like in DDA), or by using a finite element subdivision inside the blocks. Koziara and Bicanic (2008) presented the possibility to apply uniform-strain blocks. They used the term "pseudo-rigid bodies" to the approach, and suggested it as an intermediate model between perfectly rigid blocks and FEM-divided bodies. A finite element subdivision seems to be more appropriate for practical problems particularly in the case of complex block shapes or significantly varying stress fields inside the blocks.

Jean (1999) applied uniform-strain finite element subdivision as the simplest possibility for FEM meshing. A two-dimensional illustration is shown in Figure 5: a rectangular block is subdivided into eight uniform-strain triangular elements.

Figure 5. Deformable 2D block consisting of triangular finite elements: **O**: *nodes;* • : *potential contact points*

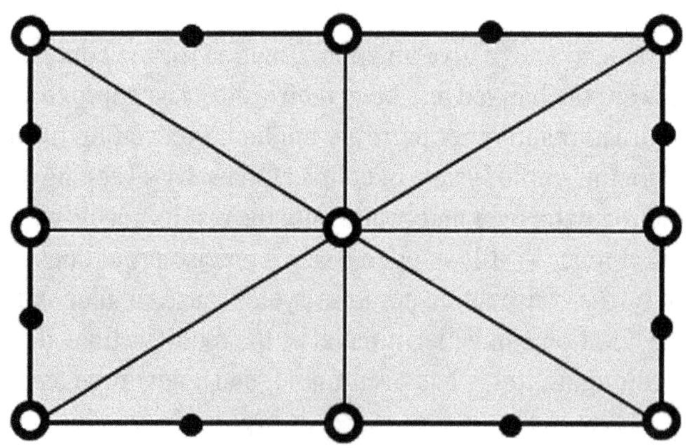

The mass of the element is distributed to the nodes (denoted by empty dots in Figure 5). The degrees of freedom are the translations of these nodes, which means that nodal rotations are not considered in the model: the usual strain field of classical continua is the basis of stress calculations (no Cosserat- or other non-classical continua are applied).

The equations of motion of a node specify the relations between the translational accelerations of the node, and between the forces reduced to the node:

- Mass-proportional forces (e.g. selfweight);
- External loads (including velocity-proportional forces like drag force);
- Contact forces acting on any face belonging to the analysed node;
- Internal forces: effect of stresses inside elements belonging to the analysed node.

The determination of the point where the contact forces act on the considered element is a crucial issue. Instead of performing a detailed analysis of the force distribution along the contact surfaces, a simplified approach is used. Dark dots in Figure 5 denote the candidates for contact points with neighbouring faces. They are chosen to be in the centre of the finite element face. Their actual position can be linearly interpolated from the nodes forming the face where the candidate contact point is located. If such a candidate touches a neighbouring face, contact forces act on the element at this contact point. The contact forces are then reduced to the nodes, according to the usual way at uniform-strain finite elements.

Note that LMGC90 offers more sophisticated FEM meshing options too (Dubois and Mozul, 2013), and in those cases the interpolation of contact point position and the reduction of the contact force become more complicated.

The Non-Uniqueness of the Solution

Similarly to the rigid-element case, the equations of motion of the whole system of deformable blocks at a time instant t can be written in the following general form:

$$\mathbf{Ma}(t) + \mathbf{Cv}(t) + \mathbf{K}(t)\Delta\mathbf{u}(t) = \mathbf{f}(t)\,.$$

In this expression the meaning of the terms are as follows:

- $\Delta\mathbf{u}(t)$ is the hypervector containing the nodal translations from an initial deformation-free configuration to the actual nodal positions at t;
- $\mathbf{v}(t)$ and $\mathbf{a}(t)$ are its first and second time derivatives (hypervectors of nodal velocities and nodal accelerations respectively);
- $\mathbf{K}(t)$ is the stiffness matrix expressing the elastic properties of the finite elements: its j-th column contains the opposite of nodal forces which arise when a unit translation is introduced at the j-th scalar of $\Delta\mathbf{u}(t)$;
- $\mathbf{C}(t)$ is the damping matrix: its j-th column contains the opposite nodal forces if a unit velocity occurs at the j-th scalar of $\mathbf{v}(t)$;

- $\mathbf{M}(t)$ is the block diagonal mass matrix that consists of as many blocks as the number of nodes, every block in it is a diagonal matrix containing three elements each of which being equal to the mass assigned to the corresponding node;
- $\mathbf{f}(t)$ is the hypervector of the external loads and contact forces reduced to the nodes.

Assuming that: (i) the structure is statically indeterminate or, at least, determinate, (ii) sliding, cracking or other abrupt changes of material behaviour can be excluded during $\Delta\mathbf{u}$, (iii) $\Delta\mathbf{u}$ is so small that $\mathbf{K}$ remains approximately the same as in the initial configuration, and (iv) loads are quasi-static, then since the stiffness matrix is invertible and constant, the iterative solver introduced in Section 2.4 corresponds to a Gauss-Seidel relaxation solution of the equilibrium position corresponding to the given loads. In this case the solution would be unique. If the structure is kinematically indeterminate but nonlinearities do not occur, then $\mathbf{K}$ is singular (though constant), and for general quasi-static loads the iterative solver does not lead to an equilibrium state but to an accelerating motion of the elements. (This phenomenon can characterize only an initial, small-displacement range of the behaviour.) Large displacements led to the gradual modification of $\mathbf{K}$. When sliding, contact cracking, or other dissipative material nonlinearities occur, $\mathbf{K}$ varies, and the solution becomes history-dependent and non-unique. In this case (similarly to the rigid-element case) repeated solutions of the same problem may differ from each other. Even if the system converges to an equilibrium, there are various paths of motions possible, and they typically lead to different equilibrium states. The user should be aware of this feature of Contact Dynamics. Acary and Jean (2000) discuss the problem and suggest a few possibilities to deal with the issue.

APPLICATIONS

The Contact Dynamics Method has been rather popular among physicists studying granular dynamics problems (e.g. Daudon et al, 1997; Radjai et al, 1998; Unger et al, 2004). In the field of masonry mechanics most applications are related to seismic simulations, though a few examples on quasi-static analysis can also be found in the literature as shown by the applications below.

Chetouane et al (2005) applied the Contact Dynamics method for the simulation of Pont Julien, a 1st-century BC roman bridge in South France, in Vaucluse. They built a 2D model with dry frictional contacts, and compared the results provided by the rigid-element and the deformable-element modelling approaches for quasi-static case. The load was the selfweight of the structure. Principal stress directions, hydrostatic stress components and contact states were compared. They found that while the computation time was definitely longer, deformable elements provided more realistic results. The 3D analysis of the same bridge under the effect of flood was published in Rafiee and Vinches (2013). The selfweight of those parts of the structure being under water level was decreased according to buoyancy, and increasing crosswise horizontal forces acting on the pillars at different levels were tested. A variety of failure mechanisms were revealed.

Raifee et al (2008a) modelled the Roman aqueduct in Arles, near Fontvielle, France. The aqueduct was collapsed, but the reasons of the failure were unknown. 3D rigid elements with dry frictional contacts were applied in the NSCD model of the aqueduct. Starting from an assumed undamaged initial geometry of the structure, selfweight and then sinusoidal seismic excitations were applied. The dynamic effects produced a cracked state of the structure, whose similarities to the in situ state suggested that a seismic event could be the reason of the destruction of the structure around 150 AD.

Rafiee et al (2008b) prepared 2D and 3D models of the amphitheatre in Nimes, one of the most beautiful and best preserved Roman arenas. The elements were rigid and deformable in their 2D models, and rigid in the 3D model, with dry contacts in all cases. In addition to selfweight, an artificial seismic vibration was simulated. The most vulnerable parts of the structure could be identified this way, so that a future restoration can take this knowledge into consideration. The amphitheatre was analysed a few years later by Bagnéris et al (2013), also with LMGC90. Rigid 3D elements were applied to model a similar part of the arena as in Rafiee et al (2008b). In Bagnéris et al (2013) the results provided by the rigid-element model were then applied to an individual block at the bottom of a pillar, and the behaviour of this block was simulated by using deformable elements in a linearly elastic FEM model. The pressure acting on this block along its boundaries was made non-uniform in the FEM analysis, in different ways (a peripheral support and surface roughness was produced by randomly translating the position of the FEM nodes perpendicularly to the surface). The results show that at some locations the magnitude of principle stresses could increase with 1 or even 2 orders of magnitude because of the contact surfaces being not perfectly planar. This phenomenon may lead to local damages that may modify the distribution of the internal forces in the structure provided by the NSCD calculation. (The authors also considered the effect f water infiltration, but that analysis is already out of the scope of an introduction to the NSCD method.)

Isfeld and Shrive (2015) modelled the cross-section of the wall of Prince of Wales Fort (built in the early 18th century) in Canada. The external part of the wall is made of cut stones lying on each other on approximately planar faces, while the core of the wall consists of ribble-like uncut stone pieces. The old mortar between the stone blocks degraded, weakened and was washed out during the centuries. In the model the external stones were represented by rigid polygons and the core consisted of rigid circular elements. The contacts were cohesive: the normal and shear strength were set to several different values in the different tests, and for every case the walls were tried to be equilibrated under selfweight. It was decided this way whether for the different joint material parameters the walls were stable or unstable. (The typical failure modes were also determined.) The authors concluded that stability could be improved by injecting grout into the walls.

Lancioni et al (2013) analysed a medieval Italian church, Santa Maria in Portuno. After its enlargement in the 11th century, the building had a nave and two aisles. For today only the nave remained, and one of the aims of the numerical analysis was to verify whether the collapse of the aisles could be caused by a 13th-century earthquake in the region. The authors applied rigid 3D polyhedral elements with dry frictional contacts. The geometry of the structure was reconstructed from the ruins that were found on the site. The model consisted of four macro-elements, corresponding to the following four main components of the structure: (1) the façade, (2) a longitudinal wall of the nave, (3) and external wall, and (4) the three apses. The 2009 L'Aquila earthquake was simulated. The analysis pointed out the most dangerous collapse mechanisms, and conclusions could be drawn regarding the expedient reinforcement of the structure.

Though the above examples demonstrate that the Contact Dynamics technique is able to simulate practical problems in an apparently realistic way, most of the applications up to now are poor in (or completely lack) a quantitative validation of the applied numerical model. An industry-inspired attempt to improve this situation can be found in the recent publications Ceh et al (2015a, 2015b). The authors conducted laboratory experiments and SOLFEC simulations on the same problem: multiple-block stacks subjected to base accelerations were analysed in both ways. Well-documented experimental and numerical tests like this would be very valuable for engineers who plan to apply a Contact Dynamics software for practical problems. With more validation studies, and with sufficiently increasing hardware capacities in the forthcoming years, the Contact Dynamics method may become a powerful tool in the everyday engineering practice in masonry analysis.

ACKNOWLEDGMENT

The above study was supported by the Hungarian National Research Fund under grant no. OTKA 100770.

REFERENCES

Acary, V., & Jean, M. (1998). Numerical simulation of monuments by the contact dynamics method. In *Procs. Monument-98 Workshop on seismic performance of monuments.*

Acary, V., & Jean, M. (2000). Numerical modeling of three-dimensional divided structures by the Non Smooth Contact Dynamics method: Application to masonry buildings In *Procs. Fifth International Conference on Computational Structures Technology.*

Bagneris, M., Dubois, F., & Martin, A. (2013). Numerical analysis of historical masonry structures for stone degradation diagnosis: An application to the Roman Amphitheater of Nîmes. In *2013 Digital Heritage International Congress (DigitalHeritage).* doi:10.1109/DigitalHeritage.2013.6743792

Camenen, J.-F., Ceh, N., Jelenic, G., Koziara, T., & Bicanic, N. (2015). Dynamic Sensitivity of a Multi-block Stack Subjected to Horizontal Harmonic Excitation.*22ème Congrès Français de Mécanique*, Lyon, France.

Ceh, N., Pellegrino, A., Camenen, J.-F., Bicanic, N., Petrinic, N., & Tuhtan, M. (2015b). *Overturning of multiple-block stack - dynamic sensitivity parameters and scaling effect.* 8th International Congress of Croatian Society of Mechanics, Opatija, Croatia.

Ceh, N., Pellegrino, A., Camenen, J.-F., Petrinic, N., Jelenic, G., Koziara, T., & Bicanic, N. (2015a). Dynamic sensitivity of multi-block stacks subjected to pulse base excitation – Experimental evidence and non-smooth contact dynamics simulations. In Procs. COMPDYN 2015.

Chetouane, B., Dubois, F., Vinches, M., & Bohatier, C. (2005). NSCD discrete element method for modelling masonry structures. *International Journal for Numerical Methods in Engineering, 64*(1), 65–94. doi:10.1002/nme.1358

Cundall, P. A. (1988). Formulation of a three-dimensional distinct element model - Part I: A scheme to detect and represent contacts in a system composed of many polyhedral blocks. *International Journal of Rock Mechanics and Mining Sciences, 25*(3), 107–116. doi:10.1016/0148-9062(88)92293-0

Cundall, P. A., & Strack, O. D. L. (1979). A discrete numerical model for granular assemblies. *Geotechnique, 29*(1), 47–65. doi:10.1680/geot.1979.29.1.47

Daudon, D., Lanier, J., & Jean, M. (1997). A micromechanical comparison between experimental results and numerical simulation of a biaxial test on 2D granular material. In Behringer et al. (Eds.), *Powders and Grains 97* (pp. 219–222). Balkema.

Dubois, F., & Jean, M. (2006). The non smooth contact dynamic method: recent LMGC90 software developments and application. In P. Wriggers & U. Nackenhorst (Eds.), *Analysis and Simulation of Contact Problems* (Vol. 27, pp. 375–378). Lecture Notes in Applied and Computational Mechanics. doi:10.1007/3-540-31761-9_44

Dubois, F., & Mozul, R. (2013). LMGC90. In *CSMA 2013, 11e Colloque National en Calcul des Structures.*

Isfeld, A., & Shrive, N. (2015). Discrete Element Modeling of Stone Masonry Walls With Varying Core Conditions: Prince of Wales Fort Case Study. *International Journal of Architectural Heritage*, *9*(5), 564–580. doi:10.1080/15583058.2013.819135

Jean, M. (1999). The non-smooth contact dynamics method. *Computer Methods in Applied Mechanics and Engineering*, *177*(3-4), 235–257. doi:10.1016/S0045-7825(98)00383-1

Jean, M., & Moreau, J. J. (1992). Unilaterality and dry friction in the dynamics of rigid body collections. In *Procs. Contact Mechanics International Symposium.* Presses Polytechniques et Universitaires Romandes.

Koziara, T. & Bićanić, N. (2008). Semismooth Newton method for frictional contact between pseudo-rigid bodies. *Computer Methods in Applied Mechanics and Engineering, 197*(33-40), 2763–2777.

Lancioni, G., Lenci, S., Piattoni, Q., & Quagliarini, E. (2013). Dynamics and failure mechanisms of ancient masonry churches. *Engineering Structures*, *56*, 1527–1546. doi:10.1016/j.engstruct.2013.07.027

Moreau, J. J. (1988). Unilateral contact and dry friction in finite freedom dynamics. In J. J. Moreau & P. D. Panagiotopoulos (Eds.), *CISM Courses and Lectures* (Vol. 302). Vienna: Springer. doi:10.1007/978-3-7091-2624-0_1

Moreau, J. J. (2006). Facing the plurality of solutions in nonsmooth mechanics. In C.C. Baniotopoulos (Ed.), *Nonsmooth/Nonconvex Mechanics with Applications in Engineering, II. NNMAE 2006 (Proc. International Conference in Memoriam of P. D. Panagiotopoulos).*

Radjai, F., & Richefeu, V. (2009). Contact dynamics as a nonsmooth discrete element method. *Mechanics of Materials*, *41*(6), 715–728. doi:10.1016/j.mechmat.2009.01.028

Radjai, F., Wolf, D. E., Jean, M., & Moreau, J. J. (1998). Bimodal character of stress transmission in granular packings. *Physical Review Letters*, *80*(1), 61–64. doi:10.1103/PhysRevLett.80.61

Rafiee, A., & Vinches, M. (2013). Mechanical behaviour of a stone masonry bridge assessed using an implicit discrete element method. *Engineering Structures*, *48*, 739–749. doi:10.1016/j.engstruct.2012.11.035

Rafiee, A., Vinches, M., & Bohatier, C. (2008a). Application of the NSCD method to analyse the dynamic behaviour of stone arched structures. *International Journal of Solids and Structures*, *45*(25-26), 6269–6283. doi:10.1016/j.ijsolstr.2008.07.034

Rafiee, A., Vinches, M., & Bohatier, C. (2008b). Modelling and analysis of the Nimes arena and the Arles aqueduct subjected to a seismic loading, using the Non-Smooth Contact Dynamics method. *Engineering Structures*, *30*(12), 3457–3467. doi:10.1016/j.engstruct.2008.05.018

Shi, G.-H. (1992). Discontinuous deformation analysis: A new numerical model for the statics and dynamics of deformable block structures. *Engineering Computations*, *9*(4), 157–168. doi:10.1108/eb023855

Unger, T., & Kertész, J. (2003). The contact dynamics method for granular media. In *Modeling of Complex Systems* (pp. 116–138). Melville, NY: American Institute of Physics. doi:10.1063/1.1571300

Unger, T., Wolf, D. E., & Kertész, J. (2004). *Force indeterminacy in the jammed state of hard disks*. cond-mat/0403089

KEY TERMS AND DEFINITIONS

Common Plane: The plane between two arbitrarily-shaped convex bodies which maximizes the gap between the plane and the closest points on the two elements.

Implicit Method: A time integration method is implicit if the approximated numerical solution of the differential equation of the analyzed initial value problem is calculated at the end of a time interval from the numerical solution at the beginning of the interval in such a way that the differential equation is satisfied at the end of the time interval according to any required exactness.

Iterative Solver: That part of an NSCD code which iteratively finds the approximate solution for contact forces at a given time instant.

Singular Stiffness Matrix: Structures that are not supported sufficiently, either internally or externally, are kinematically indeterminate. The stiffness matrix of such a structure is singular i.e. it has at least one zero eigenvalue, so that it cannot be inverted.

Time Integration: A procedure to quantitatively approximate the solution of an initial value problem described by a differential equation. The time integration provides quantitative approximations of the state variables of the modelled system at a series of separate time instants.

Chapter 6
Introduction to the Combined Finite–Discrete Element Method

Máté Hazay
Budapest University of Technology and Economics, Hungary

Ante Munjiza
Queen Mary University of London, UK

ABSTRACT

This chapter presents a general overview of the combined Finite-Discrete Element Method (FEM/DEM) which is considered as a state-of-the-art technique for the mechanical analysis of masonry structures. In a FEM/DEM simulation each discrete element representing a stone block is discretized into finite elements in order to describe the deformability of the blocks. This chapter deals with the main steps of the FEM/DEM including contact detection, contact interaction, fracture and fragmentation algorithms, calculation of deformations and the time integration of the equation of motion. The FEM/DEM is advantageously used to simulate transition from continua to discontinua processes which may lead to the collapse of the structure. Some examples for practical applications found in the literature are mentioned.

INTRODUCTION

The two basic types of mechanical models are the classical models of continuum mechanics and the models of discrete elements. The finite element method (FEM) has been the most widely used technique to model continuum mechanical problems for decades. On the other hand, the discrete element method (DEM) is capable of describing discontinuum-based phenomena including the motion and interaction of individual particles. DEM was originally introduced by Cundall (Cundall, 1971) for modelling rock mechanical problems with the assumption that infinitely rigid blocks comprise the problem. Later, this pioneering model have been improved and modified by taking into account the deformability of the blocks. Such an alternative model is the discontinuous deformation analysis (DDA) introduced by Shi (Shi, 1988) where the blocks were able to deform uniformly. Intensive research work resulted in the appearance of several DEM codes, i.e. UDEC, YADE, etc. However in certain situations the continuum- and discontinuum-like phenomena arise at the same time, thus the development of a coupled numerical

DOI: 10.4018/978-1-5225-0231-9.ch006

technique was required. Therefore, in the early 1990s the two above mentioned methods were combined and the resulting method was termed the combined finite-discrete element method (FEM/DEM). A typical combined finite-discrete element simulation comprises large number – thousands or even millions – of particles which are represented by a single discrete element. In FEM/DEM each discrete element is discretized into finite elements in order to describe the deformations of the blocks. Furthermore, the classical steps of discrete element method, including contact detection, contact interaction and time integration are applied to follow the motion and interaction of the individual particles. This method was implemented mainly to simulate so-called transition from continua to discontinua problems, including failure, fracture and fragmentation processes. Therefore, detailed structural collapse simulations can be performed with the help of the FEM/DEM. These simulations can play a vital role in the design of structures against hazardous loading conditions. In this case the load bearing capacity of the structure can be determined, and the progressive collapse modes can be identified as well. Some examples for modelling transition from continua to discontinua problems can be found at the end of this chapter and in (Munjiza, 2004, pp. 30-32).

Masonry is a heterogeneous structural material consisting of bricks joined by mortar layers. Numerical modelling of masonry structures - especially in the nonlinear range - represents an interesting challenge due to the extremely complex behaviour deriving from their semi-discrete, composite nature. Two different approaches exist for mechanical modelling of masonry structures, these are the micro modelling and macro modelling of masonry structures. In case of the macro modelling approach, the structure is handled as an orthotropic continuum and the constitutive relations are described with the help of experiments and homogenization techniques. The serious disadvantage of this continuum-based approach is its inability to describe large discontinuities which occur during the collapse process. In order to ease this restriction contact elements may be implemented into the finite element mesh. This approach enables to consider the material nonlinearity where the constitutive behaviour of the contact elements are prescribed according to the theory of plasticity or damage mechanics. Unfortunately, even with the application of contact elements, the finite element method is unable to handle the mechanical interaction of several blocks. Due to this fact, reliable collapse simulations including progressive failure modes cannot be carried out with the application of the FEM. However FEM/DEM includes contact detection and contact interaction algorithms, thus this technique is capable of modelling the mechanical interaction of several bodies. Therefore, numerical models based on the combined finite-discrete element method may be developed for analysing such problems.

Appropriate computational modelling of the response of masonry structures must involve all of the fundamental mechanisms which affect the failure process, like sliding and splitting of the blocks along the mortar joints and cracking of individual units under certain stress conditions. The combined finite-discrete element method proved to be a very effective tool for the analysis of the non-linear behaviour of masonry structures. The FEM/DEM uses this previously mentioned joint element approach to describe the fracturing behaviour of an initially continuum-like solid block represented by a single discrete element. Therefore, in case of combined finite-discrete element simulations, the number of discrete elements may vary during the collapse process. The advantage of this method derives from the possibility to model the whole process of the structural response starting from the linear-elastic phase, followed by the non-linear range including crack initiation, crack propagation, and sliding effects. The motion and interaction of the particles are followed considering inertia effects until the final, state of rest situation is reached due to the energy dissipation of the structure. The FEM/DEM can be a very advantageous technique, particularly in case of cyclic loading situations, where the appropriate modelling of the creation of new contacts due to the redistribution of the system topology is a fundamental demand. At the end of this chapter an example for the analysis of structural response under cyclic loading conditions is shown.

In the following sub-chapters the main processes of the combined finite-discrete element method are discussed including contact detection, contact interaction, fracture and fragmentation algorithms, calculation of deformations, parallelization and the time integration of the equations of motion. Different modelling approaches and examples connected to the structural analysis of masonry structures are mentioned which were found in the literature. Afterwards a brief comparison was made between FEM/DEM and other techniques belonging to the DEM.

BASIC CHARACTERISTICS OF THE MECHANICAL MODEL

This sub-chapter focuses on the basic characteristics of the mechanical model applied in a combined finite-discrete element simulation. In general, a FEM/DEM simulation may comprise a large number of arbitrary shaped deformable discrete elements. Each discrete element represents a continuum, therefore the deformations of solid blocks are followed by classical continuum mechanical description (i.e. each material point has translational freedoms). Since the finite element method is applied here to follow the continuum-like behaviour, each discrete element is discretized into finite elements (mentioning that usually constant strain triangle or tetrahedron elements are used). Therefore, the shape of discrete elements and their position in space at any time instant are given by the current coordinates of the finite element nodes:

$$\boldsymbol{x} = \begin{bmatrix} x_1 \\ x_2 \\ x_3 \\ \dots \\ x_i \\ \dots \\ x_N \end{bmatrix}, \tag{1}$$

where N is the total number of degrees of freedom of a particular discrete element. It means that N (i.e. the total number of the components of vector $\boldsymbol{x}$) equals the product of the number of finite element nodes and the number of translational freedoms corresponding to the nodes. Stresses can be calculated in a typical continuum mechanical approach, by using strain tensor and some kind of constitutive equation. During the calculation of the strains, usually small strains but large displacements and large rotations must be considered. After evaluating the stress field, the $\boldsymbol{F}_{\text{int}}$ vector of internal forces can be compiled by reducing the distributed internal forces to the finite element nodes following the finite element discretization. The velocity and acceleration fields over the discrete elements are described by nodal velocities $\boldsymbol{v}$ and nodal accelerations $\boldsymbol{a}$ given by:

$$\boldsymbol{v} = \dot{\boldsymbol{x}} = \begin{bmatrix} \dot{x}_1 \\ \dot{x}_2 \\ \dot{x}_3 \\ \ldots \\ \dot{x}_i \\ \ldots \\ \dot{x}_N \end{bmatrix} \text{ and } \boldsymbol{a} = \dot{\boldsymbol{v}} = \ddot{\boldsymbol{x}} = \begin{bmatrix} \ddot{x}_1 \\ \ddot{x}_2 \\ \ddot{x}_3 \\ \ldots \\ \ddot{x}_i \\ \ldots \\ \ddot{x}_N \end{bmatrix}. \tag{2}$$

The interaction of neighbouring discrete elements is described by the contact detection and contact interaction algorithms. As a result of the mechanical interaction, contact forces are evaluated between discrete elements. These contact forces can be assembled in $\boldsymbol{F}_{\mathbf{ext}}$ vector of external forces, which also includes the external forces acting on discrete elements directly. Again, the number of components of this vector equals *N*, since concentrated forces may act at finite element nodes in the direction of any degree of freedom. Additionally, the combined finite-discrete element method has the speciality that discrete elements can fracture and fragment, thus the number of discrete elements comprising the problem may vary during the analysis. This fracturing behaviour is described by joint elements implemented into the finite element mesh. The mechanical behaviour of these joint elements is defined by the fracture and fragmentation algorithm. The forces transmitted through joint elements can be assembled in $\boldsymbol{F}_{\mathbf{joint}}$ vector.

The applied mechanical model may include any kind of external or internal damping effects considered in damping matrix $\boldsymbol{C}$. Additionally, since the transient dynamic motion of discrete elements is analyzed, the equation of motion must include inertia forces as well. Due to the finite element discretization, the mass is also discretized. To reduce the computational challenge, the most convenient way to discretize the mass is the so-called lumped mass approach. Here it is assumed that the mass is lumped into the nodes of the finite element mesh. Hence, the mass associated with each degree of freedom is given by:

$$\boldsymbol{m} = \begin{bmatrix} m_1 \\ m_2 \\ m_3 \\ \ldots \\ m_i \\ \ldots \\ m_N \end{bmatrix}. \tag{3}$$

Following the lumped mass approach, after the compilation of equation of motion, a diagonal lumped mass matrix $\boldsymbol{M}$ is obtained. When combined finite-discrete element simulations comprise thousands or millions of discrete elements, then thousands or millions of separate finite element meshes are included as well. Referring to the computational challenge associated with large scale FEM/DEM problems, no stiffness matrices are calculated, and an explicit time integration scheme is applied on element-by-element, node-by-node and degree of freedom by degree of freedom.

Therefore, Equation 4 describes the equation of motion which is solved for each element separately:

$$\boldsymbol{Ma} + \boldsymbol{Cv} + \boldsymbol{F}_{\text{int}} - \boldsymbol{F}_{\text{ext}} - \boldsymbol{F}_{\text{joint}} = \boldsymbol{0}. \tag{4}$$

After prescribing the initial conditions, the following computational tasks are performed in each time step:

1. Evaluation of internal forces based on the deformation of solid blocks,
2. Evaluation of joint forces based on the deformation of joint elements,
3. Fracture of joints,
4. Contact detection,
5. Contact interaction (i.e. evaluation of contact forces),
6. Application of external forces,
7. Solution of the equation of motion.

After the discussion of the basic characteristics of the applied mechanical model, the further subchapters deal with the most important steps of a FEM/DEM simulation in a more detailed way. First the main aspects of the applied contact detection algorithms are considered.

CONTACT DETECTION ALGORITHM

In case of a combined finite-discrete element analysis, the transient dynamics of a large number of deformable discrete elements are simulated. One of the key issues in the development of a FEM/DEM code is the realistic and efficient handling of mechanical contacts formed between the discrete elements. The treatment how the mechanical behaviour of contacts are simulated is based on the contact detection and contact interaction algorithms.

In general, processing contact interaction of all possible contacts would involve a total number of computational operations proportional to the square of the total number of discrete elements comprising the problem. In case of large-scale FEM/DEM simulations this would represent a huge - probably unsolvable - computational challenge, and would limit the application possibilities of FEM/DEM to the simulation of relatively small problems. Therefore, a contact detection algorithm is applied which is responsible for reducing the CPU requirements of processing contact interaction, while contact detection itself must demand as small RAM and CPU requirements as possible. Contact detection algorithms reduce CPU requirements via two main steps, namely these have to eliminate those couples of discrete elements that are far from each other and are surely not in contact and these have to detect all couples of discrete elements that are actually in contact. In case of a contact detection algorithm, beside its robustness, the CPU efficiency, RAM efficiency and its easy implementation are the most important requirements.

Emphasizing the importance of CPU efficiency, contact detection algorithms can be classified according to the proportion how the CPU time depends on the size of the problem. Generally speaking, classical contact detection algorithms of discrete element method (for instance the body-based search technique) belong to the group of Hyper-linear contact detection algorithms meaning that the necessary CPU time increases faster than the size of the problem (i.e. number of discrete elements). Nowadays linear contact detection algorithms are used in combined finite discrete element simulations meaning that the CPU time is a linear function of the number of discrete elements comprising the problem. The first such an algorithm was implemented in 1995 and called Munjiza-NBS contact detection algorithm.

The NBS contact detection algorithm is based on space decomposition, where the space is subdivided into identical square cells as it is shown in Figure 1.

Discrete elements are assigned an integer identification number and similarly, each cell is assigned an identification couple of integer numbers. Afterwards each discrete element is mapped onto cells, meaning that each element is assigned to one and only one cell. This can be seen in Figure 1, where each dot represents the centre of a discrete element. Additionally, mapping of discrete elements onto columns and rows are performed. A discrete element is said to be mapped to a particular row/column of cells if it is mapped to any cell from that row/column. The CPU and RAM efficiency of Munjiza-NBS contact detection algorithm derives mainly from the way how it represents the mapping between discrete elements and cells, namely singly connected linked lists are used for this purpose. The representation of mapping is performed in two stages, starting with mapping of all discrete elements onto the rows of cells, when a singly connected list is formed for each row as it is shown in Figure 2.

Authors note that the digital representation of each list are achieved through two integer arrays. Further details can be found in (Munjiza, 2004, pp. 73-129) and (Munjiza & Andrews, 1998). In the second stage, mapping of discrete elements to individual cells are performed. After the elements were mapped onto cells, contact detection is accomplished by checking all discrete elements mapped to a particular cell against all discrete elements situated in neighbouring cells. It is important that only discrete elements from neighbouring cells can touch each other, thus for each element it is enough to check against discrete elements belonging to neighbouring - non-empty - cells. Contact detection is performed by employing a loop over discrete elements. Number of operations carried out inside the loops does not depend on the number of discrete elements, therefore the total CPU time necessary to detect every

Figure 1. Space decomposition in Munjiza-NBS contact detection algorithm
Source: Munjiza, 2004

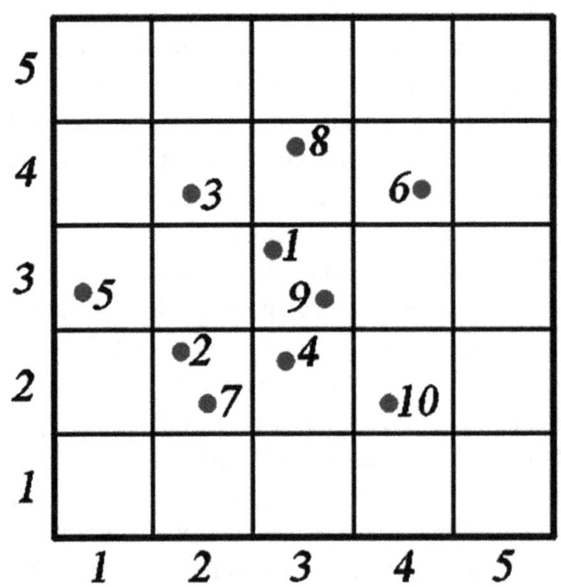

Figure 2. Singly connected lists for rows of cells
Source: Munjiza, 2004

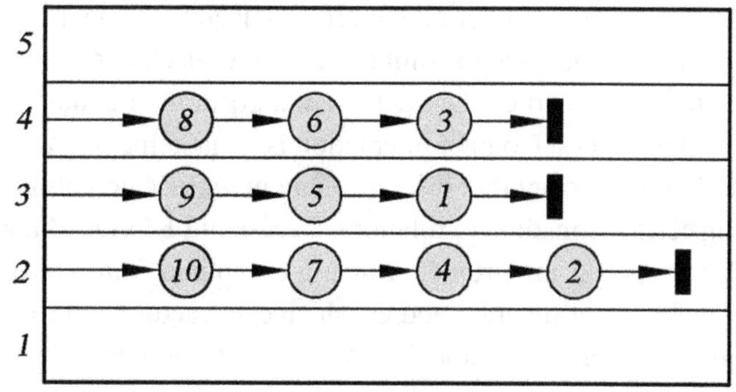

contact is proportional to the total number of discrete elements. In the Munjiza-NBS contact detection algorithm no loop over cells is involved, which causes that the required total CPU time is independent of the total number of cells. It means that the necessary CPU time does not depend either on the size of the finite space where the discrete elements are distributed or on the packing density. The Munjiza-NBS contact detection algorithm can be generalized to 3D space as well. This step requires the mapping of all discrete elements onto layers of cells in the third orthogonal direction too. Further details can be found in (Munjiza, 2004, pp. 118-128).

It should be mentioned that even though the contact detection operations could be done in each time step, it would be very expensive for CPU time point of view. Therefore, a so-called buffer can be introduced. The frequency of contact detection analyses can be controlled through the size of the buffer, because the contact detection process is performed only if the maximum travelled distance of elements exceeds the size of this buffer.

Nowadays a whole range of contact detection algorithms is available for large scale combined finite-discrete element simulations. In general, there is not an exact answer to which is the best contact detection algorithm, hence further consideration about the selection of the applied contact detection algorithm may be necessary. The screening array based contact detection algorithm is one of the most efficient algorithm from CPU time point of view, however it demands enormous RAM requirements. The sorting contact detection algorithm is very efficient in terms of RAM requirements, but this algorithm belongs to the group of hyper-linear contact detection algorithms, meaning that for a very large-scale problem CPU time may be prohibitive. In terms of CPU requirements, the Munjiza-NBS contact detection algorithm is more efficient than the binary search based contact detection algorithms or the sorting contact detection algorithm. Although it uses less RAM space than the binary search based algorithms, the sorting contact detection algorithm is a little bit more efficient from this point of view. The Munjiza-NBS contact detection algorithm and the sorting contact detection algorithm have RAM requirements proportional to the number of discrete elements. Eventually, the authors mention that further alternative contact detection algorithms can be found in the literature, referring to papers (Munjiza, Rougier, & John, 2006), (Schiava D'Albano, Munjiza, & Lukas, 2013), (Munjiza, Knight, & Rougier, 2012), (Perkins & Williams, 2002) and (Zang, Gao, & Lei, 2011).

CONTACT INTERACTION ALGORITHMS

As it has been mentioned earlier, in case of combined finite-discrete element simulations it is absolutely essential to have an efficient and robust algorithm for handling mechanical contacts. Once elements in contact are detected due to the applied contact detection algorithm, a contact interaction algorithm is applied to evaluate the contact forces. Contact interaction is a mathematical model which computes the penetration of a discrete element into another discrete element. After calculating the penetration, contact forces are evaluated using certain constitutive relations. In case of FEM/DEM applications the treatment of contact interaction has a special importance. Since a combined finite-discrete element simulation may contain millions of deformable separate bodies which can fracture and fragment, the topology and even the size of the problem change continuously. Therefore, the process how the contacts are handled defines the constitutive behaviour of the system. Due to this fact, special attention must be paid on contact kinematics and robustness in order to obtain energy balance and realistic distribution of contact forces.

In case of algorithms based on so-called concept of the contact element described in (Munjiza, Owen, & Bicanic, 1995) these requirements are not satisfied entirely. When the overlap of discrete elements in contact exceeds the contact layer, energy balance is not preserved and the same is true when new surfaces are created due to fracture and fragmentation processes. Furthermore, in this concept concentrated contact forces are considered, which causes stress and strain concentrations near the boundary of elements. These stress concentrations may result in significantly unrealistic fracture and fragmentation behaviour especially in case of brittle materials.

Recognizing these phenomena the so-called potential field-based penalty function method was introduced (Munjiza & Andrews, 2000). Here the basic assumption is that when two bodies in contact penetrate each other, this penetration results in a contact force, as it is shown in Figure 3. For further calculations, a standard contact functional U_c was implemented in the following way:

$$U_c = \int_{\Gamma} \frac{1}{2} p \left(\mathbf{r}_t - \mathbf{r}_c\right)^{\mathrm{T}} \left(\mathbf{r}_t - \mathbf{r}_c\right) \mathrm{d}\Gamma, \tag{5}$$

Figure 3. Contact force due to an infinitesimal overlap around points P_c and P_t
Source: Munjiza, 2004

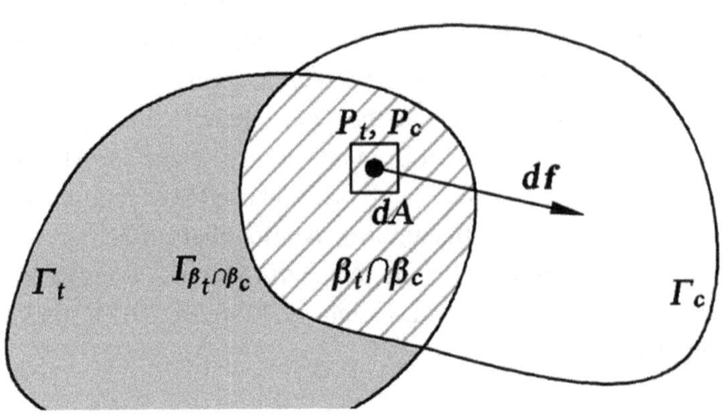

where p is the penalty term while $\boldsymbol{r}_t$ and $\boldsymbol{r}_c$ are position vectors of the points on the overlapping boundaries of the target and contactor bodies, respectively. In case of infinite penalty terms no penetration would occur, however large penalty term may cause numerical problems, thus in practical applications the penalty function method usually works with overlaps between discrete elements in contact. The penalty term can be selected as to be proportional to the modulus of elasticity E:

$$p = \alpha E, \tag{6}$$

where α is a scaling factor. In this case, the error in the displacements due to the penetration can be easily controlled. For example, if $\alpha = 100$, the local error of displacements will be less than 1%.

In this potential contact force concept distributed contact forces are considered which are evaluated from the shape and the size of overlap between the so-called contactor and target elements which form the contact. It is assumed that in the situation of elementary penetration where the overlapping area is dA, an infinitesimal contact force arises given by:

$$\mathrm{d}\boldsymbol{f} = \left[\mathrm{grad}\varphi_c\left(P_c\right) - \mathrm{grad}\varphi_t\left(P_t\right)\right]\mathrm{d}A = \mathrm{d}\boldsymbol{f}_c - \mathrm{d}\boldsymbol{f}_t, \tag{7}$$

where $\mathrm{d}\boldsymbol{f}$ is the infinitesimal contact force, due to the overlap dA defined by overlapping points P_c - belonging to the contactor - and P_t - belonging to the target, moreover φ_c and φ_t are the potential functions of the contactor and target elements, respectively. As it is shown in Equation 7, the contact can be viewed as firstly the elemental area of the contactor penetrating the target and then the elemental area of the target penetrating the contactor. The previous formula also shows that the contact force can be calculated as the gradient of the corresponding potential function, thus contact forces form a conservative vector field. Due to this fact, if point P_c of the contactor element penetrates the target through any path defined by end point A and B, the total work of the contact force does not depend on the path but on the end points only and it is given as $\varphi_t\left(A\right) - \varphi_t\left(B\right)$.

An important aspect is that when point A and B are placed on the boundary of the target discrete element, a contact-contact release situation arises, when the energy preservation law requires that no work is done by the contact force, i.e.:

$$\varphi_t\left(A\right) - \varphi_t\left(B\right) = 0 \text{ and } \varphi_c\left(A\right) - \varphi_c\left(B\right) = 0. \tag{8}$$

Therefore, if Equation 8 is valid for every arbitrary boundary points of the contactor and target elements, then the contact force given by Equation 7 preserves the energy balance regardless of the geometry and the shape of contacting elements, the size of the overlap and the value of the penalty term. Afterwards the total contact force is obtained by integrating elementary forces defined in Equation 7 over the overlapping area S as it is shown in Equation 9.

$$\boldsymbol{f} = \int_{S=\beta_t \cap \beta_c} \left[\mathrm{grad}\varphi_c - \mathrm{grad}\varphi_t\right]\mathrm{d}A. \tag{9}$$

This integral can be transformed into a line integral over the boundary of the overlapping area Γ:

$$\boldsymbol{f} = \oint_{\Gamma=\beta_t \cap \beta_c} \boldsymbol{n}_\Gamma \left(\varphi_c - \varphi_t\right) \mathrm{d}\Gamma, \tag{10}$$

where $\boldsymbol{n}_\Gamma$ is the outward unit normal to the boundary of the overlapping area.

Since in the FEM/DEM individual discrete elements are discretized into finite elements, the previous integrals can be represented by the summation over finite elements as it is shown in Equation 11 and Equation 12.

$$\boldsymbol{f} = \sum_{i=1}^{n} \sum_{j=1}^{m} \int_{\beta_{t_j} \cap \beta_{c_i}} \left[\mathrm{grad}\varphi_{c_i} - \mathrm{grad}\varphi_{t_i}\right] \mathrm{d}A, \tag{11}$$

$$\boldsymbol{f} = \sum_{i=1}^{n} \sum_{j=1}^{m} \int_{\Gamma=\beta_{t_j} \cap \beta_{c_i}} \boldsymbol{n}_{\Gamma_{\beta_{t_j} \cap \beta_{c_i}}} \left[\varphi_{c_i} - \varphi_{t_j}\right] \mathrm{d}\Gamma, \tag{12}$$

where n and m are the total number of finite elements into which the contactor and target discrete elements are discretized.

Hence, the contact force between overlapping discrete elements can be calculated by a summation over the edges of corresponding finite elements that overlap.

In case of a 3D situation, integral over the overlapping volume must be performed as it is shown in Equation 13.

$$\boldsymbol{f} = \sum_{i=1}^{n} \sum_{j=1}^{m} \int_{\beta_{t_j} \cap \beta_{c_i}} \left[\mathrm{grad}\varphi_{c_i} - \mathrm{grad}\varphi_{t_i}\right] \mathrm{d}V. \tag{13}$$

By replacing integration over finite elements by equivalent integration over finite element boundaries, the following equation for contact force is obtained:

$$\boldsymbol{f} = \sum_{i=1}^{n} \sum_{j=1}^{m} \int_{S=\beta_{t_j} \cap \beta_{c_i}} \boldsymbol{n} \left[\varphi_{c_i} - \varphi_{t_j}\right] \mathrm{d}S. \tag{14}$$

Thus, the contact force between overlapping discrete elements is calculated by a summation over the surfaces of corresponding finite elements that overlap, where $\boldsymbol{n}$ is the outward unit normal to the surface of the overlapping volume.

As a summary, it was shown that using contact interaction algorithm based on the potential field-based penalty function method has several advantages. It works with distributed contact forces, hence no artificial stress concentrations arise due to the contact, which would cause unrealistic fracturing behaviour. Furthermore, these algorithms satisfy the energy preservation which is an essential requirement for simulating physical behaviour properly. Due to this fact, the application of any kind of artificial damping is unnecessary. This potential force concept allows accurate representation of physical contact conditions including friction, sliding, plasticity, surface roughness, wet-dry conditions, etc. This technique uses only data supplied in the in-core database for contact free finite element analysis and it can be easily linked to contact detection algorithm such as Munjiza-NBS contact detection algorithm. This algorithm does not need detection of boundary surfaces and the contact force is discretized with the same algorithm regardless of the shape of discrete elements, thus algorithmic complexities are greatly reduced. All of these reasons lead to the fact that this type of contact processing is in general faster than alternative solutions. Additional details and algorithms can be found in the books (Munjiza, 2004, pp. 35-72) and (Munjiza, Knight, & Rougier, 2012).

DEFORMABILITY OF ELEMENTS

As from the basic idea of the combined finite-discrete element method follows, the continuum behaviour (i.e. the deformability) of discrete elements is described with the means of the finite element method. Each discrete element has its own finite element mesh which makes possible to follow the deformations of the discrete element. Furthermore, as it was shown earlier, the finite element discretization of individual discrete elements describes the contact conditions between discrete elements due to the discretization of distributed contact forces.

To describe the deformations of discrete elements, the classical equations and tensors of continuum mechanics are used. Thus, the current coordinates any material point having reference position vector $\boldsymbol{p}$ can be written as:

$$\boldsymbol{x} = \boldsymbol{p} + \boldsymbol{u}\left(\boldsymbol{p}\right), \tag{15}$$

where $\boldsymbol{u}\left(\boldsymbol{p}\right)$ is the displacement of the material point. Deformations are described with the help of a strain tensor calculated from the deformation gradient tensor $\boldsymbol{F}\left(\boldsymbol{p}\right)$:

$$\boldsymbol{F}\left(\boldsymbol{p}\right) = \nabla \boldsymbol{x} = \boldsymbol{I} + \nabla \boldsymbol{u}. \tag{16}$$

It should be mentioned that beside the Cartesian global coordinate system, other reference systems are also used to describe the deformations. Some of them are associated with the initial position of a particular discrete element, while the others move with the discrete element during the motion. Further details can be found about the applied reference frames and the transformations between them in (Munjiza, 2004, pp. 132-141). Typically, in a FEM/DEM simulation the well known strain measures of continuum mechanics, for instance, the right- and left Cauchy-Green, Green-St. Venant, Almansi-Hamel strain tensors, etc. are used. Although in most cases it is enough to consider small strains, it is very important to note that in FEM/DEM simulations usually large displacements must be followed. In such a case when the stretches are small but the rotations and displacements are large it is practical to decompose the deformation gradient tensor in the following way:

$$\boldsymbol{F} = \boldsymbol{R}\boldsymbol{U} = \boldsymbol{V}\boldsymbol{R}, \tag{17}$$

where $\boldsymbol{U}$ is the right stretch tensor, $\boldsymbol{V}$ is the left stretch tensor and $\boldsymbol{R}$ is the orthogonal rotation tensor. Following the polar decomposition of the deformation gradient tensor, the right Green-St. Venant strain tensor $\overline{\boldsymbol{E}}$ can be expressed as:

$$\overline{\boldsymbol{E}} = \frac{1}{2}\left(\boldsymbol{U}^{\mathrm{T}}\boldsymbol{U} - \boldsymbol{I}\right). \tag{18}$$

The application of $\boldsymbol{U}$ right stretch tensor physically means that the material is first stretched in the principal directions and these stretches are followed by the rotation. The reverse procedure can be performed by using the left stretch tensor $\boldsymbol{V}$. Here the left Green-St. Venant strain tensor is applied as:

$$\widetilde{\boldsymbol{E}} = \frac{1}{2}\left(\boldsymbol{V}\boldsymbol{V}^{\mathrm{T}} - \boldsymbol{I}\right). \tag{19}$$

In this case, the rotation occurs first, and the stretches already occur in the principal directions of the rotated configuration.

In order to measure the stress variables, for instance, the Cauchy, First- or Second Piola-Kirchhoff stress tensors can be applied. In the calculations some kind of constitutive relation is applied which makes connection between stresses and strains and contain failure criteria. Discussion of further details about the typical steps of continuum mechanics is out of the scope of this chapter. Additional details can be found, for instance, in (Munjiza, 2004, pp. 131-177), (Belytschko, Liu, & Moran, 2000, pp. 75-137) and (Munjiza, Knight, & Rougier, 2015).

Since the finite element discretization is also used to process contact interaction, in general, it is important to employ as simple geometry of finite elements as possible, in order to obtain efficient contact interaction algorithm. Due to this fact, mainly constant strain triangle finite elements are used in 2D and constant strain tetrahedron elements are used in case of 3D analyses. However in case of relatively incompressible solids, locking problems are associated with these linear finite elements which can seriously degrade the accuracy of the simulations. Therefore, in the paper (Xiang, Munjiza, & Latham, 2009), finite strain, finite rotation quadratic tetrahedral elements were implemented into the FEM/DEM.

FRACTURE AND FRAGMENTATION ALGORITHMS

The appropriate modelling of transition from continua to discontinua processes including fracture and fragmentation plays an essential role in a FEM/DEM analysis. Fracturing behaviour is described by joint elements implemented into the finite element mesh. These elements are illustrated in Figure 4.

The mechanical behaviour of these joint elements is defined by the fracture and fragmentation algorithm which handles the creation of new boundaries and new discrete elements under certain loading conditions.

Figure 4. Four-noded joint elements are situated between constant strain triangular elements

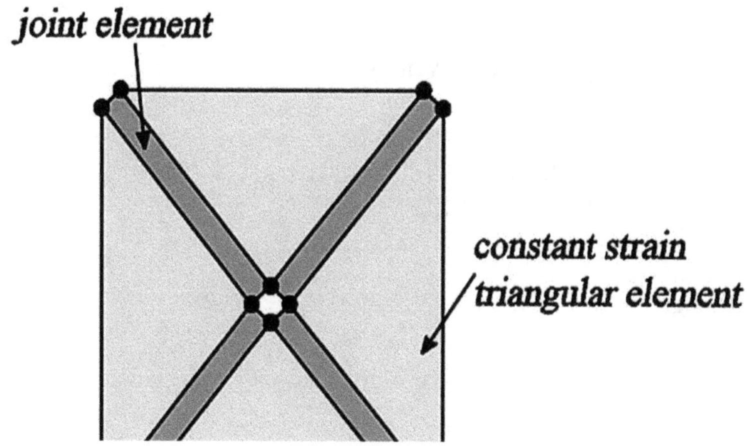

In each time step the joint forces are calculated based on the relative displacement between opposite triangle edges and then it is analyzed whether these joint forces cause the fracture of discrete elements or not. In order to proceed these steps, the fracture and fragmentation algorithm has to describe the constitutive behaviour of joint elements, including the conditions under crack initiation appears, and the way how the cracks propagate. During the crack propagation, remeshing is not necessary. Fractures can propagate along the edges of finite elements, therefore "arbitrary" (at least approximately) fracture trajectories can be obtained by applying sufficiently small element size. Several different approaches exist how to perform the analysis of cracks. The global approach is based on the representation of the stress singularity at the crack tip. This singularity can be characterized by the energy release rate *G*. The single crack approach which was used in FEM/DEM is similar to the Dugdale model (Lemaitre & Chaboche, 1990) which considers a plastic zone at the crack tip, and the plastically strained material is replaced by a zone of weakened bonds between the crack walls. On the other hand, local approaches usually apply a smeared crack approach, where the crack is replaced by a blunt crack band and the crack propagation processes are described with the help of constitutive laws or formulations of damage mechanics. In the combined finite-discrete element method a fracture based softening plasticity framework was applied, where the energy dissipation was considered through a mesh size dependent softening modulus as it is detailed in (Munjiza, Owen, & Bicanic, 1995). After further research (Munjiza, Andrews, & White, 1999), the combination of smeared and single crack approaches was implemented. The aim of this coupling was to model multiple-crack, and progressive fracture situations. In this combined single and smeared crack model the strain-hardening part of the stress-strain curve is considered through constitutive laws, while the strain-softening part is treated by the softening stress-displacement relationship being implemented through the single crack model as it can be seen in Figure 5. In this approach, the plastic zone is represented by bonding stress being the function of crack opening.

The separation begins when the bonding stress σ_b is equal to the tensile strength f_t. While the separation is increasing $\delta > \delta_t$, the bonding stress is continuously decreasing, and at separation $\delta = \delta_c$ the bonding stress drops to zero. The value of bonding stress for separation $\delta_t < \delta < \delta_c$ is given by

$$\sigma_b = z f_t, \tag{20}$$

where z is the scaling (softening) function given by

$$z = \left[1 - \frac{a+b-1}{a+b}\exp\left(D\frac{a+cb}{(a+b)(1-a-b)}\right)\right]\left[a(1-D)+b(1-D)^c\right], \tag{21}$$

where the variable D is given by

$$D = \begin{cases} 0 & \text{if } \delta \le \delta_t, \\ 1 & \text{if } \delta > \delta_c, \\ \dfrac{\delta - \delta_t}{\delta_c - \delta_t} & \text{otherwise.} \end{cases} \tag{22}$$

Using Equation 21 for the determination of scaling (softening) function z, the bonding stress will approximate the experimental stress-displacement curve, while the parameters a, b and c are obtained by curve fitting (Munjiza, Andrews, & White, 1999). The considered stress-strain diagram for tensile loading is shown in Figure 6, where the strength softening part is shaded.

The constitutive behaviour can be described similarly in case of shear loading as well. The relevant equations can be found in (Smoljanović, Živaljić, & Nikolić, 2013b). This method, when joint elements are applied to represent the fracture mechanism is termed as a combined single and smeared crack model (Munjiza, Andrews, & White, 1999). After dissipating the material fracture energy release rate

Figure 5. Plastic zone represented by bonding stress being a function of crack opening
Source: Munjiza & John, 2002

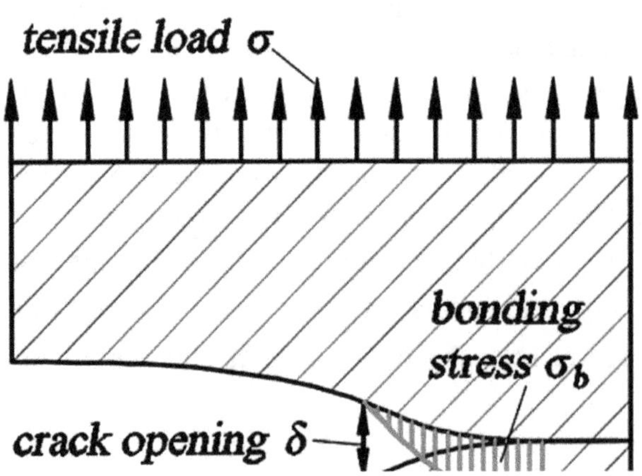

Figure 6. Constitutive behaviour of joint elements for tensile loading

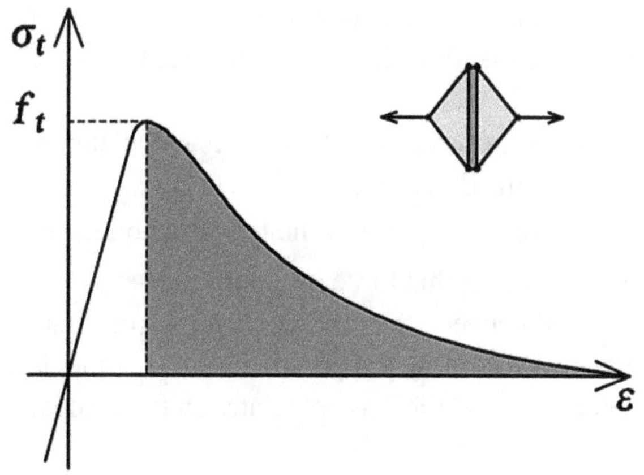

G, joint elements break and these are removed from the simulation which means that the transition from continuum to discontinuum is locally obtained.

Due to the finite element discretization of the governing equations, an approximated stress and strain field near the crack tip can be obtained. It was shown in (Munjiza, Andrews, & White, 1999), that in order to get accurate results, a very fine mesh is required near the crack tip, namely the size of the finite elements must be much smaller than the actual size of the plastic zone. It is important to note, that the combined single and smeared fracture algorithm is sensitive to the size of the finite elements situated in vicinity of the crack tip. In case of a very fine mesh, accurate results can be obtained, but if a very coarse mesh is applied, then the calculated failure load tends to the failure load belonging to the uniform stress distribution, and even in case of applying medium-sized elements, the obtained critical load will be overestimated. Beside the enormous computational challenge caused by contact detection, contact interaction and time integration, the fracture and fragmentation algorithms provides a huge computational task additionally, which – at least nowadays - may restrict the application of the combined finite-discrete element method in case of large-scale engineering problems.

TIME INTEGRATION

In order to reduce the CPU time, explicit time integration schemes may be applied. As it was mentioned earlier, the so-called lumped mass approach is used, thus diagonal mass matrix is considered during the analysis. Furthermore, the assembly of stiffness matrix can be avoided by using an explicit time integration scheme. Traditionally, the central difference method has been employed to solve the equation of motion. Since it is a conditionally stable technique, the stability and the required accuracy are achieved through reducing the size of the time step. As an alternative technique the Gear's predictor-corrector time integration schemes should be mentioned. These schemes always include three stages, namely:

- Prediction stage, where the position of the elements is calculated at $t + \Delta t$ using Taylor series based on positions and their time derivatives.
- Evaluation stage, where the forces f_{t+h} are evaluated based on the position of the elements calculated in the prediction stage. From this force, accelerations at $t + \Delta t$ can be calculated, and the difference can be determined between the predicted and the calculated accelerations.
    - Correction stage, where the predicted positions and its time derivatives are corrected using the discrepancy between accelerations calculated in the evaluation stage.

It should be mentioned that in the book (Munjiza, 2004, pp. 203-208) further alternative explicit time integration techniques, for instance CHIN, OMF30 and OMF32 integration schemes are discussed. All of these techniques can be used in a FEM/DEM simulation. The decision about which scheme should be used depends on their efficiency in terms of stability, accuracy and CPU time. In (Munjiza, 2004, pp. 208-211) and (Rougier, Munjiza, & John, 2004), a comparison analysis can be found for a single degree of freedom system. It was shown that although the higher order methods may work with larger values of critical time steps, still these methods are not necessarily faster than the lower order schemes. This conclusion is especially true in case of large scale systems, where contact detection, contact interaction, fracture and fragmentation may be involved, and higher order schemes require multiple force evaluation which is much less efficient than lower order methods.

PARALLELIZATION

As it has been mentioned, the limitation of the FEM/DEM is that it is CPU-intensive, thus it is difficult to analyze large scale problems on sequential CPU hardware. Therefore, researches (Owen, Feng, Han, & Peric, 2000), (Owen & Feng, 2001), (Wang, Feng, & Owen, 2004)] were carried out in order to implement the possibility to use high-performance parallel computers. In all those studies, a master/slave approach was implemented, meaning that one master processor performs domain decomposition and load balancing tasks, then it distributes work to slave processors. Some general approaches for parallelization of FEM/DEM are described in (Munjiza, Knight, & Rougier, 2012). Later, other versions appeared as static domain decomposition (Schiava D'Albano, 2014), hardware independent FEM/DEM parallelization framework by using virtual parallel machine (Lei, Rougier, Knight, & Munjiza, 2014), and dynamic domain decomposition based parallelization where all tasks (domain decomposition, load balancing) are performed concurrently on all processors (Lukas, Schiava D'Albano, & Munjiza, 2014). Parallelization strategies usually attempt to divide large problems (computational domains) into smaller sub-problems (sub-domains). A state-of-the-art parallelization technique has to fulfil two - often competing - requirements, namely each processor must be kept busy doing useful work while the communication between processors must be kept to a minimum. Today, parallelization algorithms are available in FEM/DEM codes, thus - at least in many cases - the solution of large scale problems became possible.

EXAMPLES FOR COMBINED FINITE-DISCRETE ELEMENT SIMULATIONS

In the following, some examples are mentioned for the application of the combined finite-discrete element method. This part focuses on the modelling of masonry structures, however other applications are shortly mentioned as well.

Firstly, referring to the paper (Smoljanović, Živaljić, & Nikolić, 2013a), seismic analysis of dry stone masonry structures were performed with the help of the FEM/DEM. In that work, each stone block was modelled as a discrete element which was discretized by constant strain triangular finite elements. The fracture and fragmentation processes were considered through contact elements which were implemented into the finite element mesh. In the applied FEM/DEM code the Munjiza-NBS contact detection algorithm was used. In the contact interaction algorithm the potential contact force concept was applied and the Coulomb-type condition was used to analyze the friction between discrete elements. The cracks were assumed to coincide with the finite element edges and the separation of these edges induced a bonding stress which was taken to be a function of the size of the separation. At first, numerical calculations for the analysis of shear behaviour of stone masonry joints were performed. Two different pre-compression stresses were applied and after shear tests were carried out under displacement control. The numerical results showed a great correspondence with the experimental results. Afterwards the behaviour of existing structures were analysed under seismic effects. The dynamic response of the structure of the Prothyron - which is a certain part of Diocletian's Palace in Split - was simulated with the combined finite-discrete element method. The FEM/DEM model of the structure can be seen in Figure 7.

The structure was subjected to gradually increasing horizontal ground acceleration until the collapse occurred. The collapse process of the structure can be seen in Figure 8.

That research (Smoljanović, Živaljić, & Nikolić, 2013a) showed the advantage of using the combined finite-discrete element method - which derives from the possibility of modelling the fragmentation of blocks - for analysis of failure modes of structures under hazardous loads. The results of the performed analyses showed good agreement with the experiments, thus this paper demonstrates the potential of the FEM/DEM for realistic modelling of the response of dry masonry structures. Further details can be found in the original paper (Smoljanović, Živaljić, & Nikolić, 2013a).

In paper (Smoljanović, Živaljić, & Nikolić, 2013b) the authors analyzed the behaviour of a dry stone masonry bell tower exposed to impact and seismic load. Again the whole structural failure (transition from continua to discontinua) process due to impact load was simulated. Using the combined finite-discrete

Figure 7. FEM/DEM model of Prothyron including discrete elements (left) and FEM mesh (right)
(© 2014, Nikolić et al. Used with permission.)

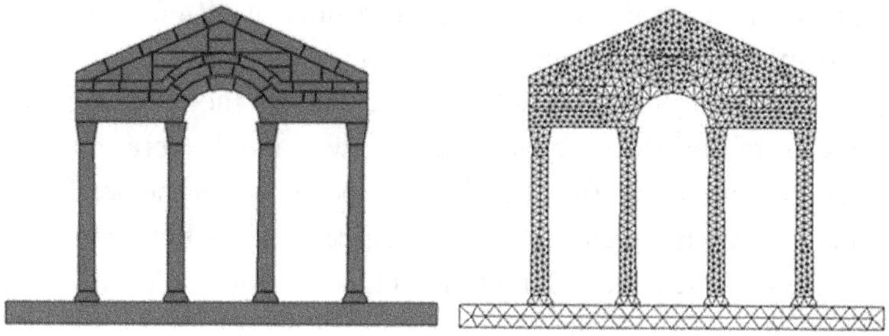

Figure 8. Collapse simulation of Prothyron due to seismic effect using FEM/DEM
(© 2014, Nikolić et al. Used with permission.)

element method the detailed analysis of failure was carried out including energy dissipation during the impact, initiation and propagation of cracks and the inertia effect of the individual parts of the structure. The seismic analysis of the structure was also performed which enables the determination of the load bearing capacity and the behaviour factor of the structure. In that paper the advantage of the combined finite-discrete element method in performing collapse simulations were shown. The applied numerical model was able to simulate the whole structural failure process including large displacements and rotations of stone blocks subjected to friction force acting between them. Further details of the analyses can be found in the original paper (Smoljanović, Živaljić, & Nikolić, 2013b) including numerical analyses of reinforced concrete structures as well.

In the paper (Baraldi, Reccia, Cazzani, & Cecchi, 2013) a comparative study is discussed. Discrete models and FEM/DEM models were used to investigate the in-plane behaviour of periodic brickwork. In the FEM/DEM simulations elastic blocks were considered by means of finite elements, but on the other hand, rigid discrete elements were applied in case of the discrete model. In the FEM/DEM model the mortar joints were idealized as elastic or elasto-plastic zero-thickness Mohr-Coulomb interfaces. Two different joint types were applied in the FEM/DEM simulations. One of these was implemented inside the blocks working with a high cohesion value in order to avoid the breaking of blocks. However, a much smaller cohesion value was applied for joints placed between the blocks in order to model the separation of blocks along the mortar joints. In the paper (Baraldi, Reccia, Cazzani, & Cecchi, 2013) masonry panels were subjected to in plane actions, like compression and shear forces in horizontal and vertical directions. Some deformed shapes of the FEM/DEM and discrete models due to horizontal shear effects can be seen in Figure 9.

The authors of paper (Baraldi, Reccia, Cazzani, & Cecchi, 2013) found that in case of the application of vertical and horizontal compressive forces, the results of different models show a very good agreement in terms of displacements and reactions. In case of compression, the influence of the texture pattern in the FEM/DEM model is less evident than in the discrete model due to the elastic behaviour of mortar joints. Even in case of horizontal shear force, the two models were in very good agreement for the running bond pattern, however significant differences occurred for the head bond pattern. In case of vertical shear force, the two models were in excellent agreement for both considered texture patterns. Further details can be found in the paper (Baraldi, Reccia, Cazzani, & Cecchi, 2013).

Figure 9. Horizontal shear deformed mesh: (a) FEM/DEM model; (b) Discrete model
(© 2013, Baraldi et al. Used with permission.)

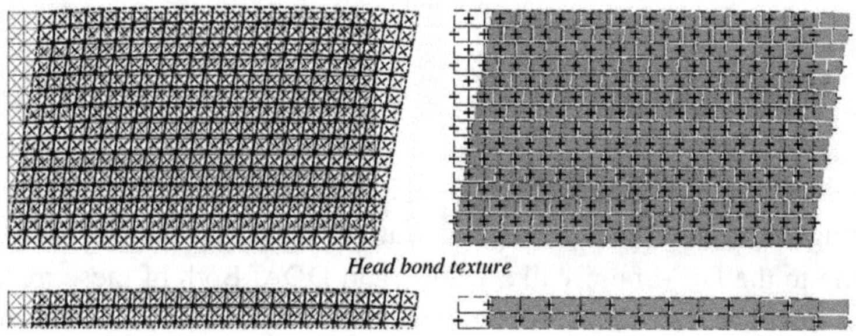

In the paper (Owen, Peric, Petrinic, Brookes, & James, 1998) a masonry arch bridge with backfill and a masonry arch bridge with anchor stitching system were analysed with the combined finite-discrete element method. The complex problem including masonry units, granular back-fill and geotechnical foundation material was considered. Here the masonry blocks were represented by deformable discrete elements in frictional contact. The fill material was modelled by spherical discrete elements and the foundation medium was considered as a Mohr-Coulomb material. In case of the strengthened bridge, the anchor system was modelled as an elasto-plastic steel bar, and the grouted bond with the masonry was simulated by prescribing a nonlinear shear stress/strain relation, whose parameters were determined from laboratory anchor pull-out tests. The FEM/DEM discretization and the deformations of the strengthened bridge just before the bottom ring fell down can be seen in Figure 10. Further information about the performed analyses can be found in the paper (Owen, Peric, Petrinic, Brookes, & James, 1998).

Further interesting examples for FEM/DEM modelling of structures can be found in (Keller et al., 2012), (De Lorenzi-Venneri et al., 2014), (Rougier et al., 2014) and (Munjiza et al., 2013). Apart from the numerical simulations of structures, the combined finite-discrete element method proved to be an effective tool in case of biomechanical applications as well. The authors refer to the paper (Xu et al.,

Figure 10. FEM/DEM discretization of the arch bridge (a); Deformations just before the collapse (b)
(© 1998, Owen et al. Used with permission.)

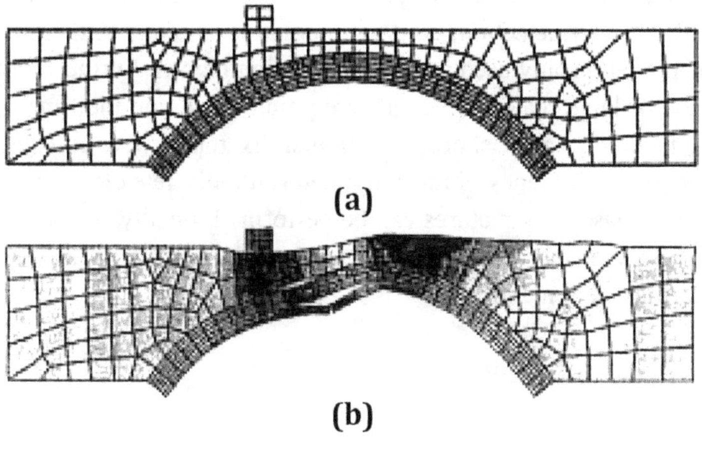

2013), where large scale simulations of red blood cell aggregation in shear flows were performed with the help of FEM/DEM.

COMPARATIVE DISCUSSION

After the introduction of the main characteristics and steps of FEM/DEM, a brief comparison can be made by summarizing the main differences and similarities between FEM/DEM and other widely used techniques belonging to the DEM family like UDEC and DDA. Both of these methods may be applicable for modelling masonry structures, since deformable polyhedron-shaped elements are available, although vital differences exist among them. It is especially true for DDA and FEM/DEM, since in DDA the deformability of blocks are handled without the finite element discretization of discrete elements. Furthermore, in DDA each element has only one reference point (at the centre of gravity), whose rotational degrees of freedom are also considered beside the translational ones. Moreover, another crucial difference is that DDA uses an implicit time integration scheme in contrast to FEM/DEM. On the other hand, some similarities can be observed between UDEC (or 3DEC) and FEM/DEM, since both methods describe deformations of blocks via finite element discretization, and usually constant strain elements are used for that purpose. Additionally, both methods work with explicit time integration scheme. In UDEC material failure may be represented either by plastic yielding (according to the prescribed failure criteria) or by relative displacements along the boundaries of discrete elements. The main difference in FEM/DEM is that fracture can occur inside discrete elements which leads to the formulation of new discrete bodies. It was shown earlier that in FEM/DEM, the need for realistic modelling of fractures led to the evolution of contact interaction algorithms where distributed contact forces are considered instead of point-like forces which are applied in UDEC. Due to these characteristics, FEM/DEM is proved to be a very effective numerical tool, especially for limit analyses of structures, where transition from continua to discontinua phenomenon is significant. A detailed comparative study, where further techniques are also discussed can be found in (Lisjak, & Grasselli, 2014).

CONCLUSION

In this chapter the main characteristics and steps of the combined finite-discrete element method were discussed. The specific contact detection, contact interaction, fracture and fragmentation algorithms were mentioned, which are typically used in FEM/DEM simulations. Furthermore, the huge computational challenge in case of large scale problems were pointed out, and the time stepping process with certain parallelization techniques were shortly mentioned. Examples were collected from the literature, where the authors proved the efficiency of the combined finite-discrete element method with which the appropriate modelling of masonry structures can be performed. Finally, a brief comparison was made between FEM/DEM and some other techniques belonging to the DEM family, where the most important differences and similarities were mentioned.

REFERENCES

Baraldi, D., Reccia, E., Cazzani, A., & Cecchi, A. (2013). Comparative analysis of numerical discrete and finite element models: the case of in-plane loaded periodic brickwork. *Composites: Mechanics, Computations, Applications: International Journal (Toronto, Ont.)*, *4*(4), 319–344.

Belytschko, T., Liu, W. K., & Moran, B. (2000). *Nonlinear finite elements for continua and structures*. John Wiley & Sons Ltd.

Cundall, P. A. (1971). A computer model for simulating progressive large scale movements in blocky rock systems.*Proc. Symp. Rock Fracture (ISRM)*.

De Lorenzi-Venneri, G., Lee, R. C., Luscher, D. J., Bronkhorst, C. A., Rougier, E., Knight, E. E., . . . Salmon, M. W. (2014). *Proceedings of the workshop on the structural cracking of the cupola of Santa Maria del Fiore*. International Report.

Keller, C.F., Jr., Salmon, M.W., Zubelewicz, A., Farrar, C.R., Luscher, D.J., Rougier, E., & De Lorenzi-Venneri, G. (2012). *Characterization and computer modeling for Brunelleschi's Dome of Santa Maria del Fiore in Florence*. International Report.

Lei, Z., Rougier, E., Knight, E. E., & Munjiza, A. (2014). A framework for grand scale parallelization of the combined finite discrete element method in 2D. *Computational Particle Mechanics*, *1*(3), 307–319. doi:10.1007/s40571-014-0026-3

Lemaitre, J., & Chaboche, J. L. (1990). *Mechanics of solid materials*. New York: Cambridge University Press. doi:10.1017/CBO9781139167970

Lisjak, A., & Grasselli, G. (2014). A review of discrete modeling techniques for fracturing processes in discontinuous rock masses. *Journal of Rock Mechanics and Geotechnical Engineering*, *6*(4), 301–314. doi:10.1016/j.jrmge.2013.12.007

Lukas, T., Schiava D'Albano, G. G., & Munjiza, A. (2014). Space decomposition based parallelization solutions for the combined finite-discrete element method in 2D. *Journal of Rock Mechanics and Geotechnical Engineering*, *6*(6), 607–615. doi:10.1016/j.jrmge.2014.10.001

Mahabadi, O. K., Grasselli, G., & Munjiza, A. (2010). A graphical user interface and pre-processor for the combined finite-discrete element code, Y2D, incorporating material heterogeneity. *Computers & Geosciences*, *36*(2), 241–252. doi:10.1016/j.cageo.2009.05.010

Munjiza, A. (2004). *The combined finite-discrete element method*. John Wiley & Sons Ltd. doi:10.1002/0470020180

Munjiza, A., & Andrews, K. R. F. (1998). NBS contact detection algorithm for bodies of similar size. *International Journal for Numerical Methods in Engineering*, *43*(1), 131–149.

Munjiza, A., & Andrews, K. R. F. (2000). Penalty function method for combined finite-discrete element systems comprising large number of separate bodies. *International Journal for Numerical Methods in Engineering*, *49*(11), 1377–1396. doi:10.1002/1097-0207(20001220)49:11<1377:AID-NME6>3.0.CO;2-B

Munjiza, A., Andrews, K. R. F., & White, J. K. (1999). Combined single and smeared crack model in combined finite-discrete element method. *International Journal for Numerical Methods in Engineering*, *44*(1), 41–57. doi:10.1002/(SICI)1097-0207(19990110)44:1<41::AID-NME487>3.0.CO;2-A

Munjiza, A., & John, N. W. M. (2002). Mesh size sensitivity of the combined FEM/DEM fracture and fragmentation algorithms. *Engineering Fracture Mechanics*, *69*(2), 281–295. doi:10.1016/S0013-7944(01)00090-X

Munjiza, A., Knight, E. E., & Rougier, E. (2012). *Computational mechanics of discontinua*. Chichester, UK: John Wiley & Sons.

Munjiza, A., Knight, E. E., & Rougier, E. (2015). *Large strain finite element method: A practical course*. Chichester, UK: John Wiley & Sons.

Munjiza, A., Lei, Z., Divic, V., & Peros, B. (2013). Fracture and fragmentation of thin shells using the combined finite-discrete element method. *International Journal for Numerical Methods in Engineering*, *95*(6), 478–498. doi:10.1002/nme.4511

Munjiza, A., Rougier, E., & John, N. W. M. (2006). MR linear contact detection algorithm. *International Journal for Numerical Methods in Engineering*, *66*(1), 46–71. doi:10.1002/nme.1538

Nikolić, Z., Smoljanović, H., & Živaljić, N. (2014). *Seismic analysis of dry stone masonry structures based on combined finite-discrete element method*. Paper presented at EURODYN, Porto, Portugal.

Owen, D. R. J., & Feng, Y. T. (2001). Parallelised finite/discrete element simulation of multi-fracturing solids and discrete systems. *Engineering Computations*, *18*(3-4), 557–576. doi:10.1108/02644400110387154

Owen, D. R. J., Feng, Y. T., Han, K., & Peric, D. (2000). *Dynamic domain decomposition and load balancing in parallel simulation of finite/discrete elements*. Paper presented at European Congress on Computational Methods in Applied Sciences and Engineering, Barcelona, Spain.

Owen, D. R. J., Peric, D., Petrinic, N., Brookes, C. L., & James, P. J. (1998). *Finite/discrete element models for assessment and repair of masonry structures*. Paper presented at Second International Arch Bridge Conference, Venice, Italy.

Perkins, E., & Williams, J. (2002). Generalized spatial binning of bodies of different sizes. *Discrete Element Methods*, *52-55*, 52–55. doi:10.1061/40647(259)10

Rougier, E., Knight, E. E., Broome, S. T., Sussman, A. J., & Munjiza, A. (2014). Validation of a three-dimensional Finite-Discrete Element Method using experimental results of the Split Hopkinson Pressure Bar test. *International Journal of Rock Mechanics and Mining Sciences*, *70*, 101–108.

Rougier, E., Munjiza, A., & John, N. W. M. (2004). Numerical comparison of some explicit time integration schemes used in DEM, FEM/DEM and molecular dynamics. *International Journal for Numerical Methods in Engineering*, *61*(6), 856–879. doi:10.1002/nme.1092

Schiava D'Albano, G. G. (2014). *Computational and algorithmic solutions for large scale combined finite-discrete elements simulations*. (Unpublished doctoral dissertation). Queen Mary, University of London.

Schiava D'Albano, G. G., Munjiza, A., & Lukas, T. (2013). *Novel MS (MunjizaSchiava) contact detection algorithm for multi-core architectures.* Paper presented at Particles, Stuttgart, Germany.

Shi, G. H. (1988). *Discontinuous deformation analysis: A new numerical model for the statics and dynamics of block systems.* (Unpublished doctoral dissertation). University of California, Berkeley, CA.

Smoljanović, H., Živaljić, N., & Nikolić, Ž. (2013). A combined finite-discrete element analysis of dry stone masonry structures. *Engineering Structures*, *52*, 89–100. doi:10.1016/j.engstruct.2013.02.010

Smoljanović, H., Živaljić, N., & Nikolić, Ž. (2013). Nonlinear analysis of engineering structures by combined finite-discrete element method. *Gradevinar*, *65*, 331–334.

Wang, F., Feng, Y. T., & Owen, D. R. J. (2004). Parallelization for finite-discrete element analysis in a distributed-memory environment. *International Journal of Computational Engineering Science*, *5*(1), 1–23. doi:10.1142/S146587630400223X

Xiang, J., Munjiza, A., & Latham, J.-P. (2009). Finite strain, finite rotation quadratic tetrahedral element for the combined finite-discrete element method. *International Journal for Numerical Methods in Engineering*, *79*(8), 946–978. doi:10.1002/nme.2599

Xu, D., Kaliviotis, E., Munjiza, A., Avital, E., Ji, C., & Williams, J. (2013). Large scale simulation of red blood cell aggregation in shear flows. *Journal of Biomechanics*, *46*, 1810–1817. PMID:23809770

Zang, M., Gao, W., & Lei, Z. (2011). A contact algorithm for 3D discrete and finite element contact problems based on penalty function method. *Computational Mechanics*, *48*(5), 541–550. doi:10.1007/s00466-011-0606-5

KEY TERMS AND DEFINITIONS

Constitutive Equation: A mathematical expression which describes the response of the material to some kind of external effects. In this context the mechanical behaviour of the material is described by establishing connection between stress and strain variables and containing failure criteria.

Contact Detection: A process, when contacting couples of discrete elements are recognized.

Contact Interaction: A mathematical model which describes the mechanical behaviour of contacts between discrete elements, including the calculation of penetration and the evaluation of contact forces.

Dry Stone Masonry: One of the oldest building techniques when the structure was constructed by assembling regular stone blocks either without the addition of mortar or using mortar having low strength which has no influence on the mechanical behaviour of the structure by now.

Explicit Time Integration Schemes: A group of time approximation techniques when the values of unknown variables at time $t + \Delta t$ are predicted from the approximated values of the variables belonging to time t. These techniques, unlike the implicit techniques, do not check whether the equations of motion are satisfied at the end point of the actual time interval.

Mohr-Coulomb Material: A material whose behaviour is described with the Mohr-Coulomb failure criterion.

Parallelization: A computational solution strategy when several processors work in parallel to solve the particular task.

Chapter 7
Discrete Element Particle Modelling of Stone Masonry

Nuno Monteiro Azevedo
National Laboratory for Civil Engineering (LNEC), Portugal

José V. Lemos
National Laboratory for Civil Engineering (LNEC), Portugal

João Rocha de Almeida
Faculty of Sciences and Technology - New University of Lisbon, Portugal

ABSTRACT

Circular Particle Models (PM) are a class of discrete elements which has been increasingly used for detailed analysis in rock and concrete structures. There have been few applications to masonry, but the potential of these techniques appears significant, due to their proven ability to simulate fracture processes through random particle assemblies representing quasi-brittle materials at the grain scale. The present chapter presents the fundamentals of this approach and reviews some previous applications of PM models to masonry. The model capabilities are first exemplified by simple models involving a few irregular blocks formed by particles. Irregular stone masonry wall specimens under compression and under in-plane shear loading are then presented. In these models both the units and the mortar are represented by circular particles, and failure processes through the joints or through joints and stones are analyzed. The main issues regarding the use of these models are finally discussed.

INTRODUCTION

Two fundamental numerical approaches towards masonry analysis are possible: macro-modelling as an equivalent homogeneous medium and micro-modelling of the individual components, joints and the units. Both approaches have specific strengths and engineering roles. While continuum models have been developed which simulate complex geometric patterns and material behaviour (e.g. Stefanou et al. 2015), the alternative discontinuum or micro-models allow a detailed representation of the internal structure, and have become a powerful tool for research on the fundamental behaviour of geo-materials.

DOI: 10.4018/978-1-5225-0231-9.ch007

The discrete element method (DEM) provides a numerical implementation framework for discontinuum modelling.The simplest element geometry that can be assumed is that of circular particles in 2D or spheres in 3D. Cundall (1971), the original paper that proposed DEM, already considered circular particle models in addition to rigid polygonal blocks. Cundall & Strack (1979) extended the circular particle approach and proposed it for the analysis of granular media, namely soils. In other fields of engineering and physics, the same concept was pursued by several researchers under different designations, namely Molecular Dynamics (e.g., Pöschel & Schwager, 2005).

Replacing the frictional contacts between particles by bonds with cohesive and tensile strength, circular particle assemblies were employed to simulate materials such as rock, concrete or asphalt pavement, in the early 1990s, Meguro and Hakuno (1989), Potyondy and Cundall (1996), Schlangen and Garboczi (1997), Chang and Meegoda (1997).

Studies of fracture propagation in these geo-materials have shown the ability of these bonded-particle models to reproduce the types of phenomena observed in laboratory experiments, such as uniaxial tension, uniaxial compression and triaxial tests. The random nature of particle assemblies allows the propagation of cracks to develop in a natural manner, replicating the irregularity of the structure of geo-materials. Circular or spherical particles may also be associated to build macro-particles with an arbitrary form. Rigid or breakable bonds can be assumed to link the component particles, thus allowing different levels of elaboration of the analysis. The interaction forces, however, are still obtained by the elementary contacts between the pairs of adjacent circular particles.

More recently, 3D rigid spherical particle models have been proposed for rock, Matsuda and Iwase (2002), Potyondy and Cundall (2004), Monteiro Azevedo and Lemos (2013), and for concrete, Lilliu and Van Mier (2003), Hentz et al. (2004). Models based on the rigid spring block method adopting 3D Voronoi shape polyhedra have also been developed for concrete, Nagai et al. (2005) and Berton and Bolander (2006). Numerical models that follow complex polyhedral based discrete element contact interaction and include the particle deformation by considering the inner block finite element mesh have been recently proposed for rock fracture, Gao and Stead (2014), Hamdi et al. (2014). A review of discrete modelling techniques for fracturing processes in rock can be found in Lisjak and Grasseli (2014).

Particle models, and more generally DEM models, embody an approach to describe the fundamental behaviour of materials by means of simple constitutive assumptions, realized by elementary contact models between the individual particles. Complexity of macroscopic behaviour arises from the combination of these mechanisms over a large random assembly. Procedures have been developed to calibrate the micro-properties to obtain the desired response, typically by comparison with the results of laboratory tests on samples. For masonry structures, particle models have the ability to represent the natural irregularity of their geometry and material structure, and therefore to simulate the complex patterns of cracking and failure development, through the units or the joints, which characterize their response.

PARTICLE MODELS FOR MASONRY

General Concepts

The fundamental concepts of particle models follow essentially the framework described in the earlier chapters for rigid block models. In the present case, the model is composed of circular particles, in 2D, or spherical particles, in 3D, which are assumed to be rigid. Explicit algorithms are typically employed to describe the motion of the particles, whether in dynamic or quasi-static processes. Some codes allow macro-particles to be built by rigidly connecting several circular or spherical particles, either adjacent or partially overlapping, and thus the motion of these macro-particles needs to be handled as a rigid body in 2D or 3D, with, respectively, 3 or 6 degrees-of-freedom.

The main difference with respect to block models is the representation of contact, which becomes much simpler in particle models. The line between the particle centres defines the contact normal, and the criterion to activate an interaction depends only on the verification of a certain distance. It is this computational simplicity that allows much larger systems to be handled with reasonable computing times. This makes possible to use much finer material representations than those that could be attained with polygonal or polyhedral block models, a point which is particularly critical in the study of localized phenomena such as crack propagation.

The constitutive assumptions that govern the response of a particle model are defined at the contact level. Given the contact normal, the relative displacements between the 2 particles can be broken into normal and shear components. The constitutive model needs to provide the normal and shear forces between the particles, as a function of the relative displacements. For each contact, it is necessary to define the deformability, in terms of normal and shear contact stiffness, typically with units corresponding to a force-displacement relationship. The non-elastic behaviour starts when the contact tensile or shear strengths are reached. In this case, either brittle or softening laws may be adopted, as described in the formulation section.

Modelling Soil and Fill Materials

The capabilities of particle models to address the behaviour of granular media have been explored by several authors to simulate the interaction between masonry structures and soil or fill materials. The representation of the fill in arch bridge analysis is one of the problems in which circular particles were applied, as done by Thavalingam et al. (2001) with the PFC code (Itasca, 2014). In this 2D model, the stone arch itself is represented by macro-particles formed by assembling clusters of circular particles joined with high strength bonds. The joints between the arch blocks are taking into account by assigning lower strength to the contacts across them.

For the same type of problem, Rouxinol et al. (2007) developed a 2D code that allows interaction between circular particles and polygonal blocks, providing a straightforward representation of the arch-fill contact. Figure 1 shows two stages of the failure mode of a model of Bridgemill bridge, for which results of an in situ load test are available. The numerical model provided a reasonable approximation to the experimental failure load.

Oetomo et al. (2014) made use of the PFC code to study the failure modes of dry-stone retaining walls. The soil is modelled as a granular material by means of circular particles, while the wall blocks are assumed to be rigid. These blocks are simulated in the PFC code by rigidly connecting a group of particles and constraining it to move as a single rigid body (named a PFC "clump").

Modelling Masonry Units With Particles

The experience gained with the application of particle models in the investigation of fracture propagation in geo-materials, such as rock or concrete, indicates the potential for their use in masonry structures. There are two options for the simulation of a masonry unit by particles: assuming rigid block behaviour,

Figure 1. Model of masonry bridge combining rigid blocks and circular particles
(Rouxinol et al., 2007)

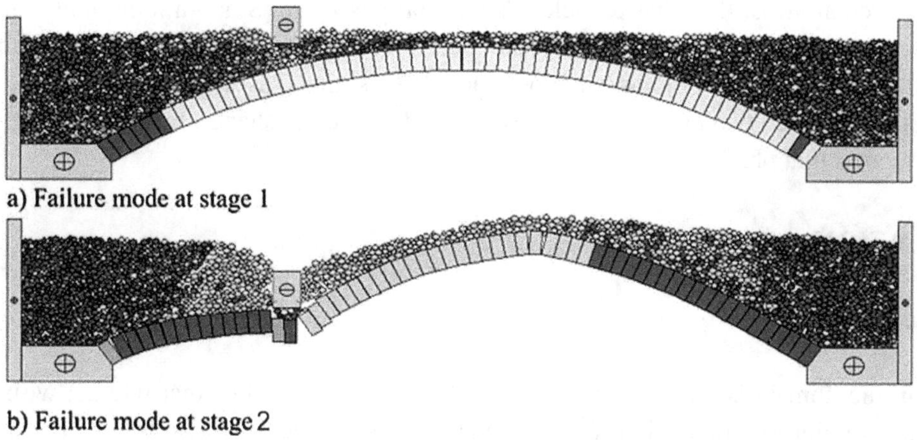

or employing breakable bonds. The former can be used if the unit material is not expected to fail under the applied loads. It requires the ability to constrain the movement of a set of particles as a single rigid body, which is available in various codes. Its main advantage is computational speed. In contrast, the use of breakable bonds to connect the particles forming a unit implies a larger computational effort, mainly due to the need of going through the force-displacement calculations for defining these intra-unit contacts. However, it provides a means to consider the fracture of the unit. Modelling the unit as a cluster of particles facilitates the representation of irregular shapes, as observed in traditional stone masonry walls, with a degree of approximation that depends on the adopted particle size.

Contact Types and Properties

For a masonry structure, several types of contact with different properties may have to be used. In the simplest case, let us consider the model of a stone masonry structure in which the blocks are represented as rigid macro-particles. In this situation only the particles on the boundary of each macro-particle would interact with the ones in the adjacent block, and a single set of contact properties would be sufficient. The contact stiffnesses need to be defined in a way that produces the correct global system deformability. In cases in which deformable, breakable bonds are adopted inside a macro-particle, there is need to distinguish between these bonded contacts and the inter-block contacts, possibly purely frictional. The behaviour of a mortared joint can be simulated by assigning the inter-block contacts with the desired deformability, as well as frictional and cohesive strengths. In the case of a very detailed model, it is possible to represent the mortar by means of small particles. If so, several types of contact and property sets must be defined: mortar-mortar; mortar-stone; intra-block stone bonds; and, possibly, inter-block stone-stone contacts.

Joint Roughness

A block composed by particles, whether internal bonds are rigid or breakable, displays a boundary formed by circular arcs, leading to a roughness scale that depends on the particle size. In some cases, this geometric roughness may correspond to a realistic frictional behaviour. However, it is more likely that the "bumpy" nature of the surface produces excessive friction and dilation, as each particle has to slide up the circular arc of the other particle. Cundall proposed the 'Smooth joint model' (e.g., Itasca, 2014) to overcome this undesirable type of behaviour, allowing the user to define friction and dilations angles as in the case of a planar joint in a block model. This model assumes that the joint corresponds to a mean plane between two blocks and ignores the actual circular shapes, taking the same normal for all contacts across that plane.

FORMULATION

Fundamentals

In the DEM, the domain is replaced by an assembly of discrete entities that interact with each other through contact points or contact interfaces. The ability to include finite displacements and rotations, including complete detachment, and to recognize new contacts as the calculation progresses are essential features. The set of forces acting on each particle are related to the relative displacements of the particle with respect to its neighbours. At each step, given the applied forces, Newton's second law of motion is invoked to obtain the new particle positions. The particle equations of motion, including local non-viscous damping, may be expressed as:

$$F_i(t) + F_i^d(t) = m\,\ddot{x}_i \tag{1}$$

$$M_3(t) + M_3^d(t) = I\,\dot{\omega}_3 \tag{2}$$

where, $F_i(t)$ and $M_3(t)$ are, respectively, the total applied force and moment at time t including the exterior contact contribution, m and I are, respectively, the particle mass and moment of inertia, $\ddot{x}_i$ is the particle acceleration, $\dot{\omega}_3$ is the particle angular acceleration. The damping forces using Cundall's (1987) local non-viscous damping formulation are given by:

$$F_i^d(t) = -\alpha \left|F_i(t)\right| sign(\dot{x}_i) \tag{3}$$

$$M_3^d(t) = -\alpha \left|M_3(t)\right| sign(\omega_3) \tag{4}$$

where, $\dot{x}_i$ is the particle velocity, ω_3 the particle angular velocity, α the local non-viscous damping and the function $sign\left(x\right)$ is given by:

$$sign\left(x\right) = \begin{cases} +1\,,\ x > 0 \\ -1\,,\ x < 0 \\ \ \ 0\,,\ x = 0 \end{cases} \tag{5}$$

The local non-viscous damping approach has, when compared with traditional viscous damping based formulations, some advantages: the amount of damping varies from particle to particle; the damping magnitude is dimensionless and frequency independent and steady state particles are not damped. In all the numerical examples that are presented a local damping coefficient of 0.7 is adopted, which is the usual value for quasi-static simulations.

Point Contact Model

The simplest contact model used in particle models (PM) is the point contact model (PCM), in which the mechanical interaction between two particles is represented by a force. Generalized contact models (GCM) have been proposed that allow transmission of forces and moments (Monteiro Azevedo and Lemos, 2005). The model naturally incorporates the force versus relative particle displacement relationships of the traditional point contact model (PCM). Further it provides both moment transmission and simple physical constitutive models based on standard force displacement relationships.

Like in the PCM and in the GCM contact models, the contact unit normal, n_i, is defined given the particles centre of gravity and the inter-particle distance, *d*, Figure 2 b):

$$n_i = \frac{x_i^{[B]} - x_i^{[A]}}{d} \tag{6}$$

Like in the PCM contact model and in the GCM contact model, the contact overlap for the reference contact point, U^n, is defined by:

$$U^n = R^{[A]} + R^{[B]} - d \tag{7}$$

Figure 2. PCM contact model

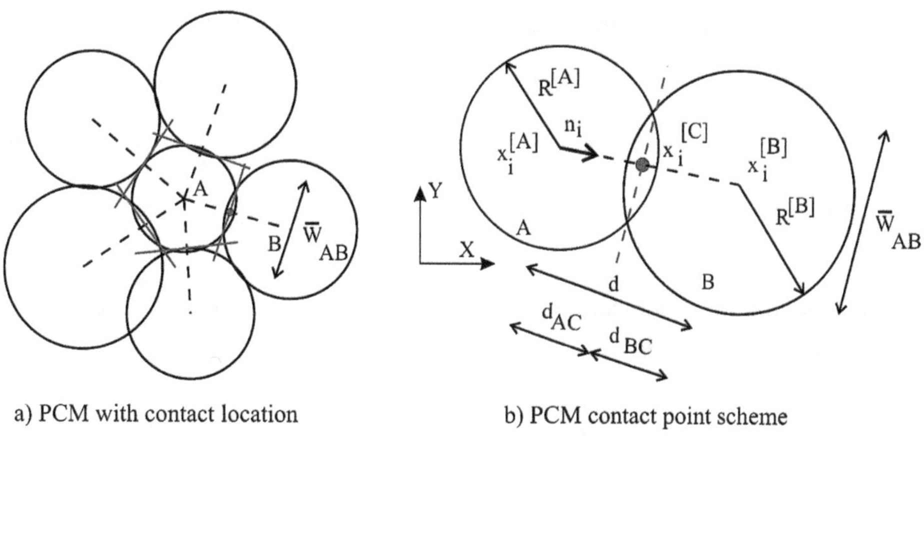

a) PCM with contact location

b) PCM contact point scheme

In the GCM contact model, the reference contact point has the same location of the traditional single PCM contact point, and the local contact points are distributed along the PCM contact plane. In the PCM contact model, the contact point, $x_i^{[C]}$, is defined by:

$$x_i^{[C]} = x_i^{[A]} + \left(R^{[A]} - \frac{1}{2} U^n \right) n_i \tag{8}$$

The contact forces acting on the contact point, $x_i^{[C]}$, can be decomposed into their normal and shear components with respect to the contact plane:

$$F_i^{[C]} = F_i^{[C,N]} + F_i^{[C,S]} \tag{9}$$

The contact velocity at the contact point, which is the velocity of particle *B* relative to particle *A*, at the contact location is given by:

$$\begin{aligned} \dot{x}_i^{[C]} &= \left(\dot{x}_i^{[C]} \right)_B - \left(\dot{x}_i^{[C]} \right)_A \\ &= \left(\dot{x}_i^{[B]} + e_{i3k}\, \omega_3^{[B]} \left(x_k^{[C]} - x_k^{[B]} \right) \right) - \left(\dot{x}_i^{[A]} + e_{i3k}\, \omega_3^{[A]} \left(x_k^{[C]} - x_k^{[A]} \right) \right) \end{aligned} \tag{10}$$

where, e_{ijk} is the permutation tensor. The contact displacement normal increment, $\Delta u^{[C,N]}$, stored as a scalar, and the contact displacement shear increment, $\Delta u_i^{[C,S]}$, stored as a vector, are given by:

$$\Delta u^{[C,N]} = \left(\dot{x}_i^{[C]} \Delta t \right) n_i \tag{11}$$

$$\Delta u_i^{[C,S]} = \left(\dot{x}_i^{[C]} \Delta t \right) - \Delta u^{[C,N]} n_i \tag{12}$$

Given the normal (k_n) and shear stiffness (k_s) of the local contact point, the normal and shear forces increments are obtained following an incremental linear law, as follows:

$$\Delta F_i^{[C,N]} = -k_n\, \Delta u_i^{[C,N]} \tag{13}$$

$$\Delta F_i^{[C,S]} = -k_s\, \Delta u_i^{[C,S]} \tag{14}$$

Due to the fact that the shear contact force is stored in global coordinates it is necessary to redefine it in the updated contact plane using:

$$F_i^{[C,S.old]} = F_i^{[C,S.old]} - e_{ij3} e_{3mn} F_j^{[C,S.old]} n_m^{old} n_n \tag{15}$$

The predicted normal and shear forces acting at the local contact point are then updated by applying the following equations:

$$F^{[C,N.new]} = F^{[C,N.old]} + \Delta F^{[C,N]} \tag{16}$$

$$F_i^{[C,S.new]} = F_i^{[C,S.old]} + \Delta F_i^{[C,S]} \tag{17}$$

Given the predictive normal and shear contact forces, the adopted constitutive model is applied. It is necessary to carry out adjustments if the predictive forces do not satisfy the constitutive model, this adjustment is model dependent. The updated contact force at the contact point is then given by:

$$F_i^{[C]} = F^{[C,N.new]}\, n_i + F_i^{[C,S.new]} \tag{18}$$

The associated particles forces and moments are updated by transferring to the centre of gravity of each particle in contact, the updated contact force through:

$$F_i^{[A]} = F_i^{[A]} - F_i^{[C]} \tag{19}$$

$$F_i^{[B]} = F_i^{[B]} + F_i^{[C]} \tag{20}$$

$$M_3^{[A]} = M_3^{[A]} - e_{3jk}\left(x_j^{[C]} - x_j^{[A]}\right) F_k^{[C]} \tag{21}$$

$$M_3^{[B]} = M_3^{[B]} + e_{3jk}\left(x_j^{[C]} - x_j^{[B]}\right) F_k^{[C]} \tag{22}$$

Numerical Stability

When only a steady state solution is sought, a mass scaling algorithm is adopted in order to reduce the number of timesteps necessary to reach the desired solution. The particle mass and inertia are scaled so that the adopted centred-difference algorithm has a higher rate of convergence for a given loading step. The particle scaled mass and inertia used in the calculations are set assuming a unit time increment, $\Delta t = 1$, given the particle stiffness at a given time through:

$$m_{scaled} = 0.25\, K_t \tag{23}$$

$$I_{scaled} = 0.25\, K_\theta \tag{24}$$

The latter equation is the result of the application of the Gershgorin's theorem (Underwood, 1983), which guarantees that the highest frequency of a structural system is less than or equal to the ratio of the sum of the absolute values of a row of the stiffness matrix and the sum of the mass matrix row. By applying Gershgorin's theorem to the PCM local contact stiffness matrix (Monteiro Azevedo, 2003), an upper bound of the total particle translation stiffness K_t and of the total rotational stiffness K_θ must be found at a given timestep:

$$K_t = \sum_{c=1}^{N} 2\left(k_n + k_s\right) \tag{25}$$

$$K_\theta = \sum_{c=1}^{N}\left(k_s\, d_{AC}^2 + k_s\, d_{AC}\, d_{BC}\right) \tag{26}$$

where, $d_{AC} = \left\| x_i^{[A]} - x_i^{[C]} \right\|$ and $d_{BC} = \left\| x_i^{[B]} - x_i^{[C]} \right\|$ are the Euclidean norms of the corresponding vectors, $\sum_{c=1}^{N}$ indicates a summation along the "*N*" contacts associated with a given particle, k_n and k_s are the contact normal and shear stiffness, respectively.

Contact Stiffness and Contact Strength

The contact model requires the user definition of the contact deformability parameters, namely the Young's modulus of the equivalent continuum material ($\bar{E}$) and the equivalent continuum Poisson's ratio (ν_c). The normal and the shear stiffness spring value are given by:

$$k_n = \frac{\bar{E}'}{d} A_c \tag{27}$$

$$k_s = \frac{\bar{E}''}{d} A_c \tag{28}$$

where, $A_c = 2.0\, R^{\min}\, t$ is the contact area associated with the contact point, being $R^{\min}$ the minimum radius of the particles in contact, t is the adopted particle out of plane thickness and d is the distance between the particles centre of gravity, and for plane stress conditions:

$$\begin{aligned} \bar{E}' &= \frac{\bar{E}}{\left(1-\nu_c^2\right)} \cap \\ \bar{E}'' &= \frac{\bar{E}}{2\left(1+\nu_c\right)} \end{aligned} \tag{29}$$

For the local inter-particle contacts, the PCM-2D model also requires the definition of the contact strength properties, the maximum contact tensile stress, $\sigma_{n.t}$, the maximum contact cohesion stress, τ, and the contact frictional term, μ_c.

The maximum contact local tensile strength, $F^J_{n.\max}$, and the maximum local contact shear strength, $F^J_{s.\max}$, are defined given the user-specified contact strength properties and the current local contact normal force, F^C_n, as follows:

$$F_{n.\max} = \sigma_{n.t} A_c \tag{30}$$

$$F_{s.\max} = \tau A_c + F^C_n \mu_c = C_{\max} + F^C_n \mu_c \tag{31}$$

where, $C_{\max}$ is the adopted maximum local contact cohesion strength. Figure 3 shows the adopted bi-linear softening contact model under tension and shear (Rokugo et al., 1989).

Figure 3. Constitutive model with bilinear softening under tension and shear

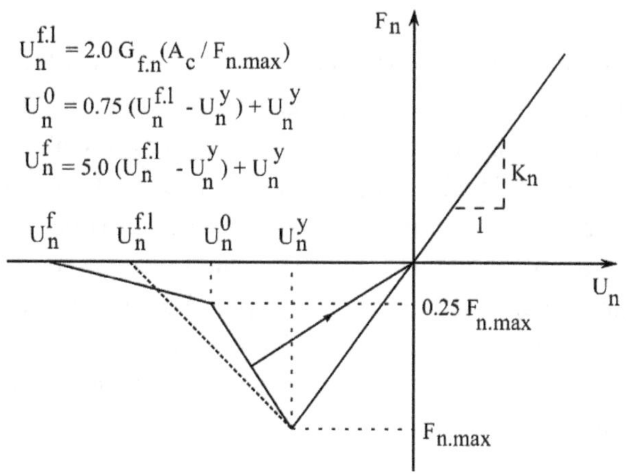

a) Normal direction

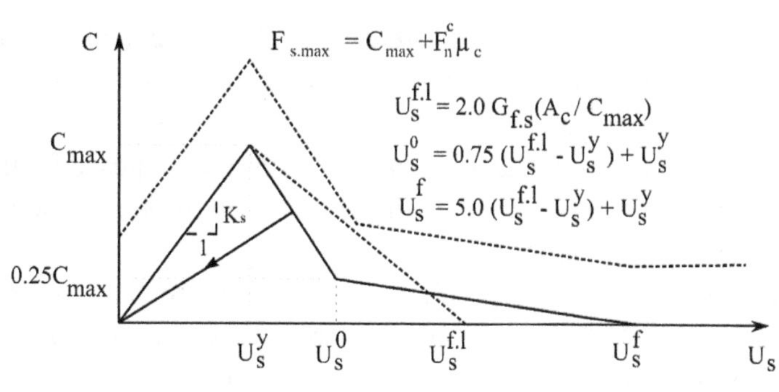

b) Shear direction

The bilinear contact model requires the definition of the contact tensile fracture energy, $G_{f.n}$, and of the contact shear fracture energy, $G_{f.s}$. As soon as the local maximum strength is reached, the local maximum normal tensile and the local maximum cohesion values are reduced based on the current contact damage value, which varies from 0, in the undamaged state, to 1, in a fully damaged state.

The tensile damage value is defined based on the current local contact normal displacement ($D_n\left(U_n\right)$) Figure 3 a), and the cohesion damage value is defined based on the current local contact shear displacement ($D_s\left(U_s\right)$), only the cohesion term is affected, Figure 3 b).

Figure 3 b) also shows the evolution of the local contact shear strength for a constant value of local contact normal force (F_n^C).

In each local contact point the contact damage, D_c , is given by the sum of the tensile and shear contact damages. Given the current local contact damage, the local maximum tensile strength and maximum local cohesion strength are updated to:

$$F_{n.\max}^{Current} = D_c\ F_{n.\max} \tag{32}$$

$$C_{\max}^{Current} = D_c\ C_{\max} \tag{33}$$

A local contact crack is considered to occur when the maximum possible damage ($D_c = 1$) is reached. At this stage the local contact point is only considered to work under pure friction. A local contact crack is considered to be a tensile crack if the local contact was under a shear/tensile loading state when the maximum damage was reached. A local contact crack is considered to be a shear crack if the local contact was under a shear/compression loading state as soon as the maximum damage was reached.

If the adopted contact fracture energy is equal to the energy corresponding to the elastic behaviour, the response of the bilinear model is the same as the response obtained using a traditional brittle Mohr-Coulomb model with tension cut-off. By using a bilinear softening model at the contact level the fracture propagation occurs in a smoother and more controlled way than the numerically observed with a brittle model, allowing a less brittle response.

Particle Generation Scheme

The modelling of a stone masonry wall with a discrete element particle model requires the generation of a particle assembly within the elements representing the stones and within the outer domain representing the mortar joints.

In Monteiro Azevedo (2003), particle generation algorithms are proposed for the generation of compact particles assemblies within a polygonal domain with a given the shape and aspect ratio, following the principles proposed by Wang et al. (1999).

In each stone unit and in the mortar it is required to set a grain size distribution. For concrete modelling, the adopted grain size distribution follows the larger size distribution given by a sieve analysis. In the stone and mortar representation, a similar principle can be followed. To comply with computational restrictions, the adopted grain structure is usually simplified adopting larger grain sizes and usually assuming a uniform radius distribution, which is the approach that is followed in the examples presented herein.

One advantage of the particle modelling, when compared with equivalent finite element techniques, is that it is easier to set the initial particle arrangement from a CAD representation or digital image representing the stone units, especially when dealing with very irregular arrangements.

Figure 4 shows the adopted particle generation scheme. At the beginning, the stone unit shapes and locations are known and a digital image of a wall structure can be adopted. Then, each stone unit is discretized with a compact particle assembly in an independent way, adopting for the stone units a uniform grain distribution, Figure 4 b). As previously mentioned, the particle assemblies with low porosity follow the void elimination procedure presented in Monteiro Azevedo (2003), in which the particle model solution scheme has a discontinuous behaviour characteristic of a granular media in order to evenly distribute the particle overlaps throughout the particle assembly.

The mortar is discretized following similar principles, Figure 4 c). For the mortar, a different grain size distribution is adopted with a lower grain size, not only to reflect the fact that a mortar has a characteristic lower grain size but also because a significant number of particles need to be adopted in order to fill narrower spaces.

In the final particle assembly representing the wall, two particles are considered to interact if the inter-particle distance is smaller than a given threshold given by the particle radius:

$$d-\left(R^{A}+R^{B}\right)\leq\gamma\ R^{min} \tag{34}$$

where, R^{min} is the particle assembly minimum particle radius and γ is a particle distance threshold which in the case studies presented here is equal to 0.25. By introducing a particle distance threshold value the number of inter-particles interactions greatly increases, making it possible to model a particle assembly with very low porosity.

Figure 4. Wall representation, from the stone unit structure to the final particle assembly

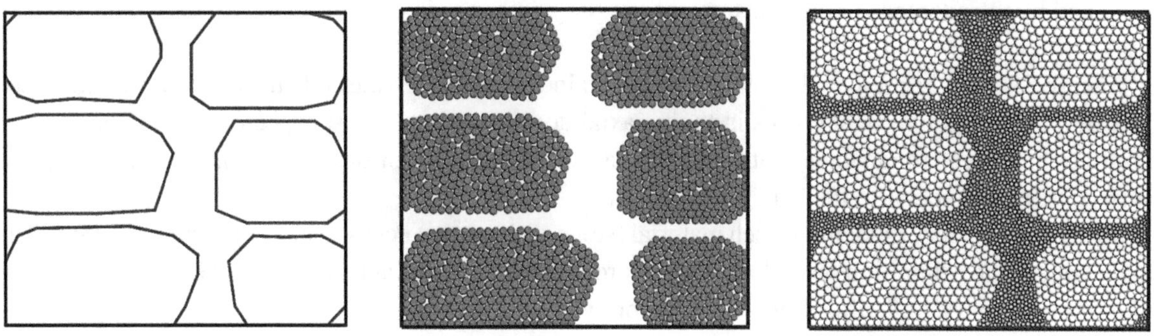

a) Stone structure b) Stone discretization c) Mortar discretization

Model Parameters and Calibration Procedure

The point contact model requires the definition of seven elastic and strength parameters. The elastic response is related with both the Young's modulus of the equivalent continuum material, $\overline{E}$, and the equivalent Poisson's ratio. The strength macroscopic response requires the definition of the maximum contact tensile stress, $\sigma_{n.t}$, the maximum contact cohesion stress, τ, the frictional coefficient, μ_c, and the contact tensile, $G_{f.n}$, and the contact shear fracture energy, $G_{f.s}$.

The properties associated with the particle generation can also be regarded as model parameters, namely the grain size distribution given by the maximum diameter (D_{max}) and the minimum diameter (D_{min}) of the circular particle assembly adopted for the stone units and for the mortar representation. As mentioned above, the particle distribution adopted should be as close as possible to the grain size distribution of the material to be represented, but for computational reasons a higher grain size is usually adopted.

Given that a direct relationship between micro-properties and macro-properties is difficult to establish, the micro-properties are traditionally defined through a calibration process in order to reproduce the known macroscopic material behavior.

Traditionally uniaxial tensile and compression experimental tests are adopted. In order to make the calibration procedure more robust, not only the peak values should be known but the full strain/stress diagram including the post-peak behaviour. The model deformability parameters are first obtained by adjusting the equivalent continuum Poisson's ratio in order to obtain the material Poisson's coefficient. Then the Young's modulus of the equivalent continuum material is adjusted in order to reproduce the macroscopic Young's modulus.

Following, the strength parameters are adjusted using an iterative procedure based on the experience gained with previous calibration procedures (Monteiro Azevedo et al. 2015):

1. The relationship between the macroscopic compression strength and the macroscopic uniaxial tensile depends fundamentally on the relationship between the cohesive term and the contact tensile strength.
2. The macroscopic compression strength depends on the cohesion, on the contact frictional term and on the adopted contact fracture energy.
3. The tensile macroscopic strength is mainly influenced by the contact maximum tensile strength and by the adopted contact fracture energy.

Figure 5 shows the predicted failure patterns obtained in direct tensile and uniaxial compression tests. The development of vertical cracks in the uniaxial compression test was expected, given that there is no friction between the loading plates and the particle assembly. For the direct tensile test, the rupture surface is perpendicular to the loading direction.

For stone masonry modelling, each material (stone and mortar) needs to be calibrated separately. The main goal of the calibration procedure is to ensure that the model is able to predict the global behaviour of a stone masonry wall, after elementary calibration procedures.

Figure 5. Particle model predicted failure pattern: Calibration procedure

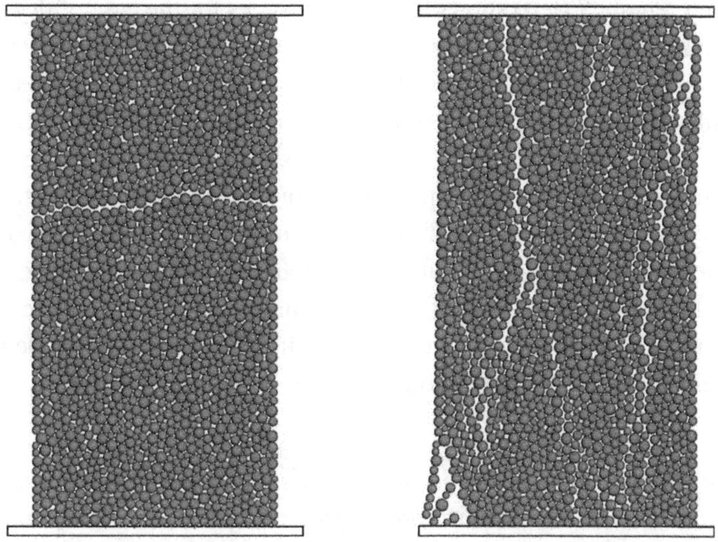

a) Direct tensile test b) Uniaxial compression test

CASE STUDIES

Analysis of Wall Segment under Compression and In-Plane Shear

A simple example of a particle model of a segment of a masonry wall is shown in Figure 6, analyzed with a model developed for concrete, Monteiro Azevedo (2003). The left picture shows the case of dry joints, with the stones represented by an assembly of bonded circular particles; on the right, is depicted the case of voids filled with smaller particles to simulate mortar. A bilinear softening constitutive model was used for the various material interfaces: contacts with strong tensile and shear bonding between particles forming each stone; purely frictional contacts between adjacent stones; and, contacts mortar-mortar and mortar-stone with appropriate weaker bond strengths.

Figure 6. Particle model of a stone masonry wall segment: (left) dry stones; (right) voids filled with mortar

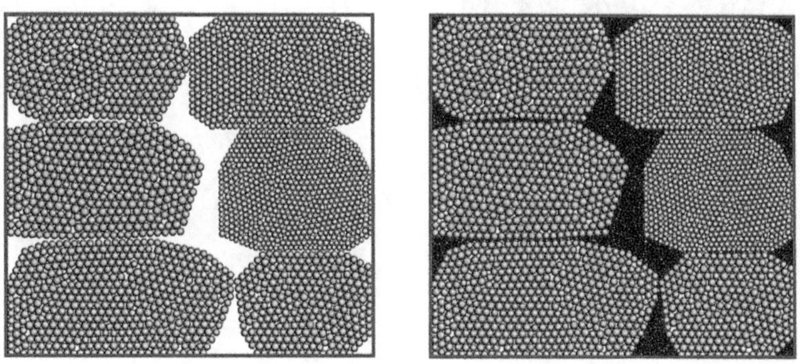

The models were loaded in compression by imposing an increasing vertical displacement to the top row of a particle clump. Figure 7 (left) shows the contact forces for the case of dry stones at the end of the test, in which cracks have developed in some stones (as plotted on the right of the same figure). The particular geometries of the stones led to more localized contact with strong stress concentrations on the left-hand side wall stones, in contrast to a more distributed loading for the more planar faces of the right-hand side ones.

The case considering voids filled with mortar led to a much more even distribution of contact forces, plotted on Figure 8. In this example, relatively high mortar strength was assumed, resulting in a distributed cracking pattern reminiscent of loading of concrete cubes between frictional plates. Note that the constitutive model assigned to the mortar particle contacts does not simulate failure by crushing, but such phenomena may also be taken into account without difficulty, by implementing more complex interface models developed for masonry (Lourenço & Rots,1997).

Figure 7 Compression of stone wall with thin mortared joints: (left) contact forces (blue denotes compression; grey denotes tension; thickness proportional to force magnitude); (right) cracked contacts (blue denotes cracks due to tensile forces; green denotes cracks due to shear forces; thickness proportional to crack opening)

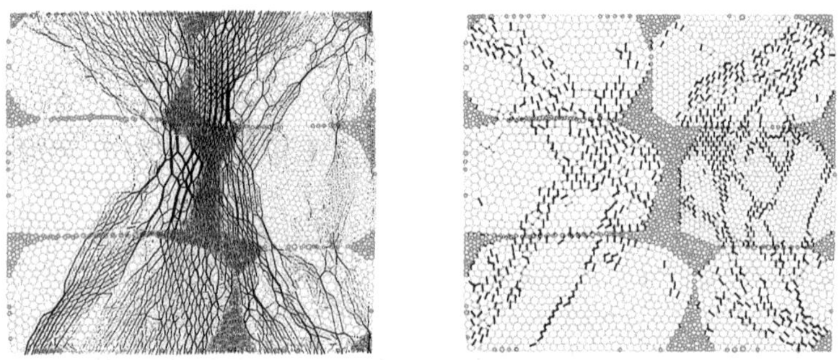

Figure 8. Compression of stone wall with thick mortared joints: (left) contact forces (blue denotes compression; grey denotes tension; thickness proportional to force magnitude); (right) cracked contacts (blue denotes cracks due to tensile forces; green denotes cracks due to shear forces; thickness proportional to crack opening

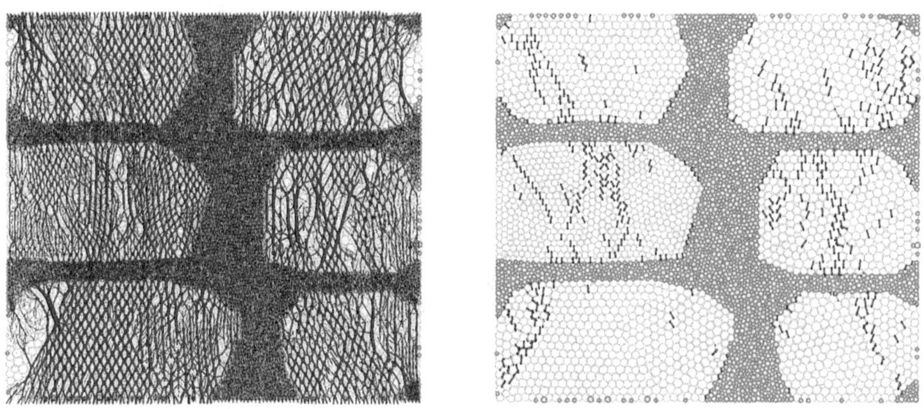

A second type of test was performed with the mortared model by loading it first in compression and then applying a progressive shear motion to the top boundary. A large diagonal crack develops, with other cracks along the joints, as may be seen in Figure 9. In the model of Figure 9, there is still direct contact between the stones at a few points, with the mortar filling essentially the inner void and connecting the two stone panes. A different particle model was also studied with thick mortared joints separating the stones (Figure 10). The same compression and shear loading was applied, and, in this case, the cracks develop fully through the weaker mortar material. The cracking pattern may be visualized in the right plot of Figure 10, which presents the cracked contacts (with line thickness proportional to the separation distance). The contact force distribution is shown on the left plot of the same figure.

In the case presented in Figure 10, a shear deformation was applied preventing the top boundary to rotate. If under shear deformation the top boundary rotation is allowed, a more localized pattern of cracking (Figure 11) is induced, with progressive opening of a mortared joint.

These analyses illustrate the potential of particle models for detailed representation of masonry components, in particular of irregular assemblages found in most historical constructions. Failure of the various types of multi-leaf walls is usually induced by loss of integrity of the assemblage and decoupling of the outer stone panes. Many experimental tests have been conducted, namely to evaluate proposed strengthening techniques, which may involve grouting and internal or external reinforcement with dif-

Figure 9. Shear with pre-compression of stone wall with thin mortared joints: (left) contact forces; (right) cracked contacts

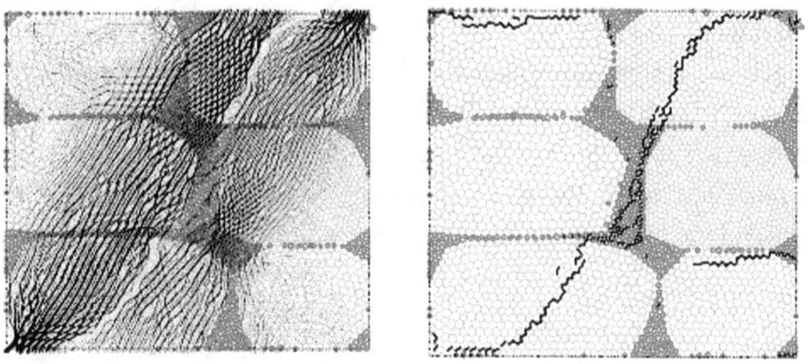

Figure 10. Shear with pre-compression of stone wall with thick mortared joints: (left) contact forces; (right) cracked contacts

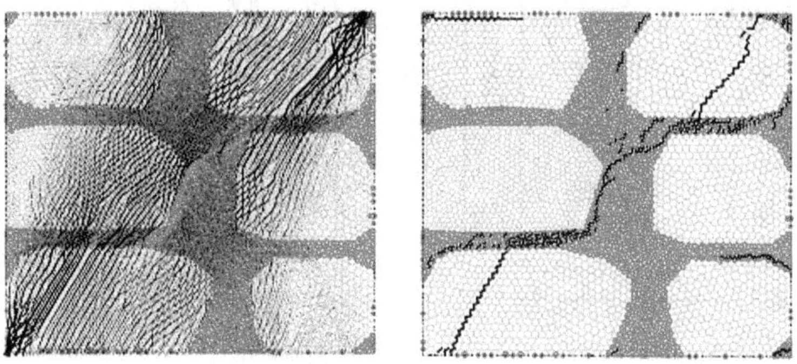

Figure 11. Shear with pre-compression of stone wall with thick mortared joints allowing top plate rotation: (left) contact forces; (right) cracked contacts

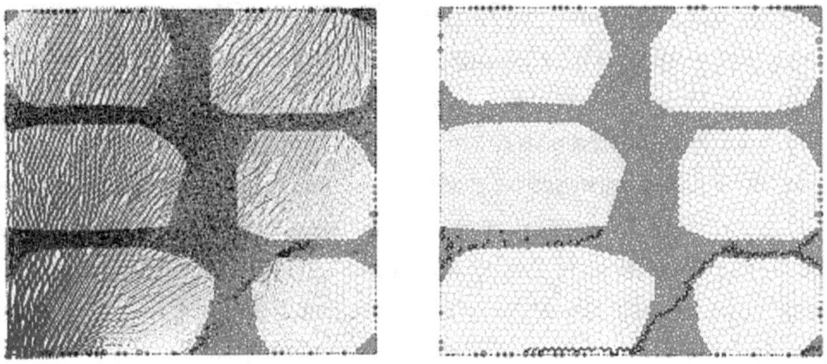

ferent materials, e.g. Valluzi et al. (2001). Numerical models capable of addressing the complexity of the mechanical response of these structures at the grain-scale are very useful in order to interpretate and generalize the experimental knowledge being gathered.

Rubble and Irregular Masonry Walls under Compression and In-Plane Shear

The capabilities of the discrete element based particle model for stone masonry analysis are assessed in uniaxial compression tests and shear tests with an initial pre-compression axial force of wall panels tested by Vasconcelos (2005) for cyclic shear loading with an initial pre-compression load; both regular and irregular mortared walls were tested.

The uniaxial monotonic compression numerical tests and monotonic numerical shear tests with an initial pre-compression loading are presented. It is intended to show that the fracture process is dependent of the adopted topology. Figure 12 shows the adopted wall geometries, which are based on the digital images of irregular mortared walls studied by Vasconcelos (2005). One wall has a stone unit arrangement lightly irregular and the other wall has a higher irregularity in the stone unit placement and shape.

Figure 12. Geometry of walls tested by Vasconcelos (2005)

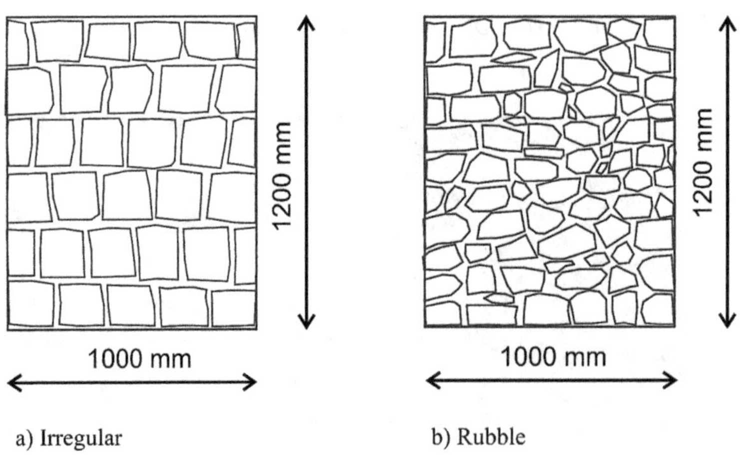

As previously mentioned, for applying the particle model it is necessary to calibrate the micro-properties of the stone units and mortar. With this purpose, uniaxial tensile and compression tests were carried out in quadrilateral specimens of 200 mm × 200 mm × 200 mm. For the mortar representation, a uniform particle distribution of circular particles with radius ranging from 2.5 mm to 4.5 mm was adopted. For the stone units, a uniform particle radius distribution from 3.0 mm to 5.0 mm was adopted. The calibrated micro-proprieties are presented in Table 1. In the mortar calibration, the particle assemblies had in average 1165 particles and in the stone unit calibration tests the particle assemblies had in average 840 particles.

Table 2 shows the numerical macroscopic properties obtained after the calibration procedure for the mortar and for the stones.

Figure 13 shows the particle assemblies considered. The particle generation procedure follows the principles discussed in the previous section. The irregular wall geometry has 35 stone units and adopts a total of 16591 particles representing the stone units and a total of 8300 particles representing the mortar. The rubble wall geometry has 80 stone units and adopts a total of 13964 particles representing the stone units and a total of 11144 particles representing the mortar.

The contacts are established adopting a particle distance threshold value of 0.25 in both wall geometries. In the irregular wall geometry, a total number of 68243 contacts are established (42594 contacts of the type stone-stone, 5720 contacts of the type stone-mortar and 19290 contacts of the type mortar-mortar). In the rubble wall geometry a total of 67271 contacts are created (33434 contacts of the type stone-stone, 8119 contacts of the type stone-mortar and 25718 contacts of the type mortar-mortar).

Table 1. Elastic and strength micro-properties

Contact Type	$\bar{E}$ [GPa]	ν_c	$\sigma_{n.t}$ [MPa]	τ [MPa]	μ_c	$G_{f.n}$ [N/m]	$G_{f.s}$ [N/m]
Stone/Stone	16.0	0.16	2.2	22.0	0.3	10.0	100.0
Stone/Mortar	6.0	0.18	1.5	1.5	0.5	0.5	0.5
Mortar/Mortar	1.6	0.20	1.5	1.5	0.5	0.5	0.5

Table 2. Elastic and strength macro-proprieties

Material	E [GPa]	ν	σ_t [MPa]	σ_c [MPa]
Stone	20.6	0.19	2.3	28.7
Mortar	2.2	0.18	1.3	5.3

Figure 13. Adopted particle assemblies

a) Irregular b) Rubble

Uniaxial Compression Test

The uniaxial compression tests are performed adopting displacement control for the upper plate represented by a polygonal type element which interacts with the particle assembly through contact interaction in a similar way to the contact formulation described in the previous section. For each wall geometry, two different simulations are carried out: in the first only normal interaction is allowed to occur between the plates and the particle assembly (low friction) and in the second both normal and shear interactions are allowed to occur, restraining the lateral deformation at the plates boundary (high friction). The upper and lower plates are not allowed to rotate.

Figure 14 shows the final crack failure patterns obtained in the irregular wall geometry. The high friction example presents the expected diagonal shear cracking. In the low friction example, the failure pattern is closer to the expected vertical cracking but the presence of the stone units makes the crack pattern more tortuous. In both cases, the macro-cracks are the result of the sum along the failure plane of several contacts that have failed mostly under tensile forces. The fracture initiates at the stone-mortar contacts and then evolves through the mortar matrix, being able also to cross through some stone units which have higher strength. Nevertheless, some cracks that propagate in the mortar matrix are in some cases restrained by the presence of the stone units.

Figure 15 shows the final crack pattern associated with the rubble wall arrangement. As in the previous example, the expected crack patterns are obtained for both the low and high plate friction examples. The fact that the stone unit positioning is more random eases the process of crack propagation through the mortar matrix. Some cracks also tend to propagate through the stone units but the failure load is less conditioned by this constraint when compared to the less irregular stone arrangement.

In Figure 16 a), it is presented the force/displacement diagram obtained for both loading plate conditions. It is shown that a higher transversal restraint increases the peak load and leads to a more ductile post-peak behaviour. It is also shown that the rubble masonry wall with higher random stone arrangement and thicker mortar joints leads to lower peak load values and to a less ductile response. A regular pattern of the stone units for this type of loading creates a mechanism that restrains the propagation of cracks through the mortar matrix.

Figure 14. Uniaxial compression test: Irregular wall

a) Low plate friction b) High plate friction

Figure 15. Uniaxial compression test: Rubble wall

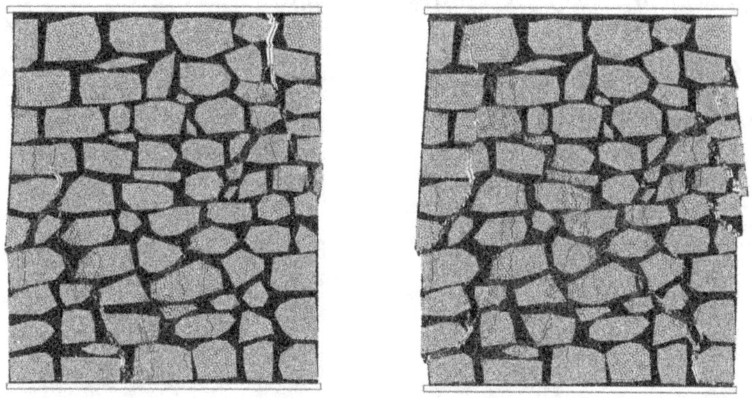

a) Low plate friction b) High plate friction

Figure 16. Force displacement diagrams

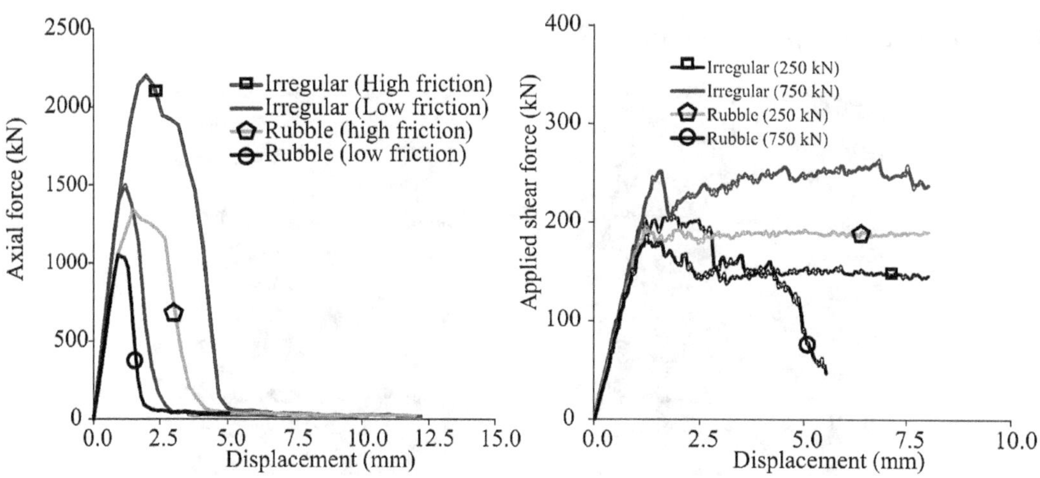

a) Uniaxial compression b)Shear with pre-compression

Shear Test with Initial Pre-Compression

In the shear tests with initial pre-compression, a high friction plate (normal and shear contact force transmission) is adopted in both the upper and the lower plates. An initial pre-compression load is adopted (250 kN and 750 kN), followed by a lateral displacement of the upper plate. On both plates, the rotation is restrained.

Figures 17 and 18 show the failure patterns obtained. It is shown that, for the high pre-compression, the shear failure type is more noticeable. Also, in the rubble geometry a more dispersed crack pattern is obtained for both initial pre-compression values. For the higher pre-compression value, it can be observed that fracture occurs at the stone units located at the comers near to the diagonal crack that is formed. Figure 16 b) shows the force/displacement diagram; it can be seen that, as expected, a higher initial pre-compression value leads to a higher peak shear load.

Figure 17. Shear test with pre-compression: Irregular wall

a) 250 kN pre-compression b) 750 kN pre-compression

Figure 18. Shear test with pre-compression: Rubble wall

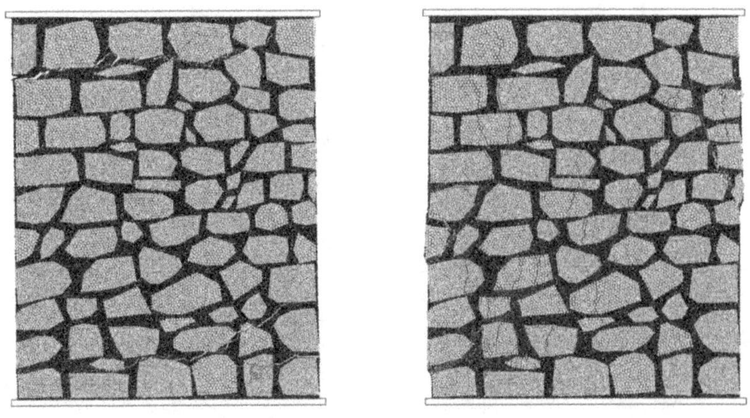

a) 250 kN pre-compression b) 750 kN pre-compression

Table 3. Run times for numerical simulations (double precision)

	Run Time [hours]	Total Number of Steps	Steps per Run Time Hour
Rubble (Compression-high friction)	15.9	476824	30017
Irregular (Compression-high friction)	15.1	486466	32306
Rubble (Shear-750 kN)	13.2	226907	17147
Irregular (Shear-750 kN)	17.1	314826	18379

Table 3 shows the numerical simulations run times in double precision for an Intel Core I7 processor with a 2,93 GHz clock, running windows 7 x64 bits professional edition. The run times that are obtained in the compression and shear tests are already within the acceptable for engineering decision making. Note that the numerical model that was adopted (Monteiro Azevedo, 2003) does not take into account any type of parallel programming optimization, both distributed memory (Wang et al., 2004) and shared memory (Holmes et al., 2010). For this reason there is still a place to further reduce the obtained computational run time and also to apply the algorithm to larger particle assemblies.

CONCLUSION

The computational resources available today allow the detailed representation of materials by means of discrete element particle models, which have become a powerful tool to investigate the fundamental behavior of materials and structures. While mainly applied for research purposes, their increased use in engineering practice can be foreseen, as well as the progressive switch from 2D to 3D representations.

For masonry structures, these models show great potential to handle particular problems, namely: study of traditional masonry structures, with irregular stones, 3-leaf walls, or rubble masonry. Generation of random particle systems that reproduce the observed patterns can be readily implemented, or the actual block configurations can be input if a detailed description is available. Particle models are also well suited for cases in which fracture of the units needs to be considered, as the experience obtained in modelling rock and concrete has shown.

After appropriate validation based on experimental data, particle models can be adopted as a numerical tool to perform parametric studies and provide benchmark solutions for complex problems, which can be used to test and calibrate simplified modelling tools, more suitable for current engineering applications.

REFERENCES

Berton, S., & Bolander, J. (2006). Crack band model of fracture in irregular lattices. *Computer Methods in Applied Mechanics and Engineering*, *195*(52), 7172–7181. doi:10.1016/j.cma.2005.04.020

Chang, K., & Meegoda, J. (1997). Micromechanical Simulations of Hot Mix Asphalt. *Journal of Engineering Mechanics*, *123*(5), 495–503. doi:10.1061/(ASCE)0733-9399(1997)123:5(495)

Cundall, P. A. (1971). A computer model for simulating progressive large scale movements in blocky rock systems. In *Proc. Int. Symposium on Rock Fracture (ISRM)*, Nancy (vol. 1, paper II-8).

Cundall, P. A. (1987). Distinct element models for rock and soil structure. In E. T. Brown (Ed.), *Analytical and computational methods in engineering rock mechanics* (pp. 129–163). London: Allen & Unwin.

Cundall, P. A., & Strack, O. D. L. (1979). A discrete numerical model for granular assemblies. *Geotechnique*, *29*(1), 47–65. doi:10.1680/geot.1979.29.1.47

Gao, F., & Stead, D. (2014). The application of a modified Voronoi logic to brittle fracture modelling at the laboratory and field scale. *International Journal of Rock Mechanics and Mining Sciences*, *68*, 1–14. doi:10.1016/j.ijrmms.2014.02.003

Hamdi, P., Stead, D., & Elmo, D. (2014). Damage characterization during laboratory strength testing: A 3D-finite-discrete element approach. *Computers and Geotechnics*, *60*, 33–46. doi:10.1016/j.compgeo.2014.03.011

Hentz, S., Daudeville, L., & Donze, V. (2004). Identification and Validation of a discrete element model for concrete. *Journal of Engineering Mechanics*, *6*(6), 709–719. doi:10.1061/(ASCE)0733-9399(2004)130:6(709)

Holmes, D., Williams, J. R., & Tilke, P. (2010). Events Based Algorithm for Distributing Concurrent Tasks on Multi-Core Architectures. *Computer Physics Communications*, *181*(2), 341–354. doi:10.1016/j.cpc.2009.10.009

Itasca (2014). *PFC (Particle Flow Code) - Version 5.0*. Minneapolis, MN: Itasca.

Lilliu, J., & Van Mier, M. (2003). 3D lattice type fracture model for concrete. *Engineering Fracture Mechanics*, *70*(7-8), 841–927. doi:10.1016/S0013-7944(02)00158-3

Lisjak, A., & Grasseli, G. (2014). A review of discrete modeling techniques for fracturing processes in discontinuous rock masses. *Journal of Rock Mechanics and Geotechnical Engineering*, *6*(4), 301–314. doi:10.1016/j.jrmge.2013.12.007

Lourenço, P. B., & Rots, J. (1997). Multisurface Interface Model for Analysis of Masonry Structures. *Journal of Engineering Mechanics*, *123*(7), 660–668. doi:10.1061/(ASCE)0733-9399(1997)123:7(660)

Matsuda, Y., & Iwase, Y. (2002). Numerical simulation of rock fracture using three-dimensional extended discrete element method. *Earth, Planets, and Space*, *54*(4), 367–378. doi:10.1186/BF03352426

Meguro, K., & Hakuno, M. (1989). Fracture analysis of concrete structures by the modified distinct element method. *Structural Engineering/Earthquake Engineering. Japanese Society for Civil Engineers*, *6*(2), 283–294.

Monteiro Azevedo, N. (2003). A rigid particle discrete element model for the fracture analysis of plain and reinforced concrete. PhD Thesis, Heriot-Watt University, UK.

Monteiro Azevedo, N., Candeias, M., & Gouveia, F. (2015). A Rigid Particle Model for Rock Fracture Following the Voronoi Tessellation of the Grain Structure: Formulation and Validation. *Rock Mechanics and Rock Engineering*, *48*(2), 535–557. doi:10.1007/s00603-014-0601-1

Monteiro Azevedo, N., & Lemos, J. V. (2005). A generalized rigid particle contact model for fracture analysis. *International Journal for Numerical and Analytical Methods in Geomechanics*, *29*(3), 269–285. doi:10.1002/nag.414

Monteiro Azevedo, N., & Lemos, J. V. (2013). A 3D generalized rigid particle contact model for rock fracture. *Engineering Computations*, *30*(2), 277–300. doi:10.1108/02644401311304890

Nagai, K., Sato, Y., & Ueda, T. (2005). Mesoscopic simulation of failure of mortar and concrete by 3D RBSM. *Journal of Advanced Concrete Technology*, *3*(3), 385–402. doi:10.3151/jact.3.385

Oetomo, J. J., Vincens, E., Dedecker, F., & Morel, J.-C. (2014). Discrete Element Method to assess the 2D failure of dry stone retaining walls, In *9th International Masonry Conference*, Guimarães, Portugal.

Pöschel, T., & Schwager, T. (2005). *Computational Granular Dynamics: Models and Algorithms*. Berlin: Springer.

Potyondy, D., Cundall, P., & Lee, C. (1996). Modelling rock using bonded assemblies of circular particles. In: M Aubertin et al. (eds). *Proceedings of the 2nd North American Rock Mechanics Symposium*, 1937-1944, Rotterdam, Balkema.

Potyondy, D., & Cundall, P. A. (2004). A bonded-particle model for rock. *International Journal of Rock Mechanics and Mining Sciences*, *41*(8), 1329–1364. doi:10.1016/j.ijrmms.2004.09.011

Rokugo, K., Iwasa, M., Susuki, T., & Koyanagi, W. (1989). Testing methods to determine tensile strain softening curve and fracture energy of concrete. In H. Mihasi et al. (Eds.), *Fracture toughness and fracture energy: Test methods for concrete and rock* (pp. 153–163). Rotterdam: Balkema.

Rouxinol, G. A. F., Providência, P., & Lemos, J. V. (2007). Bridgemill bridge bearing capacity assessment by a discrete element method. In P. B. Lourenço, D. B. Oliveira & A. Portela (Eds.), *Arch'07 - Proc. 5th International Conference on Arch Bridges* (pp. 669-676). Madeira, Portugal: Multicomp Lda Publishers.

Schlangen, E & Garboczi, E. (1997). Fracture simulations of concrete using lattice models:Computational aspects. *Engineering Fracture Mechanics*, 57(2/3), 319:332.

Stefanou, I., Sab, K., & Heck, J.-V. (2015). Three dimensional homogenization of masonry structures with building blocks of finite strength: A closed form strength domain. *International Journal of Solids and Structures*, *54*, 258–270. doi:10.1016/j.ijsolstr.2014.10.007

Thavalingam, A., Bicanic, N., Robinson, J. I., & Ponniah, D. A. (2001). Computational framework for discontinuous modelling of masonry arch bridges. *Computers & Structures*, *79*(19), 1821–1830. doi:10.1016/S0045-7949(01)00102-X

Underwood, P. (1983). Dynamic relaxation. In T. Belytschko & T. Hughes (Eds.), *Computation Methods for Transient Analysis* (pp. 246–265). New York: North-Holland.

Valluzi, M. R., da Porto, F., & Modena, C. (2001). Behaviour of multi-leaf stone masonry walls strengthened by different intervention techniques. In P. B. Lourenço & P. Roca (Eds.), *Historical Constructions* (pp. 1023–1032). Guimarães, Portugal.

Vasconcelos, G. F. M. (2205). Experimental investigations on the mechanics of stone masonry: Characterization of granites and behaviour of ancient masonry shear walls. PhD Thesis, Universidade do Minho, Portugal.

Wang, F., Feng, Y., Owen, D., Jing, Z., & Yang, L. (2004). Parallel analysis of combined finite/discrete element systems on PC cluster. *Acta Mechanica Sinica*, *20*(5), 534–540. doi:10.1007/BF02484276

Wang, Z. M., Kwan, A. K. H., & Chan, H. C. (1999). Mesoscopic study of concrete I: Generation of random structure and finite element mesh. *Computers & Structures*, *70*(5), 533–544. doi:10.1016/S0045-7949(98)00177-1

Chapter 8
Numerical Modelling of Masonry Dams Using the Discrete Element Method

Eduardo Martins Bretas
Northern Research Institute, Norway

ABSTRACT

This work concerns the numerical modelling of masonry dams using the Discrete Element Method. It begins with a review of the history of masonry dams and their behaviour. A numerical tool based on the Discrete Element Method developed specifically for the structural assessment of masonry dams is then presented. The mechanical calculations performed by the tool are discussed in detail, together with the approach used for the modelling of passive anchors and the modules for seismic analysis and hydromechanical analysis. Structural and hydraulic analyses of a diverse set of existing masonry dams conducted using the tool are then presented. The Discrete Element Method is shown to be capable of reproducing the structural behaviour of masonry dams and identifying their likely failure mechanisms as required for structural safety evaluations.

INTRODUCTION

During the second half of the 19th century, stone masonry was used in the construction of most large dams in both Europe and America. Many of these structures have since undergone rehabilitation works and remain in use. However, their continued use presents a challenge because they are required to achieve a degree of structural safety consistent with modern regulations. The structural problems of masonry dams result from aging and are different to those that occur in concrete dams. Many of them arise from a loss of masonry cohesion as a result of prolonged exposure to water seepage. The structural properties of masonry dams are also influenced by the quality of the material and the laying scheme of the stones on the dam's external faces as well as in the inner material.

Structural safety analyses of masonry dams based on numerical models require tools that can accurately represent the internal structure of the masonry. These tools and representations must be able to characterize the structure's non-linear behaviour and the hydromechanical interactions that occur when the dam is subjected to static and dynamic loads, and which lead to typical failure mechanisms such as

DOI: 10.4018/978-1-5225-0231-9.ch008

cracking, sliding and overturning. Many numerical applications that have designed for the analysis of common structures based on the Finite Element method disregard the discontinuous nature of the material, which controls the structural and hydraulic behaviour of masonry dams. To address this deficiency, models based on the Discrete Element method (DEM) have been used to analyse masonry dams. The capacity to model discontinuities explicitly and the ability to perform coupled analyses are two significant benefits of the Discrete Element Method.

This work provides a review of the literature on masonry dams. A numerical application based on the DEM and developed specifically for masonry dam modelling is presented, with a detailed description of the calculation cycle and its main modelling tools. The tool's main features are its ability to use continuous and discontinuous meshes in the same model, to recreate typical loading combinations in masonry dams, and to perform full coupled hydromechanical analyses including analyses of reinforcement elements. Studies on a set of illustrative masonry dams using continuous and discontinuous models, and different discretization techniques are presented. Static and dynamic structural analyses of the dams are discussed, considering collapse mechanisms through the dam body and the rock foundation. Hydraulic analyses that compute internal water pressures and flow rates are also performed. Modelling of rehabilitation measures like application of passive anchors, and reinforcement of drainage systems, is also discussed.

THE HISTORY OF MASONRY DAMS

The first manmade dams were built from earth and stones that were randomly deposited in order to close off a valley and thus create water reservoirs, primarily for agricultural purposes. Accidents were common and generally as resulted of flooding (International Commission on Large Dams [ICOLD], 2013). The Romans introduced important innovations in dam building, particularly by pioneering the use of new materials. Stone masonry was used in the construction of most Roman dams, which were generally built from stone blocks and hydraulic lime mortar (Schnitter, 1994). Stone masonry continued to be used for some time, but the lime was gradually replaced with cement. Subsequently, during the early twentieth century, concrete replaced stone masonry as the material of choice in dam building.

Until the mid-nineteenth century, masonry dams were designed using empirical criteria. In 1853, J.A.T. Sazilly proposed that the cross-section of a dam should be chosen using the "profile of equal resistance" method. This method computes the vertical stresses along the dam's body in order to define the cross-section's geometry based on permissible stress limits for two extreme cases: a completely full reservoir and a completely empty reservoir. This method creates a section similar to a triangle, like modern gravity sections. This criterion represented an important technological advancement because it allowed the bearing capacity of the dam's material to be fully exploited and thus enabled significant economic gains (Bretas, Lemos, & Lourenço, 2012). The first dam designed and built according to this criterion was the Furens dam in France, which is 50 m tall and 200 m long. The project, developed by F. Delocre, was begun in 1858 and the first reservoir filling was done in 1866. Figure 1a shows the Furens dam while Figure 1b depicts the Puentes dam, which was built in Spain between 1785 and 1791, before the publication of Sazilly's paper. Although both dams have similar heights, the area of the main section of the Puentes dam is around 50% greater than that of the Furens dam (Bretas, Lemos, & Lourenço, 2010).

It was also during this period that the main rules for masonry construction started to be consolidated. Most masonry dams were built from cut stones, rubble and concrete. The inner parts were built with a combination of large stones and rubble, surrounded by concrete. When laying the stones, special measures

Figure 1. Main sections of the Furens and Puentes dams

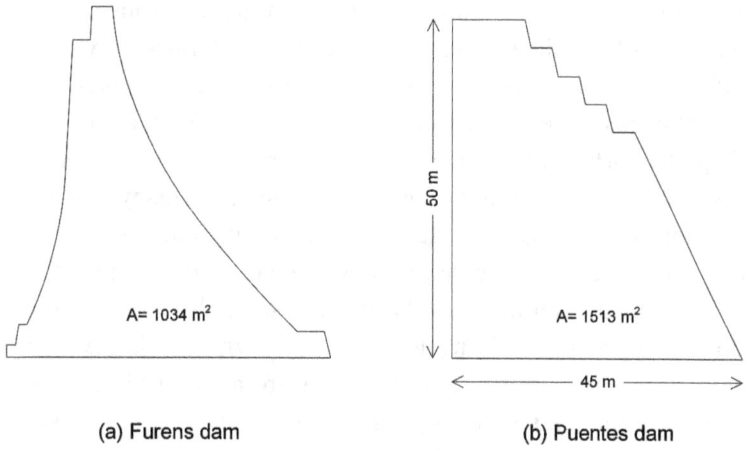

were taken to create break joints in all directions, avoiding horizontal or vertical courses (Wegmann, 1888). Because of its high cost, the cut stone was used only on the upstream and downstream faces and in ornamental works. In many cases the cut stones were laid normal to the faces, but in some cases they were laid horizontal to the faces. While dam design during this period was based on a more scientific approach than had previously been the case, some important factors such as the effect of the uplift and seismic loading remained unknown or overlooked. Uplift loading started being included in stability calculations after the failure of the Bouzey dam in France, which occurred in 1895 and whose collapse was probably related to the uplift (Smith, 1971). Uplift is caused by water seepage in the contact plane between the dam and the foundation. It effectively reduces the dam's self-loading, which reduces the structure's safety. In some regions, it is essential to take seismic loading into account when designing dams. Seismic action on dams has been considered, in a first approximation, as a static inertial load. The method for modelling the hydrodynamic reservoir effect proposed in 1933 by Westergaard is still considered valid. The development of Finite Element method in the second half of the twentieth century represented a major development in seismic calculations.

THE BEHAVIOUR OF EXISTING MASONRY DAMS

General Aspects

Most of the masonry dams that remain in use are gravity dams, which use their own weight to resist the pressure exerted by the water of the reservoir. Masonry dams are usually laid on a sound rock foundation to withstand the stresses imposed by hydrostatic pressure, which are particularly severe at downstream side. Some dams of this sort have vertical contraction joints like those found in modern concrete dams. These joints enable a degree of volumetric variation in the dam to accommodate both expansion and contraction due to thermal changes without causing cracking. The spillway is typically installed over the dam's body with the bottom outlet crossing the section at the valley's lowest point. Some masonry dams have a drainage gallery that was built with the dam or subsequently during rehabilitation works. Drainage galleries allow the installation of drains along the foundation, usually close to the upstream face, for the purpose of draining infiltrating water and thereby reducing uplift. In some cases, vertical drains through the dam body are also installed in order to drain infiltrating water and thus release the inner pressure.

Almost all masonry dams have already been subjected to structural and hydraulic rehabilitation. The main concern regarding the safety of masonry gravity dams relates to material ageing. This can be a consequence of losses of cohesion resulting from chemical and physical degradation due to water seepage. This problem is particularly severe in dams subject to harsh climatic conditions such as long and cold winters. For this reason, the rehabilitation works conducted on many masonry dams have focused on reducing the permeability of the dams' material and increasing the stiffness of the dam body. Common measures taken for this purpose include grouting the dam and its foundation, and constructing a concrete curtain or a geomembrane curtain over the dam's upstream face. Aside from material ageing, the main causes of rehabilitation works are (i) the need to comply with regulatory changes and increasingly stringent safety requirements; and (ii) deficiencies arising from the original construction that must be rectified (ICOLD, 1994). Common deficiencies further include an inadequate spillway capacity in the event of flooding and lacking structural stability due to uplift and seismic loading. The most commonly used measures for addressing problems of structural stability involve installing active and passive anchors that link the dam to its rock foundation. The rehabilitation measures are undertaken in combination with the reinforcement of the drainage system, into the dam body and the foundation, and improvement of the dam's monitoring systems (ICOLD, 2000).

Safety Assessment of Masonry Gravity Dams

Dams present a high potential risk because their failure can cause the loss of life and considerable material losses. To reduce the risk of accidents and the severity of their consequences, dams must be tested extensively to evaluate their structural behaviour in relation to the loads they experience. The assessment of potential failure scenarios is an important part of such studies. The main international regulations governing dams primarily relate to concrete and embankment dams, although some specific references to masonry dams can be found. The French guidelines are notable (Comité Français des Barrages et Réservoirs [CFBR], 2013) because they include information on the properties of masonry dams and their constituent materials based on the results of tests conducted on French dams (Table 1). The measured values vary widely because the mortar used in the dam's construction has profound effects on the properties of the masonry.

When assessing the safety and structural soundness of gravity dams in general, it is essential to consider the hydrostatic pressure and the uplift in addition to the structure's dead load. Other loads may also be considered in certain cases, such as thermal loads, ice loads, and silt loads. In some regions, the

Table 1. Properties of masonry used in masonry dams (CFBR, 2013)

Property	Values
Unit weight	22 – 24 kN/m^3
Internal angle of friction	30^0 - 46^0
Cohesion	0 – 1 MPa
Compressive strength	5 – 10 MPa
Tensile strength	0 – 1 MPa

seismic load is an important factor in determining dam safety. For example, the regulations produced by the US Bureau of Reclamation (USBR) (1976) state that the loading combinations to be considered are the "usual loading combination" (ULC), the "unusual loading combination" (UnLC) and the "extreme loading combination" (ELC). The ULC is based on the current loads for the normal design reservoir elevation while the UnLC is based on the same loads but is computed for the maximum reservoir design elevation. The ELC is based on the same conditions as the ULC but includes the maximum credible earthquake. Other loading combinations are also suggested for each case, such as those to be expected in the event of a failure of the drainage system.

The criteria for safety assessment specified in the USBR regulations are based on safety factors. The compressive stress in all points of the structure is required to be less than the material's strength (which is reduced by a safety factor of 3.0, 2.0 or 1.0 in the ULC, UnLC, and ELC cases, respectively). Furthermore, to prevent the formation of cracks and water seepage, especially at the upstream face, the compressive stress must exceed the difference between the hydrostatic pressure at the point in question and the material's tensile strength (which is reduced by the same factors mentioned above for the compressive stress in the ULC, UnLC, and ELC cases). For assessments of failure via the slide mechanism, the USBR regulations suggest that the analysis should be performed using a safety factor calculated from the equilibrium of the section. This safety factor should be greater than 3.0, 2.0 or 1.0 depending for the ULC, UnLC and ELC loading combinations, respectively. To assess failure in the foundation, safety factors of 4.0, 2.7 and 1.3 should be used for these loading combinations.

In the absence of specific criteria for evaluating the structural safety of masonry dams, it is necessary to adapt the established criteria for concrete dams. It is also important to consider that most existing masonry dams are rather old structures that are not expected to meet the same requirements as modern concrete structures. Specific aspects of masonry dams should be evaluated on the basis of the knowledge and opinions of professionals with experience in assessing the safety of these structures.

Numerical Modelling of Masonry Dams

Structural analysis using numerical models plays a central role in the design of complex structures and evaluations of their behaviour. In the case of dams, numerical analyses based on models featuring a dam, a foundation, and a reservoir are used to guide the structural design of new dams and the rehabilitation of existing ones, to investigate the structure's behaviour under normal operating conditions, and to predict its behaviour in failure scenarios involving e.g. earthquakes and floods (ICOLD, Rough copy). The numerical models should strike a balance between simplicity and reality: they must incorporate a level of detail that is sufficient to accurately represent all of the phenomena of interest without being excessively expensive in computational terms.

Masonry consists of blocks and mortar. Numerical models may include explicit representations of both elements separately, as is the case in detailed micro-modelling. In other cases involving different model sizes or studies with different objectives, it may be preferable to model the two elements together; this approach is known as macro-modelling. The models presented in this work are based on a simplified micro-modelling scheme in which both the blocks and the mortar are explicitly modelled but the joint has zero thickness and so the mortar's behaviour is represented by the properties of the joint material (Lourenço, 1996).

Bettzieche et al. (2004) described the rehabilitation works carried out on the 8 reservoirs belonging to the Ruhrverband (Ruhr River Association) in the German state of North Rhine-Westphalia. Most of the rehabilitation measures applied to the masonry dams in this group were designed on the basis of numerical models that were developed with the Finite Element method and used to analyse the seepage inside the dams, their temperature flows, and crack propagation. Crack propagation was modelled by introducing discrete crack increments into the modelled structure rather than with a smeared crack model.

Wittke et al. (2012) also described the use of an FE model to develop a rehabilitation program for the Urft dam, a masonry gravity dam located in the Eifel near Cologne. A three-dimensional analysis was performed, accounting for the interaction between the dam and the underlying foundation rock as well as the dam's anisotropic mechanical and hydraulic rock mass properties. These studies exemplify the macro-modelling approach.

Dolezalova (2004) used the simplified micro-modelling approach to analyse the stability and seepage behaviour of an almost 100-year old stone masonry dam. A safety assessment of the structure was performed using the Discrete Element Method (DEM) and treating the dam as a discontinuous medium, and its results were used to provide a basis for the dam's reconstruction.

The use of DEM seems to be acceptable in dam engineering, particularly for the numerical modelling of masonry dams. DEM approaches enable the explicit representation of discontinuities in both the dam body and its foundation, which control the dam's non-linear behaviour in failure scenarios. Furthermore, the loads acting on the dam are modelled in a consistent way, making it possible to perform a fully coupled hydromechanical analysis of the dam-foundation-reservoir system. The next section outlines a numerical application of the DEM approach for the 2D analysis of masonry gravity dams. The resulting approach is applied to a set of test cases including both static and dynamic analyses as well as hydromechanical analyses.

NUMERICAL MODELLING OF MASONRY DAMS USING THE DISCRETE ELEMENT METHOD

The tools presented in this paper are based on the DEM and were designed to meet three main requirements. Firstly, they are intended to model both the masonry dam and the rock foundation in an integrated manner as components of a blocky system. Secondly, the software tools were intended to provide a practical way of using both equivalent continuum and blocky models using the same mesh. Finally, the tool was required to include all of the features required in dam engineering analysis, including facilities for estimating the water flow and pressures at the joints, the ability to include reinforcement elements such as passive or active anchors, and the means to model the effects of the load types involved in static and seismic analyses. All of the tool's components interact through a compatible data structure. The following sections provide a detailed description of the calculations used to perform static, seismic and hydromechanical analyses. Emphasis is placed on formulating interactions on the basis of face-to-face contacts, which is more rigorous than the more common approach based on the assumption of point contacts. Additional details are available in various earlier publications (Bretas, Lemos, & Lourenço, 2013; Bretas, Lemos, & Lourenço, 2014; Bretas, Lemos, & Lourenço, 2015; Bretas, Léger, Lemos, & Lourenço, 2010).

Blocks, Macroblocks, and Contacts

The fundamental element of discretization is a block with three or four edges, which may be rigid or deformable. Both block types can be used simultaneously in the same model. The computational advantage of rigid blocks is more relevant in explicit dynamic analysis, since static solutions are usually obtained very quickly. In dam engineering, stress analyses of both the structure and foundation are usually required, so deformable blocks are preferred. The following sections deal exclusively with cases involving deformable blocks.

Blocks of arbitrary shape can be created by assembling 3- and 4-node blocks into macroblocks. This is important for the modelling of discontinuous media such as masonry dams and rock mass foundations, and makes it possible to establish an equivalent continuum representation of the whole system (or a part thereof) in which each block is just a unit of the Finite Element (FE) mesh. A macroblock is a combination of blocks that form a continuous mesh in which the vertices are coincident. Relative movement between the blocks of a single macroblock is prohibited, so there are no intra-block contact forces. The macroblock is similar to a FE mesh but with an explicit solution because no global stiffness matrix is assembled.

The mechanical interaction between two blocks is defined using a numerical contact. In this application, the fundamental contact type is face-to-face (Bretas et al., 2014). The use of face-to-face contact is unusual in DEM but it enables the use of different stress integration schemes to determine contact forces. Face-to-face contact is also advantageous because it achieves a linear distribution of stress and is compatible with the correct application of the joint constitutive model. This in turn produces a distribution of contact forces that is statically consistent with the stress diagrams and a correct bending stiffness.

Figure 2 a shows the contacts between the blocks 1-2, 1-3, 1-4, 2-3 and 2-4, which are designated respectively by $CT_{1.2}$, $CT_{1.3}$, $CT_{1.4}$, $CT_{2.3}$ and $CT_{2.4}$. The numerical implementation of face-to-face contact involves two sub-contacts whose origin points are located at the corners of the blocks. The contact $CT_{1.2}$ (Figure 2b), for example, is established numerically by the two sub-contacts $SCT_{1.2.1}$ and $SCT_{1.2.2}$. The contacts are updated in each cycle on the basis of the blocks' relative movements. The variation of the sub-contact separation is incremental and depends on the time step,

$$\Delta u_x = \dot{u}_{x,rel} \Delta t \quad (1)$$

$$\Delta u_y = \dot{u}_{y,rel} \Delta t \quad (2)$$

where Δu_x and Δu_y are the incremental variations of the contact opening, $\dot{u}_{x,rel}$ and $\dot{u}_{y,rel}$ are the relative velocities of the points that define the sub-contact; and Δt is the time step. The normal (Δu_n) and tangential) (Δu_s) components are determined according to the orientation of the edge where the sub-contact acts.

Determination of Contact Forces

The contact forces are calculated by integrating the stress diagram established for the contact. The stresses are determined based on the normal and tangential movement of the sub-contacts, updated at

Figure 2. Blocky and mechanical models

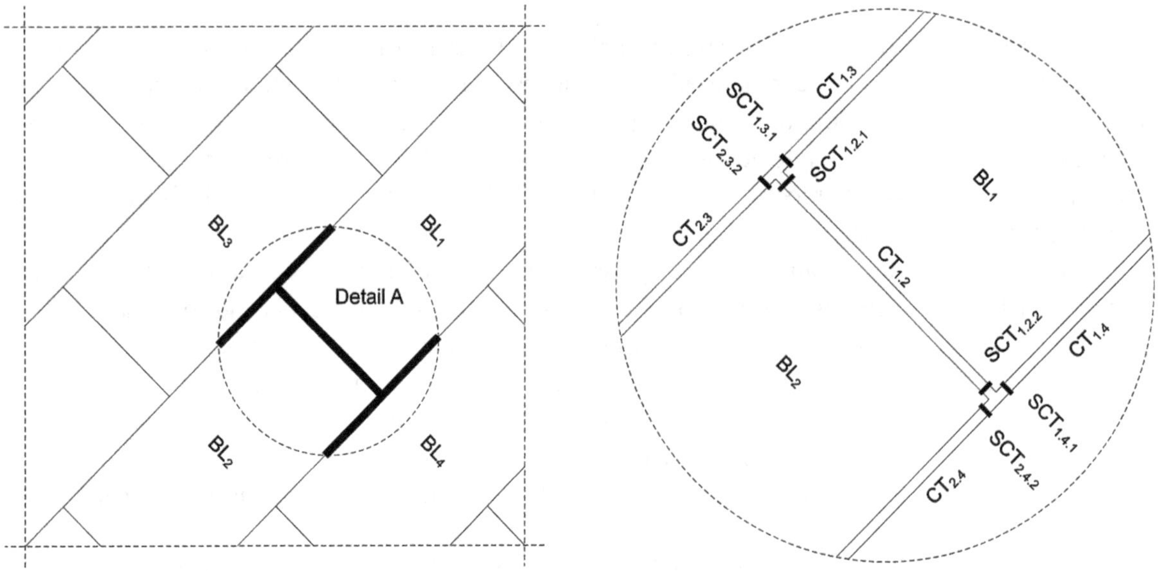

(a) Blocky model
(Blocks 1, 2, 3 and 4)

(b) Mechanical model – Contact $CT_{1,2}$
(Detail A)

each iteration, and the properties of the joint material - in particular, the normal stiffness (k_n) and tangential stiffness (k_s),

$$\sigma = \sigma_o + k_n \Delta u_n \tag{3}$$

$$\tau = \tau_o + k_s \Delta u_s \tag{4}$$

where σ and σ_o are the normal stresses of the sub-contact in the current and previous iterations, respectively; and τ and τ_o are the tangential stresses of the sub-contact in the current and previous iterations, respectively.

The stress equations must be checked against the joint constitutive model that has been adopted, which was the Mohr-Coulomb failure criterion in this work. Different contact integration procedures for the stress diagram can be used, and six solutions have been implemented in the tool: a point model (pt), a rectangular model with two control points (r2pt), a rectangular model with three control points (r3pt), a trapezoidal model with two control points (t2pt), a trapezoidal model with three control points (t3pt), and a trapezoidal model with multiple control points (tmpt). The control points are the points where the stresses are checked with respect to the constitutive model. In models with three points the third point corresponds to the midpoint of the contact, whose stress is determined by assuming a linear distribution of stresses. In the model with multiple points, the points are introduced in places where violations of the constitutive model are detected. In this model the number of control points is not predefined; instead, points are added as required to achieve a final diagram with the correct definition.

In the point contact model (pt) the forces are distributed directly at the sub-contact application point after integration, so the resultant forces are not statically compatible with the diagram from which the originated. Despite this shortcoming, the point contact model is the most widely used in existing DEM codes of those considered in this work. In the model of rectangular contact with two control points, the integration is done in the same way but the resultant forces are applied to the middle of each sub-contact's influence area and application points are allocated to the sub-contacts from these points, yielding a set of forces that are statically consistent with the diagram of stresses. The same is true for the rectangular model with three control points. In this case, the diagram is discretized using three rectangles whereas the previously discussed models use only two. All models that use trapezoidal diagrams of stresses give rise to forces that are distributed through the sub-contacts' application points in a way that is statically compatible with the diagram of stresses. In the model with three control points, the diagram is approximated by two trapezoidal stress diagrams that have one side in common. In the model with multiple points, the stress diagram is decomposed into as many trapezoids as are needed to define the diagram, after which the forces are distributed through the sub-contacts' application points.

Figure 3 shows the constitutive model test for the normal direction, for both elastic behaviour and in case of violation of the tension and compression limits. The integration of the stress diagram and the distributions of the resultant forces for all of the mentioned contact models are also shown. The method of integration controls the rotation stiffness of the contact. The point contact model is the most rigid, while the trapezoidal models are more flexible and all provide the same rotation stiffness in linear elastic analyses. The model with multiple points provides the most accurate integration scheme and should be adopted in parts of the model where stress analysis is important. In parts of the model not directly involved in the failure mechanism, the point contact model can be used since it is more robust and requires less computation time.

General Aspects of the Numerical Solution

Numerically, the incremental step involves setting up and integrating the equation of motion for each degree of freedom,

$$c\dot{u} + m\ddot{u} = f \tag{5}$$

$$f = f_{ext} + f_{\text{int}} + f_{ct} + f_m + f_{hyd} + f_r \tag{6}$$

where $\dot{u}$ is the velocity; $\ddot{u}$ is the acceleration; c is the viscous damping coefficient; m is the mass; f is the total force; f_{ext} is the external force; f_{int} is the internal force, which is equivalent to the stress field of the element (only for deformable blocks); f_{ct} is the contact force; f_m is the mass force; f_{hyd} is the hydraulic force; and f_r is the reinforcement elements force. Damping includes only the component proportional to the mass,

$$c = \alpha m \tag{7}$$

where α is the viscous damping coefficient. The internal forces are calculated according to standard FE method practice,

Figure 3. Integration of the stress diagram and distribution of the resultant forces in case of failure

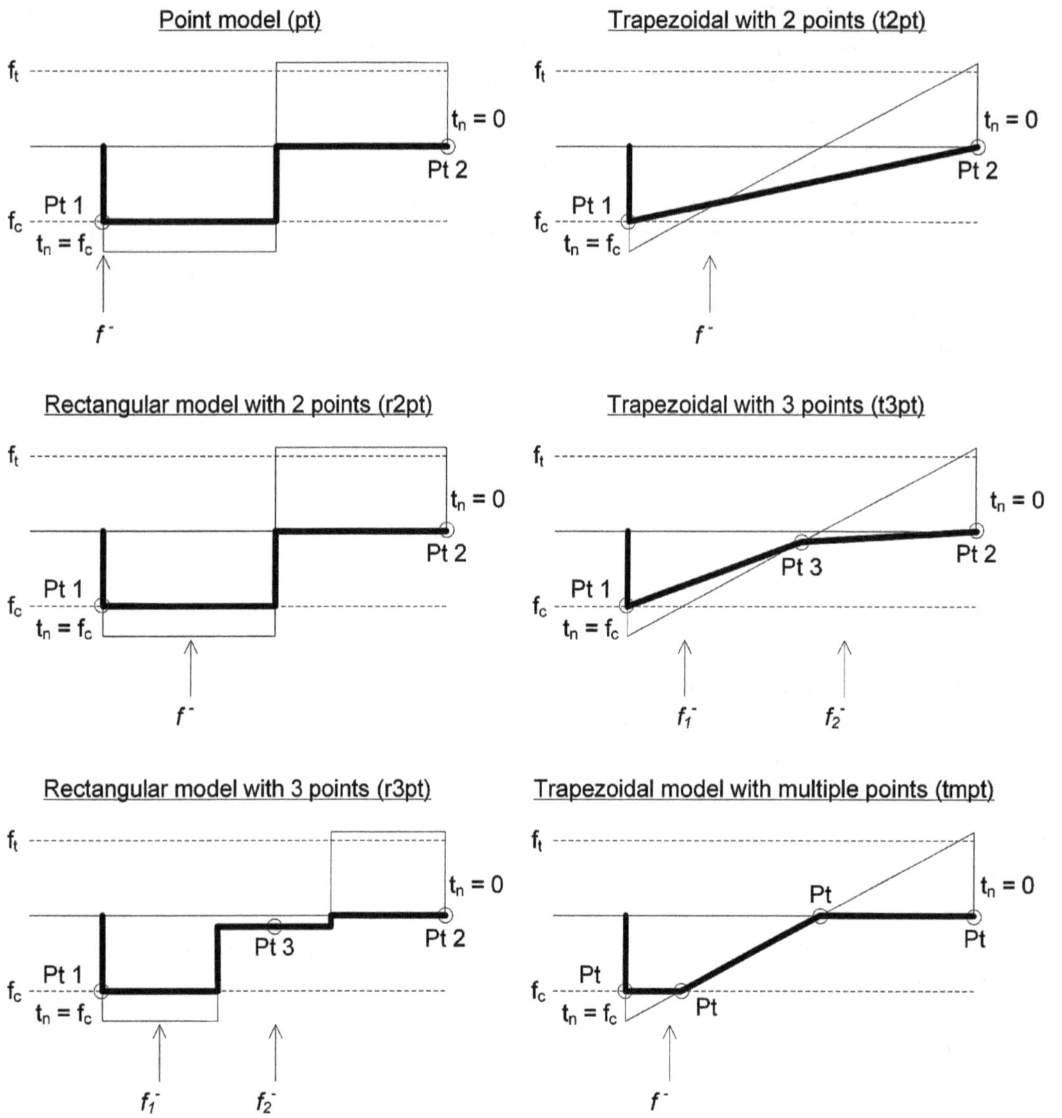

$$f_{\text{int}} = \int_A B^T \sigma dA \tag{8}$$

where B^T is the transposed matrix of deformation; σ is the stress field; and A is the area of the element. In linear deformable blocks with four corners there are eight degrees of freedom in total corresponding to two translations for each vertex, one in the horizontal direction (x) and the other in the vertical direction (y). The integration of the general equation of motion, which is a second order differential equation, is performed explicitly using the central difference method. This makes it possible to determine the velocity at intermediate instant ($t + \Delta t/2$) from the velocity at instant ($t - \Delta t/2$). The displacements are obtained in an incremental way,

$$u_{x,t+\Delta t/2} = u_{x,t-\Delta t/2} + \dot{u}_{x,t+\Delta t/2}\Delta t \tag{9}$$

$$u_{y,t+\Delta t/2} = u_{y,t-\Delta t/2} + \dot{u}_{y,t+\Delta t/2}\Delta t \tag{10}$$

where u_x is the total displacement in the x direction; u_y is the total displacement in the y direction. In each incremental step, the positions of the block's vertices are given by,

$$x_{t+\Delta t/2} = x_{t-\Delta t/2} + u_{x,t+\Delta t/2} \tag{11}$$

$$y_{t+\Delta t/2} = y_{t-\Delta t/2} + u_{y,t+\Delta t/2} \tag{12}$$

where x and y are the vertex coordinates.

The process of explicitly integrating the equation of motion using the central difference method imposes a numerical restriction on the length of the time step, which should be lower than the time required for a longitudinal wave to travel through the model's smallest elastic element. According to the Courant criterion, in an elastic system, the following equation should be observed,

$$\Delta t \leq 2/\omega \tag{13}$$

where ω is the system's highest natural frequency. An upper bound of this frequency can be estimated from the mass and stiffness of each degree of freedom of the model.

The static solutions are obtained by a process of dynamic relaxation using scaled masses and artificial damping. Viscous mass-proportional damping is used, with an adaptive scheme that updates the damping coefficient in a stepwise fashion based on the dominant frequency of the structure from the Rayleigh quotient (Sauvé & Metzger, 1995). To determine the Rayleigh quotient, it is necessary to calculate the tangent stiffness for each degree of freedom, $k_{\tan,x}$ and $k_{\tan,y}$,

$$k_{\tan,x} = \sum \left[\frac{\left(f_{\text{int},x,t+\Delta t/2} - f_{ct,x,t+\Delta t/2}\right) - \left(f_{\text{int},x,t-\Delta t/2} - f_{ct,x,t-\Delta t/2}\right)}{\dot{u}_{x,t-\Delta t/2}\Delta t} \right] \tag{14}$$

$$k_{\tan,y} = \sum \left[\frac{\left(f_{\text{int},y,t+\Delta t/2} - f_{ct,y,t+\Delta t/2}\right) - \left(f_{\text{int},y,t-\Delta t/2} - f_{ct,y,t-\Delta t/2}\right)}{\dot{u}_{y,t-\Delta t/2}\Delta t} \right] \tag{15}$$

Thereafter the Rayleigh quotient, Q_{ray} is calculated for the model,

$$Q_{ray} = \frac{\sum\left[k_{\tan,x}\left(\Delta u_x\right)^2 + k_{\tan,y}\left(\Delta u_y\right)^2\right]}{\sum\left[m_x\left(\Delta u_x\right)^2 + m_y\left(\Delta u_y\right)^2\right]} \tag{16}$$

The mass-proportional damping is given by,

$$\alpha = 2\sqrt{Q_{ray}} \tag{17}$$

Numerical models of dams may feature components with different dynamic behaviours, namely the dam and its foundation. It is therefore possible to select the blocks that will contribute to determining the overall coefficient. The adaptive scheme described above provides a smoother convergence path than alternative dissipation options such as non-viscous local damping.

Modelling Passive Anchors within the DEM

In dam engineering, passive anchors are steel bars that work essentially in the shear direction. For this reason, anchors are installed perpendicular to the plane in which the bearing capacity must be improved. The interface between the dam and the foundation as well as discontinuities extending into the dam's foundation and the dam's body are all typical failure planes that may require reinforcement. Passive anchors are unusual in that they only come into operation after an initial sliding along the plane has occurred. This initial sliding is not elastic, which is likely to adversely affect the initial properties of the rock or masonry joint. The resultant forces in the passive anchor vary according to the relative stiffness of the material, steel and concrete, and whatever sliding has occurred. A numerical model for analysing passive anchors was developed in this work (Bretas, Léger, et al., 2010). This model makes it possible to integrate the assessment of passive with the discrete element model in both static and dynamic calculations.

Axial reinforcements are accounted for in the calculation cycle when the relative displacements are determined and also during the phase when forces originating in the reinforcements are applied to the structure's other components. The creation of an axial reinforcement begins with the definition of initial (P_i) and final (P_f) points, from which the relative displacements are recorded. The P_i and P_f points are obtained with the assistance of a global reference. To this end, it is necessary to start by identifying the blocks in which P_i and P_f, are located. The points' local coordinates are then determined with respect to the local axes of the elements in which they are located. The velocities and displacements of P_i and P_f are then obtained from the element's shape functions and the velocities and displacements of the respective nodes. The P_i velocities are determined using the same procedure as for P_f,

$$\dot{u}_{x,i} = \sum_{n=1}^{4} N_{n,1}\dot{u}_{x,n} \tag{18}$$

$$\dot{u}_{y,i} = \sum_{n=1}^{4} N_{n,1}\dot{u}_{y,n} \tag{19}$$

where $\dot{u}_{x,i}$ and $\dot{u}_{y,i}$ are the velocities of P_i (the initial anchor point); $N_{n,i}$ is the shape function; and $\dot{u}_{x,i}$ and $\dot{u}_{y,i}$ are the velocities of the nodal points of the finite element where P_i is located.

In the next step, the relative velocities and the relative displacements are calculated,

$$\dot{u}_{x,rel} = \dot{u}_{x,f} - \dot{u}_{x,i} \tag{20}$$

$$\dot{u}_{y,rel} = \dot{u}_{y,f} - \dot{u}_{y,i} \tag{21}$$

$$\Delta x = \Delta x_0 + \dot{u}_{x,rel} \Delta t \tag{22}$$

$$\Delta y = \Delta y_0 + \dot{u}_{y,rel} \Delta t \tag{23}$$

where $\dot{u}_{x,f}$ and $\dot{u}_{y,f}$ are velocities of P_f (the anchor end point); $\dot{u}_{x,rel}$ and $\dot{u}_{y,rel}$ are relative velocities between P_i and P_f; Δx and Δy are the relative displacements between P_i and P_f; and Δx_0 and Δy_0 are the relative displacements computed until the previous step.

In addition to geometry-related factors, the properties of the anchor must be defined, including its axial stiffness, shear stiffness, maximum axial force, the maximum tangential force, the maximum axial strain and maximum shear strain. The forces determined with respect to the anchor's normal and tangential directions because the properties are defined with respect to these directions. The orientation of the anchor is updated at each step based on the displacements determined by the immediately preceding calculation step,

$$f_a = k_a \left(\Delta x a_0 + \Delta y a_1 \right) \tag{24}$$

$$f_s = k_s \left(\Delta x s_0 + \Delta y s_1 \right) \tag{25}$$

where f_a and f_s are the axial and shear forces; and $\left(a_0, a_1\right)$ and $\left(s_0, s_1\right)$ are the axial and shear directions of the anchor.

The values obtained must be compared to the anchor's limits of stress and deformation. The forces are transferred to the vertices that define the blocks containing the points P_i and P_f. Another aspect to consider relates to the determination of the time step. An anchor introduces an additional degree of stiffness into the structure, which may be significant and should be taken into account when calculating or updating the time step. The stiffness is distributed by the vertices that define the blocks containing the points P_i and P_f.

Extension of the DEM Model to Seismic Analysis

Three different procedures for modelling dynamic loads and defining boundary conditions were implemented (Figure 4) (Bretas, Lemos, et al., 2015). In the first scheme, the foundation is modelled as a rigid material and the dynamic load is applied using a velocity history (Figure 4a). This procedure can be used if the earthquake is assumed to act in one or two directions. For the horizontal component, the direction (x) is associated with a velocity history while the velocity in the vertical direction (y)is set null. In the general case, both horizontal and vertical components are applied, with different velocity histories prescribed in each direction.

In the second scheme, the foundation is modelled as a deformable mesh and the dynamic load is applied using a stress history (Figure 4b). In this case, it is no feasible to introduce the seismic load using a velocity history at the base of the model because it is necessary to simultaneously impose a non-reflecting boundary to absorb the waves reflected by the free surface and the structure. The viscous boundary formulation proposed by Lysmer and Kuhlemeyer (1969) is therefore used instead.

Consequently, the seismic load is applied through a shear or normal stress history that is computed from the velocity history by considering the elastic properties of the rock mass. This makes it possible to apply one of the components of the dynamic load (i.e. the horizontal or the vertical component) at a time. In the case of a vertically propagating shear wave, shear stresses are applied at the model base resulting in horizontal displacements, and the lateral boundaries of the foundation are fixed in the vertical direction to ensure that theoretical shear wave conditions obtain. In addition, a viscous boundary is applied in the horizontal direction at the base of the foundation. The model is intended to be capable of describing situations in which the foundation rock mass behaves non-elastically, so the traditional assumption of a massless and elastic medium are is not invoked. The rock mass is represented with its real mass and a constitutive model.

The third option for seismic load modelling uses the "free field" technique (Lemos & Cundall, 1999), which is applied at the lateral boundaries to simulate an infinite lateral extension of the model (Figure 4c). A complementary model mesh is computed in parallel with the main model, consisting of

Figure 4. Three different procedures for modelling dynamic loads and defining boundary conditions

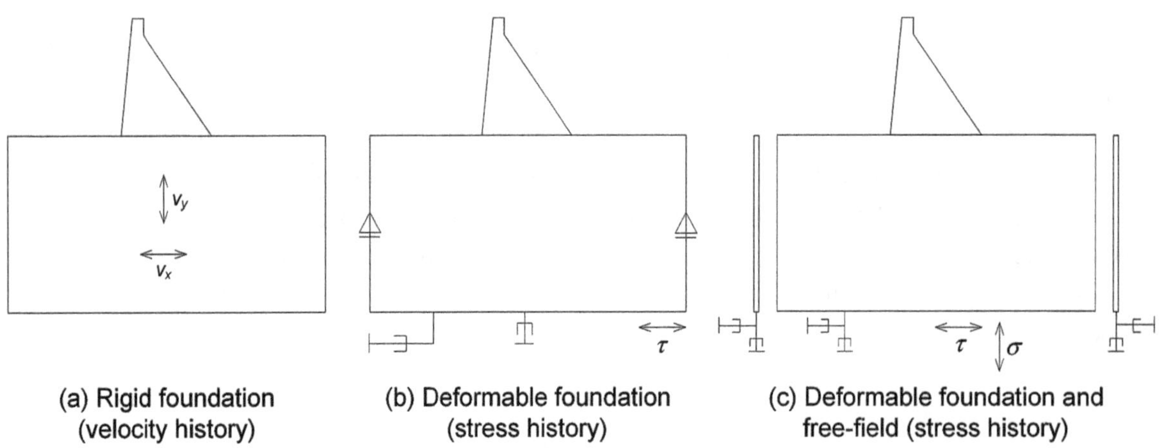

two columns (one on each side of the foundation), each with unitary width and heights coincident with the foundation's discretization. The boundary conditions and dynamic loads from the main model are applied at the bases of these columns. The velocities and stresses experienced by a horizontally infinite medium are obtained by analysing the 1D propagation of this input in the vertical direction. Each column consists of four side elements with four nodal points, and four Gauss points are used to integrate the stiffness matrix. Each free-field element is associated with the element of the foundation, inheriting its material and properties. The columns represent the dynamic behaviour of the free-field at some distance from the dam. At each time step, the horizontal and shear stresses of the free-field mesh are applied to the main model. In addition, to absorb waves reflected from the dam and foundation, viscous dampers are used between the main mesh and the free-field. These dampers are applied to the difference in velocities, following the formulation of Lysmer and Kuhlemeyer (1969). At each time step, the following forces derived from the free field are applied to the nodes of the lateral boundaries of the main mesh:

$$f_x = \left[\sigma_{xx} n - \rho C_p \left(v_x - v_{x,ff}\right)\right] L \tag{26}$$

$$f_y = \left[\sigma_{xy} n - \rho C_s \left(v_y - v_{y,ff}\right)\right] L \tag{27}$$

where: σ_{xx} is the horizontal stress of the free-field element; σ_{xy} is the shear stress of the free-field element; n is -1 or 1 for the left and right lateral boundaries, respectively; ρ is the density of the rock mass; C_s is the propagation velocity of a transverse wave in a continuous medium; C_p is the propagation velocity of a longitudinal wave in a continuous medium; v_x and v_y are the main mesh velocity in the x- and y-directions, respectively; $v_{x,ff}$ is the velocity of the free-field in the x-direction; $v_{y,ff}$ is the velocity of the free-field in the y-direction; and L is the point's area of influence.

The main advantage of using the free-field method compared to the second solution is the simultaneous application of the vertical and horizontal components of the seismic load. Another advantage is the absorption of the reflected wave which is incident on the sides of the model, which is only possible at the base.

The well-known Westergaard (1933) added mass method is used to represent the dynamic dam-reservoir interaction. Experimental evidence has shown that this solution reproduces the decrease in dams' natural frequencies due to the presence of a reservoir with reasonable accuracy (e.g. Kuo, 1982). Formulations based on fluid elements that provide more precise representations of the fluid-structure interaction have also been used, mainly for arch dams in which the hydrodynamic effects of the reservoir are more relevant. In this approach, the additional mass due to the reservoir is added to the mass of the points that define the upstream face.

Rayleigh damping proportional to the dam's mass and stiffness was used to model the dissipation of energy following a seismic event. In cases where neither the mass or stiffness components is null, the Rayleigh damping is almost constant for a frequency range around the minimum of the damping curve. The damping is adjusted to reflect the dominant frequencies associated with the problem at hand. The damping component proportional to the mass is taken into account in the equations of motion. The stiffness-proportional component is considered by adding damping forces to the normal and tangential contact forces, and to the nodal forces of each finite element. The Rayleigh damping proportional to the stiffness should only be applied in the contact forces when the joint presents an elastic behaviour, since the corrections applied by the constitutive model already include some energy dissipation. For numerical stability of the explicit algorithm, the time step should be reduced when damping proportional to the stiffness is used. According to Belytschko (1983), the adjusted time step is given by,

$$\Delta t_{dyn} = \Delta t_{sta}\left(\sqrt{1+\lambda^2} - \lambda\right) \tag{28}$$

where Δt_{dyn} is the dynamic time step when considering damping proportional to the stiffness; Δt_{sta} is the static (reference) time step; and λ is the relative damping coefficient for the maximum frequency. While reducing Δt may be very costly in terms of computational performance, the stiffness-proportional component is important in nonlinear analyses such as those required to model the failure of gravity dams, particularly via the sliding mechanism. In sliding failures, the dam fails gradually in a way that progressively reduces the structure's natural frequency. The damping proportional to the stiffness decreases when the damping proportional to the mass increases (Hall, 2006). In addition, the damping proportional to the stiffness seems to be appropriate as it is applied directly to the dam-foundation interface, while the damping proportional to the mass is applied to the entire structure.

The Fully Coupled Hydromechanical Model

The development of the hydromechanical model is based on the assembly of a flow mesh consisting of flow channels and hydraulic nodes (Bretas et al., 2013). The hydraulic model is superimposed on the mechanical model. The flow channels correspond to the face-to-face contacts, while the hydraulic nodes (which are associated with one or more element nodes) correspond to sub-contacts or a set of sub-contacts around the node. The flow rates are calculated at the flow channels and the pressures are calculated at the hydraulic nodes. If we consider contact $CT_{1,2}$ as shown in Figure 2 and Figure 5, the flow channel overlaps the mechanical model and is designated FC_1; the flow rate q_1 is calculated at this point. At the ends of the flow channel FC_1, pressures P_1 and P_2 are established in the hydraulic nodes HN_1 and HN_2, respectively. The hydraulic node HN_1 consists of the sub-contacts $SCT_{1.3.1}$, $SCT_{2.3.2}$ and $SCT_{1.2.3}$ where the flow rates q_1, q_2 and q_3 converge and will be summed to determine the pressure P_1. The pressures P_1 and P_2, will act on the faces of the channel.

The stress generated by the flow acting on the walls of the joints (whose magnitude is determined by summing all of the acting forces) will create a new opening, changing the dam's permeability and thus the flow conditions. Four different levels of coupling can be assumed. Level 0 corresponds to uncoupling, in which the hydraulic aperture is constant and the fluid does not act mechanically on the discontinuities. Level 1 (Figure 6) involves hydraulic aperture that varies according to the mechanical opening, with no evidence of fluid pressure acting on the discontinuities. Level 2 involves constant hydraulic aperture with the fluid acting mechanically on the discontinuities. Finally, Level 3 involves hydraulic aperture that varies according to the mechanical opening while the fluid acts on the discontinuities.

The hydraulic calculation begins with the determination of the hydraulic aperture for each flow channel. This parameter defines the channel's permeability and plays an important role in determining the total flow rate. The definition of the numerical hydraulic aperture (a_h) requires three reference values: a lower limit, referred to as the residual opening (a_{res}), which is the minimum permeability that exists

Figure 5. The hydromechanical model

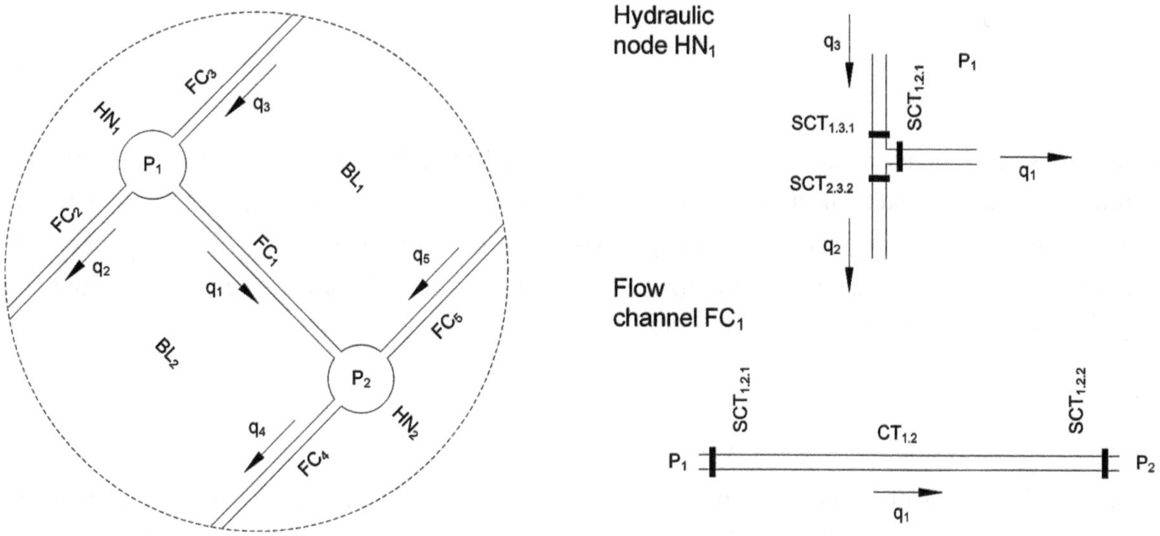

Figure 6. Hydromechanical coupling levels

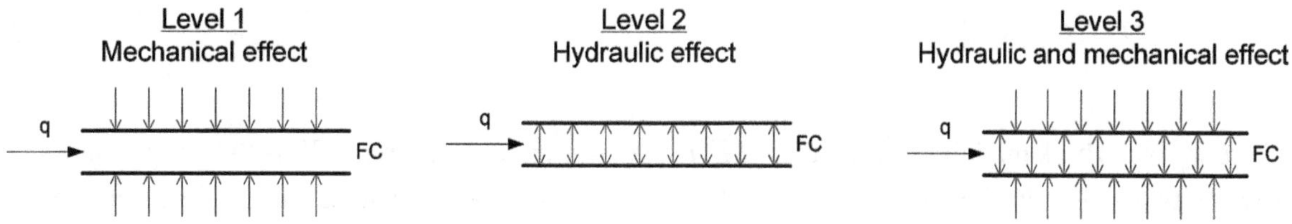

even when the joints are highly compressed; an upper limit, designated the maximum opening ($a_{\max}$), a value that cannot be exceeded for numerical stability reasons; and finally, the nominal hydraulic aperture (a_0), which reflects the flow in an initial normal stress state. The hydraulic aperture varies according to the mechanical opening; it will be smaller than the characteristic opening when the discontinuity is in compression and the mechanical opening is negative. Similarly, the hydraulic aperture will be larger than the characteristic opening when the discontinuity is in tension and the mechanical opening is positive. The size of the mechanical opening varies linearly with the normal stiffness (k_n) of the joint, which makes this an important parameter for the hydromechanical coupling. In an uncoupled calculation, the hydraulic aperture does not change, so the nominal hydraulic aperture value is used as a reference data point. The determination of the flow rate depends on the available energy established by the hydraulic gradient, in accordance with the permeability of the media. In a laminar flow model between two plates (Bear, 1988), the flow rate is given by the “cubic flow law”, as

$$Q = \frac{1}{12\mu} a_h^3 \frac{\Delta h}{L} \tag{29}$$

$$\Delta h = \left| h_1 - h_2 \right| \tag{30}$$

$$h_i = P_i + y_i \rho_w g \tag{31}$$

where μ is the dynamic viscosity of the fluid; Δh is the head difference; L is the length of the discontinuity; h_i is the head of the hydraulic node i; P_i is the pressure of the node i; y_i is the y-coordinate of the node i; ρ_w is the fluid density; and g is the acceleration of gravity. In the next step, new pressures at the hydraulic nodes are determined. For each hydraulic node, a balance of the positive and negative flow rates is computed from the sub-contacts, as

$$Q_{hn} = \sum Q_{sct} \tag{32}$$

where Q_{hn} is the sum of the input and output flows in the hydraulic node; and Q_{sct} is the flow contribution from a convergent flow channel. The pressure variation depends on the net flow and, for compressible fluids, is given by

$$\Delta P = \frac{k_w}{V}\left(Q_{hn}\Delta t + \Delta V\right) \tag{33}$$

where k_w is the bulk modulus of the fluid; ΔP is the differential pressure; ΔV is the volume variation between two consecutive cycles ; V is the volume of the flow channel; and Δt is the time step. For the steady flow analysis, the volume change between two consecutive cycles does not need to be considered. Only the final pressure, after the convergence of the model to an equilibrium state, is relevant and this equation takes the form

$$P = P_0 + k_w \frac{Q_{hn}\Delta t}{V} \tag{34}$$

where P is the pressure in the current time step; and P_0 is the pressure in the preceding time step. Finally, all hydraulic sub-contacts associated with the node under consideration inherit the same pressure, which is used to define the hydraulic forces. Based on a face-to-face contact with two sub-contacts SCT_1 and SCT_2 subject to the pressures P_1 and P_2, respectively, the pressures form a trapezoidal diagram acting on the faces of the blocks, in the normal direction, in conjunction with existing contact forces. The hydraulic forces are determined by integrating the pressure diagram, according to,

$$f_{hyd,1} = \left(P_1 + \frac{P_2}{2}\right)\frac{L}{3} \tag{35}$$

$$f_{hyd,2} = \left(\frac{P_1}{2} + P_2\right)\frac{L}{3} \tag{36}$$

where $f_{hyd,1}$ is the hydraulic force to node 1 and $f_{hyd,2}$ is the hydraulic force to node 2. In addition to the confined flow case, this formulation allows the flow analysis of the free surface condition. In order to handle this type of boundary, additional precautions to avoid negative numerical pressures that may arise are mandatory. It is necessary to progressively reduce the flow rate towards a zero pressure node to obtain convergence to a balanced equilibrium solution. Generally, if $h_1 > h_2$ and $y_1 > y_2$, but $P_1 < (y_1 - y_2)\rho_w g$, then the flow rate should be reduced by a certain factor (f_{red}) to provide a numerically smooth representation of the free-surface condition by simulating a gradual evolution to an unsaturated state,

$$f_{red} = \frac{P_1}{(y_1 - y_2)\rho_w g} \tag{37}$$

As with the mechanical calculations, the duration of the hydraulic time step must be limited to achieve numerical stabilization of the explicit algorithm. The hydraulic time step for a given hydraulic node is directly proportional to its volume and inversely proportional to the conductivity of the flow channels connected to it, and is given by

$$\Delta t_h = \min\left(\frac{\sum V_i}{k_w \sum k_i}\right) \tag{38}$$

$$k_i = \frac{1}{12\mu} a_{h,i}^3 \frac{1}{L_i} \tag{39}$$

where Δt_h is the hydraulic time step; V_i is the volume of the flow channel i; k_i is the conductivity of the flow channel i; and L_i is the length of the flow channel i. For steady state flow analyses, volumes can be scaled to obtain a given time step,

$$V_i = \Delta t_{\min} k_w \sum k_i \tag{40}$$

where V_s is the scaled volume of the flow channel and $\Delta t_{\min}$ is the minimum time step of all hydraulic nodes in the system. In steady state hydromechanical analysis, hydraulic calculations are carried out simultaneously with the mechanical calculation. In such cases it is convenient to adopt a common time step, which means that the volumes should be scaled in relation to the minimum time step, which may be either the mechanical or the hydraulic time step. For true transient analysis, different time steps must be used. Hydraulic boundary conditions and other restraints must be adequately modelled. The lateral and basal boundaries of the foundation are generally considered impermeable. A fixed pressure equivalent to the level of the reservoir is applied on the surface of the foundation upstream and downstream of the dam. Triangular or trapezoidal pressure diagrams are then applied on the dam's upstream and downstream faces, while a free surface flow occurs inside the masonry dam. Therefore, the top face of the crest and the downstream face of the dam are assumed to have permeable boundaries. Another important factor is the ability to model drainage systems and grout curtains. A drainage system is represented by fixing the hydraulic pressure in the nodes located in the alignment of the drains, varying in depth according to the hydrostatic gradient. It is also possible to account for the discharge flowing out of the drains. In such analyses, the pressures are obtained via flow analysis. To model the grout curtain, the conductivity of the hydraulic channels within the grouted region is reduced. The final important aspect to consider in the hydromechanical modelling of masonry gravity dams relates to earthquake analysis. During an earthquake, the water pressure in pre-existing water-filled cracks can be assumed to remain constant (Ebeling, Nuss, & Tracy, 2000). In practice the flow calculation is interrupted and the hydraulic pressures already in place remain unchanged.

NUMERICAL MODELLING OF REAL MASONRY DAMS USING THE DISCRETE ELEMENT METHOD

Five case studies on gravity masonry dams are presented, comprising eight numerical analyses. Three of the numerical analyses examine the discontinuous nature of the dam body, focusing on the loss of cohesion scenario, the crack propagation scenario and the seismic performance of a masonry dam. The remaining four analyses deal with stress analysis, global stability, the permanent sliding resulting from a seismic event and hydromechanical analysis.

Static Analysis of a Masonry Gravity Dam using Continuous and Discontinuous Models

The actual failure of a gravity dam is a complex process involving overstress, sliding and overturning scenarios. For the sake of simplicity, these scenarios are tested separately. Three models were developed for the static analysis of a masonry gravity dam with the aim of identifying local and global failure modes (Bretas, Lemos, & Lourenço, 2014). The modelled dam is about 31 m high, with a base 23 m wide. The first analysis to be performed was a stress analysis. A macroblock-based equivalent continuous model of the dam and its foundation was adopted (Figure 7a), with the assumption of an elastic joint between the dam and the foundation. The foundation was represented by a mesh 47.4 m high and 86.5 m wide, in which a box boundary condition was applied, i.e. at the foundation base, the model is free to move in the horizontal direction, and at the lateral faces of the foundation, the model is free to move in the vertical direction. The loads considered in this model were the self-weight and the hydrostatic pressure. The stress field was represented through the main stresses. These loads place the dam under compression, which peaks at -0.96 MPa in the downstream toe (Figure 7b). The safety factor is defined on the basis of the main stresses and the material properties according to the Mohr-Coulomb criterion. Figure 7c shows the dam's safety factors based on an assumed friction angle of 55°, compressive strength of 10 MPa, tensile strength of 0.25 MPa, and cohesion of 0.79 MPa. The main failure mode is shear failure, and the lowest safety factor (1.9) occurs in the downstream toe. Failure by overstress is not a major concern when considering a gravity masonry dam under normal conditions.

Figure 7. Stress analysis of a masonry gravity dam

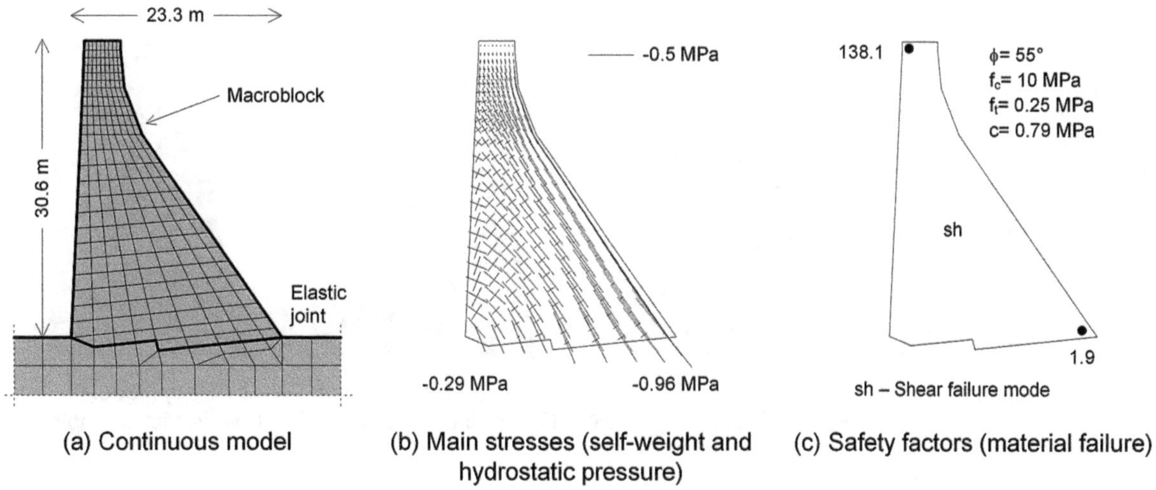

The structural problems of masonry dams result from aging and are different to those that occur in concrete dams. Most of the problems relate to a loss of cohesion in the masonry due to the chemical and physical consequences of exposure to water. The second model was developed to examine this problem (Figure 8a). It featured two block types: quadrangular blocks, which were used to model the upstream and downstream faces; and Voronoi blocks, which were used to model the masonry structure in the dam's body. Each Voronoi block consists of triangular elements with one common point that is coincident with the centroid of the Voronoi element. The dam is subject to its own self-weight and hydrostatic pressure, as well as the internal pressure due to seepage through the masonry joints. The internal pressure was established by means of a hydromechanical analysis. The boundary condition was applied directly at the base of the dam, without including a foundation block. A non-linear constitutive model was used to describe the interactions between the masonry blocks. A parametric study was conducted in which the friction angle was reduced and the consequences of setting the cohesion and tensile strength to zero were considered. Three different discretization schemes were tested, referred to as Models A, B and C. For Model A, a failure on the crest occurred at a friction angle of 18° (Figure 8b). A similar failure mechanism occurred under Model B at a friction angle of 17° (Figure 8c). Under Model C, a global collapse of the dam occurred at a friction angle of 15° (Figure 8d). It should be noted that the discretization scheme used can influence the failure mechanism observed in non-linear analyses of this sort. It is therefore advisable to conduct multiple analyses using different meshes to assess the robustness of one's findings.

The importance of considering global dam sliding and overturning mechanisms during safety assessments of gravity dams is widely acknowledged. These failure mechanisms are induced by discontinuities that extend through the dam body into the foundation and the underlying rock mass. The model used to assess failures of this sort was similar to the first model (Figure 7a) except that it assumed non-linear behaviour in the joint between the dam and the foundation. Both dam and foundation were modelled using a continuous macroblock approach. For the sliding scenario, two failure surfaces were considered. In the first case, sliding occurs at the interface between the dam and the foundation (Figure 9a). In the second, the failure involves part of the foundation close to downstream toe (Figure 9b). The conditions of null tensile strength and null cohesion were adopted, and the friction angle was reduced stepwise from an initial value of 45°. In the first case, the dam started to slide when the friction angle reached

Figure 8. Assessment of a loss of cohesion scenario in a masonry dam

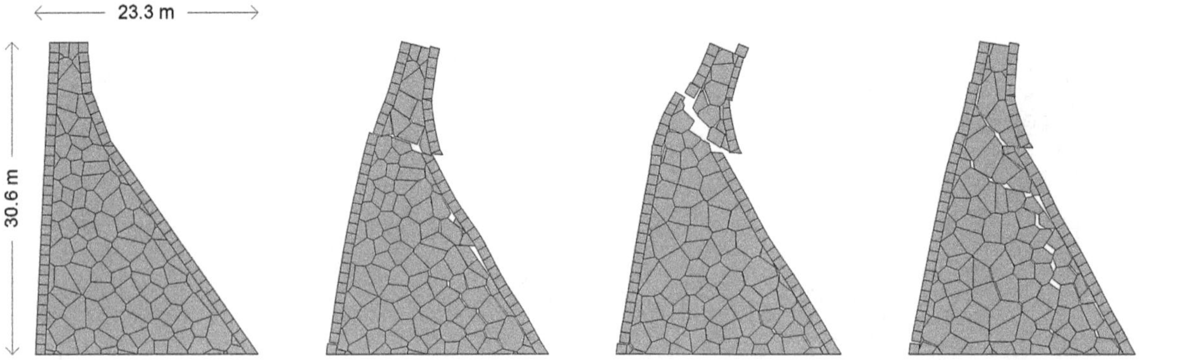

(a) Discontinuous model (b) Model A Failure friction angle: 18º (c) Model B Failure friction angle: 17º (d) Model C Failure friction angle: 15º

27°, yielding a sliding safety factor (SSF) of 2.0. In the second case, the dam started sliding at a friction angle of 25°, giving an SSF of 2.1. The potential for the dam to overturn around its downstream toe was analysed by considering a flood scenario. The dam proved to be stable even when the water level was raised over 5 m above the normal design reservoir elevation. The final crack in the upstream heel and the compressive stress at the downstream toe under these circumstances are illustrated in Figure 9c.

Analysis of Potential Failures Mechanisms in an Old Masonry Gravity Dam

The failure of a masonry gravity dam through the dam body was investigated (Bretas, Batista, & Lemos, 2015). The dam was located at a high elevation (ca. 1575 m) with harsh winters. It is a gravity dam with a maximum height of 32.5 m above the foundation. The upstream and downstream faces were made

Figure 9. Static analysis of a masonry gravity dam assessing the failure mechanisms of global sliding and overturning

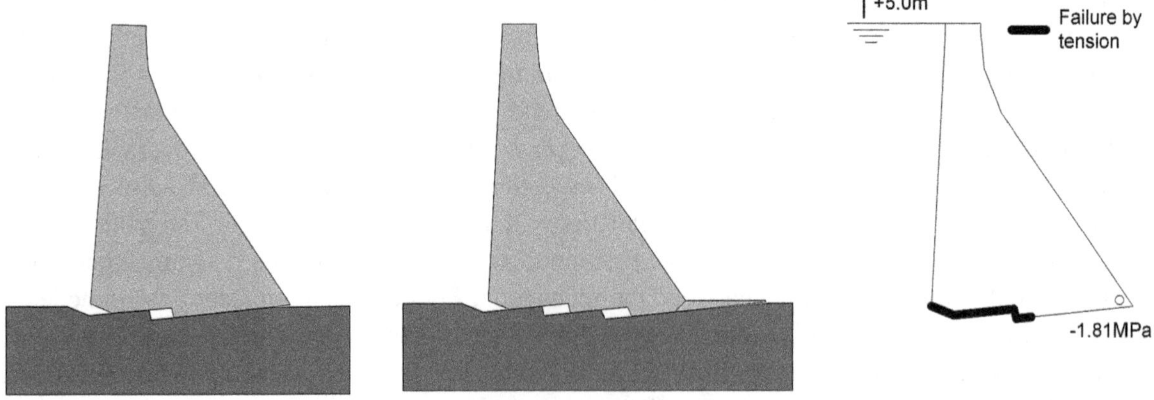

(a) Sliding of the dam (b) Sliding of the dam and part of the foundation (c) Stress state in case of flooding of rectangular masonry blocks connected with cement mortar, while the inside is filled with irregular masonry blocks and hydraulic lime mortar. The analysis was intended to identify potential dam failure mechanisms arising from a loss of masonry cohesion. The model of the dam's highest section is shown in Figure 10a. The upstream and downstream faces were discretized using quadrangular blocks, while the interior of the dam body was represented by Voronoi elements. The size of these elements does not match that of the actual blocks but was sufficiently close to enable investigation of key failure modes. The foundation is represented by a single rigid block, since the analysis focused on failure mechanisms occurring in the dam body and the foundation plane. Three crack types in the upstream face were considered, at 1/4, 1/2 and 3/4 of the dam's total height; each crack was initially 2.5 m long (Figure 10a). A friction angle of 35° was adopted for all joints. A null tensile strength and cohesion was assumed for all joints other than those between the blocks in the upstream and downstream faces, where a tensile strength of 1 MPa and cohesion of 2 MPa was adopted. The loads were applied sequentially - first the self-weight, then the hydrostatic load, and finally the internal pressure and the uplift. The internal pressure and the uplift were obtained from the hydromechanical analysis of the dam. For these loads, the model was stable, so the dam was considered safe. Failure mechanisms would only be expected in the event of a change in loading conditions. If the hydrostatic pressure were to increase as a consequence of flooding, two modes of failure are possible. The first was observed when the dam-foundation joint was assumed to be nonlinear. In this case, the dam slid through the foundation plane in the downstream direction (Figure 10b). The second failure mechanism was observed when the joint between the dam and the foundation was assumed to be elastic (Figure 10c). In this case the failure originated from the first crack, close to the upstream heel. Additional calculations were performed to evaluate the effect of the ice load using this model, with an elastic joint between the dam and the foundation. The ice load was modelled as a distributed horizontal force based on the reservoir level. In this case, the failure originated from the uppermost crack, close to the crest (Figure 10d).

Analysis of Crack Propagation in the Dam Body of a High Masonry Gravity Dam

One important objective in the design of a gravity dam is to avoid tensile stresses in any region of the dam body or at the contact surface between the dam and the foundation. The origin of a crack can

Figure 10. Potential failure mechanisms of an old masonry gravity dam

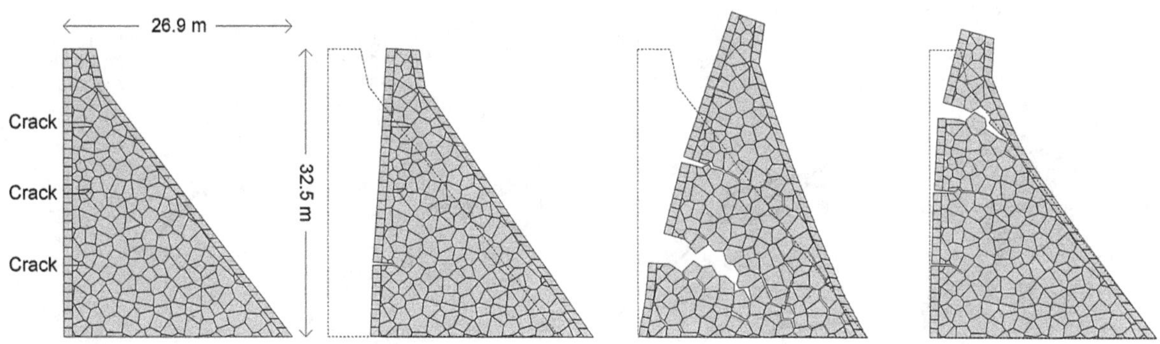

(a) Discontinuous model (b) Sliding (flooding) (c) Cracking (flooding) (d) Cracking (ice load) generally be linked to unexpected loads such as earthquakes, or construction deficiencies. The crack's subsequent propagation is typically due to water seepage and incremental uplift inside the discontinuity. Figure 11 shows the results of a crack propagation analysis for a large (82 m tall) masonry gravity dam with a severe crack extending from the upstream face to the downstream toe (Bretas, Lemos, Léger, & Lourenço, 2011). The purpose of this study is reproduce the behaviour of the existing dam, and simulate the cracking process, testing different material properties. The dam model is based on triangular elements laid (Figure 11a) over a foundation that was modelled as a continuous elastic macroblock. The dam's loading includes its self-weight and the hydrostatic pressure on the upstream side. An elastic joint at the dam-foundation interface was assumed. Three different models (A, B, and C) governing the interactions between the triangular elements of the dam model were considered, each of which yielded distinct failure mechanisms as shown in Figure 11, according to the joint properties. In all models, the crack was initiated on the upstream face as a consequence of reductions in properties relating to the first numerical contact, and developed as the internal pressure between the blocks was updated to simulate the effects of water seepage. This ultimately led to complete failure. Model C seemed to give the best description of the phenomenon, giving a simulated crack path propagation that closely resembled that seen in the real dam.

Seismic Analysis of a Masonry Gravity Dam

Ensuring dam safety in the face of seismic events is a major concern in many countries. It is essential to ensure that dams are resistant to both global and local failures. A standard test for resistance to global failure involves assessment of permanent sliding. Therefore, a model for the seismic analysis of a masonry gravity dam focusing on permanent sliding was developed (Bretas, Lemos, et al., 2015). The model is based on two continuous elastic macroblocks representing the dam body and the foundation rock mass (Figure 12a). A nonlinear connection between the dam and the foundation was assumed, with null cohesion and null tensile strength, and a friction angle of 45°. The dynamic load was imposed using a free-field boundary condition. The dynamic load was applied by means of shear and vertical stress histories acting directly on the base of the block foundation and the bases of the free-field columns. It should be

Figure 11. Analysis of crack propagation on the dam body of a tall masonry gravity dam

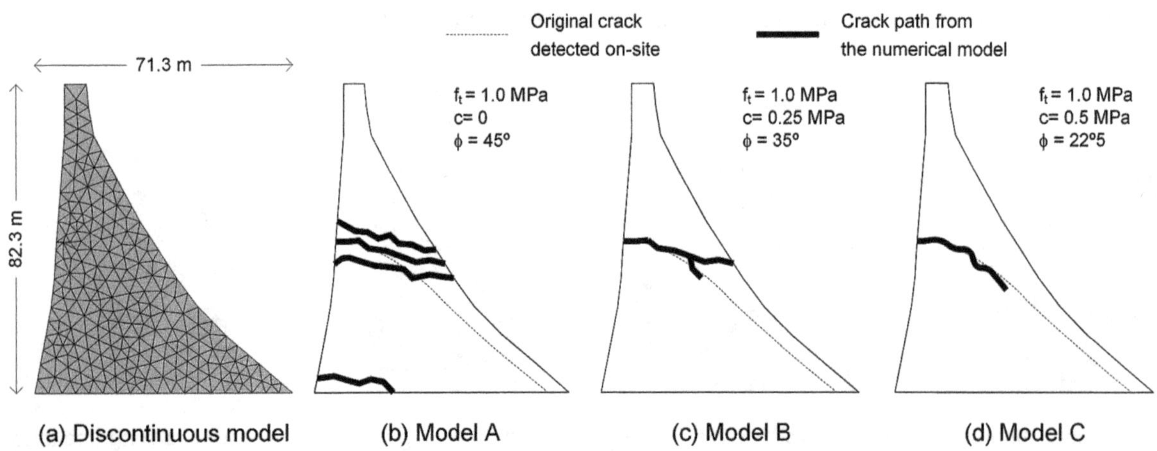

noted that there are various oversimplified solutions that could be used instead of the free-field scheme. For example, a velocity history could be applied directly on the nodes of the foundation block, which assumes a rigid behaviour; or a stress history could be applied at the base of a deformable foundation, in combination with viscous boundaries. In addition to the seismic loads, the model accounted for static loads including the self-weight, the hydrostatic pressure, and the uplift. The dam's modelled displacement history is plotted in Figure 12 for various assumptions including a rigid foundation, a deformable foundation, and the free-field boundary conditions with acceleration in both horizontal and vertical directions. The maximum permanent displacement of around 10 cm was predicted with the rigid foundation model.

Pre-existing discontinuities within the body of the dam should be considered when evaluating seismic stability, especially in the case of masonry dams (Bretas, Lemos, et al., 2015). To this end, the analysis developed in the previous section was repeated, replacing the dam mesh with a discontinuous and non-linear model that allowed the joint between the dam and the foundation to behave elastically. The model was based on quadrangular elements arranged in vertical layers (Figure 13a). This approach was intended to reflect the structural features of a dam that had undergone a series of heightening works. In the dam's most recent rehabilitation programme, a new concrete plate was erected on the upstream face, which was also considered in the model. The load combination included the self-weight, the hydrostatic pressure, the uplift and the seismic load (which was modelled using the free-field boundary condition). During the seismic analysis sliding movements were detected between the horizontal masonry layers, as shown in Figure 13b. No internal pressure was applied within the body of the dam, which would reduce the safety implications of the observations. Many regulatory bodies strongly recommend the performance of post-seismic analyses to identify damage that may have been caused by an earthquake. In the studied case, the simulated earthquake was found to have increased the uplift on the dam body as a result of seepage through masonry discontinuities and cracks.

Figure 12. Dynamic analysis of a masonry gravity dam focusing on potential permanent sliding

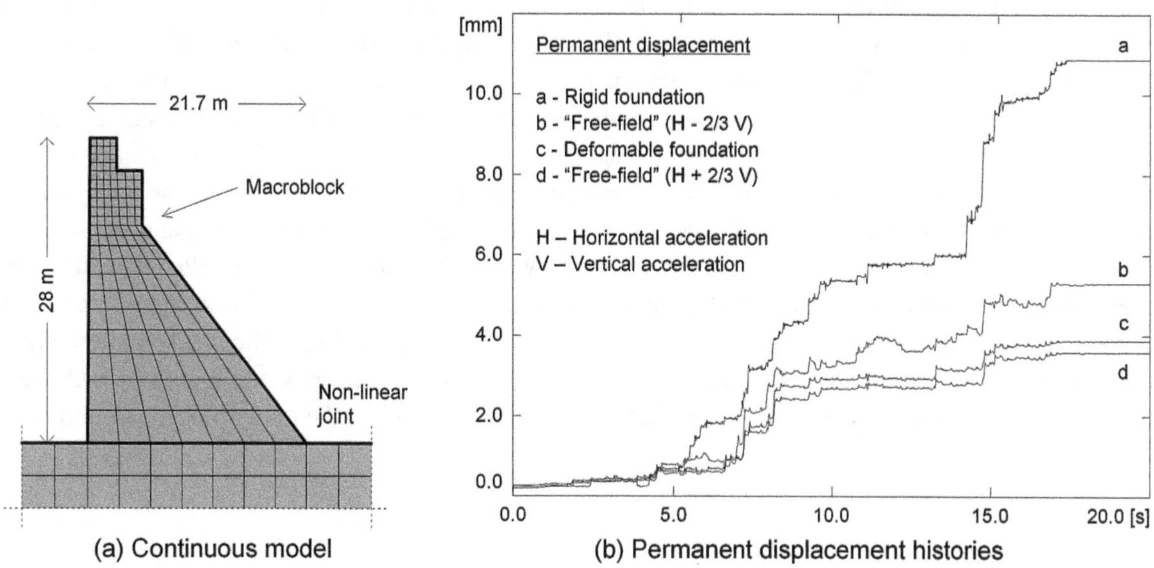

Figure 13. Seismic analysis of a masonry gravity dam considering the discontinuities on the dam body

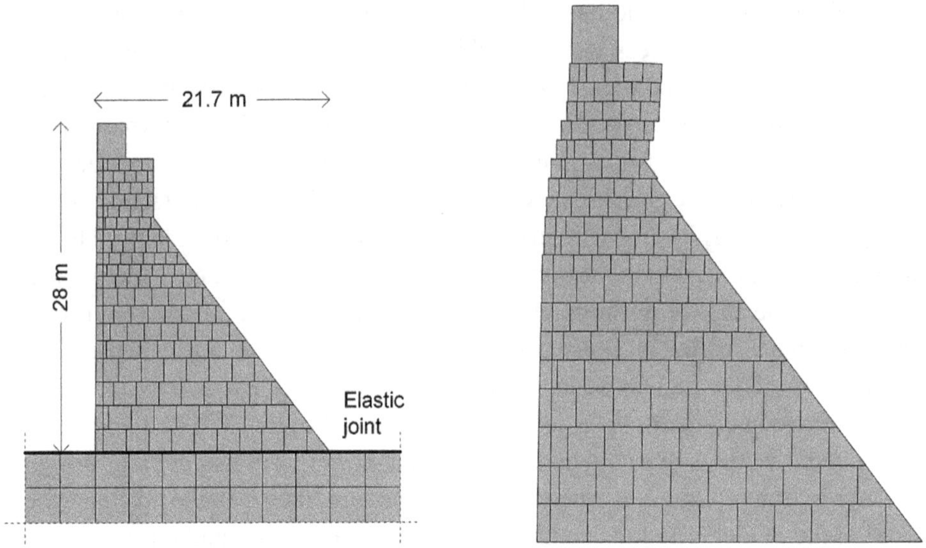

(a) Discontinuous model (b) Failure mechanism

Assessment of a Rehabilitation Project to Improve the Hydraulic Behaviour of a Masonry Dam and its Foundation

Many rehabilitation works carried out on masonry dams are prompted by seepage problems. As mentioned previously, reducing the dam's permeability on the water-facing side tends to be the primary goal of such efforts. The seepage of water from the reservoir has short term consequences for stability because of the internal pressure of the water on the dam body. However, it also has long-term consequences that reduce the dam's structural integrity. For this reason, studies on these phenomena must include coupled analyses with linked mechanical and hydraulic calculations. Figure 14 presents an example of this type of analysis (Bretas et al., 2013). It shows results obtained during the assessment of a rehabilitation project in which grouting was applied to the dam's foundation and through the dam's body together with the opening of new drains in the foundation and the dam, as well as the excavation of a new drainage gallery. A continuum elastic model approach was used to model the dam and its foundation, and its hydraulic properties were simulated on the basis of both its known pre-rehabilitation permeability and its predicted permeability after the completion of the rehabilitation works. The drains were modelled by setting the pressures at the nodes along their development. The pressure inside each drain is assumed to be equal to the hydrostatic pressure. The changes in the hydraulic performance due to the rehabilitation works are indicated by the equipotential lines (Figure 14a). Uplift diagrams are shown in Figure 14b. Overall, the results obtained indicate that the rehabilitation works should reduce the total uplift by around 60%.

Figure 14. Hydromechanical analysis of a masonry gravity dam showing the predicted consequences of rehabilitation measures

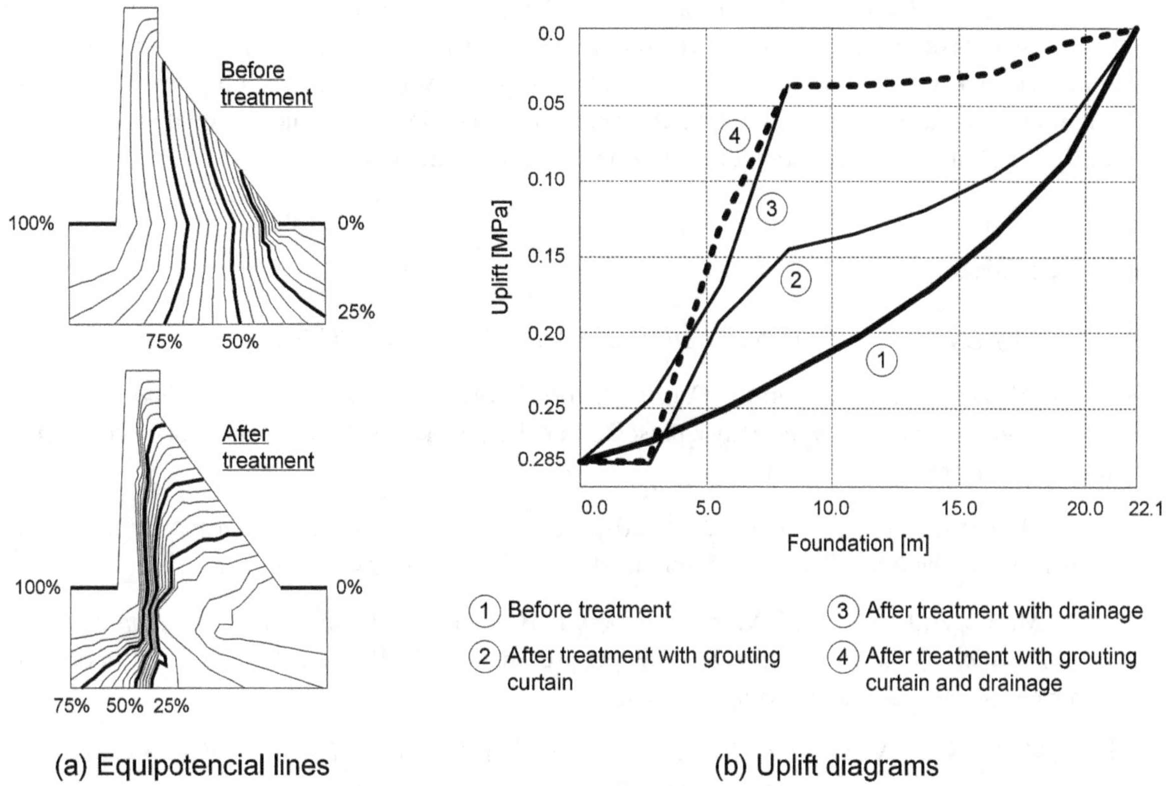

CONCLUSION

This work focuses on the numerical modelling of masonry dams using the Discrete Element Method. A numerical tool for the structural analysis of masonry dams based on this method was developed and applied in five case studies of existing masonry dams. The tool has specific modules for modelling static, seismic, and hydraulic loads; determining boundary conditions; and creating graphs and other means of visualizing numerical results. It is capable of modelling dams and their behaviour using discontinuous meshes, as demonstrated in the case studies. Voronoi elements were used to study loss of cohesion scenarios, while triangular elements were used in the crack propagation model. In addition, a discontinuous model featuring rectangular elements was used to conduct seismic analyses. The results of the analyses demonstrate that it is important to tailor the representation of the masonry dam being studied to the purpose of the analysis. When studying failure mechanisms involving masonry discontinuities, it is clearly essential to use a discontinuous model. However, continuum models could be viable alternatives for some purposes, such as analyses of global stability problems. The tool presented herein has the unique and valuable ability to integrate continuous and discontinuous meshes in a single model. In the case studies, static, hydromechanical and seismic calculations were executed in conjunction with evaluations of structural reinforcements. These calculations were performed sequentially, using a single numerical model. The tool enabled all of the calculations to be performed using the same data model in an integrated and consistent manner, which is a clear advantage relative to existing traditional methods.

The results presented herein demonstrate that the Discrete Element Method is a powerful tool for safety assessments of masonry dams and can provide direct insight into their main failure mechanisms. In future, the tool will be developed further to allow for other ways of representing masonry and its discontinuities, the use of different failure criteria for the material under study (including alternative fracture criteria), and the application of alternative constitutive models.

REFERENCES

Bear, J. (1988). *Dynamics of fluids in porous media*. New York: Dover Publications.

Bettzieche, V., Deutsch, R., & Heitfuss, C. (2004). *100 Years of experience in ageing of masonry dams and life-time-based rehabilitation*. Paper presented at the 2nd International Conference on Lifetime Oriented Design Concept, Bochum.

Bretas, E. M., Batista, A. L., & Lemos, J. V. (2015). *Stability analysis of an old masonry gravity dam*. Paper presented at the XXX Seminário Nacional de Grandes Barragens, Foz do Iguaçu.

Bretas, E. M., Léger, P., Lemos, J. V., & Lourenço, P. B. (2010). *Analysis of a gravity dam considering the application of passive reinforcement*. Paper presented at the II International Congress on Dam Maintenance and Rehabilitation, Zaragoza, Spain.

Bretas, E. M., Lemos, J. V., Léger, P., & Lourenço, P. B. (2011). *Structural analysis of the rehabilitation works of Bhandardara dam*. Paper presented at the XXVIII Seminário Nacional de Grandes Barragens, Rio de Janeiro.

Bretas, E. M., Lemos, J. V., & Lourenço, P. B. (2010). *Masonry Gravity Dams - A numerical application for stability analysis*. Paper presented at the 8th International Masonry Conference, Dresden, Germany.

Bretas, E. M., Lemos, J. V., & Lourenço, P. B. (2012). Masonry dams: Analysis of the historical profiles of Sazilly, Delocre, and Rankine. *International Journal of Architectural Heritage*, *6*(1), 19–45. doi:10.1080/15583058.2010.501399

Bretas, E. M., Lemos, J. V., & Lourenço, P. B. (2013). Hydromechanical analysis of masonry gravity dams and their foundations. *Rock Mechanics and Rock Engineering*, *46*(2), 327–339. doi:10.1007/s00603-012-0305-3

Bretas, E. M., Lemos, J. V., & Lourenço, P. B. (2014). A DEM based tool for the safety analysis of masonry gravity dams. *Engineering Structures*, *59*(Feb), 248–260. doi:10.1016/j.engstruct.2013.10.044

Bretas, E. M., Lemos, J. V., & Lourenço, P. B. (2015). Seismic analysis of masonry gravity dams using the Discrete Element Method: Implementation and application. *Journal of Earthquake Engineering*.

Comité Français des Barrages et Réservoirs (CFBR). (2013). *Guidelines for the justification of the stability of gravity dams*. French Dams and Reservoirs Committee.

Dolezalova, M. (2004). Numerical analysis of an old masonry dam using UDEC. In Konietzky (Ed.), Numerical Modeling of Discrete Materials (pp. 269-277). London: Taylor & Francis Group.

Ebeling, R. M., Nuss, L. K., & Tracy, F. T. (2000). *Evaluation and comparison of stability analysis and uplift criteria for concrete gravity dams by three federal agencies*. Washington, DC: Academic Press.

Hall, J. F. (2006). Problems encountered from the use (or misuse) of Rayleigh damping. *Earthquake Engineering & Structural Dynamics*, *35*(5), 525–545. doi:10.1002/eqe.541

International Commission on Large Dams (ICOLD). (1994). *Ageing of dams and appurtenant works: review and recommendations (Bulletin 93)*. Paris: International Commission on Large Dams.

International Commission on Large Dams (ICOLD). (2000). *Rehabilitation of dams and appurtenant works: State of the art and case histories (Bulletin 119)*. Paris: International Commission on Large Dams.

International Commission on Large Dams (ICOLD). (2013). *Historical review on ancient dams (Bulletin 143)*. Paris: International Commission on Large Dams.

International Commission on Large Dams (ICOLD). (n.d.). *Dam Safety Management: operational phase of the dam life cycle (Bulletin 154)*. France: International Commission on Large Dams.

Kuo, J. S. H. (1982). *Fluid-structure interactions: added mass computations for incompressible fluid (UCB/EERC-82/09)*. Berkeley, CA: Earthquake Engineering Research Centre, University of California.

Lemos, J. V., & Cundall, P. A. (1999). Earthquake analysis of concrete gravity dams on jointed rock foundations. In V. M. Sharma, K. R. Saxena, & R. D. Woods (Eds.), Distinct element modelling in geomechanics (pp. 117-143). New Delhi: Oxford & IBH Publishing CO. PVT. LTD.

Lourenço, P. B. (1996). *Computational strategies for masonry structures. (PhD)*. Delft University of Technology.

Lysmer, J., & Kuhlemeyer, R. L. (1969). Finite dynamic model for infinite media. *Journal of the Engineering Mechacnics Division*, 859-877.

Sauvé, R. G., & Metzger, D. R. (1995). Advances in dynamic relaxation techniques for nonlinear finite element analysis. *Journal of Pressure Vessel Technology*, *117*(2), 170–176. doi:10.1115/1.2842106

Schnitter, N. J. (1994). *A history of Dams – The useful pyramids*. Rotterdam: A.A. Balkema.

Smith, N. (1971). *A history of dams*. London: Peter Davies.

US Bureau of Reclamation (USBR). (1976). *Design of gravity dams*. Denver, CO: Bureau of Reclamation, US Department of the Interior.

Wegmann, E. (1888). *The desing and construction of dams*. New York: John Wiley & Sons.

Westergaard, H. M. (1933). Water pressure on dams during earthquakes. *Transactions of the American Society of Civil Engineers*, *98*, 418–433.

Wittke, W., Wittke, M., & Kiehl, J. R. (2012). Interaction of a masonry dam and the rock foundation. *Geotechnical and Geological Engineering*, *30*(3), 581–601. doi:10.1007/s10706-012-9493-6

Chapter 9
Discrete Element Modeling of Masonry-Infilled Frames

A. Mohebkhah
Malayer University, Iran

Vasilis Sarhosis
Newcastle University, UK

ABSTRACT

In this chapter, the different modeling strategies for simulating the behavior of masonry infilled frames are investigated. Particular emphasis is given on the suitability of the Discrete Element Method (DEM) to accurately represent the mechanical behavior, strength and ductility of concrete and brickwork masonry infilled frames. Within DEM, masonry infill panels are represented by individual bricks and blocks separated by zero thickness interfaces representing mortar joints. The assumptions adopted, the numerical implementation and the advantages and disadvantages of modeling masonry infilled frames using the discrete element method are discussed. This 'discontinuum' approach, an alternative to modeling masonry as a homogenized continuum, is particularly suited for studying the mechanical behavior and interaction between the individual masonry brick/blocks and their interaction with the framed structure.

INTRODUCTION

Building frames are usually infilled with masonry walls. Infilled frames have high in-plane strength and stiffness compared to the corresponding bare frames (no infill). At low levels of lateral forces, the surrounding frame and infill panel behave as a monolithic composite frame. However, as the levels of lateral load increases, an interaction between the surrounding frame and infill panel occurs. The frame behaves in flexural mode and the infill panel deforms in a shear mode. Also, the provision of the infill increases the later stiffness and strength of the frame. However, in current practice the structural effect of infill walls is not taken into account during the analysis of building frames and leads to inaccurate predictions of the building's stiffness, strength and ductility.

Over the last 60 years, extensive numerical and experimental studies have been performed on lateral load behavior of masonry-infilled frames. Experimental and analytical investigations on the lateral stiff-

DOI: 10.4018/978-1-5225-0231-9.ch009

ness and strength of steel frames infilled with masonry panels have been carried out by Stafford-Smith (1962 and 1966), Mallick and Garg (1971), Riddington and Stafford-Smith (1977), Liauw and Kwan (1985) and Moghadam et al (2006). Also, Dawe and Seah (1989), Mosalam et al. (1997), Schneider et al. (1998), Flanagan and Bennett (1999) and El-Dakhakhni et al. (2002) have studied the behavior of masonry-infilled steel frames under lateral in-plane loads. Saneinejad and Hobbs (1995) developed a method for infilled steel frames inelastic analysis and design subjected to in-plane forces. The model was later adopted by Madan et al. (1997) and implemented in software IDARC for dynamic analysis of infilled frames. A smeared-crack finite element model to study the nonlinear behavior of infilled reinforced concrete frames have been proposed by Mehrabi and Shing (1997). Based on minimizing the factor of safety with reference to the failure surfaces in a masonry infill, Moghadam (2004, 2006) proposed a new analytical approach for the evaluation of shear strength and cracking pattern of masonry infill panels.

Stavridis (2009) and Stavridis and Shing (2010) by combining the smeared and discrete crack approaches, proposed a modeling strategy to capture different failure modes of infilled frames. A parametric study has also performed to evaluate the sensitivity of the numerical results to the modeling input parameters. Koutromanos et al. (2011) extended the work of Stavridis and Shing (2010) and used a newly developed cohesive crack interface model and an improved smeared-crack model to capture the cyclic behavior of infilled frames. Recently, Chen and Lui (2015) developed a finite element model adopting the micro-modeling strategy proposed by Dawe and Seah (2001) to study the effect of the size and location of openings on the stiffness and strength of infilled frames.

As an alternative to the finite element-based approaches, a discrete element method (DEM) can be employed to simulate the nonlinear behavior of masonry-infilled steel frames. The DEM was originally used in rock engineering projects where continuity between the separate blocks of rock does not exist (Cundall, 1971). Today, applications of the DEM found application for the understanding of the mechanical behavior of blocky masonry structures (Toth, 2009; Sarhosis et al., 2008; Sarhosis & Sheng, 2014; Sarhosis, 2015; Giamundo, 2014; Lemos, 1997, 2007; Sarhosis et al., 2014a). Within discrete element method, sliding, large displacements and rotations between blocks, and even the complete detachment of the blocks are allowed with the automatically detection of new contacts as the calculations proceeds. Mohebkhah and Tasnimi (2007) and Mohebkhah et al. (2008) developed a 2D micro-model using the discrete element method to investigate the nonlinear lateral load behavior of confined masonry walls and masonry-infilled frames under monotonic loading. The model was later extended by Sarhosis et al. (2014b) to investigate the influence of the openings on the shear resistance of masonry infilled steel frames. In the following section, a review of the current strategies for modeling masonry-infilled frames will be given with an emphasis on those proposed by the discrete element method.

MECHANICAL BEHAVIOR OF MASONRY-INFILLED FRAMES

In the early 1950's, for the first time Polyakov (1956) studied experimentally the behavior of masonry-infilled steel and reinforced concrete frames subjected to lateral loads. In order to investigate the effects of parameters such as masonry units, mortar strength, loading type (monotonic or cyclic), panel-to-frame interaction conditions, Polyakov (1956) conducted some large-scale experiments and found that prior to the occurrence of cracks around the perimeter of panel-to-frame interface, the infill panel and the surrounding frame is acting as a monolithic unit. However, as the lateral load increases, the infill panel undergoes some step-wise cracks along the compression diagonal. It was also observed that cracks in the masonry infill wall panel pass through mortar head joints and bed joints. Increasing the magnitude of the lateral load, the cracks widened while new cracks appeared until failure of masonry units along the compression diagonal observed. Based on these observations, Polyakov (1956) proposed that masonry-infilled frames can be analyzed as an equivalent braced frame with a compression diagonal strut replacing the infill panel.

Stafford-Smith (1962) conducted some tests on steel frames infilled with concrete masonry blocks. The results showed that the separation between the infill and the frame occurs over three quarters of the length of the surrounding beams and columns. It was also observed that the infill increases the lateral stiffness and strength of the steel frame. Based on further experimental tests and numerical analyses using the finite difference method, Stafford-Smith (1966) proposed the concept of effective width of the infill panel diagonal strut to be used in the analysis of the equivalent braced frame. Later, Mallick and Garg (1971) investigated the effect of openings on the lateral stiffness of infilled frames by testing a series of small-scale steel frames infilled with cement mortar. Mallick and Garg (1971) observed that as the opening position moves towards the compression diagonal, the frame-infill composite action is adversely affected.

Extensive experimental studies conducted at the University of New Brunswick in the early1980's by Dawe to investigate the effect of panel-to-column ties, mortar strength, panel-to-frame friction and bond, gap between roof beam and panel and panel openings on the lateral load behavior of concrete masonry-infilled steel frames. A summary of these studies presented in Dawe and Seah (1989). Dawe and Seah (1989) showed the provision of a gap between the floor beam and panel interface reduces significantly the cracking and ultimate capacity of infilled frames. The study also showed that due to the variable nature of panel-to-frame interface conditions and variable behavior of concrete masonry infill panel, predicting the lateral load capacity of infilled frames is difficult. Comprehensive description of the studies performed until 1987 has been reported in the state-of-the-art report on infilled frames by Moghadam and Dowling (1987).

Mosalam et al. (1997) conducted a quasi-static cyclic experimental program to investigate the behavior of gravity-load designed steel frames infilled with masonry walls. Based on the observed crack patterns, they concluded that the stressed part of the infill (equivalent diagonal strut) has a non-prismatic cross section with large width in the center of the infill panel. Schneider et al. (1998) investigated the in-plane behavior of steel frames with masonry infills having large window openings. Test parameters included the masonry pier width and the number of spans. They found that narrow piers and double span infills tend to be more ductile than wide piers. Flanagan and Bennett (1999) tested some large-scale structural clay tile infilled steel frames under in-plane loading. The results revealed that the failure load of these frames was relatively insensitive to the surrounding frame characteristics. Moghadam (2004) presented the results of an experimental program on retrofitting brick masonry infilled steel frames. El-Dakhakhni et al. (2002) conducted an experimental investigation to study the effect of retrofitting unreinforced concrete masonry-infilled steel frame structures using GFRP laminates. Moghadam et al. (2006) reported the results of an experimental investigation on small and medium scale masonry and concrete infilled frames with and without horizontal reinforcement as well as bond beams under in-plane cyclic loading.

Recently, Tasnimi and Mohebkhah (2011) conducted an experimental program to investigate the in-plane seismic behavior of steel frames with clay brick masonry infills having openings. The experimental results indicate that infill panels with and without openings can improve the seismic performance of steel frames. Also, the amount of cumulative dissipated energy of the infill panels with and without openings at ultimate state is almost identical. Furthermore, the results indicated that infilled frames with openings are not always more ductile than the one with solid infill. It appears that, the ductility of such frames depends on the failure mode of infill piers. This experimental investigation showed that the infilled frames with openings experienced pier diagonal tension or toe crushing failure presenting smaller ductility factor than those frames with solid infill.

Markulak et al. (2013) conducted an experimental program to investigate the in-plane cyclic behavior of steel frames with commonly used masonry infills in Croatia. They studied three types of masonry infills: a) perforated clay blocks (C); b) lightweight AAC blocks (A); and c) a newly proposed combination of these materials (CA). They observed that by partial separation of the infill panel from the steel frame at certain drift levels and using the proposed combined masonry infill (CA), it was possible to eliminate the infill's detrimental effects.

Liu and Manesh (2013) tested 14 concrete masonry-infilled steel frames subjected to combined in-plane axial and lateral loading. They observed that for the relatively low axial load levels, the presence of axial loading increased the lateral load capacity of the infilled steel frames.

Based on experimental and numerical studies performed in the literature on the behavior of infilled frames, five different in-plane failure modes have been identified for such frames as follows:

1. **Corner Crushing (CC):** This failure mode can occur in case of rather weak infills as local crushing of masonry units at the ends of diagonal compression struts as shown in Figure 1(a).
2. **Horizontal Sliding Shear (HSS):** This failure mode can occur in case of very weak infills with very low bed joints shear resistance as a horizontal sliding plane along a bed joint as shown in Figure 1(b).
3. **Diagonal Tension Cracking (DTC):** This failure mode can occur in case of strong frames with weak connections or weak frames infilled with masonry panels having strong mortar. This failure mode takes place in the form of a step-wise diagonal crack through the bed and head joints or an inclined crack through masonry units as shown in Figure 1(c).
4. **Column Flexural Failure (CFF):** This failure mode can occur in case of strong frames with weak connections or weak frames infilled with very strong masonry. This failure mode takes place in the form of premature flexural failure of columns or of beam-column connections as shown in Figure 1(d).
5. **Column Shear Failure (CSF):** This failure mode may occur in infilled reinforced concrete frames' columns after formation of HSS failure mode (Paulay & Priestley, 1992). In this case, the two masonry sub-panles above and below the horizontal sliding plane in HSS mode would act as two knee bracing members resulting in column shear failure as shown in Figure 1(e).

NUMERICAL MODELING STRATEGIES FOR MASONRY-INFILLED FRAMES

The abovementioned literature review on the behavior of masonry infilled frames reveals that simulation of masonry-infilled frames is one of the most complicated problems in structural engineering field. This is due to the fact that an infilled frame is a composite frame comprising of two interacting components of different material (e.g. the surrounding frame and masonry infill panel) with different structural behavior and failure mechanisms. Masonry infill panel itself is a heterogeneous material with different failure modes and complicated behavior depending on its constituent elements' characteristics and the applied loads intensity and type. This complexity is mainly attributed to the discontinuum nature of

Figure 1. In-plane failure modes of infilled frames: (a) corner crushing mode; (b) horizontal sliding shear mode; (c) diagonal tension cracking model; (d) column flexural failure mode (El-Dakhakhni et al., 2003); and (e) column shear failure mode (Paulay & Priestley, 1992)

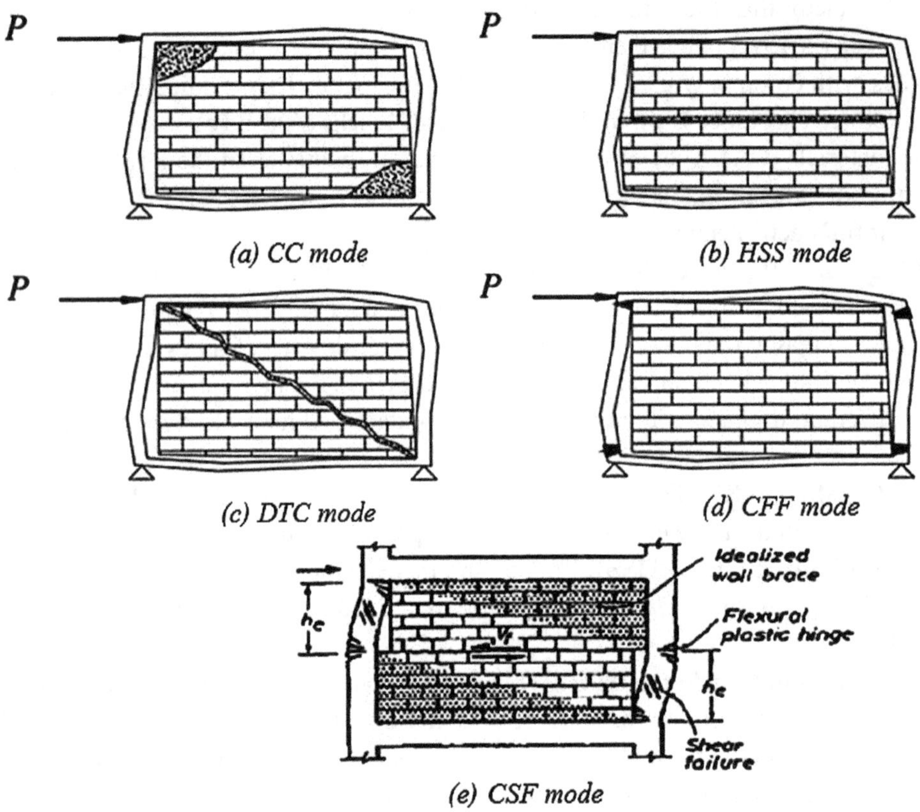

masonry panel and the existence of joints between masonry units as the major source of weakness and material nonlinearities. Another important factor that increases the complexity of infilled frames' behavior simulation is the interaction between the infill-to-frame. The infill-to-frame interaction depends on infill-frame relative stiffness, infill-frame friction and bond as well as gap between roof beam and panel. Therefore, a suitable numerical modeling strategy for the 2D analysis of infilled frames is to be capable of simulating the following phenomena:

1. Mortar joints sliding and opening.
2. Compressive and shear failure of masonry units (blocks).
3. Block-mortar or block-block interaction.
4. Infill-frame interaction.

Numerical modeling strategies of infilled frames are divided into two distinct categories (see Figure 2). These are: a) macro-modeling; and b) micro-modeling.

In macro-modeling strategy for modelling the mechanical behavior of masonry-infilled frames, the solid masonry infill panel is replaced by an equivalent system. The simplest model in this category was

Figure 2. Micro-modeling strategies for masonry walls: (a) detailed; and (b) simplified (Lourenco 1996)

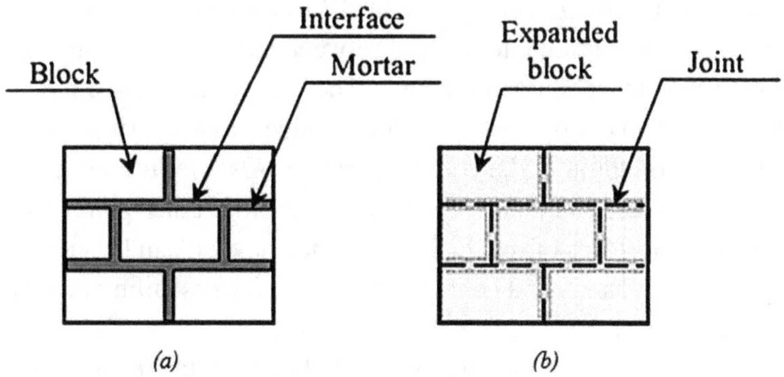

proposed by Stafford-Smith (1966) and then adopted and developed by Mainstone (1971). According to this model, an equivalent pin-jointed diagonal strut is substituted for the infill panel. In the literature, a number of macro-models have been proposed by several researchers. Among them, the models proposed by Chrysostomou et al. (2002), Mosalam et al. (1997*a*, 1998), Saneinejad and Hobbs (1995), and El-Dakhakhni et al. (2003) are distinguished. In the model suggested by Chrysostomou et al. (2003), the infill is idealized with three compression only inclined struts in each direction, which follow the behavior defined by the strength envelope and hysteretic loop equations. The off-diagonal struts are located to represent the interaction between the infill and confining steel frame. Mosalam et al. (1998) proposed a macro-model in which the infill panel is represented by an equivalent nonlinear truss with contact and tie (tension) elements. Saneinejad and Hobbs (1995) developed an inelastic analysis and design method for infilled steel frames subjected to in-plane forces. This method was later adopted and implemented in software IDARC for dynamic analysis of such frames by Madan et al. (1997). El-Dakhakhni et al. (2003) adopting the analysis methodology and concept of Saneinejad and Hobbs' model (1995), proposed a simple nonlinear macro-model to estimate the stiffness and the lateral load capacity of solid masonry-infilled steel frames failing in corner crushing (CC) mode. In this method, each masonry panel is replaced by three struts (one diagonal and two off-diagonal) with nonlinear force-deformation characteristics. The abovementioned macro-models are capable of predicting the global behavior of infilled frames failing in CC mode. However, these models cannot capture most of the other complicated failure modes and local effects of infill-frame interaction mentioned in Sec. 2 and Figure 1.

In *micro-modeling* strategy for modeling the mechanical behavior of masonry-infilled frames, both the surrounding frame and the infill panel components details (i.e. masonry units and mortar joints) can be simulated using a numerical method such as finite element method or discrete element method. In this strategy, the interaction between masonry units and mortar along the joints as well as the frame-infill interaction is taken into account ("*detailed micro-modeling method*"). There is also a "*simplified micro-modeling method*" in which each joint is modeled as an interface with zero thickness. In this method, fictitious expanded block dimensions are used that are of the same size as the original dimensions plus the real joint thickness as shown in Figure 2(b). The interface's stiffness is represented numerically by the stiffness of the real joint. The inelastic, isotropic model is used for the behavior of the blocks. The blocks can be considered rigid or fully deformable; thus allowing deformation to occur both in the blocks and joints and a better simulation of blocks crushing, crack propagation and sliding in the joints. Deformable blocks are internally discretized into Finite Difference triangular zones and each element responds according to a prescribed linear or non-linear stress-strain law. These zones are continuum elements as they occur in the Finite Element Method (FEM). Mortar joints between masonry blocks can be viewed as the interfaces representing interactions between the blocks. These are simulated employing the appropriate stress-displacement constitutive laws. The interaction between the blocks is represented either by a set of "point" contacts or by a set of "edge-to-edge" contacts (without any attempt to obtain a continuous stress distribution through the contact surface). Also, failure in the masonry block can be simulated by introducing interface elements (or otherwise fictitious crack patters) within the block. Such interface will have the characteristics of the bock. From the above, it can be seen that micro-models are capable of predicting both global and local behavior of infilled frames failing in different in-plane modes introduced in Sec. 2 and Figure 1.

Achyutha et al. (1986) developed a two dimensional elastic finite element model to investigate the effect of size of opening and types of stiffeners on the lateral stiffness of infilled frames with openings. Mehrabi and Shing (1997) have proposed a smeared-crack nonlinear finite element model to study the nonlinear behavior of infilled reinforced concrete frames. Simplified modeling strategies have also been investigated using finite element method by Mosalam (1997), Dawe et al. (2001), Ghosh and Amde (2002), Asteris (2003), Mohebkhah (2007), Mohebkhah et al. (2008), Stavridis (2009), Stavridis and Shing (2010), Koutromanos (2011), Koutromanos et al. (2011) and Sarhosis et al. (2014b). Extensive reviews on numerical micro-modeling aspects of masonry-infilled frames and masonry structures have been presented in Asteris et al. (2013) and Asteris et al. (2015).

DISCRETE/DISTINCT ELEMENT METHOD

Discrete/distinct element methods (DEMs) were originally developed for the study of jointed and fractured rocks due to the capability of the method to explicitly represent the motion of multiple, intersecting discontinuities and fractures (Sharma et al. 1999). More recently, discrete element methods have been used for the analysis of the masonry structures in which a significant part of the deformation is due to relative motion between the blocks and cracking tends to be along the brick/mortar interfaces and failure usually results from de-bonding of the bricks. So far, various discrete element applications to masonry shear walls (Mohebkhah & Tasnimi, 2007; Sarhosis & Sheng, 2014), unreinforced masonry buildings (Sincraian & Azevedo, 1998; Mohebkhah & Sarv-Cheragi, 2015), stone masonry arches (Toth et al., 2009), ancient structures (Sincraian et al., 1998; Azevedo et al., 2000; Psycharis et al., 2000; Psycharis et al., 2003; Giordano et al., 2002), and dry-stone retaining walls (Claxton et al.; 2005) have been reported for both static and dynamic analysis in the literature. Mohebkhah et al. (2008) developed a 2D distinct/discrete element model using the Universal Discrete Element Software UDEC (Itasca, 2004) for the inelastic analysis of concrete masonry-infilled steel frames which considers both geometric and material nonlinearities. The Universal Distinct Element Code (UDEC) is a two-dimensional numerical program based on the distinct element method for discontinuum modeling (Cundall, 1971). UDEC simulates the response of discontinuous media subjected to either static or dynamic loading. The discontinuous medium is represented as an assemblage of discrete blocks. In DEM, the discontinuities are treated as boundary conditions between blocks in which large displacements along discontinuities and rotations of blocks are allowed. In other words, the joints are viewed as interfaces between distinct bodies. Therefore, the starting point of a discrete element model is a discontinuous system, in contrast with the FE method, which evolved from the representation of a continuum media (Lemos, 2007).

Individual blocks can be represented as either rigid or deformable material. Rigid blocks when masonry units are subjected to low intensity stresses and the deformations in them are relatively small compared to their total strength. Alternatively, masonry unit blocks can be considered as deformable. Deformable blocks are subdivided into a mesh of finite-difference triangular elements, where each element responds according to a prescribed linear or nonlinear stress-strain law. The formulation of these elements is similar to the constant strain triangle (CST) finite element formulation. The complexity of the deformation of the bricks depends on the number of zone elements into which they are divided. Zones obey to the constitutive model assigned to them and, for each separate block the strain can be estimated. In UDEC, the deformable block zones can be assumed to be linear elastic or non-linear according to the Mohr-Coulomb criterion (Itasca, 2004a). The disadvantage of this element is that, in case of complicated deformation problems such as beam-to-column behavior in an infilled frame, the number of triangular elements in which the area has to be discretized may become very large.

Mechanical interaction between blocks is considered by normal and shear elastic springs connected at the blocks. The relative motion of the discontinuities is also governed by linear or nonlinear (as appropriate) force-displacement relations for movement in both the normal and transverse (tangential) directions. The formulation used in the program permits both geometric and physical non-linearities of the intact material to be modeled. UDEC has several built-in material behavior models (i.e. linear elastic, Mohr-Coulomb, Mohr-Coulomb with residual strength, etc.), for both intact blocks and the discontinuities, which permit the simulation of response representative of discontinuous materials. The program uses "Lagrangian" calculation scheme that is well-suited to model the large movements and deformations of a blocky system (Itasca, 2004). In this program, the explicit solution procedure is used, in which motion equations are set up for each time increment. This takes place for each block sequentially. The behaviour of the masonry is based on the knowledge of the properties of each constituent and the interface. Material properties are normally obtained from experimental tests on the masonry units, the mortar and on masonry samples.

The original UDEC program was based on the plane strain situation. In the UDEC 4.0 version it is possible to give the stress perpendicular to the plane of the structure a constant value. The plane stress situation is then obtained by giving the stress perpendicular to the plane of the structure a value of zero. This is encountered, for example, in masonry structures loaded only in the plane of the structure.

The calculations performed in the discrete element method alternate between the application of a *force-displacement law* at all contacts and *Newton's second law* at all blocks or nodes (Itasca, 2004). The force-displacement law is used to find contact forces from known displacements. Newton's second law gives the motion of the blocks resulting from the known forces acting on them. Figure 3 demonstrates the calculation cycle for the distinct element method (Itasca, 2004). Mechanical damping is used in the DEM to solve both static and dynamic solutions. For each case, a different type of damping is used. For static analysis, an approach similar to dynamic relaxation technique is employed and the equations of motion are damped to reach the equilibrium state.

Discontinuous deformation analysis (DDA) is an alternative method for the analysis of jointed and blocky systems. The method proposed by Shi and Goodman (1988). DDA is an implicit discrete element method for the analysis of jointed rock mass problems. To determine the variables in DDA (i.e. the displacements), the equations of equilibrium are solved similar to the finite element method. On the contrary to the distinct element method where a "soft contact" approach is used and blocks can overlap when they are in compression, contacts in DDA are considered to be rigid (the "hard contact approach"). The main differences between the DEM and DDA have been given in Bićanić et al. (2001);

Figure 3. Calculation cycle for the discrete element method
(Itasca, 2004)

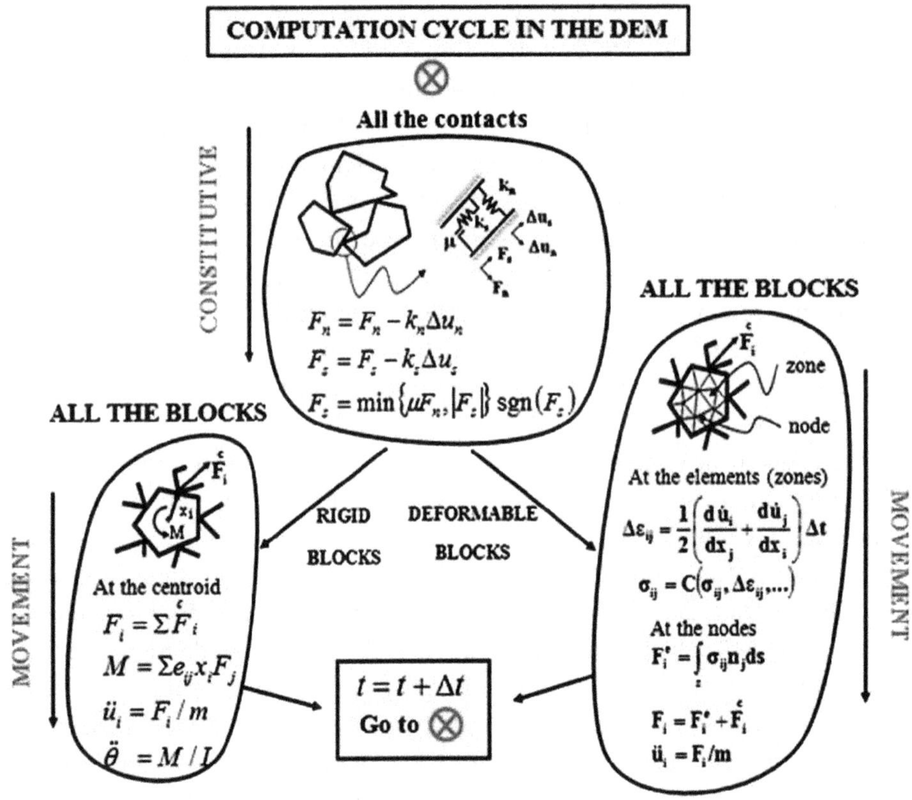

Bobet et al. (2009); and Lisjak and Grasselli (2014). DDA has two advantages over the DEM. These are: a) permission for relatively larger time steps; and b) closed-form integrations for the stiffness matrices of elements (Jing, 2003).

There are numerous simplified micro-modeling strategies in the literature which can be utilized using the FEM for the analysis of masonry-infilled frames. However, these procedures generally fail to simulate the response of infilled frames when large displacements and rotations between masonry units or complete detachment of the units occur. Nevertheless, there are some advanced discontinuum finite element programs such as DIANA (developed by Lourenco, 1996) that include interface elements allowing the user to analyze properly masonry panels with large displacements and rotations between their constituents. As an alternative, the DEM as pointed out in previously has the capability of simulating all failure modes of masonry-infilled frames even complete and large detachment of masonry units. The DEM also has the capability of detecting new contact points as the analysis proceeds.

PROCEDURE USED FOR MODELLING MASONRY INFILL WALL PANELS

The procedure used to model masonry structures using the discrete element method is discussed in more detail below. This methodology is based on the authors' personal experience of using UDEC.

Step 1: Understand the objectives of the model analysis.

The objectives of the analysis are important in determining the level of the detail to be included in a model. Often, it is not necessary to produce a complex model with excessive details just because it is like that in reality. Complicated features should be omitted if they are likely to have little influence or they are irrelevant to the purpose of modelling. A simple model should be built first for the feasibility study and validation purposes and later refinements can be added as (and if) necessary. For example, it is usually more appropriate to model plain masonry first and then add elements such as reinforcement.

Step 2: Create a conceptual picture of the physical system.

It is very important to have a conceptual picture of the problem in order to provide an initial estimate of the physical characteristics of the system. The modeller should decide upon the suitability and applicability of the model's geometry, constitutive laws, the boundary conditions and the initial equilibrium state.

Step 3: Generate the model.

After deciding upon the objectives of the modelling analysis and the physical characteristics of the system, it is time to generate the initial model. The first task involves the generation of the model's geometry. The model should begin with a single block which spans the physical region being analysed. By cutting this block into smaller blocks, the detailed features (i.e. mortar joints, windows/doors openings etc.) are introduced. Also, a choice between rigid or deformable blocks should be made by the modeller. Next, the material behaviour and properties should be assigned. The constitutive behaviour and the associated material properties will dictate the response of the model to external loading. As described previously, the user has a number of choices for the constitutive model that can be used in the analysis. Also, material properties can be obtained from the literature or from testing. The boundary and initial conditions that represent the in-situ state (i.e. the condition before a change or disturbance such as external loading is introduced) are then defined. Finally, if required by the user, structural members such as reinforcement can also be included.

Step 4: Bring the model into equilibrium.

The next step is to bring the model into a state of equilibrium otherwise alternations cannot be performed. At this state only the initial boundary conditions, including the self weight effects, are applied to the model. As UDEC uses an explicit solution process, the solution is reached after a series of computational steps. The user is required to specify a number of calculation steps to bring the model into a state of equilibrium. According to Itasca (2004), the model is in equilibrium only when "the net nodal vector at each grid point of deformable block is zero". However, for a numerical analysis, the out of balance force will never reach exactly zero. In practice, it is assumed that the model is in equilibrium only when the maximum unbalanced force is relatively small compared to the force applied to the model (i.e. a difference of less than 1%). The force imbalance in the model is caused by the slight difference in the location of contact forces and grid point forces at block corners (Itasca, 2004a & b). This is related to corner rounding (Section 3.3).

Step 5: Perform alterations.

Once the model is in equilibrium, alterations such as external loading and thermal changes can be applied and new structural elements (such as reinforcement) can be introduced.

Step 6: Step to solution.

UDEC uses an explicit time-machine method to solve the algebraic equations, described earlier in Section 3.4. A solution is reached after a series of computational steps have been executed. The number of steps is controlled manually by the user who must determine if the number of steps is sufficient to reach the solved state required.

Step 7: Examine model response.

When modelling masonry in order to assess the failure load, it is recommended that loading is applied monotonically to the model until failure occurs. According to Itasca (2004), in order to determine a collapse load, it is often appropriate to use "strain-controlled" boundary conditions rather than "stress-controlled" – i.e., a constant velocity is applied and the boundary reaction forces are measured rather than applying forces and measuring displacements. According to Itasca (2004), "a system that collapses becomes difficult to control as the applied load approaches the collapse load". Collapse can occur when the system is incapable of carrying any more load; this condition is judged by the user rather than being identified by the software. The model's response can be examined against theoretical/analytical methods of analysis or from similar laboratory tests.

Step 8: Present results for interpretation.

The final stage of problem solving involves the presentation of the results for a clear interpretation and analysis. In UDEC the results can be presented graphically either electronically or as a hard copy for easy comparison with field or laboratory measurements and observations. Various results can be generated by the software including: block velocity; displacements; forces; grid point stresses and strains; and contact detachment/crack patterns. In addition, history plots of load against displacement, crack widths and crack lengths at each time step of the analysis can be obtained. History plots are especially helpful to monitor whether a state of equilibrium or failure has been reached.

DEM MODELING ASPECTS OF MASONRY-INFILLED STEEL FRAMES

For the analysis of the nonlinear lateral load behavior of masonry-infilled steel frames using a simplified DEM micro-modeling method, simulations should be carried out sequentially. First each model is brought to equilibrium under its own dead weight and then external load is applied to it. In order to determine a collapse load and simulate the post-peak behavior of masonry-infilled steel frames, it is often better to use displacement-controlled boundary conditions rather than force-controlled. Therefore, incremental horizontal displacement is usually applied to each model/domain to be studied.

Elastic Parameters

As pointed out earlier, individual masonry units can be modeled as either rigid or deformable material. To capture the linear-elastic behavior of masonry units, a linear-elastic constitutive model is sufficient. In this case, deformable units are subdivided automatically into a mesh of triangular elements in which each element responds according to a prescribed linear or nonlinear stress–strain law as it will be discussed later in Sec. 5.2. To define the constitutive model of interface joints, the following stress-displacement relationship between the interface stress vector $\mathbf{t}$ and the interface displacement vector δ is adopted:

$$\mathbf{t} = \mathbf{K}' \tag{1}$$

in which

$$\mathbf{t} = \begin{Bmatrix} \sigma \\ \tau \end{Bmatrix}, \ \delta = \begin{Bmatrix} u \\ v \end{Bmatrix}, \ \mathbf{K} = \begin{bmatrix} k_n & 0 \\ 0 & k_s \end{bmatrix} \tag{2}$$

and k_n and k_s are the joint normal and shear stiffnesses, respectively. In order to generate a simplified DEM micro-model based on interface elements with zero thickness, the size of the blocks has to be expanded by the mortar thickness (h_m) in both directions. It follows that the elastic properties of the "*expanded*" *block* and the "*interface*" *joint* must be adjusted to yield correct results. Contact points in UDEC are assumed to be "soft". This means that contact forces are generated as a result of interpenetration of adjacent blocks at their contacts. The joint stiffness parameters (k_n and k_s) describe the stress deformation characteristics of the joints. According to Itasca (2004), the joint stiffnesses do not represent actual masonry parameters but they are mathematical values required by the 'soft' contact model used in UDEC which control the interpenetration of adjacent blocks in the model. For example, if the block stiffnesses are high then small amounts of interpenetration will occur. The amount of interpenetration is controlled by the user and does not influence the calculation time. Also, due to the zero thickness of the interfaces, the geometry of the bricks has to be expanded to incorporate the thickness of the joint in both directions. Due to the relative dimensions of mortar and block, it is assumed that the elastic properties of the blocks remain unchanged. Then, under the assumption of stack bond and uniform stress distributions both in the block and mortar as shown in Figure 4, the following relation can be written between the equivalent axial stiffness of the real (detailed) and the simplified situations, which is a function of the Young's Modulus and thickness of the joint:

$$\sum \left(\frac{1}{k_i} \right)_{detailed} = \sum \left(\frac{1}{k_i} \right)_{simplified} \tag{3}$$

then

$$\frac{h_b}{E_b} + \frac{h_m}{E_m} = \frac{h_b + h_m}{E_b} + \frac{1}{k_n} \tag{4}$$

Figure 4. Scheme to determine joint normal stiffness: (a) real masonry prism; (b) expanded block with zero-thickness interface element (springs in series analogy)

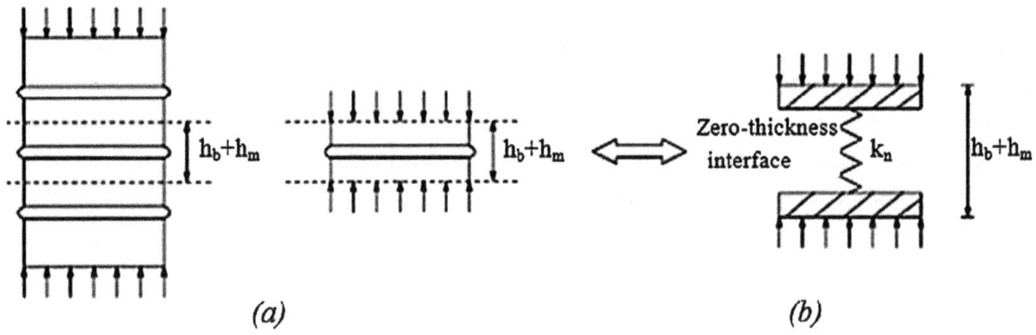

or

$$k_n = \frac{E_b E_m}{h_m(E_b - E_m)} \tag{5}$$

k_n is the stiffness of the equivalent spring as shown in Figure 4(b). A similar expression can be derived for joint shear stiffness as follows:

$$k_s = \frac{G_b G_m}{h_m(G_b - G_m)} \tag{6}$$

In the above relations, h_b = height of the block; E_b and E_m are the Young's modulus, G_b and G_m are the shear modulus, respectively, for block and mortar. For the analysis of dry joint stone masonry walls using the simplified micro-modeling method, Lourenco et al. (2005) proposed the following expressions to determine the joint normal and shear stiffnesses:

$$k_n = \frac{1}{h_s\left(\frac{1}{E_{wall}} - \frac{1}{E_{stone}}\right)} \tag{7}$$

$$k_s = \frac{k_n}{2(1+\nu)} \tag{8}$$

where h_s = height of the stone unit; ν =Poisson's ratio of wall; and E_{wall} and E_{stone} are the Young's modulus of the wall and the stone unit, respectively. Using advanced detailed discontinuum finite element analyses, the accuracy of simplified micro-modeling strategy has been verified by Lourenco (1996).

Apart from the aforementioned spring elements to connect masonry units to each other and the infill panel boundary nodes to the surrounding frame, there are various infill-frame interface elements in the literature (King & Pandey, 1978; Dhanasekar & Page, 1986; Mosalam et al., 1997; Dawe et al., 2001). Also, Koutromanos (2011) and Koutromanos et al. (2011) extended the work in Stavridis and Shing (2010) by using a combined discrete and smeared crack approach to investigate the behavior of masonry-infilled, non-ductile RC frames subjected to lateral cyclic, static and earthquake loads. The main novelty in this research is the use of a newly developed cohesive crack interface model and an improved smeared-crack model to capture the cyclic behavior of such structures.

INELASTIC PARAMETERS AND CONSTITUTIVE CRITERIA

Infill Panel Material

A characteristic feature of UDEC is the geometric non-linearity of the intact bricks/blocks to be modelled. The displacement of the bricks due to shear and opening up of the interfaces can be immediately included in the calculations. So, it is possible that masonry units which were originally adjacent to each other to become partially or entirely detached from each other, or new contact points to be formed. This is significant feature of UDEC software when modelling problems involving discontinuities such as low bond strength masonry, where the predominant failure mechanism is due to the de-bonding of the bricks or blocks from the mortar, as the location and the magnitude of surface crack widths within a masonry structure can be determined realistically. Detecting and updating brick/block contacts is the most time consuming part of the software's calculation.

There are many parameters that affect the behavior of the model. The most fundamental parameter is the constitutive model chosen to represent the material behavior. UDEC version 4.0 has some built-in constitutive material models such as: isotropic elastic, Drucker-Prager, Mohr-Coulomb and strain-hardening/softening. In the case of very low stress levels, a linear-elastic model is sufficient. However, for high stress levels a nonlinear model which can simulate crack formation, shear or crushing is needed. The nonlinear post-peak behavior of stone-like materials such as masonry is characterized by softening which determines the way in which crack formation propagates within a block (Rots, 1997). It has been shown that when the shear displacement increases, the masonry block cohesion does not suddenly, but more or less gradually decreases to zero (Rots, 1997). This quasi-brittle behavior (known as softening) has been observed in tension, compression and shear. This softening behavior in masonry units is attributed to the gradual break down of the weakest links followed by formation of micro-cracks and finally linking up the cracks to form a macro-crack or shear plane (Rots, 1997). Hence, masonry units softening behavior (e.g. concrete blocks or clay bricks) is possible to be simulated using a strain-hardening/softening material model such as UDEC Mohr-Coulomb model with tension cut-off in conjunction with non-associated shear and associated tension flow rules. The difference, however, lies in the possibility that the cohesion, friction, dilation and tensile strength may harden or soften after the onset of plastic yield. In the Mohr-Coulomb model those properties are assumed to remain constant (Itasca 2004). However, in the strain-hardening/softening material model, the user can define the cohesion, friction and dilation as piecewise-linear functions of a hardening parameter measuring the plastic shear strain. In other words, the post-failure behavior in a strain-softening model is dictated by values for angle of internal friction and cohesion, which decrease with increasing percent strain after failure. Cohesive strength of masonry units can be calculated using uniaxial and triaxial tests. However, as an approximation, cohesion strength of units can be estimated based on their internal friction angle (φ) and compressive strength (f_b'). To this end, utilizing the Mohr-Coulomb model as shown in Figure 5(a), the following relation is derived:

$$c = \frac{f_b'}{2\,\mathrm{Tan}\left(45 + \frac{\varphi}{2}\right)} = \frac{f_b'}{2}\left(\frac{1-\sin\varphi}{\cos\varphi}\right) \tag{9}$$

Strain-softening model for the blocks is described by two parameters. These are the cohesion (c) and the angle of internal friction (φ). The angle of internal friction and cohesive strength of masonry blocks can be calculated based on uniaxial and triaxial tests which are performed on representative sample blocks. For the case of concrete-masonry infills, the angle of internal friction and cohesive strength of blocks/units can be calculated using the relations given by Tasnimi and Farzin (2006). They developed a 3D finite-element model for nonlinear analysis of plain concrete columns. The model was validated with the results of some experimental tests. Then, the FEM model was used for a parametric study to obtain two relations for the angle of internal friction and cohesion of concrete in terms of concrete compressive strength (f_c'). In the parametric study, bearing capacity of a plain concrete column varying

Figure 5. (a) Mohr-Coulomb failure model with tension cut-off, (b) tension strength softening; and (c) cohesion strength softening

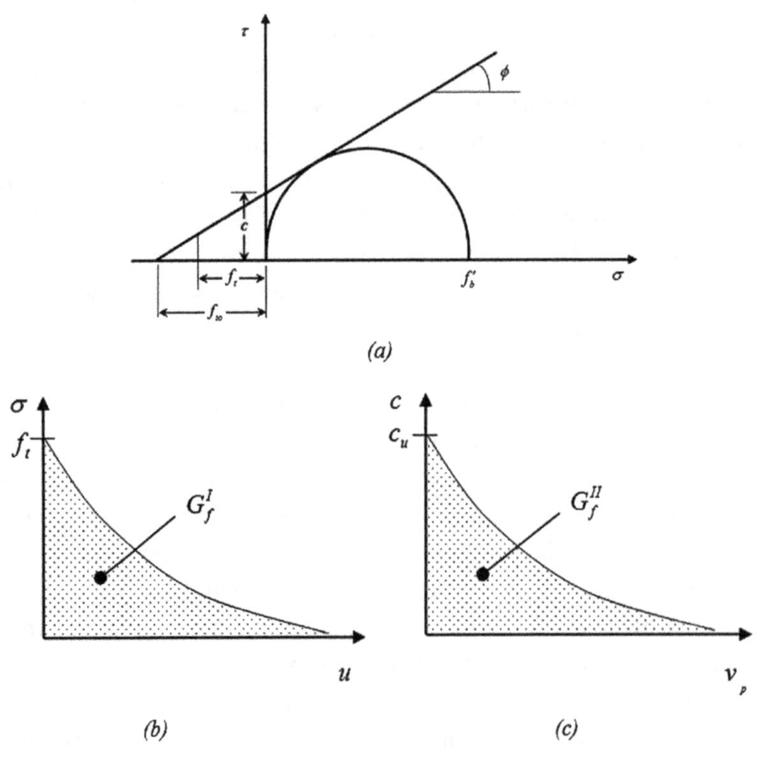

the values of concrete cohesion and internal friction angle was determined. Finally, using the statistical curve fitting methods, the following equations were proposed:

$$\varphi^{\circ} = 0.145 f_c'(MPa) + 49.71 \tag{10}$$

$$c\,(MPa) = 0.1065 f_c'(MPa) + 0.531 \tag{11}$$

Softening of masonry units can be allowed in DEM. Softening is described using softening rate of tensile strength and cohesion as shown in equation 12 and 13 below (Rots, 1997):

$$\sigma = f_t\, e^{-\frac{f_t}{G_f^I} u_p} \tag{12}$$

$$c = c_u\, e^{-\frac{c_u}{G_f^{II}} v_p} \tag{13}$$

where G_f^I denotes the tensile fracture energy (generally named Mode-I fracture energy) and is defined as the amount of energy to create a unitary area of a tension crack, Figure 5(b). Also, G_f^{II} stands for the shear fracture energy (generally named Mode-II fracture energy) and is defined as the amount of energy to create a unitary area of a shear crack, Figure 5(c). According to Lourenco (1996), parameter G_f^I for clay bricks ranges from 0.005 to 0.02 *N.mm/mm*^{2} for a tensile strength ranging from 0.3 to 0.9 *MPa* and parameter G_f^{II} ranges from 0.01 to 0.25 *N.mm/mm*^{2} for a cohesion strength ranging from 0.1 to 1.8 *MPa*. These parameters are required to capture the inelastic behavior of concrete masonry blocks using the strain-softening model.

Joints

For the joints, mortar mechanical behavior can be simulated suing the Mohr-Coulomb slip constitutive law model. For such constitutive law, the inelastic parameter related to the dilation angle is also required. Dilatancy is a measure of the change in volume that occurs when shear stress is applied to a material (Atkinson et al., 1989). This change is characterized by a dilation angle, ψ_j, which measures the uplift upon shearing. It has been shown (Atkinson et al., 1989; Mosalam et al., 1997) that dilatancy must be taken into account in the analysis of confined masonry structures such as infilled frames. According to Itasca (2004), for concrete material, a typical dilation angle value is that of 12°. In addition, another inelastic material parameter is that of the coefficient of friction angle. For the simulation of steel masonry infill frames, the coefficient of friction between steel components and masonry panel has an influence on the behavior of the model and must be defined. Based on the available literature, a coefficient angle of friction equal to 0.25 can be used for the frame-to-panel interfaces (Dawe &Seah, 2001). However, it is highly recommended that further experimental studies should be carried out to define such value.

A problem often encountered when modelling masonry is the unrealistic response when brick interaction occurs close to or at two opposing brick contacts. At this point, numerically, blocks may be locked or hung-up due to the modelling assumptions that brick corners are sharp or have infinite strength. In a real masonry structure this will only occur as a result of stress concentrations. However, simulating such a phenomenon is impractical. A realistic representation can be achieved by rounding the corners of the blocks so that they can smoothly slide past one another when two opposing corners interact. According to Itasca (2004), a short corner rounding length (e.g. 1% of the block's length) gives a good level of accuracy.

Sarhosis et al. 2015, carried out a series of sensitivity studies supported with regression analysis to investigate the significance of the brick–mortar interface properties (normal and shear stiffnesses, tensile strength, cohesive strength and frictional resistance) on the load at first cracking and ultimate load that a masonry wall panel containing an opening can carry. Computational results were also compared against full scale experimental tests carried out in the laboratory. From the sensitivity analyses it was found that the joint tensile strength is the predominant factor that influences the occurrence of first cracking in the panel, while the cohesive strength and friction angle of the interface influence the behaviour of the panel from the onset of cracking up to collapse.

Steel Frame Material

Since the steel frame components in a masonry-infilled steel frame are expected to behave in-elastically at ultimate state of loading, a Von-Mises material model must be chosen to represent the steel frame real behavior. The Von-Mises criterion is not available in UDEC. However, according to Itasca (2004), the Drucker-Prager criterion can be degenerated into the Von-Mises criterion when friction angle is equal to zero (i.e. $\varphi = 0$). Although the steel frame components are made up of steel I-sections, they can be modeled as solid blocks of steel with equivalent elastic and inelastic mechanical properties.

DEM STUDIES ON MASONRY-INFILLED FRAMES

Although there are many DEM studies in the literature on masonry shear walls, unreinforced masonry buildings, stone masonry arches, ancient structures, colonnades and monuments analysis (Sarhosis et al., 2016), only a few studies have been done on DEM analysis of masonry-infilled RC and steel frames. Some of these few studies are presented in this section.

Masonry-Infilled RC Frames

In order to evaluate the seismic behavior of masonry walls in rural areas surrounded by reinforced concrete frames (in which the RC frames only provide some binding effect), Sincraian and Azevedo (1998) for the first time analyzed some one-bay and two-story RC frames with different masonry infill panels' configurations using DEM. Elastic isotropic materials were used for both masonry blocks and frame elements. All the frames were subjected to an artificially generated accelerogram matching the elastic response spectrum for soil class A (hard soil) presented in Eurocode 8. The accelerogram were used with different PGA intensities for each model to simulate different severities of the seismic motion. They found that the Discrete Element Method can be used to capture the collapse mechanisms of the infilled RC frames as shown in Figure 6. This figure shows the capability of the DEM in capturing the mixed failure mode of masonry panels as joints separation and sliding.

Lang and Benzoni (2014) developed a DEM model of a confined masonry wall using a micro-modeling strategy. The model could simulate the linear and nonlinear behavior of a confined masonry shear wall subjected to in-plane cyclic loading up to 1% drift (Figure 7). From Figure 7, the capability of the DEM in capturing the DTC failure mode of the infill panel which is in the form of a step-wise diagonal crack through the bed and head joints is demonstrated.

Chiou et al. (1998, 1999) studied the static lateral load behavior of masonry-infilled RC frames using a refined discontinuous deformation analysis (DDA). In those studies, the masonry infill wall and RC frame were cut into sub-blocks by virtual joints with finite tension and shear strength of mortar and concrete materials, respectively as shown in Figure 8. It was found that the DDA results are in good agreement with experimental results. However, the proposed model has some disadvantages such as:(1) ignoring the bond slip of reinforcement in the RC columns and beam; (2) ignoring mixed tension-shear failure of mortar joints and concrete joints between concrete sub-blocks; (3) not taking into account the possibility of bricks' tensile failure in the infill wall regions with high normal stresses; and (4) considering linear constitutive relations for the bricks and concrete material (Chiou et al. 1998).

Figure 6. Deformed geometry of the masonry infilled RC frames
(Sincraian and Azevedo, 1998)

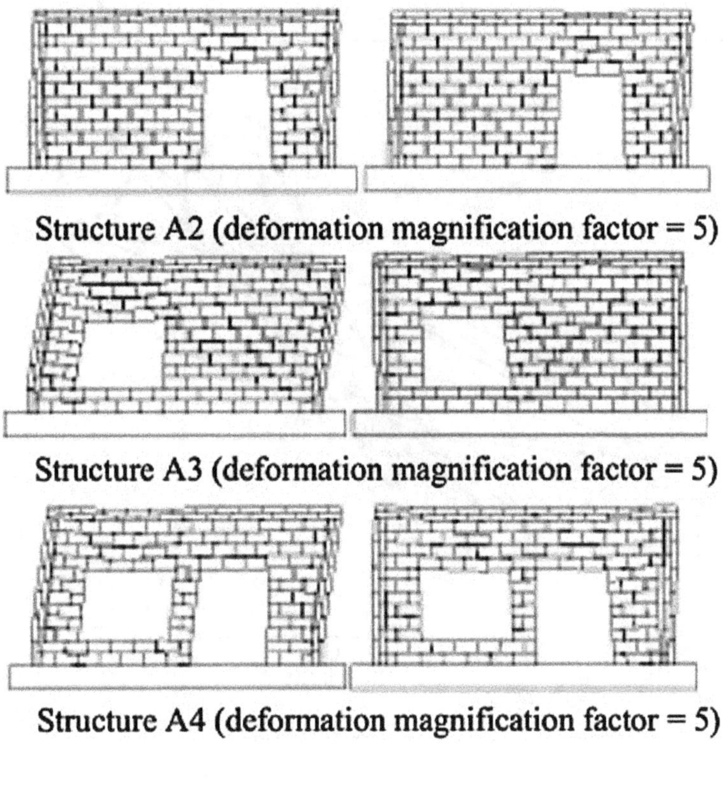

Figure 7. (a) DE model damage pattern and (b) Force-displacement results of experimental and numerical model of a confined masonry wall subjected to cyclic loading
(after Lang and Benzoni 2014)

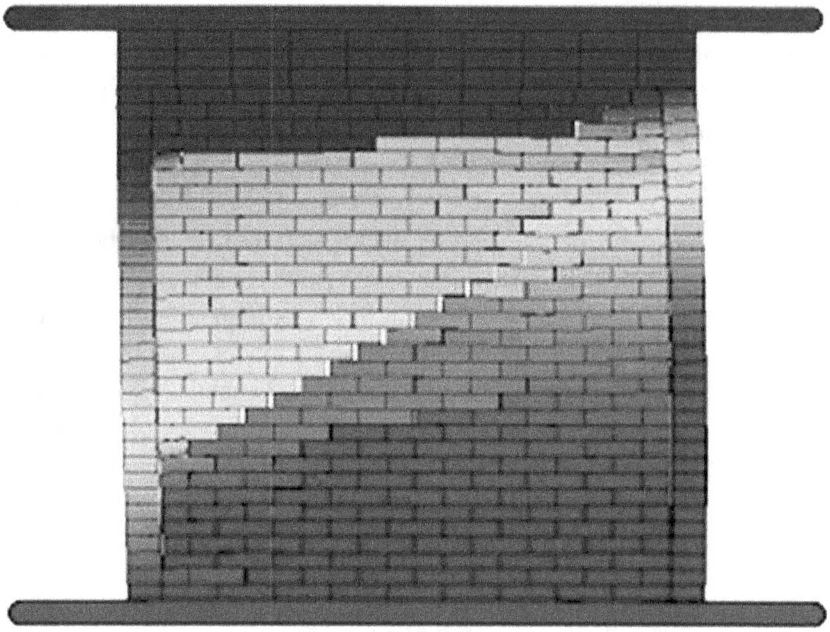

(a)

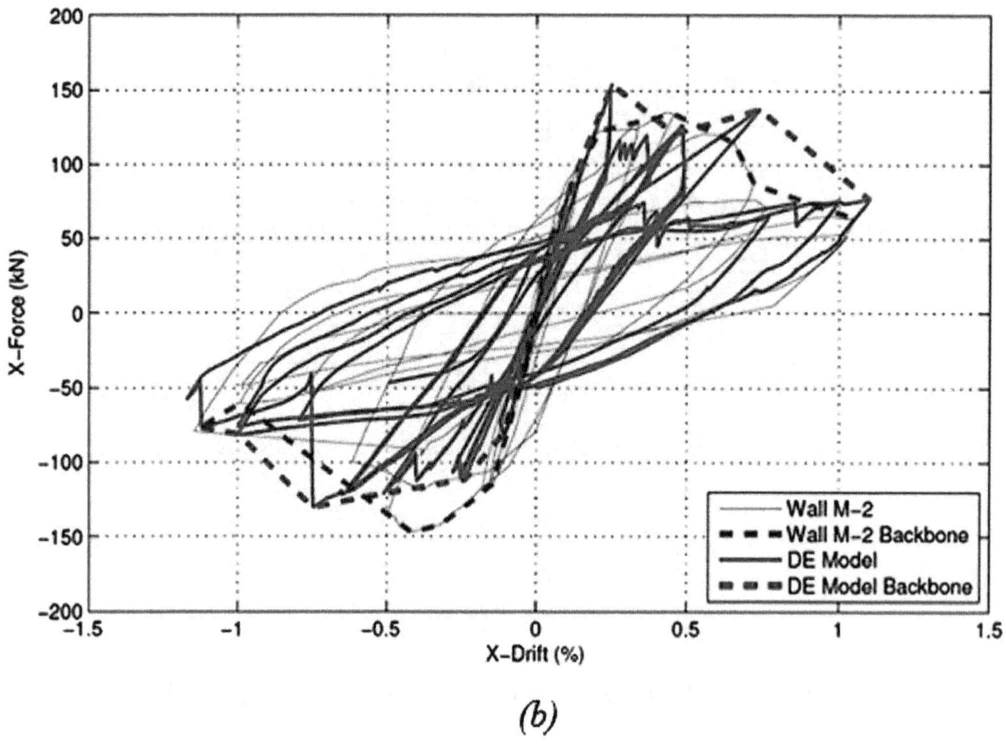

(b)

Figure 8. Principal stress distribution of a masonry-infilled RC frame using DDA
(after Chiou et al. 1998)

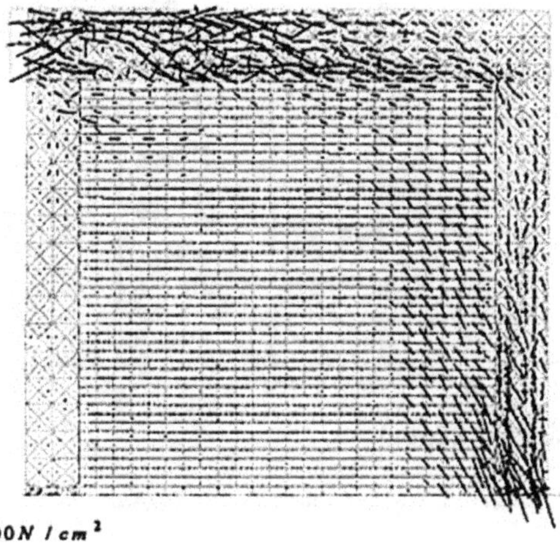

Masonry-Infilled Steel Frames

Opening Size and Location Effects

Mohebkhah et al. (2008) developed a two-dimensional numerical model using the specialized discrete element method (DEM) software UDEC (2004) for the nonlinear static analysis of masonry-infilled steel frames with openings subjected to in-plane monotonic loading. It was found that the numerical model is applicable to a simplified micro-modeling simulation of the response of concrete masonry-infilled steel frames with and without openings throughout the loading process leading to failure (as shown in Figure 9) and to predict their failure mechanisms based on joint separation and sliding.

Furthermore, the DEM model was used to investigate the effect of door frame on lateral load capacity and stiffness of infilled frames with a central opening as shown in Figure 10. In addition, Figure 10 shows the capability of the DEM in capturing the DTC and CC failure modes and diagonal step-wise cracking which is consistent reasonably with laboratory experimental observations presented in Dawe and Seah (1989). Mohebkhah and Tasnimi (2012) also used the developed DEM model to simulate the lateral load behavior of some tested clay brick masonry-infilled steel frames in Mohebkhah (2007).

In order to study the effect of non-standard case of positions of openings (as shown for example in Figure 11) on the behavior of steel in-filled frames, Sarhosis et al. (2014b) developed a 2D numerical model using the specialized discrete element method (DEM) software UDEC. The model was used for a series of parametric studies to investigate the significance of the size and location of the openings on the lateral load capacity, as well as the stiffness and failure mechanisms of the infilled steel frames. The effect of the actual loading type on the steel frame has also been investigated by evaluating the responses of two lateral load conditions. The first case was the typical point load on the top left hand corner of the steel frame (Figure 12a), as used in all previous analyses presented in this chapter, while the second case considered a lateral uniformly distributed load at the left hand column of the steel frame (Figure 12b). Comparisons made with respect to the strength and stiffness of the infilled system as well as their

Figure 9. Lateral load-displacement response of masonry-infilled steel frames
Source: Mohebkhah et al. (2008)

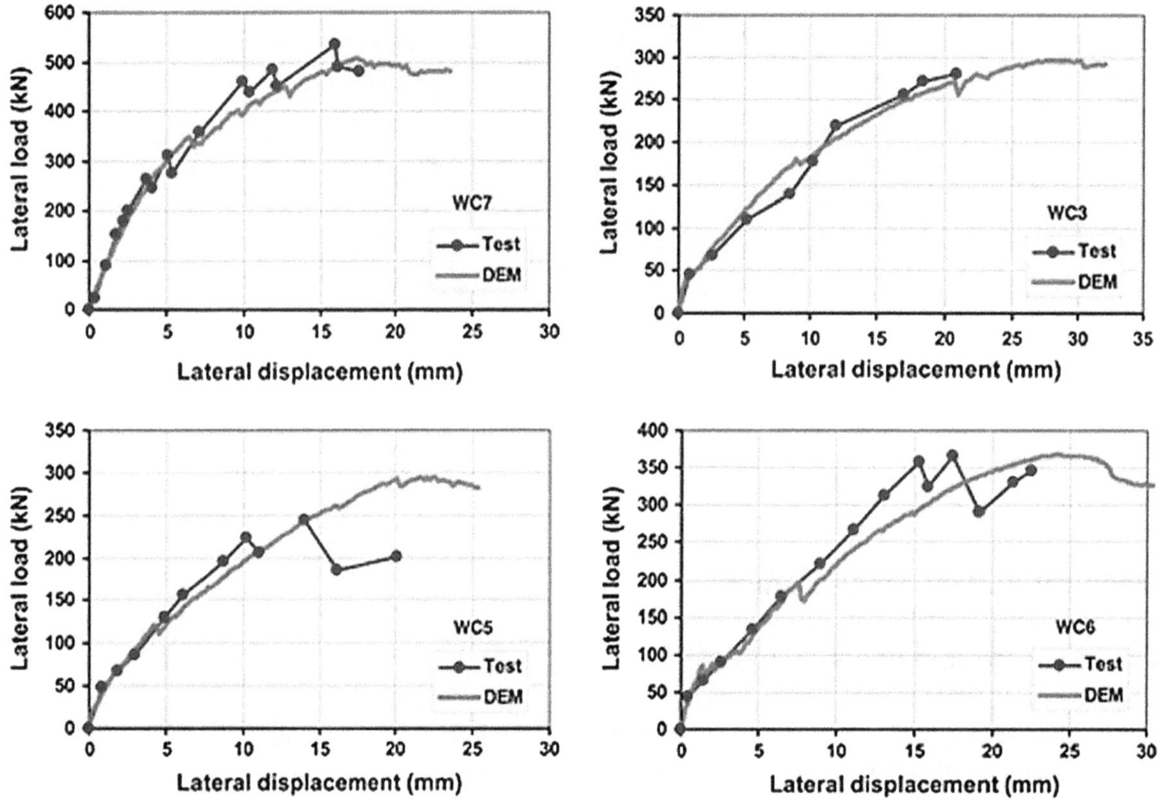

failure mechanisms, in order to draw the main differences and provide better understanding for such cases where, for instance, another frame is in full contact and distributes lateral loads.

From the results analysis, it was found that when the point load was applied on the steel frame, the initial stiffness was relatively high. However, after a certain load (approx. 125 kN) the stiffness reduced and kept constant up until the failure of the masonry infill panel. On the other hand, the strength of the masonry infill steel frame is lower when the distributed load applied on the steel frame. The main reason resulted this behaviour is that the distributed load applied on the entire column, hence it deflects more affecting the bottom part of the infill, which now deforms severely, leading to higher overall displacements. When a point load applied to the masonry panel, the corresponding load capacity observed was 535 kN; the corresponding capacity for the case with the distributed load was 450 kN (Figure 13).

Effect of Gravity Load

In order to study the effect of gravity loading on the behavior of infilled frames, Mohebkhah (2007) using the developed DEM model, analyzed some tested clay masonry-infilled steel frames. The specimens included one solid infilled frame (SW) and four infilled frames with central openings of different relative openings area (PW1, PW2, PW3 and PW4). The DEM lateral load-displacement curves of the frames as shown in Figure 14 indicated that the presence of gravity loads on columns and top beam

Figure 10. DEM results of specimen WC3 with a door frame at a horizontal displacement of 23 mm: (a) blocks' failure points and crack patterns of the joints; (b) magnified principal stress tensors; (c) horizontal displacement contours (m); (d) deformed geometry (magnification factor=10); (e) principal finite strain tensors; and (f) bending moment of door frame
Source: Mohebkhah et al. (2008)

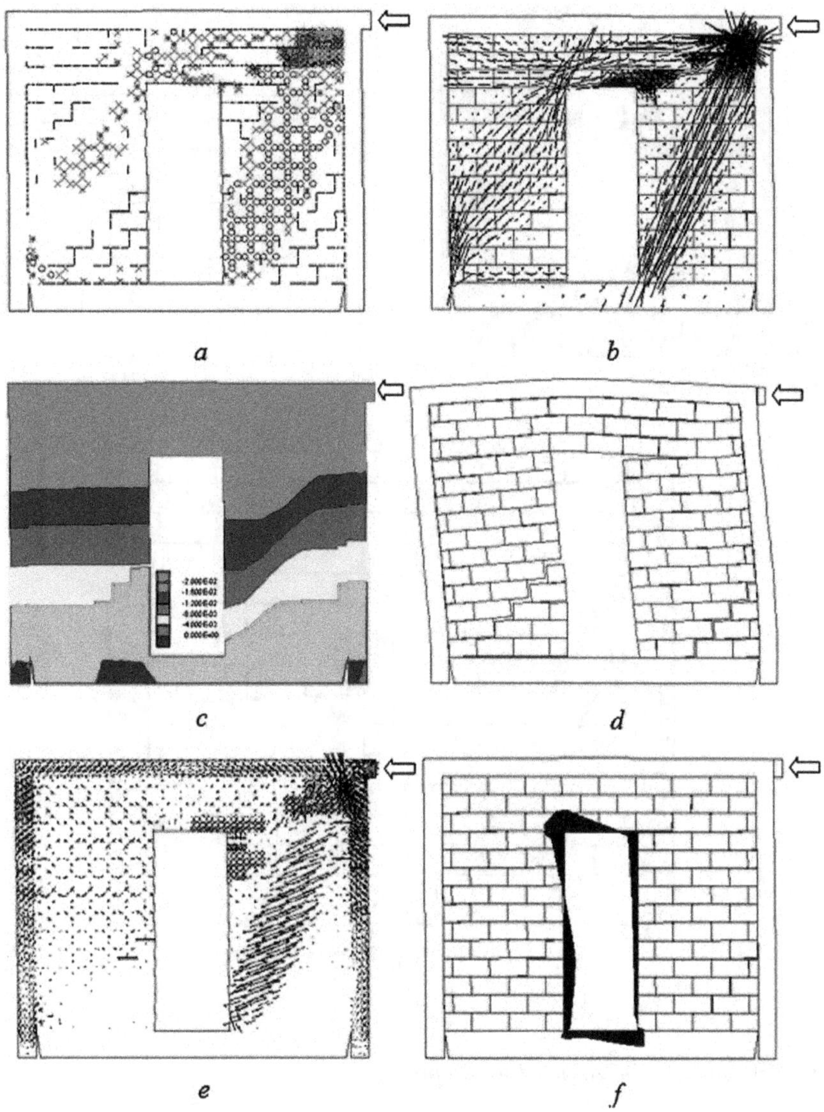

notably reduces the infilled frames lateral stiffness degradation compared to the frames without gravity loads (i.e. curves designated as W/O GL in the figure). It was also observed that applying gravity loads (resulted from the upper stories gravitational loads of a typical three-story infilled frame) resulted in an increase in lateral load capacity of 16% for specimen SW. However, the increase in lateral capacity for other specimens with openings depends on the opening relative length; as the opening length increases the capacity increase reduces.

Figure 11. DEM analysis of a masonry-infilled steel frame with multiple openings
Source: Sarhosis et al. (2014b)

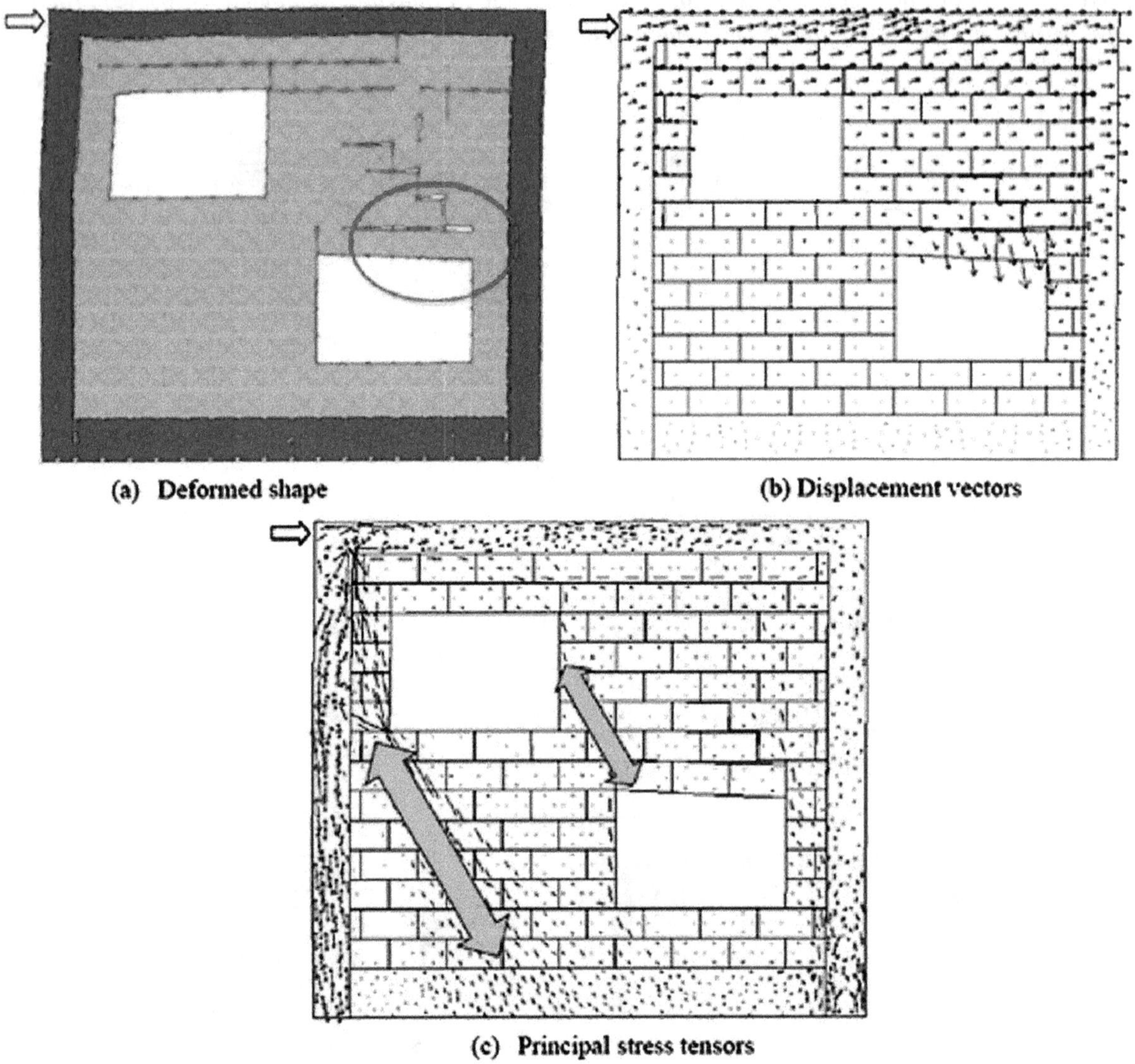

Figure 12. Configurations of load applied to the panel

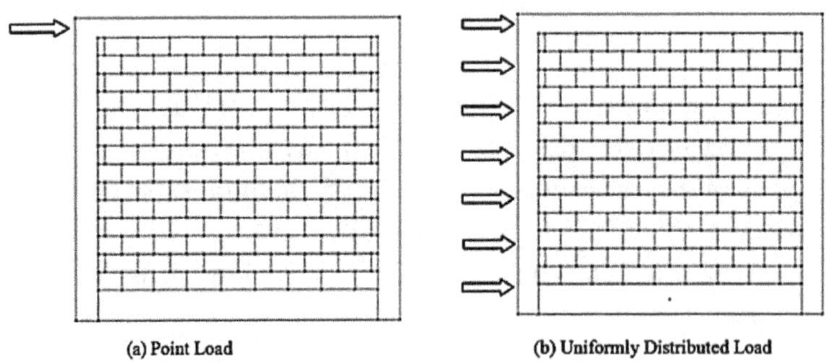

Figure 13. Load displacement behaviour for both loading conditions

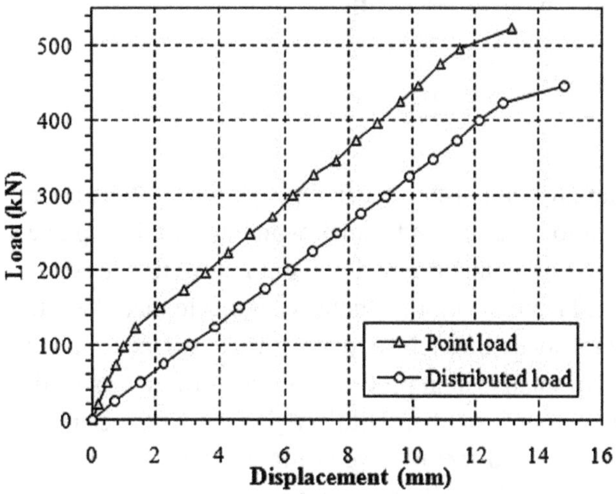

Figure 14. Effect of gravity loads on lateral load-displacement response of masonry-infilled steel frames using DEM
Source: Mohebkhah (2007)

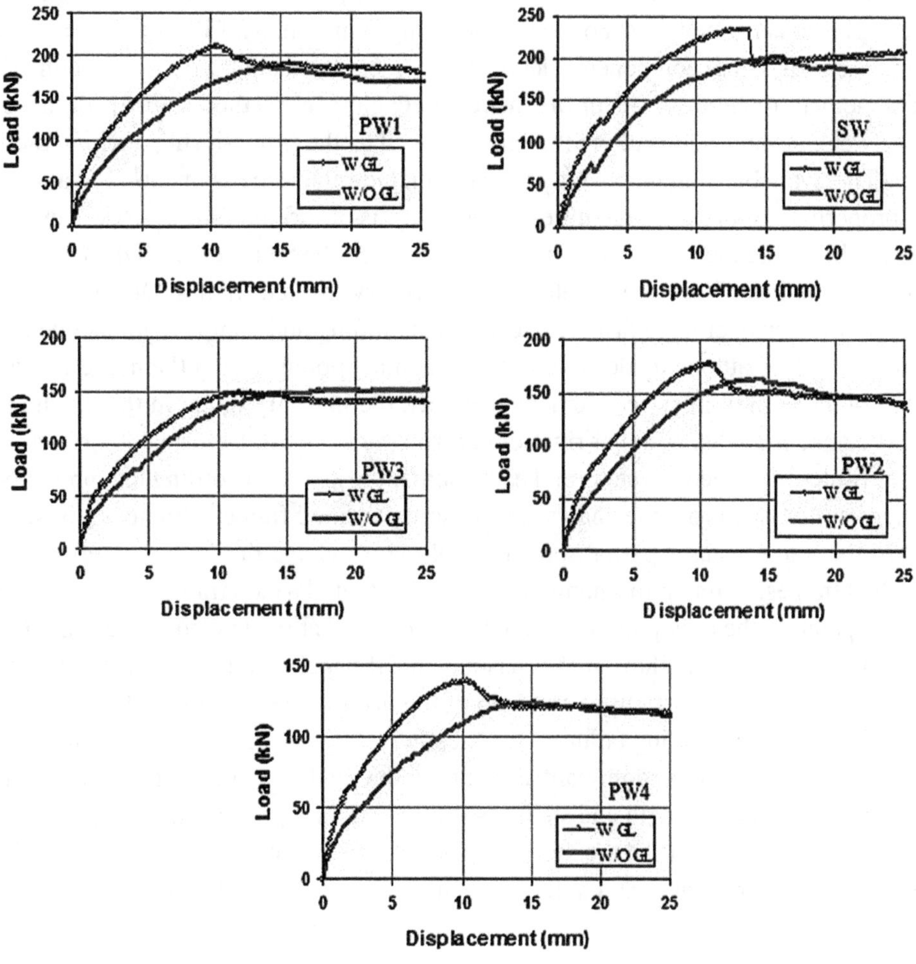

It should also be noted that, although applying gravity loads increases the frames' capacity, however, the frames behavior becomes more or less brittle.

CONCLUSION

Framed buildings in urban areas are usually infilled with masonry as interior and exterior walls. The resulting system is referred to as an infilled frame, which has high in-plane stiffness and strength. Although, such frames are analyzed and designed without taking into account the effects of infill panels, however, the lessons learned from the past earthquakes indicate that often, the infill panels can stabilize the buildings against lateral loads through increasing their stiffness and strength. Hence, the effects of infill panels in the analysis and design processes of such frames as well as seismic vulnerability of existing infilled frames is of major concern for the engineers. As experimental research is prohibitively expensive, it is fundamentally important to have available computational tools that can be used to predict the mechanical behaviour of the infill panels with sufficient reliability. Once such tools have been established, they can be used to investigate a range of complex problems and scenarios that would not, otherwise, be possible.

Numerical modeling of masonry-infilled frames presents many difficulties. These arise from the existence of joints as the major source of weakness and material nonlinearities as well as the infill-frame interaction. Numerical modeling strategies of infilled frames are divided into two distinct categories, micro-modeling and macro-modeling. For micro-modeling of masonry-infilled frames, both the surrounding frame and the infill panel components details are modeled and analyzed using a numerical method such as finite element method (FEM) or distinct/discrete element method (DEM). In this strategy, the interaction between masonry units along the joints as well as the frame-infill panel interaction is taken into account. The available finite element programs include interface elements that allow the user to incorporate properly masonry discontinuities in the analysis of infilled frames. However, as an alternative to the available finite element methods, discrete/distinct element method (DEM) can also be used to investigate the nonlinear lateral load behavior of masonry-infilled frames. Discrete element method has the capability to consider large displacements, shear sliding and complete joints openings between masonry units as well as automatic detection of new contact points during the numerical analysis process considering no explicit interface elements between the masonry units. In this method, individual masonry units can be modeled as either rigid or deformable material. Deformable units are subdivided into a mesh of finite-difference elements and each element responds according to a prescribed linear or nonlinear stress-strain law. From the analysis of masonry-infilled framed structures, it was shown that the swaying of the frames causes partial separation of the panel and the frame. In terms of the sway of masonry-infilled frames, a similar mechanism is noticed to that of a bare frame however to a much lesser degree. Once separated, the only point of contact can be found at the loaded corner and the diagonally opposite corner of the frame. Failure of the masonry infill panel could be described by the following four modes: diagonal cracking; localised crushing of the bricks/ blocks; sliding through the mortar or a combination of diagonal and sliding failure. Current studies suggest that localising crushing is the most common failure mode. Failure predominantly occurs in the loaded corner of the frame and diagonally opposite corner. The overall strength and stiffness is affected through variation in connections between the column and the panel, strength of the mortar and blocks and the rigidity of the frame. Also, cracking occurs mainly within the mortar through diagonal and sliding failure. The diagonal cracks propagated along the compression diagonal which runs from the loaded corner down to the opposite corner. Different models and experiments have resulted varying results in terms of the compression areas once the load is applied. Compression strut models have been carried out since the 1960s. In relation to masonry infilled frames containing openings, the openings within the masonry panel affect the lateral resistance of the frame, the failure mode and the ultimate load carrying capacity of the frame. From both experimental and analytical studies, it has been shown that changing the position of a door opening within the panel affects the lateral ultimate load. Also, the further away the opening is from the location of the application of the external load, the higher the load that the structure can resist. As the size of the opening increases the overall stiffness and strength of the system reduces. When the opening is located towards the compression diagonal the overall strength and stiffness is adversely affected. Stiffness has been monitored through stiffness factor values and interpreting load-displacement profiles of the frames. Failure modes similar to solid infills occur with corner crushing of the bricks close to the applied load and diagonal stepped cracking was found.

The discrete element method and in particular the UDEC software has been proved to accurately represent the experimental test results and predict the failure mechanisms and the strength and stiffness of the masonry infilled panels. According to the experience of the authors, the advantage of discrete element method for modeling masonry infill steel framed structures lies in the fact that each brick is considered individually and large displacement and rotations of masonry units are allowed. Also, within DEM the geometrical, physical and mechanical characteristics of a masonry structure can be represented realistically. This is opposed to other numerical methods where continuity theories are used to simplify the masonry structure into an unrealistic continuum medium. Moreover, the use of DEM for modelling infill wall panels can be used to predict accurately the location and magnitude of surface crack widths between adjacent masonry blocks. In this way, the serviceability limit state performance of masonry structures can be estimated accurately. However, it is worth highlighting that the size of the simulation is limited by the number of masonry units which is a function of the available computing power and that the accuracy of the results is highly dependent on the reliability of the material parameters. However, nowadays, with the fast development of fast processors and parallel programming, computational issues related to the analysis and execution of numerical simulations are becoming less of an issue.

REFERENCES

Achyutha, H., Jagadish, R., Rao, P. S., & Shakeebur, R. S. (1986). Finite element simulation of the elastic behavior of infilled frames with openings. *Computers & Structures*, *23*(5), 685–696. doi:10.1016/0045-7949(86)90077-5

Asteris, P. G. (2003). Lateral stiffness of brick masonry infilled plane frames. *Journal of Structural Engineering*, *129*(8), 1071–1079. doi:10.1061/(ASCE)0733-9445(2003)129:8(1071)

Asteris, P. G., Cotsovos, D. M., Chrysostomou, C. Z., Mohebkhah, A., & Al-Chaar, G. K. (2013). Mathematical micromodeling of infilled frames: State-of-the-art. *Engineering Structures*, *56*, 1905–1921. doi:10.1016/j.engstruct.2013.08.010

Asteris, P. G., Sarhosis, V., Mohebkhah, A., Plevris, V., Papaloizou, P., Komodromos, P., & Lemos, J. V. (2015). *Numerical Modeling of Historic Masonry Structures. In Seismic Assessment and Rehabilitation of Historic Structures*. IGI Global. doi:10.4018/978-1-4666-8286-3

Atkinson, R. H., Amadei, B. P., Saeb, S., & Sture, S. (1989). Response of masonry bed joints in direct shear. *Journal of Structural Engineering*, *115*(9), 2276–2296. doi:10.1061/(ASCE)0733-9445(1989)115:9(2276)

Azevedo, J. J., Sincraian, G. E., & Lemos, J. V. (2000). Seismic behavior of blocky masonry structures. *Earthquake Spectra*, *16*(2), 337–365. doi:10.1193/1.1586116

Bićanić, N., Ponniah, D., & Robinson, J. (2001). Computational modelling of masonry, brickwork and blockwork structures. In J. W. Bull (Ed.), *Discontinuous Deformation Analysis of Masonry Bridges*. Edinburgh, UK: Saxe-Coburg Publications.

Bobet, A., Fakhimi, A., Johnson, S., Morris, J., Tonon, F., & Yeung, M. R. (2009). Numerical models in discontinuous media, review of advances for rock mechanics applications. *Journal of Geotechnical and Geoenvironmental Engineering*, *135*(11), 1547–1561. doi:10.1061/(ASCE)GT.1943-5606.0000133

Chen, W. F. (1982). *Plasticity in reinforced concrete*. New York: McGraw-Hill.

Chen, X., & Liu, Y. (2015). Numerical study of in-plane behaviour and strength of concrete masonry infills with openings. *Engineering Structures*, *82*, 226–235. doi:10.1016/j.engstruct.2014.10.042

Claxton, M., Hart, R. A., McCombie, P. F., & Walker, P. J. (2005). Rigid block distinct-element modeling of dry-stone retaining walls in plane strain. *Journal of Geotechnical and Geoenvironmental Engineering*, *131*(3), 381–389. doi:10.1061/(ASCE)1090-0241(2005)131:3(381)

Cundall, P. A. (1971). *A Computer Model for Simulating Progressive Large Scale Movements in Blocky Rock Systems*. In *Proceedings of the Symposium of the International Society for Rock Mechanics*.

Dawe, J. L., & Seah, C. K. (1989). Behavior of masonry infilled steel frames. *Canadian Journal of Civil Engineering*, *16*(6), 865–876. doi:10.1139/l89-129

Dawe, J. L., Seah, C. K., & Liu, Y. (2001). A computer model for predicting infilled frame behavior. *Canadian Journal of Civil Engineering*, *28*(1), 133–148. doi:10.1139/l00-083

Dhanasekar, M., & Page, A. W. (1986). Influence of brick masonry nfill properties on the behaviour of infilled frames. Proc, Inst Civil Engrs. *Lond*, *81*(Part 2), 593–605.

El-Dakhakhni, W. W. (2002). *Experimental and analytical seismic evaluation of concrete masonry-infilled steel frames retrofitted using GFRP laminates*. (Ph.D. thesis). Drexel University, Philadelphia, PA.

El-Dakhakhni, W. W., Elgaaly, M., & Hamid, A. A. (2003). Three-strut model for concrete masonry-infilled steel frames. *Journal of Structural Engineering*, *129*(2), 177–185. doi:10.1061/(ASCE)0733-9445(2003)129:2(177)

Flanagan, R. D., & Bennett, R. M. (1999). In-plane behavior of structural clay tile infilled frames. *Journal of Structural Engineering*, *125*(6), 590–599. doi:10.1061/(ASCE)0733-9445(1999)125:6(590)

Gambarotta, L., & Lagomarsino, S. (1997). Damage models for the seismic response of brick masonry shear walls. Part I: The mortar joint model and its applications. *Journal of Earthquake Engineering and Structural Dynamics*, *26*(4), 423–439. doi:10.1002/(SICI)1096-9845(199704)26:4<423::AID-EQE650>3.0.CO;2-#

Ghosh, A. K., & Made, A. M. (2002). Finite element analysis of infilled frames. *Journal of Structural Engineering*, *128*(7), 881–889. doi:10.1061/(ASCE)0733-9445(2002)128:7(881)

Giamundo, V., Sarhosis, V., Lignola, G. P., Sheng, Y., & Manfredi, G. (2014). Evaluation of different computational strategies for modelling low strength masonry. *Engineering Structures*, *73*, 160–169. doi:10.1016/j.engstruct.2014.05.007

Giordano, A., Mele, E., & Luca, A. (2002). Modelling of historical masonry structures.Comparison of different approaches through a case study. *Engineering Structures*, *24*(8), 1057–1069. doi:10.1016/S0141-0296(02)00033-0

Itasca Consulting Group, Inc. (2004). UDEC- Universal Distinct Element Code, version 4.0- User's manual. Minneapolis, MN: Author.

Jing, L. (2003). A review of techniques, advances and outstanding issues in numerical modelling for rock mechanics and rock engineering. *International Journal of Rock Mechanics and Mining Sciences*, *40*(3), 283–353. doi:10.1016/S1365-1609(03)00013-3

King, G. J. W., & Pandey, P. C. (1978). The analysis of infilled frames using finite elements. *Proceedings - Institution of Civil Engineers*, *65*(Part 2), 749–760. doi:10.1680/iicep.1978.2707

Koutromanos, I. (2011). *Numerical analysis of masonry-infilled reinforced concrete frames subjected to seismic loads and experimental evaluation of retrofit techniques*. (Ph.D. thesis). University of California, San Diego, CA.

Koutromanos, I., Stavridis, A., Shing, P. B., & Willam, K. (2011). Numerical modeling of masonry-infilled RC frames subjected to seismic loads. *ComputStruct.*, *89*, 1026–1037.

Lang, A. F., & Benzoni, G. (2014). Modeling of nonlinear behavior of confined masonry using discrete elements. In *Proceedings of the 10th National Conf. on Earthquake Engineering*. Frontier of Earthquake Engineering.

Lemos, J. V. (1997). Discrete element modeling of the seismic behavior of stone masonry arches. Computer Methods in Structural Masonry- 4.

Lemos, J. V. (2007). Discrete element modeling of masonry structures. *International Journal of Architectural Heritage*, *1*(2), 190–213. doi:10.1080/15583050601176868

Liauw, T. C., & Kwan, K. H. (1985). Unified plastic analysis for infilled frames. *Journal of Structural Engineering*, *111*(7), 1427–1448. doi:10.1061/(ASCE)0733-9445(1985)111:7(1427)

Liu, Y., & Manesh, P. (2013). Concrete masonry infilled steel frames subjected to combined in-plane lateral and axial loading – An experimental study. *Engineering Structures*, *52*, 331–339. doi:10.1016/j.engstruct.2013.02.038

Lofti, H. R., & Shing, P. B. (1994). Interface model applied to fracture of masonry structures. *Journal of Structural Engineering*, *120*(1), 63–80. doi:10.1061/(ASCE)0733-9445(1994)120:1(63)

Lourenco, P. B. (1996). *Computational strategies for masonry structures.* (Ph.D. thesis). Delft University of Technology, Delft, The Netherlands.

Lourenco, P. B., Oliveira, D. V., Roca, P., & Orduna, A. (2005). Dry joint masonry walls subjected to in-plane combined loading. *Journal of Structural Engineering*, *131*(11), 1665–1673. doi:10.1061/(ASCE)0733-9445(2005)131:11(1665)

Lourenço, P. B., & Rots, J. G. (1997). A multi-surface interface model for the analysis of masonry structures. *Journal of Engineering Mechanics*, *123*(7), 660–668. doi:10.1061/(ASCE)0733-9399(1997)123:7(660)

Madan, A., Reinhorn, A. M., Mander, J. B., & Valles, R. E. (1997). Modeling of masonry infill panels for structural analysiss. *Journal of Structural Engineering*, *123*(10), 1295–1302. doi:10.1061/(ASCE)0733-9445(1997)123:10(1295)

Markulak, D., Radic´, I., & Sigmund, V. (2013). Cyclic testing of single bay steel frames with various types of masonry infill. *Engineering Structures*, *51*, 167–277. doi:10.1016/j.engstruct.2013.01.026

Mehrabi, A. B., & Shing, P. B. (1997). Finite element modeling of masonry-infilled RC frames. *Journal of Structural Engineering*, *123*(5), 604–613. doi:10.1061/(ASCE)0733-9445(1997)123:5(604)

Moghaddam, H. A. (2004). Lateral load behavior of masonry infilled steel frames with repair and retrofit. *Journal of Structural Engineering*, *130*(1), 56–63. doi:10.1061/(ASCE)0733-9445(2004)130:1(56)

Moghaddam, H. A., & Dowling, P. J. (1987). *The state-of-the-art in infilled frames. ESEE Res. Rep. No. 87-2, Imperial Coll. of Sci. and Technol.* London: Civ. Eng. Dept.

Moghaddam, H. A., Mohammadi, M. G. H., & Ghaemian, M. (2006). Experimental and analytical investigation into crack strength determination of infilled steel frames. *Journal of Constructional Steel Research*, *62*(12), 1341–1352. doi:10.1016/j.jcsr.2006.01.002

Mohebkhah, A. (2007). *A nonlinear-seismic model for brick masonry-infill panels with openings in steel frames*. (Ph.D. Dissertation). TarbiatModares University, Tehran, Iran.

Mohebkhah, A., & Sarv-Cheraghi, A. A. (2015). Nonlinear analysis of unreinforced masonry buildings using distinct element method[in Persian]. *Modares Civil Engineering Journal,*15(3), 85-91, *Tehran, Iran.*

Mohebkhah, A., & Tasnimi, A. A. (2007). Seismic behavior of brick masonry walls recommended by IS-2800: experimental and numerical approaches. In *Proceedings of the 5thConf. on Seismology and Earthquake Engineering*. International Institute of Earthquake Engineering and Seismology.

Mohebkhah, A. and Tasnimi, A.A. (2012). Distinct element modeling of masonry-infilled steel frames with openings. *The Open Construction and Building Technology Journal*, *6*(Suppl 1-M2), 42-49.

Mohebkhah, A., Tasnimi, A. A., & Moghadam, H. A. (2008). Non-linear analysis of masonry in-filled frames with opening using discrete element method. *Journal of Constructional Steel Research*, *64*(21), 1463–1472. doi:10.1016/j.jcsr.2008.01.016

Mosalam, K. M., White, R. N., & Gergely, P. (1997). Static response of infilled frames using quasi-static experimentation. *Journal of Structural Engineering*, *123*(11), 1462–1469. doi:10.1061/(ASCE)0733-9445(1997)123:11(1462)

Mosalam, K. M., White, R. N., & Gergely, P. (1997). *Computational strategies for frames with infill walls: discrete and smeared crack analysis and seismic fragility.* Technical Report NCEER-97-0021. National Center for Earthquake Engineering Research, State Univ. of N.Y. at Buffalo.

Paulay, T., & Priestley, M. J. N. (1992). *Seismic design of reinforced concrete and masonry buildings.* New York, NY: John Wiley & Sons, Inc. doi:10.1002/9780470172841

Polyakov, S. V. (1956). Masonry in framed buildings. In *Translated by G. L. Cairns in 1963.* Boston Spa, UK: National Lending Library for Science and Technology.

Psycharis, I. N., Lemos, J. V., Papastamatiou, D. Y., Zambas, C., & Papantonopoulos, C. (2003). Numerical study of the seismic behavior of a part of the Parthenon Pronaos. *Earthquake Engineering & Structural Dynamics*, *32*(13), 2063–2084. doi:10.1002/eqe.315

Psycharis, I. N., Papastamatiou, D. Y., & Alexandris, A. P. (2000). Parametric investigation of the stability of classical columns under harmonic and earthquake excitations. *Earthquake Engineering & Structural Dynamics*, *29*(8), 1093–1109. doi:10.1002/1096-9845(200008)29:8<1093::AID-EQE953>3.0.CO;2-S

Riddington, J. R., & Stafford-Smith, B. (1977). Analysis of infilled frames subjected to racking with design recommendations. *The Structural Engineer*, *55*(6), 263–268.

Rots, J. G. (1997). *Structural masonry: an experimental/numerical basis for practical design rules. Center for Civil Engineering Research and Codes (CUR).* Balkema.

Saneinejad, A., & Hobbs, B. (1995). Inelastic design of infield frames. *Journal of Structural Engineering*, *121*(4), 634–650. doi:10.1061/(ASCE)0733-9445(1995)121:4(634)

Sarhosis, V., Garrity, S. W., & Sheng, Y. (2008). *Distinct element modelling of masonry wall panels with openings.* 9th International Conference on Computational Structures Technology, Athens, Greece.

Sarhosis, V., Garrity, S. W., & Sheng, Y. (2015). Influence of the brick-mortar interface on the mechanical response of low bond strength masonry lintels. *Engineering Structures*, *88*, 1–11. doi:10.1016/j.engstruct.2014.12.014

Sarhosis, V., Lignola, G. P., & Asteris, P. (2016). Seismic Vulnerability of Ancient Colonnade: Numerical analysis of the two storey colonnade of the Forum in Pompeii. In Asteris & Repapis (Eds.), Seismic Assessment and Rehabilitation of Historic Structures. IGI Global.

Sarhosis, V., Oliveira, D. V., Lemos, J. V., & Lourenco, P. (2014a). The effect of the angle of skew on the mechanical behaviour of arches. *Journal of Mechanics Research Communications*, *61*, 53–49. doi:10.1016/j.mechrescom.2014.07.008

Sarhosis, V., & Sheng, Y. (2014). Identification of material parameters for low bond strength masonry. *Engineering Structures*, *60*, 100–110. doi:10.1016/j.engstruct.2013.12.013

Sarhosis, V., Tsavdaridis, K., & Giannopoulos, G. (2014b). Discrete Element Modelling of masonry in-filled steel frames with multiple window openings subjected to lateral load variations. *The Open Construction and Building Technology Journal*, *8*(1), 93–103. doi:10.2174/1874836801408010093

Schlegel, R., & Rautenstrauch, K. (2004). Failure analyses of masonry shear walls. In H. Konietzky (Ed.), *Numerical modelling of discrete materials in geotechnical engineering, civil engineering and earth sciences* (pp. 15–20). Taylor and Francis Group London.

Schneider, S. P., Zagers, B. R., & Abrams, D. P. (1998). Lateral strength of steel frames with masonry infills having large openings. *Journal of Structural Engineering*, *124*(8), 896–904. doi:10.1061/(ASCE)0733-9445(1998)124:8(896)

Sharma, V. M., Saxena, K. R., & Woods, R. D. (1999). *Distinct Element Modeling in Geomechanics*. Rotterdam: A. A. Balkema.

Shi, G. H., & Goodman, R. E. (1988). Discontinuous deformation analysis – a new method for computing stress, strain and sliding of block systems. In P.A. Cundall, R. Sterling, & A. Starfield (Eds.), *Proceedings of the 29th U.S. Symposium: Key Questions in Rock Mechanics.*

Sincraian, G. E., & Azevedo, J. J. (1998). Numerical simulation of the seismic behavior of stone and brick masonry structures.*Proc. of the 11th European Conference on Earthquake Engineering*. Balkema.

Sincraian, G. E., Lemos, J. V., & Oliveira, C. S. (1998). Assessment of the seismic behavior of stone masonry aqueduct using the discrete element method.*Proc. 11th European Conference on Earthquake Engineering*. Balkema.

Stafford-Smith, B. (1962). Lateral stiffness of infilled frames. *Journal of the Structural Division*, *88*(6), 183–199.

Stafford-Smith, B. (1966). Behavior of square infilled frames. *Journal of the Structural Division*, *92*(1), 381–403.

Stavridis, A. (2009). *Analytical and experimental study of seismic performance of reinforced concrete frames infilled with masonry walls.* (PhD thesis). Department of Structural Engineering, University of California, San Diego, CA.

Stavridis, A., & Shing, P. B. (2010). Finite-element modeling of nonlinear behavior of masonry-infilled RC frames. *Journal of Structural Engineering*, *136*(3), 285–296. doi:10.1061/(ASCE)ST.1943-541X.116

Tasnimi, A. A., & Farzin, M. (2006). Inelastic behavior of RC columns under cyclic loads based on cohesion and internal friction angle of concrete. *Modarres Technical and Engineering Journal*, *23*, 29–40.

Tasnimi, A. A., & Mohebkhah, A. (2011). Investigation on the behavior of brick-infilled steel frames with openings, experimental and analytical approaches. *Engineering Structures*, *33*(3), 968–980. doi:10.1016/j.engstruct.2010.12.018

Toth, A. R., Orban, Z., & Bagi, K. (2009). Discrete element analysis of stone masonry arch. *Mechanics Research Communications*, *36*(4), 469–480. doi:10.1016/j.mechrescom.2009.01.001

ADDITIONAL READING

Al-Chaar, G., Issa, M., & Sweeney, S. (2002). Behavior of masonry-infilled nonductile reinforced concrete frames. *Journal of Structural Engineering*, *128*(8), 1055–1063. doi:10.1061/(ASCE)0733-9445(2002)128:8(1055)

Al-Chaar, G. L., & Mehrabi, A. (2008). *Constitutive models for nonlinear finite element analysis of masonry prisms and infill walls (No. ERDC/CERL TR- 08- 19)*. US Army Corps of Engineering.

Almeida, J. C., Lourenco, P. B., & Barros, J. A. (2002). Characterization of brick and brick–mortar interface under uniaxial tension, in: Proceedings of 7th Int. Seminar on Structural Masonry, Brazil: CEFET-MG. 67–76.

Amato, G., Cavaleri, L., & Fossetti, M. (2008). Infilled frames: Influence of vertical load on the equivalent diagonal strut model. Presented at the 14 WCEE, Beijing, China.

Braga, F., Gigliotti, R., Laterza, M., D'Amato, M., & Kunnath, S. (2012). Modified steel bar model incorporating bond-slip for seismic assessment of concrete structures. *Journal of Structural Engineering*, *138*(11), 1342–1350. doi:10.1061/(ASCE)ST.1943-541X.0000587

Buonopane, S., & White, R. (1999). Pseudodynamic testing of masonry infilled reinforced concrete frame. *Journal of Structural Engineering*, *125*(6), 578–589. doi:10.1061/(ASCE)0733-9445(1999)125:6(578)

Calvi, G. M., & Bolognini, D. (2001). Seismic response of reinforced concrete frames infilled with weakly reinforced masonry panels. *Journal of Earthquake Engineering*, *5*(2), 153–185. doi:10.1080/13632460109350390

Calvi, G.M., Bolognini, D., Penna, A., (2004). Seismic performance of masonry-infilled RC frames: benefits of slight reinforcement. SÍSMICA 2004 - 6 o Congreso Nacional de Sismología e Engeñaría Sísmica.

Cavaleri, L., Fossetti, M., & Papia, M. (2004). Effect of vertical loads on lateral response of infilled frames. Presented at the 13WCEE, Vancouver, B.C., Canada.

Fiorato, A. E., Sozen, M. A., & Gamble, W. L. (1970). *Behavior of five-story reinforced concrete frames with filler walls, Interim report to the Department of Defense Office of Secretary of the Army*. Urbana, Illinois: Office of Civil Defense.

Fiorato, A. E., Sozen, M. A., & Gamble, W. L. (1970). *An investigation of the interaction of reinforced concrete frames with masonry filler walls (No. UILU-ENG-70-100)*. Urbana- Champaign, IL: Dept. of Civil Engineering, University of Illinois.

Flanagan, R. D., & Bennett, R. M. (1999). Arching of masonry infilled frames: Comparison of analytical methods. *Practice Periodical on Structural Design and Construction*, *4*(3), 105–110. doi:10.1061/(ASCE)1084-0680(1999)4:3(105)

Gabrielsen, B. L., & Kaplan, K. (1997). Arching in masonry walls subjected to out-of-plane forces. *Earthquake Resistant Masonry Construction. NBS Building Science Series*, *106*, 283–313.

Gumaste, K. S., Nanjunda Rao, K. S., Venkatarama Reddy, B. V., & Jagadish, K. S. (2006). Strength and elasticity of brick masonry prisms and wallets under compression. *Materials and Structures*, *40*(2), 241–253. doi:10.1617/s11527-006-9141-9

Hamburger, R. O., & Meyer, J. D. (2006). The performance of steel-frame buildings with infill masonry walls in the 1906 San Francisco earthquake. *Earthquake Spectra*, *22*(1), 43–67. doi:10.1193/1.2185656

Haselton, C.B., Goulet, C.A., Mitrani-Reiser, J., Beck, J.L., Deierlein, G.G., Porter, K.A., (2007). An assessment to benchmark the seismic performance of a code-conforming reinforced concrete moment-frame building (No. PEER 2007/12).

Haselton, C. B., Liel, A. B., Lange, S. T., & Deierlein, G. G. (2008). Beam-column element model calibrated for predicting flexural response leading to global collapse of RC frame buildings (No. PEER 2007/03). CA, Berkeley.

Hashemi, A., & Mosalam, K. (2007). *Seismic evaluation of reinforced concrete buildings including effects of masonry infill walls, PEER Technical Report*. Pacific Earthquake Engineering Research Center.

Lepage, A. (1997) A Method for Drift-Control in Earthquake-Resistant Design of RC Building Structures. PhD Thesis, University of Illinois, Urbana, 251.

Matjaz, D., & Fajfar, P. (2005). Simplified Non-linear Seismic Analysis of Infilled Reinforced Concrete Frames. *Earthquake Engineering & Structural Dynamics*, *34*(1), 49–66. doi:10.1002/eqe.411

Negro, P., & Verzeletti, G. (1996). Effect of Infills on the Global Behaviour of Frames: Energy Considerations from Pseudo-dynamic Tests. *Earthquake Engineering & Structural Dynamics*, *25*(8), 753–773. doi:10.1002/(SICI)1096-9845(199608)25:8<753::AID-EQE578>3.0.CO;2-Q

Park, Y. J., Reinhorn, A. M., & Kunnath, S. K. (1987). IDARC: Inelastic Damage Analysis of Reinforced Concrete Frame – Shear-Wall Structures. Technical Report NCEER-87-0008, State University of New York at Buffalo.

Rodriguez, M. E. (2008) Selecting Earthquake Records for Nonlinear Dynamic Analysis of Structures, *Proc Seismic Engineering International Conference Commemorating the 1908 Messina and Reggio Calabria Earthquake*. doi:10.1063/1.2963824

Saneinejad, A., & Hobbs, B. (1995). Inelastic Design of Infilled Frames. *Journal of Structural Engineering*, *121*(4), 634–650. doi:10.1061/(ASCE)0733-9445(1995)121:4(634)

Wood, S. (1991). Performance of Reinforced Concrete Buildings Duringthe 1985 Chile Earthquake: Implications for the Design of Structural Walls. *Earthquake Spectra*, *7*(4), 607–639. doi:10.1193/1.1585645

Zarnic, R., Gosti, S., Crewe, A. J., & Taylor, C. A. (2001). Shaking Table Tests of 1:4 Reduced-Scale Models of Masonry Infilled Reinforced Concrete Frame Buildings. *Earthquake Engineering & Structural Dynamics*, *30*(6), 819–834. doi:10.1002/eqe.39

KEY TERMS AND DEFINITIONS

Computer Simulation: Simulation run on a computer to reproduce behavior of a system.

Discrete Element Method: Is any of a family of numerical methods for computing the motion and effect of a large number of small particles.

Infill Wall: A panel that is built in between the beam and column components of the primary structural frame of a building.

Loads: Forces which are applied to a structure.

Masonry: Construction material composed of individual units laid in and bound together with or without mortar.

Steel Frame: Is a building technique which includes vertical steel columns and horizontal beams, constructed in a rectangular grid to support the floors, roof and walls of a building which are all attached to the frame.

Stiffness: Refers to the rigidity of a structural element and the extent to which it resists deformation with the application of external load.

APPENDIX: LIST OF SYMBOLS AND ACRONYMS

δ : Interface displacement vector.
σ : Joint normal stress.
τ : Joint shear stress.
ν = Poisson's ratio.
ϕ = Internal friction angle.
ψ_j = Dilatancy angle.
c: Cohesion strength.
c_u: Cohesion, ultimate value bond shear strength.
E_b: Block Young's modulus.
E_m: Mortar Young's modulus.
E_{wall}: Wall Young's modulus.
E_s: Stone unit Young's modulus.
f_b': Block compressive strength.
f_c': Concrete compressive strength.
f_t: Tensile strength.
G_f^I : Mode-I fracture energy for tensile fracture.
G_f^{II} : Mode-II fracture energy for shear fracture.
G_b: Block shear modulus.
G_m: Mortar shear modulus.
h_b: Block height.
h_m: Mortar thickness.
h_s: Stone unit height.
K: Joint stiffness vector.
k: Axial stiffness of masonry prism.
k_n: Joint normal stiffness.
k_s: Joint shear stiffness.
t: Interface stress vector.
u: Relative displacement at normal to interface.
u_p: Crack opening.
v: Relative shear displacement at interface.
v_p: Crack slip.
CC: Corner Crushing failure mode.
CFF: Column Flexural Failure mode.
CSF: Column Shear Failure mode.
DDA: Discontinuous Deformation Analysis.
DEM: Discrete/Distinct Element Method.
DTC: Diagonal Tension Cracking.
FEM: Finite Element Method.
HSS: Horizontal Sliding Shear failure mode.
UDEC: Universal Distinct Element Code.

Chapter 10
Vulnerability Assessment of Damaged Classical Multidrum Columns

Michalis Fragiadakis
National Technical University of Athens, Greece

Ioannis Stefanou
Université Paris-Est, France

Ioannis N. Psycharis
National Technical University of Athens, Greece

ABSTRACT

Multi-drum columns are articulated structures, made of several discrete bulgy stone blocks (drums) placed one on top of the other without mortar. The multi-drum column is a typical structural element of temples of the Classical, Hellenistic and earlier Roman period. Despite the lack of any lateral load resisting mechanism, these columns have survived several strong earthquakes over the centuries. The Chapter focuses on the effect of past drum dislocations on the vulnerability of classical columns and presents a performance-based framework for their seismic risk assessment. The vulnerability is numerically calculated through response estimations using detailed three-dimensional models based on the Discrete Element Method. Conditional limit-state probabilities are calculated and appropriate performance criteria are suggested. The proposed methodology is able to pinpoint cases where past damage affects the vulnerability of such structures and can serve as a valuable decision-making tool.

INTRODUCTION

Classical monuments are particular masonry structures made of bulgy stone blocks. Major structural members of these ancient structures are their columns, which consist of bulgy discrete drums that lie one on top of the other without mortar (Figure 1). During earthquakes, the columns respond with intense rocking, wobbling and, depending on the magnitude of the induced accelerations, sliding of the drums. In rare cases, steel connections (dowels) are provided at the joints, which restrict sliding without, usually, affecting rocking.

DOI: 10.4018/978-1-5225-0231-9.ch010

Figure 1. Olympieion of Athens, Greece: (a) standing South East corner; (b) single standing and fallen column showing the multi-drum construction of ancient columns

Several investigators have examined the seismic response of classical monuments and, in general, of stacks of rigid bodies analytically, numerically or experimentally, mostly using two-dimensional models (Allen et al., 1986; Sinopoli, 1991; Psycharis, 1990; Konstantinidis & Makris, 2005; Lemos, 2007; Papaloizos & Komodromos, 2009; Sarhosis V., Lignola G.P. & Asteris P., 2014). Three-dimensional studies are less, e.g. Psycharis et al. (2003), Dasiou et al. (2009b), Stefanou et al. (2011a, 2011b). These studies have shown that:

- Owing to rocking and sliding, the response is nonlinear. The nonlinear nature of the response is pronounced even for the simplest case of a rocking single block (Housner, 1963; Makris & Zhang, 2001). In addition, multidrum columns can rock in various 'modes', which might alternate during the response increasing thus the complexity of the problem (Psycharis, 1990). The word 'mode' denotes the pattern of rocking motion rather than a natural mode in the classical sense, since rocking structures do not possess such modes and periods of oscillation.
- The dynamic behaviour is sensitive to even trivial changes in the geometry of the structure or in the base-motion characteristics. The sensitivity of the response has been verified experimentally, since 'identical' experiments produced significantly different results in some cases (Yim *et al,.* 1980; Mouzakis *et al.,* 2002; Dasiou *et al.,* 2009a). The sensitivity of the response is responsible for the significant out-of-plane motion observed during shaking table experiments for purely planar excitations (Mouzakis *et al.,* 2002).
- The vulnerability of the structure greatly depends on the predominant period of the ground motion, with earthquakes containing low-frequency pulses being in general much more dangerous than high-frequency ones (Makris & Roussos, 2000; Psycharis *et al.,* 2000). The former pulses force the structure to respond with intensive rocking, whereas the latter produce significant sliding of the drums, especially at the upper part of the structure.
- The size of the structure affects significantly the stability (Psycharis, 1985; Makris & Roussos, 2000; Psycharis *et al.* 2000), with larger structures being much more stable than smaller ones of the same slenderness.

Previous analyses of the seismic response of classical columns have shown that these structures, despite their apparent instability to horizontal loads, are, in general, earthquake-resistant (Psycharis et al. 2000). This is also proven by the fact that many classical monuments built in seismic prone areas have survived for almost 2500 years. However, many others have collapsed.

The assessment of the seismic reliability of a monument is a prerequisite for the correct decision-making during a restoration process. The seismic vulnerability of the column, not only in what concerns the collapse risk, but also the magnitude of the expected maximum and the additional residual displacements of the drums, is vital information that can help the authorities decide and prioritize the necessary interventions. This assessment is not straightforward, since fully detailed analyses for the near-collapse state are practically impossible due to the sensitivity of the response to small changes in the geometry and the difficulty in modeling accurately the existing imperfections. The results of such analyses depend highly on the ground motion characteristics, which introduces additional uncertainty on the problem.

On top of the above uncertainties, one must realize that, in their current condition, ruins of ancient structures present many different types of damage. Most common are: missing pieces (cut-offs) that reduce the horizontal sections in contact, foundation problems resulting in tilting of the columns, dislocated drums from previous earthquakes or due to the collapse of the main structure (Figure 2), and cracks in the structural elements that in some cases split the block in two. Such imperfections affect significantly the stability of the columns which, thus, are much more vulnerable to earthquake excitations compared with the original intact structures. Therefore, the above-mentioned impressive stability of ancient monuments against earthquakes might not be still present, if significant damage is present.

An example of the effect of existing imperfections on the stability of ancient columns is shown in Figure 3 for a column of the Parthenon in Athens (Psycharis *et al.* 2003). The maximum permanent displacement of the column is plotted versus the peak ground acceleration (*pga*) of the ground motion and it is seen that the presence of the imperfections shown in the left drawing of Figure 3 leads to larger displacements and the column collapses for a smaller pga.

Figure 2. (a) Columns at Propylaia of Acropolis hill in Athens, Greece; (b) Remaining standing column of the temple of Hera in Samos, Greece. Significant drum dislocations are observed in both cases

Figure 3. Maximum permanent displacements of a column of the Parthenon under the Aigion, Greece (1995) earthquake amplified to several values of pga without and with the imperfections shown in the left diagram (Psycharis et al. 2003)

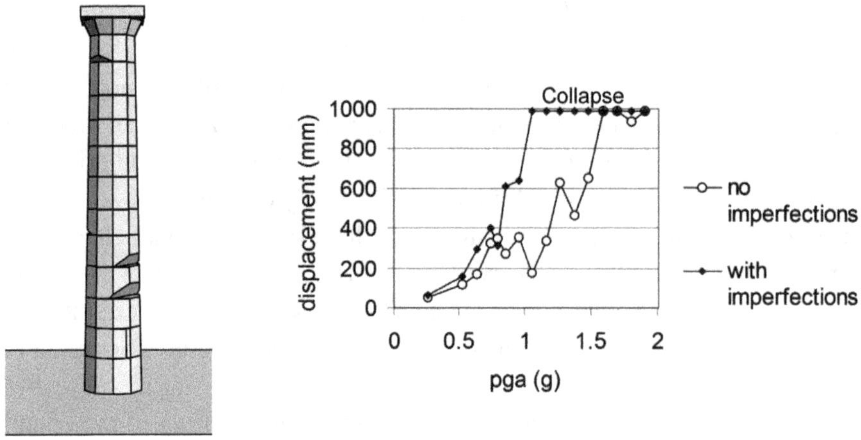

The effect of drum dislocations on the vulnerability of these systems is studied in this chapter. For this purpose, we introduce a performance-based seismic risk assessment framework for developing seismic fragility surfaces of damaged and undamaged classical multidrum columns. Performance-Based Earthquake Engineering (PBEE) concepts are tailored to these unique structures, while appropriate limit states are proposed. Limit-state exceedance probabilities are subsequently calculated in order to develop fragility surfaces, i.e. surfaces that provide the limit-state exceedance probability as function of magnitude and distance. Such fragility surfaces will serve as a valuable decision-making tool.

Modeling of the Columns

In order to explore the effect of drums' dislocations on the seismic performance of the classical columns, we study a simplified model of the column of Propylaia (Figure 2a) considering two different configurations (Figure 4). The first configuration refers to an "intact" column geometry (Figure 4, left) and the second to a "damaged" column of the same size and number of drums (Figure 4, right). In the latter column, above the first drum a dislocation of 15% of the drum's diameter, equal to 15 cm, is today present. This dislocation may be attributed to human activity, to previous earthquake events, or to the collapse of the main structure. Both damage and intact columns consist of seven drums of equal diameter of 1.00 m and height equal to 0.85 m.

All three-dimensional numerical simulations of this Chapter were performed using the Discrete Element Method (DEM) code 3DEC developed by Itasca (Itasca) corporation. A critical factor for the numerical analysis is the choice of appropriate constitutive laws for modeling the mechanical behavior of the joints. Joint elements are placed between the drums. A Coulomb failure criterion is here adopted, while the values assumed for the friction angle, the cohesion, the ultimate tensile strength and the stiffness of the joints are listed in Table 1. Notice that the stiffness might affect considerably the results of the analysis. A parametric investigation performed by Toumbakari and Psycharis (2010) showed that stiff joints might lead to larger permanent dislocations of all drums for strong ground motions compared with joints of soft stiffness. The values of Table 1 correspond to marble columns and were calibrated

Figure 4. On the left: geometry of the intact column. On the right: geometry of damaged column where the column over the first drum has been dislocated for 15cm (15% of the drums' diameter)

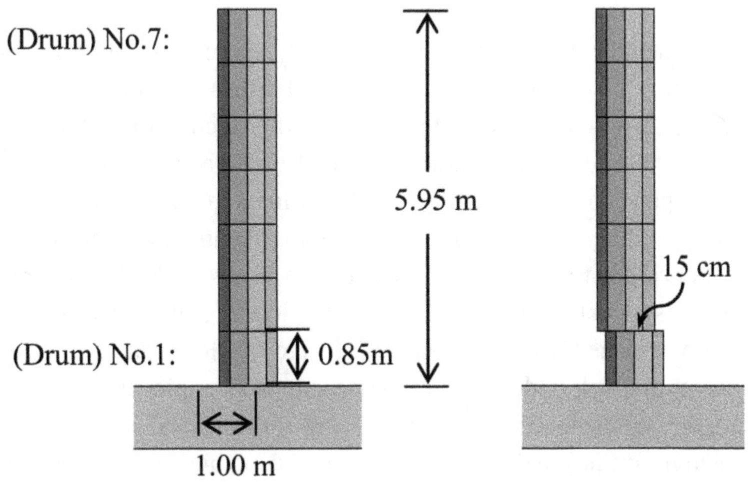

against shaking table experiments on the seismic response of free-standing columns (Papantonopoulos et al. 2002). Using these values, good agreement was achieved for both the maximum displacement of the top drum and also for the residual drum displacements. It must be pointed out, however, that different values should be assigned to the stiffness parameters for material other than marble of good quality. One way to choose the appropriate value for a specific column is through calibrating the stiffness using ambient vibration measurements (see Ambraseys & Psycharis 2012).

No connections were considered between the drums, as the only connectors that are present in such structures are wooden dowels, known as 'empolia'. Empolia were used to center the drums during the erection of the column and not to provide a shear resistance mechanism. Moreover, according to Konstantinidis and Makris (2005), the shear strength of the wooden dowels is small and has only marginal effect to the response of the column. For this reason, the wooden dowels were not considered in our numerical model. Furthermore, no artificial (numerical) damping was introduced to the system. According to the results of a previous investigation (Papantonopoulos et al. 2002), damping is set to zero only during the intense rocking response. Non-zero damping is considered at the end of the seismic excitation in order to dissipate the free vibrations.

Table 1. Constitutive parameters for the Coulomb frictional model considered for the mechanical behaviour of the joints between the drums

Parameter	Value
Normal Stiffness	1 GPa/m
Shear Stiffness	1 GPa/m
Friction Angle	37°
Cohesion	0 MPa
Tensile strength	0 MPa

PERFORMANCE-BASED RELIABILITY ASSESSMENT OF CLASSICAL MONUMENTS

Performance-Based Earthquake Engineering (PBEE) and seismic risk assessment combine computational tools and reliability assessment procedures in order to obtain the system fragility for a wide range of limit states. The seismic risk assessment requires the calculation of the failure probabilities of a pre-set number of performance levels. According to PBEE, the acceptable level of damage sustained by a structural system depends on the level of ground shaking and its significance. For example, under a frequent earthquake a building structure should be able to tolerate minor, non-structural, damage, but a critical facility (e.g. a bridge or a hospital) should remain intact and fully operable. Thus, the goal in seismic risk assessment is to obtain the probabilities of violating limit-state performance objectives, ranging from little or no damage for frequent earthquakes to severe damage for rare events.

Today, these concepts are well understood among earthquake engineers, but, when classical monuments are considered, the performance-based criteria may differ considerably. For example, in order to retrofit an ancient column, one has to decide what is the 'acceptable level' of damage for a given intensity level. The approach for making such decisions is not straightforward. A consensus among various experts in archaeology and monument preservation is necessary, while a number of non-engineering decisions must be made.

Performance Levels

In order to assess the risk of a monument, the performance levels of interest and corresponding limit-state capacity thresholds need first to be decided. Demand and capacity should be measured with appropriate parameters (e.g. stresses, strains, displacements) at critical locations, in accordance to the different damage (or failure) modes of the structure. Subsequently, this information has to be translated into one or a combination of engineering demand parameters (*EDP*s), e.g., maximum column deformation, drum dislocation, foundation rotation or maximum axial and shear stresses. For the *EDP*s chosen, appropriate threshold values that define the various performance objectives, e.g. light damage, collapse prevention etc. have to be established. Since such threshold values are not always directly related to visible damage, the *EDP*s should be related to damage that is expressed in simple terms, e.g., crack width, crack density or exfoliation surface area. In all, this is a challenging, multi-disciplinary task that requires experimental verification, expert opinion and a rigorous formulation.

For the assessment of the seismic vulnerability of intact and damaged classical columns a single engineering demand parameters (*EDP*) is here adopted. This is the maximum displacement of the top drum (drum no. 7 for the column of Figure 4) over the whole time history normalized by the base diameter (D), i.e. $u_{top} = \max(u(t))/D$, where $u(t)$ is the displacement of the top drum measured from the reference axis of the column (see Figure 5). Note that the top displacement usually corresponds to the maximum displacement among all drums. This is a parameter that provides a measure of the amount that a column has been deformed during ground shaking and also shows how close to collapse the column was brought during the earthquake. For example, for a cylindrical column that responds as a monolithic block with a pivot point at the corner of its base (Figure 5a), collapse is probable to occur for $u_{top} > 1$, as the weight of the column switches from a restoring ($u_{top}<1$) to an overturning force. But, if the same column responds as a multidrum spinal system with rocking at all joints (Figure 5b), a larger value of u_{top} can be attained without threatening the overall stability of the column. In fact, the top displacement can be larger than the base diameter without collapse, as long as the weight of each part of the column above an opening joint gives a restoring moment about the pole of rotation of the specific part. In the numerical analyses presented here, the maximum value of u_{top} that was attained without collapse was in the order of 1.15. Clearly, it is not straightforward to assign a value of u_{top} that corresponds to collapse, as collapse depends on the 'mode' of deformation, which in turn depends on the ground motion characteristics.

Based on the above *EDP*, the performance criteria of Table 2 are proposed. Three performance levels are recommended, similarly to those typically chosen for modern structures. The first performance level (*damage limitation*) corresponds to weak shaking of the column with very small or no rocking. At this level of shaking, no damage, or any severe additional residual deformations are expected. The second level (*significant damage*) corresponds to intense shaking with significant rocking/wobbling and evident residual deformation of the column after the earthquake; however, the column is not brought close to collapse. The third performance level (*near collapse*) corresponds to very intense shaking with significant rocking and probably sliding of the drums. The column does not collapse at this level, as $u_{top}<1$, but has been brought close to collapse. In most cases, collapse occurred when this performance level was exceeded.

Figure 5. Top displacement for two extreme modes of deformation of a classical column: (a) as a monolithic block; (b) with opening of all joints (displacements are shown exaggerated)

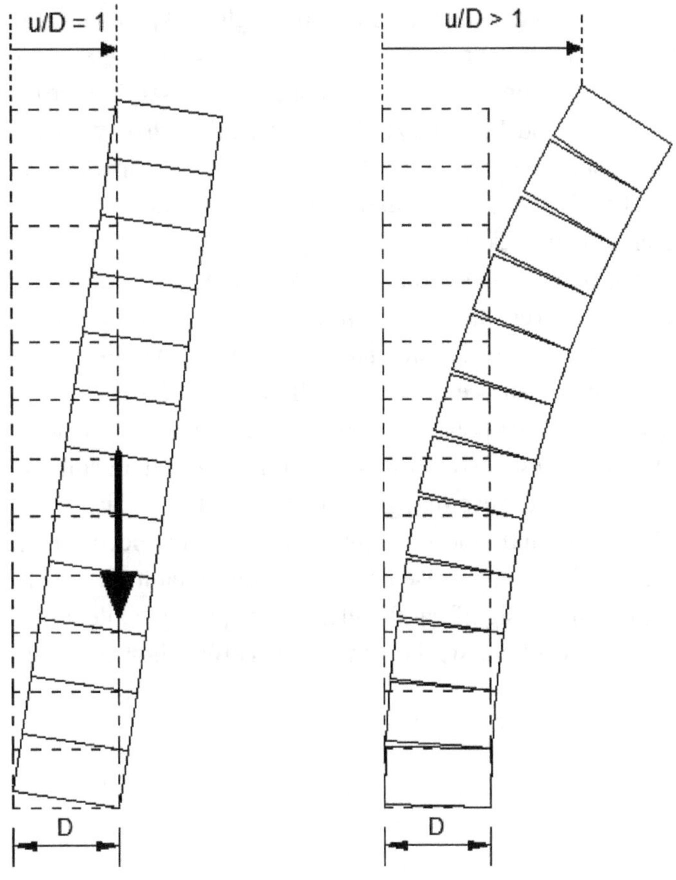

Table 2. Proposed performance criteria concerning the risk of collapse

u_{top}	Performance Level	Description
0.15	Damage limitation	No danger for the column. No significant permanent drum dislocations expected.
0.35	Significant damage	Large opening of the joints with probable damage due to impacts and considerable residual dislocation of the drums. No serious danger of collapse.
1.00	Near collapse	Very large opening of the joints, close to partial or total collapse.

GENERATION OF SYNTHETIC, HAZARD–CONSISTENT GROUND MOTIONS

The assessment of the seismic reliability that is presented herein is based on synthetic ground motions, representative of near-field sites. The reason of using synthetic instead of natural ground motions, is the limited number of the latter for the range of pairs M_w–R (magnitude-distance) that are examined, especially for stiff soil conditions on which monuments are typically founded. The synthetic records were generated using the process that has been proposed by Mavroeidis and Papageorgiou (2003), which allows for the combination of independent models that describe the low-frequency (coherent) component of the directivity pulse, with models that describe the high-frequency component of an acceleration time history. In the present paper, the generation of the high-frequency component was based on the stochastic (or engineering) approach discussed in detail in (Boore, 2003). Based on a given magnitude-distance scenario (M_w–R) and depending on a number of site characteristics, the stochastic approach produces synthetic ground motions. It must be noted that, due to the high nonlinear nature of the rocking/wobbling response and the existence of a minimum value of the peak ground acceleration that is required for the initiation of rocking, the high frequency part of the records is necessary for the correct simulation of surrogate ground motions. Long-period directivity pulses alone, although they generally produce devastating effects to classical monuments (Psycharis et al. 2000), might not be capable to produce intense shaking and collapse (Psycharis et al. 2013), as the maximum acceleration of such pulses is usually small and not strong enough to even initiate rocking.

It is noted that the use of point-source models, inherent in the above approach, is not appropriate for near-fault ground motions; however, this approach is adopted here for simplicity and is not expected to qualitatively affect the risk assessment calculations. For more details regarding the creation of the synthetic ground motions used herein we refer to Psycharis *et al.* (2013).

Classical monuments were usually constructed on the Acropolis of ancient cities, i.e. on top of cliffs; thus, most of them are founded on stiff soil or rock, and only few on soft soil. For this reason, the effect of the soil on the characteristics of the exciting ground motion was not considered in the present analysis, assuming that any significant ground motion amplification does not occur due to the local topography. It is noted that, although the directivity pulse contained in near-fault records is not generally affected by the soil conditions, soft soil can significantly alter the frequency content of the ground motion and, consequently, affect the response of classical columns. This effect, however, is beyond the scope of this Chapter.

Directivity Pulse

For the long-period component of the synthetic ground motions we applied the pulse wavelet proposed by Mavroeidis and Papageorgiou (2003). This wavelet has been calibrated using actual near-field ground motions from a wide database. The velocity pulse is given by the expression:

$$V\left(t\right) = 0.5V_{p}\left[1+\cos\left(\frac{2\pi f_{p}}{\gamma_{p}}\left(t-t_{0}\right)\right)\right]\cos\left[2\pi f_{p}\left(t-t_{0}\right)+\nu_{p}\right], \quad t\in\left[t_{0}-\frac{\gamma_{p}}{2f_{p}}, t_{0}+\frac{\gamma_{p}}{2f_{p}}\right]$$

where A_p, f_p, v_p, γ_p and t_0 describe the amplitude of the envelope of the pulse, the prevailing frequency, the phase angle, the oscillatory character (i.e., number of half cycles) and the time shift to specify the epoch of the envelope's peak, respectively. For every magnitude – distance scenario (M_w–R), the velocity amplitude of the directivity pulse, V_p, and the frequency f_p were obtained using the expressions produced by Rupakhety *et al.* (2011). Note that V_p is not in general equal to the envelope amplitude A_p, but one can be calculated from the other if the phase angle v_p is known. Specifically, the mean value of V_p was obtained by:

$$\log\left(V_{p}\right) = -5.17 + 1.98\cdot M_{w} - 0.14\cdot M_{w}^{2} - 0.10\cdot\log\left(R^{2} + 0.562\right)$$

where M_w cannot exceed M_{sat}, which is considered equal to 7.0. Thus, for magnitude values above M_{sat}, we set $M_w = M_{sat}$ to obtain V_p using Eq. . Similarly, the mean pulse frequency f_p is obtained as:

$$\log(1/f_{p}) = -2.87 + 0.47M_{w}$$

Low-frequency pulse-like ground motions were constructed using Eq. and random sampling V_p, f_p, ν_p and γ for every M_w–R combination using Latin Hypercube Sampling. We assumed that the logarithms of V_p and f_p follow the normal distribution with standard deviations equal to 0.16 and 0.18, respectively (Rupakhety, *et al.* 2011). The phase angle v_p was randomly chosen in the [–π/2, π/2] range. Moreover, being consistent with the data of Mavroeidis and Papageorgiou (2003), the number of half cycles γ_p was assumed to follow a normal distribution with mean and standard deviation equal to 1.8 and 0.4, respectively. The distribution of γ_p was left-truncated to one, while V_p and f_p were also left-truncated to zero, ensuring that no negative values were sampled. Figure 6 shows the histogram of the four random parameters used for creating pulses for the $M_w = 7$ and $R = 5$ km case.

High Frequency Component Using the Stochastic Approach

The stochastic approach was selected for modeling the high-frequency component of the ground motions. The stochastic method is discussed in detail in Boore (2003) and is based on the ground motion radiation spectrum $Y(M_w, R, f)$, which is the product of quantities that consider the effect of source, path, site and instrument (or type) of motion. By separating the spectrum to its contributing components, the models based on the stochastic method can be easily modified to account for different problem characteristics. The shape and the duration of the ground motions depend on an envelope function $w(M_w, R, t)$.

Figure 6. Histogram of the random parameters that describe the low-frequency plot ($M_w = 7$ and $R = 5$ km)

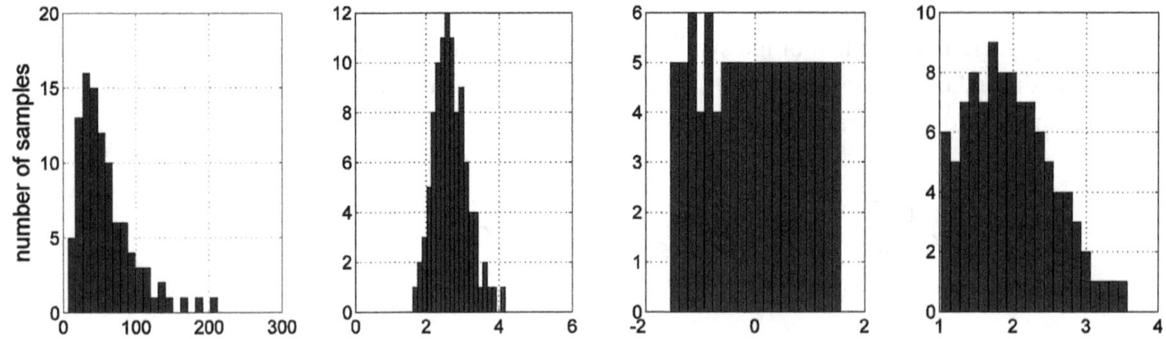

The steps followed to generate the high frequency component are briefly summarized as follows.

- First generate white noise (Gaussian or uniform) for a duration given by the duration of the motion as predicted by an appropriate ground motion prediction equation.
- The noise is then windowed and transformed into the frequency domain using the envelope function $w(M_{w,}R,t)$.
- The spectrum is normalized by the square root of the mean square amplitude spectrum and multiplied by the ground motion spectrum $Y(M_{w,}R,f)$.
- The resulting spectrum is transformed back to the time domain.

The $Y(M_{w,}R,f)$ spectrum and the model parameters adopted in our study are these of Atkinson and Silva (2000). All simulations have been performed using the SMSIM program, freely available from http://www.daveboore.com.

Combination of Low- and High-Frequency Components

The procedure suggested by Mavroeides and Papageorgiou (2003) was adopted in order to combine the low- and the high-frequency ground motion components. The procedure consists of first obtaining the Fourier transform of both the high- and the low-frequency components. Subsequently, the Fourier amplitude of the pulse is subtracted from that of the high-frequency component of the ground motion and a synthetic acceleration time-history is constructed so that its Fourier amplitude is that of the previous step and its phase angle is that of the high-frequency record. The final synthetic record is obtained by adding the pulse time-history.

FRAGILITY ANALYSIS

Fragility (or vulnerability) curves are a valuable tool for the seismic risk assessment of a system. Fragility analysis was initially developed for the reliability analysis of nuclear plants in an effort to separate the structural analysis part from hazard analysis which, typically, is performed by engineering seismologists. Vulnerability analysis requires the calculation of the exceedance probabilities of several monotonically increasing limit-states. The seismic fragility F_R is defined as the limit-state probability conditioned on seismic intensity. The seismic intensity is here expressed in terms of magnitude M_w and distance R, resulting to a surface $F_R(M_w,R)$. Therefore, the fragility of a system is the probability that an engineering demand parameter (EDP) exceeds a threshold value edp and is defined as:

$$F_R\left(M_w, R\right) = P\left(EDP \geq edp \left|M_w, R\right.\right)$$

Eq. provides a single-point of a limit-state fragility surface, while engineering demand parameters (EDPs) are quantities that characterize the system response, e.g., permanent or maximum deformation, drum dislocation, etc. To calculate F_R we performed Monte Carlo Simulation (MCS) using Latin Hypercube Sampling (LHS) for a range of magnitude and distance (M_w, R) scenarios. For this purpose, a large number of nonlinear response history analyses for every M_w–R pair is needed, especially when small probabilities are sought. Therefore, suites of records that correspond to the same M_w and R value must be compiled. Since it is very difficult to come up with such suites of natural ground motion records, we produced synthetic ground motions following the procedure discussed in the previous section.

Assuming that seismic data are lognormally distributed, $F_R(M_w,R)$ can be calculated analytically once the mean and the standard deviation of the logs of the EDP are calculated, which are denoted as $\mu_{\ln EDP}$ and $\beta_{\ln EDP}$, respectively. Once they are known, they can be used to calculate F_R using the normal distribution:

$$F_R = P\left(EDP \geq edp \left|M_w,R\right.\right) = 1 - \Phi\left(\frac{\ln\left(edp\right) - \mu_{\ln EDP}}{\beta_{\ln EDP}}\right)$$

where edp is the EDP's threshold value that denotes that the limit-state examined is violated and Φ denotes the standard normal distribution. For example, if we are calculating the fragility surface that corresponds to normalized displacement of the column's capital u_{top} (defined previously) larger than 0.3, ln(edp) would be equal to ln(0.3). Alternatively, a good approximation of Eq. can be obtained by the ratio of successful simulations over the total number of simulations performed, thus bypassing the assumption of lognormality. For the case study examined in this paper, the two approaches give similar results.

As the ground motion intensity increases, some records may result in collapse of the structure. When collapsed simulations exist, Eq. is not accurate, since the EDP takes an infinite or a very large value that cannot be used to calculate $\mu_{\ln EDP}$ and $\beta_{\ln EDP}$. To handle such cases, Eq. is modified by separating the data to collapsed and to non-collapsed. The conditional probability of collapse is calculated as:

$$P\left(C\left|M_w, R\right.\right) = \frac{\text{number of simulations collapsed}}{\text{total number of simulations}}$$

If $\mu_{\ln EDP}$ and $\beta_{\ln EDP}$ are the mean and the dispersion of the non-collapsed data respectively, Eq. is modified as follows:

$$F_R = P\left(EDP \geq edp \left|M_w,R\right.\right) = \\ = P\left(C\left|M_w,R\right.\right) + \left(1 - P\left(C\left|M_w,R\right.\right)\right)\cdot\left(1 - \Phi\left(\frac{\ln\left(edp\right) - \mu_{\ln EDP}}{\beta_{\ln EDP}}\right)\right)$$

NUMERICAL RESULTS

Fragility Assessment

The proposed fragility assessment methodology was applied to the damaged and the undamaged column shown in Figure 4. The response of the columns was calculated for 35 M_w–R scenarios. For every M_w–R scenario 100 Monte Carlo Simulations (MCS) were performed, thus resulting to 3500 simulations for every column.

In Figure 7 we present the collapse probabilities of the intact and the damaged column as function of magnitude and distance. Collapse is considered independently of whether it is local (collapse of a few drums on the top) or total (collapse of the whole column). Apparently, the number of collapses is larger for smaller fault distances and larger magnitudes. For instance, for $M_w = 7.5$ and $R = 5$ km, 70% of the simulations of the intact column and 95% (almost all) of the simulations of the damaged column caused collapse. However, practically zero collapses occurred for both columns for magnitudes smaller than 6.0. As expected, the damaged column is clearly more prone to collapse compared to the one that is intact.

Concerning the normalized mean top displacement during the seismic motion, u_{top}, which defines the performance of the columns according to the criteria of Table 2, Figure 8 shows that for both columns and for small distances from the fault, up to approximately 7.5 km, the mean value of u_{top} increases monotonically with the magnitude. However, for larger fault distances, the maximum u_{top} occurs for magnitude $M_w = 6.5$ and for larger M_w the top displacement decreases.

Figure 7. Collapse probabilities for the intact and the damaged column. The damaged column is evidently more prone to collapse

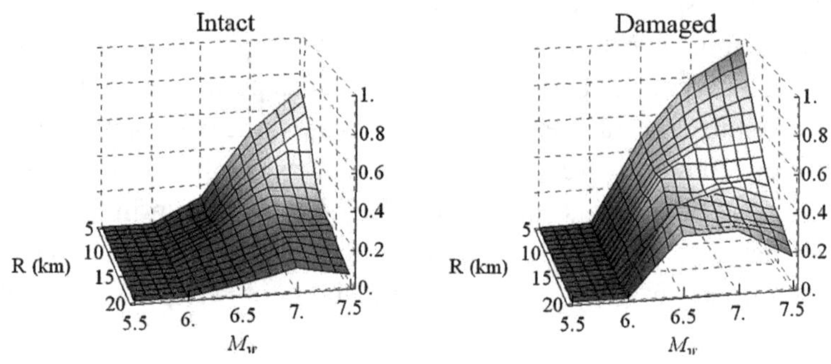

Figure 8. Mean maximum normalized top displacements, u_{top}, for the intact and for the damaged column

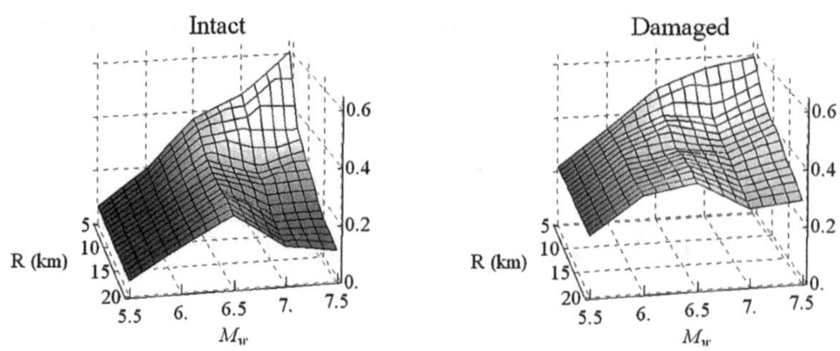

This counter-intuitive response is attributed to the saturation of *PGV* for earthquakes with magnitude larger than $M_{sat} = 7.0$, while the period of the pulse is increasing exponentially with the magnitude. As a result, the directivity pulse has small acceleration amplitude for large magnitudes, which is not capable to produce intense rocking. For the intact column, the validity of this 'strange' observation was verified from results obtained using 30 natural ground motions from the NGA PEER database recorded in distances ranging from 17 to 23 km (Psycharis *et al.* 2013): The same behaviour was observed for the natural ground motions, i.e. the maximum u_{top} demand occurred for $M_w = 6.5$, while for higher magnitudes the demand gradually decreased as in the case of the synthetic records. For more details on this issue the reader may also consult Psycharis *et al.* (2013).

Comparing the response between the intact and the damaged column, Figure 8 shows that the maximum normalized displacements, u_{top}, follow the same trend for both columns. Nevertheless, for weaker earthquakes the maximum displacements of the damaged column are somehow lower. This is due to the initial drum dislocation of 15% that was considered herein. It has to be mentioned that the simulations that lead to column collapse were excluded from these plots and therefore, Figure 8 has to be cross-examined together with Figure 7.

Figure 9 shows the fragility surfaces of the columns for the three performance levels of Table 2. The performance levels range from *significant damage* ($u_{top} > 1$) to *damage limitation* ($u_{top} > 0.15$). It is reminded that $u_{top} > 0.15$ means that the maximum top displacement during ground shaking is larger than 15% of the base diameter and $u_{top} > 1$ corresponds to intense rocking that could collapse the column, or bring it on the verge of collapse. When *damage limitation* is examined (Figure 9c), the exceedance probability for the intact column is high, exceeding 0.6 for M_w=6 and further increases for larger magnitudes. Looking at the worst-case scenario among those examined, M_w=7.5, R=5 km, the probability that the top displacement is larger than 15% of D is equal to one, while in the range $M_w = 6.5 - 7.5$ and $R > 15$ km the probability decreases. For the damaged column, the probability of exceedance of the *damage limitation* ($u_{top} > 0.15$) is equal to one, since the column has already failed as the initial dislocation is 15 cm and D =1.00 m.

Similar observations hold for the *significant damage* limit state ($u_{top} > 0.35$), but the probabilities obtained are considerably smaller. For the *near collapse* limit state ($u_{top} > 1.0$), the probability of exceedance reduces significantly as the distance increases, even for large magnitudes. Notice that the $u_{top} > 1.0$ fragility surface is quite similar to the probability of collapse of Figure 7, which indicates that if the top displacement becomes equal to the base diameter, then there is a high possibility that the column will collapse a little later. The fragility curves of Figure 9 show that, in general, the damaged column exhibits larger displacements than the intact and hence is more likely to collapse.

Comparing the behavior of the intact column studied here (Figure 4) with the behavior of a typical column of the Parthenon Pronaos, which was studied following the same methodology in Psycharis *et al.* (2013), we observe that, considering the same *EDP*, the collapse and the exceedance probabilities that are obtained here are larger than those obtained for a column of the Parthenon Pronaos. Since the Parthenon Pronaos column is larger, this is in accordance with the observation that larger columns are more stable than smaller ones of the same aspect ratio of dimensions (cf. also Housner 1963).

RESIDUAL DISPLACEMENTS AND DAMAGE BUILD-UP

The ability of the spinal systems under consideration to wobble and dissipate energy through sliding and block impacts during rocking, results to residual displacements at the end of the seismic motion ($t=t_{end}$).

Figure 9. Fragility surfaces with respect to the performance levels of Table 2 for the intact and the damaged column: (a) $P(u_{top} > 1.0)$; (b) $P(u_{top} > 0.35)$; (c) $P(u_{top} > 0.15)$

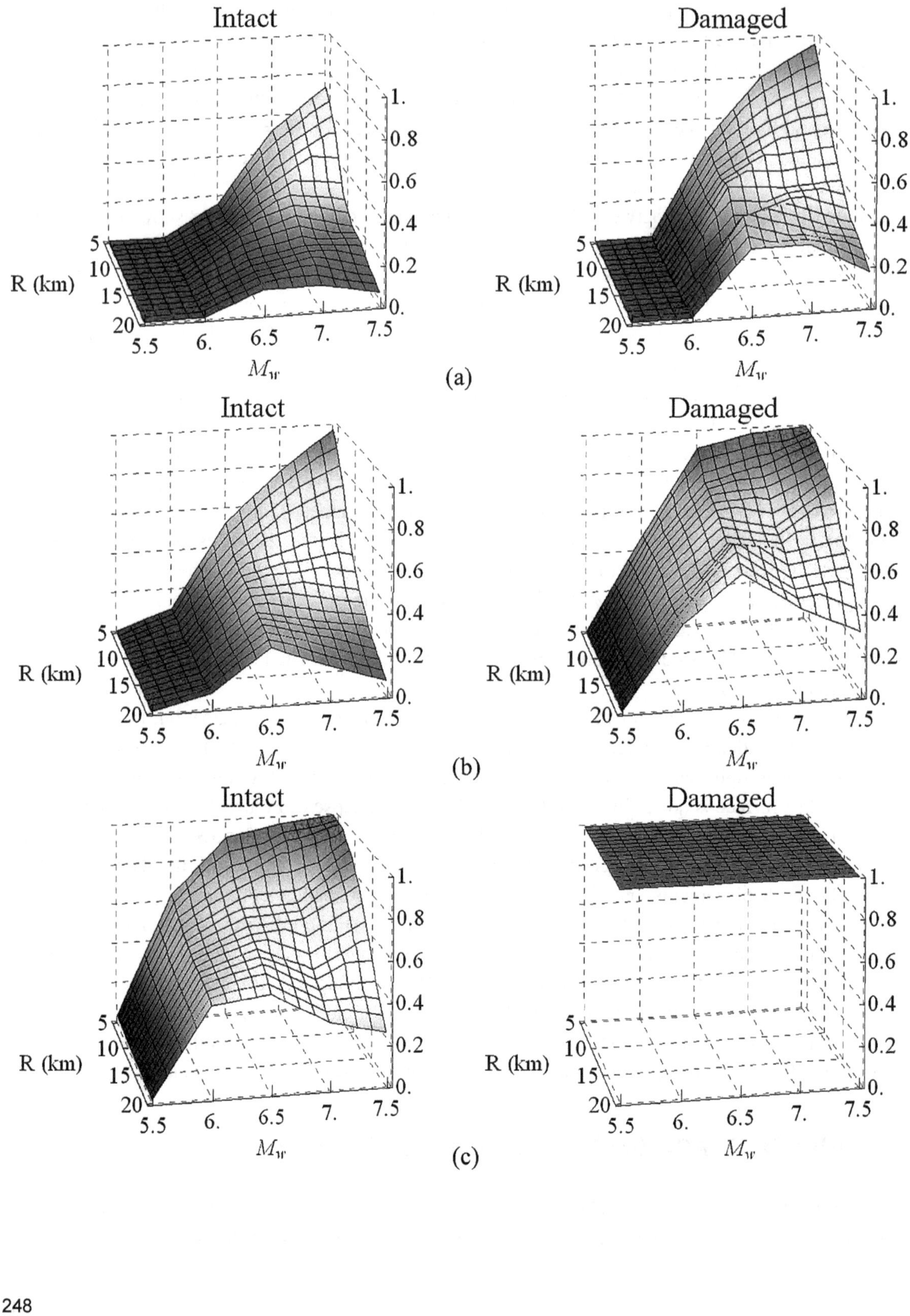

This damage build-up is apparent in classical monuments, though usually of small extent, i.e. of some centimeters (1-2% of the drums' diameter). It is also noted that, given the long history of these structures, it is not evident whether drum dislocations observed today should be attributed to seismic actions or to other events that may have taken place during the life of such monumental structures. This holds also for the damaged column considered here, which was considered with excessive damage (permanent dislocation equal to 15% of the drum diameter).

In order to quantify the permanent dislocations caused by a single earthquake, we introduce a new parameter, u_d, which provides a measure of how much the geometry of the column has been altered after a seismic event. In other words, u_d expresses the damage build-up because of one earthquake. The drum dislocations are measured from the initial ($t=t_0$) deformed/damaged shape of the column, and u_d is defined as $u_d = \max[\Delta(\mathrm{res}u_i)]/D_i$, where $\Delta(\mathrm{res}u_i)$ is the *additional* residual dislocation of drum *i* caused by the earthquake compared with the initial damaged state ($t=t_0$) and D_i is the corresponding drum diameter.

Figure 10 shows the mean maximum additional permanent drum dislocations u_d with respect to M_w and *R* for the intact and for the damaged column. We observe that the maximum mean residual displacement between the drums of both columns is approximately 2% of the base diameter (1 cm) which is in accordance with field observations. Consequently, the initially considered dislocation of 15% (15 cm) could hardly be the result of only one earthquake. Therefore, as there is no evidence that several strong earthquakes have hit Athens in the past (Ambraseys & Psycharis 2012), it seems that in the case of the column of Propylaia that was studied herein, the dislocation of Figure 4 should be attributed to factors other than the seismic activity (e.g. human activity, collapse of adjacent parts of the monument, impacts etc.).

Figure 11 presents the probabilities of exceedance of certain earthquake damage build-up thresholds. For instance, the probability of having a residual damage of 1.5% (Figure 11a) after an earthquake of magnitude M_w=6 is quite small (of the order of 0.1), thus showing that only strong earthquakes are able to cause significant residual displacements.

In the case study examined, the differences between the damaged and the intact column, when u_d is the *EDP*, are not significant. However, due to the sensitivity of the response, this conclusion cannot be generalized. In other cases, existing imperfections might have a pronounced effect on the permanent displacements after an earthquake and on the stability of the column to future ones, as, for example, in case of Figure 3.

Figure 10. Mean maximum normalized residual deformations, u_d for the intact column (left) and for the damaged column (right). In the latter case, the initial dislocation of the upper part of the column (see Figure 4) is not included in the value of u_d. It is noted that the damaged column cannot develop significant additional permanent displacements since it is already on the verge of collapse

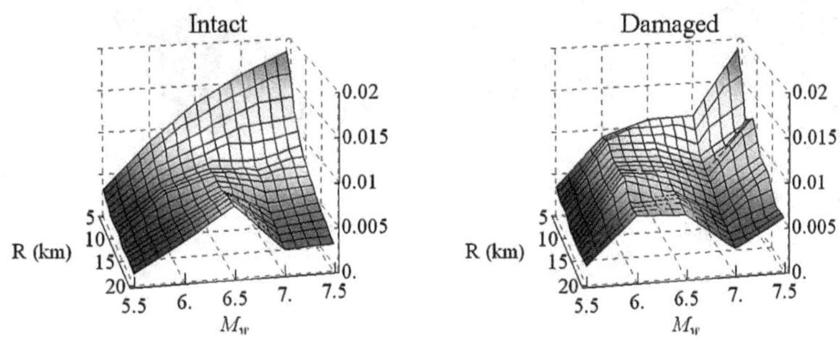

Figure 11. Probabilities of exceedance of certain damage build-up thresholds, for the intact and the damaged column: (a) $P(u_d > 0.015)$, (b) $P(u_d > 0.01)$, (c) $P(u_d > 0.005)$

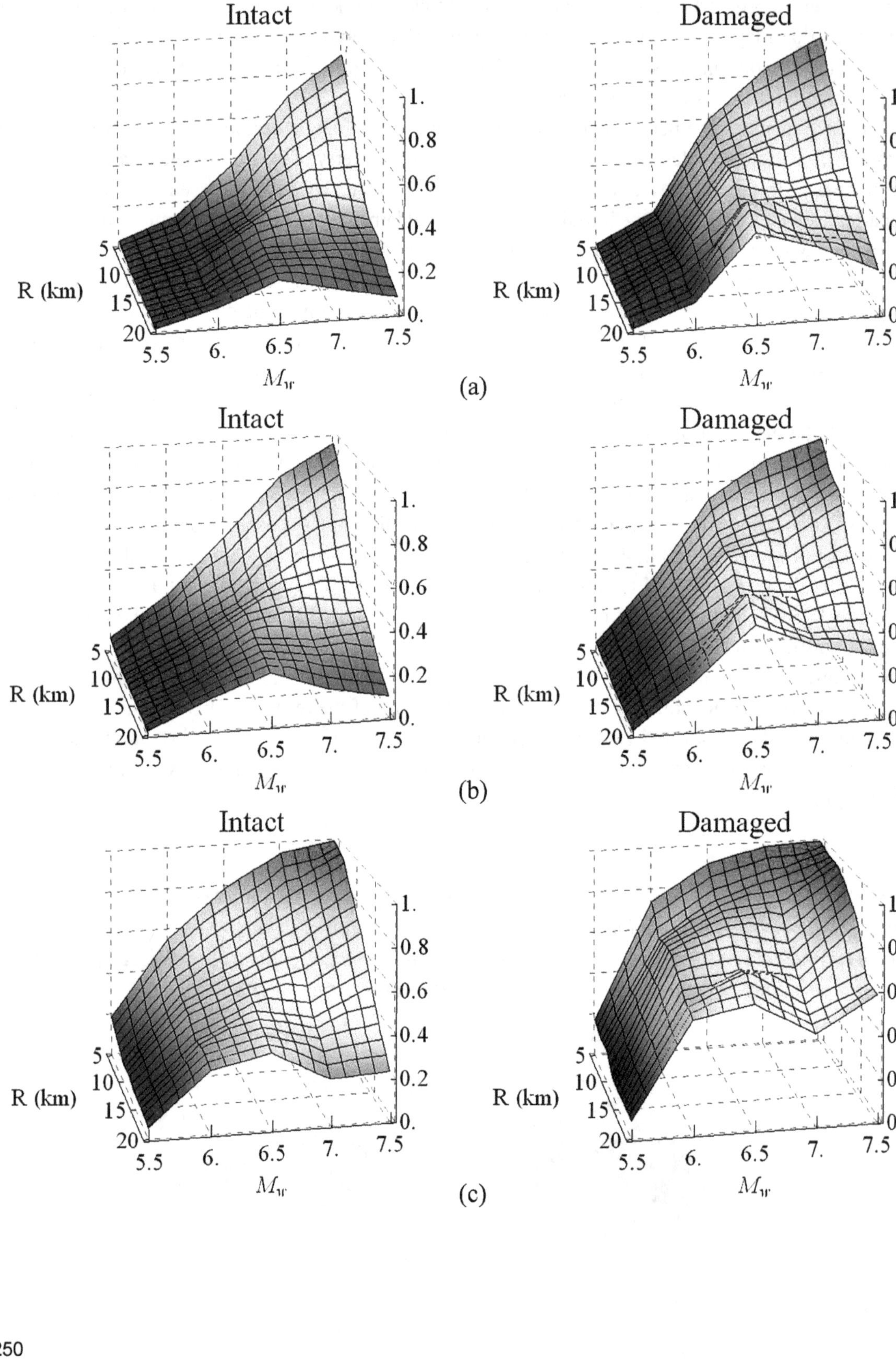

CONCLUSION

A methodology for the seismic risk assessment of damaged multi-drum columns is presented. Existing damage is common in monuments and threatens their stability under seismic actions. In this investigation, only damage in the form of pre-existing drum dislocations to classical columns is examined. The dislocations may have been caused either by past seismic events or by several other factors, including human activity. The proposed methodology, however, can be extended to cover other types of damage as well.

An engineering demand parameter (*EDP*) related to the column collapse risk was adopted for the assessment of the vulnerability of the classical columns at hand. The fragility analysis was performed for the case study of a column at the Propylaia on the Acropolis of Athens, Greece which, in its current state shows a drum dislocation of 15% of the drum diameter. The seismic performance of the "damaged" column was compared to the performance of its "intact" geometry, i.e. the same column without the pre-existing dislocations. In order to account for the probabilistic nature of the seismic events and the strong non-linearities of the dynamical system at hand, the Monte Carlo method was applied using synthetic ground motions, which contain a high- and a low- frequency component. The response of both columns was calculated and compared for 35 M_w–R seismic scenarios.

The analyses showed that the risk of collapse is quite important and higher for the damaged compared to the undamaged column (95% vs 70% collapse probability). Nevertheless, the damage build-up (residual displacements) because of a single seismic event was similar between the two columns and in the order of 2% of the drum's diameter. Therefore, as there is no evidence that several strong earthquakes have occurred in Athens in the past (Ambraseys and Psycharis 2012), it seems that, in the case of the column of Propylaia, the observed excessive dislocation should be rather attributed to factors other than seismic activity (e.g. the explosion that occurred in 1687, human activity, collapse of adjacent parts of the monument –most probable–, impacts, among others). Finally, comparing the results of the present analysis to those of previous studies (Psycharis *et al.* 2013) we corroborate the fact that larger columns are more stable than columns of a smaller aspect ratio.

REFERENCES

Allen, R. H., Oppenheim, I. J., Parker, A. P., & Bielak, J. (1986). On the dynamic response of rigid body assemblies. *Earthquake Engineering & Structural Dynamics*, *14*(6), 861–876. doi:10.1002/eqe.4290140604

Ambraseys, N., & Psycharis, I. N. (2012). Assessment of the long-term seismicity of Athens from two classical columns. *Bulletin of Earthquake Engineering*, *10*(6), 1635–1666. doi:10.1007/s10518-012-9388-1

Atkinson, G. W., & Silva, W. (2000). Stochastic modeling of California ground motions. *Bulletin of the Seismological Society of America*, *90*(2), 255–274. doi:10.1785/0119990064

Boore, D. M. (2003). Simulation of Ground Motion Using the Stochastic Method. *Pure and Applied Geophysics*, *160*(3), 653–676. doi:10.1007/PL00012553

Dasiou, M.-E., Mouzakis, H. P., Psycharis, I. N., Papantonopoulos, C., & Vayas, I. (2009a). Experimental investigation of the seismic response of parts of ancient temples. In *Prohitech Conference*.

Dasiou, M.-E., Psycharis, I. N., & Vayas, I. (2009b). Verification of numerical models used for the analysis of ancient temples. In *Prohitech Conference*.

Housner, G. W. (1963). The behavior of inverted pendulum structures during earthquakes. *Bulletin of the Seismological Society of America*, *53*, 403–417.

Itasca Consulting Group, Inc. (n.d.). *3DEC – Three-dimensional distinct element code*. Minneapolis, MN: Itasca

Konstantinidis, D., & Makris, N. (2005). Seismic response analysis of multidrum classical columns. *Earthquake Engineering & Structural Dynamics*, *34*(10), 1243–1270. doi:10.1002/eqe.478

Lemos, J. V. (2007). Discrete Element Modeling of Masonry Structures. *International Journal of Architectural Heritage*, *1*(2), 190–213. doi:10.1080/15583050601176868

Makris, N., & Roussos, Y. (2000). Rocking response of rigid blocks under near-source ground motions. *Geotechnique*, *50*(3), 243–262. doi:10.1680/geot.2000.50.3.243

Makris, N., & Zhang, J. (2001). Rocking Response of Anchored Blocks under Pulse-Type Motions. *Journal of Engineering Mechanics*, *127*(5), 484–493. doi:10.1061/(ASCE)0733-9399(2001)127:5(484)

Mavroeidis, G. P., & Papageorgiou, A. S. (2003). A Mathematical Representation of Near-Fault Ground Motions. *Bulletin of the Seismological Society of America*, *93*(3), 1099–1131. doi:10.1785/0120020100

Mouzakis, H., Psycharis, I. N., Papastamatiou, D. Y., Carydis, P. G., Papantonopulos, C., & Zambas, C. (2002). Experimental investigation of the earthquake response of a model of a marble classical column. *Earthquake Engineering & Structural Dynamics*, *31*(9), 1681–1698. doi:10.1002/eqe.184

Papaloizou, L., & Komodromos, P. (2009). Planar investigation of the seismic response of ancient columns and colonnades with epistyles using a custom-made software. *Soil Dynamics and Earthquake Engineering*, *29*(11-12), 1437–1454. doi:10.1016/j.soildyn.2009.06.001

Papantonopoulos, C., Psycharis, I. N., Papastamatiou, D. Y., Lemos, J. V., & Mouzakis, H. P. (2002). Numerical prediction of the earthquake response of classical columns using the distinct element method. *Earthquake Engineering & Structural Dynamics*, *31*(9), 1699–1717. doi:10.1002/eqe.185

Psycharis, I. N. (1990). Dynamic behaviour of rocking two-block assemblies. *Earthquake Engineering & Structural Dynamics*, *19*(4), 555–575. doi:10.1002/eqe.4290190407

Psycharis, I. N., Fragiadakis, M., & Stefanou, I. (2013). Seismic reliability assessment of classical columns subjected to near-fault ground motions. *Earthquake Engineering & Structural Dynamics*, *42*(14), 2061–2079.

Psycharis, I. N., & Jennings, P. C. (1985). Upthrow of objects due to horizontal impulse excitation. *Bulletin of the Seismological Society of America*, *75*(2), 543–561.

Psycharis, I. N., Lemos, J. V., Papastamatiou, D. Y., Zambas, C., & Papantonopoulos, C. (2003). Numerical study of the seismic behaviour of a part of the Parthenon Pronaos. *Earthquake Engineering & Structural Dynamics*, *32*(13), 2063–2084. doi:10.1002/eqe.315

Psycharis, I. N., Papastamatiou, D. Y., & Alexandris, A. P. (2000). Parametric investigation of the stability of classical columns under harmonic and earthquake excitations. *Earthquake Engineering & Structural Dynamics*, *29*(8), 1093–1109. doi:10.1002/1096-9845(200008)29:8<1093::AID-EQE953>3.0.CO;2-S

Rupakhety, E., Sigurdsson, S. U., Papageorgiou, A. S., & Sigbjörnsson, R. (2011). Quantification of ground-motion parameters and response spectra in the near-fault region. *Bulletin of Earthquake Engineering*, *9*(4), 893–930. doi:10.1007/s10518-011-9255-5

Sarhosis, V., Lignola, G. P., & Asteris, P. (2014). Seismic Vulnerability of Ancient Colonnade: Numerical analysis of the two storey colonnade of the Forum in Pompeii. In Seismic Assessment and Rehabilitation of Historic Structures. IGI Global.

Sinopoli, A. (1991). Dynamic Analysis of a Stone Column Excited by a Sine Wave Ground Motion. *Applied Mechanics Reviews*, *44*(11S), 246–256. doi:10.1115/1.3121361

Stefanou, I., Psycharis, I. N., & Georgopoulos, I. (2011a). Dynamic response of reinforced masonry columns in classical monuments. *Construction & Building Materials*, *25*(12), 4325–4337. doi:10.1016/j.conbuildmat.2010.12.042

Stefanou, I., Vardoulakis, I., & Mavraganis, A. (2011b). Dynamic motion of a conical frustum over a rough horizontal plane. *International Journal of Non-linear Mechanics*, *46*(1), 114–124. doi:10.1016/j.ijnonlinmec.2010.07.008

Toumbakari, E., & Psycharis, I. N. (2010). Parametric investigation of the seismic response of a column of the Aphrodite Temple in Amathus. In *14th European Conference on Earthquake Engineering (ECEE)*. Ohrid: FYROM.

Yim, C. S., Chopra, A. K., & Penzien, J. (1980). Rocking response of rigid blocks to earthquakes. *Earthquake Engineering & Structural Dynamics*, *8*(6), 565–587. doi:10.1002/eqe.4290080606

Chapter 11
Numerical Study of Discrete Masonry Structures under Static and Dynamic Loading

Rossana Dimitri
Università del Salento, Italy

Giorgio Zavarise
Università del Salento, Italy

ABSTRACT

Much of the world's architectural heritage consists of Unreinforced Masonry (URM) structures whose preservation is a topical subject. To prevent possible collapse of multi-block systems in hazardous conditions, a promising tool to investigate their structural response is represented by numerical modelling with the Discrete Element Method (DEM). Gothic buttresses of trapezoidal and stepped shapes are first analysed comparatively under static loading, defining the optimal configurations. Numerical results are verified against the analytical predictions of overturning and sliding resistances, based on a continuum approximation of masonry. The DEM is then successfully adopted to assess the first-order seismic behavior of arches and buttressed arches with different shapes as compared to predictions based on limit analysis. A systematic investigation on dynamic behavior failure domains and on modes of collapse of URM structures is finally performed for varying input parameters, as needed to gain more confidence on the numerical results.

INTRODUCTION

Among the structural components in masonry building, buttresses, arches, and portals (i.e. buttressed arches) deserve particular attention. They are very widespread in European historical centers, and their preservation as part of the cultural heritage is a topical subject. Because of their ages, or for accidental events as earthquakes, these structural elements can suffer several types of damage. Therefore, a fundamental understanding of their structural behavior has to be evaluated in terms of stability, in order to prevent their brittle collapse in possible future hazardous conditions. In this perspective, the limitations

DOI: 10.4018/978-1-5225-0231-9.ch011

of sophisticated analytical modelling and experimental investigations for complex geometries have increased the use of numerical modelling for structural analyses (Fanning, Boothby, & Roberts, 2001; Raman, 2004; Viola, Panzacchi, & Tornabene, 2005; Gonen, Dogan, Karacasu, Ozbasaran, & Gokdemir, 2013). Although the Finite Element Method (FEM) is widely used for the analysis of problems with discontinuities, this method is not particularly suitable to analyze discrete multi-block systems characterized by continuous changes of their geometry and contact conditions among bodies.

An alternative approach is represented by the Discrete Element Method (DEM), as developed by Cundall (1971), where discontinuous bodies are able to move freely in space and interact reciprocally with contact forces, leading to an automatic updating of contact detection. Based on this approach, a system of multiple bodies may change its position in a time-stepping scheme under the action of external and interaction forces between bodies, which in turn lead to a steady-state configuration when the static equilibrium is reached. For rigid masonry blocks, the contact interaction law is therefore the only constitutive law considered, without including any continuum constitutive law (e.g. elasticity, plasticity, damage, fracturing), as differently occurs for deformable bodies. The DEM features many advantages, e.g. the low storage, simplicity of coding, suitability for parallel processing, adoption of the same algorithm for statics and dynamics. A systematic numerical analysis based on the DEM is herein performed in three parts, in order to evaluate the static, quasi-static, and dynamic response of some bi-dimensional URM structures (see more details in Dimitri, 2009), as implemented in the code UDEC (Universal Distinct Element Code) by Itasca (2004).

More specifically, the first part of the chapter is concerned with the evaluation of the stability of buttresses of different shapes subjected to concentrated lateral loads, accounting for fracturing and sliding before collapse. Gothic buttresses of trapezoidal and stepped shapes, are analyzed comparatively, where numerical results are verified against some theoretical predictions of overturning and sliding resistances, obtained considering masonry as a continuum body without tension resistance.

In the second part of the chapter, the first-order seismic assessment of masonry structures is studied with the discrete modelling in order to estimate the lateral loading they can withstand before collapse. The numerical results are compared with theoretical predictions based on limit analysis. Due to a lack of information about the relative efficiency of different arch shapes, for any possible mechanism of collapse, circular, pointed or basket-handle shapes are analyzed comparatively. This makes clearly interesting a parametric assessment of masonry structure for varying geometry parameters, as herein performed with the DEM.

The last part of the chapter is concerned with the evaluation of the dynamic behavior and resistance of multi-drum columns subjected to simple base excitations. A detailed investigation on failure domains and on modes of collapse is discussed, and the simulations are performed with varying geometry and mechanical parameters. Besides the parametric evaluation of the sensitivity of the response to the input parameters, (i.e. the input properties of the joints, and the damping ratio), a novel aspect of this DEM-based research is pointed out with respect to the available literature, such as the size effect on failure domains.

BACKGROUND

In a context where researchers and engineers differently emphasize the importance of strength or stability in a static or dynamic assessment of masonry structures, the combination of the two aspects should be certainly considered depending on nature and geometry of the specific structure. The structural studies of historical buildings usually emphasize strength and neglect stability, although, in most cases, their failure is due to instability instead of a lack of material strength. This results in a general misunderstanding of the structural behavior of URM, leading to useless or destructive repair interventions to be avoided.

Stresses in masonry structures are only a fraction of the loading capacity of the stone, such that rigid block models with the same proportions as in real structures represent a valid way to understand their stability. Differently from elastic problems, stability problems can be scaled, and an equilibrium method is most appropriate for a structural assessment of masonry constructions. The sizing of many structural elements has been of paramount importance in the traditional theories of construction based on geometrical rules. Ancient builders were traditionally aware that the design of a structural element affected the safety of the whole structure. However, many historical buildings are still standing in a state of equilibrium. This means that the stability or equilibrium approach need a better consideration, and the DEM can provide a useful numerical framework.

This numerical technique was initially developed by Cundall (1971) to model blocky-rock systems and sliding along rock mass. The approach has been later applied to evaluate the statics (Baggio & Trovalusci, 1995; Mamaghani, Aydan, & Kajikawa, 1999; Bićanić, Stirling, & Pearce, 2003; Lemos, 2007; Halabian, Mirshahzadeh, & Hashemol-Hosseini, 2014) or dynamics (DeJong & Ochsendorf, 2006; Winkler, Meguro, & Yamazaki, 1995; Papantonopoulos, Psycharis, Papastamatiou, Lemos, & Mouzakis, 2002; Psycharis, Papastamatiou, & Alexandris, 2000; Giordano, Mele, & De Luca, 2002; Meyer, Ochsendorf, Germaine, & Kausel, 2007; DeJong, De Lorenzis, & Ochsendorf, 2008; Sarhosis, Oliveira, Lemos, & Lourenco, 2014; among others) of URM structures including arches, buttresses and buttressed arches, with fracturing occurring along mortar joints. Many of these works demonstrate the suitability of DEM in predicting results and describing realistically the collapse load and mechanisms. In most cases, however, a methodical way of defining modelling parameters before numerical analyses is missing, as performed in this chapter, in order to give more confidence to DEM predictions.

MAIN FOCUS OF THE CHAPTER

Statics

A realistic collapse evaluation of a discrete masonry buttress in static conditions must account for fracturing before overturning or possible sliding between blocks. An ideal geometrical design can be planned adapting the masonry shape to the trajectory of the line of thrust (Heyman, 1992), and minimizing the amount of material for a required level of resistance. Different mathematical problems associated with the design of buttresses have been first solved analytically by Mosely (1843), Rankine (1858), Milankovitch (1910), Huerta (2004), assuming a monolithic behavior up to failure. The effect of fracture propagation on the global resistance of buttresses has been later studied by Ochsendorf et al. (2004) and De Lorenzis et al. (2012a,b) for different shapes, assuming the masonry as a continuous medium with no tensile strength, infinite compressive strength and a linear distribution of compressive stresses in the effective portions of the buttress (middle-third rule by Heyman, 1992). These studies have clearly verified the unsafe assumption of the monolithic behavior, due to reaction of the fractured buttresses against the applied force with only a portion of its mass. The reliability of the analytical predictions obtained in De Lorenzis et al. (2012a,b) is herein confirmed in the first part of the chapter with a numerical discrete approach both in terms of overturning and sliding resistances of trapezoidal and stepped buttresses.

The discrete approach is adopted to study realistically the failure capacity and modes of Gothic trapezoidal and stepped buttresses under static loading, looking into possible fracturing and sliding. Numerical results based on DEM are compared to theoretical predictions based on assumptions of masonry as continuous medium with no tensile strength and infinite compressive strength (more details about the theoretical modelling can be found in De Lorenzis et al., 2012a,b). Hence the minimum value between the computed overturning and sliding loads defines the critical failure mode and controls the capacity of the buttresses. Within a discrete setting, the Gothic buttresses are numerically considered as an assemblage of rigid blocks with frictional joints, whereas solid buttresses are modelled as single block jointed directly at the support. The contact forces between blocks are assumed proportional to the entity of their relative interpenetration, considered by means of the normal and shear penalty parameters defined in input. For a question of consistency with the analytical modelling, the properties of the joints, like the tensile strength, cohesion and dilatancy angle are all assumed equal to zero, whereas large values are considered for the elastic normal and shear stiffnesses (i.e. 10^5 and $5x10^4$, respectively) to limit possible interpenetrations. The geometries of the buttresses to model numerically are chosen from the most significant parameter ranges examined analytically in De Lorenzis et al. (2012a,b).

Trapezoidal Buttresses

In particular for trapezoidal buttresses the analyses consider geometries with top width $b = 1$ m, bottom width $B = 2$ m, height $h_b = 5$ m, subjected to an inclined lateral force with a constant vertical component V, and a variable horizontal component H, applied at a distance $h = 4.5$ m from the base (Figure 1). A weight $\gamma = 30$ kN/m^2 per unit area of the elevation is also assumed for the material, accounting for the material density, ρ and the through-thickness depth, s. The problem has been defined by means of some dimensionless parameters as follows: $\beta = B/b = 2$, $h_b/b = 5$, $\mu = h/h_b = 0.9$. The buttress is subjected to a vertical force: $V = 11.25$ kN, which corresponds to a dimensionless load ratio $\psi = 0.05$ relating the vertical load, V, and the weight of the buttress, W_b. Two values are assigned to the friction coefficient, i.e. $f_s = 1.2$ and 0.7, respectively, in order to evaluate both the overturning or sliding failures, respectively.

Figure 1. Trapezoidal buttress geometry

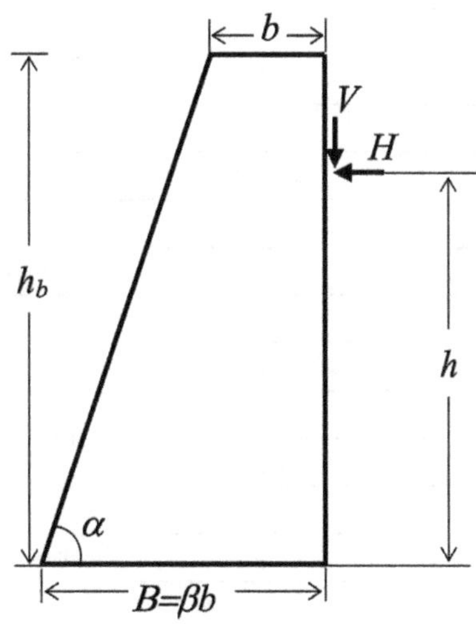

Different sizes and shapes of blocks are also considered for a parametric investigation of numerical results. For this reason, decreasing length, *l*, and height, *t*, of the blocks are considered (1x0.5, 0.6x0.3, 0.5x0.25, respectively) for a fixed *l/t* ratio (*l/t* = 2); whereas decreasing *l/t* ratios are considered (*l/t* = 3, 2.5, 2, 1.5, respectively) for a fixed value of *t* (*t* = 0.3). Table 1 summarizes all the results obtained for different sizes and shapes of the blocks. The ultimate failure load, H_u, of the buttress (reported in bold) represents the minimum value between the overturning load of the fractured structure, H_{ov}, and the sliding load, H_{sl}. It is worth noticing as numerical overturning loads match rather well the analytical results found in De Lorenzis et al. (2012a). The numerical results are clearly larger than the analytical ones, since the fracture path is not exactly linear and straight as assumed analytically.

Table 1. Overturning and sliding load for a trapezoidal buttress: effect of the block size and shape

EFFECT OF THE BLOCK SIZE			
	Block Dimensions *lxt* (m)	**Analytical Model**	**DEM**
$H_{ov,s}$ (kN)	-	66.1	65.3
H_{ov} (kN)	1x0.5	**30.8**	**36.2**
	0.6x0.3		**33.2**
	0.5x0.25		**32.5**
$f_s = 1.2$			
H_{sl} (kN)		32.4	-
Failure mode		**Overturning**	**Overturning**
$f_s = 0.7$			
H_{sl} (kN)	1x0.5	**18.9**	**18.8**
	0.6x0.3		**18.8**
	0.5x0.25		**18.8**
Failure mode		**Sliding**	**Sliding**
EFFECT OF THE BLOCK SHAPE			
$H_{ov,s}$ (kN)	-	66.1	65.3
H_{ov} (kN)	0.9x0.3	**30.8**	**36.1**
	0.75x0.3		**35.6**
	0.6x0.3		**33.2**
	0.45x0.3		**30.9**
$f_s = 1.2$			
H_{sl} (kN)		32.4	-
Failure mode		**Overturning**	**Overturning**
$f_s = 0.7$			
H_{sl} (kN)	0.9x0.3	**18.9**	**18.8**
	0.75x0.3		**18.8**
	0.6x0.3		**18.8**
	0.45x0.3		**18.8**
Failure mode		**Sliding**	**Sliding**

As shown in Figures 2 and 3, the buttress fracturing features a zigzag pattern, as obtained numerically (see the red double lines) instead of a linear pattern (see the black lines), as predicted analytically. This leads to a certain sensitivity of numerical results to the size and shape of the blocks. As visible in Figures 2 and 3, indeed, this pattern becomes gradually closer to the analytical straight fracture for smaller block sizes and *l/t* ratios, together with the overturning load. The sliding loads perfectly agree with the analytical response, since sliding failures are not affected by the shape and the size of the blocks, as shown in Figures 4 and 5. The vectors in Figures 2-5 refer to the motion of the blocks in terms of displacement, and the vectors in bold represent the motion of the points where the force is applied, located at a distance *h* from the support.

Figure 2. Overturning failure of the trapezoidal buttress with blocks of different sizes, for a fixed l/t ratio and f_s = 1.2, as given by the DEM

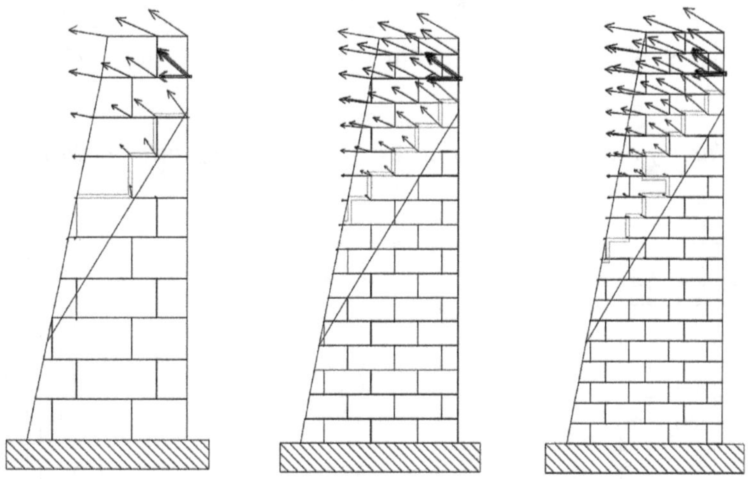

(a) 1x0.5 blocks (b) 0.6x0.3 blocks (c) 0.5x0.25 blocks

Figure 3. Overturning failure of the trapezoidal buttress with blocks of different l/t ratios and f_s = 1.2, as given by the DEM

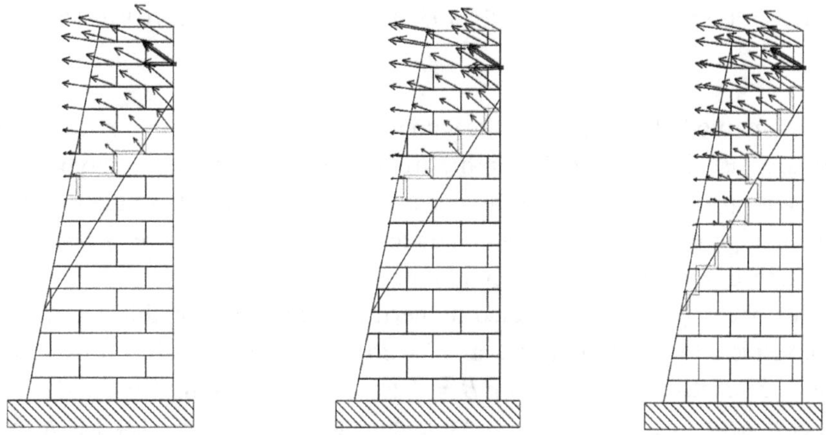

(a) 0.45x0.3 blocks (b) 0.75x0.3 blocks (c) 0.9x0.3 blocks

Figure 4. Sliding failure of the trapezoidal buttress with blocks of different sizes for a fixed l/t ratio and f_s = 0.7, as given by the DEM

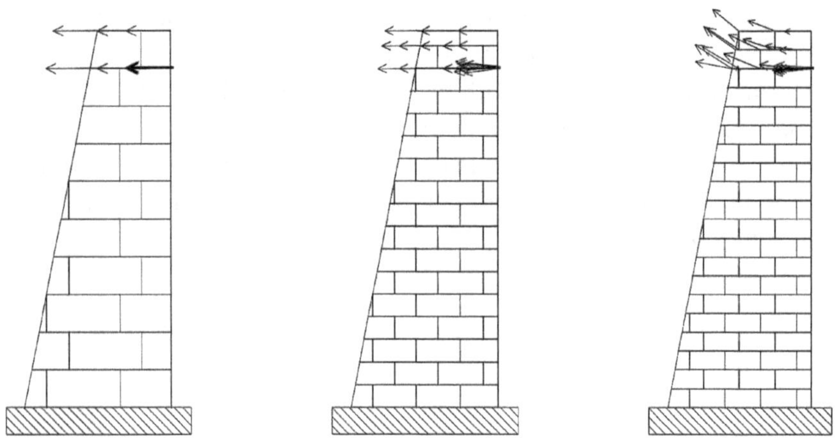

(a) 1x0.5 blocks (b) 0.6x0.3 blocks (c) 0.5x0.25 blocks

Figure 5. Sliding failure of the trapezoidal buttress with blocks of different l/t ratios and f_s = 0.7, as given by the DEM

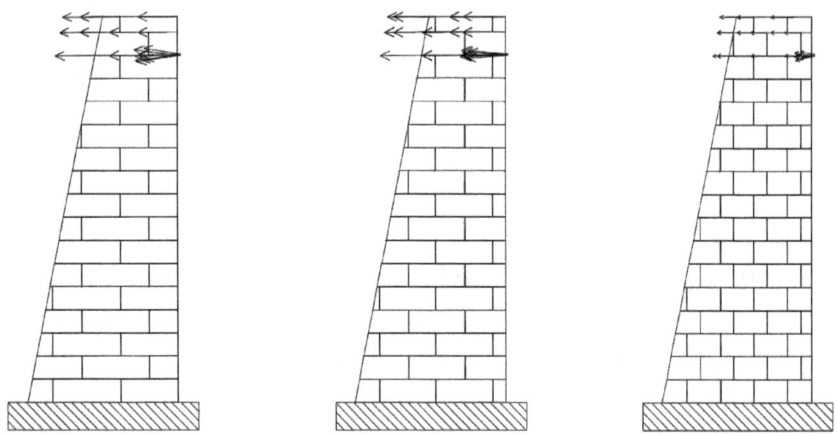

(a) 0.45x0.3 blocks (b) 0.75x0.3 blocks (c) 0.9x0.3 blocks

Based on a the same modelling approach, the numerical investigation is also extended to more complex stepped buttresses, as usually found in the Gothic architecture. As predicted analytically in De Lorenzis et al. (2012b), several patterns of fracture can occur for stepped buttresses, due to changes in profiles. This is illustrated in Figures 6 and 7 for two- and three-step buttresses, where different fracturing modes are assumed, contained or not within the height of each single step.

The geometries of the buttresses modeled numerically are chosen within the most significant parameter ranges considered in De Lorenzis et al. (2012b). More specifically, for two-step buttresses the analyses examines geometries with b = 1 m, B = 2-1.4-1.2 m, h_b = 4 m, h = 3.5-1.7 m, γ = 30 kN/m^2 (Figure 6). The corresponding dimensionless parameters are: $\beta = B/b$ = 2-1.4-1.2, h_b/b = 4, $\mu = h/h_b$ = 0.875-0.425. The buttresses are subjected to a vertical force V = 15 kN, which yields $V^* = V/(bh_b\gamma)$ =

Figure 6. Two-step buttress geometry and possible fracture patterns

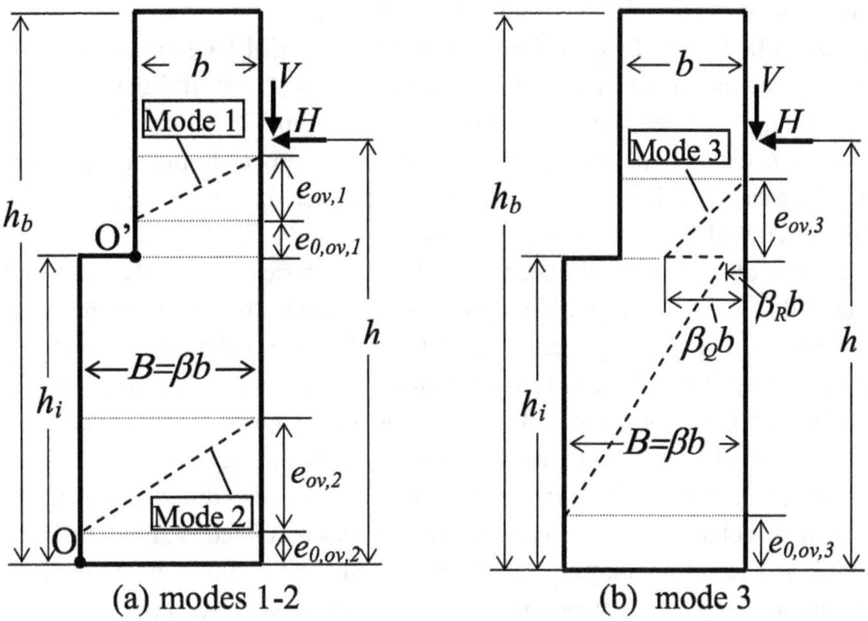

Figure 7. Three-step buttress geometry and possible fracture patterns

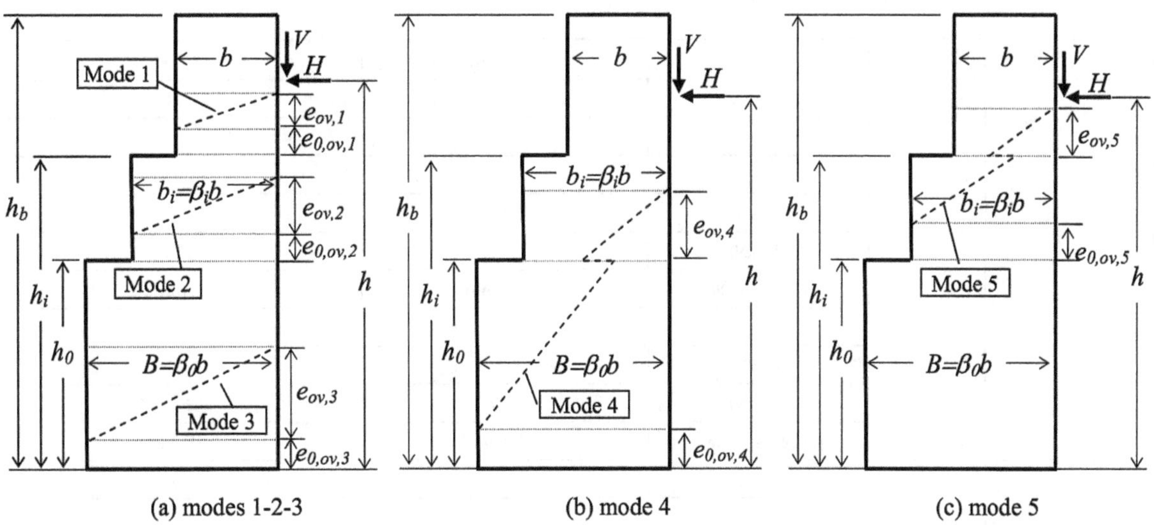

0.1. For three-step buttresses the analyses consider geometries with $b = 1$ m, $b_i = 1.5$ m, $B = 2$ m, $h_b = 5$ m, $h_i = 2.4$ m, $h_o = 1.2$ m, $h = 3.6$-2.88-0.96 m, $\gamma = 30$ kN/m^2 (Figure 7) . The dimensionless parameters result as follows: $\beta_i = b_i/b = 1.5$, $\beta_o = B/b = 2$, $h_b/b = 5$, $\mu_i = h_i/h_b = 0.6$, $\mu_o = h_o/h_b = 0.3$, $\mu = h/h_b = 0.9$-0.72-0.24. In each case, the buttresses are discretized with blocks 0.6 x 0.3 m, except for solid buttresses modeled with single blocks jointed at the bottom of the buttress or along each step interface, in order to study both global and single-step overturning without fracturing. In order to compare both overturning and sliding loads, once again the friction coefficient is set to two different values: 1.2 (to ensure overturning failure) and 0.7 (to ensure sliding failure).

As summarized in Tables 2 and 3, a good agreement between DEM-based results and analytical predictions is obtained for each analyzed case. Once more, the numerical overturning load seems to be slightly higher or equal than the expected analytical value, while the numerical sliding load is sometimes lower than the analytical value due to the combined failure modes based on sliding and overturning instead of a pure sliding. These cases lead to a lower capacity than predicted analytically (see the sliding collapse for two-step buttresses with $\beta = 1.4 - \mu = 0.875$, $\beta = 1.2 - \mu = 0.875$, $\beta = 1.2 - \mu = 0.425$, and a static coefficient of friction of 0.7). A good matching between numerical and analytical results is generally reflected also on failure modes, except for two cases of discrepancy between the approaches. For a fixed $f_s = 1.2$, the failure mode of two-step buttresses with $\beta = 2 - \mu = 0.425$ and three-step buttresses with $\mu = 0.24$ is controlled by overturning instead of sliding, as predicted analytically (see Tables 2 and 3).

Figures 8 and 9 show the failure modes of two-step and three-step buttresses, respectively, as provided numerically and analytically, along with the black vectors representing the motion of the blocks. In both figures, the double black and grey lines refer to the overturning and sliding paths as given by the DEM, whereas the continuous black lines refer to the fracture shapes predicted analytically. When overturning controls, the numerical results display a good matching with the fracture location predicted analytically under the assumption of a linear distribution of compressive stresses (see Figures 8a-f and 9a,c,e).

Table 2. Overturning and sliding load for 2-step buttresses

β	μ		Analytical Model	DEM
2	0.875	$H_{ov,s}$ (kN)	27.3	30.0
		H_{ov} (kN)	22.1	27.2
			$f_s = 1.2$	
		H_{sl} (kN)	32.4	-
		H_u (kN)	**22.1**	**27.2**
		Failure mode	**Overturning**	**Overturning**
			$f_s = 0.7$	
		H_{sl} (kN)	18.9	20.8
		H_u (kN)	**18.9**	**20.8**
		Failure mode	**Sliding**	**Sliding**
2	0.425	$H_{ov,s}$ (kN)	137.7	139.3
		H_{ov} (kN)	115.4	115.5
			$f_s = 1.2$	
		H_{sl} (kN)	108.0	-
		H_u (kN)	**108.0**	**115.5**
		Failure mode	**Sliding**	**Overturning**
			$f_s = 0.7$	
		H_{sl} (kN)	63.0	64.2
		H_u (kN)	**63.0**	**64.2**
		Failure mode	**Sliding**	**Sliding**

continued on following page

Table 2. Continued

β	μ		Analytical Model	DEM
		$H_{ov,s}$ (kN)	28.4	29.9
		H_{ov} (kN)	21.6	23.3
		$f_s = 1.2$		
		H_{sl} (kN)	33.1	-
	0.875	**H_u (kN)**	**21.6**	**23.3**
		Failure mode	**Overturning**	**Overturning**
		$f_s = 0.7$		
		H_{sl} (kN)	19.3	10.3 (Slid.+Overt.)
		H_u (kN)	**19.3**	**10.3**
1.4		**Failure mode**	**Sliding**	**Sliding + Overturning**
		$H_{ov,s}$ (kN)	77.1	77.8
		H_{ov} (kN)	65.9	70.8
		$f_s = 1.2$		
		H_{sl} (kN)	102.5	-
	0.425	**H_u (kN)**	**65.9**	**70.8**
		Failure mode	**Overturning**	**Overturning**
		$f_s = 0.7$		
		H_{sl} (kN)	59.8	60.8
		H_u (kN)	**59.8**	**60.8**
		Failure mode	**Sliding**	**Sliding**
		$H_{ov,s}$ (kN)	28.5	29.2
		H_{ov} (kN)	18.1	23.2
1.2	0.875	$f_s = 1.2$		
		H_{sl} (kN)	32.4	-
		H_u (kN)	**18.1**	**23.2**
		Failure mode	**Overturning**	**Overturning**
		$f_s = 0.7$		
		H_{sl} (kN)	18.9	10.4 (Slid.+Overt.)
		H_u (kN)	**18.1**	**10.4**
		Failure mode	**Overturning**	**Sliding + Overturning**
		Hov,s (kN)	58.6	*60.0*
		Hov (kN)	50.1	*54.6*
1.2	0.875	$f_s = 1.2$		
		H_{sl} (kN)	99.3	-
		H_u (kN)	**50.1**	**54.6**
		Failure mode	**Overturning**	**Overturning**
		$f_s = 0.7$		
		H_{sl} (kN)	57.9	53.5 (Overt.+Slid.)
		H_u (kN)	**50.1**	**53.5**
		Failure mode	**Overturning**	**Overturning + Sliding**

Table 3. Overturning and sliding load for 3-step buttresses

μ		Analytical Model	Discrete Element Model
0.9	$H_{ov,s}$ (kN)	30.0	30.0
	H_{ov} (kN)	24.3	24.3
		$f_s = 1.2$	
	H_{sl} (kN)	36.0	-
	H_u (kN)	**24.3**	**24.3**
	Failure mode	**Overturning**	**Overturning**
		$f_s = 0.7$	
	H_{sl} (kN)	21.0	20.9
	H_u (kN)	**21.0**	**20.9**
	Failure mode	**Sliding**	**Sliding**
0.72	$H_{ov,s}$ (kN)	63.4	62.8
	H_{ov} (kN)	51.0	51.0
		$f_s = 1.2$	
	H_{sl} (kN)	68.4	-
	H_u (kN)	**51.0**	**51.0**
	Failure mode	**Overturning**	**Overturning**
		$f_s = 0.7$	
	H_{sl} (kN)	39.9	39.3
	H_u (kN)	**39.9**	**39.3**
	Failure mode	**Sliding**	**Sliding**
0.24	$H_{ov,s}$ (kN)	245.3	243.3
	H_{ov} (kN)	224.2	229.0
		$f_s = 1.2$	
	H_{sl} (kN)	192.6	-
	H_u (kN)	**192.6**	**229.0**
	Failure mode	**Sliding**	**Overturning**
		$f_s = 0.7$	
	H_{sl} (kN)	112.3	110.4
	H_u (kN)	**112.3**	**110.4**
	Failure mode	**Sliding**	**Sliding**

For $f_s = 1.2$ and $\mu = 0.425$, overturning collapse of two-step buttresses is always controlled by mode 2, independently of β, (see Figures 8d-f), while mode 1 (for $\beta = 2$), or mode 3 (for $\beta = 1.4$-1.2) control if $\mu = 0.875$ (see Figures 8a-c). Differently, when $f_s = 0.7$ is assumed, two-step buttresses collapse under pure sliding on a horizontal plane at the level of the applied load (Figures 8g,l,m), or by a combined effect of sliding and overturning. In the last case, overturning can predominate over sliding (Figure 8n) or vice versa (Figures 8h,i). The three-step buttresses analyzed in Figure 9 collapse under pure sliding for $f_s = 0.7$, whereby overturning always controls for $f_s = 1.2$ with different modes. By the way, the numeri-

Figure 8. Failure modes for two-step buttresses

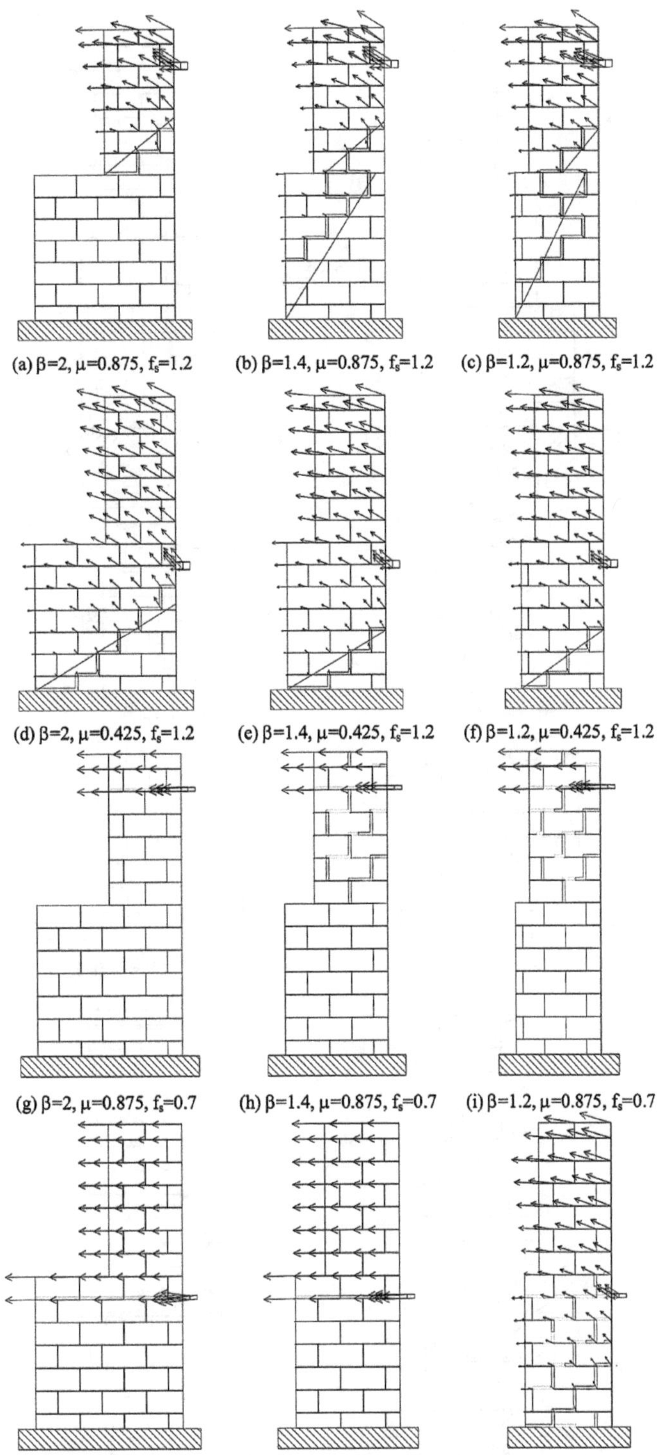

(a) β=2, μ=0.875, f_s=1.2 (b) β=1.4, μ=0.875, f_s=1.2 (c) β=1.2, μ=0.875, f_s=1.2

(d) β=2, μ=0.425, f_s=1.2 (e) β=1.4, μ=0.425, f_s=1.2 (f) β=1.2, μ=0.425, f_s=1.2

(g) β=2, μ=0.875, f_s=0.7 (h) β=1.4, μ=0.875, f_s=0.7 (i) β=1.2, μ=0.875, f_s=0.7

(l) β=2, μ=0.425, fs=0.7 (m) β=1.4, μ=0.425, fs=0.7 (n) β=1.2, μ=0.425, fs=0.7

Figure 9. Failure modes for three-step buttresses

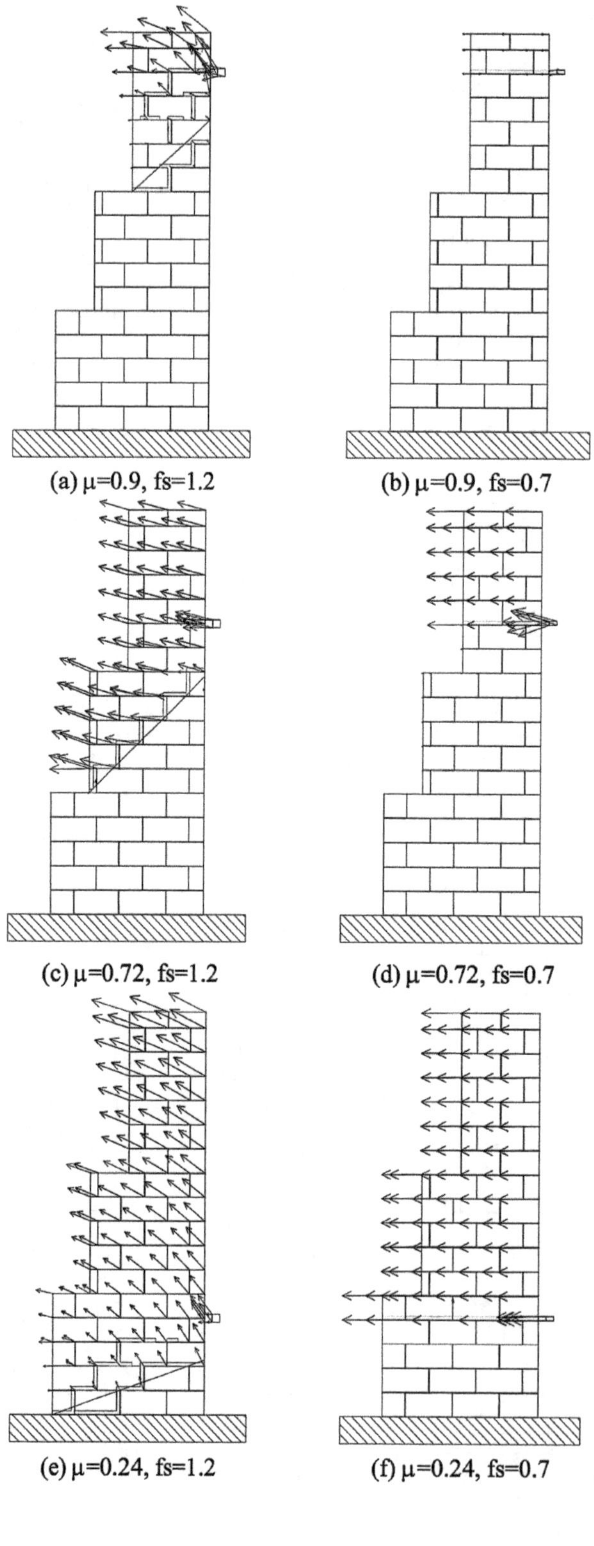

(a) μ=0.9, fs=1.2 (b) μ=0.9, fs=0.7

(c) μ=0.72, fs=1.2 (d) μ=0.72, fs=0.7

(e) μ=0.24, fs=1.2 (f) μ=0.24, fs=0.7

cal fracturing shape follows zigzag lines along the joint interfaces which could be approximated with simplified straight lines as considered analytically.

Quasi Statics

A DEM-based numerical analysis is performed in the second part of the chapter as validation tool for the first-order seismic assessment of complex unreinforced masonry structures (URM), under a constant acceleration, combined with the vertical acceleration due to gravity. The studies conducted thus far in this sense were mainly focused on the structural behavior of semicircular geometries (see, i.e. Giordano, Mele, & De Luca, 2002; Baggio & Trovalusci, 1995; Baratta, Zuccaro, & Binetti, 2004). On the contrary, there is a lack of information about the relative efficiency of different arch shapes, for any possible mechanism of collapse, such as pointed or basket-handle arches. Pointed arches are typical structural elements of the Gothic architecture. They allowed the Gothic cathedrals to reach larger heights than the Romanesque ones, while bearing lower thrusts for given loads and spans and reducing the weight on the lateral walls. It is well known however, as the ancient master builders were capable to use geometrical rules, developed through centuries of trial and error, to build complex arches and vaults. They usually scaled up the same proportions for new larger elements, without knowing anything else about the material properties or allowable stresses. The ideal geometries and proportions of arches and vaults were typically related to the in-plane dimensions and/or height of the space to be built, and in many cases dictated by some practical difficulties in the building site. This makes clearly interesting a parametric assessment of masonry structure for varying geometry parameters, as herein performed.

Within a wider analytical and numerical research project on the structural behavior and collapse mechanisms of masonry arches and portals of different shapes, pointed and basket-handle arches are comparatively studied with respect to the circular shape (for more details see Dimitri & Tornabene, 2015). Theoretical predictions were based on the usual hypotheses of limit analysis of infinite compressive strength, no tensile strength, and absence of sliding failure, such that masonry becomes an assemblage of rigid ashlars unable to slide or crush, but able only to disconnect forming non-dissipative hinges (Heyman, 1982). For a comparative study of lateral stability of pointed and basket-handle shapes, the main geometry parameters were theoretically defined in dimensionless form as: the t/R_{circ} ratio between the arch thickness t and the centerline radius of the reference circular arch R_{circ}; the angle of embrace 2α; for pointed arches, the e/R_{circ} ratio between the eccentricity of the arch center e and R_{circ}; for basket-handle arches, the ratios r/R_{circ} and d/R_{circ}, r being the smaller radius and d the distance of the middle center from the horizontal springline of the reference circular arch (Figure 10). Collapse of the arch requires four non-dissipative hinges to form, two at the extrados (B and D in Figure 10), and two at the intrados (A and C in Figure 10). The locations of the four hinges were found by iterative computations based on the equilibrium application in the form of the principle of virtual work, where the minimum constant lateral acceleration (indicated as λg, λ being the multiplier and g the base acceleration) was computed within all kinematically admissible positions of hinges.

A similar analysis was then extended to portal shapes considering the additional parameters B/R_{circ} (B being the width of the buttress) and h/R_{circ} (h being the height of the buttress from the base to the horizontal springline of the reference circular arch), as illustrated in Figure 11. In the last cases, three possible types of mechanism can be activated within portals, as already mentioned by De Luca et al. (2004): a local mechanism involving the only arch, without any hinge occurring in the buttresses; a global mechanism, featuring two hinges at the base of the buttresses and two hinges in the arch; and a

Figure 10. Geometry and hinges location for a pointed (a) and basket-handle (b) arch

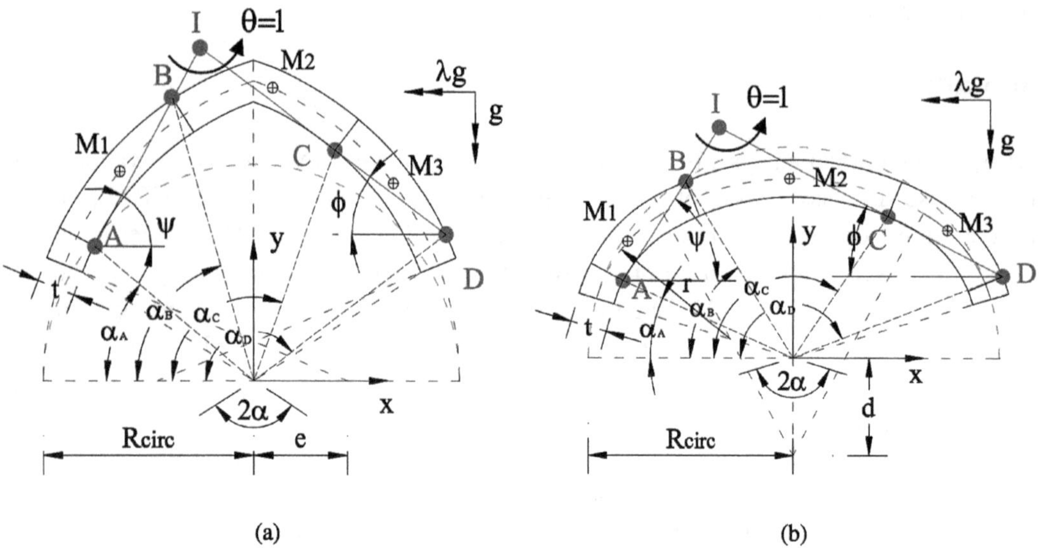

semi-global mechanism with one hinge at the base of one buttress and three hinges in the arch. While global mechanisms interest portals subjected to the only ash load or low seismic actions and prevailing ash load ones, the local or semi-global mechanisms interest stocky and massive portals, respectively, under the only seismic actions or prevailing seismic actions plus low ash load actions (Baratta, Zuccaro, & Binetti, 2004). In all cases considered in this parametric study, the hinge locations corresponding to the minimum lateral stability of portals are those represented in Figure 11, i.e. a semi-global mechanism.

Figure 11. Geometry and hinges location for a portal with pointed (a) and basket-handle (b) arch

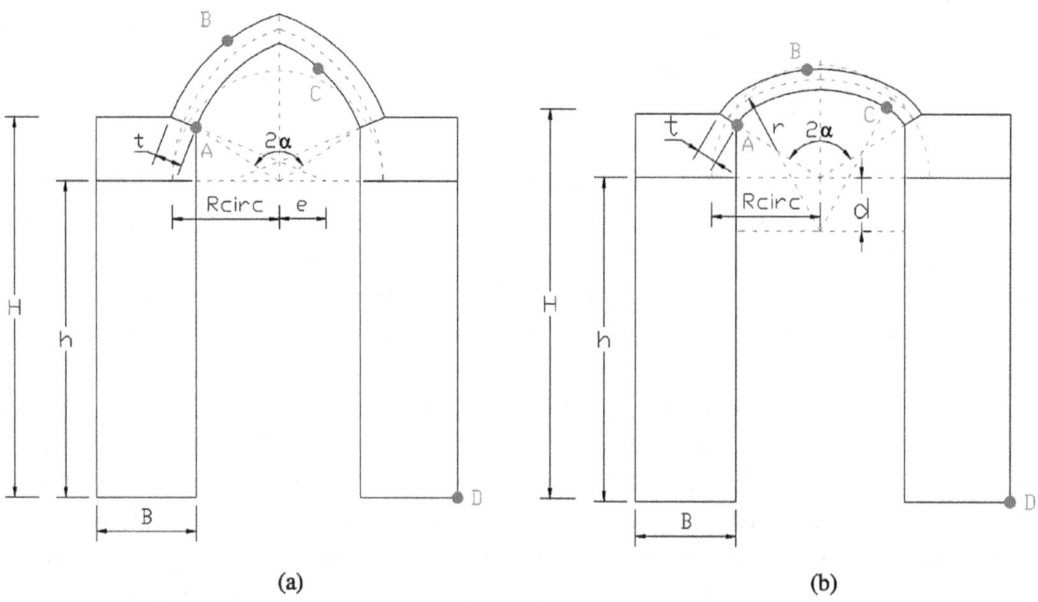

For the numerical investigation, masonry arches and portals are considered as an assemblage of rigid blocks with frictional joints, where the following properties are required: the elastic shear and normal stiffnesses, cohesion and friction angle, tensile strength and dilatancy angle. For consistency with the assumptions of the analytical investigation, the tensile strength and cohesion of the joints as well as the dilatancy angle are herein taken as zero, whereas the elastic normal and shear stiffnesses are arbitrarily set as $5x10^3$ GPa/m. It is, however, verified that these values do not affect quasi-static results. An high value of friction coefficient is also assumed (i.e. f_s=1.2) to ensure an overturning collapse, as predicted analytically.

Arches

The analyses consider first circular arches with R_{circ}=1, pointed arches with e/R_{circ}=0.5 and basket-handle arches with d/R_{circ}= r/R_{circ} =0.5, all characterized by a thickness ratio t/R_{circ}=0.2 and a density of 2000 kg/m^3. Three angles of embrace are taken into account (2α=120°, 150°, 180°), as already done in the analytical investigation in Dimitri & Tornabene, (2015). Table 4 summarizes comparatively the analytical and numerical results obtained in terms of onset acceleration for the three different shapes under consideration, while Figures 12-14 represent their corresponding kinematics at collapse.

As expected analytically with limit analysis, the analyzed pointed arches are always less stable than the reference circular ones, and their stability decreases as the angle of embrace increases. On the contrary, basket-handle arches seem to be always more stable compared to the circular counterpart with a maximum lateral stability reached for an angle of embrace 2α=120°. The last geometry would represent the optimal configuration herein analyzed. Furthermore, the mechanism of collapse is always characterized by the formation of two hinges at the supports, and two other hinges within the span, for angles of embrace 2α=120°, 150°, i.e. for angles of embrace under a certain critical value. Differently, three hinges form within the span, with only one hinge forming at the support, for 2α=180°, i.e. for angles of embrace above this critical value. In addition, the numerical results are in perfect agreement with the analytical predictions both in terms of the onset accelerations (λg) and mechanisms of collapse, confirming the efficiency of the DEM for an equivalent static study of generally shaped arches under lateral accelerations.

Table 4. Analytical and numerical onset acceleration for circular, pointed and basket-handle arches

Geometrical Shape	Angle of Embrace (°)	Analytical Results (λg)	Numerical Results (λg)
CIRCULAR	120	1.27g	1.28g
	150	0.62g	0.64g
	180	0.29g	0.31g
POINTED	120	0.53g	0.53g
	150	0.38g	0.39g
	180	0.24g	0.25g
BASKET-HANDLE	120	2.35g	2.38g
	150	1.06g	1.10g
	180	0.54g	0.56g

Figure 12. Mechanisms of collapse for circular arches with $R_{circ}=1$, and $t/R_{circ}=0.2$: analytical (a, c, e) and numerical (b, d. f) configurations

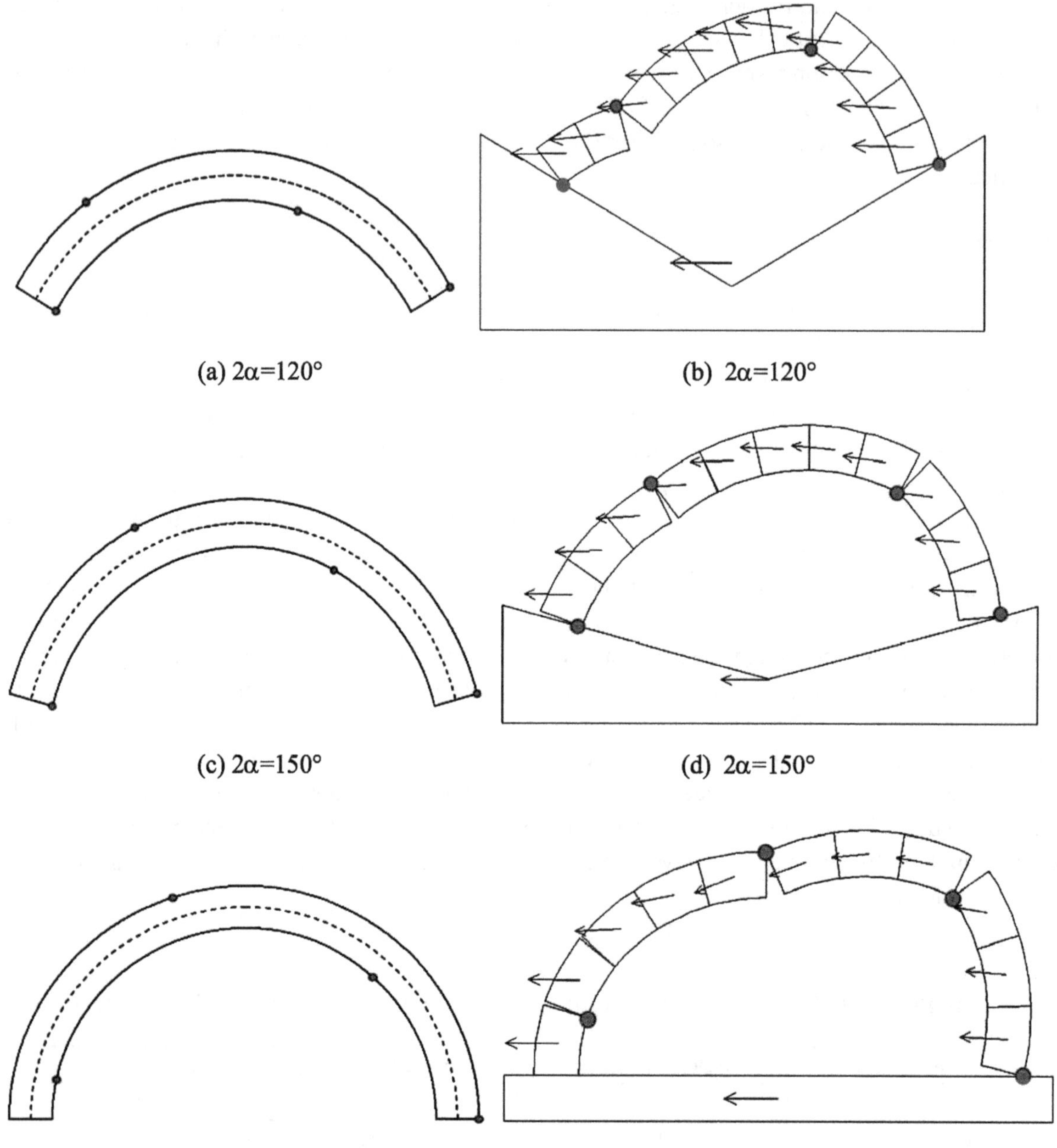

(a) 2α=120° (b) 2α=120°

(c) 2α=150° (d) 2α=150°

(e) 2α=180° (f) 2α=180°

Figure 13. Mechanisms of collapse for pointed arches with $R_{circ}=1$, $e/R_{circ}=0.5$, and $t/R_{circ}=0.2$: analytical (a, c, e) and numerical (b, d. f) configurations

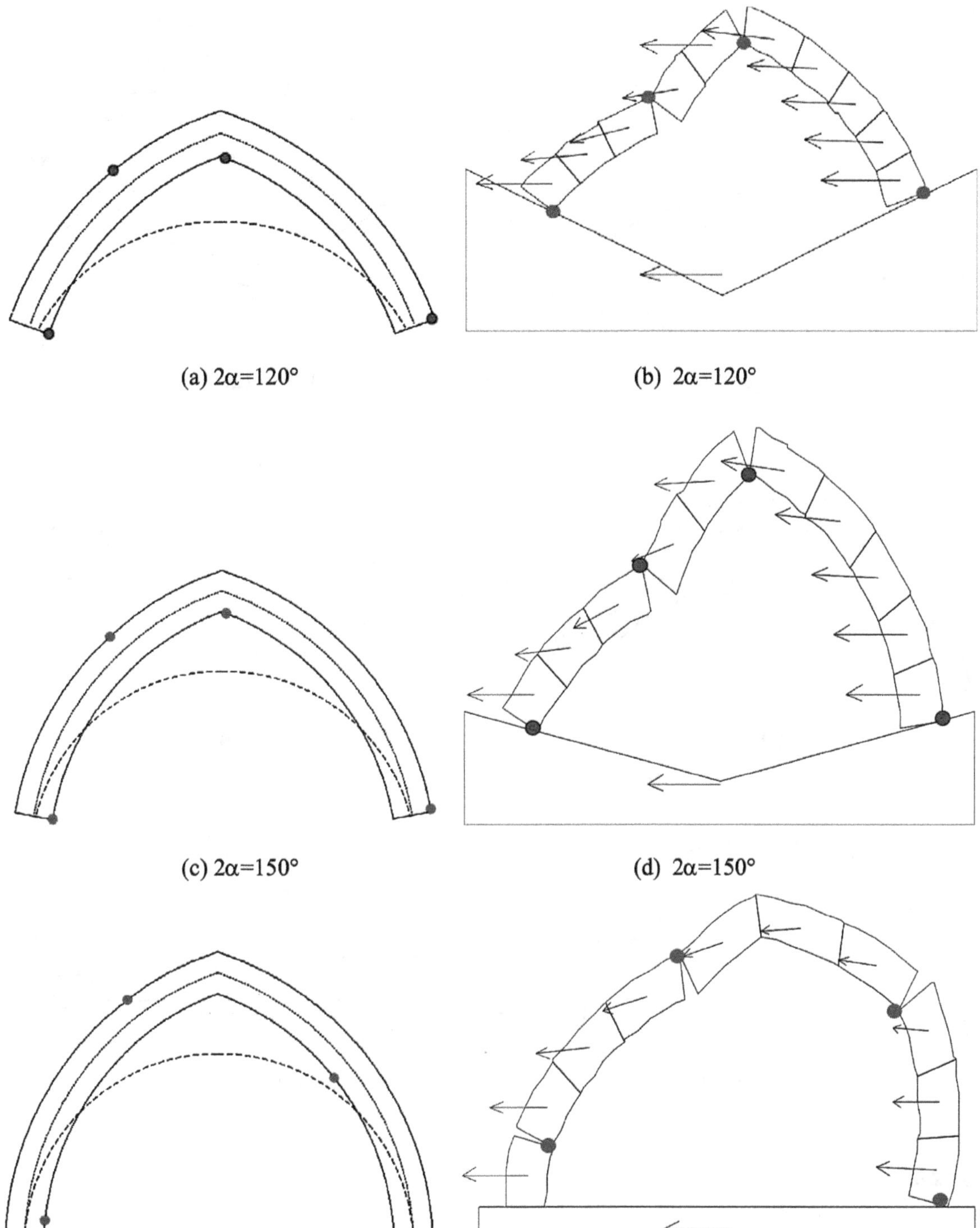

Figure 14. Mechanisms of collapse for basket-handle arches with $R_{circ}=1$, $d/R_{circ}=0.5$, $r/R_{circ}=0.5$ and $t/R_{circ}=0.2$: analytical (a, c, e) and numerical (b, d, f) configurations

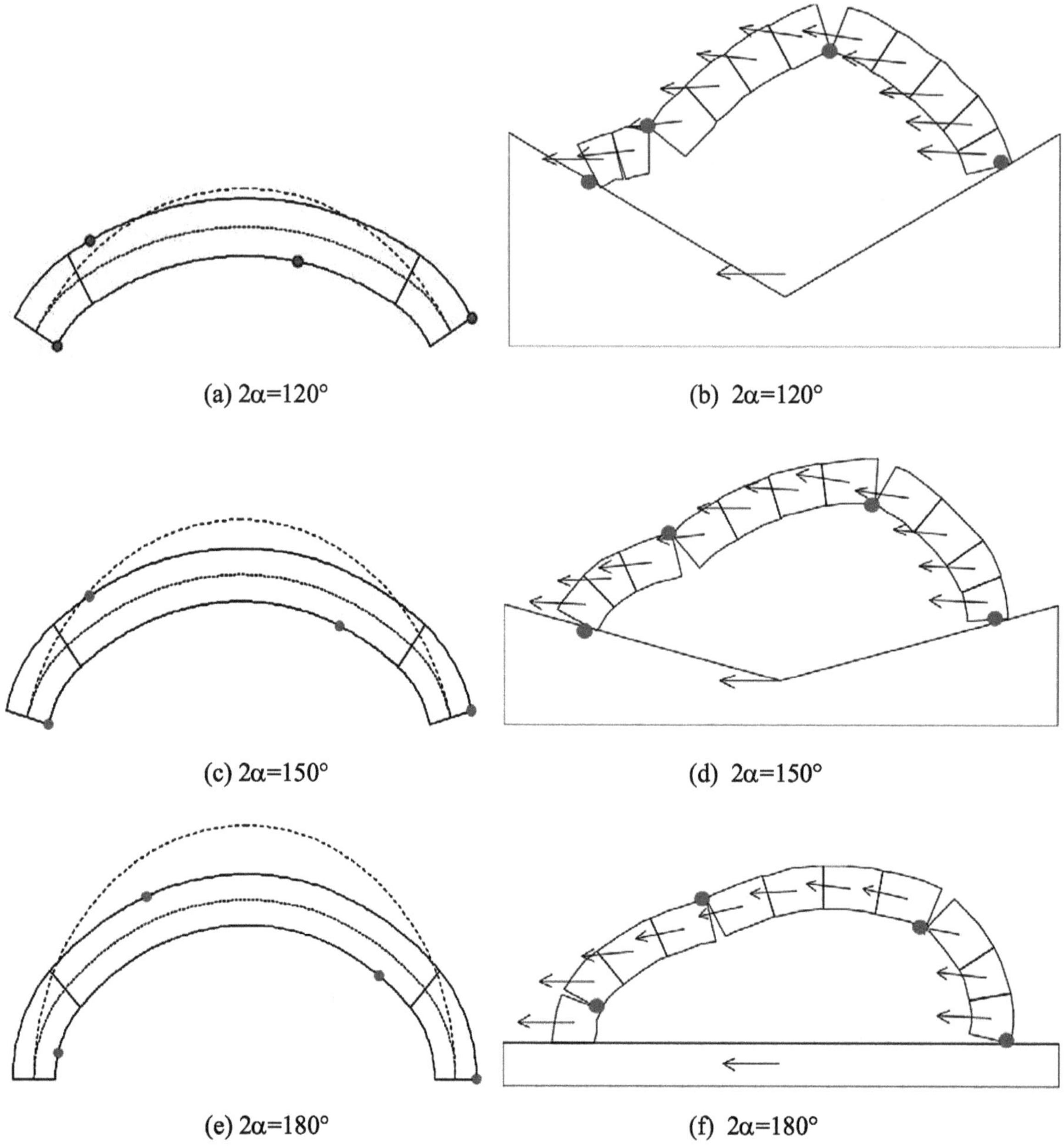

(a) 2α=120° (b) 2α=120°

(c) 2α=150° (d) 2α=150°

(e) 2α=180° (f) 2α=180°

Portals

Similar analyses are then repeated for circular, pointed and basket-handle portals with the same material and geometry parameters as considered for the only arches, with the additional parameters $B/R_{circ}=0.8$ and $h/R_{circ}=3$ related to the dimensions of the buttresses. Three angles of embrace are considered again, for comparison purposes with the analytical investigation in Dimitri & Tornabene, (2015) (2α=120°, 150°,

180°). Results are summarized in Table 5, in terms of lateral resistance, together with the mechanisms of collapse as shown in Figures 15-17, for circular, pointed and basket-handle portals, respectively. As visible in Table 5, it is worth noting as portals with pointed arches are more stable than their circular counterparts independently of the angles of embrace, or equally stable, as found numerically for the only angle of embrace $2\alpha=180°$. On the contrary, portals with basket-handle arches are always less stable than their corresponding circular counterparts for the given geometry ratios. The lateral stability anyway tends to increase for increasing angles of embrace, as found both theoretically and numerically.

From a kinematical point of view, it seems also that a semi-global mechanism always controls, independently of the arch shape, which yields three hinges forming into the arch and one hinge fixed at the extrados side of the support (see Figures 15-17). A comparison between analytical and numerical results reveals, once more, a perfect agreement between the two approaches both in terms of dynamic resistance and mechanism of collapse.

Benchmark for a Comparative Assessment of DEM, FEM and Experimental Results

The suitability of the DEM approach for an approximated dynamic study of masonry structures is finally assessed in comparison to FEM outcomes and results from an experimental campaign. A standard circular arch is therefore chosen as benchmark, as already considered in the literature by DeJong et al. (2008), with a scaled geometry defined by a dimensionless thickness ratio $t/R_{circ}=0.15$, an angle of embrace $2\alpha=152°$, and a radius $R_{circ}=20$cm.

The discrete arch, consisting of 13 blocks and subjected to a constant horizontal base motion is assessed with DEM and compared to a classical procedure based on the non-linear FEM, as implemented in the computer software STRAND7. An equivalent frame approach, known as ''macro-element approach'' is adopted in the last case to study the in-plane behavior of the arch in question. Based on this approach, the masonry is modeled as a homogeneous material with equivalent mechanical properties.

The FEM arch is discretized by 60 x 5 plane-stress 9-node isoparametric elements in the x-y plane (see Figure 18a), which gives a sufficiently accurate approximation of the bending behavior of the model

Table 5. Analytical and numerical onset acceleration for circular, pointed and basket-handle portals

Geometrical Shape	Angle of Embrace (°)	Analytical Results (λg)	Numerical Results (λg)
CIRCULAR	120	0.16g	0.17g
	150	0.18g	0.18g
	180	0.19g	0.21g
POINTED	120	0.17g	0.18g
	150	0.19g	0.20g
	180	0.20g	0.21g
BASKET-HANDLE	120	0.14g	0.15g
	150	0.15g	0.16g
	180	0.16g	0.17g

Figure 15. Mechanisms of collapse for circular portals with $R_{circ}=1$, $H/R_{circ}=3$, $B/R_{circ}=0.8$, and $t/R_{circ}=0.2$: analytical (a, c, e) and numerical (b, d. f) configurations

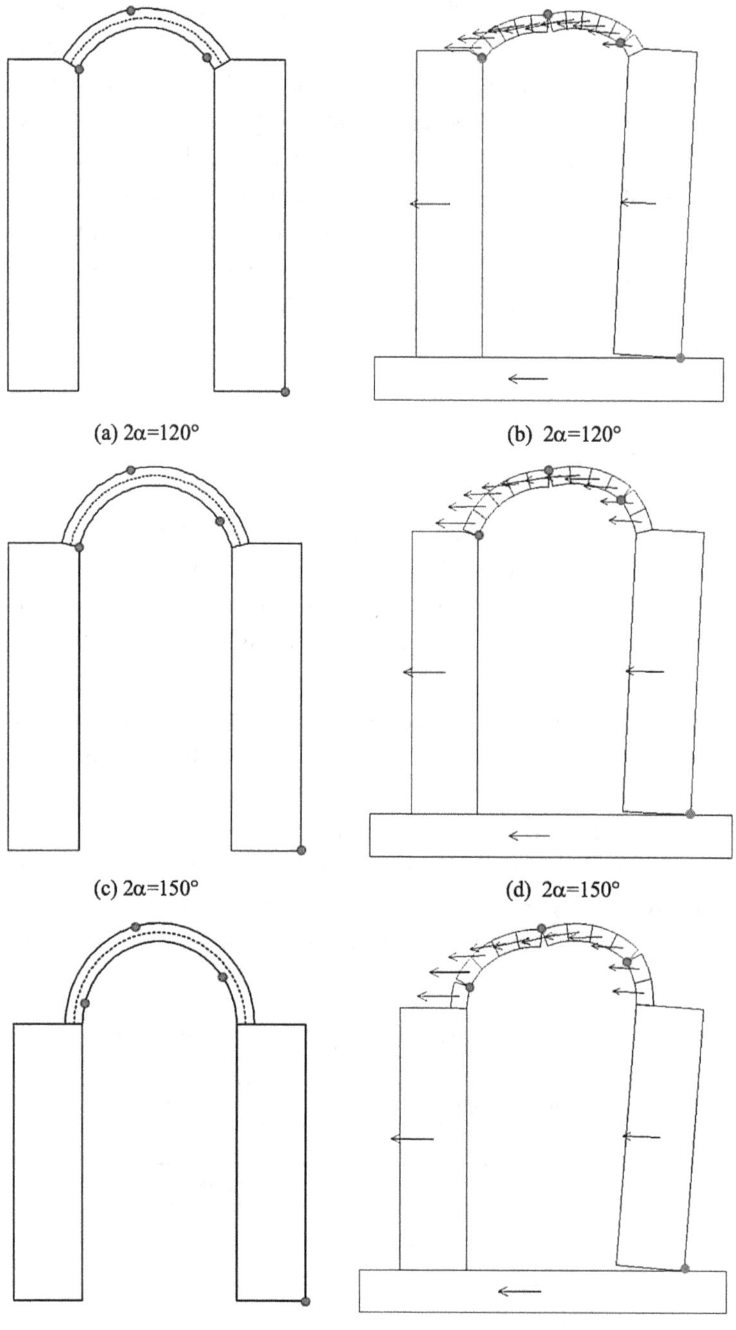

(a) 2α=120° (b) 2α=120°

(c) 2α=150° (d) 2α=150°

(e) 2α=180° (f) 2α=180°

Figure 16. Mechanisms of collapse for pointed portals with R_{circ}=1, H/R_{circ}=3, B/R_{circ}=0.8, e/R_{circ}=0.5, and t/R_{circ}=0.2: analytical (a, c, e) and numerical (b, d. f) configurations

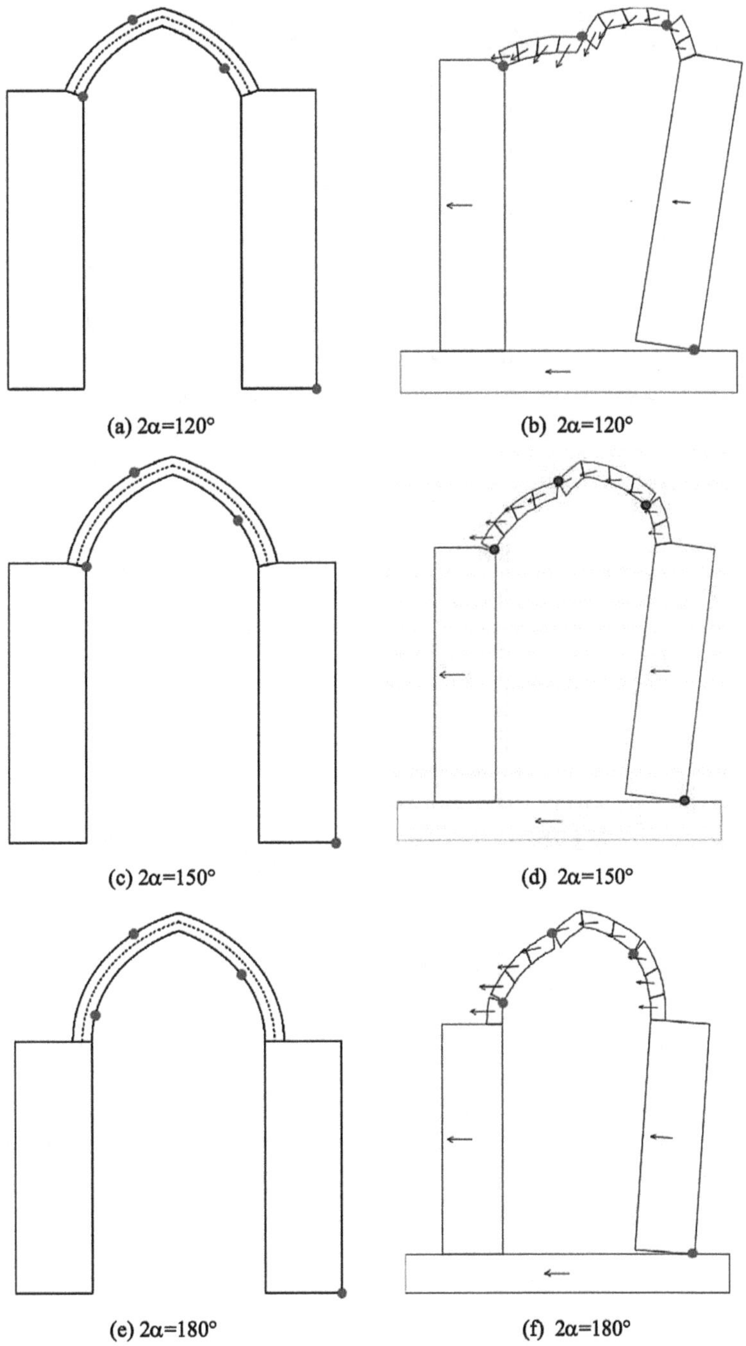

Figure 17. Mechanisms of collapse for basket-handle portals with R_{circ}=1, H/R_{circ}=3, B/R_{circ}=0.8, d/R_{circ}=0.5, r/R_{circ}=0.5 and t/R_{circ}=0.2: analytical (a, c, e) and numerical (b, d, f) configurations

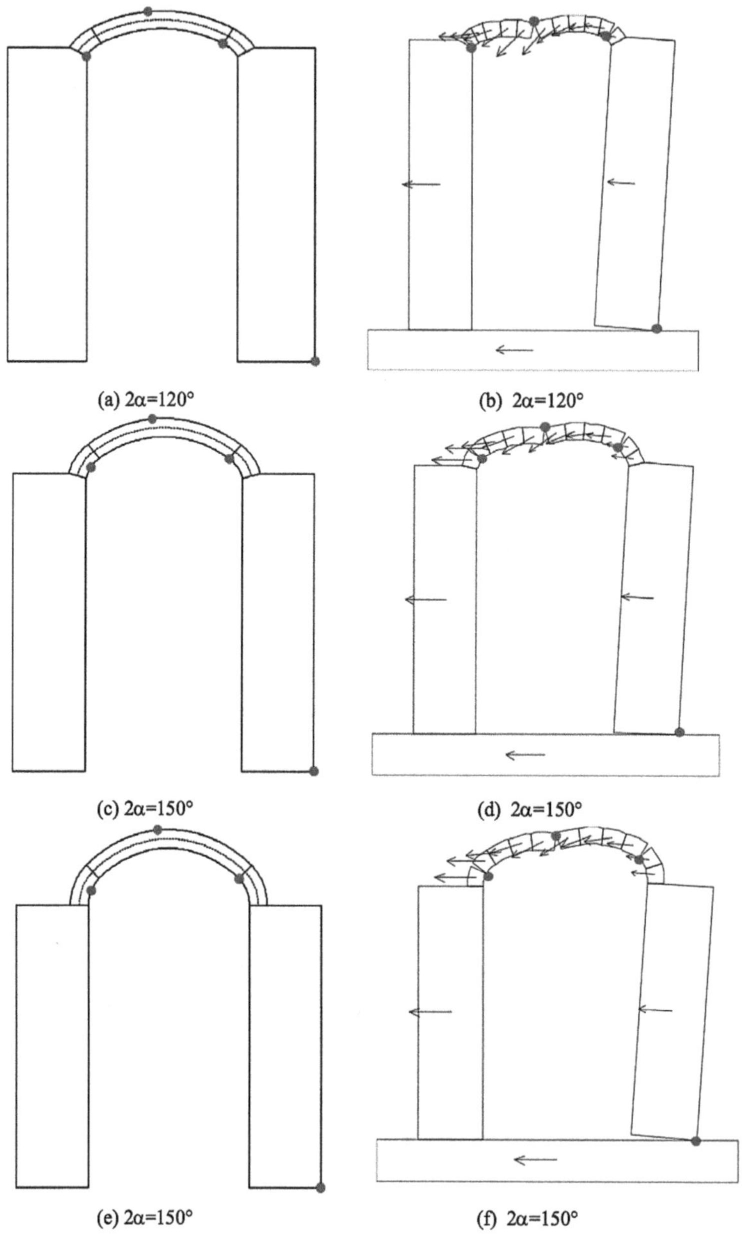

during the loading process. An elasto-plastic ''tuff masonry model'' is assigned to the body, with a density of 2000kg/m^3, an elastic modulus E=1.1GPa, a Poisson's ratio ν=0.2, a compressive strength f_c=3.3 MPa, and a tensile strength f_t=0.33 MPa (see De Luca et al., 2004).

Figure 18a illustrates the deformed configuration at the last load step of the analysis, along with the horizontal stress contour imposed. The stress state resulting from the finite element approach is correctly representative of the collapse mechanism, due to an adequate identification of the zones within the arch where hinges are most likely to occur. The loading capacity obtained via FEM results to be equal to 0.32, and lower than analytical or DEM-based predictions (i.e. λ= 0.37 and λ=0.38, respectively). In other words, FEM yields to a lower bound limit for strength capacity, in agreement with findings of De

Figure 18. Quasi-static response for a circular arch: FEM (a, b) vs. limit analysis (c), DEM (d), and experimental results (e)

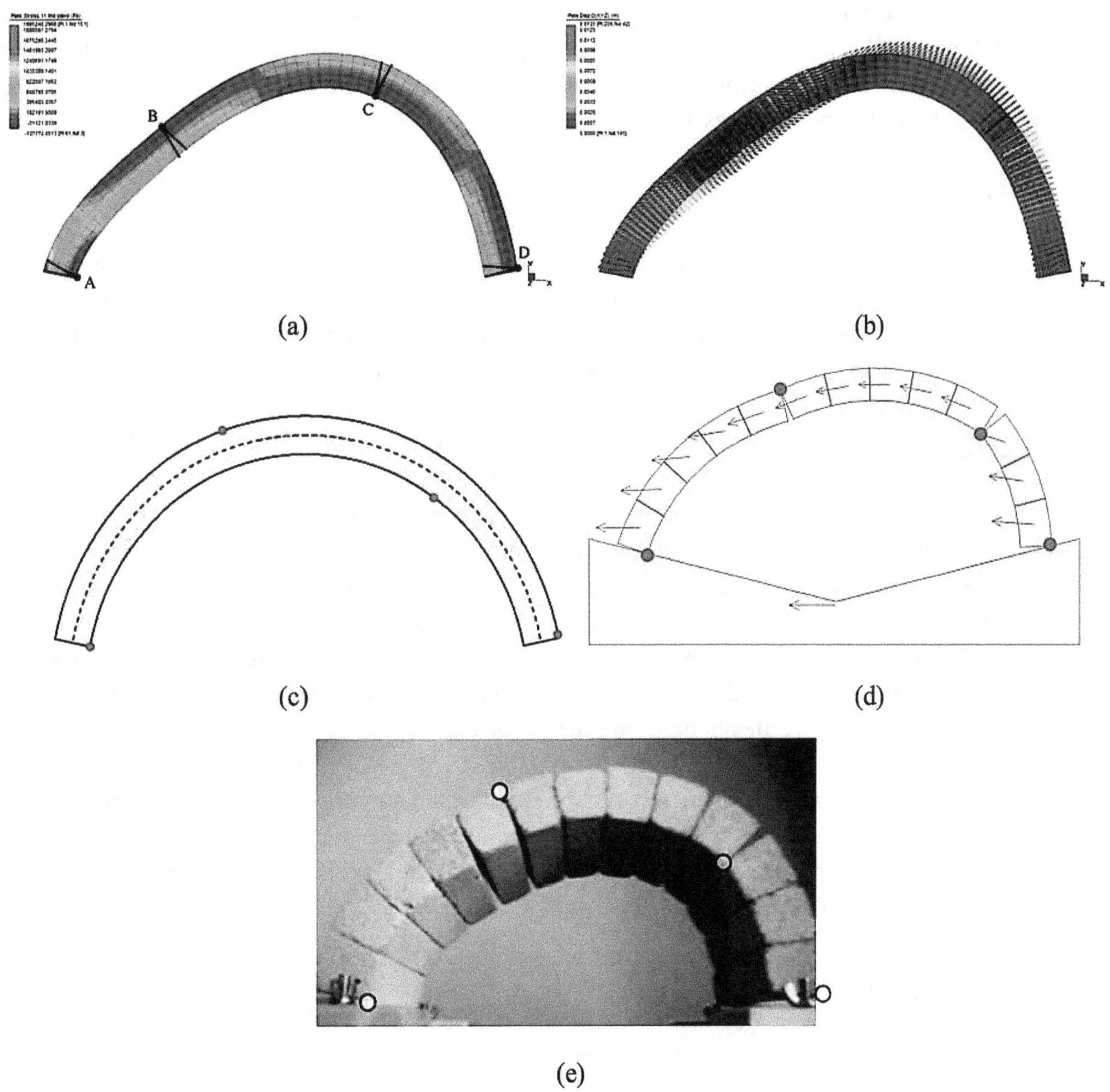

Luca et al. (2004). A vector representation of the FEM-based displacement field is provided in Figure 18b, which clearly features a 4-hinge mechanism at collapse, as predicted with limit analysis and DEM (compare Figures 18a,b with Figures 18c,d). Despite the different assumptions of FEM and DEM, both the approaches provide useful indications about the approximated seismic capacity of discrete masonry.

The quasi-static analysis is finally completed with a qualitative comparison with results from an experimental campaign available from the literature (DeJong et al., 2008) including some seismic tests performed on scaled prototypes of autoclaved aerated concrete, under different earthquakes with varying frequency content and maximum amplitude (i.e. earthquakes occurred in Parkfield 1966, El Centro 1940, Golden Gate 1957, Northridge 1994, and Helena 1935). While the quasi-static analysis gives only an information about the onset acceleration leading to a mechanism, the actual dynamic strength of an arch is expected to be higher due to the sliding and/or rocking motions of the blocks. The analytical and numerical collapse multiplier (i.e. $\lambda= 0.37$ and $\lambda= 0.38$), in fact, represents only a lower bound for the dynamic strength of the arch, as its value lies always below the experimental failure domain computed experimentally by DeJong et al. (2008), with magnitudes varying between $\lambda=0.4$ and $\lambda=1.3$. Regardless, a good matching between theoretical, numerical and experimental results is observed, from a kinematical point of view, with two hinges forming at the support and the other two ones throughout the span of the arch, as visible by comparing Figures 18a,c,d,e.

Dynamics

Although the available literature has already proved the capability of the DEM to predict the static or dynamic response of multi-block structures, a limited attention is given to the sensitivity of numerical results to the input parameters. This sensitivity would be necessary at least to gain a certain confidence of numerical results, especially in view of the uncertainty and difficulty of some experimental investigations in terms of practical realization and interpretation of results. A real case could be, for example, the definition of an elastic joint stiffness between blocks. The geometric enforcement of contact constraints, indeed, would require an approximate definition of the joint stiffness values both in the normal and tangential direction as penalty parameters, which should be as large as possible to limit the interpenetrations and relative displacements in the stick phase. At the same time, a refined expression of these parameters may be possible, when defining the mortar properties of joints, or the elastic deformability of blocks by means of constitutive laws, as well as the real contacting surfaces with given roughness profiles. In addition, the joint stiffness is adopted sometimes to consider the elastic deformations of blocks which are assumed to be rigid. Based on the aforementioned uncertainties, a parametric study on the dynamic behavior of URM masonry elements based on the DEM for varying input parameters is required for completeness (see also Dimitri, De Lorenzis, & Zavarise, 2011).

A dynamic behavior of masonry structures is now represented numerically with DEM by a timestepping algorithm where the size of the timestep is limited assuming a constant value for velocities and accelerations within the timestep. The required timestep is defined by the mass of the blocks and the stiffness contact interfaces, and is estimated as

$$\Delta t = \left(frac\right)2\left(\frac{M_{\min}}{k_{\max}}\right)^{1/2} \tag{1}$$

where M_{min} is the mass of the smallest block in the masonry system, and k_{max} is the maximum contact stiffness. The term *frac* is a coefficient that accounts for the simultaneous contact between a single block and several other ones, and is usually set to 0.1. In dynamic modelling, the choice of the damping properties is generally a key point. In the analyses presented herein, Rayleigh damping is applied, i.e.

$$\boldsymbol{C} = \alpha_R \boldsymbol{M} + \beta_R \boldsymbol{K} \tag{2}$$

$\boldsymbol{C}$, $\boldsymbol{M}$, and $\boldsymbol{K}$ being the damping, mass and stiffness matrices, respectively, α_R the mass-proportional damping constant, and β_R the stiffness-proportional damping constant. Mass-proportional damping is dominant at lower frequencies, and responsible of the movement in unison of several blocks ("sloshing"), while stiffness-proportional damping is dominant at higher frequencies and responsible of the inter-blocks vibrations ("rattling") (Housner, 1963). Rayleigh damping can be defined with the critical damping ratio for mode i, ζ_i which depends on the frequency of a multiple-degree-of-freedom system, f_i, as in the following (Housner, 1963)

$$\zeta_i(f_i) = \frac{1}{4\pi}\left(\frac{\alpha_R}{f_i} + 4\pi^2 \beta_R f_i\right) \tag{3}$$

whose minimum point of the curve gives two parameters ζ_{min}, and f_{min}, defined as

$$\zeta_{\min} = \left(\alpha_R \beta_R\right)^{1/2} \tag{4}$$

$$f_{\min} = \frac{1}{2\pi}\left(\alpha_R / \beta_R\right)^{1/2} \tag{5}$$

The last two equations express the input parameters adopted to define Rayleigh damping.

The ninth column of the eastern side of the Temple of Apollo at Bassae in the Peloponnese, Greece, is chosen for numerical modelling, due to the availability of its geometrical properties, as given in Psycharis et al. (2000). The Doric column is characterized by seven drums and one capital, with a total height of 5.95 m, a base of 1.11 m in diameter, and a taper angle of 1°. The input model is constituted by the assemblage of eight rigid marble blocks of density equal to 2620 kg/m^3, interacting through contact elements with a reference normal and shear stiffness, j_{kn} and j_{ks} respectively, set to 5x10^{10} Pa/m for both, and a reference value of the friction angle, α, of 35° corresponding to a friction coefficient f_s=0.7. The stiffness and mass damping are initially set to zero. A parametric investigation on the dynamic response and failure domain of the column under harmonic impulses (Figure 19) is performed for varying input properties of joints (friction angle, normal and shear stiffness), and damping ratios, together with the size effect and the effect of the number of drums. The impulse magnitude required to initiate rocking, equal to 0.19g, is first determined with an equivalent static analysis performed on the monolithic structure under a constant horizontal acceleration in addition to a vertical acceleration of gravity. The model is then subjected to a range of harmonic impulses with periods T varying from 0.25 to 2 s, for a total duration of each analysis of 20 s, in order to be sure of capturing the dynamic behavior of the column during and after application of the base excitation. For each impulse period the analysis is repeated increasing

Figure 19. Harmonic time history

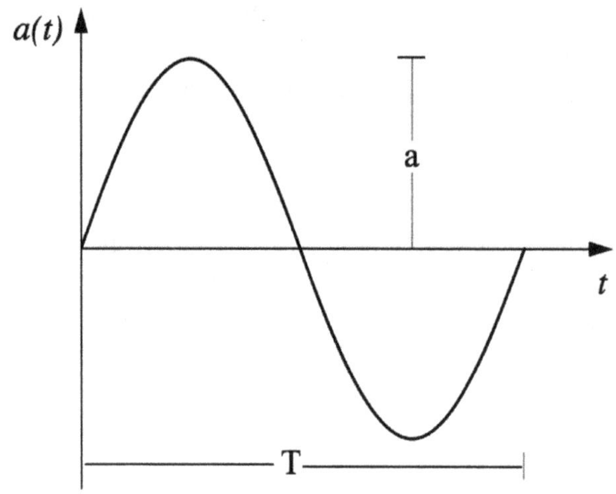

by small steps the magnitude of the base acceleration until collapse of the structure occurs. The effect of the number of cycles of the excitations on the failure domain is also analyzed.

As shown in Figure 20, failure domains of columns feature a chaotic response since columns can collapse at a certain acceleration level while remaining stable for a larger magnitude input. This agrees with findings of Psycharis et al. (2000). The failure domains can be, therefore, determined by the smallest acceleration causing collapse, which explains their irregular profile. For high frequency excitations the seismic resistance is highly affected by the number of cycles (see Figure 21), i.e. it decreases considerably for increasing number of cycles until it stabilizes for 10-cycles excitations, herein adopted in what follows. The effect of the number of cycles seems to decrease for decreasing frequency excitations, where the structure collapses during the first cycle of excitation or soon after it.

Figure 20. The chaotic behavior of the column subjected to the 10-cycle harmonic impulse

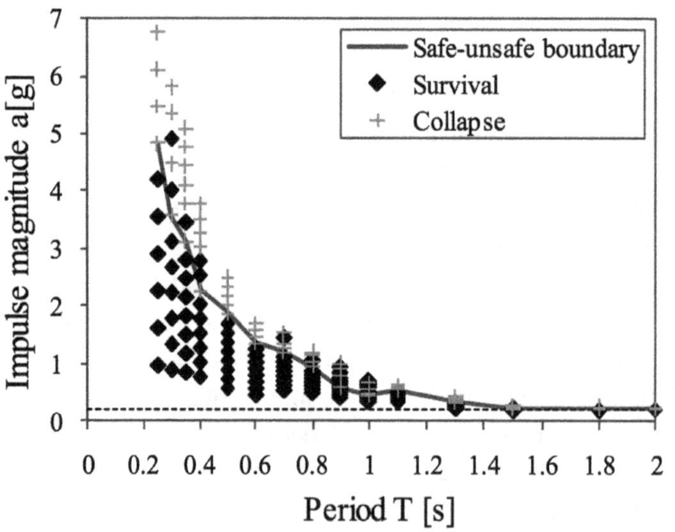

Figure 21. Failure domains of the column subjected to harmonic impulses

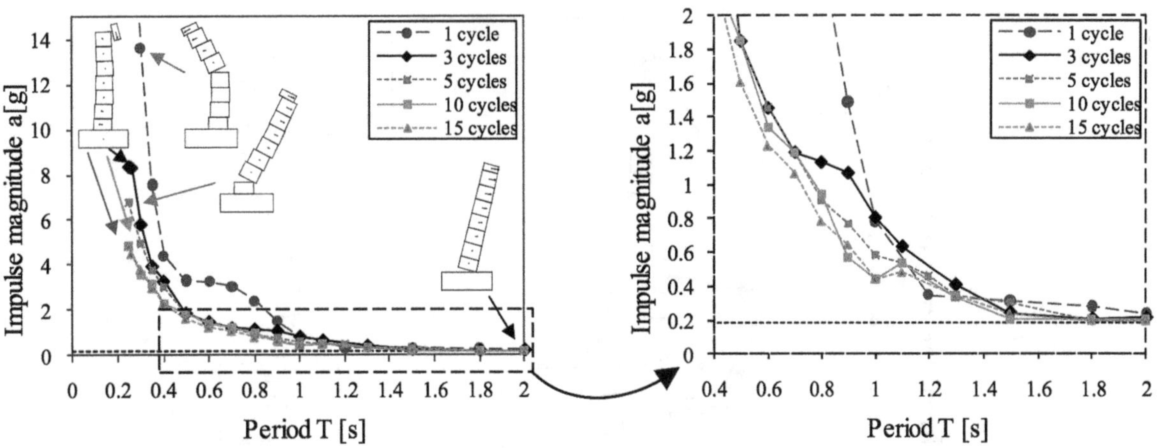

In the magnitude-period space, the failure domains gradually decrease for increasing periods T. In other words, the rocking column collapses more easily for long period base motions than for short period ones. For 1-cycle base excitations, however, the dynamic resistance keeps approximately constant at a value of 3.2g for a period varying between 0.5 and 0.7 s, where the evolution of the structure to collapse remains almost the same, before declining to the static solution. Except for some cases where high frequency excitations lead to a local instability of the top capital falling down after bouncing many times, large rocking, sliding and/or inter-stone vibrations along joints interest the column for high frequency excitations. A monolithic overturning differently occurs for low frequency excitations. More specifically, the response of the structure is dominated by sliding phenomena and inter-stone vibrations for $T < 0.5$s, as well as by rocking for $0.5 < T < 1$s, which becomes the only motion of the structure for $T > 1$s.

Effect of Friction Coefficient

The mechanical behavior of the interfaces between blocks is expected to be affected by the friction coefficient, which varies considerably for varying materials and environmental conditions. A reference value $\alpha = 35°$ (f_s=0.7) is considered for the friction angle, as determined from some conventional shear tests for marble (Psycharis et al., 2000; Papantonopoulos et al., 2002), together with two additional values $\alpha = 25°, 45°$ (i.e. $f_s = 0.47, 1.0$), here chosen for comparative purposes. As shown in Figure 22, the dynamic resistance of the column increases for decreasing friction angles and becomes quite pronounced in the period interval with significant sliding phenomena between drums, i.e. for low periods where relative inter-stone vibrations cause irreversible displacements of the drums. This also occurs for the largest value of coefficient of friction herein examined.

Effect of Damping Ratio

Another important parameter affecting the dynamic response of masonry structures is the damping coefficient, which accounts for the energy dissipation due to the internal friction of materials. In order to quantify the sensitivity of the failure domain of the column to the specified damping ratio, the domain

Figure 22. Effect of friction on the failure domain

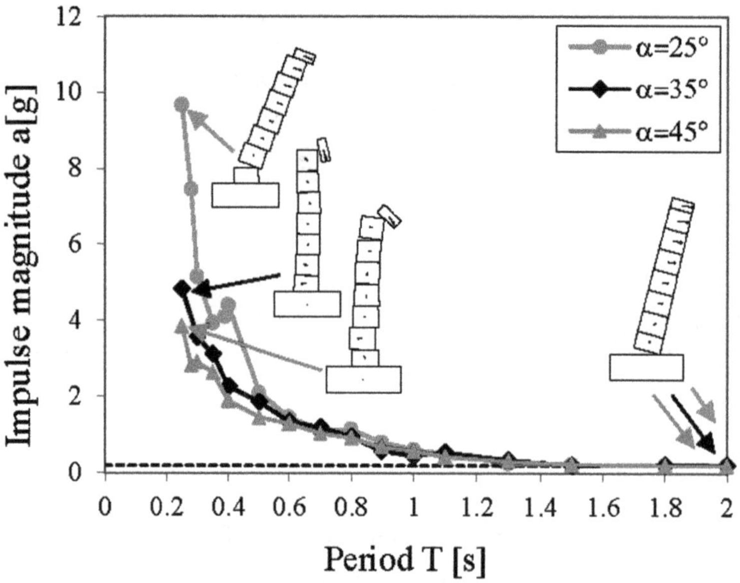

in Figure 20 is computed again for a wide range of values of the mass- and stiffness-proportional damping constants, α_R and β_R respectively, as reported in Table 6. Some of these values were adopted in the literature for dynamic investigations of rocking blocks and arches (De Lorenzis, DeJong, & Ochsendorf, 2007; DeJong, 2009). The dynamic response is evaluated for increasing stiffness-proportional damping, α_R, and mass-proportional damping, β_R, where results are compared to the reference ones obtained for the undamped structure (α_R=0, β_R=0).

Except for the highest considered level of mass damping (i.e. α_R=6.28x10^{-1}, β_R=1.59x10^{-4}), the failure domain, for harmonic base excitations, seems to be quite unaffected by the increase in mass and stiffness damping. As shown in Figure 23, only a slight variation can be observed in the dynamic resistance of damped columns with respect to the initial undamped structure. Some differences can be also highlighted in the kinematics of the structure, especially for low period impulses. More specifically, the increase in stiffness-proportional damping does not affect the low frequency damping but it increases the

Table 6. Damping parameters adopted for numerical simulations

	Damping Constants		ξ_{min}	f_{min}
	α_R	β_R	[%]	[Hz]
Stiffness-proportional effect	6.28x10^{-5}	6.37x10^{-6}	0.002	0.5
	6.28x10^{-5}	3.98x10^{-5}	0.005	0.2
	6.28x10^{-5}	1.59x10^{-4}	0.01	0.1
	6.28x10^{-5}	3.98x10^{-3}	0.05	0.02
Mass-proportional effect	6.28x10^{-5}	1.59x10^{-4}	0.01	0.1
	6.28x10^{-3}	1.59x10^{-4}	0.1	1
	6.28x10^{-1}	1.59x10^{-4}	1	10

high frequency damping. The results show as bouncing phenomena of the top drums can be completely prevented for increasing values of stiffness-proportional damping, while the response of the structure features a dominant rocking motion, with an increasing number of drums involved (see Figure 23a). On the contrary, the increase in mass-proportional damping increases the low frequency damping but it leaves the high frequency damping almost unaffected. For the highest level of mass-proportional damping, a substantial increase in the dynamic resistance of the structure is noticed, especially for medium and high impulse periods. Moreover, sliding phenomena occur instead of a pure rocking for medium and high periods (Figure 23b).

Effect of Joint Stiffness

The joint stiffnesses are considered as penalty parameters, whose values do not consider the roughness characteristics of the stone surfaces and are chosen as large as possible to minimize the interpenetration between contacting surfaces in a context where pure contact non-penetration conditions are geometrically

Figure 23. Effect of damping on the failure domain

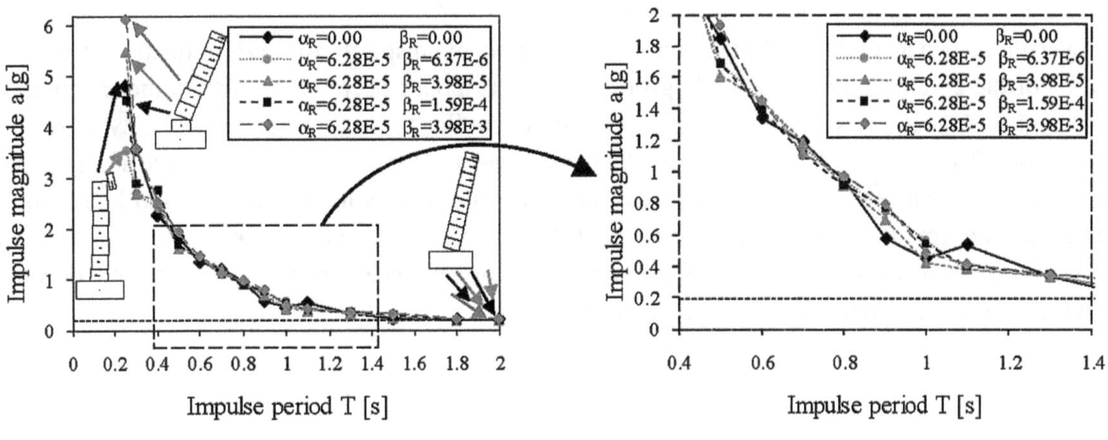

(a) Stiffness-proportional damping

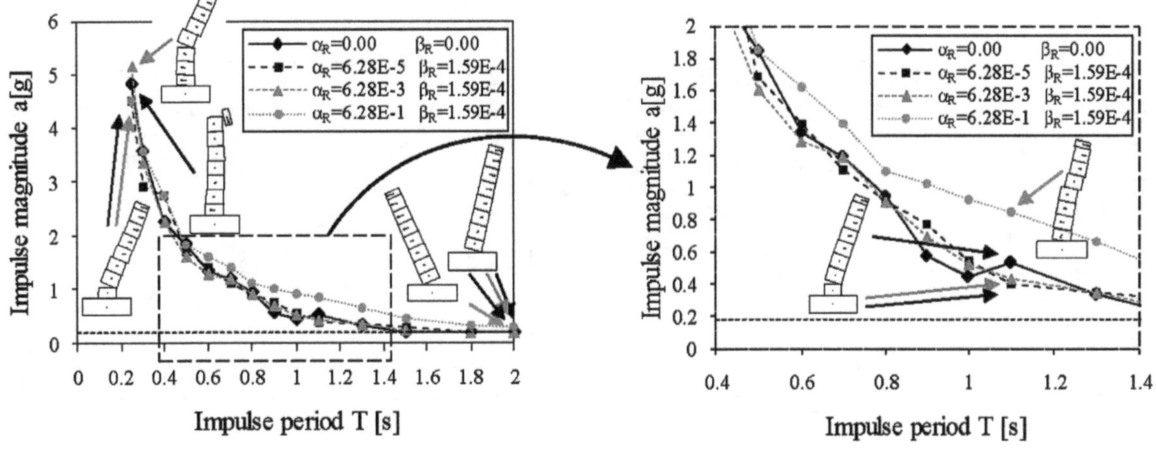

(b) Mass-proportional damping

enforced between interfaces. In order to evaluate the effect of the joint properties on the dynamic response of the column, the normal and shear joint stiffnesses are varied within a range of $5x10^9 \div 5x10^{12}$ Pa/m, for fixed values of damping parameters, α_R and β_R, equal to $6.28x10^{-5}$ and $1.59x10^{-4}$, respectively. A slight increase in the dynamic resistance of the structure results from high frequency impulses (Figures 24). In such cases, indeed, higher stiffness values lead to a slight increase of the natural frequencies; higher natural frequencies are critically damped, causing some variations into the kinematics of the structure where vibration displacements give space to rocking and sliding movements. This emphasizes as joint stiffnesses are interrelated to frequency dependent damping parameters.

Comparison with the Monolithic Column

As expected, the behavior of multi-drum column is quite different in comparison to a monolithic one, both in terms of dynamic resistance and mode of collapse. Based on numerical results, it seems that the multi-drum column is always more stable than the monolithic one with the same dimensions, as visible in Figure 25. When the dynamic behavior involves both rocking and sliding phenomena (i.e. for low frequency impulses), the discrete column is always more resistant than the monolithic one as the inter-stone displacements due to sliding and the impacts due to "partial rocking" of smaller groups of drums lead to an increase of kinetic energy dissipation.

Conversely, when a "global rocking" dominates the dynamic response for high frequency impulses, all drums of the column rotate in a single group, and the response becomes similar to the monolithic structure. In this range the safe-unsafe boundary is quite similar for both models, as clearly shown in Figure 25 for 1.3 s $\leq T \leq$ 2 s. A monolithic model is therefore too conservative for practical assessments of the dynamic response of discrete structures, since the only dissipation of energy would be associated to impacts during the global rocking motion of the monolithic structure before collapse. Based on these results, it is expected that possible devices of connection between drums, as, i.e. internal dowels, effectively decrease the dynamic resistance of the structure.

Figure 24. Effect of joint stiffness on the failure domain

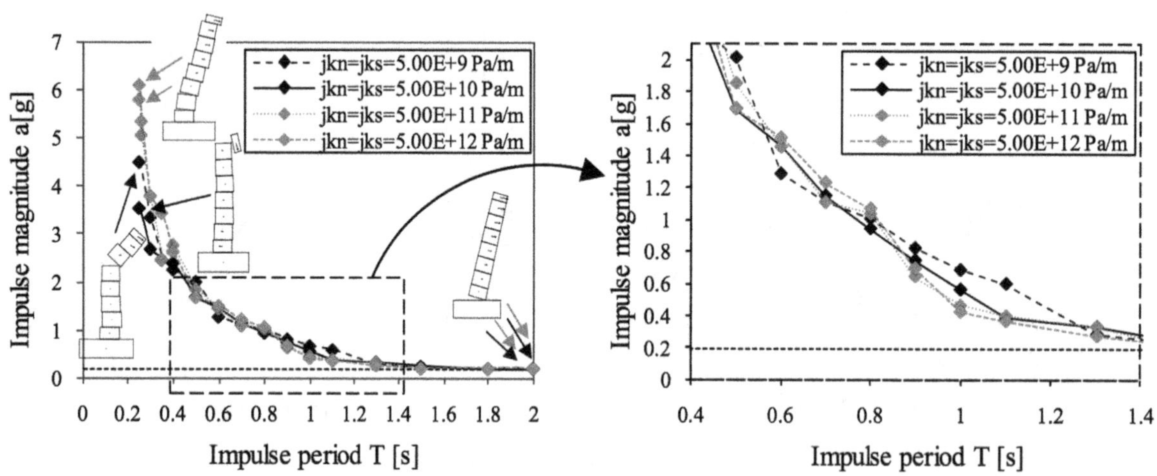

Figure 25. Effect of drums on the failure domain

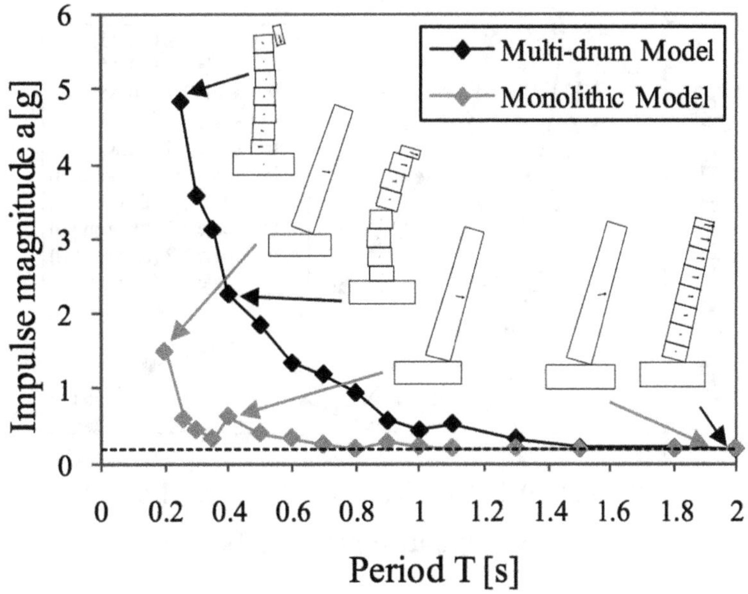

Size Effect

Another parameter affecting the dynamic response of multi-drum columns is represented by their size, i.e. columns with different size and the same proportions show a different dynamic behavior and resistance. As visible from failure domains of Figure 26, the domain curves translate upwards for increasing dimensional scales, i.e. bigger columns are more stable than smaller ones of the same geometric proportions. Furthermore, the curves are almost scalable with the square root of the length R, independently of the collapse mode, where R is the radial distance between the center of gravity and the instantaneous center of rotation O during rocking (Figure 27). This is visible in Figure 26 where the theoretical curves are computed by means of the aforementioned size effect law with respect to the reference case ("scale 1:1"), in agreement with findings of Housner (1963) for a single rocking block.

Despite the complex kinematics of multi-drum columns, which features inter-stone vibrations and sliding together with rocking motions, the scale effect found by Housner remains almost well reproduced along the whole range of periods. This means that also for multi-drum columns it is sufficient to evaluate the failure domain for an arbitrary reference length, and the other domain curves can be obtained by multiplying the impulse period corresponding to each given magnitude by the square root of the ratio of the requested length R to the reference length.

FUTURE RESEARCH DIRECTIONS

A possible further research could consider a static, quasi-static, or dynamic numerical investigation of strengthened masonry arches or portals with fiber reinforced polymer composites (FRP), applied at the intrados, and/or extrados sides of the structure. FRP composites represent an attractive solution for

Figure 26. Size effect on the failure domain for columns

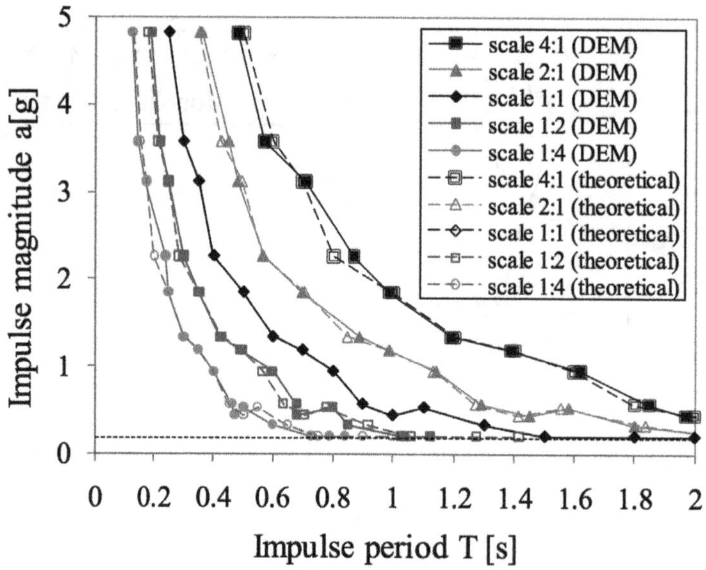

Figure 27. Multi-drum column rocking monolithically at its base

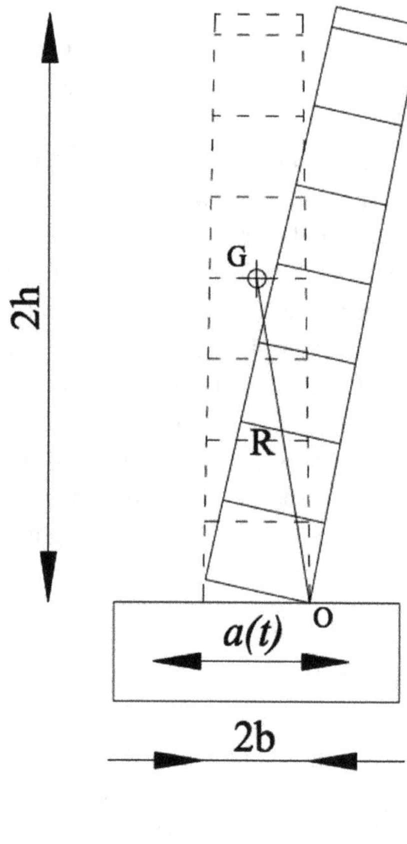

strengthening and rehabilitation of masonry structures within the respect of the original characteristics in terms of their architectural and functional aspects. A parametric analysis could be performed for different arrangements of the reinforcement (e.g. different locations, spacing and width), as well as for a different number of blocks where the strip is bonded. A systematic investigation on the dynamic behavior of masonry structures is additionally suggested for more complex base excitations, considering some real earthquake motions. An extension of the dynamic investigation with DEM to 3D is suggested by using the 3DEC code. In this case, all three components of the base excitations, (two components in the horizontal direction and one in the vertical direction) could be applied simultaneously on the structures to capture more complex aspects of the real response, as for example, the main rotations around the vertical axis due to the simultaneous rocking in two normal directions. The sensitivity of the 3D rocking response to small geometrical perturbations as well as to mechanical input properties of joints and damping could be similarly analyzed and discussed.

CONCLUSION

This chapter provides an increased understanding of the stability and the global behavior of some masonry structures under static, quasi-static, and dynamic loading and validate the efficiency of the DEM as numerical tool to model the discontinuous nature of URM structures also for complex geometries, for which analytical treatments could be unfeasible or difficult. The numerical results obtained by means of the universal discrete element code (UDEC) are in good qualitative and quantitative agreement with analytical predictions, where available. More in detail, the primary general contributions are drawn as follows:

- Overturning or sliding failure typically govern the collapse of masonry buttresses of non rectangular geometries under lateral loads, as trapezoidal or stepped buttresses. For overturning-controlled failure modes, a fracture develops at collapse, thus reducing the capacity of the buttress to resist horizontal loads. The numerical predictions of fracturing and capacity within buttresses agree quite well with the analytical ones based on a straight fracturing assumption at failure, independently of the exact knowledge of the joint parameters expressed in terms of elastic normal and shear stiffnesses. Analytical results are approached more closely as the dimensions of the blocks are reduced, for which the numerical discrete nature of masonry would tend to the analytical continuum description of the medium.
- DEM is useful in predicting the first-order seismic behavior of masonry structures under constant lateral accelerations both in terms of mechanism of collapse and collapse load. Results from the numerical approach, indeed, are in perfect agreement with analytical predictions based on an equivalent static analysis. Masonry arches generally collapse due to the formation of a four-hinge mechanism with a predictable pattern of hinge locations for different geometries, and one hinge forming always at the extrados side of the support. Masonry portals, instead, can collapse with a semi-global or local mechanism depending on their shape and geometric parameters. When semi-global mechanisms control, three hinges form into the arch, and one forms at the base of one buttress, whereby four hinges form within the arch for local mechanisms. The stability of the arches and portals, however, highly depends on the geometric shape and parameters. The analyzed pointed arches are always less stable than the reference circular ones, and their stabil-

ity decreases as the angle of embrace increases. Basket-handle arches, instead, are always more stable compared to the circular counterpart with a maximum lateral stability reached for an angle of embrace 2α=120°, corresponding to the optimal geometrical configuration herein analyzed. By repeating the same analyses for buttressed arches, portals with pointed arches are more stable than their circular counterparts independently of the angles of embrace, or equally stable, as found for 2α=180°. The contrary occurs for portals with basket-handle arches which are always less stable than their corresponding circular counterparts for the given geometry ratios herein considered. The lateral stability anyway tends to increase for increasing angles of embrace, as found both theoretically and numerically.

- The DEM correctly predict the dynamic response of masonry structures of complex geometries. Through a parametric study performed by varying the excitation frequencies and accelerations, as well as the mechanical and geometrical parameters, it can be concluded as the safe-unsafe boundary follows, in all cases, a decreasing trend. This means that base motions containing long-period impulses are much more threatening than short-period ones. The period of excitation plays a significant role also in the kinematic response of the structures, i.e. short-period impulses impose substantial inter-stone vibrations and sliding, while the exhibited response is dominated by rocking for long-periods. Anyway, a deep investigation on the sensitivity of numerical results to the assigned mechanical and geometrical parameters is required, before nothing can be predicted a priori, and the damping parameters as well as the friction coefficients and joint stiffnesses must be defined appropriately by engineers for planning purposes. Based on the numerical results, it seems that failure domains are slightly sensitive to the analyzed mechanical parameters especially for low period impulses where sliding and inter-stone vibrations control the structural response. More specifically, the effects of joint stiffnesses and damping parameters are correlated. Higher stiffnesses increase the natural frequencies. Higher natural frequencies are critically damped, causing some variations into the kinematics of the structure with prevailing rocking and sliding motions instead of vibrations. A varying stiffness- and mass-proportional damping also leads to some variations in the kinematics. Mass-proportional damping causes an increase in low frequency damping where rocking occurs, while bouncing phenomena of the top drums can be prevented for increasing values of stiffness-proportional damping for low impulse periods. At the same time, the friction coefficient affects the response especially for short periods. Lower values of the friction coefficient increase the incidence of sliding, which results in a beneficial effect for the structure, due to an increasing amount of dissipated energy.

REFERENCES

Baggio, C., & Trovalusci, P. (1995). Stone assemblies under in-plane actions. Comparison between nonlinear discrete approaches, Computer methods in structural masonry. In *Proceedings of the 3rd International Symposium* (pp. 184-193). Swansea, UK: Academic Press.

Baratta, A., Zuccaro, G., & Binetti, A. (2004). Strength capacity of a no tension portal arch-frame under combined seismic and ash loads. *Journal of Volcanology and Geothermal Research*, *133*(1-4), 369–376. doi:10.1016/S0377-0273(03)00408-6

Bićanić, N., Stirling, C., & Pearce, C. J. (2003). Discontinuous modelling of masonry bridges. *Computational Mechanics*, *31*(1-2), 60–68. doi:10.1007/s00466-002-0393-0

Cundall, P. A. (1971). *A computer model for simulating progressive large scale movements in blocky rock systems*. Paper presented at the International symposium, international society of rock mechanics, Nancy, France.

De Lorenzis, L., DeJong, M. J., & Ochsendorf, J. (2007). Failure of masonry arches under impulse base motion. *Earthquake Engineering & Structural Dynamics*, *36*(14), 2119–2136. doi:10.1002/eqe.719

De Lorenzis, L., Dimitri, R., & Ochsendorf, J. (2012a). Structural study of masonry buttresses: The trapezoidal form. *Proceedings of the Institution of Civil Engineers. Structures and Buildings*, *165*(9), 483–498. doi:10.1680/stbu.10.00042

De Lorenzis, L., Dimitri, R., & Ochsendorf, J. (2012b). Structural study of masonry buttresses: The stepped form. *Proceedings of the Institution of Civil Engineers. Structures and Buildings*, *165*(9), 499–521. doi:10.1680/stbu.10.00043

De Luca, A., Giordano, A., & Mele, E. (2004). A simplified procedure for assessing the seismic capacity of masonry arches. *Engineering Structures*, *26*(13), 1915–1929. doi:10.1016/j.engstruct.2004.07.003

DeJong, M. I. (2009). *Seismic assessment strategies for masonry structures*. (Doctoral Dissertation). Massachusetts Institute of Technology, Cambridge, UK.

DeJong, M. J., De Lorenzis, L., & Ochsendorf, J. (2008). Rocking stability of masonry arches in seismic regions. *Earthquake Spectra*, *24*(4), 847–865. doi:10.1193/1.2985763

DeJong, M. J., & Ochsendorf, J. (2006). *Analysis of vaulted masonry structures subjected to horizontal ground motion*. Paper presented at the 5th International conference on structural analysis of historical constructions, New Delhi, India.

Dimitri, R. (2009). *Stability of Masonry Structures under Static and Dynamic*. (Doctoral Dissertation). Università del Salento, Lecce, Italy.

Dimitri, R., De Lorenzis, L., & Zavarise, G. (2011). Numerical study on the dynamic behavior of masonry columns and arches on buttresses with the discrete element method. *Engineering Structures*, *33*(12), 3172–3188. doi:10.1016/j.engstruct.2011.08.018

Dimitri, R., & Tornabene, F. (2015). A parametric investigation of the seismic capacity for masonry arches and portals of different shapes. *Engineering Failure Analysis*, *52*, 1–34. doi:10.1016/j.engfailanal.2015.02.021

Fanning, P. J., Boothby, T. E., & Roberts, B. J. (2001). Longitudinal and transverse effects in masonry arch assessment. *Construction & Building Materials*, *15*(1), 51–60. doi:10.1016/S0950-0618(00)00069-6

Giordano, A., Mele, E., & De Luca, A. (2002). Modeling of historical masonry structures: Comparison of different approaches through a case study. *Engineering Structures*, *24*(8), 1057–1069. doi:10.1016/S0141-0296(02)00033-0

Gonen, H., Dogan, M., Karacasu, M., Ozbasaran, H., & Gokdemir, H. (2013). Structural failures in refrofit historical murat masonry arch bridge. *Engineering Failure Analysis*, *35*, 334–342. doi:10.1016/j.engfailanal.2013.02.024

Halabian, A. M., Mirshahzadeh, L., & Hashemol-Hosseini, H. (2014). Non-linear behavior of unreinforced masonry walls with different Iranian traditional brick-work settings. *Engineering Failure Analysis*, *44*, 46–65. doi:10.1016/j.engfailanal.2014.04.018

Heyman, J. (1982). *The masonry arch*. West Sussex, UK: Ellis Horwood-Wiley.

Heyman, J. (1992). Leaning towers. *Meccanica*, *27*(3), 153–159. doi:10.1007/BF00430041

Housner, G. W. (1963). The Behavior of Inverted Pendulum Structures During Earthquakes. *Bulletin of the Seismological Society of America*, *53*(2), 403–417.

Huerta, S. (2004). *Arcos, Bóvedas y Cúpulas – Geometria y Equilibrio en el Cálculo Tradicional de Estructuras de Fábrica*. Spain: Instituto Juan de Herrera, Escuela Técnica Superior de Arquitectura de Madrid.

Itasca Consulting Group. (2004). UDEC Version 4.0 User's Guide. Minneapolis, MN: Author.

Lemos, J. V. (2007). *Numerical issues in the representation of masonry structural dynamics with discrete elements*. Paper presented at the ECCOMAS Thematic conference on computational methods in structural dynamics and earthquake engineering, Crete, Greece.

Mamaghani, I. H. P., Aydan, Ö., & Kajikawa, Y. (1999). Analysis of masonry structures under static and dynamic loading by discrete finite element method. Journal of Structural Mechanics Earthquake Engineering, 626(16), 75-86.

Meyer, P., Ochsendorf, J., Germaine, J., & Kausel, E. (2007). The impact of high frequency/low energy seismic waves on unreinforced masonry. *Earthquake Spectra*, *23*(1), 77–94. doi:10.1193/1.2431211

Milankovitch, M. (1910). Zur Statik der massive Widerlager. *Zeitschrift für Mathematik und Physik*, *58*, 120–128.

Moseley, H. (1843). *The Mechanical Principles of Engineering and Architecture*. London, UK: Longman, Brown, Green and Longmans.

Ochsendorf, J., Huerta, S., & Hernando, J. I. (2004). Collapse of masonry buttresses. *Journal of Architectural Engineering*, *10*(3), 88–97. doi:10.1061/(ASCE)1076-0431(2004)10:3(88)

Papantonopoulos, C., Psycharis, I. N., Papastamatiou, D. Y., Lemos, J. V., & Mouzakis, H. (2002). Numerical prediction of the earthquake response of classical columns using the distinct element method. *Earthquake Engineering & Structural Dynamics*, *31*(9), 1699–1717. doi:10.1002/eqe.185

Psycharis, I. N., Papastamatiou, D. Y., & Alexandris, A. P. (2000). Parametric investigation of the stability of classical column under harmonic and earthquake excitations. *Earthquake Engineering & Structural Dynamics*, *29*(8), 1093–1109. doi:10.1002/1096-9845(200008)29:8<1093::AID-EQE953>3.0.CO;2-S

Raman, P. G. (2004). Structural masonry and architectural expression. *Construction & Building Materials*, *18*(2), 133–139. doi:10.1016/j.conbuildmat.2003.08.016

Rankine, W. J. M. (1858). *A Manual of Applied Mechanics*. London, UK: Charles Griffin.

Sarhosis, V., Oliveira, D. V., Lemos, J. V., & Lourenco, P. B. (2014). The effect of skew angle on the mechanical behavior of masonry arches. *Mechanics Research Communications*, *61*, 53–59. doi:10.1016/j.mechrescom.2014.07.008

Viola, E., Panzacchi, L., & Tornabene, F. (2005). General analysis and application to redundant arches under static loading. *Construction & Building Materials*, *21*(5), 1129–1143. doi:10.1016/j.conbuildmat.2005.10.001

Winkler, T., Meguro, K., & Yamazaki, F. (1995). Response of rigid body assemblies to dynamic excitation. *Earthquake Engineering & Structural Dynamics*, *24*(10), 1389–1408. doi:10.1002/eqe.4290241008

Chapter 12
Validation of the Discrete Element Method for the Limit Stability Analysis of Masonry Arches

Haris Alexakis
University of Patras, Greece

Nicos Makris
University of Central Florida, USA

ABSTRACT

This chapter revisits the limit equilibrium analysis of masonry arches when subjected to gravity and lateral loads. Firstly, the major contributions during the last three centuries either with geometric or energy formulations are discussed, and subsequently, the performance of the Distinct Element Method (DEM) is examined against rigorous solutions. Analytical solutions with the use of energy methods are presented for the assessment of the stability of masonry arches with circular or elliptical shapes under various load conditions, including gravity, lateral inertial loading or earth pressures. The DEM is implemented in all loading cases and reproduces the analytical results with remarkable accuracy. The DEM is used either for a direct correlation with the classic limit analysis that assumes that the joints of the masonry blocks do not transmit tension, masonry blocks are rigid and incompressible and do not slide at the joints, or by permitting sliding with the adoption of Coulomb sliding failure between the joints.

INTRODUCTION

During the last three decades we have witnessed a growing interest regarding the preservation of our cultural heritage and the retrofit of historic structures. The long history of these structures, most of them built out of masonry, is partly due to their sound structural configuration, the advanced level of construction practices and the quality of building materials. At the same time, the long loading history on these structures ranging from ground settlements to earthquake excitations has resulted in the accumulation

DOI: 10.4018/978-1-5225-0231-9.ch012

of a variety of deformations which are responsible for a prevailing stress field that its current state may be challenging to assess.

In view of the challenge to assess the actual local stresses in a historic structure, which through the centuries has suffered a variety of deformation, an "elastic stress" analysis becomes less attractive; whereas, the pre-twentieth century "limit equilibrium" analysis has become again increasingly popular (Sinopoli et al., 1997; Foce & Aita, 2003; De Luca et al., 2004; among others).

Figure 1 summarizes the modelling strategies for masonry structures. Continuous modelling is mainly represented with Finite Element Analysis (linear and nonlinear), while discontinuous modelling is represented with Limit Analysis (classic and advanced formulation) and Discrete Element Method (DEM).

Below we present an overview of the major contributions for the analysis of limit stability of masonry arches during the last three decades together with the latest developments of limit analysis, including DEM.

BACKGROUND

Stability Analysis of Masonry Arches: The Early Concepts

Robert Hooke (1675) was apparently the first to propose a rational rule for sizing masonry arches by describing the analogy in the load path between a "hanging chain", which forms a catenary in tension under its own weight, and a masonry arch which stands under compression. This analogy conceived by Hooke is expressed in the literature "As hangs the flexible line, so but inverted will stand the rigid arch" (Heyman, 1998; O'Dwyer, 1999; Block et al., 2006).

Due to its zero tensile strength, the masonry arch acts only in compression; therefore, according to Hooke's analogy, it is expected to satisfy equilibrium if its finite thickness can accommodate the shape of an inverted "hanging chain" which acts only in tension (Gregory, 1697). The shape of the catenary (the "hanging chain") is a physically realizable load path which keeps the constituents of the one-dimensional chain (rings) in equilibrium. In the two-dimensional masonry arch, the physically realizable load path is

Figure 1. Modelling strategies for masonry structures

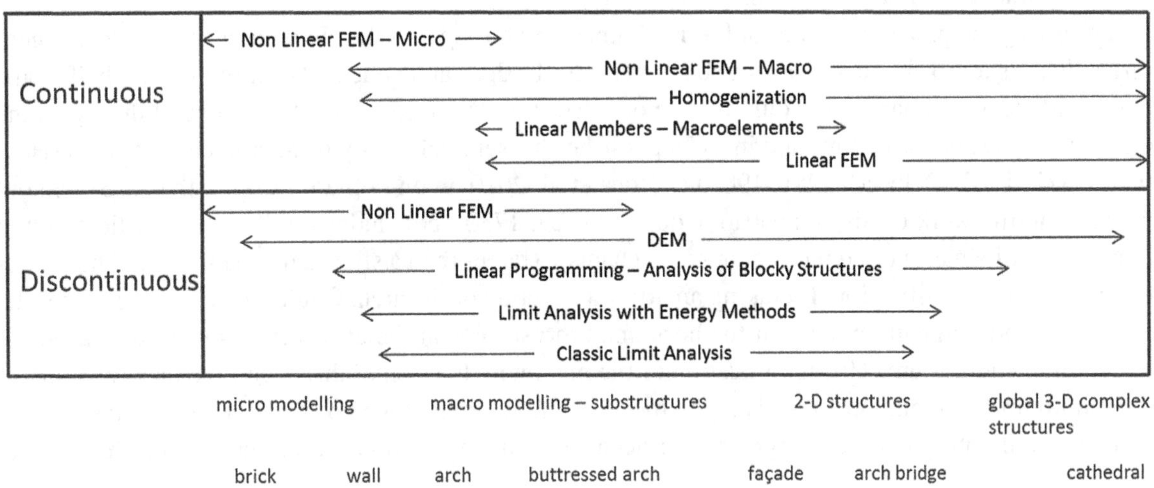

the one defined by the points of application of the resultant forces that keep the voussoirs of the arch in equilibrium. This load path has been traditionally called the thrust-line (or the line of resistance, Moseley 1843, or the druckkurve, Milankovitch 1904, 1907) and is defined as the geometrical locus of the application points of the resultant thrust-force that develops at any cross-section of the arch. Recently, it was shown that the shape of the thrust line of a masonry arch depends on the stereotomy exercised and Hooke's "hanging chain" (the catenary) cannot be a physically admissible thrust line for the masonry arch with finite thickness regardless the stereotomy exercised on the arch, although is a neighboring line (Makris & Alexakis, 2012, 2013; Alexakis, 2013; Alexakis & Makris, 2013b, 2014a).

Limit Equilibrium Analysis of Arches in France during the 18th Century: The Collapse Mechanism

As early as in the first half of the eighteenth century, Pierre Couplet (1729, 1730), based on the work of La Hire (1695, 1712), presented a remarkably complete solution to the problem of computing the minimum thickness of a semicircular masonry arch with mid-thickness radius, R, and thickness, t, capable of supporting its own weight. His work brings forward in a most lucid way the two key concepts needed in the limit equilibrium analysis of masonry arches that do not sustain tension: (a) the limiting thrust line; and (b) the imminent mechanism of collapse. Couplet (1730) recognized that when the thickness of the semicircular arch is sufficiently small, the arch develops a five-hinge symmetric mechanism. The five hinges of the symmetric mechanism develop at the extrados of the crown of the arch, at the extrados of the springer and at an unknown location along the intrados of the arch. In order to determine the intrados hinge position, Couplet needed a third equation other than the two equations offered by moment equilibrium. At the limit equilibrium state of the arch, this third equation is offered from the application of the principle of stationary potential energy that matured upon Lagrange (1788) developed his variational calculus some 50 years after Couplet's work. Pierre Couplet by-passes this impasse by merely assuming (perhaps after being influenced by Bélidor, 1729) that the location of the rupture that forms the intrados hinge is at 45°. In this way, Couplet forces the selection of the failure mechanism; therefore, his solution for the minimum thickness, t/R=0.101, is unconservative, given that a neighboring mechanism might have developed at a larger thickness (see also Alexakis & Makris, 2014b). Nevertheless, in the absence of variational calculus back in 1730, Pierre Couplet did as well as one could possibly do to estimate the minimum thickness of a semicircular arch.

Following the pioneering work of Pierre Couplet and throughout the eighteenth and nineteenth centuries there is a growing need in Europe for masonry bridges and aqueducts. In the second half of the eighteenth century Charles A. Coulomb is motivated from the need to develop a general theory on the stability of masonry structures and in 1773 publishes his seminal theory of maxima and minima (Heyman, 1972, 1982; Delbecq, 1981, 1982a,b; Roca et al. 2010 among others). Coulomb (1773) without referencing the work of his compatriot Couplet (1729, 1730), concludes that the most critical failure mechanism of a masonry arch is the hinge mechanism (rather than a sliding mechanism) and he is concerned with its identification. By taking an arbitrary section of the arch, Coulomb (1773) calculates the maximum and minimum bounds of the horizontal thrust force at the crown that is needed to keep the structure in equilibrium. The minimization of the maximum horizontal thrust and the maximization of the minimum horizontal thrust leads to the most critical internal hinge position. The main contribution of C.A. Coulomb is that he recognizes the need to examine neighboring admissible mechanisms in structural mechanics and his entire reasoning is in-line with the newly developed theory in mathematics known as calculus of variations (Lagrange, 1788).

The First Energy Approach

While Coulomb's (1773) theory of maxima and minima derives from static equilibrium, Mascheroni (1785) states that the arch has reached its limit equilibrium state when during an infinitesimal variation of the configuration of the imminent mechanism, the work of the weights of the articulated portions of the arch equals zero. Mascheroni (1785) considers two distinct mechanisms: (a) that corresponding to the five hinge mechanism as identified by Couplet (1730); and (b) a sliding mechanism where the center portion of the arch slides downwards without friction at the imminent rupture joints and infinite friction at the springings of the arch. Clearly, Mascheroni's (1785) statement that at the limit equilibrium state the work of the corresponding weights shall be zero during an infinitesimal displacement is a clear statement of the principle of stationary potential energy—an early application of the principle of virtual work (see also Sinopoli et al. 1997).

19th and 20th Century

In the early nineteenth century, the Spanish engineer Joaquin Monasterio works on the problem of determining the minimum thickness of masonry arches. His in-depth study that was unknown until recently (Huerta & Foce 2003; Foce 2005; Albuerne & Huerta, 2010) builds upon the earlier static theory of maxima and minima developed by Coulomb (1773) and he concludes that the minimum thickness of the semicircular arch needs to satisfy: $1/9<t/R_{in}<1/8$, where $R_{in}=R-(t/2)$. Monasterio's inequality for the minimum thickness of semicircular arches, $0.1053<t/R<0.1176$, improves Couplet's unconservative result, $t/R=0.101$.

In the early nineteenth century Lamé and Clapeyron (1823) make an important contribution by noticing that for symmetric arches of any shape the calculation of the location of the intrados hinge can be greatly simplified if instead of radial-sections, vertical cross-sections are contemplated (see also Timoshenko, 1953; Alexakis & Makris, 2014b). This work had a profound influence on the books of "graphic statics" that were most popular during the end of 19th and at the beginning of 20th century (Culmann 1866; Lauenstein & Bastine 1913; Wolfe, 1921).

In his work of unparallel quality and rigor, the Serbian engineer and astronomer Milutin Milankovitch (1904, 1907) presents a general formulation for the thrust line of curved masonry structures. For the specific case of symmetric circular arches Milankovitch's formulation employs radial cuts, and after adopting a polar coordinate system, it yields a closed-form expression for the radius of curvature of the thrust line (see also Foce, 2007). The limiting thrust line must pass by the extrados points of the springing and the crown and be tangent to the intrados. This geometric condition led for the first time to the exact and complete solution to the problem of the minimum thickness of the semicircular masonry arch under gravity loads that is $t/R=0.1075$.

Application of the Theorems of Plastic Analysis

Almost sixty years after Milankovitch, Jacques Heyman (1966), based on the work of Kooharian (1952), proposes that the theorems of plastic analysis of steel structures could also be applied to masonry structures under certain simplifications; that (a) the joints of the masonry do not transmit tension, (b) the compression strength is infinite and (c) the masonry blocks do not slide at the joints. These assumptions are compatible with the traditional analysis techniques developed by the Heyman's predecessors mentioned earlier. The compact theoretical background offered by Heyman (1966), together with the unique advantages of the limit analysis that when the structure is considered in relation to its ultimate state prior knowledge of initial stresses and material parameters are not necessarily required, gave a new impetus to the research for the stability of masonry structures and new advanced techniques were emerged. Current studies have revealed some limitation regarding the limit state theorems presented by Heyman and through rigorous proofs offer more complete statements for the stability of structures considered as no-tension continua (Como 2012) or discrete systems (Bagi 2014).

LATER DEVELOPMENTS AND THE DISCRETE ELEMENT METHOD

Rigid Block Limit Analysis

Livesley (1978) was the first to propose an advanced formulation of limit stability analysis that was reported later as Discrete Limit Analysis (Gilbert et al. 2007), Rigid Block Limit Analysis (Block, 2009) or Analysis of Blocky Structures (Roca et al. 2010). Using linear programming based on kinematic approach, in relation with an optimization technique that Livesley (1978) developed, the method calculates the maximum load of any structure composed by rigid blocks at its limit stability state. This work was later followed by several authors (Gilbert & Melbourne, 1994; Baggio & Trovalusci, 1998; Ferris & Tin-Loi, 2001; Casapulla & D'Ayala, 2001; Orduña & Lourenço, 2003, 2005a,b, among others) that showed that it is possible to relax the simplifying assumptions of classic limit analysis by taking into account the finite compressive strength of masonry or the possibility of sliding failure between the joints.

Force/Thrust Network Method and Catenary Models

O'Dwyer (1999), based on the original idea of the funicular polygon of forces that was used in the books of "graphic statics" (Culmann, 1866; Lauenstein & Bastine, 1913; Wolfe, 1921, among others) proposes an advanced method for the analysis of 3d-vaults and domes. O'Dwyer (1999) developed an optimization process that is in search for a network of the resultant compressive forces – in analogy with the thrust line of the 2d-arch – that must be contained within the thickness of the vault, satisfying the lower bound theorem. The external loading is gradually increased until the moment of limit stability, when the structure becomes statically determinate. In this way, it is possible to estimate the collapse load factor for various load patterns. The method was recently extended by Block and Ochsendorf (2007) and Block (2009). An alternative computer technique for the analysis of the stability of complex masonry structures was developed by Andreu et al. (2007, 2010) that is in search for a network of hanging chains (catenary curves) that must satisfy the lower bound theorem.

Discrete Element Method

While Finite Element Method (FEM) is the usual choice for the analysis of the majority of structures, modelling the discontinuous nature of masonry and estimating the stability of articulated structures is still a challenging task. To exemplify, high tensile stresses from static or dynamic FEM analysis indicate a material failure and, by extension, a structure failure, while in reality the structure may be perfectly stable without relying on its tensile strength (DeJong, 2009). This is the main reason why the two last decades an increasing number of engineers are using the Discrete Element Method (DEM) for the analysis of masonry structures (Cundall, 1971; Pagnoni, 1994; Pagnoni & Vanzi, 1995; Lemos, 1995, 2007; Papantonopoulos et al., 2002, among others) where discontinuous modelling is inherent. Recently, results from theoretical studies for the minimum thickness of masonry arches were compared with numerical results from DEM, using the DDA—Discontinuous Deformation Analysis formulation (Cocchetti et al.. 2012; Rizzi et al.. 2014) and results were found to be in a good agreement.

One of the most representative commercial software of DEM is UDEC—Universal Distinct Element Code (Itasca Consulting Group, Inc., 2004). The software UDEC simulates the response of discontinuous media (such as a jointed rock mass) subjected to either static or dynamic loading. In the DEM, a structural system is represented as an assembly of discrete blocks. The method works in terms of forces and displacements and each block is treated as a separate entity, allowed to undergo unlimited translation and rotation, and the calculations must satisfy equilibrium at every time step. The method adopts an elastic behavior which leads to the need to define the normal-to-the-contact surface, K_n (normal stiffness), and the tangential-to-the contact surface, K_s (shear stiffness), equivalent linear elastic constants. The discontinuities are treated as boundary conditions between blocks; whereas large displacements along discontinuities and rotations of blocks are allowed.

The software UDEC may be used for direct correlation with the classic limit state analysis that uses the three aforementioned assumptions: (a) joints do not transmit tension, (b) compression strength is infinite and (c) blocks do not slide at the joints, or with a more advanced version of limit state analysis (e.g. by permitting sliding between the joints). Every stone of the masonry arches that are analyzed in the next sections has been considered as a rigid block with sufficient compressive strength (second assumption of limit state analysis). This rigid body analysis adopts a linear Mohr-Coulomb sliding model between the block joints that demands the determination of the cohesion, the friction angle, the tensile strength and the dilatancy angle. In this work the effect of the mortar shall be neglected and the tensile strength, the cohesion and the dilatancy angle will be zero (first assumption of limit state analysis). Classic limit state analysis assumes that only a collapse hinging mechanism of the arch is possible and sliding failure cannot occur (third assumption of limit state analysis). This assumption is satisfied by just giving a very large value for the friction angle φ that would prevent sliding. However, the effect of a possible sliding failure in the overall stability of the arch may be also studied by giving a reasonable value of the friction angle.

Joints are represented numerically in UDEC as contact surfaces formed between two block edges. For rigid blocks, a contact is created at each corner interacting with a corner or edge of the opposing block. Numerically, blocks may become locked or hung-up for very sharp corners. The singularities and abrupt changes are avoided in UDEC by rounding the corners so that blocks can smoothly slide past one another. Rounding the corners may also serve to simulate the loss of material at the corners due to crashing, affecting the result of the minimum thickness at limit equilibrium state. However, in order to be consistent with classic limit state analysis that does not take into account crashing at the corners, the rounding length, which is equivalent to the radius of the rounded corner, was chosen to be sufficiently small. For the following analysis, the value of the rounding length is 0.001m. In most cases, this value is below the 1% of the average block edge length, as it is recommended in the manual of the software

(Itasca 2004). For the cases where the size of the stone blocks has been decreased appreciably, the value of the rounding length is adjusted below the 1% boundary.

UDEC is an explicit time marching model and the choice of damping parameters is crucial, especially for dynamic problems. Rayleigh damping has been widely used for the dynamic response of masonry structures and the optimum choice of the mass-proportional and stiffness-proportional constants has been investigated by several authors (De Lorenzis et al. 2007; Peña et al. 2007; DeJong, 2009; Dimitri et al. 2011, among others). On the other hand, searching for the limit state condition of a masonry arch is purely a static problem. In static analysis the equations of motions are damped to reach a force equilibrium state as quickly as possible under the applied initial and boundary conditions. For this reason, a special form of damping, referred to as "local damping", is provided in UDEC in which the damping force on a node is proportional to the magnitude of the unbalanced force, in a way that energy is always dissipated (see also Cundall, 1987; Itasca, 2004). By this selection, the numerical analysis converges in few time steps and with reliable results.

In the analysis that follows, all arches are modelled with rigid elements (stone blocks) and the structure stands free on an inelastic base, which is modelled as a massive rigid block with fixed boundaries. The size and the self-weight of this block do not affect the results of the analysis. UDEC considers the density of the material and the geometry of every block to automatically calculate and apply the resultant gravity forces at the center of gravity of every rigid block, from the first step of the analysis. In the case where external forces should be applied (e.g. lateral inertial loading or soil pressure), the structure is first considered as free-standing (self-weight only), and when stability is achieved in few time steps (unbalance force approaches zero), then all the external loads are released together in the nest time step. In rigid block analysis, the forces can only be applied at the center of gravity of the blocks.

The results obtained by DEM are in remarkable agreement with the analytical results presented in the next sections, which indicates that DEM may present dependable results for the stability analysis of complex masonry structures with large number of elements that it would be most challenging to assess with analytical solutions.

MASONRY ARCHES UNDER GRAVITY LOADING

During a limit state analysis the various collapse mechanisms are first identified and with simple calculations which derive primarily from energy methods one can conclude whether the various rigid portions of the structure are in equilibrium Lagomarsino (2006). The masonry arch stands out as the structural system that has challenged several of the most eminent engineers and mathematicians from the second half of the seventeenth century to the beginning of the twentieth century during their efforts to develop and refine the analysis at the limit equilibrium state (Hooke, 1675; La Hire, 1712; Couplet, 1729, 1730; Coulomb, 1773; Mascheroni, 1785; Lamé & Clapeyron, 1823; Persy, 1825; Navier, 1826; Culmann, 1866; Durand-Claye, 1867; Milankovitch, 1904, 1907 among others). Interestingly, the majority of studies on the limit equilibrium analysis of the masonry arch adopt a geometric formulation which asks for the identification of a limiting thrust line and only a handful of studies use energy methods which are liberated from the concept of the thrust line (Mascheroni, 1785; Clemente, 1995; Ochsendorf, 2002; Sinopoli, 2003; Makris & Alexakis, 2012, 2013; Alexakis, 2013; Alexakis & Makris 2013b, 2014a,b). The problem of determining the minimum thickness of semicircular masonry arches has challenged the engineering community since the early eighteenth century (Couplet, 1729, 1730), was tackled with remarkable ingenuity by Monasterio in the early nineteen century (Albuerne & Huerta, 2010), was addressed rigorously in the early twentieth century in the nearly unknown work by Milankovitch (1904, 1907), was revisited in the late 1960s by Heyman (1969) in association with the stability of historic bridges, and remains worth discussing until today (Ochsendorf, 2002; Heyman, 2009; Makris & Alexakis, 2012, 2013; Alexakis, 2013; Alexakis & Makris, 2013b, 2014a,b).

This section shows that the answer for the minimum thickness of symmetric elliptical masonry arches is not unique and that it depends on the stereotomy exercised. This result is confirmed by developing a variational formulation after selecting the appropriate directions of the rupture that initiates at the intrados hinge.

Thrust Lines of an Arched Monolith with Zero Tensile Strength

Consider the monolithic semicircular arch shown in Figure 2 which has zero tensile strength and therefore acts only in compression. Any point O that is located on the mid-thickness line of the arch (dashed line) has coordinates $x=R\sin\varphi$, $(0\leq x\leq R)$, and $y=R\cos\varphi$ $(0\leq y\leq R)$ since the angular coordinate φ starts measuring from the vertical axis of symmetry, y. We are interested in computing the resultant thrust force acting within the arch at location O with coordinates $x=R\sin\varphi$ and $y=R\cos\varphi$ $(0\leq\varphi\leq\pi/2)$. The standard way to proceed is to make a cut passing through point O and work the force and moment equilibrium. One way to cut is along the radial direction after adopting a polar coordinate system as shown in Figure 2 (left). With this choice we isolate the portion of the arch ABCOD. Another way to cut is along the vertical direction after adopting a cartesian coordinate system as shown in Figure 2 (right). With this choice we isolate the portion of the arch ABC´OD´, which has different size and weight than the portion ABCOD; therefore, the thrust T_p along the radial cut COD is different both in magnitude and direction than the thrust T_c along the vertical cut C´OD´. As this exercise is repeated for different points O along the mid-thickness line of the arch, one obtains two neighboring; yet different thrust lines——one thrust line that results from radial cuts after adopting a polar coordinate system and a different thrust line that results from vertical cuts after adopting a cartesian coordinate system.

Figure 3 (left) plots these two different minimum thrust lines within a monolithic semicircular arch with t/R=0.12. They have been constructed with a custom-made computer code which repeats the force equilibrium shown in Figure 2 as point O runs from the crown to the springing. These two physically admissible thrust lines are not distinguishable in the scale of the arch; yet, if one zooms in the neighborhood isolated with the dashed parallelogram, the two thrust lines are clearly different as shown in Figure 3 (right). Both minimum thrust lines are equally correct and the fact that they lie within the physical boundaries of the arch ensures that the arch is stable. In the event that one adopts a different stereotomy (other than the radial or vertical) then a third physically admissible thrust line will emerge.

Interestingly, the inverted "hanging chain" (catenary) that passes from the extreme points A and F of the extrados of the arch shown in Figure 3 offers a third line (not shown) that is different from the two minimum thrust lines——the one computed by taking radial cuts after adopting a polar coordinate system (see Figure 2-left) and the other computed with vertical cuts after adopting a cartesian coordinate system (see Figure 2-right). Recently, Makris and Alexakis (2012, 2013) showed by reducing to the absurd that the catenary curve is not a physically acceptable thrust line of the semicircular arch.

Figure 2. Monolithic semicircular arch with zero tensile strength and t/R=0.12. At any point O along the center line of the arch (dashed line) with polar coordinates R, φ or cartesian coordinates x=Rsinφ, y=Rcosφ, the thrust force T_p along a radial cut after adopting a polar coordinate system (left) is different than the thrust force T_c along a vertical cut after adopting a cartesian coordinate system (right). Accordingly, one obtains two distinguishable physically admissible minimum thrust lines (see Figure 3) depending on the coordinate system that is selected.

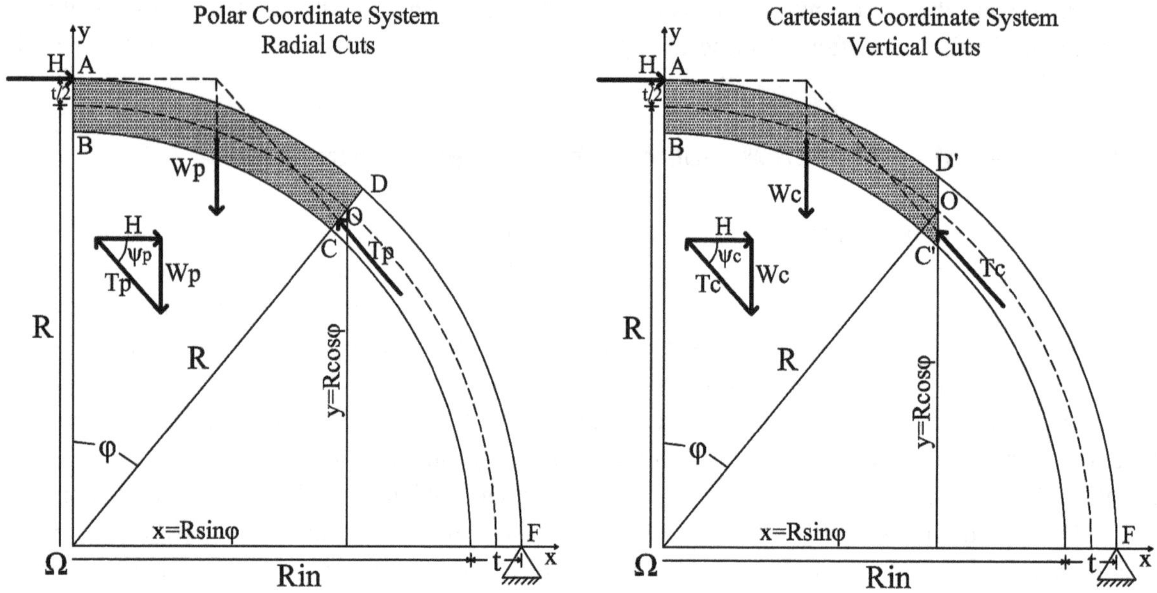

Figure 3. Monolithic semicircular arch with the two different physically admissible minimum thrust lines (left) and t/R=0.12. The two physically admissible thrust lines (ρ(φ) obtained with successive radial cuts and η(x) obtained with successive vertical cuts) are distinguishable in the enlarged parallelogram (right).

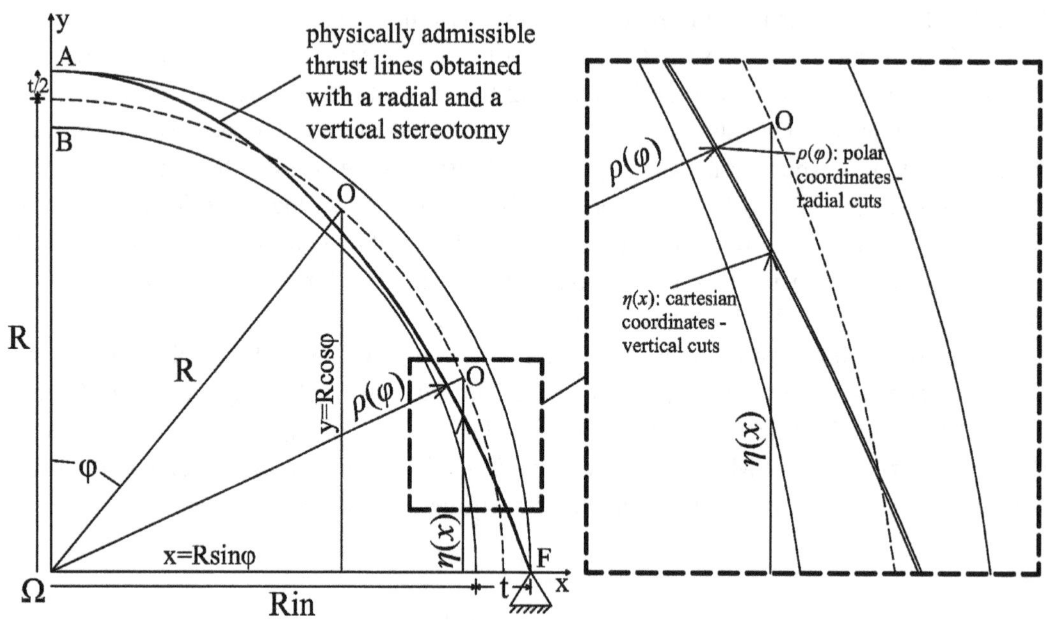

Limit Equilibrium Analysis of Masonry Arches with Energy Methods

Milutin Milankovitch (1904, 1907) computed for the first time the correct value of the minimum thickness of the semicircular arched monolith (t/R=0.1075) by assuming radial ruptures (and the associated polar coordinate system). In his analysis he developed a geometric formulation in which the major achievement is the construction of a closed-form expression of the minimum thrust line equation. Recently, Makris and Alexakis (2012, 2013) showed that Milankovitch's solution, t/R=0.1075, is not unique and that it depends on the stereotomy exercised. The adoption of vertical cuts as suggested by Lamé and Clapeyron (1823) in association with a cartesian coordinate system yields a neighboring minimum thrust line and a different, slightly higher value for the minimum thickness, t/R=0.1095, than the value computed by Milankovitch. While Makris and Alexakis (2012, 2013) succeeded in deriving the closed-form expression of the minimum thrust line when vertical cuts are assumed, the algebra and calculus involved is complicated to the extent that the geometric formulation becomes less attractive as the shape of the arch departs from the geometry of the circle. For instance, when radial ruptures are assumed, a closed-form expression for the minimum thrust line of elliptical arches is not known to the authors (Alexakis & Makris, 2013b).

In parallel with the geometric formulation, Makris and Alexakis (2012, 2013) and Alexakis and Makris (2013b, 2014a,b) formulated the limit equilibrium analysis of masonry arches by employing the principle of stationary potential energy. The main attraction of a variational formulation is that the limit equilibrium analysis is liberated from the concept of the limiting thrust line and the result depends solely on the direction of the rupture as the collapse mechanism initiates.

The fundamental concept of a variational formulation is that among all geometric admissible states (all admissible symmetric five-hinge mechanisms), a natural state – that is a state which also satisfies equilibrium – can be detected by observing its behavior under infinitesimal geometrically admissible variations of the state. With reference to Figure 4, the symmetric arched monolith is subjected only to its

Figure 4. Formation of a hinge mechanism in an elliptical arch where the rupture that initiates the intrados hinge, K, happens along a direction that is at an angle ω with the horizontal

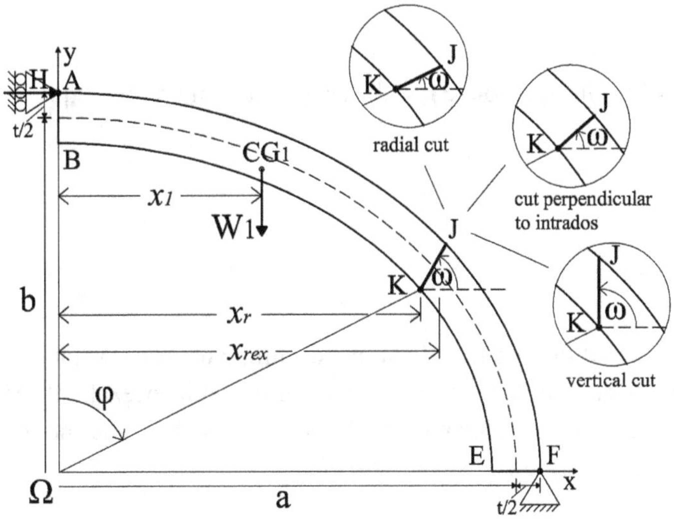

own weight (no external forces); while, when it assumes its minimum thickness each of the two ruptured portions of the arch hinging at point K, neither deform elastically nor dissipate energy at the hinges.

The multiplicity in the answers for the minimum thickness of the arched monolith which does not sustain tension originates from the infinite many ways that the arch may rupture when is about to form an internal hinge at point K.

For instance, Figure 4 shows the formation of a mechanism in an elliptical masonry arch with semi-span, a, and height, b, when developing a rupture that is at an angle ω with the horizontal. The rupture might be radial, vertical or perpendicular to intrados (see Figure 4) and so forth.

The structural system shown in Figure 4 is a typical case where the only forces (weights) acting in the system are conservative and where the work of all forces is accounted by the potential energy $V(\varphi_r, b/a)$ given by (Alexakis & Makris, 2013b)

$$V(\varphi_r, \frac{b}{a}) = a^3 f(\varphi_r, \frac{b}{a})[\frac{1}{3}\frac{b}{a}(2 + \frac{b}{a}) + \frac{1}{12} f(\varphi_r, \frac{b}{a})^2] \tag{1}$$

in which $f(\varphi_r, b/a) = t/a$ is a transcendental equation which derives solely by moment equilibrium of the hinged portions of the arch and relates the unknown rupture location φ_r with the unknown minimum thickness t/a of any elliptical arch described with a given height-to-semispan ratio b/a.

According to the principle of stationary potential energy, the geometric admissible hinged mechanism shown in Figure 4 is in an equilibrium state if and only if the total potential energy of the arch is stationary, i.e., (Alexakis & Makris, 2013b)

$$\delta V(\varphi_r, \frac{b}{a}) = \frac{dV(\varphi_r, \frac{b}{a})}{d\varphi_r} \delta\varphi_r = 0 \tag{2}$$

Substitution of equation (1) into equation (2) gives

$$\frac{df(\varphi_r, \frac{b}{a})}{d\varphi_r}[1 + \frac{1}{4} f(\varphi_r, \frac{b}{a})^2] = 0 \tag{3}$$

The quantity in brackets in equation (3) is always positive; therefore, equation (3) is satisfied when $df(\varphi_r, b/a)/d\varphi_r = 0$

$$\frac{df(\varphi_r, \frac{b}{a})}{d\varphi_r} = \frac{d(t/a)}{d\varphi_r} = 0 \tag{4}$$

where the minimum required t/a reaches a local maximum. Numerical solution of the system of equations from moment equilibrium of the hinged portions of the arch (Alexakis & Makris, 2013b) together with equation (4) gives the unknown location of the rupture angle φ_r and the minimum width t/a of an elliptical arch for a given height-to-semispan ratio b/a.

Table 1 (left columns) shows the minimum thickness t/a and rupture angle φ_r=BΩK for various ratios of b/a as computed with the variational formulation after adopting a radial rupture; whereas, Table 1 (right columns) shows the slightly different values of the minimum thickness t/a and rupture angle φ_r=BΩK as computed with the variational formulation (and also confirmed with a geometric formulation) after adopting a vertical rupture (Alexakis & Makris, 2013b). Table 1 (center columns) also offers the value of the minimum thickness of an elliptical arch where the rupture is assumed perpendicular to the intrados.

Table 1 shows that the minimum thickness of an elliptical arch that is needed to support its own weight is slightly larger when one assumes vertical ruptures rather than radial ruptures or ruptures perpendicular to the intrados, implying that the solution that corresponds to radial ruptures or ruptures perpendicular to the intrados is unconservative. In view of this result the question that arises is what is the direction of rupture that results to the larger value of the minimum thickness (given that the minimum thickness values that result from any other rupture direction are unconservative values).

This problem is addressed by expressing the direction of rupturing with the angle, ω, as shown in Figure 4 and applying the principle of stationary potential energy for the entire range of ω. When $\omega=(\pi/2)-\varphi$ the rupture is radial; whereas, when $\omega=\pi/2$, the rupture is vertical. For the case where the rupture is perpendicular to the intrados $\omega=\arctan(-(dy_{in}(x_r)/dx_r)^{-1})$, where $y_{in}(x)$ is the ordinate of the intrados of the arch expressed in cartesian coordinates and x_r is the abscissa of the rupture point K (Alexakis and Makris, 2014b). In the analysis we also consider values of $\omega>\pi/2$ in an effort to identify the most critical direction of rupture. This problem involves three unknowns:

1. The horizontal thrust force, H, at the crown,
2. The minimum thickness, t/a, of the arch; and
3. The unknown location, φ_r, of the formation of the hinge K.

At the same time, there are three equations available—that is moment equilibrium about hinge F and hinge K, together with equation (4) which originates from the application of the principle of stationary potential energy. The analytical expression of these equations has been presented by Alexakis and Makris (2014b) and the result from the solution of the system of the three equations is illustrated in Figure 5.

Table 1. Minimum thickness t/a and radial locations φ_r=BΩK of the intrados hinge K of elliptical arches with semispan=a and height=b

b/a	**Radial Cuts**		**Cuts Perpendicular to Intrados**		**Vertical Cuts**	
	Minimum Thickness: *t/a*	**Rupture Angle:** φ_r**(°)=BΩK**	**Minimum Thickness:** *t/a*	**Rupture Angle:** φ_r**(°)=BΩK**	**Minimum Thickness:** *t/a*	**Rupture Angle:** φ_r**(°)=BΩK**
0.25	0.08273	84.123	0.08413	84.088	0.08422	84.101
0.5	0.10127	74.442	0.10278	74.395	0.10340	74.533
0.75	0.10709	64.015	0.10788	63.974	0.10925	64.299
1	**0.10748**	**54.484**	**0.10748**	**54.484**	**0.10946**	**54.923**
1.25	0.10513	46.567	0.10454	46.625	0.10689	47.090
1.5	0.10135	40.323	0.10038	40.430	0.10291	40.875
1.75	0.09684	35.516	0.09565	35.658	0.09821	36.067
2	0.09200	31.857	0.09074	32.018	0.09324	32.395

Figure 5. Minimum thickness values, t/a (top), and radial locations, φ_r (bottom), of the hinge K shown in Figure 4 as a function of the assumed angle of rupturing direction ω of elliptical arches with semispan, a, and height b

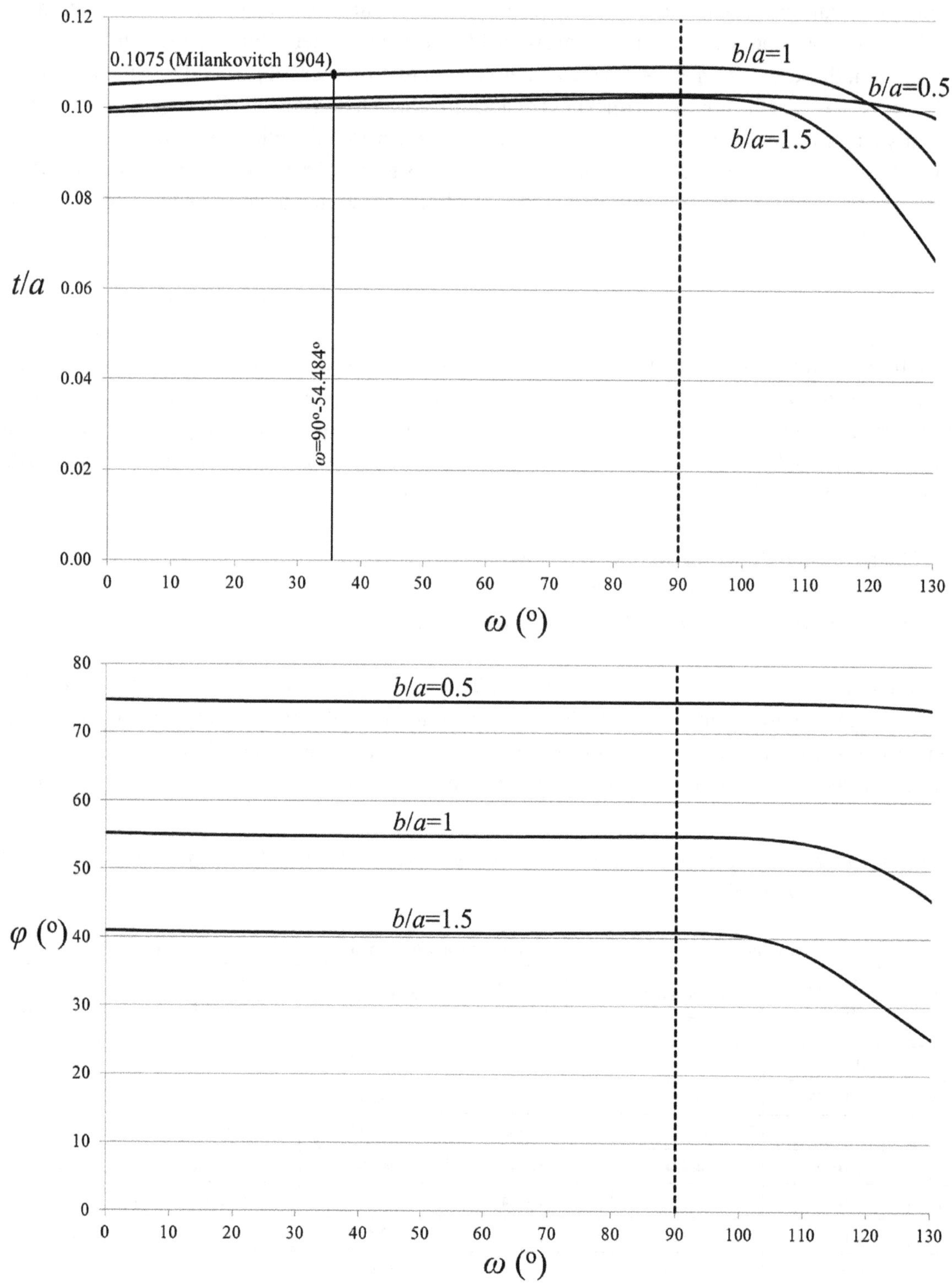

Figure 5 plots the minimum thickness values, t/R (top) and the radial locations, φ_r, of the hinge K as the angle of the rupturing direction, ω, increases. Clearly, for all three configurations (b/a=0.5, 1.0, and 1.5) presented in Figure 5, the minimum thickness value is largest when $\omega=\pi/2$, that is for vertical ruptures. Interestingly, in the range $0\leq\omega\leq\pi/2$, the minimum thickness value, t/R, exhibits a mild increase; whereas, the radial location, φ_r, of the corresponding hinge K remains almost the same (see also Alexakis and Makris, 2014b).

VALIDATION OF DEM RESULTS

The results presented in Table 1 for the case of radial rupture or rupture perpendicular to intrados are compared with the results obtained from Discrete Element Method (DEM) analysis, where the commercially available software UDEC (Universal Distinct Element Code – Itasca, 2004) has been used.

As it was mentioned before, in order for the result of DEM to be comparable with those of limit equilibrium analysis, the stone blocks must be considered as rigid; whereas, the angle of friction between the joints, φ, must be large enough (e.g. $\varphi\geq 80^{o}$) to prevent sliding. In addition, the effect of the mortar has been neglected and the tensile strength, the cohesion and the dilatancy angle shall be zero. De Lorenzis et al. (2007) and DeJong (2009) used this approach for the dynamic study of rocking arches.

The method adopts an elastic behavior which leads to the need to define the normal-to-the-contact surface, K_n (normal stiffness), and the tangential-to-the contact surface, K_s (shear stiffness), equivalent linear elastic constants. The variation of these values had infinitesimal effect for the assessment of the limit state that is the minimum thickness and the imminent mechanism of collapse. By using values between 10^7 Pa/m and 10^9 Pa/m the algorithm converges fast and with reliable results. In this analysis $K_n=K_s=10^8$ Pa/m.

The material density does not appear in equations of limit analysis, which is governed by geometry. The DEM confirms this behavior since very small or very large values of density have been tested without affecting the limit equilibrium state. In this analysis the density was chosen to be 2000 kg/m^3.

Similar is the conclusion for the size of the structure. Very small or very large arches where tested without affecting the analogy of the geometrical parameters at the limit equilibrium state (i.e. the minimum thickness t/a for every b/a). In this analysis the semispan, a, was chosen to be unity (1m).

Crucial is the selection of the stereotomy of the arch, i.e. the number of stone blocks and the direction of the joints. For the case of radial stereotomy arches with 18 blocks (joint every 10°) and 54 blocks (joint every 3.33^{o}) have been considered. Figure 6 shows the discretization of an elliptical masonry arch with radial joints and height-to-semispan ratio b/a=0.75. For the case where the joints are perpendicular to the intrados the same pattern of discretization has been considered. While these joints start from the same intrados points – as in the case of radial stereotomy – they evolve perpendicular to the intrados (not radial) and they meet the extrados at different points. Figure 7 shows the discretization of an elliptical masonry arch with joints perpendicular to intrados and height-to-semispan ratio b/a=1.25. In Figures 6 and 7 every stone block is a quadrilateral where its four edges are points of the intrados and extrados curves of the elliptical arch. The same discretization was used for all the values of b/a given in the left column of Table 1.

Our DEM analysis is in search for the minimum thickness, t/a, for arches with 18 or 54 blocks, b/a=0.25 to 2 and stereotomy with radial joints or joints perpendicular to the intrados; that is 32 cases in total (3 cases are repeated given that for the case of the semicircular arch, b/a=1, the radial joints are

Figure 6. Left: Elliptical masonry arches with radial joints, height-to-semispan ratio b/a=0.75 and 18 (top) and 54 (bottom) discrete elements that were used for the DEM analysis. Right: The collapse mechanism is always the symmetric 5-hinge-mechanism (Couplet 1730). The models experience only gravity loads.

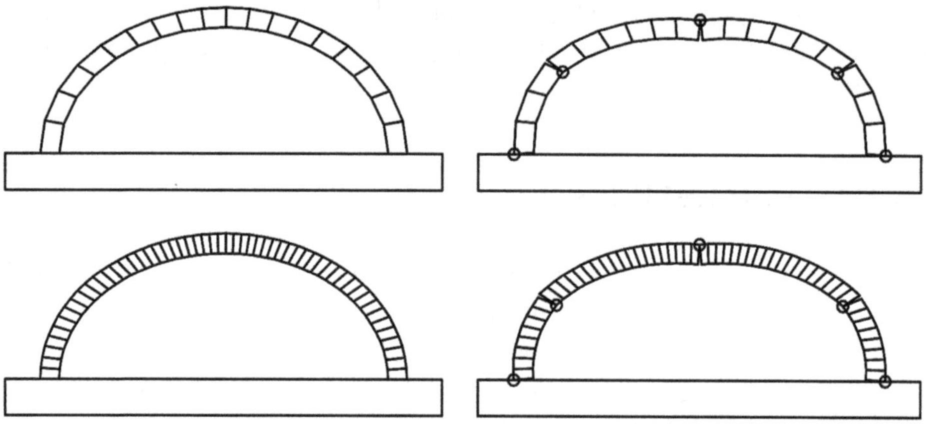

Figure 7. Left: Elliptical masonry arches with joints perpendicular to the intrados, height-to-semispan ratio b/a=1.25 and 18 (top) and 54 (bottom) discrete elements that were used for the DEM analysis. Right: The collapse mechanism is always the symmetric 5-hinge-mechanism (Couplet 1730). The models experience only gravity loads.

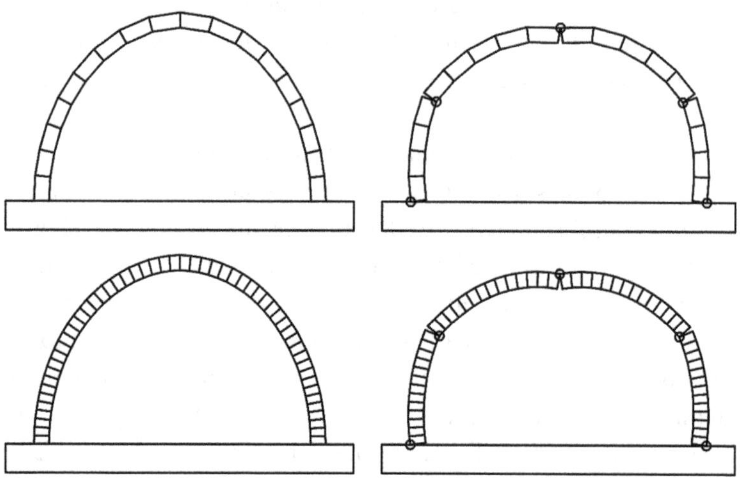

also perpendicular to the intrados). An automatic process was developed where the cartesian coordinates of the edges of every block were expressed as a function of *t/a* and *b/a*. For every case the thickness of the arch, *t/a*, was gradually decreased until the moment of limit state. Stability in UDEC is understood as the moment where the algorithm converges to a static solution, which means that the variation of the kinetic energy of the system approaches zero (Itasca, 2004).

The only forces applying on the structure are the gravity loads acting in the center of gravity of every discrete element. As the analysis is static, the value of damping was chosen to be the default proposed by the manual of UDEC for static problems ("damp local", Itasca 2004). The analysis was performed until we reach the minimum thickness, *t/a*, with five decimal digits accuracy.

Table 2 and Table 3 present the values of the minimum thickness of elliptical masonry arches with radial joints or joints perpendicular to the intrados respectively derived from DEM analysis. These values are compared with the values obtained from analytical solution (variational formulation). The indication "err (%)" expresses the percentage of the deviation between the analytical solution and the numerical solution from DEM. If the number is positive, the result from DEM is conservative (it works towards the safety).

Figures 8 and 9 are the graphical representation of Tables 2 and 3 respectively. The black continuous line represents the results of the analytical solution, while the grey lines represent the numerical results from DEM (dashed line: 18-segments-arch, continuous line: 54-segments-arch).

Table 2. Values of the minimum thickness of elliptical masonry arches with radial joints derived from DEM analysis are compared with the values obtained from analytical solution (variational formulation)

Radial Joints	t/a				
b/a	Analytical Solution	UDEC			
		18 Segments	Err (%)	54 Segments	Err (%)
0.25	0.08273	0.07214	-12.80	0.08374	1.22
0.50	0.10127	0.09890	-2.34	0.10290	1.61
0.75	0.10709	0.10713	0.04	0.10909	1.87
1.00	0.10748	0.10781	0.31	0.10964	2.01
1.25	0.10513	0.10638	1.19	0.10713	1.90
1.50	0.10135	0.10356	2.18	0.10332	1.94
1.75	0.09684	0.09810	1.30	0.09871	1.93
2.00	0.09200	0.09990	8.58	0.09388	2.04

Table 3. Values of the minimum thickness of elliptical masonry arches with joints perpendicular to intrados derived from DEM analysis are compared with the values obtained from analytical solution (variational formulation)

Joints Perpendicular to Intrados	t/a				
b/a	Analytical Solution	UDEC			
		18 Segments	Err (%)	54 Segments	Err (%)
0.25	0.08413	0.07640	-9.19	0.08634	2.63
0.50	0.10278	0.10098	-1.75	0.10472	1.89
0.75	0.10788	0.10804	0.15	0.10993	1.90
1.00	0.10748	0.10781	0.31	0.10964	2.01
1.25	0.10454	0.10560	1.01	0.10659	1.96
1.50	0.10038	0.10184	1.45	0.10240	2.01
1.75	0.09565	0.09549	-0.17	0.09756	2.00
2.00	0.09074	0.09039	-0.39	0.09252	1.96

Figure 8. Minimum thickness t/a of elliptical masonry arches with radial joints and height-to-semispan ratio b/a. The black continuous line represents the results of the analytical solution, while the grey lines represent the numerical results from DEM (dashed line: 18-segments-arch, continuous line: 54-segments-arch).

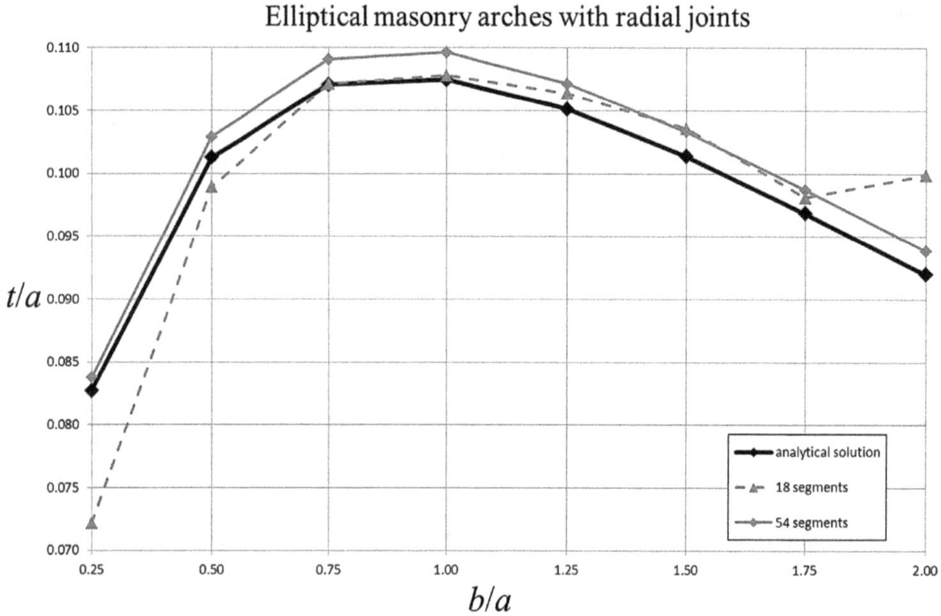

Figure 9. Minimum thickness t/a of elliptical masonry arches with joints perpendicular to the intrados and height-to-semispan ratio b/a. The black continuous line represents the results of the analytical solution, while the grey lines represent the numerical results from DEM (dashed line: 18-segments-arch, continuous line: 54-segments-arch).

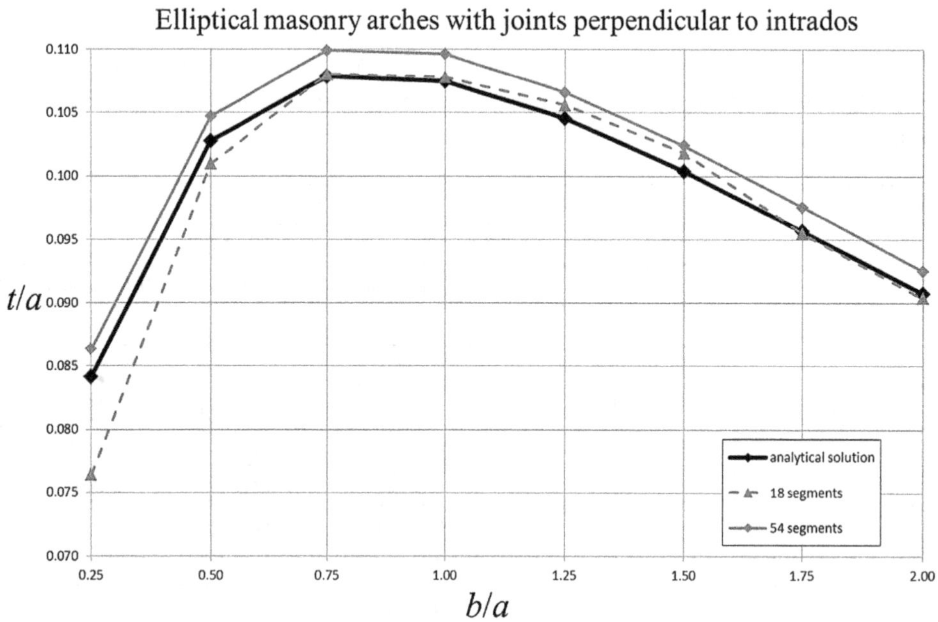

If one uses a poor discretization he may approach the theoretical correct minimum thickness t/a for moderate values of the height-to-semispan ratio b/a. However, the results are not realistic for extreme low or extreme high values of b/a. The reason is that for extreme values of b/a the geometry of the model deviates from the exact shape of the ellipse, while the theoretical position of the intrados hinge is drawn away from the closest joint. For the 18-segments-arch, the results are approaching significantly the analytical solution; whereas, for extreme values of b/a the deviation is in the order of 10%. When the discretization is fine, the 54-segments-arch simulates better the geometry of the ellipse. In addition, there are joints closer to the theoretical positions of the hinges of the collapse mechanism and the results deviate only an order of 2% (towards the safety) for any value of the height-to-semispan ratio b/a.

It is important to note that DEM analysis showed that when elliptical arches of any shape are in a limit equilibrium state under their own weight, they will always form the symmetric five-hinge-mechanism presented in the analytical solution.

MASONRY ARCHES UNDER GRAVITY AND LATERAL LOADING

While the analysis of masonry arches under static gravity loads has a long history and a rich literature, the analysis of masonry arches when subjected to lateral dynamic loads essentially starts with the seminal paper by Oppenheim (1992) (see also Clemente, 1998; Ochsendorf, 2002; De Luca et al., 2004; Alexakis & Makris, 2014a).

The planar dynamic response of masonry arches that do not sustain tension can be analyzed in two steps. Step (a) is a static-equilibrium stability analysis and deals with the calculation of the minimum horizontal acceleration, $\ddot{u}_g^{up}$, that is needed to rupture an arch at four locations (hinge formation) and induce rocking. Upon uplift has been induced, step (b) deals with the dynamic response analysis of the rocking arch by solving the nonlinear equations of motion; while, addressing satisfactorily the impact which happens whenever the motion reverses.

This section presents a rigorous solution for step (a) that computes the value of the rupture/uplift acceleration of circular masonry arches with embrace angle β, mid-thickness radius R and thickness t after identifying the exact locations of the four imminent hinges by adopting a variational formulation via the application of the principle of stationary potential energy and further confirms that the results obtained from DEM analysis are in very good agreement with the rigorous solution.

Figure 10 shows the formation of a four-hinge mechanism in a circular arch when subjected to a horizontal ground acceleration $\ddot{u}_g=\varepsilon g$. In this study we assume that the direction of the ruptures is radial.

Clemente (1998) is the first to recognize that when a masonry arch is subjected to a lateral load (say to the right), the extreme right hinge will happen at the right springing A; while, the extreme left hinge D may happen within the arch at a location above the left springing. Accordingly, depending on the slenderness of the arch (t/R, β) and the level of the horizontal ground acceleration the circular arch may develop a one-springing or a two springing, 4-hinge mechanism.

Substitution of the three equations that result after taking moment equilibrium of the combined segment 1-2 (segment ABC) about hinge A, moment equilibrium of segment 2 (segment BC) about hinge B and of segment 3 (segment CD) about hinge D eliminates the unknown force components T_{Cx} and T_{Cy} of hinge C (see Figure 10) and yields a transcendental equation which involves five quantities: the level of the horizontal ground acceleration $\ddot{u}_g=\varepsilon g$ expressed by the coefficient ε, the minimum thickness t/R

Figure 10. Formation of a four-hinge mechanism in a circular arch (R, t, β) when subjected to a horizontal ground acceleration $\ddot{u}_g = \varepsilon g$. The direction of the ruptures at the hinge locations is assumed radial.

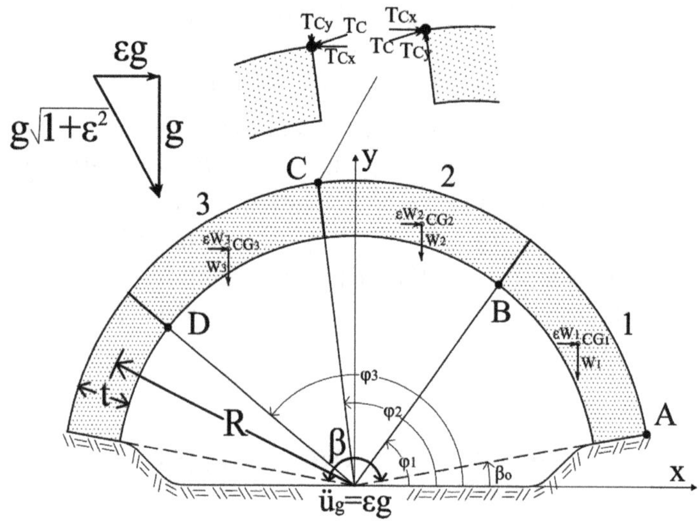

and the unknown locations of hinge B, φ_1, of hinge C, φ_2, and of hinge D, φ_3. The solution of this equation can be expressed in the form (Alexakis & Makris, 2014a)

$$t/R = f(\varphi_1, \varphi_2, \varphi_3, \varepsilon) \tag{5}$$

Alexakis and Makris (2014a) have shown that the work of all forces may be accounted by the potential energy $V(\varphi_1, \varphi_2, \varphi_3, \varepsilon)$ given by equation

$$V(\varphi_1, \varphi_2, \varphi_3, \varepsilon) = 2\sin(\frac{\beta}{2})R^3 f(\varphi_1, \varphi_2, \varphi_3, \varepsilon)[1 + \frac{1}{12} f(\varphi_1, \varphi_2, \varphi_3, \varepsilon)^2] \tag{6}$$

According to the principle of stationary potential energy, the geometrically admissible hinged mechanism shown in Figure 10 is in an equilibrium state if and only if the total potential energy of the arch is stationary ($\delta V=0$). Since in this case the potential energy $V(\varphi_1, \varphi_2, \varphi_3, \varepsilon)$ is a function of three variables φ_1, φ_2 and φ_3, the potential energy is stationary when (Shames & Dym, 1985, among others)

$$\frac{\partial V(\varphi_1, \varphi_2, \varphi_3, \varepsilon)}{\partial \varphi_1} = 0 \tag{7}$$

$$\frac{\partial V(\varphi_1, \varphi_2, \varphi_3, \varepsilon)}{\partial \varphi_2} = 0 \tag{8}$$

and

$$\frac{\partial V(\varphi_1, \varphi_2, \varphi_3, \varepsilon)}{\partial \varphi_3} = 0 \tag{9}$$

Substitution of equation (6) to any of the equations (7) to (9) ($j\in\{1,2,3\}$) gives

$$\frac{\partial V(\varphi_j,\varepsilon)}{\partial \varphi_j} = 2\sin(\frac{\beta}{2})R^3\,\frac{\partial f(\varphi_j,\varepsilon)}{\partial \varphi_j}[1+\frac{1}{4}f(\varphi_j,\varepsilon)^2]=0\,,\; j\in\{1,2,3\} \tag{10}$$

The quantity in brackets in equation (10) is always positive; therefore, equation (10) is satisfied only when $\frac{\partial f(\varphi_j,\varepsilon)}{\partial \varphi_j}=0$. Accordingly, the arch is in limiting equilibrium state ($\delta V(\varphi_1,\varphi_2,\varphi_3,\varepsilon)=0$) if and only if all partial derivatives of the function $f(\varphi_1,\varphi_2,\varphi_3,\varepsilon)$ vanish

$$\frac{\partial f(\varphi_1,\varphi_2,\varphi_3,\varepsilon)}{\partial \varphi_1}=\frac{\partial(t/R)}{\partial \varphi_1}=0 \tag{11}$$

$$\frac{\partial f(\varphi_1,\varphi_2,\varphi_3,\varepsilon)}{\partial \varphi_2}=\frac{\partial(t/R)}{\partial \varphi_2}=0 \tag{12}$$

$$\frac{\partial f(\varphi_1,\varphi_2,\varphi_3,\varepsilon)}{\partial \varphi_3}=\frac{\partial(t/R)}{\partial \varphi_3}=0 \tag{13}$$

Numerical solution of the system of equations (5) and (11)-(13) gives the unknown locations φ_1,φ_2 and φ_3 of the three hinges (the location of the fourth hinge is known at the downstream springing A) together with the minimum thickness t/R of a circular arch with a given angle of embrace β.

Table 4 presents values of t/R for arches with β=180°, 155°, 125° and 90° for different levels of horizontal acceleration ε. The bolded numbers correspond to the development of a two-springing, four-hinge mechanism (2SM). Note that the minimum thickness needed reduces appreciably as the embrace angle reduces.

The results from the rigorous solution for arches with β=180°, 155°, 125° and 90° are compared with the results obtained from DEM analysis, where the commercially available software UDEC (Itasca, 2004) has been used. The spring constants assume the value $K_n=K_s=10^9$ Pa/m, while the angle of internal friction has been assumed large enough to not allow any sliding. Upon sliding has been prevented, our analysis concludes that the results are insensitive to the exact values of K_n and K_s. Every voussoir has a size of 5° as shown in Figure 11. Accordingly, the semicircular arch has 36 voussoirs; whereas the arch with embracing angle β=90° has 18 voussoirs. The gravity load and horizontal inertia load corresponding to each voussoir have been applied as static loads at the center of gravity of each voussoir. Figure 11 shows snapshots from the initiation of collapse when the arches assume their critical thickness, t, when the seismic coefficient, ε=0.3 ($\ddot{u}_g$=0.3g). The two top arches with β=180° and β=155° develop a one-springing, 4-hinge mechanism; whereas, the two bottom arches with β=125° and β=90° develop a two-springing, 4-hinge mechanism.

The right-hand columns of Table 4 present the deviation between the analytical solution and the numerical solution from DEM. The values of the dimensionless minimum thickness, t/R, obtained with

Table 4. Values of the minimum thickness t/R and locations of the imminent hinges B, C and D of a circular arch with embrace angle β to withstand a constant horizontal acceleration $\ddot{u}_g = \varepsilon g$. The bolded numbers correspond to the development of a two-springing, four-hinge mechanism (2SM). The values of the minimum thickness obtained from analytical solution (variational formulation) are compared with DEM results. The direction of the ruptures at the hinge locations is assumed radial.

	Analytical Solution				UDEC		
	β=180°						
ε	t/R_{anal}	φ_1 (°)	φ_2 (°)	φ_3 (°)	t/R_{UDEC}	t/R_{UDEC} - t/R_{anal}	Err(%)
0.0	0.10748	35.516	90.000	144.484	0.10947	*0.00199*	*1.85*
0.1	0.13590	38.894	95.710	152.528	0.13786	*0.00196*	*1.44*
0.2	0.16897	42.412	101.310	160.208	0.17057	*0.00160*	*0.95*
0.3	0.20636	46.025	106.699	167.373	0.20809	*0.00173*	*0.84*
0.4	0.24752	49.693	111.801	173.910	0.24928	*0.00176*	*0.71*
0.5	0.29175	53.389	116.565	179.741	0.29334	*0.00159*	*0.55*
0.6	**0.33788**	**57.063**	**120.852**	**180 (2SM)**	0.33946	*0.00158*	*0.47*
0.7	**0.38495**	**60.705**	**124.617**	**180 (2SM)**	0.38687	*0.00192*	*0.50*
	β=155°						
ε	t/R_{anal}	φ_1 (°)	φ_2 (°)	φ_3 (°)	t/R_{UDEC}	t/R_{UDEC} - t/R_{anal}	Err(%)
0.0	0.06083	41.257	90.000	138.743	0.06255	*0.00172*	2.82
0.1	0.07974	44.244	95.711	147.177	0.08215	*0.00241*	3.02
0.2	0.10217	47.337	101.310	155.283	0.10421	*0.00204*	2.00
0.3	0.12789	50.480	106.699	162.918	0.12960	*0.00171*	*1.34*
0.4	**0.15638**	**53.611**	**111.761**	**167.5 (2SM)**	0.15821	*0.00183*	*1.17*
0.5	**0.18607**	**56.567**	**116.083**	**167.5 (2SM)**	0.18780	*0.00173*	*0.93*
0.6	**0.21627**	**59.343**	**119.737**	**167.5 (2SM)**	0.21761	*0.00134*	*0.62*
0.7	**0.24671**	**61.983**	**122.900**	**167.5 (2SM)**	0.24852	*0.00181*	*0.74*
	β=125°						
ε	t/R_{anal}	φ_1 (°)	φ_2 (°)	φ_3 (°)	t/R_{UDEC}	t/R_{UDEC} - t/R_{anal}	Err(%)
0.0	0.02673	49.133	90.000	130.867	0.02871	*0.00198*	*7.40*
0.1	0.03737	51.728	95.711	139.693	0.03976	*0.00239*	*6.39*
0.2	0.05051	54.411	101.310	148.209	0.05251	*0.00200*	*3.95*
0.3	**0.06592**	**57.093**	**106.578**	**152.5 (2SM)**	0.06789	*0.00197*	*3.00*
0.4	**0.08225**	**59.522**	**110.792**	**152.5 (2SM)**	0.08382	*0.00157*	*1.91*
0.5	**0.09899**	**61.707**	**114.136**	**152.5 (2SM)**	0.10075	*0.00176*	*1.78*
0.6	**0.11588**	**63.702**	**116.882**	**152.5 (2SM)**	0.11765	*0.00177*	*1.53*
0.7	**0.13277**	**65.548**	**119.215**	**152.5 (2SM)**	0.13419	*0.00142*	*1.07*

continued on following page

Table 4. Continued

	Analytical Solution				UDEC		
	β=90°						
ε	t/R_{anal}	φ_1 (°)	φ_2 (°)	φ_3 (°)	t/R_{UDEC}	t/R_{UDEC} - t/R_{anal}	Err(%)
0.0	0.00750	59.485	90.000	120.515	0.00964	*0.00214*	*28.60*
0.1	0.01193	61.692	95.711	129.729	0.01437	*0.00244*	*20.46*
0.2	**0.01781**	**63.928**	**101.145**	**135 (2SM)**	0.01988	*0.00207*	*11.64*
0.3	**0.02435**	**65.827**	**105.050**	**135 (2SM)**	0.02639	*0.00204*	*8.36*
0.4	**0.03116**	**67.406**	**107.758**	**135 (2SM)**	0.03287	*0.00171*	*5.50*
0.5	**0.03807**	**68.763**	**109.771**	**135 (2SM)**	0.04004	*0.00197*	*5.19*
0.6	**0.04500**	**69.965**	**111.372**	**135 (2SM)**	0.04698	*0.00198*	*4.40*
0.7	**0.05192**	**71.056**	**112.717**	**135 (2SM)**	0.05368	*0.00176*	*3.40*

Figure 11. Initiation of collapse of circular arches with 5° voussoirs and various embracing angles β=180°, 155°, 125°, 90° subjected to a constant horizontal acceleration $\ddot{u}_g$=0.3g (ε=0.3) in association with their gravity load as predicted with the software UDEC. The gravity load (1g) and the horizontal inertia load (0.3g) are applied as static loads at the center of gravity of each voussoir.

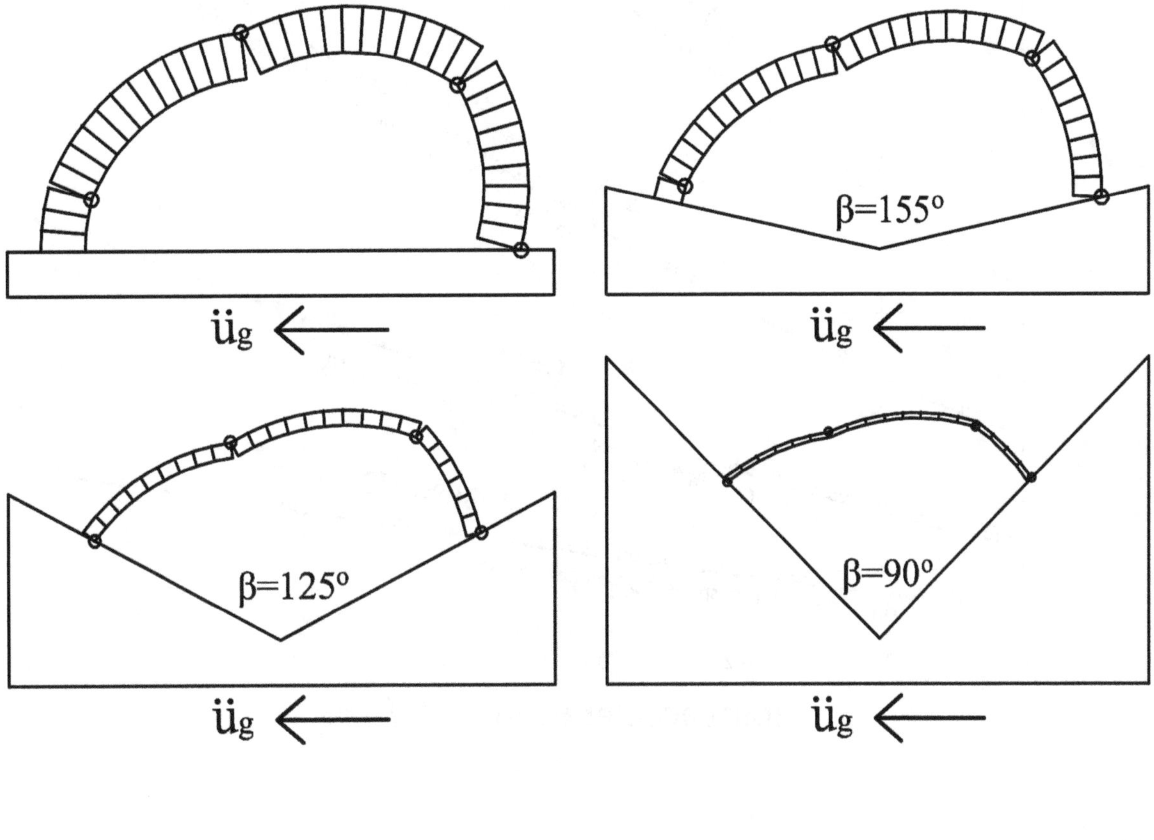

DEM are slightly conservative, and the minor difference between the analytical and the numerical solution is approximately 0.002 for all cases. Figure 12 plots with solid lines the minimum thickness values t/R from the rigorous solution and with dotted lines from the DEM analysis of a circular arch with a given angle of embrace β when subjected to various levels of horizontal acceleration $\ddot{u}_g = \varepsilon g$. The numerical solution of UDEC is in excellent agreement with the theoretical solution from the variational formulation.

To this end, Alexakis and Makris (2014a) have shown that the minimum thickness of the arch and the locations of the imminent hinges depend on the directions of the ruptures that the arch with finite thickness develops when it converts into a mechanism.

Figure 12. Minimum thickness values, t/R, of a circular arch with embrace angle, β, needed to withstand a constant horizontal acceleration $\ddot{u}_g = \varepsilon g$. The rigorous solution derived with a variational formulation (solid lines) is compared with the results obtained from DEM analysis (dotted lines). The direction of the ruptures at the hinge locations is assumed radial.

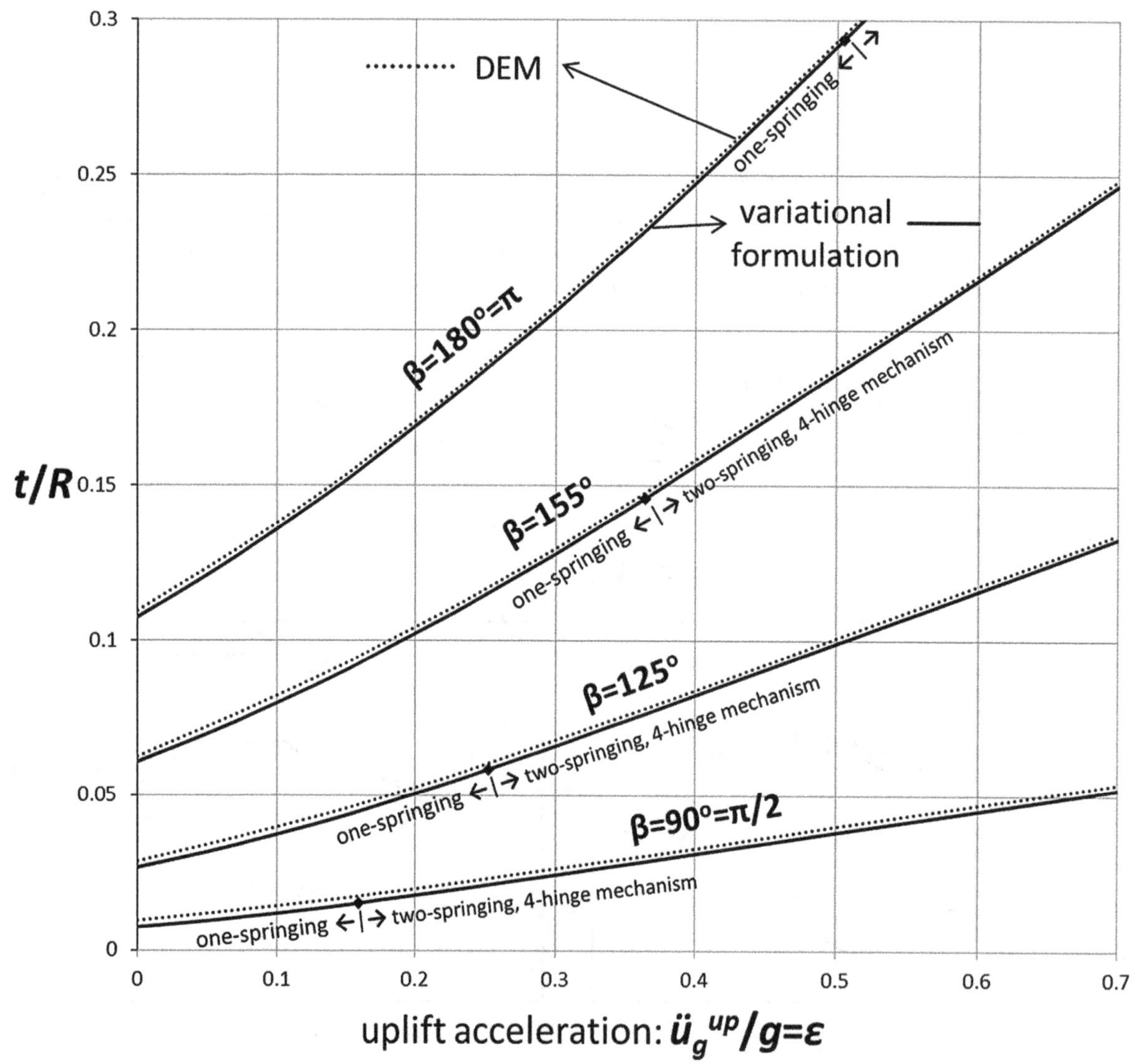

STABILITY ANALYSIS OF THE TUNNEL-ENTRANCE TO THE STADIUM OF ANCIENT NEMEA USING DEM

In the archaeological site of Ancient Nemea, Greece, there is an ancient stadium in which the athletes of the past entered through a 36 m-long tunnel——a cut-and-cover vaulted structure, constructed of limestone. Both ends of the tunnel were discovered in 1978. They were completely filled with and covered over by silt soil. The interior however, had not completely filled. Figure 13 shows a section of the tunnel along its longitudinal axis, together with the overburden earth, as was measured in 1978 (Miller, 2001).

The tunnel was initially constructed in a straight trench that was excavated in a firm soil/soft rock. It consists of two vertical stone walls which support the cylindrical vault. Figure 14 shows a drawing with the average dimensions of the stone blocks at a given cross section of the tunnel. At present, some limestone blocks of the tunnel show appreciable damage mainly due to the combined action of humidity fluctuation within the tunnel and concentration of compressive stresses at selective stone blocks (Alexakis & Makris 2013a). Figure 15 is a photograph of the interior of the tunnel that shows clearly the loss of material at the bottom of the walls and in the vicinity of the springer stone. No excavation over the tunnel has been attempted until now, because there was the fear that this could affect the stability of the structure with destructive consequences.

This section presents a parametric DEM analysis in an effort to estimate the influence of this absence of stone material (exfoliation) to the overall stability of the tunnel. In the case of the vaulted tunnel of Nemea, the first two classic assumptions of limit analysis that the joints of the masonry blocks do not transmit tension and that the masonry blocks are rigid and incompressible apply since the working stress along the tunnel (~1.0 MPa) are more than an order of magnitude less than the average strength of the limestone (~10 MPa) (Makris & Psychogios, 2004; Alexakis & Makris, 2013a). Moreover, given that there is no mortar at the stone block interfaces in association with the relative high value of the modulus of elasticity (~E=6.5 GPa) to the working stresses (~1.0 MPa), the stone blocks may considered as rigid. The third assumption that sliding cannot occur between the joints is relaxed by adopting a linear Mohr-Coulomb sliding model between the block joints. The coefficient of friction along the limestone surface of Nemea ancient stone blocks μ_{LS}>0.7 (Konstantinidis & Makris, 2005) so the friction angle assumes the value ψ=35°. The mass density of limestone is 2200 Kg/m³.

The backfill-overburden soil also assumes a Mohr–Coulomb failure law with friction angle (angle of shearing resistance) φ=30°. Given the lack of information regarding the exact behavior of the back-

Figure 13. Schematic section of the tunnel along its longitudinal axis showing the levels of earth fill inside and above the tunnel

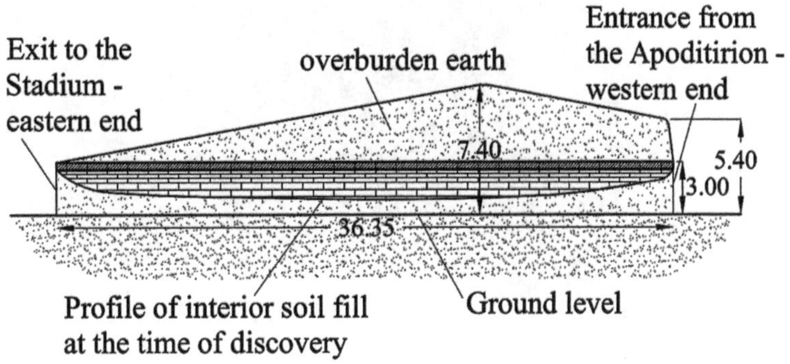

Figure 14. Schematic of a tunnel section showing the dimensions of the stone blocks used in the analysis

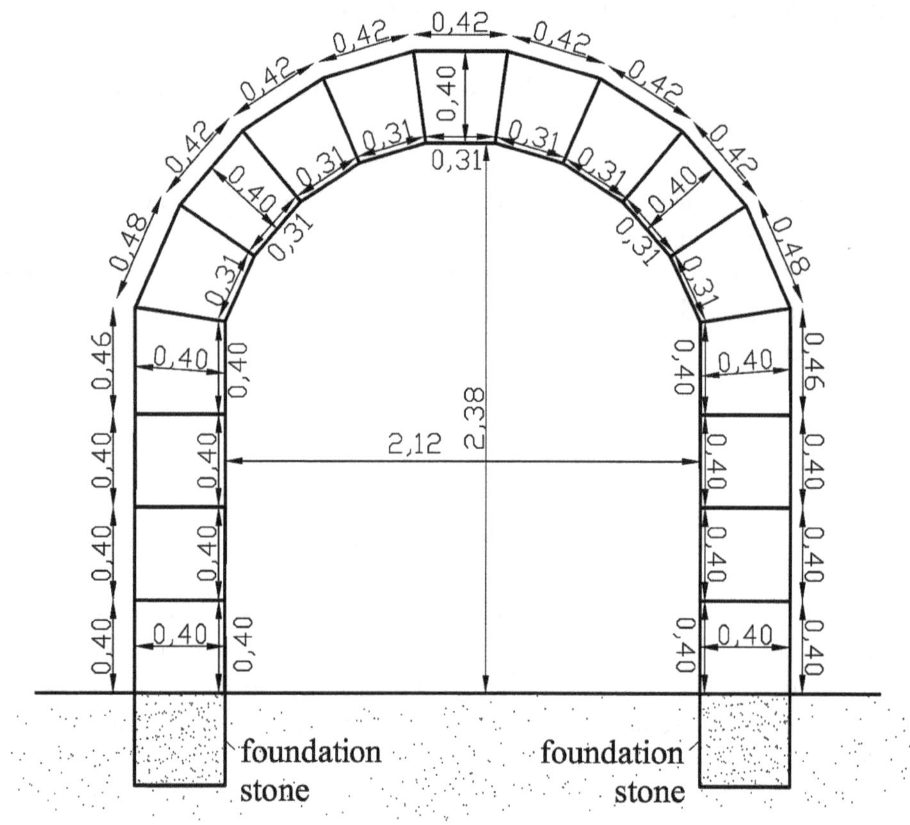

Figure 15. Current view of the interior of the tunnel

fill soil at failure, we assume that the cohesion intercept c=0 (that works towards the safety). The mass density of soil is 1800 Kg/m^3.

A thrust line limit analysis presented by Alexakis and Makris (2013a) showed that the tunnel is a well engineered "self-locking" structure which does not allow any movements of its constituent stones either to the outside or to the inside. Additionally, dense laser pointer measurements of the width of the tunnel across its length excluded any possibility that the vault might have opened during the years (Alexakis & Makris 2013a). These findings imply that the soil next and above the tunnel remained through the years "at rest" and that the possibility that it deteriorated to an active limit state is remote, while the development of passive soil stresses is not physically realizable. Consequently, the coefficient of lateral earth pressure assumes the value K_o=0.5 (soil-structure condition "at-rest", Jaky, 1944).

According to Figure 13, the level of the surrounding soil varies from H=3 m (at the eastern end of the tunnel) to H=7.4 m (at a middle section of the tunnel) measured from the ground level. Figure 16 shows the idealization of gravity loads and soil pressure of the 1.0m thick section of the tunnel with e.g. 5m backfill-overburden soil. The vertical stresses of the overburden soil, $\sigma_v=\gamma z$ (z is the vertical distance measured from the overburden soil surface), are idealized as vertical concentrated static loads at the center of gravity of the voussoir stones, while the lateral earth pressure, $\sigma_h=K_o\sigma_v$, as horizontal concentrated static loads at the center of gravity of the voussoir and wall stones. Note that when rigid-block analysis with the UDEC is performed, the loads can only be applied as concentrated forces at the center of gravity of every block. In this analysis, the case where the tunnel is free-standing (H=0 m, no backfill - there is no vertical or lateral earth pressure) has been also considered in an effort to estimate the stability of the tunnel during the time of construction.

Figure 16. Idealization of gravity loads and soil pressure of the 1.0m thick section of the tunnel with e.g. 5m backfill-overburden soil, together with the additional load needed to initiate failure. Ko is the coefficient of lateral earth pressure with soil-structure condition "at-rest".

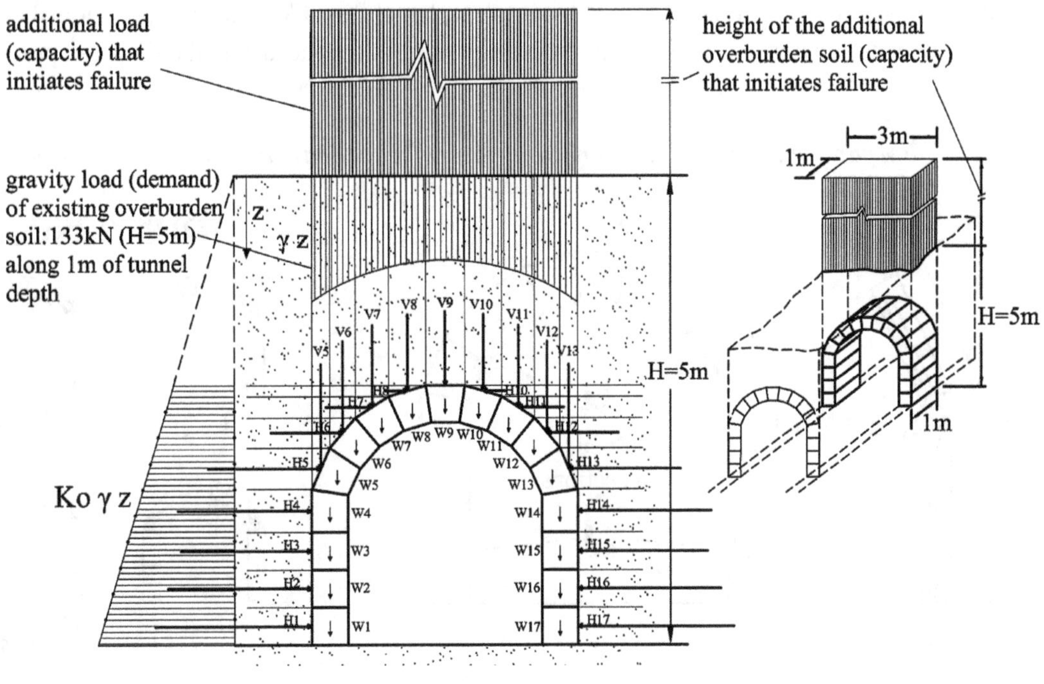

The normal-to-the contact surface K_n assumes values in the range $5\bullet10^9$ Pa/m $\leq K_n \leq 150\bullet10^9$ Pa/m and the tangential-to-the contact surface K_s assumes values in the range 10^9 Pa/m $\leq K_s \leq 50\bullet10^9$ Pa/m (Bandis et al. 1983). However, it was concluded that the results remained indifferent to wide variations of K_n and K_s.

Our DEM analysis is performed for a slice of the tunnel that is 1.0m thick and with variable backfill height H (see Figure 16). If the structure with the surrounding soil is computed as stable, the analysis proceeds by further loading the structure with a vertical distributed load that ranges over the 3 m width of the tunnel until the limit equilibrium state. In order to estimate the influence of the absence of stone material (exfoliation) to the stability of the tunnel, we gradually reduce the width of the stone in the neighbourhood of the bottom stones and in the neighbourhood of the springer voussoir. Figure 17 shows schematically a section of the tunnel and the location of the reduced stone blocks. The reduction of the stone material ranges from 2 cm up to 16 cm at a stone block with width = 40 cm. Our parametric analysis examines four cases: (a) the stones in the neighbourhood of the springer at one side have been reduced - case "u", (b) the stones in the neighbourhood of the springer and the base at one side have been reduced - case "ud", (c) the stones in the neighbourhood of the springer at both sides have been reduced - case "uu" and (d) the stones in all four critical locations have been reduced - case "uudd". The failure mechanism for all cases is the five-hinge symmetric mechanism. The hinges are located at the extrados at the base of the walls, at the extrados at the vicinity of the keystone, and at the intrados at the vicinity of the springing stones (n. 5 and n. 13, see Figure 16).

Figure 18 presents the vertical load capacity of the tunnel as a function of H. All lines except the demand line (load demand from the existing backfill) show the entire vertical load that a one-meter-thick slice of the tunnel can sustain. The bold dashed line represents the limit stability of the idealized non-damaged tunnel (no reduction of the stone material), while the continuous lines represent the limit stability of the tunnel with the reduced stone blocks. Figure 18 shows that even when the width of the stones of the tunnel has been reduced at the critical locations, the tunnel with its overburden soil exhibits appreciable stability. Nevertheless, when the reduction of the stone material is more than 8 cm the tunnel can no longer stand free and needs backfill soil to reach stability. Consequently, caution should be exercised before excavate above the tunnel given that the present overburden soil enhances appreciably

Figure 17. The DEM parametric analysis examines four cases of reduction of the stone material (exfoliation)

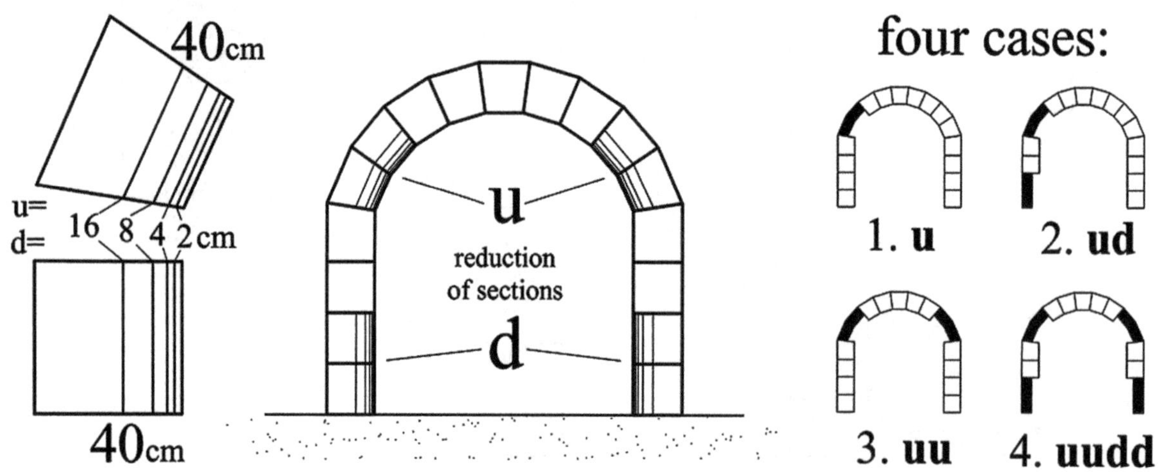

Figure 18. Variation of the vertical load carrying capacity of the tunnel due to the severity of reduction of the stone material at four critical locations (shown in Figure 17). Ko is the coefficient of lateral earth pressure with soil-structure condition "at-rest".

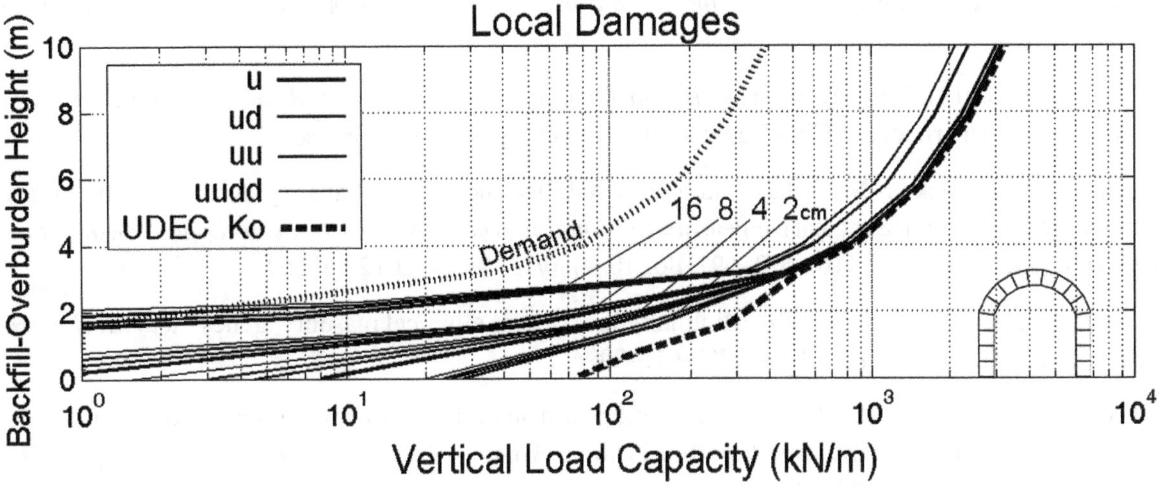

the stability of the exfoliated tunnel. Furthermore, selective stones of the structure need to be retrofitted in order to avoid further local failures.

Nemea tunnel is a 3d structure and our 2d inelastic limit state analysis does not account for longitudinal compression that may influence the in-plane behavior because of the Poisson effect, which is beyond the scope of this study. A more detailed 3d DEM analysis with or without deformable blocks could be interesting for future studies.

FUTURE RESEARCH DIRECTIONS

The most dependable performance of the discrete element method established in this work indicates that the DEM is a valuable tool for the limit analysis of monuments and is being currently used for the limit stability analysis of masonry buttresses and towers (Makris & Alexakis 2015).

CONCLUSION

This chapter shows that the DEM offers most dependable results for the limit equilibrium state of masonry arches that have been validated against closed form expressions obtained either with geometrical or variational formulation.

ACKNOWLEDGMENT

This work was funded by the research project "SeismoRockBridge" No 2295 which is implemented under the "ARISTEIA" Action of the "OPERATIONAL PROGRAMME EDUCATION AND LIFELONG LEARNING" and is co-funded by the European Social Fund (ESF) and Greek National Resources.

REFERENCES

Albuerne, A., & Huerta, S. (2010). Coulomb's theory of arches in Spain ca. 1800: the manuscript of Joaquín Monasterio. In *Proceedings of the 6th international conference on Arch Bridges* (pp. 354-362). Fuzhou, China: Academic Press.

Alexakis, H. (2013). *Limit State Analysis and Earthquake Resistance of Masonry Arches.* (Doctoral dissertation). Department of Civil Engineering, University of Patras, Greece.

Alexakis, H., & Makris, N. (2013a). Structural stability and bearing capacity analysis of the tunnel-entrance to the stadium of ancient Nemea. *International Journal of Architectural Heritage: Conservation, Analysis, and Restoration*, *7*(6), 673–692. doi:10.1080/15583058.2012.662262

Alexakis, H., & Makris, N. (2013b). Minimum thickness of elliptical masonry arches. *Acta Mechanica*, *224*(12), 2977–2991. doi:10.1007/s00707-013-0906-2

Alexakis, H., & Makris, N. (2014a). Limit equilibrium analysis and the minimum thickness of circular masonry arches to withstand lateral inertial loading. *Archive of Applied Mechanics*, *84*(5), 757–772. doi:10.1007/s00419-014-0831-4

Alexakis, H., & Makris, N. (2014b). Limit equilibrium analysis of masonry arches. *Archive of Applied Mechanics*. doi: 10.1007/s00419-014-0963-6

Andreu, A., Gil, L., & Roca, P. (2007). Computational analysis of masonry structures with a funicular model. *Journal of Engineering Mechanics*, *133*(4), 473–480. doi:10.1061/(ASCE)0733-9399(2007)133:4(473)

Andreu, A., Gil, L., & Roca, P. (2010). Analysis of masonry structures by funicular networks. *Engineering and Computational Mechanics*, *163*(3), 147–154.

Baggio, C., & Trovalusci, P. (1998). Limit Analysis for No-Tension and Frictional Three-Dimensional Discrete Systems. *Mechanics of Structures and Machines*, *26*(3), 287–304. doi:10.1080/08905459708945496

Bagi, K. (2014). When Heyman's Safe Theorem of rigid block systems fails: Non-Heymanian collapse modes of masonry structures. *International Journal of Solids and Structures*, *51*(14), 2696–2705. doi:10.1016/j.ijsolstr.2014.03.041

Bandis, S. C., Lumsden, A. C., & Barton, N. R. (1983). Fundamentals of rock joint deformation. *International Journal of Rock Mechanics and Mining Sciences & Geomechanics Abstracts*, *20*(6), 249–268. doi:10.1016/0148-9062(83)90595-8

Bélidor, B. F. (1729). *La science des ingénieurs dans la conduite des travaux de fortification et d'architecture civile.* Paris: C-A Jombert.

Block, P. (2009). *Thrust Network Analysis: Exploring Three-dimensional Equilibrium.* (Doctoral dissertation). Department of Architecture, Massachusetts Institute of Technology, Cambridge, MA.

Block, P., DeJong, M., & Ochsendorf, J. (2006). As Hangs the Flexible Line: Equilibrium of Masonry Arches. *Nexus Network Journal*, *8*(2), 13–24. doi:10.1007/s00004-006-0015-9

Block, P., & Ochsendorf, J. (2007). Thrust network analysis: a new methodology for three-dimensional equilibrium. *Journal of the International Association for Shell and Spatial Structures, 48*(3).

Casapulla, C., & D'Ayala, D. (2001). Lower bound approach to the limit analysis of 3D vaulted block masonry structures. In *Proceedings of the 5th International Symposium on Computer Methods in Structural Masonry* (pp. 28-36). Rome: STRUMAS V.

Clemente, P. (1998). Introduction to dynamics of stone arches. *Earthquake Engineering & Structural Dynamics, 27*(5), 513–522. doi:10.1002/(SICI)1096-9845(199805)27:5<513::AID-EQE740>3.0.CO;2-O

Clemente, P., Occhiuzzi, A., & Raithel, A. (1995). Limit behavior of stone arch bridges. *Journal of Structural Engineering, 121*(7), 1045–1050. doi:10.1061/(ASCE)0733-9445(1995)121:7(1045)

Cocchetti, G., Colasante, G., & Rizzi, E. (2012). On the Analysis of Minimum Thickness in Circular Masonry Arches. *Applied Mechanics Reviews, 64*(5).

Como, M. (2012). Statics of bodies made of a compressionally rigid no tension material. In M. Frémond & F. Maceri (Eds.), *Mechanics, Models and Methods in Civil Engineering, LNACM* (Vol. 61, pp. 61–78). Springer Berlin Heidelberg. doi:10.1007/978-3-642-24638-8_2

Coulomb, C. A. (1773). Essai sur une application des regles des maximis et minimis à quelques problémes de statique relativs a l'arquitecture. In Mémoires de mathematique et de physique présentés à l'académie royal des sciences per divers savants et lus dans ses assemblées (vol. 7, pp. 343-382).

Couplet, P. (1729, 1730). De la poussée des voûtes. In Histoire de l'Académie Royale des Sciences (pp. 79-117 & pp. 117-141). Paris.

Culmann, K. (1866). *Die graphische Statik*. Zurich: Verlag von Meyer und Zeller.

Cundall, P. A. (1971). A computer model for simulating progressive large-scale movements in blocky rock system. In *Proceedings of the Symposium of the International Society for Rock Mechanics* (*vol. 1*, paper no II-8). Nancy, France: Academic Press.

Cundall, P. A. (1987). Distinct element models of rock and soil structure. In E. T. Brown (Ed.), *Analytical and Computational Methods in Engineering Rock Mechanics*. London: Allen and Unwin.

De Lorenzis, L., DeJong, M., & Ochsendorf, J. (2007). Failure of masonry arches under impulse base motion. *Earthquake Engineering & Structural Dynamics, 36*(14), 2119–2136. doi:10.1002/eqe.719

De Luca, A., Giordano, A., & Mele, E. (2004). A simplified procedure for assessing the seismic capacity of masonry arches. *Engineering Structures, 26*(13), 1915–1929. doi:10.1016/j.engstruct.2004.07.003

DeJong, M. (2009). *Seismic Assessment Strategies for Masonry Structures*. (Doctoral dissertation). Department of Architecture, Massachusetts Institute of Technology, Cambridge, MA.

Delbecq, J. M. (1981). Analyse de la stabilité des voûtes en maçonnerie de Charles Augustin Coulomb à nos jours. *Annales des Ponts et Chaussées, 19*, 36–43.

Delbecq, J. M. (1982a). Analyse de la stabilité des voûtes en maçonnerie par le calcul à la rupture. *Journal de Mécanique théorique et appliquée, 1*, 91-121.

Delbecq, J. M. (1982b). *Les Ponts en Maçonnerie – Constitution et Stabilité.* Paris: Ministère des Transports, SETRA.

Dimitri, R., De Lorenzis, L., & Zavarise, G. (2011). Numerical study on the dynamic behavior of masonry columns and arches on buttresses with the discrete element method. *Engineering Structures, 33*(12), 3172–3188. doi:10.1016/j.engstruct.2011.08.018

Durand-Claye, A. (1867). Note sur la vérification de la stabilité des voûtes en maçonnerie et sur l'emploi des courbes de pression. *Annales des Ponts et Chaussées, 13*, 63–93.

Ferris, M. C., & Tin-Loi, F. (2001). Limit analysis of frictional block assemblies as a mathematical program with complementarity constraints. *International Journal of Mechanical Sciences, 43*(1), 209–224. doi:10.1016/S0020-7403(99)00111-3

Foce, F. (2005). On the safety of the masonry arch. Different formulations from the history of structural mechanics. In S. Huerta (Ed.), *Essays in the history of theory of structures* (pp. 117–142). Madrid: Instituto Juan de Herrera.

Foce, F. (2007). Milankovitch's Theorie der Druckkurven: Good mechanics for masonry architecture. *Nexus Network Journal, 9*(2), 185–210. doi:10.1007/s00004-007-0039-9

Foce, F., & Aita, D. (2003). The masonry arch between «limit» and «elastic» analysis. A critical reexamination of Durand-Claye's method. In S. Huerta (Ed.), *Proceedings of the 1st International Congress on Construction History* (*vol. 2*, pp. 895-905). Madrid: Instituto Juan de Herrera.

Gilbert, M. (2007). Limit analysis applied to masonry arch bridges: state-of-the-art and recent developments. In P.B. Lourenço, D.B. Oliveira & A. Portela (Eds.), *Proceedings of the 5th International Conference on Arch Bridges* (pp.13-28). Funchal, Madeira: Multicomp, Lda Publishers.

Gilbert, M., & Melbourne, C. (1994). Rigid-block analysis of masonry structures. *Structural Engineer, 72*(21), 356–361.

Gregory, D. (1697). Catenaria. *Philosophical Transactions of the Royal Society, 19*, 637–652.

Heyman, J. (1966). The Stone Skeleton. *International Journal of Solids and Structures, 2*(2), 249–279. doi:10.1016/0020-7683(66)90018-7

Heyman, J. (1969). The safety of masonry arches. *International Journal of Mechanical Sciences, 11*(4), 363–385. doi:10.1016/0020-7403(69)90070-8

Heyman, J. (1972). *Coulomb's Memoir on Statics: An Essay in the History of Civil Engineering*. Cambridge, UK: Cambridge University Press.

Heyman, J. (1982). *The Masonry Arch.* Chichester, UK: Ellis Horwood.

Heyman, J. (1998). *Structural Analysis: A Historical Approach.* Cambridge, UK: Cambridge University Press. doi:10.1017/CBO9780511529580

Heyman, J. (2009). La Coupe des Pierres. In K.E. Kurrer, W. Lorenz, & V. Wetzk (Eds.), *Proceedings of the 3rd International Congress on Construction History* (*vol. 2*, pp. 807-812). Berlin: Neunplus1.

Hooke, R. (1675). *A description of helioscopes, and some other instruments*. London.

Huerta, S., & Foce, F. (2003). Vault theory in Spain between XVIIIth and XIXth century: Monasterio's unpublished manuscript Nueva teorica sobre el empuje de bóvedas. In S. Huerta (Ed.), *Proceedings of the 1st International Congress on Construction History* (vol. 2, pp. 1155-1166). Madrid: Instituto Juan de Herrera.

Itasca Consulting Group, Inc. (2004). UDEC: Universal Distinct Element Code. Version 4.0. Minneapolis: Itasca.

Jaky, J. (1944). The coefficient of earth pressure at rest. *Journal of Society of Hungarian Architects and Engineers*, *78*(22), 355–358.

Konstantinidis, D., & Makris, N. (2005). Seismic Response Analysis of Multidrum Classical Columns. *Earthquake Engineering & Structural Dynamics*, *34*(10), 1243–1270. doi:10.1002/eqe.478

Kooharian, A. (1952). Limit Analysis of Voussoir (Segmental) and Concrete Arches. *Journal Proceedings of American Concrete Institute*, *49*(12), 317–328.

La Hire, P. (1695). *Traité de mécanique, où l'on explique tout ce qui est nécessaire dans la pratique des Arts, et les proprietés des corps pesants lesquelles ont eu plus grand usage dans la Physique*. Paris: Imprimerie Royale.

La Hire, P. (1712). Sur la construction des voûtes dans les édifices. In Mémoires de l'Académie Royale des Sciences (pp. 69-77). Paris: Academic Press.

Lagomarsino, S. (2006). On the vulnerability assessment of monumental buildings. *Bulletin of Earthquake Engineering*, *4*(4), 445–463. doi:10.1007/s10518-006-9025-y

Lagrange, J. L. (1788). *Mécanique analytique*. Paris: Chez La Veuve Desaint, Libraire.

Lamé, M. G., & Clapeyron, E. (1823). Mémoire sur la stabilité des voûtes. *Annales des mines*, *8*, 789-836.

Lauenstein, R., & Bastine, P. (1913). *Die graphische Statik: Elementares Lehrbuch fur den Schul- und Selbstunterricht sowie zum Gebrauch in der Praxis*. Leipzig: Alfred Kruner.

Lemos, J. V. (1995). Assessment of the ultimate load of a masonry arch using discrete elements. In J. Middleton & G. N. Pande (Eds.), *Computer Methods in Structural Masonry* (pp. 294–302). Swansea: Books & Journals International.

Lemos, J. V. (2007). Discrete element modelling of masonry structures. *International Journal of Architectural Heritage*, *1*(2), 190–213. doi:10.1080/15583050601176868

Livesley, R. K. (1978). Limit analysis of structures formed from rigid blocks. *International Journal for Numerical Methods in Engineering*, *12*(12), 1853–1871. doi:10.1002/nme.1620121207

Makris, N., & Alexakis, H. (2012). *From Hooke's "Hanging Chain" and Milankovitch's "Druckkurven" to a variational formulation: The Adventure of the Thrust-line of Masonry Arches. Report series in EEAM 2012-02*. University of Patras.

Makris, N., & Alexakis, H. (2013). The effect of stereotomy on the shape of the thrust-line and the minimum thickness of semicircular masonry arches. *Archive of Applied Mechanics*, *83*(10), 1511–1533. doi:10.1007/s00419-013-0763-4

Makris, N., & Alexakis, H. (2015). Limit equilibrium analysis of masonry buttresses and towers under lateral and gravity loads. *Archive of Applied Mechanics*. doi: 10.1007/s00419-015-1027-2

Makris, N., & Psychogios, T. (2004). Static and dynamic analysis of columns and part of the entablature of the NE corner of the temple of Zeus at Nemea. Report to American School of Classical Studies, Athens.

Mascheroni, L. (1785). *Nuove ricerche sull'equilibrio delle volte*. Bergamo.

Milankovitch, M. (1904). *Beitrag zur Theorie der Druckkurven*. Dissertation zur Erlangung der Doktorwürde. Vienna: K.K. technische Hochschule.

Milankovitch, M. (1907). Theorie der Druckkurven. *Zeitschrift für Mathematik und Physik*, *55*, 1–27.

Miller, S. G. (2001). *Excavations at Nemea II: The early Hellenistic Stadium*. Berkeley, CA: University of California Press.

Moseley, H. (1843). *The mechanical principles of engineering and architecture*. London: Longman.

Navier, C. L. M. H. (1826). *Résumé des Leçons données à l'École des Ponts et Chaussées sur l'application de la mécanique à l'établissement des constructions et des machines*. Paris: Dibot.

O'Dwyer, D. (1999). Funicular analysis of masonry vaults. *Computers & Structures*, *73*(1-5), 187–197. doi:10.1016/S0045-7949(98)00279-X

Ochsendorf, J. (2002). *Collapse of masonry structures*. (Doctoral dissertation). Department of Engineering, University of Cambridge, Cambridge, UK.

Oppenheim, I. J. (1992). The masonry arch as a four-link mechanism under base motion. *Earthquake Engineering & Structural Dynamics*, *21*(11), 1005–1017. doi:10.1002/eqe.4290211105

Orduña, A., & Lourenço, P. B. (2003). Cap Model for Limit Analysis and Strengthening of Masonry Structures. *Journal of Structural Engineering*, *129*(10), 1367–1375. doi:10.1061/(ASCE)0733-9445(2003)129:10(1367)

Orduña, A., & Lourenço, P. B. (2005). Three-dimensional limit analysis of rigid block assemblages. Part I: Torsion failure on frictional interfaces and limit analysis formulation. *International Journal of Solids and Structures*, *42*(18-19), 5140–5160. doi:10.1016/j.ijsolstr.2005.02.010

Orduña, A., & Lourenço, P. B. (2005). Three-dimensional limit analysis of rigid blocks assemblages. Part II: Load-path following solution procedure and validation. *International Journal of Solids and Structures*, *42*(18-19), 5161–5180. doi:10.1016/j.ijsolstr.2005.02.011

Pagnoni, T. (1994). Seismic analysis of masonry and block structures with the discrete element method. In G. Duma (Ed.), *Proceedings of the 10th European Conference on Earthquake Engineering* (vol. 3, pp. 1669-1674). Rotterdam: Balkema.

Pagnoni, T., & Vanzi, I. (1995). Experimental and numerical study of the seismic response of block structures. In J. Middleton & G. N. Pande (Eds.), *Computer Methods in Structural Masonry* (pp. 213–222). Swansea, UK: Books & Journals International.

Papantonopoulos, C., Psycharis, I. N., Papastamatiou, D. Y., Lemos, J. V., & Mouzakis, H. P. (2002). Numerical prediction of the earthquake response of classical columns using the distinct element method. *Earthquake Engineering & Structural Dynamics*, *31*(9), 1699–1717. doi:10.1002/eqe.185

Peña, F., Prieto, F., Lourenço, P. B., Campos Costa, A., & Lemos, J. V. (2007). On the dynamics of rocking motion of single rigid-block structures. *Earthquake Engineering & Structural Dynamics*, *36*(15), 2383–2399. doi:10.1002/eqe.739

Persy, N. (1825). *Cours sur la stabilité des constructions, à l'usage des élèves de l'école royale de l'Artillerie et du Génie*. Metz: Lithographie de l'École Royale de l'Artillerie et du Génie.

Rizzi, E., Rusconi, F., & Cocchetti, G. (2014). Analytical and numerical DDA analysis on the collapse mode of circular masonry arches. *Engineering Structures*, *60*, 241–257. doi:10.1016/j.engstruct.2013.12.023

Roca, P., Cervera, M., Gariup, G., & Pela', L. (2010). Structural Analysis of Masonry Historical Constructions. Classical and Advanced Approaches. *Archives of Computational Methods in Engineering*, *17*(3), 299–325. doi:10.1007/s11831-010-9046-1

Shames, I. H., & Dym, C. L. (1985). *Energy and Finite Elements Methods in Structural Mechanics*. New York: Hemisphere Publishing Corporation.

Sinopoli, A. (2003). The role of geometry in the theories on vaulted structures by Lorenzo Mascheroni. In S. Huerta (Ed.), *Proceedings of the 1st International Congress on Construction History* (pp. 1865-1873). Madrid: Instituto Juan de Herrera.

Sinopoli, A., Corradi, M., & Foce, F. (1997). Modern formulation for preelastic theories on masonry arches. *Journal of Engineering Mechanics*, *123*(3), 204–213. doi:10.1061/(ASCE)0733-9399(1997)123:3(204)

Timoshenko, S. P. (1953). *History of Strength of Materials*. New York: McGraw-Hill Book Company, Inc.

Wolfe, W. S. (1921). *Graphical Analysis: a text book on Graphic Statics*. New York: McGraw-Hill Book Company, Inc.

Chapter 13
Application of DEM to Historic Masonries, Two Case-Studies in Portugal and Italy:
Aguas Livres Aqueduct and Arch-Tympana of a Church

Alberto Drei
Technical University of Milan, Italy

Gabriele Milani
Technical University of Milan, Italy

Gabriela Sincraian
Vancouver Coastal Health, Canada

ABSTRACT

Two engineering applications of the Distinct Element Method to the analysis of historic masonries are presented. In particular, the commercial software UDEC, which implements DEM in a variety of engineering problems, is here used to analyze the Águas Livres aqueduct in Lisbon (Portugal) and multi-leaf masonry arch-tympana carrying systems of a basilica in Como (Italy). When dealing with the aqueduct, the most important portion of the structure is modeled and loaded with some accelerograms to evaluate its seismic vulnerability as well as the most critical zones. The second example analyzes the arch-tympana carrying system of a church in Italy. They present an unusual building technology, relying into a multiple-leaf arch, and a tympanum, made by a mixture of bad quality mortar and small stones. Again the structure is discretized into distinct elements and the load carrying capacity under dynamic excitation is evaluated, discussing the role played by the infill.

DOI: 10.4018/978-1-5225-0231-9.ch013

INTRODUCTION

A major part of the worldwide built stock is formed by old masonry buildings, made of stone or brick blocks. They also constitute the old urban nuclei in many towns, especially in Europe, and many of them are relevant part of the architectural and historical heritage. As they are representative of an earlier building tradition, there is a strong need of preserving them for future generations. Such structures, of great cultural significance, are generally vulnerable to seismic events, which have often caused massive damage or even their destruction. Examples can be found in past earthquakes, and similarly, in recent events, earthquakes have caused extensive damage and the destruction of both monumental buildings and traditional masonry houses.

Given the urgent need to protect the built cultural heritage from destruction caused by external agents, among which, especially in southern Europe, the seismic action is probably the most threatening, the effort to understand the behavior of historical structures has significantly increased.

Despite the fact that masonry is one of the oldest building materials, the assessment of the seismic behavior of old masonry structures often lacks of a scientific background. Their seismic vulnerability cannot be inferred by means of existing codes and certified analysis methodologies as those used for today's constructions. Special difficulties arise for these traditional structures, with structural typologies, geometry, mechanical characteristics and detailing which are very different from the ones assumed for modern structures. Moreover, as historical monuments are several centuries old, many of them have already survived important earthquakes that led to damage and changes in the mechanical behavior of the materials, expressed by stiffness degradation and overall weakening. The analysis of their behavior under dynamic loading is essential for their safety assessment and the eventual application of retrofitting systems. The low tensile strength of the masonry components and its discontinuous nature are features that dominate the complex mechanical behavior displayed by this material, non-linear in essence. These unreinforced blocky structures cannot be considered a continuum, but rather an assemblage of compact stone or brick elements linked by means of mortar joints. The joints constitute discontinuities between distinct bodies and can alter the response of the system, compromise its stability and introduce irreversible displacements.

The evaluation of the seismic vulnerability of such structures, as for other types of constructions, depends on a good understanding of their mechanical behavior as well as on analysis and prediction tools, namely reliable numerical simulations of their seismic response. Numerical modelling of the seismic behavior of masonry structures represents a very challenging problem due to the constitutive characteristics of the structural material and its highly physical and geometrical non-linear behavior when subjected to strong ground motions. Therefore, a numerical idealization should be capable of predicting the behavior of a masonry structure from the linear stage, through cracking and degradation until the complete loss of strength, since this is a key step towards a full understanding of the fundamental physical phenomena and the reliable assessment of the structural safety of a masonry structure.

The main objective of the Chapter is to present some numerical results in the dynamic non-linear range obtained for two meaningful blocky masonry structures basing on the Discrete Element software called UDEC (Lemos, 1997; Lemos et al., 1998; Itasca, 1998, 2000; Lemos, 2007). This methodology is based on the Discrete Element Method (DEM), which is a discontinuous analysis technique originally developed to model jointed and fractured rock masses (Cundall, 1971, 1987, 1988; Hart et al., 1988;

O'Connor et al., 1992; Kuhn & Bagi, 2009, etc.). Discrete element methods are especially suitable for problems in which a significant part of the deformation is accounted for by relative motion between distinct bodies. Masonry provides a natural field of application for such techniques (Sincraian & Azevedo, 1997, 1998; Sincraian & Lemos, 1998; Giamundo et al., 2014; Sarhosis et al., 2008, 2014, 2015; Tóth et al., 2009; Tóth & Bagi, 2011; Lengyel & Bagi, 2015), as the deformation and failure modes of these structures are mainly dependent on the joint behavior. Whilst not specifically identified into DEM, but rather into Rigid Body and Spring Mass models RBSM, other approaches that essentially correspond to DEM have been presented in the recent literature for the analysis of a variety of masonry problems in the static and dynamic range (Casolo & Milani, 2010, 2013; Casolo et al., 2013; Gilbert, 2006; Gilbert & Melbourne, 1994; Milani et al., 2008; Milani, 2008, 2011; Minghini et al., 2014; Orduna, 2006; Orduna & Lourenço, 2005). These methodologies allow, on one hand the modelling of the response up to stages of collapse corresponding to large displacements and highly geometrical non-linear behavior, and on the other hand the modelling of the physically non-linear behavior of both the at structural blocks and their interfaces, the identification of failure patterns and the control of internal stresses and deformations as a function of the seismic input.

The principal goals pursued by the Chapter are the following:

- An evaluation of the applicability of the discrete element method to simulate the structural behaviour of two historic large scale blocky masonry structures, especially in what regards their seismic behaviour. In absence of experimental data available, this can be done by simulating the observed behaviour of this type of structures in past earthquakes;
- An analysis of what are the most suitable input parameters to be used within DEM applied to masonry structures. The objectives are to provide guidelines on the use of the DEM (particularly some already available structural analysis DEM packages (e.g. UDEC), to address the discretisation issue, mainly in terms of contact between blocks, the problem of damping in dynamic analysis, the application of different kinds of loads, especially earthquakes, to analyse the problem of boundary conditions, to show the possibility of sequential modelling (by applying different model loading conditions at different stages of an analysis in order to simulate changes in physical loading);
- A critical assessment of two typical masonry structures exhibiting high complexity, in terms of their structural response, failure modes, damage indicators obtained from the program results.

In particular the chapter presents two engineering applications of the Distinct Element Method to the analysis of historic masonries.

The approach is intrinsically heterogeneous, so the structure is discretized by rigid or deformable blocks connected by joints. The numerical solution well adapts to the dynamic range, because it consists into the integration of block motion equations, with geometric and material non-linearity embedded. The interaction among blocks takes place in correspondence of the joints, taking into account their frictional behavior.

The commercial software UDEC, which implements DEM for a variety of engineering problems, is here used to analyze two historic masonry structures: Águas Livres aqueduct in Lisbon, Portugal (Oliveira et al., 1991, 1995; Azevedo & Drei, 1995; Sincraian, 2001; Sincraian & Lemos, 1999; Sincraian et al.

1998) and multi-leaf masonry arch-tympana carrying systems of the San Fedele Romanesque basilica in Como, Italy (Drei & Fontana, 2001).

When dealing with the aqueduct, the most important portion of the structure is modelled, subjected to a series of accelerograms in order to evaluate its seismic vulnerability as well as the most critical zones, thus giving suitable information on its dynamic behavior and rehabilitation techniques to be adopted.

As second example, arch-tympana carrying systems of the San Fedele Romanesque basilica is analyzed. Such structural elements present an unusual building technology, constituted by two main parts, the first relying into a multiple-leaf arch that is the effective bearing element, and the tympanum, made by a mixture of bad quality mortar and stones of small dimensions. Again the structure is discretized into distinct elements and the load carrying capacity under dynamic excitation is evaluated, discussing the role played by the mechanical properties of the infill.

THE DISCRETE ELEMENT METHOD, BASIC PRINCIPLES AND APPLICATIONS TO THE STRUCTURAL ANALYSIS OF MASONRY STRUCTURES

The discrete element methods fall within the general classification of discontinuous analysis techniques. Originally used to model jointed and fractured rock masses, they were adopted for the analysis of structures composed of particles or blocks. Joints are viewed as interfaces between blocks, i.e. the discontinuity is treated as a boundary condition.

The distinct element method (DEM) was introduced by Cundall in the early '70s (Cundall, 1971) for study of jointed rock, idealised as an assemblage of rigid blocks. The most important aspects of any discrete element program are: the representation of solid material, the representation of contacts and the scheme used to detect and revise the set of contacts (Cundall, 1987, 1988).

The material may be assumed rigid or deformable. The assumption of material rigidity is a good one when most of the deformation in a physical system is accounted for by movement on discontinuities. In this case, the movements consist mainly of sliding and rotation of blocks and of opening and interlocking of interfaces.

If the deformation of the solid material cannot be neglected, deformable blocks can be used. To introduce deformability, the body is divided into internal elements, i.e., a finite element (or finite difference) mesh is created for each block, in order to increase the number of degrees-of-freedom. The complexity of the deformation depends on the number of elements into which the body is divided. For example, in the program *UDEC* (used in this work) (Itasca, 2000), the formulation of these zones is similar to the constant-strain finite element formulation. The disadvantage of this technique is that, in case of complicated shape bodies, the number of zones in which the block has to be discretised may become very large, even for a simple deformation pattern. The formulation used in *UDEC* permits both geometric and physical non-linearities of the intact material to be modelled.

One of the most important components of any discrete element method is the formulation to represent contact between blocks. All methods allow contacts to open and slide, but the methods are divided into two groups by the way they treat behaviour in the normal direction of motion at the contacts. In the first group sometimes designated as the "soft contact" approach, a finite normal stiffness is taken to represent the measurable stiffness that exists at a contact or joint. A small overlap is present in the case of compression forces. The contact force is assumed to start from zero at the point where the bodies first touch. Thus, one body must interpenetrate the other to produce any finite value of contact force, since contact forces are related to contact displacements.

In the second group, the "hard contact" approach, interpenetration is regarded as non-physical, and algorithms are used to prevent any interpenetration of the two bodies that form a contact.

In most distinct element models, the representation of the interface between blocks relies on sets of point contacts. Adjacent blocks can touch along a common edge segment or at discrete points where a corner meets an edge (face) or another corner. No special joint element or interface element is defined. Joint elements are not practical in discrete element methods, since these methods are intended to simulate large displacements and special types of contact involving edges or single vertices must be allowed. The remarkable generality and robustness under complicated load paths of the point contact approach make it ideal for simulation of the dynamics of blocky systems where large displacements between blocks may occur. Also, the point contact hypothesis leads to a great computational efficiency.

The formulation described in the following sections follows essentially the one implemented in the programs *UDEC* (Itasca, 2000) and *3DEC* (Itasca, 1998), respectively.

The UDEC Program

UDEC (Universal Distinct Element Code) is a very general code, capable of modelling many types of jointed systems, ranging from assemblies of many discrete blocks to extended continua crossed by a few major fractures (Lemos, 1987).

Among the most important capabilities of *UDEC* that make it very suitable for masonry, we could mention the possibility to choose rigid or deformable blocks, the ability to simulate progressive failure associated with crack propagation, the capability of simulating large displacements/rotations between blocks, the possibility to use rounded corners to overcome interlocking, etc.

Each deformable block is independently discretised into an internal element mesh. An explicit, large deformation, Lagrangian formulation with constant-strain triangles is used. While the original formulation follows the finite-difference approach, these triangular zones are equivalent to finite elements. General constitutive relations can be assumed for the block material.

The soft contact approach is used, so a finite normal stiffness is taken to represent the measurable stiffness that exists at a contact or joint. A joint is represented numerically as a contact surface formed between two block edges. The representation of the interface between blocks relies on sets of point contacts. For each pair of blocks that touch (or are separated by a small gap), data elements are created to represent point contacts. Adjacent blocks can touch along a common edge segment or at discrete points where a corner meets an edge or another corner. No special joint element or interface element is defined.

A point contact hypothesis is used. In the point contact approach, the interaction force at each contact is a function of solely the relative displacement between blocks at that location. This assumption implies that a larger number of contact points is needed to get an accurate contact stress distribution on the joint surface.

When two blocks come into contact, a force develops between them which can be resolved into normal and shear components, as shown in Figure 1 (left).

The simplest model of mechanical interaction is to assume that the blocks are connected by normal and shear elastic springs, see Figure 1 (center and right), i.e. interaction forces are proportional to the relative displacement between the two blocks.

Figure 1. Forces between blocks (left) and Representation of joints within DEM (blocks are in contact, separation is shown for clarity)

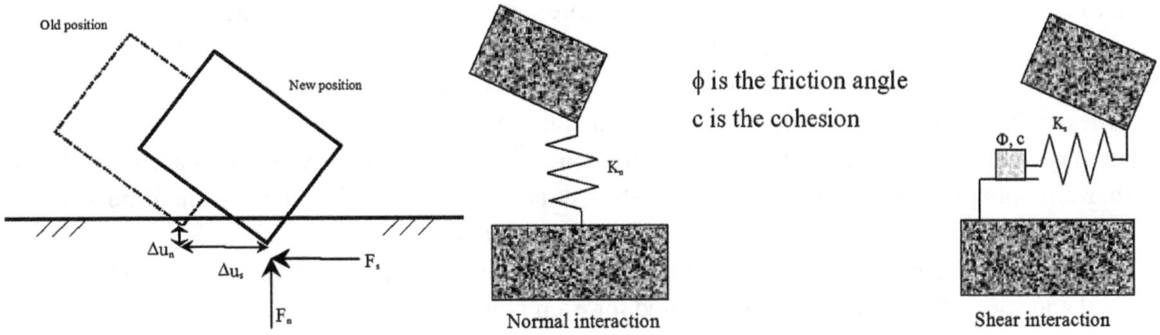

This force-displacement law at the contacts is expressed in incremental form as:

$$\Delta F_n = K_n \, \Delta u_n$$

$$\Delta F_s = K_s \, \Delta u_s \tag{1}$$

where ΔF_n and ΔF_s are the normal and shear force increments, Δu_n and Δu_s are the normal and shear displacement increments, K_n and K_s are the contact normal and shear stiffnesses.

Contacts between two block edges (Figure 2-a) can be represented by two corner-edge contacts. The contact length, l, allows contact stresses to be calculated as (assuming a unit thickness):

$$\sigma_n = F_n / l$$

$$\sigma_s = F_s / l \tag{2}$$

and stress increments to be expressed in terms of the usual joint normal (k_n) and shear (k_s) stiffnesses (stress/length) as:

$$\Delta \sigma_n = k_n \, \Delta u_n$$

$$\Delta \sigma_s = k_s \, \Delta u_s \tag{3}$$

Figure 2. Contact between blocks

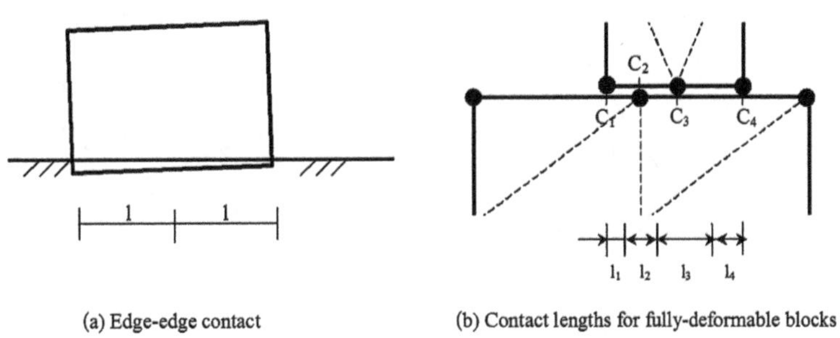

When blocks are discretised into a fine internal mesh (deformable blocks), grid-points may be placed along the original edges (Figure 2-b). These grid-points are treated as new corners, since the edge is now able to deform into a polygonal line. The same expressions are used, with contact lengths defined as shown in Figure 2-b, and where the length associated with each grid-point is equal to half the distance between the two closest grid-points located to each side of the edge it contacts.

The overlaps displayed in these figures represent only a mathematically convenient way of measuring relative normal displacements. In finite element models, joints are similarly assigned a zero thickness, with overlapping indicating compressive joint stresses and separation indicating tension. If normal joint stiffness is increased, overlaps can be made as small as desired.

A force-displacement law is used to find contact forces from known displacements. Incremental normal and shear displacements are calculated for each point contact.

The basic joint model is the Coulomb slip, see Figure 3, capable of capturing several of the features that are representative of the physical response of joints. The necessary parameters to be defined are the normal and shear stiffnesses (k_n and k_s), the friction angle (Φ) the cohesion (c) and the tensile strength (σ_t).

For the joints simulating the characteristics of the mortar, a Coulomb slip model (linear elastic with damage and residual strength) is usually sufficient. The graphic representation of the model constitutive behaviour is given in Figure 3, where the force-displacement relationships are presented both for the shear and normal components.

SOLUTION ALGORITHM

The selection of a solution algorithm for the discrete element method must take into account the fact that the geometry of the system, as well as the number and type of contacts between the discrete bodies, may change during the analysis.

In the discrete element method the structural analysis, both static and dynamic, is based on explicit time-stepping algorithms. The size of the time step is limited by the numerical stability requirements of the central difference method. For rigid block models, the central difference method is used to integrate the equations of motion of each block. For deformable block models, it is applied to the equations of motion of each node, when block discretisation into elements is performed.

Figure 3. Joint constitutive laws (normal and shear)

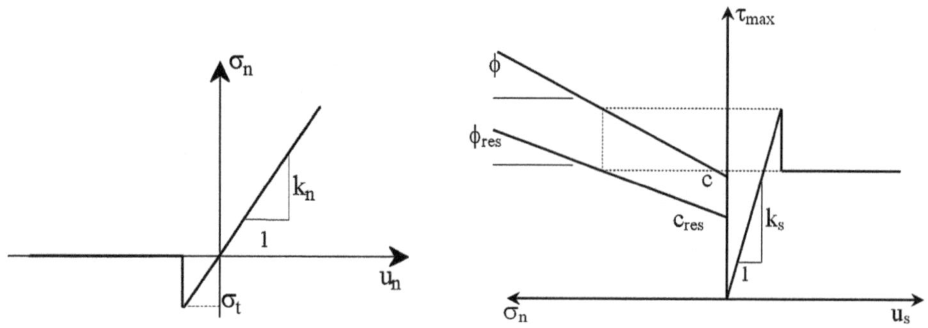

Static analysis can be performed by the same time-stepping algorithm, by introducing viscous damping to dissipate the kinetic energy, and thus make the system converge to a static solution or to a failure mechanism. This approach is conceptually similar to dynamic relaxation technique (Underwood, 1983).

The calculations performed in the Distinct Element Method at each time step alternate between the application of Newton's second law at all blocks and a force-displacement law at all contacts. The force-displacement law, the contact constitutive model, is used to determine contact forces from known displacements. Newton's second law evaluates the motion of the blocks resulting from the known forces acting on them. If the blocks are deformable, motion is calculated at the grid-points of the triangular elements within the block. Then, the application of the block material constitutive relations allows the evaluation of new stresses within the elements.

In summary, the distinct element solution algorithm is based on two sets of computations performed at each time step. First, kinematic quantities are updated. Then, the application of constitutive models provides new internal stresses and interaction forces.

New grid-point velocities are obtained from the equations of motion as:

$$\frac{d\dot{u}_i}{dt} + \alpha\dot{u}_i = \frac{F_i}{m} + g_i \tag{4}$$

where u_i are the components of the grid-point velocity (i=1,2), α is the viscous damping constant, m the grid-point mass, F_i the grid-point forces and g_i the gravity acceleration components. The grid-point forces result from the external applied loads, the contact forces, and the nodal forces derived from the element stresses.

Displacement increments are calculated as $\Delta u_i = \dot{u}_i \Delta t$. The location of the block boundary is updated, and contact displacement increments are calculated. Then the application of the joint constitutive relations provides the new contact stresses and forces. Zone strain increments are calculated as $\Delta\varepsilon_{ij} = 1/2\left(\partial\dot{u}_i / \partial x_j + \partial\dot{u}_j / \partial x_i\right)\Delta t$ and the application of the block material constitutive relations gives the new zone stresses. Finally, contact forces, zone stresses and external loads are assembled into the new force vector F_i for the next time step.

The motion of an individual block is determined by the magnitude and direction of resultant out-of-balance moment and forces acting on it. Considering the one-dimensional motion of a single mass acted on by a force *F(t)*, without damping (for clarity), Newton's second law of motion can be written as $d\dot{u} / dt = F / m$, where $\dot{u}$ = velocity, t = time, and m = mass.

The central difference scheme for the left-hand side of equation (1) at time t can be written as $d\dot{u} / dt = \left(\dot{u}^{t+\Delta t/2} - \dot{u}^{t-\Delta t/2}\right) / \Delta t$ providing the following formula:

$$\dot{u}^{t+\Delta t/2} = \dot{u}^{t-\Delta t/2} + \frac{F(t)}{m}\Delta t \tag{5}$$

With velocities stored at half-time step point, the displacement can be expressed as $u^{t+\Delta t/2} = u^t + \dot{u}^{t+\Delta t/2}\Delta t$.

As the force depends on displacement, the force-displacement calculation is done at one time instant.

The central difference scheme is second order accurate, i.e. first-order error terms vanish from the solution. This important characteristic prevents long-term drift in a distinct element simulation.

For two dimensional blocks acted upon by several forces as well as gravity, the equations of motion for the three degrees of freedom (two translations and one rotation) are as follows:

$$\begin{array}{ll} m\ddot{u}_i + \alpha m\dot{u}_i = \sum F_i + mg_i & \sum F_i = F_i^E + F_i^C \\ I\ddot{\theta} + \alpha I\dot{\theta} = \sum M & \sum M = M^E + M^C \end{array} \tag{6}$$

where $\dot{\theta}$ is the angular velocity of a block about centroid, I is the mass polar moment of inertia of the block, $\dot{u}_i$ are velocity components of block centroid (i=1,2), g_i are the components of gravitational acceleration (body forces), $\sum F_i$ are the components of the resultant force in the center of gravity of the block due to the action of exterior loads (F_i^E) and contact forces (F_i^C), $\sum M$ is the resultant moment in the centre of gravity of the block due to exterior loads (M^E) and contact forces (M^C). Index i denotes components in a Cartesian coordinate frame.

The central difference method gives the following expressions for the accelerations and velocities:

$$\ddot{u}_i^t = \frac{\dot{u}_i^{t+\Delta t/2} - \dot{u}_i^{t-\Delta t/2}}{\Delta t}$$

$$\dot{u}_i^t = \frac{1}{2}\left(\dot{u}_i^{t+\Delta t/2} + \dot{u}_i^{t-\Delta t/2}\right) \tag{7}$$

By substituting these velocities and accelerations in (6) we obtain:

$$\dot{u}_i^{t+\Delta t/2} = \dot{u}_i^{t-\Delta t/2}\frac{1-\frac{\alpha}{2}\Delta t}{1+\frac{\alpha}{2}\Delta t} + \frac{\sum F_i^t}{m} + g_i\frac{\Delta t}{1+\frac{\alpha}{2}\Delta t}$$

$$\dot{\theta}_i^{t+\Delta t/2} = \dot{\theta}_i^{t-\Delta t/2}\frac{1-\frac{\alpha}{2}\Delta t}{1+\frac{\alpha}{2}\Delta t} + \frac{\sum M^t}{I}\frac{\Delta t}{1+\frac{\alpha}{2}\Delta t} \tag{8}$$

In this scheme, velocities are stored at the mid-points of the time intervals, while forces or displacements are calculated at the interval endpoints.

The new velocities are used to determine the new block location according to:

$$x_i^{t+\Delta t} = x_i^t + u_i^{t+\Delta t/2}\Delta t$$

$$\theta^{t+\Delta t} = \theta^t + \dot{\theta}^{t+\Delta t/2}\Delta t \tag{9}$$

where θ is the rotation of block about centroid and x_i are the coordinates of block centroid.

Thus, each time step produces new block positions that generate new contact forces. Resultant forces and moments are used to calculate linear and angular accelerations of each block. Block velocities and displacements are determined by integration over increments in time. Mechanical damping is used in the equations of motion to provide both static and dynamic solutions.

For a static analysis, an approach similar to dynamic relaxation technique is employed. The equations of motion are damped to reach the equilibrium state as soon as possible. Damping is velocity-proportional, i.e. the magnitude of the damping force is proportional to the velocity of the blocks. In order to overcome some difficulties caused by this damping, an adaptive damping scheme is used to achieve faster convergence. Adaptive damping continuously adjusts the viscosity such that the power absorbed by damping is a constant proportion of the rate of change of kinetic energy in the system. Therefore, as the kinetic energy approaches zero, the damping power also tends to zero.

For dynamic analysis, Rayleigh damping (mass- and stiffness-proportional) may be used to provide damping that is approximately frequency-independent over a restricted range of frequencies. The main inconvenient in using the stiffness proportional component of Rayleigh damping is the extremely low time step required in some cases in order to reach the convergence and stability of the solution.

The mass-proportional component of Rayleigh damping is introduced in the equations of motion by the term corresponding to a damping force $F_i^V = -\alpha m \dot{u}_i$ where F_i^V is the damping force. The stiffness-proportional component is introduced in the contact forces and zone stresses, e.g. for rigid blocks, the viscous contact forces are $F_{nv} = -\beta k_n \dot{u}_n$ and $F_{sv} = -\beta k_s \dot{u}_s$.

"ÁGUAS LIVRES" AQUEDUCT IN LISBON

The Águas Livres Aqueduct in Lisbon, built in the 18th century, survived the 1755 earthquake with minor damage, in spite of the great losses produced in the surrounding region (Oliveira et al. 1995). This structure, composed of 35 arches, extends about 940 m over the Alcântara Valley, with a maximum height of about 65 m (Figure 4). Previous numerical studies of its seismic behavior were performed with finite element models, assuming continuous idealizations (Oliveira et al. 1991, Azevedo and Drei 1995). In the present analysis, an evaluation is made of the ability of the discrete element approach to provide insight into the behavior of this type of structure, particularly in view the assessment of ultimate loading scenarios. In the first phase of the analysis, reported herein, two-dimensional models were used to analyze independently the in-plane and out-of-plane seismic responses.

Figure 4. The "Águas Livres" Aqueduct in Lisbon

Only the central part of the "Águas Livres" aqueduct was analyzed. The length of the modelled region is 280 m out of 370 m of the arched section that extends over the Alcântara Valley. The lowest part of the valley is crossed by the largest and tallest arch – about 40 m wide and 65 m high (58 m of free height). The other arches are about 26 m wide. The numerical model includes ten pillars with gothic arches (Figure 5.2). The pillars have different heights ranging from 30 to 65 m, and, in their slender part, an average section of 7 x 9 m. Some have a visible enlargement at the base, with a section of 12 x 9 m; for others, the enlargement is buried. At the top of the structure there is the water channel that runs along the axis of the aqueduct, protected by an accessible gallery. There is also an aeration tower every two pillars.

The aqueduct was built, according to the 18th century usual technology, with two different materials (Oliveira et al. 1991). As in many other historical structures, the pillars of the aqueduct have cross-sections composed by an outer hard stone shell and some in-fill material. The outer shell is a regular wall made with good limestone. The stones are well cut having approximately similar dimensions. The joints between stones are very thin (about 4 mm). The mortar seems to be compact and of good quality. The inner part of the structure is filled with a heterogeneous material made up by mortared rubble and irregular stones, with highly variable properties.

The aqueduct is a landmark for the city of Lisbon. Given the importance of this historical structure, it has been the object of several experimental and analytical studies (Oliveira et al. 1991). *In situ* vibration measurements provided the characterisation of its dynamic behaviour. More recently, sample cores extracted from this structure supplied the basic information on the mechanical properties of the limestone blocks and the fill material (Schiappa de Azevedo et al. 1996).

Figure 5. Discrete Element models of the Águas Livres aqueduct

a) Different material regions

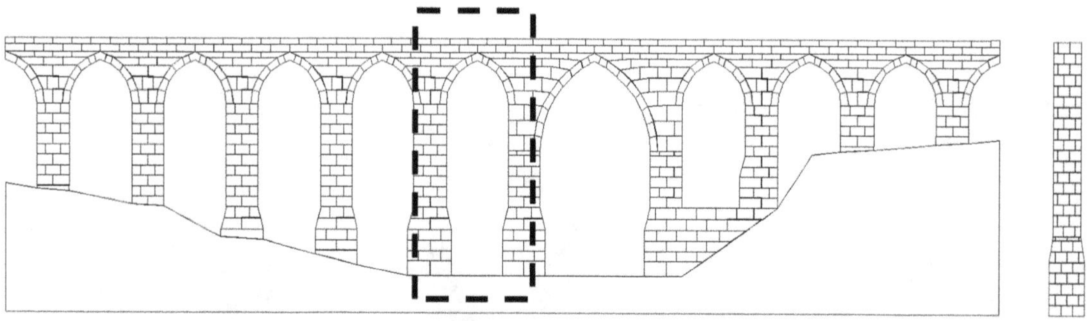

(1) Longitudinal model

(2) Transverse model

b) Block structure

Modelling Assumptions and Mechanical Properties

The numerical representation of such a large and complex structure is not an easy task and involves many simplifications that must be critically analyzed. An important decision in the modelling process is related to the dimensionality of the model used in the numerical analyses. Obviously, a three-dimensional discrete element model would be the best choice in assessing the seismic behavior of the aqueduct, since the movement in the transverse direction of a structure like this is of crucial importance. However, the complexity of a 3D model could render the analysis excessively expensive due to the very large number of blocks and joints employed in the model.

A better choice, with a great practical benefit, at least for early stage runs, may be an analysis conducted with two-dimensional models. It will be shown that 2D discrete element models can provide important insight into the structural behaviour of this very complex historical construction. For this purpose two-dimensional models were created for independent analyses in the longitudinal and transverse directions.

In a discontinuum numerical model joint spacing is a geometrical parameter rarely possible to reproduce. As a rule, the real spacing would create a very large number of blocks leading to excessive run times and inconvenient delays in obtaining results.

The aim of a discrete element model is not to include every block, but only enough joints to allow the essential deformation and failure modes to develop. For closely spaced joints, as is the case of this particular structure, a multiple of the real spacing can be safely employed. Therefore, only some joints are selected, and the block size in the numerical model is much larger than in reality. However, the simplified representation must respect the overall moduli of the masonry mass, i.e. its deformability. There are basically two ways to proceed: (1) reduce the stiffness of the joints by the ratio between real spacing and numerical spacing; and (2) keep the joint stiffness unchanged, and reduce the block moduli to obtain the same effect. The first option tends to be more straitforward for simple joint models, or when a non-linear constitutive model is assumed for the blocks. For complex joints, it is better to rely on the second option, shifting the correction to the block material. For the models representing the aqueduct, approach (1) was followed.

The approximate values for the joint normal and shear stiffnesses were back calculated from information on the deformability and joint structure in the jointed masonry mass and the deformability of the intact blocks (limestone). If the jointed masonry mass is assumed to have the same deformational response as an equivalent elastic continuum, then relations can be derived between jointed masonry properties and equivalent continuum properties. For example, for uniaxial loading of masonry containing a single set of uniformly spaced joints oriented normal to the direction of loading, the following relation applies (Itasca, 2000; O'Connor & Dowding, 1992):

$$k_n = \frac{E_m E_b}{s\left(E_b - E_m\right)}$$

$$k_s = \frac{E_s E_s}{s\left(E_s - E_s\right)} \quad (10)$$

E_m (G_m) is the mortar Young's (shear) modulus, E_b (G_b) is the intact block Young's (shear) modulus, k_n (k_s) is the joint normal (shear) stiffness (per unit area), and s is the joint spacing.

The values of the Young's and shear modulus for both the jointed masonry mass and the limestone blocks were assumed equal to the ones used in previous studies carried out on the aqueduct (Oliveira et al., 1991; Schiappa de Azevedo et al., 1996).

After the approximate stiffness values were obtained, the model was further calibrated in order to match the natural frequencies measured *in situ*. For this, the values for the joint normal and shear stiffness obtained with the previous formulas had to be slightly modified, while the blocks' mechanical properties were kept constant. Since there is very little information about the joint normal and shear stiffness for masonry, some preliminary analyses were carried out in order to understand the true influence of these parameters in the results. Changes of one order of magnitude in the joint stiffnesses had no important influence on the results. Also, some runs were performed with equal values for joint normal and shear stiffness and the results did not significantly change. Therefore, for the sake of simplicity, this solution was adopted in the analyses reported.

Based on the available information, the thickness of the external wall was assumed to be equal to one meter in both directions, while higher values were assumed for the pillars' enlargements (about 1.8 m in the transverse direction and 2.8 m in the longitudinal one).

The different material regions used in the longitudinal model are shown in Figure 5a, in different colors, and the block structure in Figure 5b. The 2 m thick pointed arches are assigned limestone properties. However, for the spandrels, only the resistant contribution of the walls was included, the fill being represented only as dead weight. In a 2D model, it is impossible to represent the real structure of the pillars, with the outer limestone shell and fill, so a blocky pattern was created, with equivalent material properties, listed below, to account for the global behaviour of the pillar section. For the wider pillar section at the base, properties were scaled as a function of the out-of-plane thickness.

The transverse section model (Figure 5b, second frame) corresponds to the tallest pillar. Equivalent pillar properties were also used in the slender part, which were scaled to simulate the thicker base and upper sections. In the upper region, thickness accounts for the half-bay arches on either side of the pillar. Deformable blocks with a coarse internal mesh were used (Figure 6).

Blocks were assumed to be elastic with the mechanical properties shown in Table 1. The basic joint model assumed constant elastic stiffnesses, zero tensile strength and a Mohr-Coulomb failure criterion in shear, with the mechanical properties also shown in Table 1.

It should be mentioned that, since blocks are assumed elastic, the failure modes may depend on the particular joint pattern employed, and it is always advisable to perform more than one analysis with models with different joint patterns. This approach was adopted here only for the analyses carried out in the transverse direction of the structure, since this is the direction for which the structure is most vulnerable to the earthquake loading.

The pillar's equivalent Young's modulus is similar to the one used in previous studies in order to match the natural frequencies measured *in situ* (Oliveira et al. 1991). In the transverse direction the UDEC model lowest frequency is 0.9 Hz, as obtained in the tests; in the longitudinal model, it is close to the 2.9 Hz, value measured for the 7th mode which involves essentially longitudinal motion.

In the dynamic analyses, only the mass-proportional component of Rayleigh damping was used, corresponding to 1.0% of critical damping at 0.9 Hz for the transversal model and to 1.0% of critical

Figure 6. The triangular element mesh of deformable blocks

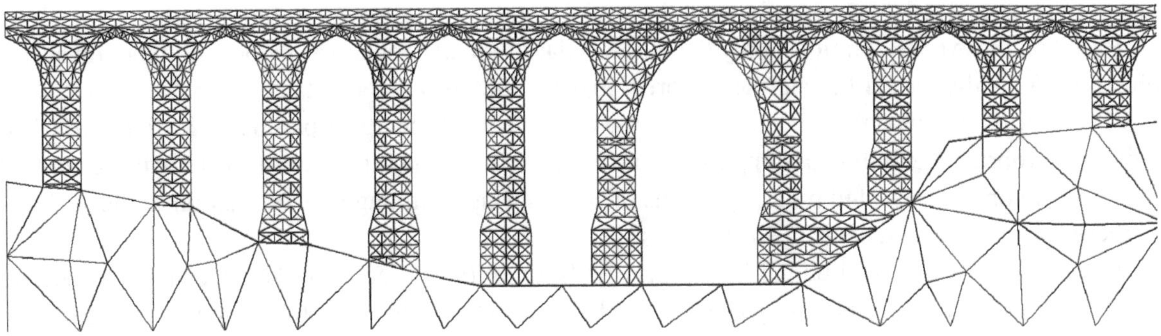

Table 1. Adopted mechanical properties for the blocks and the joints (aqueduct)

Blocks	Young's Modulus (GPa)	Poisson's Ratio	Density (kg/m³)	Mortar Joints				
Limestone blocks	50	0.24	2700	Normal stiffness (GPa/m)	Shear stiffness (GPa/m)	Friction angle	Cohesion (MPa)	Tensile strength (MPa)
Pillars	20	0.24	2500	100	100	30°	1.0	0

damping at 2.9 Hz in case of the longitudinal model. The stiffness-proportional component was not used to avoid the need for reduction of the time step. Since no stiffness-proportional damping was introduced, higher frequencies have less damping. However, joint slip produces additional energy dissipation in the structure. Once the response to the earthquake record was completed, and if the system was stable, damping was increased to a high value (50%) and the run continued for about 10 sec., in order to get a faster convergence to equilibrium and evaluate permanent displacements.

Boundary Conditions

The selection of boundary conditions, as those resulting from interconnections with adjacent structures, is always a matter that requires some effort, particularly in the case of a large and complex structure such as the Lisbon Aqueduct. As specified before, only a limited part of the structure was studied herein, i.e. the one comprising the tallest pillars of the aqueduct, over the Alcântara Valley.

The campaign of *in situ* experimental tests, based on environmental vibration and excitation of the structures, carried out on the aqueduct (Oliveira et al. 1991), showed that the first 6 modes of vibration are almost all in the out-of-plane direction (transversal). The results of the Finite Element analysis performed in the same study, considering a linear elastic model with beam elements, were in a very good agreement with the experimental ones. This shows that the bending of the tallest pillars must be approximately two-dimensional. Also, since the tallest pillars have the lowest vibration frequencies, they should be more affected by earthquakes with low frequency content, such as two of the accelerograms considered in this study. In fact, given the huge mass of the structure and despite its apparently quite high longitudinal stiffness, the first frequencies measured were very low (in the range of 0.9 to 1.5 Hz), just because the aqueduct is very high and the stiffness in the transverse direction is rather low.

Based on these considerations, only the tallest pillar of the aqueduct was modelled in the transverse direction. As in the case of FE analyses reported in other studies (Oliveira et al., 1991; Azevedo & Drei, 1995), the pillar was modelled as a cantilever, the movements of the upper part being let free. Probably, a more appropriate modelling assumption would be to consider the influence of the remaining parts of the structure on the truncated pillar, by means of some springs, although the estimation of the stiffness of these springs could create some difficulties.

In what regards the longitudinal DE model, the fact that the structure of the aqueduct is truncated, was taken into account by assigning an elastic behaviour to the joints of both end arches of the model, so no sliding or separation was permitted to occur in these particular joints (they do not break in tension or shear). This modelling assumption, in spite of being simple and convenient, was considered good enough to compensate the influence of the remaining parts of the aqueduct.

The same remarks are valid if a three-dimensional model is to be used for the analysis of this complex structure, with the additional observation that torsional movements could have some effect on the overall seismic response, although their consideration in a discrete element model is not an easy task.

Loading

The analyses were carried out in two stages: first, each model was brought to equilibrium under its own weight; then, an accelerogram was applied at all the nodes along the model base. The accelerograms used are displayed in Figure 7.

The main seismic input used for both longitudinal and transversal model was an artificially generated accelerogram according to the Portuguese code (I.N.C.M., 1983), for a long-distance earthquake in a hard soil site (type 2, soil type I), with a peak ground acceleration equal to 0.11g, shown in Figure 7-a, which is assumed to represent an event of the type of the 1755 Lisbon earthquake.

For each type of model a series of runs was performed, with the input record scaled by factors ranging from 1.0 to 3.0. For the transverse model a study of sensitivity to the input motion was undertaken by using two other accelerograms with different frequency content, also displayed in Figure 7.

These accelerograms were: first, a type 1 earthquake, soil type I (according to the Portuguese code), that corresponds to a close and shallow event with a high frequency content and a shorter duration period (10 seconds), and with a peak ground acceleration equal to 0.15g (Figure 7b); and, second, the NS component of the 1977 Vrancea earthquake (Romania), corresponding to a deep earthquake with a low frequency content, having one big pulse and a peak ground acceleration equal to 0.21g (Figure 7c).

DISCUSSION OF RESULTS

As already indicated in previous studies, the collapse of the aqueduct would most certainly involve an out-of-plane mode. For safety evaluation, the 2D longitudinal model has little practical interest and the analysis is computationally expensive. However, it allows a further check on the elastic properties, by comparison with the measured frequency, and provides some insight into failure mechanisms involving the arches.

Figure 7. The dynamic loading used for this study

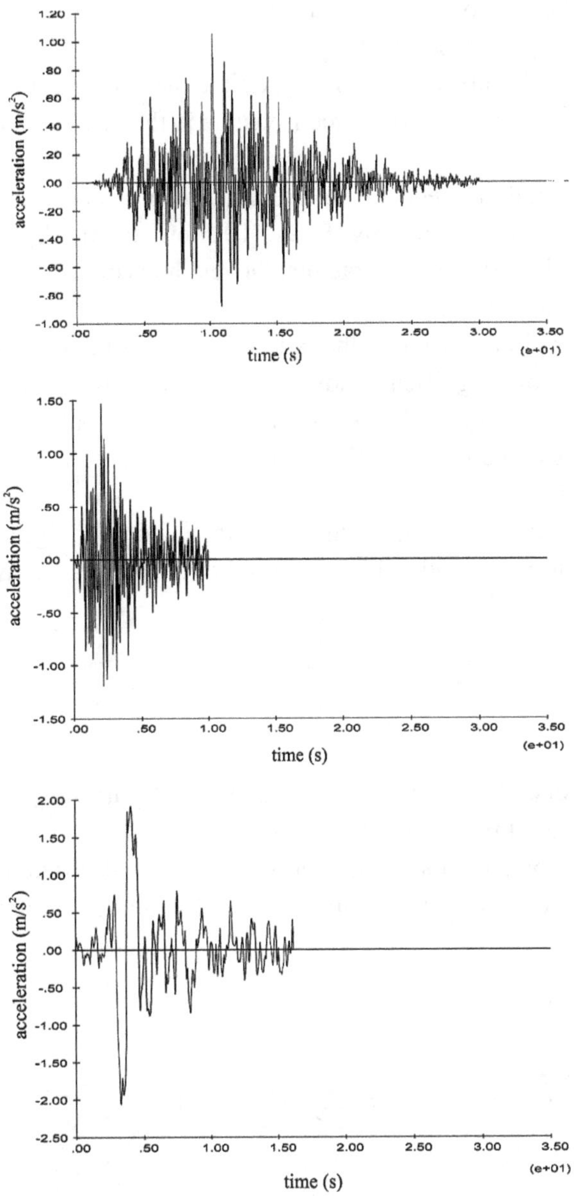

a) Portuguese code – Earthquake type 2, Soil type I

b) Portuguese code – Earthquake type 1, Soil type I

c) N-S component of the 1977 Vrancea earthquake

This model was subjected only to the Portuguese code, type 2 earthquake. As stated in the previous studies (Oliveira et al. 1991), the aqueduct did not suffer significant damage during the 1755 earthquake. This may have been due to the fact that the predominant direction of the seismic action was longitudinal with respect to the aqueduct. Its frequency in this direction is 2.9 Hz, so it is a very stiff structure not particularly affected by the predominant low frequency content of that seismic motion.

The present analyses lead to the same conclusion. For a peak acceleration equal to 0.11g, a maximum absolute top horizontal displacement equal to 10 cm was obtained, which is in a good agreement with the one obtained in the previous work.

Figure 8 shows the residual displacements observed at the end of the run. It can be seen that maximum permanent displacements, in the order of about 4 mm, occur in the left upper part of the structure, where pillars are taller. There was no important joint opening or sliding.

Principal stresses at the end of the run are shown in Figure 9a. It is clearly visible the concentration and flow of stresses in the arches and columns. The effect of the larger arch is visible in the detail given in Figure 9b.

In order to get an insight about the initiation of a failure mechanism, the initial record was scaled by a factor equal to 10, thus leading to peak acceleration equal to 1.0g. Figure 10 shows the damaged structure at the end of the input (displacements magnified 10 times). Note that the restraints of both end arches were assumed elastic, given the truncation of the structure.

The effect of shear is much stronger in the shorter pillars, while in the tallest ones the bending moment is predominant. The main arch of the aqueduct shows significant movement in the abutments. It is interesting to notice the effect of the main arch, which transmits a thrust to the pillar that is not balanced by the smaller arch, causing thus an important sliding and opening of the joints.

The distribution of joint separation is shown in Figure 11.

The analysis was repeated with the record scaled to an extremely high PGA value, equal to 3g, in order to obtain a failure mechanism with this longitudinal model, which occurred in the main arch, as shown in Figure 12 and Figure 13.

TRANSVERSE MODELS

Since the aqueduct is most vulnerable in the transverse direction, for safety assessment purposes, an extensive out-of-plane analysis of the seismic response of this structure is necessary.

The basic transverse section model (Figure 14a) corresponds to the tallest pillar. Several different alternative joint patterns were analysed for the transverse models shown in Figure 14b and c. Failure

Figure 8. Grid-point displacements at the end of the run

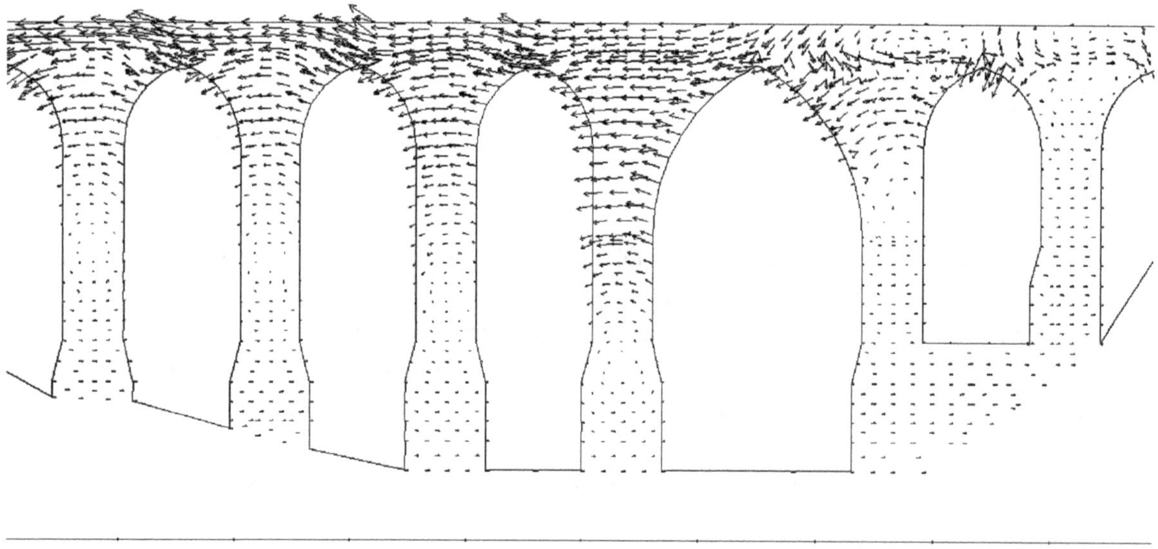

Figure 9. Principal stresses in the model (at the end of the run) (Maximum stress = 2.7 MPa)

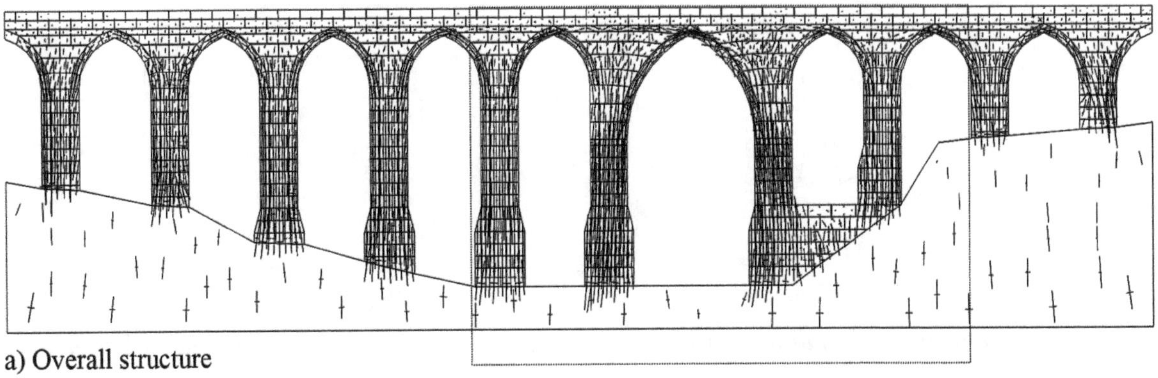

a) Overall structure

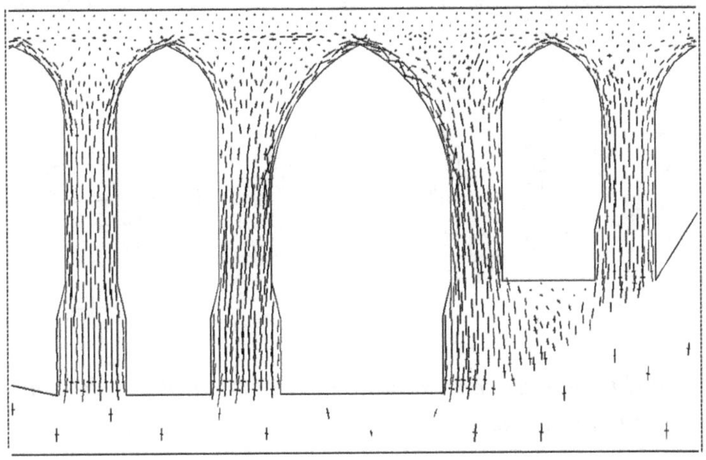

b) Detail – Zone of the tallest arches

Figure 10. Deformed pattern for 1.0g (deformation magnification factor equal to 10)

Figure 11. Joint opening (PGA = 1.0g)

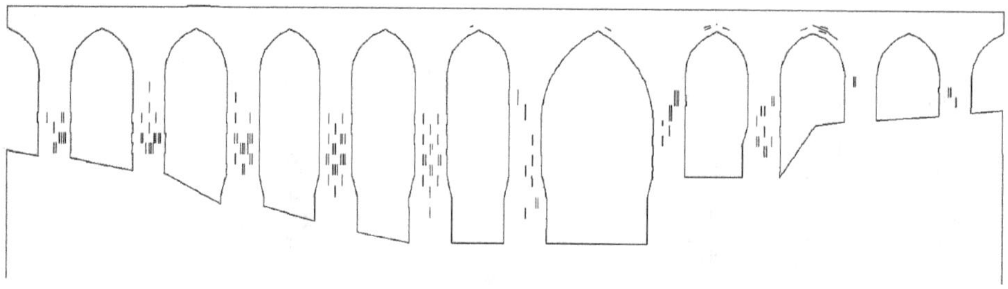

Figure 12. Development of a failure process for the "Águas Livres" aqueduct (longitudinal direction)

Figure 13. Details of two different sequences of the development of a failure process for the "Águas Livres" aqueduct (longitudinal direction)

mechanisms involving joint opening or sliding were investigated. Meanwhile, some initial cracks in the blocks were considered to accommodate possible fracturing and detachment of the limestone shell (Figure 14d and e).

For the mortar joints, the model employed assumed constant elastic stiffnesses, zero tensile strength and a Mohr-Coulomb failure criterion in shear, with the properties shown in Table 2. For the joints simulating initial cracks in the limestone blocks, the Coulomb slip model with residual strength was used in order to approximate a displacement-weakening response. This is accomplished by setting the joint friction, cohesion and tensile strength to reduced values (usually zero) whenever either the tensile or shear strength is exceeded (Itasca, 2000). However, in order to avoid some intermediate state during gravity

Figure 14. The transverse models developed in the study

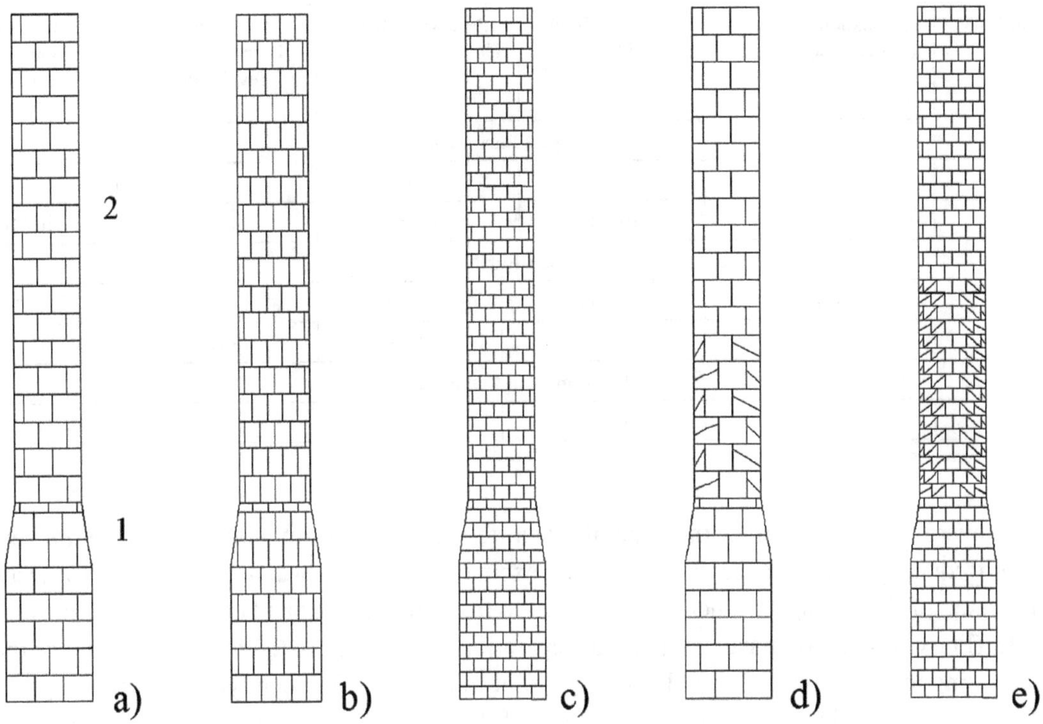

application that could cause fracturing of the joints, high values for the cohesion and tensile strength were used for the static stages (when the model was brought to equilibrium under the self-weight). Then, the real mechanical characteristics of the joints were used for the dynamic runs.

Table 3 summarizes the results of the runs performed with the basic transverse model (model a) developed for one pillar of the aqueduct using as input the three accelerograms mentioned above, multiplied by various scaling factors.

For each case, the table shows: the input multiplier factor; the maximum input acceleration at the base; the peak base horizontal displacement; the maximum top absolute horizontal displacement (corresponding to the input); the maximum top permanent displacement at the end of the run; and the maximum permanent joint shear displacement. One can notice that, even for similar input acceleration values,

Table 2. Adopted mechanical properties for the joints (aqueduct pillar)

	Normal Stiffness (GPa/m)	Shear Stiffness (GPa/m)	Friction Angle	Cohesion (MPa)	Tensile Strength (MPa)
			Residual Friction Angle	Residual Cohesion (MPa)	Residual Tensile Strength (MPa)
Mortar joints	100	100	30º	1.0	0
			-	-	-
Joints in the intact limestone	100	100	40º	7.0	2.0
			40º	0	0

Table 3. Results for the transverse model (a)

E/Q Multiplier	Maximum Input Acceleration (g)	Peak Base Horiz. Displ. (mm)	Peak Top Absol. Horiz. Displ. (mm)	Maximum Permanent Top Displ. (mm)	Maximum Permanent Joint Shear Displ. (mm)
Type 2 Earthquake					
1.0	0.1	51.00	156.00	4.97	8.70
2.0	0.2	102.00	300.00	49.16	107.10
Type 1 Earthquake					
1.0	0.15	42.60	65.20	4.35	6.79
1.5	0.22	63.90	109.00	75.06	63.40
1977 Romanian Earthquake					
1.0	0.21	387.00	726.00	142.90	144.50

the most severe seismic action for this type of structure is the 1977 Romanian earthquake. Its very low dominant frequency leads to a particularly strong excitation and to substantial damage in the structure. A comparison of the damaged structures subjected to the above-mentioned earthquakes is presented in Figure 15 (PGA ≅ 0.2g), by means of plots of separation and sliding in the joints at the end of the run. It is interesting to notice the different location of joint movements for each type of input.

Figure 16 presents three stages of the response of the structure, with extensive joint damage, when subjected to the 1977 Romanian record (PGA = 0.5g).

Some time histories of displacements and stresses for the type 2 earthquake with a PGA equal to 0.22g are presented in what follows.

Figure 17 shows the evolution of the relative horizontal displacements at the top and mid-height of the pillar. The maximum displacement at the top of the structure is about 15 cm, while at the mid-height it is about five times smaller, reaching a maximum value equal to 3 cm. One can also notice that both displacements are almost always in phase, showing that the first mode of vibration is the prevailing one. The equivalent frequency of vibration is around 0.7 Hz, which already shows some degradation of the structure, but still does not indicate highly non-linear overall behaviour.

Joint normal stress histories are illustrated in Figure 18, for the horizontal joints located 19.0m (where the section narrows) and 57.0m high (marked as 1 and 2 in (Figure 14). It can be noticed the difference in the magnitude of the stresses in these two particular joints: while for the joint located in the lower part of the pillar the normal stress reaches a value equal to 7.5 MPa, the value of the normal stress in the joint located at the top is about three times lower (about 2.5 MPa). This suggests that the section located above the enlargement of the pillar is a very vulnerable one and an eventual collapse process could originate here (as it will be presented in more detail later).

In the graphs shown in Figure 18, one can notice very clearly the opening of the joint (zero normal stress). It can also be seen that the top joints open much more often than the lower ones. The high frequency content of the response is related to the impact between blocks. The corresponding joint normal and shear displacement histories are represented in Figure 19. Interesting to observe is the fact that each time a large opening occurs there is also a large amount of sliding, since the shear resistance is lost.

Figure 15. Damaged structure (model a) for different earthquakes (PGA ≅ 0.2g)

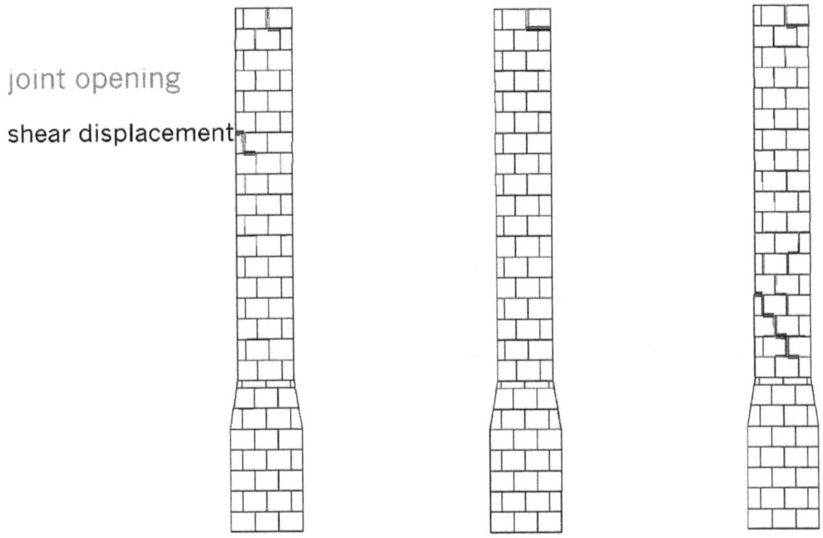

(1) type 2 earthquake (2) type 1 earthquake (3) Romanian earthquake

Figure 16. Stages of damage process – Romanian earthquake (PGA = 0.5g)

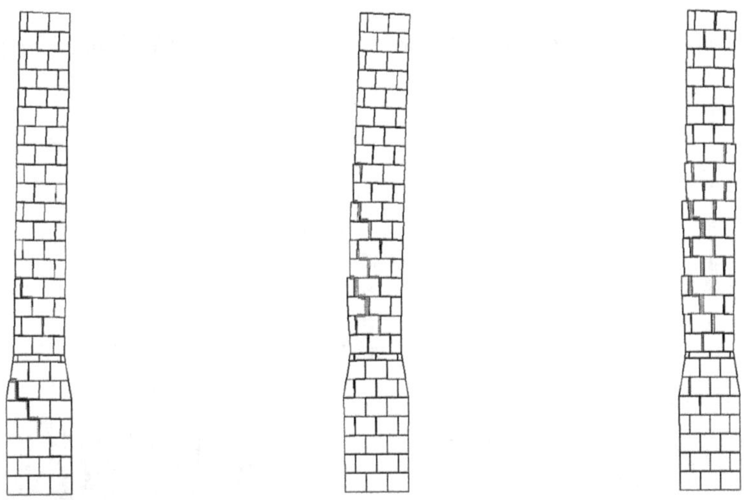

The peak normal stress, as displayed in Figure 18, is about 7.5 MPa, still below the limestone strength, however, for the Romanian record it reaches 40 MPa at the same joint. Therefore, the results for the runs with PGA levels equal to 0.2g and higher must be viewed with some caution, as the blocks were assumed elastic. In reality, collapse may occur due to brittle compression failure in the blocks. The possible fracturing and detachment of the limestone shell were not accounted for in the model a).

In order to solve the problem of handling the high compressive stresses observed in the previous analyses, causing compression failure in the blocks and mortar joints, several different joint patterns based on the initial transverse model were developed. The chosen joint patterns were obtained by halving the block sides, both vertically and horizontally.

Figure 17. Histories of relative horizontal displacements at the top (continuous line) and mid-height (dashed line) of the structure

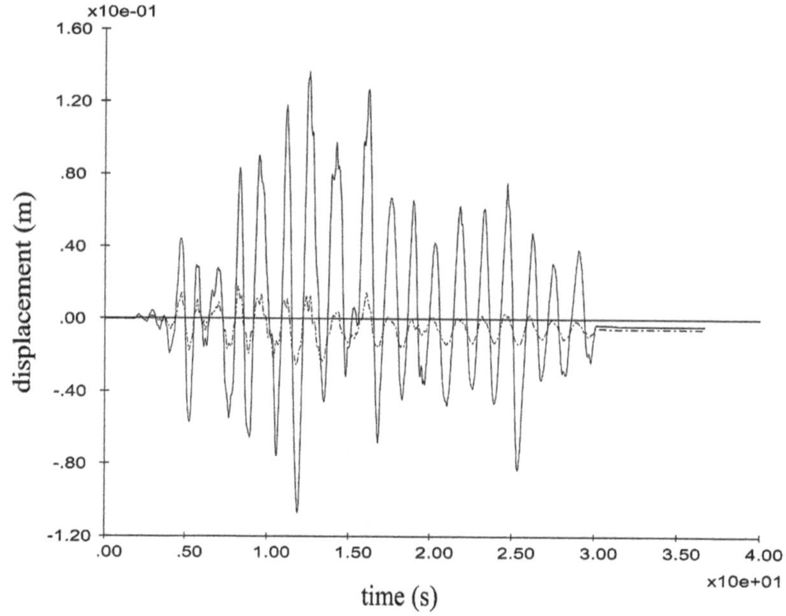

Figure 18. Joint normal stress histories for the joins located at 19m (left) and 57m (right)

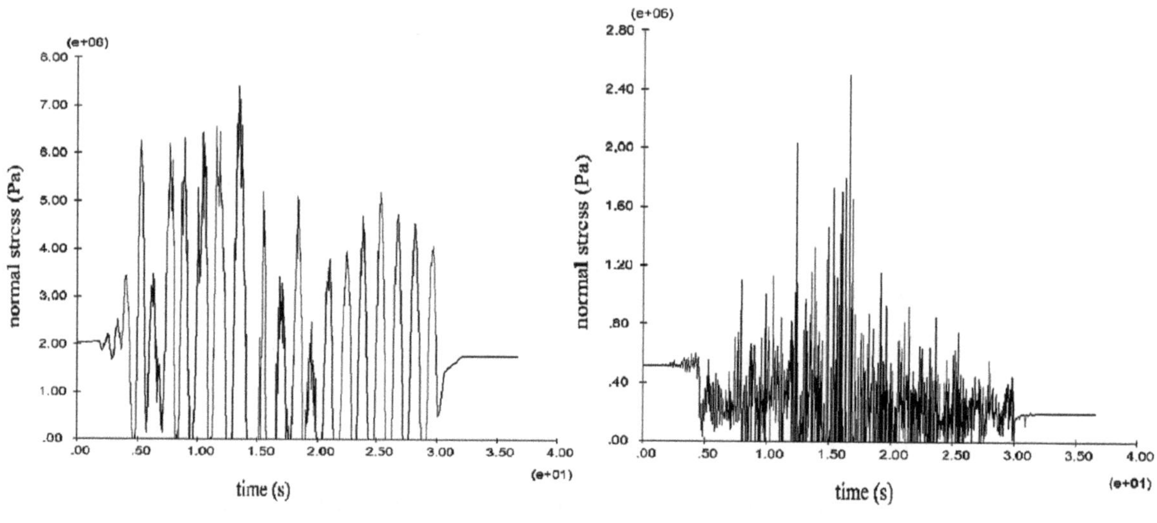

As mentioned before, the possible fracturing and detachment of the limestone shell were also accounted for by considering some initial possible cracks in the blocks. For this sensitivity study, the 1977 Romanian earthquake record was used for the input.

Figure 20 displays the damaged structure at the end of the run (input seismic motion magnified 1.5 times, PGA = 0.315g) for two different joint patterns (b and c in Figure 14.). It can be noticed that the structure is already severely damaged, some blocks are falling and the joint shear and separation are out of an acceptable range. The maximum relative top displacement is higher than 1 m. Also, the initia-

Figure 19. Joint normal and shear displacement histories for the joins located at 19m and 57m

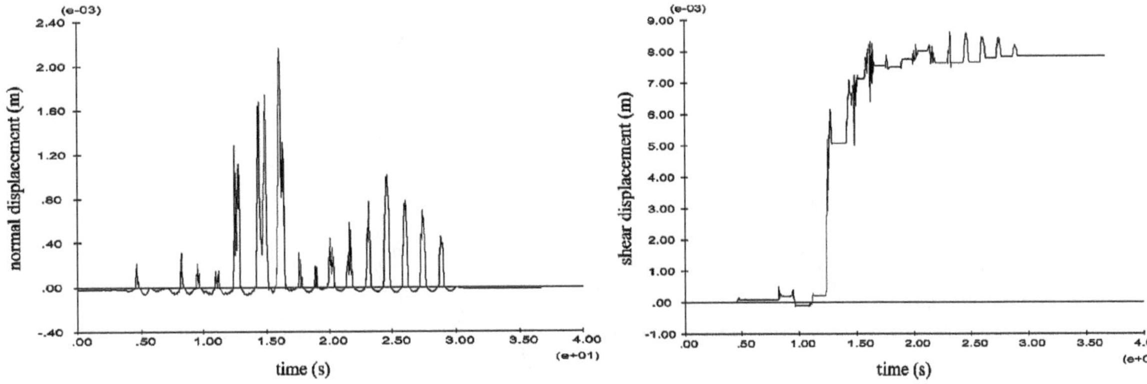

a) Joint located 19.0 m high

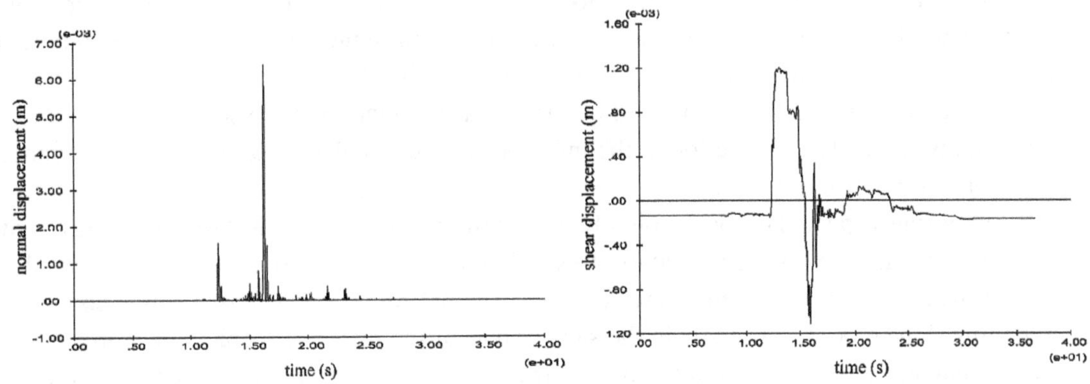

b) Joint located 57.0 m high

tion of the failure mechanism is clearly observed. Obviously, it starts with the weakest section, above the enlargement of the pillar. It can be seen that the model with larger blocks is more vulnerable to the seismic motion, probably due to the slenderness of the blocks. The most slender blocks start falling first, followed by the others. The second model behaves more similar to the initial one (shown in Figure 14a). They have the same ratio between block height and width and are more stable, although the permanent relative top displacement (at the end of the run) is in the order of 80 cm. It should be mentioned that the normal stress peaks in the section above the pillar enlargement dropped to a value of about 14 MPa, which is well below the limestone strength. Also, important to mention is the fact that the ratio between the dimensions of the blocks in the models presented in Figure 14a and c, are more similar to the real structure, while for the model shown in Figure 14b, the blocks are too slender compared to the ones in the real structure.

For this model, two runs were performed with the initial record multiplied by factors equal to 3 and 4, in order to induce structural failure. The damaged structure at the end of the run is shown in Figure 21, for both PGA levels, being clear that collapse occurs for PGA = 0.8g, but for a PGA value equal to 0.6g the structure is already very much damaged and almost close to collapse.

The 1977 Romanian earthquake with a PGA equal to 0.3g led to considerable damage in the structure. A second earthquake of a reduced size could have more serious consequences. In order to study this effect, the same seismic input, unmagnified (PGA = 0.2g), was applied to the damaged structure.

Figure 20. The damaged structure for PGA = 0.315g

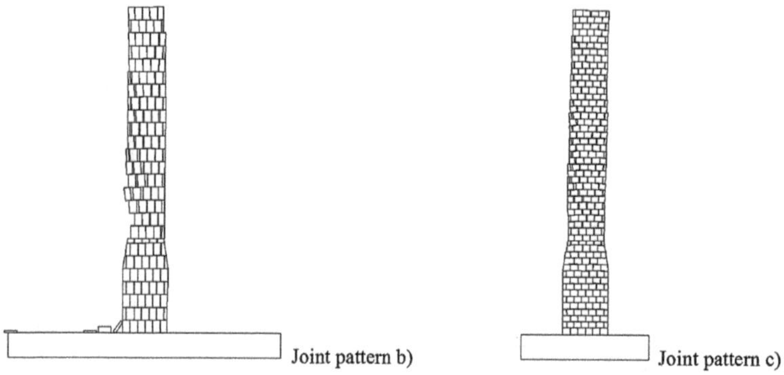

Only the model with larger blocks was used. This second earthquake motion caused the collapse of the already damaged pillar. The evolution of the failure process is presented in Figure 22. The initiation of the failure mechanism is clearly visible in the slender mid-height of the structure.

This observation indicates the importance of damage accumulation in the seismic response of this type of structures and calls the attention to the influence of earthquake duration and/or occurrence of multiple events on the structural safety.

Another important aspect of the behaviour of block masonry structures is the block splitting or crushing of block corners due to very high compressive stresses. Usually, the mortar joints are destroyed first. The joints of the Águas Livres aqueduct are very thin (about 4 mm), so it is possible that some cracks exist in the limestone blocks, especially in the slender part of the pillars. This phenomenon could lead to significant damage or even the collapse of the structure in the case of a severe seismic event. Two different models were developed (displayed in Figure 14d and e) in order to take into account this unfavourable condition. Some initial cracks were considered in the blocks situated in the weakest part of the analyzed pillar, with mechanical properties simulating potential joints that have the strength of the intact limestone.

Figure 21. The damaged pillar for two different PGA levels

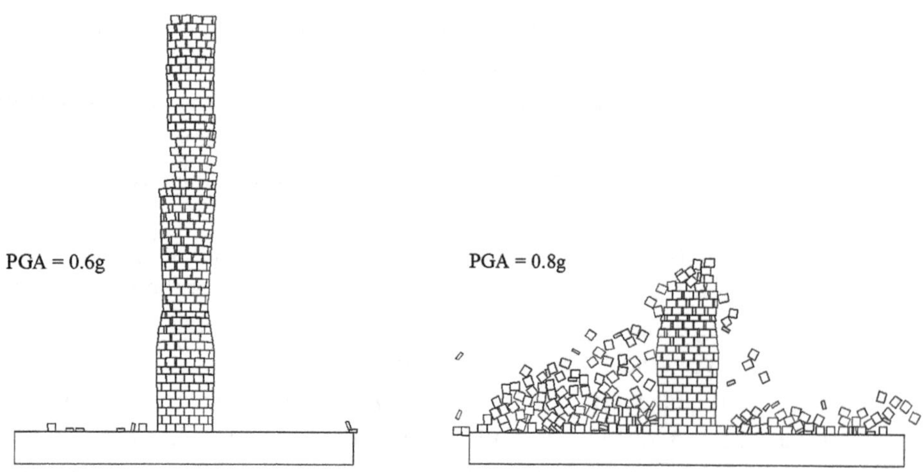

Figure 22. Different stages of the failure process of an already damaged pillar

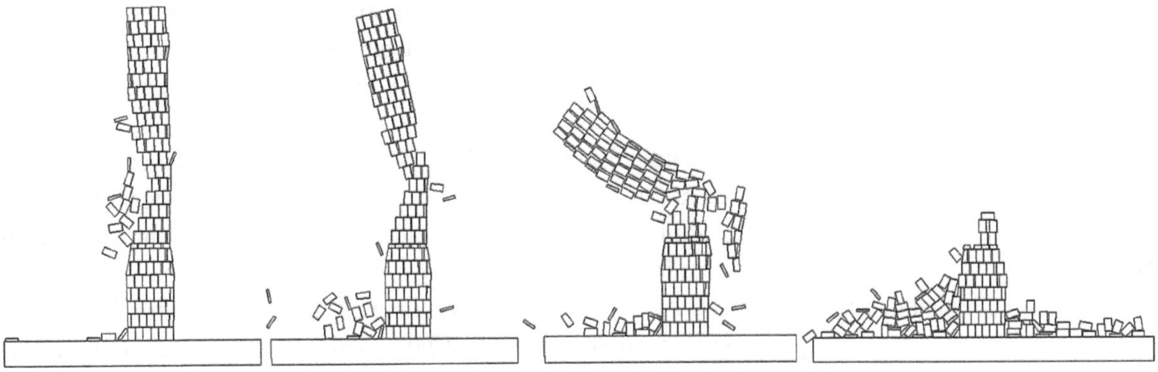

A study of sensitivity to the input motion was first undertaken. The model shown in Figure 14d was subject to the Romanian record scaled to a PGA of 0.3g and to the artificially generated accelerogram based on the Portuguese seismic code (I.N.C-M, 1983), simulating an event similar to the 1755 Lisbon earthquake, also scaled to a PGA equal to 0.3g. The results are presented in Figure 23.

As it can be seen, the damage is much larger in the case of the first record. For this earthquake two different stages of the damage process are illustrated in Figure 23a. The cracks in the stones yield to large shear displacements, opening of the joints and detachment of some blocks. Although the pillar did not fall, it can be considered as in structural failure. In what regards the second record, it was thought that its longer strong motion could cause the collapse of the structure. On the contrary, the structure seems to be more stable in this case and has less damage than in the case of the Romanian record. The damaged structure (first frame in Figure 23b) and the separation and sliding in the joints (darker lines in Figure 23b, second frame) at the end of the run are presented in the figure.

The influence of a second earthquake simulating an after shock or a later event was also investigated. For this, the model shown in Figure 14e was employed. The 1977 Romanian record scaled to a PGA

Figure 23. Structural sensitivity to the input motion

a) Romanian earthquake (PGA = 0.3g) **b) Portuguese code (Type 2, soil I) (PGA = 0.3g)**

equal to 0.3g was applied as a main shock and then the same earthquake reduced to a PGA equal to 0.2g was used as a following shock. It was found, as expected, that the structure with initial cracks in the blocks is more vulnerable than the one with intact blocks. Some images of the failure process during the main seismic shock are displayed in Figure 24. This motion rendered the structure with unrecoverable structural damage.

The second earthquake caused the complete failure of the structure. The evolution of collapse is depicted in Figure 25. It is interesting to notice the rigid body like movement of the upper thicker part of the pillar. The blocks move together with little separation or sliding. This phenomenon was already observed in real cases of very tall and slender structures acted by strong seismic motions.

It should be added that for the last two models analysed here, the very high normal stress peaks obtained in the previous study dropped to values in the range of 12 MPa, which is within acceptable values and shows the effectiveness of adopting potential crack surfaces as a means of coping with block failure.

Another possibility of simulating damage induced in the structure by block failure would be the use of an elasto-plastic model for the limestone blocks. There are several plastic models available in *UDEC*. The basic failure model for blocks in *UDEC*, used here, is the Mohr-Coulomb failure criterion (shear yield function) with tension cut-off (tensile yield function). The shear flow rule is non-associated (Itasca, 2000). This model includes brittle collapse under tension and shear, not a gradual tension- or cohesion-softening. Besides the mechanical characteristics for the linear-elastic model, the following block parameters are needed for the non-linear model: the cohesion, the angle of internal friction, the angle of dilation and the tensile strength. Table 4 presents the assumed values.

Only the transverse model presented in Figure 14a was analysed. Again, the 1977 Romanian earthquake record was used for the input. Two runs were performed with this seismic motion scaled to PGA levels equal to 0.2g and 0.3g, respectively. Figure 26a shows the deformed structure at the end of the run, for a PGA value equal to 0.2g. Fracturing of the limestone shell is represented by means of plastic points in the blocks (crosses). It can be observed that the deformed shape is pretty much the same as in the case where elastic blocks were used (Figure 15). The largest opening and sliding occur in the more slender part of the pillar, right above the enlargement (Figure 26b). Failure in the blocks occurs mainly

Figure 24. Stages of the damage process for the main seismic shock (Romanian earthquake – PGA = 0.3 g)

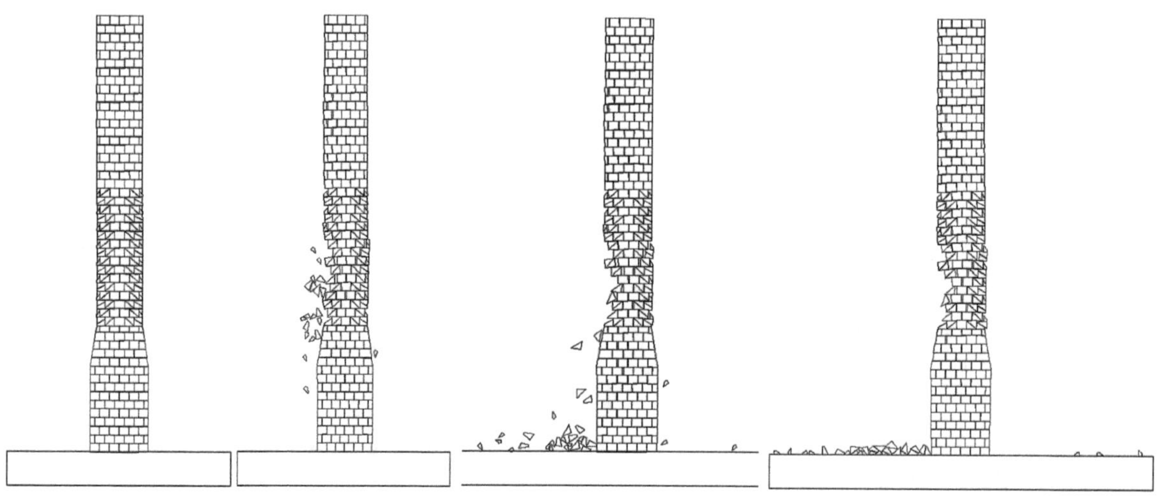

Figure 25. Stages of the collapse process for the second seismic shock (Romanian earthquake – PGA = 0.2g)

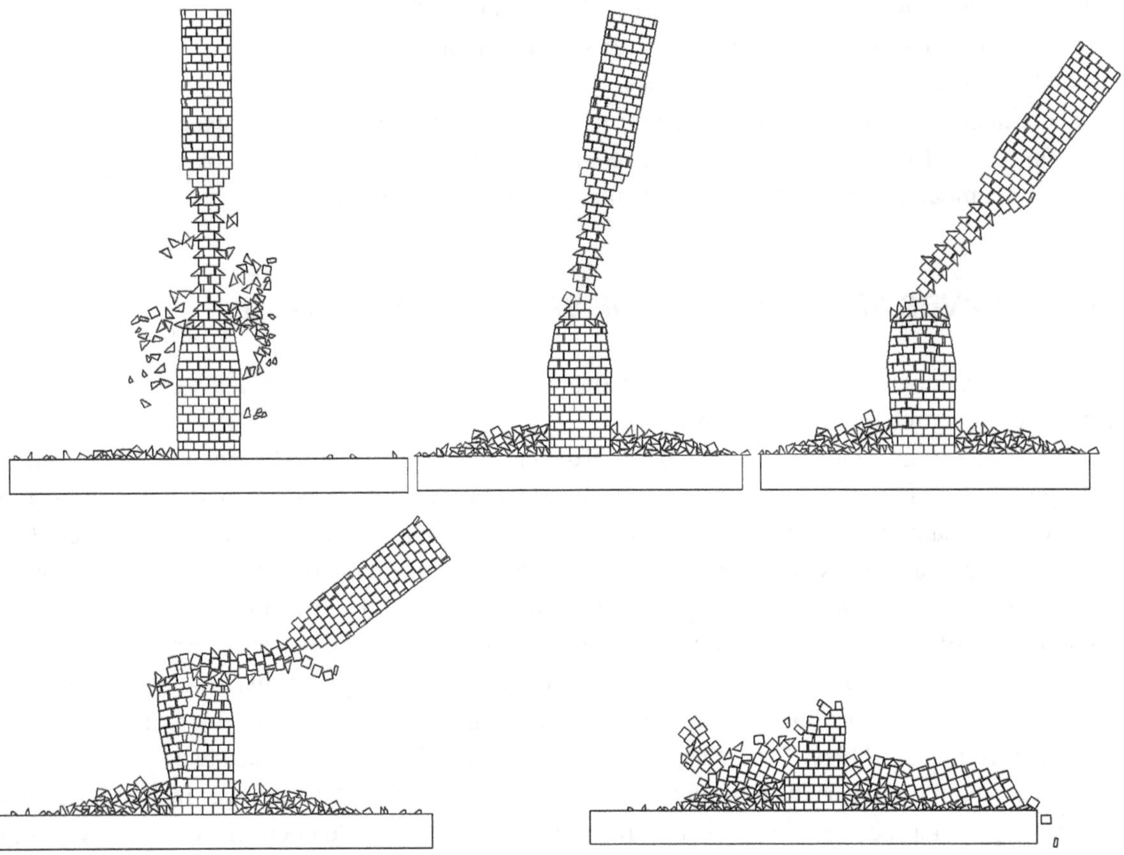

Table 4. Adopted mechanical properties for the blocks (elasto-plastic)

Cohesion (MPa)	Friction Angle	Dilation Angle	Tensile Strength(MPa)
3.0	40°	0	1.5

in the lower part of the structure where the stresses are higher. The very high normal stress (about 40 MPa), obtained in the case of elastic blocks, dropped to a value of about 16 MPa, showing thus the importance of considering the possible fracturing of the limestone shell for this type of structure. Probably the aqueduct would not withstand an earthquake similar to the 1977 Romanian earthquake, without major damage that could lead to collapse. The permanent displacements at the end of the run are in the range of 40 cm, as shown in Figure 26c.

For a PGA level equal to 0.3g, almost all the blocks have failed, as well as the joints and extremely large permanent displacements (about 1 m) were noticeable.

To conclude, the numerical analyses performed indicate that the type of block masonry pillar under study is able to withstand a significant seismic action, if the blocks remain intact. A possible mechanism of block failure was investigated, considering the existence of incipient fractures within some blocks, which led to structural collapse under a lower seismic input. The same conclusion can be drawn from the analysis done using elasto-plastic blocks.

The vulnerability of a damaged structure to a second earthquake was also examined, stressing the importance of defining damage indicators to characterise the progressive effects of seismic loading, and to allow a comparative evaluation of the results of numerical simulations.

MULTI-LEAVES ARCH-TYMPANA OF SAN FEDELE IN COMO

The second example of the UDEC utilization deals with the analysis of multi-leaves arch tympana systems used for church roofs in the Alps region.

As a matter of fact, the arch timpani of the Romanesque basilicas of the alpine region of Lombardy (Northern Italy) often present an unusual building technology, whose behavior is investigated in this Section. Two main parts can be distinguished in the structure: a multiple-leaf arch that is the effective bearing element, and the tympanum. They are made by means of quite different types of masonry.

The multiple-leaf arch has generally a strong stone masonry arch at the intrados, which can be seen from the nave of the church. Coupled to this stone arch, there is a brick masonry arch, connected to the stone arch by means of a thick (3-5 cm) mortar layer, usually a bad quality material. The tympanum is made by a poor masonry, that is a matrix of bad quality mortar including stones of small dimensions, sometimes rounded pebbles. The arch-tympani, spaced about 5 – 6 m along the nave, carry the longitudinal wood main girders of the roof, in this case study in five points. The cover is made by stone slabs 3-4 cm thick, on a secondary wood beam frame; therefore the final load on the arch-tympani is quite

Figure 26. The damaged structure (PGA = 0.2g)

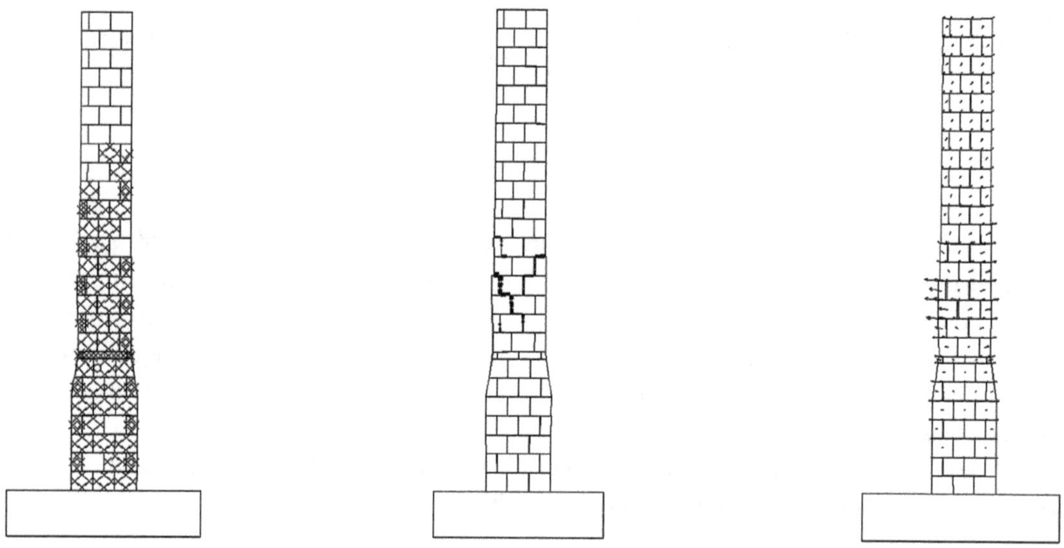

a) Deformed structure b) Opening and sliding c) Max. displ. 37.83 cm large. A false barrel vault covers the nave, so the covering system is not visible from inside, apart from the stone masonry arches. As a consequence, the direct observation of damages in the arch- tympani is not possible.

The mechanical behaviour of the structural element, particularly when subjected to dynamic loads, depends obviously from that of the different parts and on the characteristics of the connecting joints.

In a recent restoration activity of the arch-tympani of the Romanesque Basilica of San Fedele in Como (Northern Lombardy) an absence of contact in the central part of the arches was observed, between the stone masonry leaf and the brick masonry leaf, with significant opening of the joints.

Moreover the structure was weakened by the presence in tympani of openings (catwalks) for inspection and maintenance of the garret.

The chain restraints in the photograph of the inside of the church (Figure 27) regard only the false barrel vault; they were added in XIX century.

Assuming a small increase of the global bearing capacity of the system due to the presence of the masonry tympanum, a bad quality material, the resisting behaviour of the arch-tympanum can be considered that of the two layers round arch. It depends mainly on the interface characteristics, because of the lack of chains in the round arches of Romanesque Basilicas.

The absence of a specific system of absorption of the horizontal forces of the arches, which are resisted only by the large masses involved in the construction, makes the structural system more sensitive to base displacements, and the presence of different types of masonries more sensitive to season thermal gradient.

Figure 27. Inside of Basilica di San Fedele, in Como

In this research, the geometric shape of the structure taken into account is: 5 m internal arch radius, stone arch thickness 30 cm, brick arch thickness 25 cm.

Two main joint characteristics were investigated, namely the effective joint, modelled with a Coulomb friction behaviour and a no tangential stress joint, with only radial stress transfer.

The assumed loading condition and mechanical characteristics are aimed also to model the local separation between the arch layers. This kind of damage, particularly dangerous, was observed in real structural elements, mainly near to the keystones.

Probably this situation is due to the combined effect of the loads and of the ground displacements at the buttresses, resisting against the horizontal thrust of the arches.

In the analysis performed, the arch-tympani are assumed braced in the out-of-plane direction, for the presence of lateral walls and of the girders of the cover, so the static and dynamic loads act in the plane of the structure.

The analysis was aimed also to evaluate the influence of the weak mechanical characteristics of the tympanum masonry, which transmits the load of the roof girders to the multiple arches.

The tympanum wall behaves mainly as an additional load to the arch, which is the real and effective bearing structure. Its contribution to the global load carrying capacity is negligible, but it permits a better distribution of the loads from the girders on the multiple arches. Nevertheless from the tympanum masonry originates an additional horizontal thrust, resisted at each side by a stone masonry wall.

Moreover the mass of the tympanum wall is large, and surely affects the dynamic response of the system.

The geometry of the model, specific for San Fedele, is representative of the shape and dimensions of many Romanesque churches of the Alpine region: the central naves are generally about 10 - 11 m wide. Indeed at that time the master masons (the fairly famous *Magistri Cumacini*, just from the town of Como) respected the proportion and the dimensions of the various structural elements indicated by the traditional building experience as the best values with regards both to static and aesthetic reasons.

During the Romanesque period the mortar was composed by fine sand, whose granulometry was rather uniform, in a matrix of lime putty.

Its bad quality and degrade have a large influence on the resistance of the walls and of the joints between different kinds of masonry.

Numerical Modelling and Results

The study was carried out focusing the attention mainly on the influence of the mechanical properties of the joint between the two leaves. The analysis, both static and dynamic, was performed by means of *UDEC*.

The use of a large number of blocks, more than 1700, gives a quite refined model: the number and the dimensions of the blocks used are comparable with the real number and dimensions of the stones and bricks in the real structure.

The uncertain values of the model are the mechanical characteristics of the joints: therefore different models were performed, assuming sets of different joint properties. The results here presented give a comparison between a weak and a stronger material condition, which could be obtained, as an example, by injections. In the "weak" model the stone masonry joints cohesion was set equal to 0.50 MPa and the friction angle equal to 30 degrees; for the brick masonry, the intermediate arch mortar layer and the tympanum masonry the cohesion is respectively 0.3, 0.20, 0.25 MPa, and the friction angle is 25 degrees. A small tension resistance was considered, from 0.20 to 0.10 MPa.

The stronger model adopted has the same stone masonry characteristics, but larger values of strength for the other materials (cohesion 0.40 MPa for brick masonry, 0.30 MPa for the others elements, friction angle 30 degrees for all, a small increase in joint tensile resistance).

Weak Model

The results are represented in Figure 28 and Figure 29, where the opening of cracks is indicated.

Assuming for the exercise load of the roof on an arch-tympanum the value of 50 - 60 KN, as a rough estimate, transmitted by the five longitudinal wood girders, in the "weak" structural system at a static load about 1.5 times there are fractures in the element between the arch layers, at the imposts of the arch. Increasing the load, a limit condition is reached at about a 1.7 multiplier. In this situation there are fractures between the stone and brick masonry in the upper part of the arch (Figure 28). Figure 29 represents the initial collapse of the arch-tympanum at a load about 2.5 the assumed exercise load.

It should be pointed out that these results, that are a plausible evaluation of a safety coefficient, depend on the assumed mechanical characteristics of the joints; therefore they cannot be simply applied to real situations, without a proper material calibration. The "weak" model refers to a damaged situation due to a possible degrade of the materials, mainly the mortar between blocks, caused by aging, pollution, or other factors.

Figure 28. Fracture between stone and masonry arches

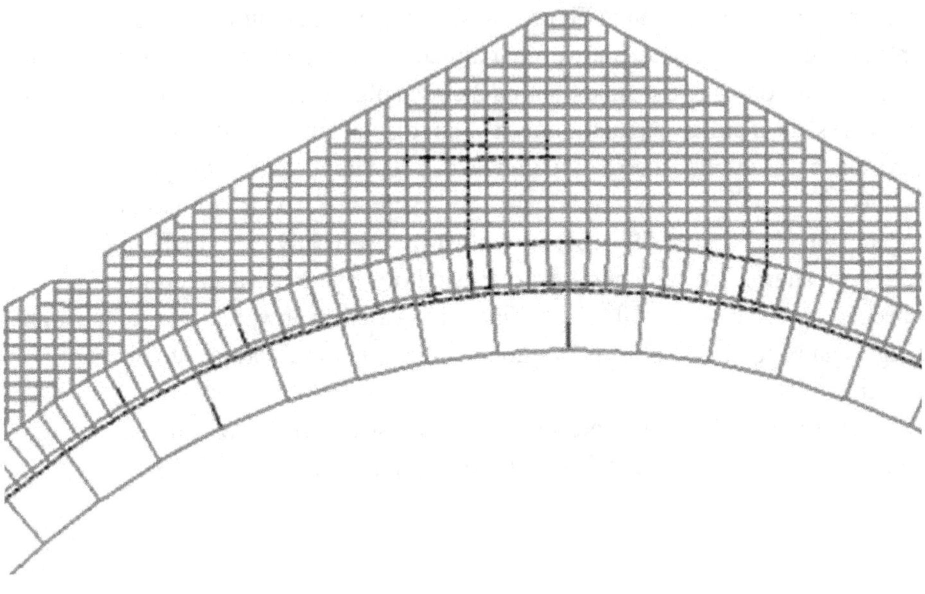

Figure 29. Collapse of the arch-tympanum due to static load

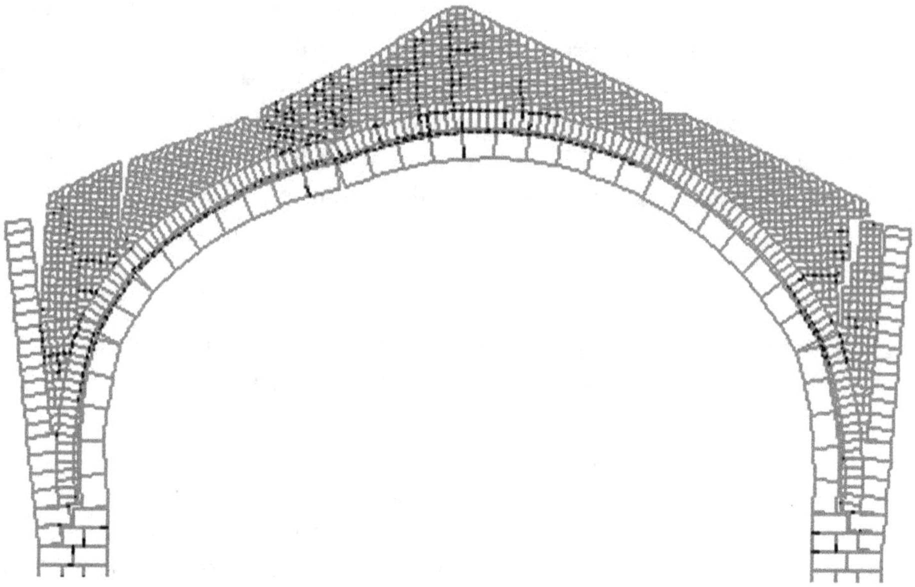

Main Model

The other set of parameters probably is more realistic for the description of the usual good conditions of the structure: at a load corresponding to a 1.5 load multiplier, no crack is detected, differently from the previous model.

In order to investigate the effect of differential base displacements, assuming a load multiplier equal to 1.25, the right buttress of the arch was subjected to a settlement. The structure appears quite sensitive to this kind of stress, as it can be seen in Figure 30, showing a limit equilibrium situation. The impressed displacements are 1cm in horizontal and vertical direction. A large fracture between the arch layers appears. A similar condition can be reached with a 2 cm displacement only in vertical direction.

Figure 31 represents a collapse situation for a 4 cm displacement in the two directions.

To arch-tympana was applied also a dynamic load, to get an insight in the possible behaviour in case of an earthquake. Nevertheless the numerical experimentation performed is not an effective seismic modelling of the church, because the acceleration was applied only to a small part of the global structure, the arch- tympanum. The dynamic load consisted of a synthetic accelerogram with PGA 14% of the gravity acceleration. The regional seismic hazard in Como is low, but this value can be considered corresponding to an adequate level of protection for the structure, taking into account the importance of the monument.

Though the external static loads were neglected, the structure shows almost no resistance to this dynamic loading, as it can be seen at different times in Figure 32 and Figure 33.

Figure 30. Cracks generated by right base displacement (detachments indicated in cyan color)

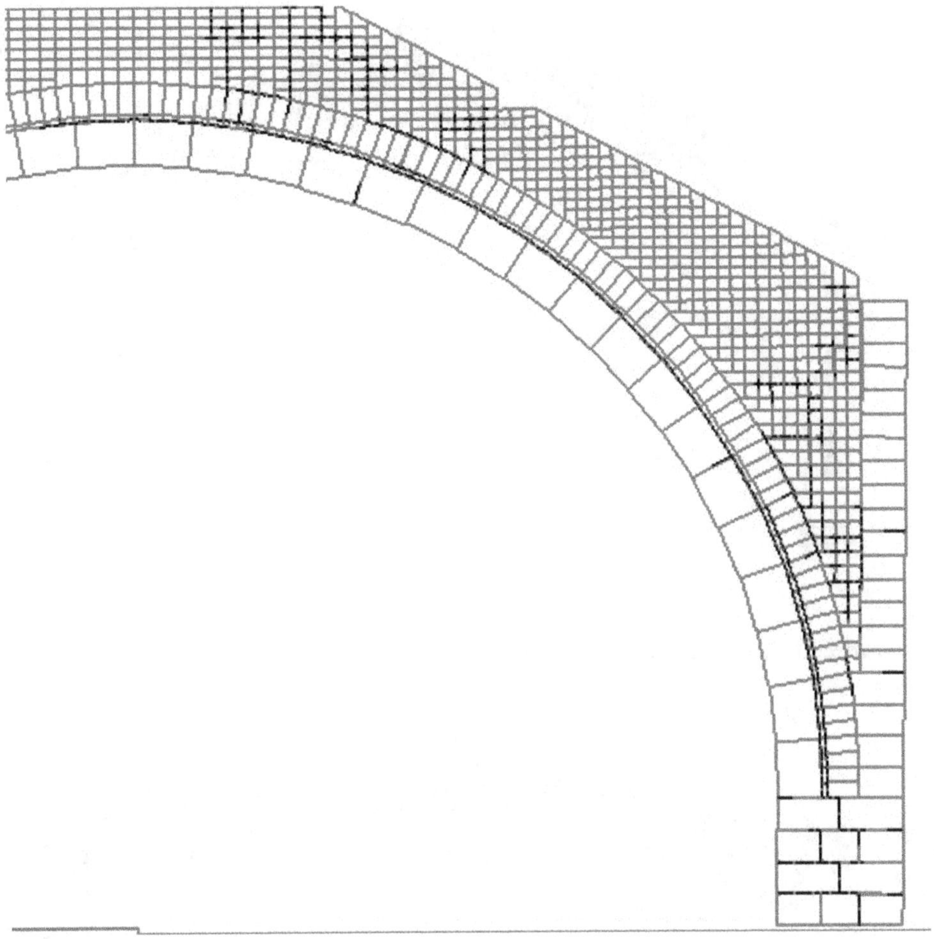

Figure 31. Collapse due to right base displacement

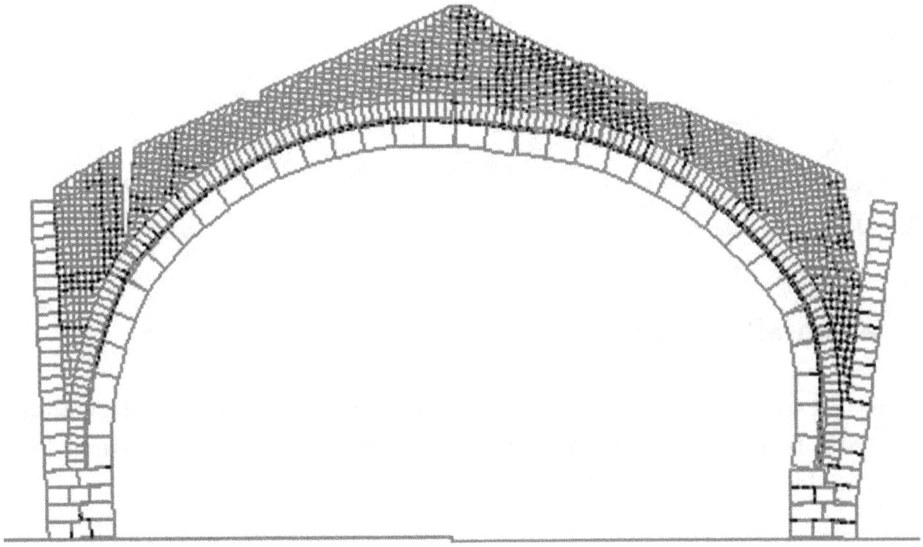

Figure 32. Collapse due to dynamic load, configuration at t = 3.4 sec

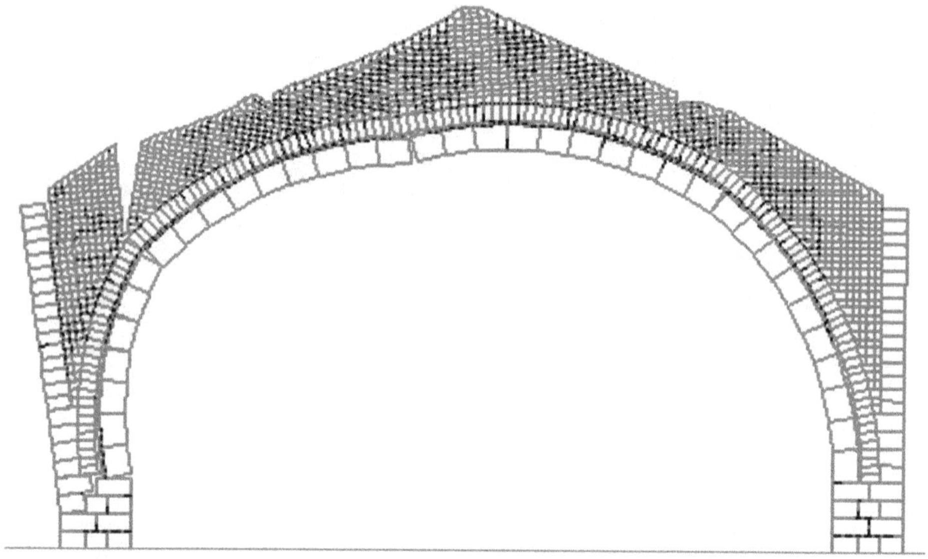

Figure 33. Collapse due to dynamic load, configuration at t = 4.0 sec

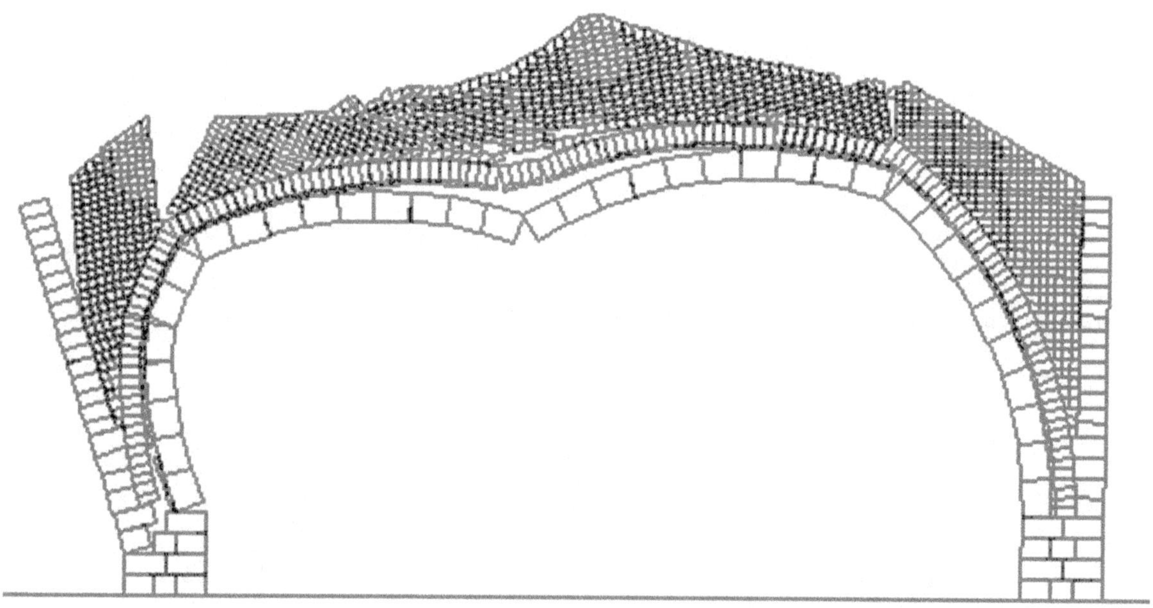

Arch-Tympanum with Catwalks

The effect of the presence of catwalks was modelled assuming the set of mechanical properties for the regular joint. A significant weakening of the structure can be underlined; at a 1.5 static load multiplier, there are fractures at the keystone and at the reins.

This structure with catwalk appears even more damaged than the corresponding "weak" model, without catwalk. Figure 34 reports the configuration generated by a 2 cm vertical displacement at the right impost. It must be pointed out that this is not an equilibrium condition, according to the model performed, the structure will undergo a collapse.

Non-Effective Intermediate Arch Layer

Generally the quality of the mortar used in medieval times was poor from the beginning; in time the effectiveness of the bound became very low. In a multiple-leaf arch an important consequence of this behaviour is the loss of tangential stress transfer between the two arches. A numerical model was performed to test this condition, assuming the joint mechanical properties of the weak model, and a very small (almost null) friction resistance at the interface between the arches.

A limit load condition is reached at a load multiplier about 0.5. In order to obtain an effective structure with a small increase of the external load (0.6 multiplier) it is necessary to provide a strong restraint of horizontal displacements, that is to apply a system of chains. Figure 35 shows a condition of an insufficient horizontal restraint, which causes displacements and separation of the layers, particularly close to the imposts of the arches.

It is important to notice that the presented results do not limit to this particular monument, but they are intended as a general survey of the expected behaviour of typical structural elements, quite uniform for the materials, the dimensions and the building technology, common in the churches of that historical period in northern Lombardy. For every single case the real behaviour depends naturally from the history of the monument, from the suffered degrade, from loading and restraint conditions. Nevertheless a comparison of the models performed and the observed configuration of damages in the real arch-tympani considered, suggests a good reliability of the obtained indication. The use of a distinct element program allows a quite realistic model of block structures like stone masonry monuments, and a plausible evaluation of the load multipliers for their collapse conditions.

Figure 34. Right base displacement in arch-tympanum with catwalk

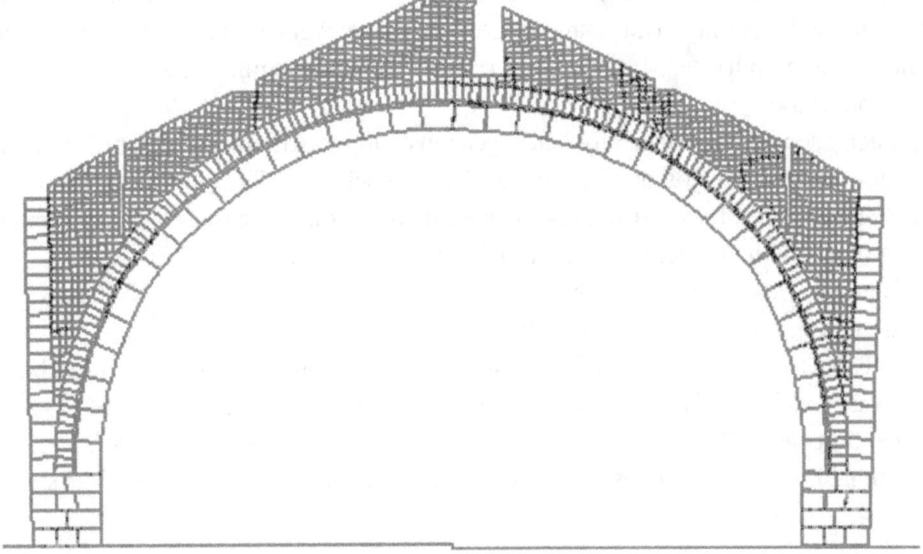

Figure 35. Non-effective intermediate layer

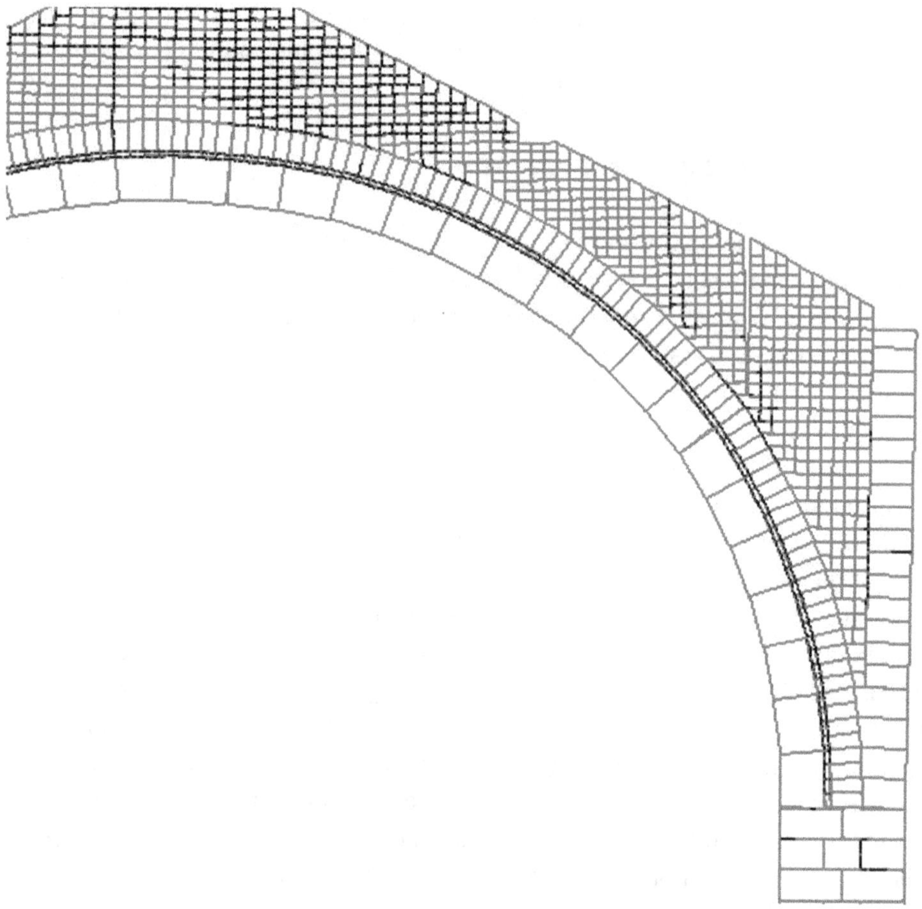

CONCLUSION

The Chapter presents two meaningful applications of the Distinct Element Method (DEM) code to real complex case studies belonging to the European built heritage: Águas Livres aqueduct in Lisbon (Portugal) and a multi-leaf masonry arch-tympanum carrying system belonging to the San Fedele Romanesque basilica in Como (Italy).

The approach adopted is intrinsically heterogeneous and the structure is discretized by rigid or deformable blocks connected by joints exhibiting a frictional behavior. The numerical solution well adapts to the dynamic analysis of blocky structures, because it consists into the integration of motion equations of each single block, with geometric and material non-linearity embedded.

The commercial software UDEC, which is a specific 2D implementation of the DEM theoretical formalization in a variety of engineering problems, is here used.

For Águas Livres, a meaningful portion of the structure –thought by the authors as being the most representative- is modelled and loaded with a series of accelerograms. The aim is an evaluation of both the seismic vulnerability as well as the most critical zones, thus giving suitable information on the dynamic behavior and rehabilitation techniques to be adopted. The out-of-plane behaviour of the slenderest pillar is also investigated under different hypothesis on the texture and loads applied. It is found, in agreement with intuition, that the out-of-plane strength of the system is much lower.

The second example is the arch-tympana carrying system belonging to the the San Fedele Romanesque basilica in Como (Italy). Arch-tympana carrying structural elements present an unusual building technology. Indeed, they are constituted by two main parts, the first is a multiple-leaf arch that is the effective bearing element, whereas the second is the tympanum, made by a mixture of bad quality mortar and stones of small dimensions. Again the structure is discretized into distinct elements and the load carrying capacity under dynamic excitation is evaluated, discussing the role played by the mechanical properties of the infill, eventual presence of catwalk, weak mechanical properties of the intermediate layer, and other structural and material characteristics.

REFERENCES

Azevedo, J., & Drei, A. (1995). *Characterisation of Typical Historical Constructions from a Structural Point of View: Application to the Seismic Behaviour assessment of the Águas Livres aqueduct in Lisbon*. Report IC-IST AI 8/95. Instituto Superior Técnico, Lisboa.

Casolo, S., & Milani, G. (2010). A simplified homogenization-discrete element model for the non-linear static analysis of masonry walls out-of-plane loaded. *Engineering Structures*, *32*(8), 2352–2366. doi:10.1016/j.engstruct.2010.04.010

Casolo, S., & Milani, G. (2013). Simplified out-of-plane modeling of three-leaf masonry walls accounting for the material texture. *Construction & Building Materials*, *40*, 330–351. doi:10.1016/j.conbuildmat.2012.09.090

Casolo, S., Milani, G., Uva, G., & Alessandri, C. (2013). Comparative seismic vulnerability analysis on ten masonry towers in the coastal Po Valley in Italy. *Engineering Structures*, *49*, 465–490. doi:10.1016/j.engstruct.2012.11.033

Cundall, P. A. (1971). A computer model for simulating progressive large scale movements in blocky rock systems. In *Proc. of the Symposium of the International Society of Rock Mechanics*.

Cundall, P. A. (1987). Distinct element models of rock and soil structure. In E. T. Brown (Ed.), *Analytical and Computational Methods in Engineering Rock Mechanics* (pp. 129–163). London: George Allen and Unwin.

Cundall, P. A. (1988). Formulation of a three-dimensional distinct element model – Part I: A scheme to detect and represent contacts in a system composed of many polyedral blocks'. *International Journal of Rock Mechanics and Mining Sciences*, *25*(3), 107–116. doi:10.1016/0148-9062(88)92293-0

Drei, A., & Fontana, A. (2001). Load carrying capacity of multiple-leaf masonry arches. In *Proc. 3rd Historical Construction Conference*.

Giamundo, V., Sarhosis, V., Lignola, G. P., Sheng, Y., & Manfredi, G. (2014). Evaluation of different computational modelling strategies for the analysis of low strength masonry structures. *Engineering Structures*, *73*, 160–169. doi:10.1016/j.engstruct.2014.05.007

Gilbert, M., Casapulla, C., & Ahmed, H. M. (2006). Limit analysis of masonry block structures with non-associative frictional joints using linear programming. *Computers & Structures*, *84*(13-14), 873–887. doi:10.1016/j.compstruc.2006.02.005

Gilbert, M., & Melbourne, C. (1994). Rigid-block analysis to masonry arches. *Structural Engineering*, *72*, 356–361.

Hart, R., Cundall, P. A., & Lemos, J. (1988). Formulation of a three-dimensional distinct element model-part II. mechanical calculations for motion and interaction of a system composed of many polyhedral blocks. *International Journal of Rock Mechanics and Mining Sciences*, *25*(3), 117–125. doi:10.1016/0148-9062(88)92294-2

I.N.C.M. (1983). *Regulamento de segurança e acções para estruturas de edifícios e pontes*. Lisbon.

Itasca Consulting Group, Inc. (1998). Three-Dimensional Distinct Element Code, version 2.0 – User's manual. Minneapolis, MN: Author.

Itasca Consulting Group, Inc. (2000). Universal Distinct Element Code– User's manual. Minneapolis, MN: Author.

Kuhn, M. R., & Bagi, K. (2009). Specimen size effect in discrete element simulations of granular assemblies. *Journal of Engineering Mechanics*, *135*(6), 485–492. doi:10.1061/(ASCE)0733-9399(2009)135:6(485)

Lemos, J. V. (1997). *Discrete element modelling of the seismic behaviour of stone masonry arches.*In G. N. Pande, J. Middleton, & B. Kralj (Eds.), *Computer Methods in Structural Masonry - 4* (pp. 220–227). London: E&FN Spon.

Lemos, J. V. (2007). Discrete element modelling of masonry structures. *International Journal of Architectural Heritage*, *1*(2), 190–213. doi:10.1080/15583050601176868

Lemos, J. V., Azevedo, F. S., Oliveira, C. S., & Sincraian, G. (1998). Three dimensional analysis of a block masonry pillar using discrete elements. In *Proc. Monument 98, Workshop on Seismic Performance of Monuments*.

Lengyel, G., & Bagi, K. (2015). Numerical analysis of the mechanical role of the ribs in groin vaults. *Computers & Structures*, *158*, 42–60. doi:10.1016/j.compstruc.2015.05.032

Milani, E., Milani, G., & Tralli, A. (2008). Limit analysis of masonry vaults by means of curved shell Finite Elements and homogenization. *International Journal of Solids and Structures*, *45*(20), 5258–5288. doi:10.1016/j.ijsolstr.2008.05.019

Milani, G. (2008). 3D upper bound limit analysis of multi-leaf masonry walls. *International Journal of Mechanical Sciences*, *50*(4), 817–836. doi:10.1016/j.ijmecsci.2007.11.003

Milani, G. (2011). Simple homogenization model for the non-linear analysis of in-plane loaded masonry walls. *Computers & Structures*, *89*(17-18), 1586–1601. doi:10.1016/j.compstruc.2011.05.004

Minghini, F., Milani, G., & Tralli, A. (2014). Seismic risk assessment of a 50m high masonry chimney using advanced analysis techniques. *Engineering Structures*, *69*, 255–270. doi:10.1016/j.engstruct.2014.03.028

O'Connor, K. M., & Dowding, C. M. (1992). Distinct Element Modeling and Analysis of Mining-induced Subsidence. *Rock Mechanics and Rock Engineering*, *25*(1), 1–24. doi:10.1007/BF01041873

Oliveira, C. S., Martins, A., & Lopes, M. S. (1995). Seismic studies for the 'Águas Livres' Aqueduct in Lisbon. In *Proc. 10th European Conf. on Earthquake Engineering*.

Oliveira, C. S., Portugal, A., Lopes, M. S., & Martins, A. (1991). *Comportamento sísmico do Aqueduto das Águas Livres, Report 202/91 NDA*. Lisboa, Portugal: Laboratório National de Engenharia Civil.

Orduna, A. (2006). Three-dimensional limit analysis of ancient masonry buildings with rigid block models. *WIT Transactions on the Built Environment*, *85*, 673–682.

Orduna, A., & Lourenço, P. B. (2005). Three-dimensional limit analysis of rigid blocks assemblages. Part II: Load-path following solution procedure and validation. *International Journal of Solids and Structures*, *42*(18-19), 5161–5180. doi:10.1016/j.ijsolstr.2005.02.011

Sarhosis, V., Garrity, S. W., & Sheng, Y. (2008). Distinct element modelling of masonry wall panels with openings.*9th International Conference on Computational Structures Technology, CST 2008*.

Sarhosis, V., Garrity, S. W., & Sheng, Y. (2015). Influence of brick-mortar interface on the mechanical behaviour of low bond strength masonry brickwork lintels. *Engineering Structures*, *88*, 1–11. doi:10.1016/j.engstruct.2014.12.014

Sarhosis, V., Tsavdaridis, K. D., & Giannopoulos, I. (2014). Discrete element modelling of masonry infilled steel frames with multiple window openings subjected to lateral load variations. *Open Construction and Building Technology Journal*, *8*(1), 93–103. doi:10.2174/1874836801408010093

Schiappa de Azevedo, F., Sousa, M. R., Reis, M. O. B., Casaca, J. M. M., Rodrigues, J. D., Veiga, M. R., & Silva, A. S. (1996). *Estudo dos efeitos das vibrações no Aqueduto das Águas Livres, produzidas pelas construção e tráfego do eixo Norte-Sul. Relatório 207/96 - NDA*. Lisboa: Laboratório Nacional de Engenharia Civil.

Sincraian, G. E. (2001). *Seismic behaviour of block masonry structures – A discrete element method approach*. (PhD Thesis). Instituto Superior Técnico, Lisbon, Portugal.

Sincraian, G. E., & Azevedo, J. J. (1997). Earthquake analysis of structural topologies with masonry infill panels using the distinct element method. In Proc. 3rd Encontro Sobre Sismologia e Engenharia Sismica.

Sincraian, G. E., & Azevedo, J. J. (1998). Numerical simulation of the seismic behavior of stone and brick masonry structures using the discrete element method. In *Proc. 11th European Conference on Earthquake Engineering*. Balkema.

Sincraian, G. E., & Lemos, J. V. (1998). *A discrete element program based on a rigid block formulation. Report 40/98- NDE/NEE*. Lisbon: Laboratorio National de Engenharia Civil.

Sincraian, G. E., & Lemos, J. V. (1999). Seismic analysis of a stone masonry acqueduct using discrete elements. In *Proc. 8th Canadian Conference on Earthquake Engineering*.

Sincraian, G. E., Lemos, J. V., & Oliveira, C. S. (1998). Assessment of the seismic behavior of a stone masonry acqueduct using the discrete element method. In *Proc. 11th European Conference on Earthquake Engineering*. Balkema.

Tóth, A. R., & Bagi, K. (2011).Analysis of a Lunar Base Structure Using the Discrete-Element Method. *Journal of Aerospace Engineering*, *24*(3), 397–401.

Tóth, A. R., Orbán, Z., & Bagi, K. (2009). Discrete element analysis of a stone masonry arch. *Mechanics Research Communications*, *36*(4), 469–480. doi:10.1016/j.mechrescom.2009.01.001

Underwood, P. (1983). Dynamic relaxation. In T. Belytschko & T. J. R. Hughes (Eds.), *Computational Methods for Transient Analysis* (pp. 1–65). North Holland.

Chapter 14
FEM/DEM Approach for the Analysis of Masonry Arch Bridges

Emanuele Reccia
University IUAV of Venice, Italy

Antonella Cecchi
University IUAV of Venice, Italy

Gabriele Milani
Technical University of Milan, Italy

ABSTRACT

The problem of masonry arch bridges load carrying capacity is studied by means of a coupled FEM/DEM 2D approach. The numerical model relies into a triangular discretization of the domain with embedded crack elements that activate whenever the peak strength is reached. The proposed approach can be regarded as a combination between Finite Elements allowing for the reproduction of elastic strain into continuum and DEM, suitable to model frictional cohesive behavior exhibited by masonry structures even at very low levels of external loads. The aforementioned numerical approach is applied to masonry arch bridges interacting with infill. A preliminary validation of the procedure is addressed for the prediction of the masonry arches limit state behavior where the stones are supposed infinite resistant and plastic hinges can occur exclusively on mortar joints, modeled as cohesive frictional interfaces. The sensitivity of the infill role varying mechanical properties of the infill is extensively discussed.

INTRODUCTION

In this Chapter, the feasibility of the utilization of a combined Finite Element/Discrete Element (FEM/DEM) approach to investigate the behavior of masonry arch bridges is assessed. In particular, the Chapter proposes and discusses a possible approach to FEM/DEM modelling of two existing masonry arch bridges. Attention is paid to the assessment of the load carrying capacity of the structures by means of a suitable coupled FEM/DEM 2D approach.

DOI: 10.4018/978-1-5225-0231-9.ch014

As a matter of fact, Finite Element Method (FEM) is the most widely used numerical tool in computational solid mechanics. However, some decades ago, ad-hoc computational methods have been developed to deal with particulates, jointed rock, granular flows and problems where the so called emergent properties of a system are a result of interaction between large numbers of individual solid particles: the most widely used method for a large class of these problems is the so called Discrete Element Method (DEM). This latter approach well adapts to all those non-linear problems characterized by the mutual movement of rigid bodies, eventually interacting by means of both contact and friction, also under large displacement hypotheses. It is therefore quite intuitive to understand why DEM received growing attention in the field of both soil mechanics and masonry modelling, due to the intrinsic capability of the procedure to tackle the most important and distinctive aspects of their behavior, especially in the non-linear dynamic range, but also when limit analysis computations are carried out.

In the early 1990s FEM and DEM methods were combined and the resulting method was termed as combined FEM/DEM method. The first attempts are due, among the others, by Munjiza (2004), who shown how to join the advantages of FEM and the ones of DEM into a single numerical tool. FEM/DEM is in its essence a Discrete Element Method where the Discrete Elements are meshed into Finite Elements. Finite Elements allow to model elastic deformation (but also crushing on blocks when required), while Discrete Element algorithms allow modeling interaction, fracture and fragmentation processes in a very straightforward fashion.

FEM/DEM provides a consistent procedure to study masonry structures, thanks to the possibility of creating models made by separated blocks. In particular, these models result particularly suited to analyze quickly but accurately the behavior of historical masonry constructions (Reccia, Cecchi & Cazzani, 2012; Smoljanović, Živaljić & Nikolić, 2013; Baraldi, Reccia, Cazzani, & Cecchi, 2013; Baraldi, Reccia & Cecchi, 2015).

In the present Chapter, the combination of DEM and FEM provided by the open source code Y2D, developed again by Munjiza (2004), is utilized to study masonry arch bridges behavior interacting with the infill and subjected to increasing static loads up to failure. Typically, the problems show non-linearity at early stages of the application of the external loads. Blocks are usually either clay bricks or stones, showing excellent compression strength and high elastic modulus. As a consequence, they can be conveniently modeled as very stiff elastic bodies, while mortar joints might be idealized as elastic or elastic-plastic zero-thickness Mohr-Coulomb interfaces.

Briefly, the advantages of the coupled FEM/DEM method are:

1. It allows modelling both deformable or rigid bodies – in this chapter this aspect is highlighted modelling the arch voussoirs as a rigid bodies while backfill as a deformable material;
2. Cracks elements are embedded between all the Finite Element of the mesh, therefore joints may be modelled in order to prefigure fixed crack patterns – like in the case of masonry arches, where cracking may occur only between voussoirs – or random crack patterns – like in the case of backfill, where cracks may occur everywhere, according with the mesh geometry and with any orientation.

Masonry arch bridges are among the most ancient and best-preserved historical structures. Their construction dates back to the dawn of the history and their development has gone hand in hand with the technological advancement. Masonry arch bridges are a remarkable evidence of the engineering progress and the technological achievement and skills the mankind has developed over the centuries: they are an essential part of the architectural historical heritage. Their presence is a characteristic feature of the European landscape.

Moreover the European railroad networks include thousands of masonry arch bridges, mainly built during the XIX century, that are still in exercise, as reported by UIC (2015). The present rail traffic is heavier than in the past. While in case of monumental bridges, which its historic value is recognized, the performance requirements may be sacrificed in order to ensure the conservation, in case of masonry arch railway bridges their conservation is guaranteed by their functioning. The capacity of historical masonry arch rail-bridges to carry the actual traffic must be verified. Considering the great number of those bridges, it needs to develop methodologies of analysis that must be reliable and fast.

In this view, applicability of FEM/DEM method to the study of masonry arch bridge may be an important tool for the assessment of the load carrying capacity. It may provide a reliable evaluation of safety of masonry arch bridges, by providing kinematic mechanisms of collapse taking into account the mechanical properties of joints and the deformability of the elements, and not only the geometry, like standard limit analysis. Moreover, respect to standard FEM analysis, FEM/DEM may be more suitable to catch the intrinsic behavior of masonry material, thanks to the possibility to model units and joints.

The chapter aims at outlining fields and limits of applicability of the FEM/DEM method to the study of masonry arch bridges. The main target is to evaluate the applicability of FEM/DEM, in particular its reliability to describe the nonlinear behavior of masonry arch bridges under increasing static loads, to catch kinematic failure mechanisms and collapse load multipliers, as well as to evaluate the role played by the backfill.

FEM/DEM models of two masonry arch bridges have been made:

1. The Prestwood Bridge (UK), which have been tested up to collapse at the beginning of 90's, and which is recognized as a standard benchmark (Page, 1993; Cavicchi & Gambarotta, 2005, 2006, 2007) to test any numerical model;
2. The Venice trans-Lagoon Bridge, which has been already studied by the authors (Reccia, Cecchi & Tralli, 2013; Reccia, Milani, Cecchi, & Tralli, 2014), which represents an important piece of heritage in the European panorama of arch bridges.

Whit such a purpose, a series of parametric analyses has been conducted in order to evaluate the influence of the different parameters involved on the behavior of the bridges. Pushover analyses have been performed to investigate the nonlinear behavior up to the collapse and a up to a clear formation of a failure mechanism in the model. Attention has been paid to the mechanical properties of joints, which have the strongest influence on the analyses results. Several numerical simulations have been carried out varying the parameters of joints, mainly cohesion and friction.

When fulfilling the main target, some additional goals have been also reached. The models have been initially tuned studying the single masonry arches without infill. Afterwards, models have been made more and more complex studying the backfill influence in the most realistic manner, i.e. with a 2D discretization, and finding, in agreement with consolidated literature in the field, the relevant role played on the global behavior and on the load carrying capacity increase, also when horizontal loads simulating a seismic action are applied.

SCIENTIFIC BACKGROUND

A so called masonry like-material is a heterogeneous and anisotropic material exhibiting non-linear behavior. Masonry may be formed by an ordered set of interconnected blocks, joined together by means of dry or mortar joints. The properties of masonry are strongly dependent on the properties of its individual constituents: blocks and mortar joints and their arrangement.

Briefly it is possible to point out the main mechanical characteristics of masonry:

1. Low and uncertain resistance to tensile stress;
2. Quite good compressive strength;
3. Shear strength depending by compression, on the base of Coulomb's law on friction;
4. Diagram load/displacement elastic-linear for very low loads, but that turns to nonlinear behavior just for loads that are far away from the last ones.

All these aspects can considerably change in base of specific characteristics of each masonry, such as quality of components, dimension of units, thickness of joints, arrangement of units. And the characteristics of masonry have a strong influence on the structural behavior of masonry buildings.

The difficulty in modelling such structures depends on three fundamental problems: the composite nature of masonry, made up of a complex system of blocks and joints; the size of heterogeneity respect to the size of masonry structure; several geometric complexities of masonry constructions, which impose the adoption of 2D and 3D modelling approaches.

In literature, many models and tools of analysis have been developed. They may be distinguished by the scale of the problem faced, constructive features, masonry type, acting forces.

At micro-scale, masonry is modelled as a discrete system of elements: blocks, joints and/or interfaces. Attention is devoted to interactions between blocks through mortar joints that may be modelled as interface or continuum material characterized by either a linear or a nonlinear response (Lourenço & Rots, 1997; Alpa & Monetto, 1994; Lofti & Benson Shing, 1994; Masian & Trovalusci, 1996; de Buhan & de Felice, 1997; Cecchi & Sab, 2004).

At the meso-scale masonry is modelled as an equivalent continuum, usually obtained by homogenization procedures: Cauchy continuum, Cosserat continuum or higher order continua (Anthoine, 1995; Masiani, Rizzi & Trovalusci, 1995; Massart, Peerlings & Geers, 2007). In this way linear analysis (Cecchi & Sab, 2002a, 2002b, 2004; Salerno & de Felice, 2009; Baraldi & Cecchi, 2014; Baraldi, Cecchi & Tralli, 2015), non-linear analysis (Luciano & Sacco, 1997; Shieh-Beygi & Pietruszczak, 2008) or limit analysis (Milani, Lourenço & Tralli, 2006; Sab, 2003; Cecchi & Milani, 2008) have been proposed.

At the macro-scale level the characteristic length is represented by macro-elements like masonry panels or portion of structure (Turnšek & Cačovič, 1971; Tomaževic & Turnšek, 1980; Turnšek & Sheppard, 1980; Braga & Dolce, 1982; D'Asdia & Viskovi, 1994; Gambarotta & Lagomarsino, 1997a, 1997b; Magenes & Calvi, 1997).

In particular, at micro-scale, some approaches are based on the Discrete Element Method, like Cundall (1976), therefore require a dynamic incremental analysis performed through the explicit integration of the equations of motion (Azevedo, Sincraian & Lemos, 2000; de Felice & Giannini, 2001). Most of these approaches have been developed to model periodic regular masonries. A complete and exhaustive overview of the different possible approaches to the modelling of masonry structures by means of Discrete Elements has been provided by Lemos (2007).

In recent times an increasing number of models attempted to combine the advantages of Finite and Discrete Element methods. Cundall (1988) and Hart, Cundall and Lemos (1988) adopted deformable blocks with an internally Finite Element mesh with 2D triangular plane strain elements. Barbosa (1996) proposed a Discrete-Finite Element model where deformable blocks are meshed by quadrilateral isoparametric Finite Element. Shi and Goodman (1988) developed a discontinous deformation analysis method where deformable blocks are assumed to be in a state of uniform strain and stress.

One of the approaches which combines the advantages of the Finite and Discrete Element method is a combined Finite-Discrete Element method (FEM/DEM) proposed by Munjiza (Munjiza, Owen & Bicanic, 1995; Munjiza, 2004). FEM/DEM Y2D code has been recently updated by the Geo Group of the Toronto University (Mahabadi, Grasselli, & Munjiza, 2010; Mahabadi, Lisjak, Munjiza, & Grasselli, 2012). In the last years, FEM/DEM method has been successfully adopted to study the behavior of historical masonry structures (Reccia, Cecchi & Cazzani, 2012; Smoljanović, Živaljić & Nikolić, 2013; Baraldi, Reccia, Cazzani & Cecchi, 2013; Baraldi, Reccia & Cecchi, 2015).

At present, a large amount of literature regarding the analysis up to collapse of masonry arch bridges and masonry arches in general is present.

The masonry arch is the main element of the historic bridge, which connotes its form and structure and defines its architectural and engineering characteristics. The structural behavior of the whole bridge is strongly dependent by the behavior of the masonry arch, which provide the main contribution to the load-bearing capacity, although not the only one.

The masonry arch is a structure made of wedge shaped blocks – stones or bricks – called voussoirs, placed one next to the others, with or without mortar joints, in order to precisely create an arch ring. In large-span arch voussoirs are usually stones accurately cut and assembled without mortar, or just with a minimum of it. Ancient arches built by Romans were usually made like this, such as many bridges built from renaissance to the XIX century. Instead small arches may be realized using stones roughly cut or bricks assembled with mortar joints.

Many authors dealt with structural behavior of masonry arch. Besides the historic rules (Brencich & Morbiducci, 2007), the classic approach to determine the stability of arch bridges is probably due to Pippard and Ashby (1936), Pippard (1948) and Heyman (1969). Heyman (1969) was the first to extend in a clear and explicit way both the kinematic and static theorems of limit analysis to masonry arches, according to which the structure is safe if a thrust line inner to the arch depth can be determined in equilibrium with the external loads.

As previously stated, the most common idealizations of masonry material behavior are elastic, non-linear elastic and elastic plastic (Lourenço, Milani, Tralli & Zucchini, 2007), but in the case of masonry arch bridges and curved structures in general the most diffused approach still remains limit analysis (Gilbert, 2006; Drosopoulos, Stavroulakis & Massalas, 2006). Several rigid blocks analysis methods have been developed to study the behavior at collapse of masonry arch (Gilbert & Melbourne, 1994; Hughes & Blackler, 1997; Boothby, 1995). This approach is based on a rigid block discretization of the arches within limit analysis concepts coupled with Finite Elements. While such an approach is very appealing because it provides failure mechanisms and load multipliers for a variety of different 2D geometries and loading conditions, still it is based on strong simplifications and consider the role played by the backfill only in an approximate way.

To rigorously investigate the role played by the backfill in the determination of the actual load carrying capacity of 2D bridges, a discretization with plane strain rigid-plastic elements and interfaces is needed, as recently proposed by Cavicchi and Gambarotta (2005, 2006, 2007). The role of backfill respect to both service and ultimate loads and the transversal effect of load have been studied by the authors (Reccia, Milani, Cecchi & Tralli, 2014).

Discrete models and FEM/DEM model have been proposed for stone arch and masonry arch bridges by several authors (Mirabella Roberti & Calvetti, 1998; Owen, Peric, Petrinic, Brookes & James, 1998; Audenaert, Fanning, Sobczak & Peremans, 2008; Tòth, Orbàn & Bagi, 2009).

FEM/DEM BASIC ASSUMPTIONS FOR MASONRY ARCH BRIDGES

The approach to FEM/DEM analysis of masonry arch bridges is here presented. The numerical model relies into a triangular discretization of the domain with embedded crack elements that activate whenever the peak strength is reached.

The analyses have been performed by means of the FEM/DEM Y2D code (Munjiza, 2004), in particular using the Y-GUI (Mahabadi, Grasselli & Munjiza, 2010) to prepare the input file and the Y-Geo code developed by the Geo Group of the Toronto University (Mahabadi, Lisjak, Munjiza & Grasselli, 2012) to run the analyses. Analyses are performed under 2D plane stress conditions.

Bridges have been modelled adopting Finite Element meshes. The meshes have been realized by means of a FEM commercial code, namely Straus7, and then imported in the input file. Results have been processed by means of ad-hoc MatLab scripts and by means of spreadsheets.

The properties adopted for Finite Elements are Young's modulus E^B and Poisson's coefficient ν, density ρ and viscous damping μ. The viscous damping μ depends by the mechanical properties – Young's modulus E and density ρ – and by the dimension of finite elements. On the base of these values the viscous damping adopted is equal to:

$$\mu = 2\xi\left(\frac{1}{k}\right)\sqrt{E^B\rho} \tag{1}$$

where ξ is the damping ratio, which for critical damping is equal to 1, and k is the inverse of the wave length, which has the same order of the height of elements h.

In order to avoid compenetration of blocks, a penalty contact parameter is adopted. The penalty contact is equal to the Young's modulus E^B, and the tangential penalty adopted is equal to its half.

Time step size, which is a fundamental parameter of the analysis depends both by mechanical properties and density and by the mesh, in particular the dimension of finite elements. Critical time step T_c has been calculated as:

$$T_c = \frac{2}{\omega}\left[\sqrt{\left(1+\xi^2\right)} - \xi\right] \tag{2}$$

where ω is the angular frequency, which is given by:

$$\omega = \frac{1}{h}\sqrt{\frac{E^B}{\rho}} \tag{3}$$

where $\sqrt{E / \rho}$ is the wave propagation speed inside the elements and h the minimum height of elements.

A series of parametric analyses have been performed to evaluate the influence of the different mechanical parameters adopted for the joints respect to the behavior of the bridge. Joints are modelled as elastic-plastic Mohr-Coulomb interfaces. Parametric analyses have been carried out varying the parameters involved: cohesion c and friction φ, and then the other related parameters, tensile strength τ and fracture energy GI_C (first mode) and GII_C (second mode) which have been calculated on the base of the values of cohesion and friction adopted (Carpinteri, 1992).

Tensile strength has been calculated on the base of Mohr-Coulomb criterion as

$$\tau = c \,/\, \tan\varphi \tag{4}$$

The fracture energy G_C characterizes the softening behavior, in particular its value governs the non-linear behavior the crack elements embedded between each finite element of the mesh. Fracture energy of first mode GI_C concerns de-cohesion mechanism of joint crack, it has been calculated as

$$GI_C = \left(a \cdot \pi \cdot c^2\right) / \, E^M \tag{5}$$

where a is the dimension of the interface and E^M the Young's modulus of the joints. In the model the joints are modeled as zero-thickness interfaces, therefore the young's modulus E^M that appears in the formula adopted to calculate fracture energy represents the hypothetic modulus of the mortar joints.

Fracture energy of second mode GII_C concerns the slippage mechanism of joint crack, therefore has been calculated in the same way as first mode fracture energy but using the value of tensile strength instead that of cohesion

$$GII_C = \left(a \cdot \pi \cdot \tau^2\right) / \, E^M \tag{6}$$

A first parametric analysis has been performed on the arch without taking into account the backfill.

The arch has been divided in wedges, called voussoirs, separated by joints modeled as elastic-plastic Mohr-Coulomb interface. Voussoirs are considered infinite rigid, by adopting a very high value of Young's Modulus E^B –1000 time higher then the value of Young's modulus adopted for mortar joints – a Poisson's coefficient ν equal to zero and a very strong internal joints, in order to avoid cracking inside the voussirs, that may occur only in the joints between them. This assumption is suitable to describe the behavior of historical masonry arch.

The parametric analysis has been performed varying the mechanical properties of the joints between the voussoirs; therefore different values of cohesion c and friction φ have been adopted. The purpose is to assess the value of the collapse load multiplier and the nonlinear behavior of the arch and to evaluate how they change at the increasing of the mechanical parameters of the joints. Therefore, a pushover analysis has been performed for each value of cohesion and friction adopted and then the results of the different analyses have been compared.

The pushover analysis has been performed increasing the value of the horizontal load applied until the collapse of the arch. The horizontal load has been considered as an increasing ratio of the vertical load applied, which represents the self-weight of the arch. Therefore, several incremental nonlinear dynamic analyses have been performed for each increasing value of horizontal loads.

Afterwards, a parametric analysis has been carried out in order to evaluate the backfill role on the global behavior of masonry arch bridges.

In backfill, cracking may occur everywhere: crack elements are embedded at the interface of all elements of backfill and only one set of mechanical properties for the backfill joint is considered (differently from the arch, in which internal joints of voussoirs and joints between voussoirs are modelled with different mechanical properties, in order to allow cracking only between voussoirs). The position of crack depends on the mesh: fine meshes should be preferred, but taking into account also the computational costs needed.

A very high value of friction angle has been adopted for backfill, $\varphi = 50°$: the adoption of a high angle of friction provides a value of tensile strength lower than cohesion, which seems to be appropriate to simulate the behavior of incoherent filling. The properties of backfill internal joints have been adopted also to model arch-fill interaction.

Parametric analyses have been performed on two study cases, namely the Prestwood Bridge and the Venice trans-Lagoon Bridge.

PRESTWOOD BRIDGE

In order to evaluate the applicability and reliability of the FEM/DEM method to study the behavior of masonry arch bridges, a first analysis has been performed on the Prestwood Bridge, Figure 1.

The Prestwood Bridge, located at Staffordshire UK, is a single span bridge that has been tested up to collapse within an experimental research on masonry bridge supported by the Transport Research Laboratory (Page, 1993). The load was applied across the bridge at quarter of the span. The configuration of the bridge just before collapse is shown in Figure 2. The experimental collapse load was equal to 228kN and the collapse occurred exhibiting a four hinges mechanism. The test on full-scale bridge

Figure 1. The Prestwood Bridge, Staffordshire, UK

has highlighted the strong influence of fill and spandrels on the collapse mechanisms and the load carrying capacity. For this reason, the Prestwood Bridge has become a benchmark and many researchers have studied it, in particular in order to assess the effect of backfill to the global behavior of the bridge (Cavicchi & Gambarotta, 2005, 2006, 2007).

On the base of such a premise, the Prestwood Bridge has been chosen as a first case of study to evaluate the applicability of FEM/DEM method to the study of masonry arch bridges and to qualitatively validate it. In particular, FEM/DEM analyses have been performed on the bridge in order to verify the reliability of the method to catch the behavior at collapse of the bridge, but in presence of increasing horizontal loads representing a seismic excitation in a static fashion. As a consequence, the comparison with the experimental collapse mechanism may be only qualitative, because the load system is different. Initially, an analysis has been performed only on the arch, while in the sequel other analyses have been done taking into account also the backfill, in order to evaluate its influence on the global behavior of the bridge.

The geometry of the bridge is shown in Figure 3. The span is 6.55 m and the rise is 1.43 m, the arch has a span/rise $R_{s/r}$ ratio equal to about 1/5. The vault has a curvature radius of 4.69 m and a thickness of 0.22 m and it is made by a single ring of bricks, laid as headers. The joints between the bricks are made of mortar. The ratio thickness/span $R_{t/s}$ is about 1/30. The backfill height at the crown is 0.17 m. The bridge has no piers: the arch rests directly on abutments.

Figure 2. Collapse of Prestwood Bridge
(Page, 1993)

Figure 3. Geometry of the Prestwood Bridge, dimensions in meters

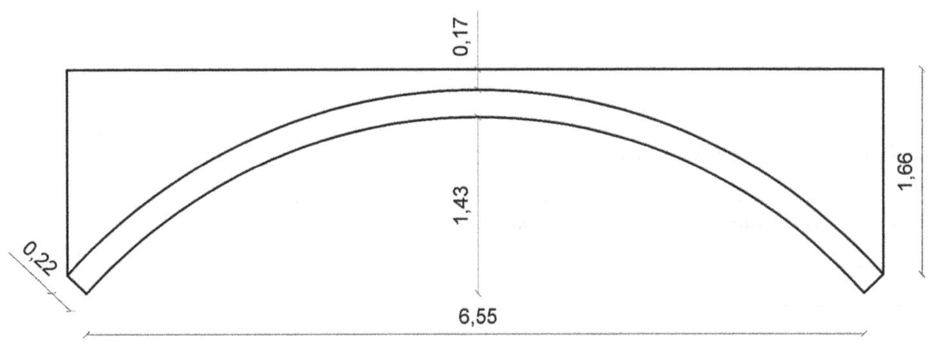

In the model the arch has been divided into 50 wedges having the same dimension. The depth considered is equal to 1 m. Load is applied in the center of blocks, each one meshed by 4 CST triangular bi-dimensional finite elements, in the nodes at the cross of the four elements. The mesh adopted is reported in Figure 4.

Thus in each block both the self-weight and the horizontal load on the base of the multiplier used are applied. The density adopted for blocks is $\rho = 20$ kN/m^3. Being the volume of each block V_B equal to (0.22 * 0.15 * 1) = 0.0331 m^3, the vertical load applied is equal to $P_B = 0.662$ kN. The whole vertical load applied is equal to $P = 33.1$ kN. The horizontal load applied is equal to $H_B = \lambda P_B$, where λ increases up to the value needed to activate the collapse mechanism of the bridge.

Two additional blocks are added at the abutments, that are restrained by 2 hinges each one in the external nodes, in order to simulate a pinned connection to the ground. Therefore the arch is connected to the abutments by the same joint adopted between the arch wedges: it allows the possibility of developing kinematic mechanisms showing hinges at abutments, as often occurs in reality. The forces applied and the boundary conditions have been reported in Figure 4.

A previously said, in order to simulate rigid blocks, a very high value of Young's modulus E^B has been adopted with Poisson's ratio ν=0. Viscous damping has been calculated by Equation 1. All the parameters adopted for the blocks are reported in Table 1.

The adopted time step size is $\delta t = 1/10$ of the critical time step T_c, therefore $\delta t = 2.26$E-07 s. Analysis ran for 1E+06 time steps.

As previously said, internal joints inside blocks are modeled to avoid cracking, while the joints between blocks are modeled varying cohesion and friction. Three values of cohesion have been used: $c = 0.10$ MPa, $c = 0.25$ MPa and c = 0.50 MPa. Two values of angle of friction have been adopted: $\varphi = 30°$ and $\varphi = 37°$. All the other parameters of joint vary accordingly with cohesion and friction and have been calculated using Equations 4, 5 and 6. All the mechanical parameters adopted for joints are reported

Figure 4. FEM/DEM model of the Prestwood Bridge arch (a) and loads applied (b)

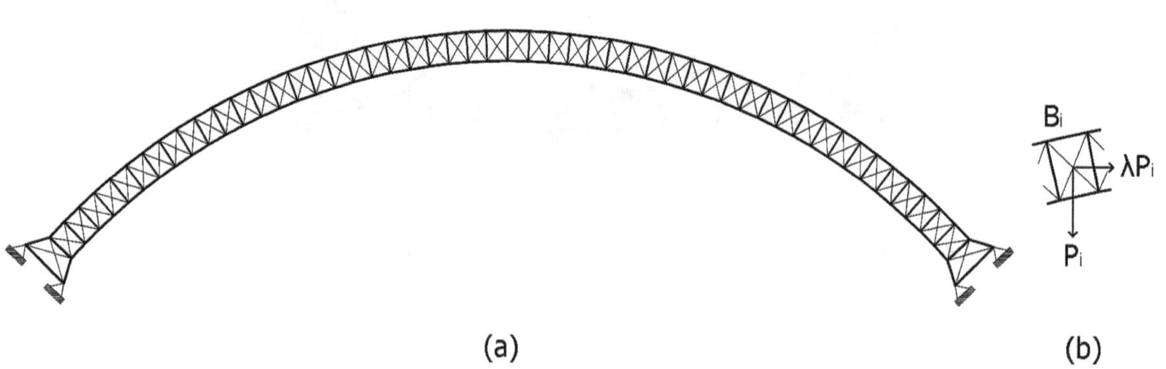

Table 1. Mechanical properties of blocks

Young Modulus E^B (MPa)	Poisson Ratio ν	Viscous Damping μ	Density ρ (kN/m^3)	Contact Penalty (MPa)	Tangential Penalty (MPa)
2E+06	0	1.26E+04	20	2E+06	1E+05

Table 2. Mechanical properties of joints adopted for Prestwood Bridge arch

Cohesion *c* (MPa)	Friction *φ* (°)	Tensile strength *τ* (MPa)	Fracture Energy I GI_c	Fracture Energy II GII_c
0.10	30	0.17	6.91E-06	2.07E-05
0.25		0.43	4.32E-05	1.29E-04
0.50		0.86	1.72E-04	5.18E-04
0.10	37	0.13	6.91E-06	1.27E-05
0.25		0.33	4.32E-05	7.60E-05
0.50		0.66	1.72E-04	3.04E-04

in Table 2. In particular, the dimension a of the interface is equal to the arch thickness, therefore a = 0.22 m. The value adopted for the elastic modulus is E^M = 1000MPa, which is a typical value for mortar (Baraldi, Reccia & Cecchi, 2015).

The parametric analysis shows that increasing the mechanical properties of joints, the load carrying capacity of the arch increases, as expected. Graphs of the load-displacement curves for the different values of cohesion and friction are reported in Figure 5 and Figure 6. Vertical and horizontal displacements are calculated at crown. They clearly show that the arch begins to behave non-linearly later and the collapse load multiplier increases as cohesion increases. Being the blocks infinite rigid, so having a very high value of Young's modulus E^B, the displacements obtained are very small.

A comparison between the results obtained with different values of friction φ = 30° and φ = 37° keeping the same value of cohesion c = 0.25 MPa is reported in Figure 7.

For all the values of cohesion and friction, the collapse mechanism occurred exhibits 4 hinges, alternatively disposed at intrados and extrados of the arch, with first hinge at extrados of left abutment and fourth hinge at intrados of right abutment. Mechanism obtained is in good agreement with the ones that usually occur in masonry arch. The deformed shape at collapse obtained with c = 0.25 MPa and φ = *37°* is reported in Figure 8.

Figure 5. Vertical (a) and horizontal (b) displacements at crown of the Prestwood Bridge arch for different values of cohesion c and constant friction φ = 30°

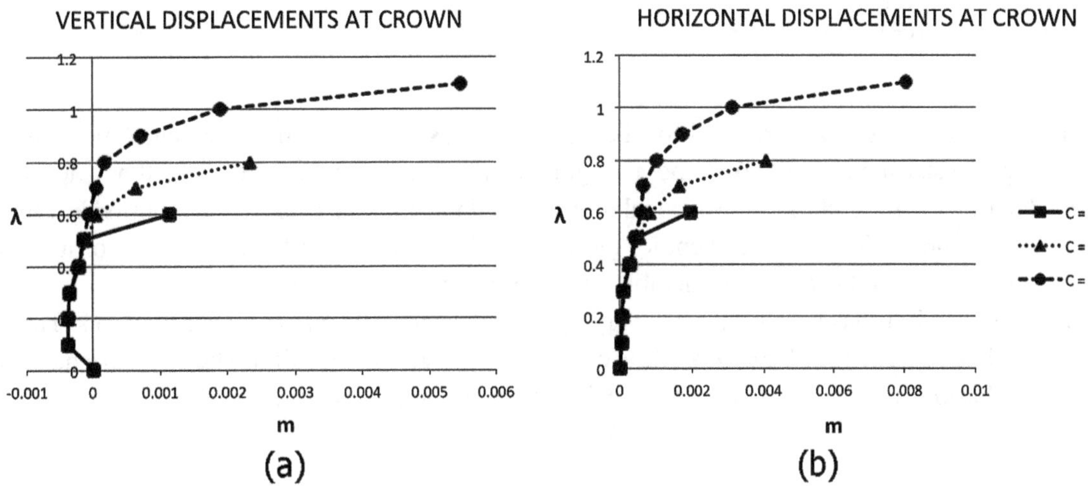

Figure 6. Vertical (a) and horizontal (b) displacements at crown of the Prestwood Bridge arch for different values of cohesion c and constant friction φ = 37°

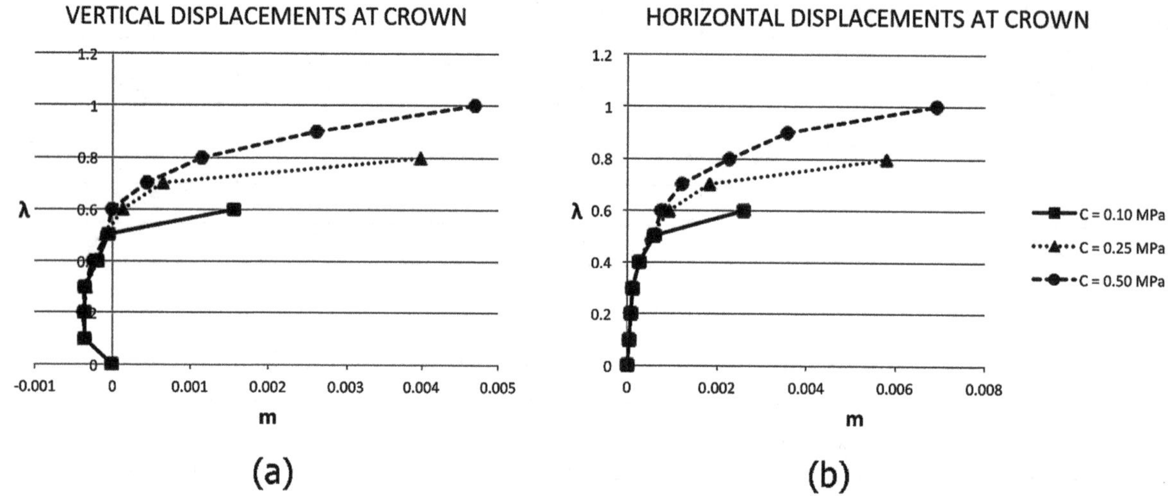

Figure 7. Vertical (a) and horizontal (b) displacements at crown of the Prestwood Bridge arch for different values of friction φ and cohesion constant c = 0.25 MPa

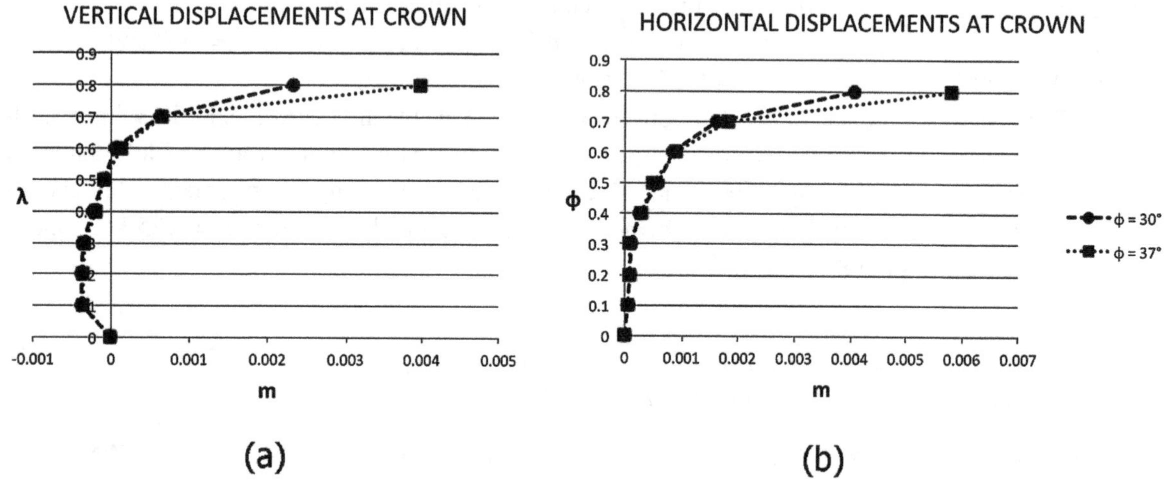

Afterwards to the first parametric analysis, backfill effect has been taken into account. A new mesh has been prepared starting from the first one: triangular elements representing the backfill profile have been added to the original arch. The mechanical properties adopted for the backfill are reported in Table 3, while the same properties have been kept for the arch and its joints. Backfill has not been considered rigid, so its Young's modulus E^F is considerably lower respect to the arch, $E^F = 1400$ MPa.

A parametric analysis has been performed varying the cohesion of the internal backfill joints. Three value of cohesion have been used: $c = 0.04$ MPa, c = 0.08 MPa and c = 0.12 MPa. The other mechanical parameters of backfill internal joints have been calculated using Equations 4, 5 and 6.

Figure 8. Collapse mechanism of Preston Bridge arch, $c = 0.25$ MPa - $\varphi = 37°$

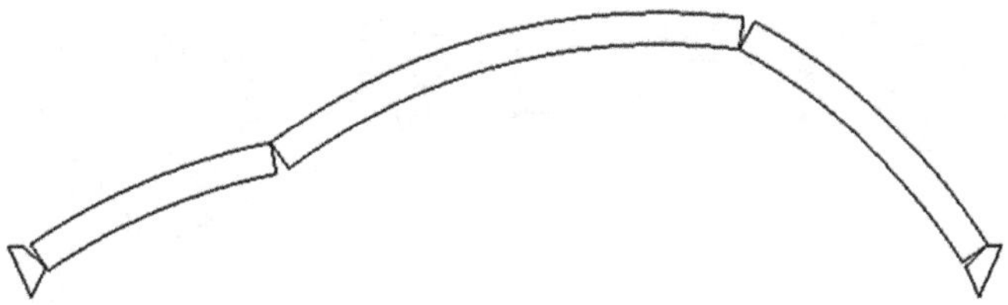

Table 3. Mechanical properties of Prestwood Bridge backfill

Young Modulus E^F (MPa)	Poisson Coefficient ν	Viscous Damping μ	Density ρ (kN/m³)	Contact Penalty (MPa)	Tangential Penalty (MPa)
1.4E+3	0	1.34E+02	20	1.4E+03	7E+02

The fracture energies GI_C and GII_C have been calculated adopting $E^M = E^F$ and $a = 0.04$ m, which is the smallest size of backfill elements. The properties adopted for backfill are reported in table 4. The properties of the arch have been kept constant: $c = 0.25$ MPa and $\varphi = 37°$ have been used.

Vertical and horizontal loads of backfill are distributed along the mesh, on the nodes belonging to each column of backfill over the arch. Thus vertical loads P_{Fi} represent the weight of each portion of backfill supported by the several wedges in which the arch has been divided, and horizontal loads are $H_{Fi} = \lambda P_{Fi}$, as in the previous analysis. Boundary conditions applied are the same for the abutments: they are fixed; while the external sides of backfill have horizontal restraints, in order to avoid contraction. The FEM/DEM model of the Prestwood Bridge and the portions of backfill considered for self-weight are reported in Figure 9.

The model is appropriate to take into account the stabilizing effect of backfill. In fact the multipliers obtained are higher than the ones obtained for the single arch. The parametric analysis shows that increasing the cohesion of backfill, an increase of the load carrying capacity of the arch is obtained, as expected. Graphs of the load-displacement curves for the different values of backfill cohesion are reported in Figure 10, with a comparison between the results obtained with or without backfill. Vertical and horizontal displacements are calculated at crown, in the same node of the previous analysis.

In all the cases, the collapse mechanism of the arch exhibits 4 hinges, alternatively disposed at intrados and extrados of the arch, with the first hinge at extrados of the left abutment and the fourth at intrados of the right abutment. Diffuse cracking occurs in backfill. However, the presence of backfill provides some change in the mechanism of the single arch. The mechanism obtained is in good agreement with

Table 4. Mechanical properties of internal backfill joints

Cohesion c (MPa)	Friction φ (°)	Tensile strength τ (MPa)	Fracture Energy I GI_c	Fracture Energy II GII_c
0.04	50	0.033	1.44E-07	9.77E-08
0.08		0.067	5.47E-07	4.03E-07
0.12		0.10	1.29E-06	8.97E-07

Figure 9. FEM/DEM model of the Preston Bridge (a) and loads applied (b)

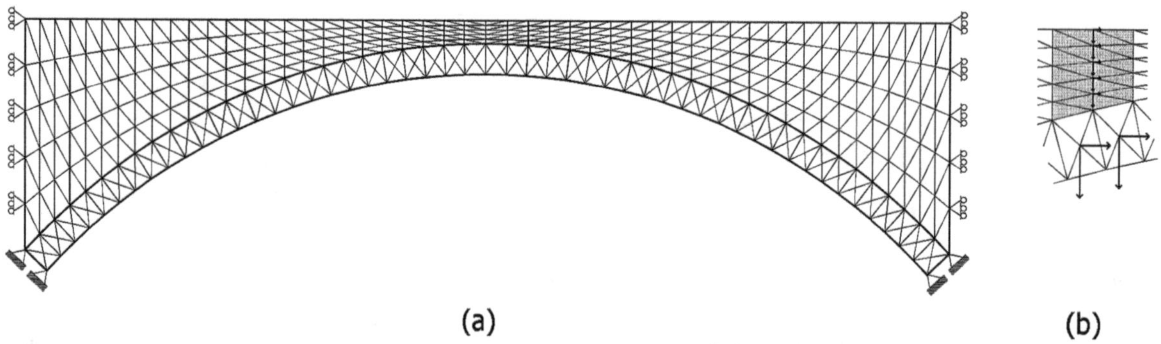

Figure 10. Vertical (a) and horizontal (b) displacements at crown for different values of backfill cohesion (c)

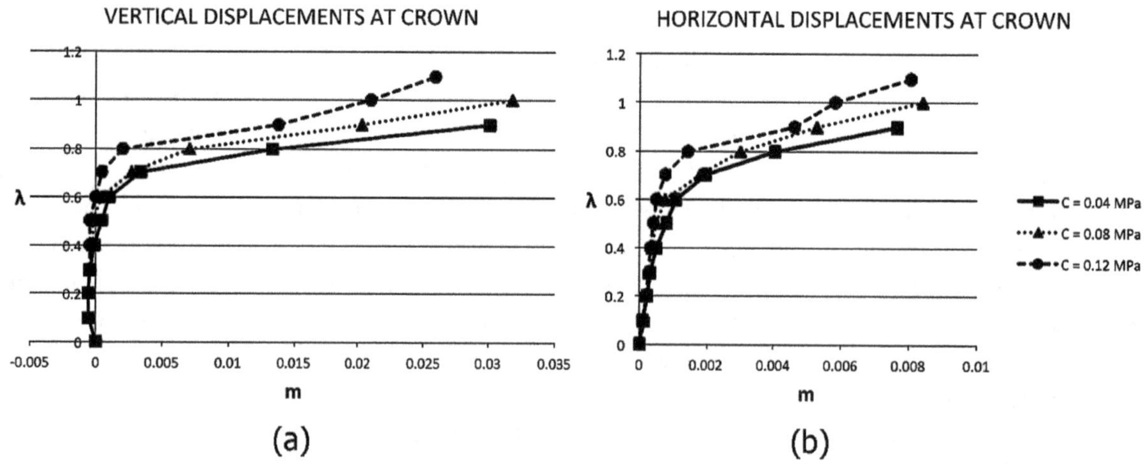

the real collapse occurred during the test carried out on the Prestwood Bridge, see Figure 2. The collapse mechanism obtained with FEM/DEM model is reported in Figure 11.

From simulations results, it can be affirmed that FEM/DEM seems to be a reliable tool to study the nonlinear behavior of masonry arch bridge and to assess its safety. The method provides and affordable description of the collapse mechanism both for the arch and for the whole bridge.

VENICE TRANS-LAGOON BRIDGE

The second case- study that is presented in this chapter is the Venice trans-Lagoon rail bridge, that connects Venice to its mainland, in Italy. The bridge was built in 1846 and during its life was subjected to several interventions and enlargements. At the moment it consists of three different bridges coupled: the historical masonry arch bridge, the roadway bridge built in 1933 and the new rail bridge, built in 1973. The historical bridge is hidden by the new bridges that have been built, and it is partially connected to them. However here only the historical masonry arch bridge is considered. An image of the historical

Figure 11. Collapse mechanism of Preston Bridge with backfill cohesion c = 0.08 MPa and friction φ = 50°

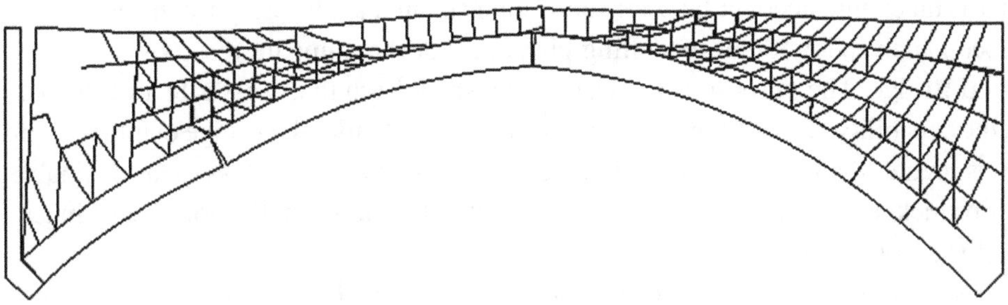

bridge is shown in figure 12. Previous studies have been conducted on the bridge by means of continuous FEM analysis with 3D homogenization procedure, (Reccia, 2013; Reccia, Cecchi & Tralli, 2013; Reccia, Milani, Cecchi & Tralli, 2014), here 2D FEM/DEM analyses are performed.

The bridge carries two rails belonging to the railway Milan-Venice, which is a main railroad in Italy, highly congested by traffic.

The bridge has a total length of 3600 m, it consists of 222 arches divided in 6 modules of 37 arches each one, named *stadii*, which are separated by artificial islands. Each *stadio* is divided in 7 sequences of 5 arches, except the central one consisting of 7 arches: between each sequence there is a big pier in order to prevent a global collapse due to the fall down of a single arch. For this reason the bridge could be considered as a sum of minor bridges made of 5 or 7 arches. Each arch has a span of 10 m and a rise of 1.73 m, with a ratio span/rise $R_{s/r}$ equal to about 1:5.8. The vault, made by bricks and mortar joints, has a curvature radius of 8.80 m at intrados and a transversal depth of 9 m. The thickness changes: 0.65 m at the crown, 0.80 m in the half of middle span, 0.94 at the abutments. However here its thickness is considered constant equal to 0.80 m, the ratio thickness/span is $R_{t/s}$ = 1/12.5. The barrel vault is completely made by bricks and mortar joints. Abutments and piers are made of Istrian stones. Backfill is made by heterogeneous incoherent materials.

Figure 12. The Venice trans-Lagoon Bridge, Italy, during the XIX century

Here an analysis is performed on a single arch. The first simulation deals with the arch without backfill, that is taken into account later for more sophisticated analyses. Only the arch and backfill systems have been analyzed, without modelling piers, in order to compare results with the ones obtained for Prestwood Bridge. The purpose is to evaluate the shape/arch thickness effect of the behavior of the bridge and the capability of FEM/DEM to reproduce it. In particular the bridge exhibits an arch thickness about 4 times greater than the Prestwood Bridge one, while the span is less than the double, leading to a higher ratio thickness/span $R_{t/s}$. A collapse constituted by the formation of shear hinges is therefore possible in this second case.

The first parametric analysis has been performed on the arch without backfill. As in the previous case, the arch has been divided in wedges separated by joints modeled as elastic-plastic Mohr-Coulomb interfaces. Blocks are considered infinitely rigid and cracks may occur only in the joints between the blocks. A parametric analysis has been performed varying cohesion: values $c = 0.10$ MPa, c = 0.20 MPa and $c = 0.25$ MPa have been adopted, while friction has been kept constant $\varphi = 37°$.

A pushover analysis has been performed by increasing the value of horizontal load up to the collapse of the bridge. Vertical and horizontal loads are applied at the center of blocks. A depth of 1 m is considered. The volume of blocks V_B is (0.80 * 0.55 * 1) = 0.46 m^3 and the density is $\rho = 20$ kN/m^3, therefore the vertical load applied in each block is $P_B = 8.556$ kN. The horizontal load applied is equal to $H_B = \lambda\, P_B$, where λ increases up to the value needed to activate the mechanism of collapse of the arch. As before, two additional blocks are added at the abutments to allow possible cracking. The geometry of the bridge, considering only arch and backfill, is shown in Figure 13, the FEM/DEM model of the Venice trans-Lagoon Bridge arch and the loads applied are reported in Figure 14.

The mechanical properties of blocks and the parameters of the joints between them are reported in Tables 5 and 6. The same properties adopted for the Prestwood Bridge arch have been used. However dimension of the elements are bigger with respect to the mesh adopted for Prestwood Bridge, therefore higher values of viscous damping μ and fracture energies GI_C and GII_C have been adopted. In particular the dimension of the interface has been considered equal to the arch thickness, $a = 0.80$ m. Such a choice implies considerably bigger values of fracture energy, both for the first and second mode, in comparison with the Prestwood Bridge case.

Results of the parametric analysis are reported in Figure 15. As expected, the larger thickness increases the load bearing capacity of the arch.

Figure 13. Geometry of the Venice trans-Lagoon Bridge, dimension in meters

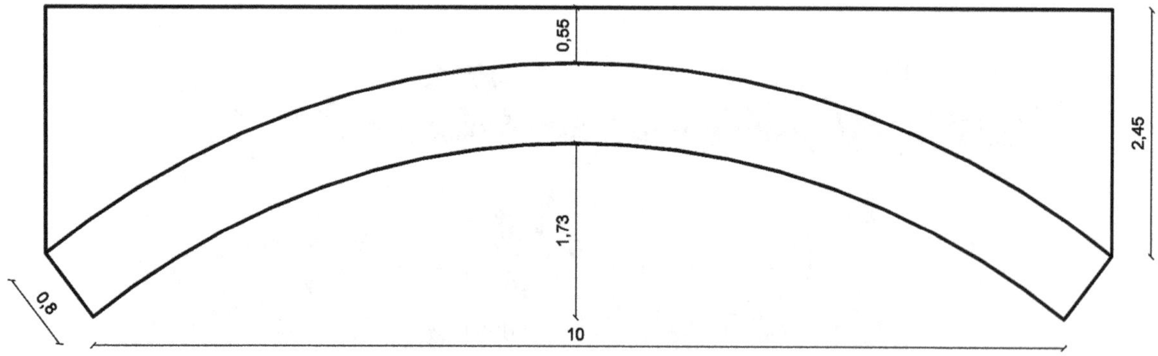

Figure 14. FEM/DEM model of the Venice trans-Lagoon Bridge arch

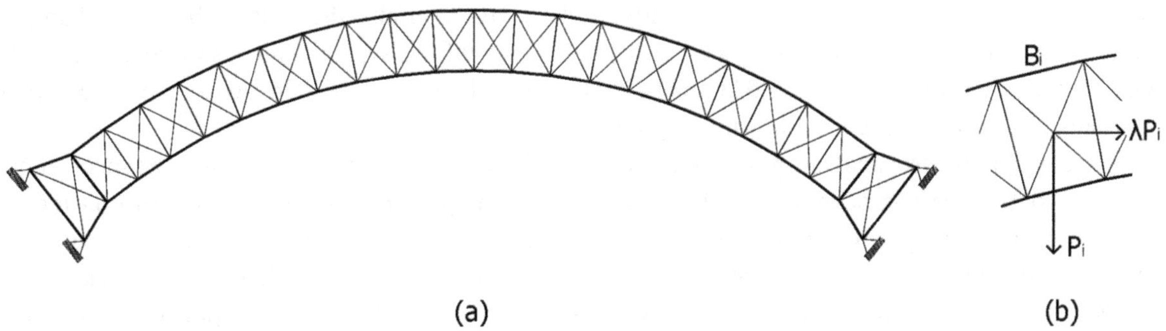

Table 5. Mechanical properties of blocks

Young Modulus E^B (MPa)	Poisson Ratio ν	Viscous Damping μ	Density ρ (kN/m³)	Contact Penalty (MPa)	Tangential Penalty (MPa)
2E+06	0	3.16E+4	20	2E+06	1E+05

Table 6. Mechanical properties of joints

Cohesion c (MPa)	Friction φ (°)	Tensile strength τ (MPa)	Fracture Energy I GI_c	Fracture Energy II GII_c
0.10	37	0.13	2.51E-05	4.25E-05
0.20		0.26	1.00E-04	1.70E-04
0.25		0.33	1.57E-04	2.74E-04

Figure 15. Vertical (a) and horizontal (b) displacements at crown for different values of cohesion c and constant friction φ = 37°

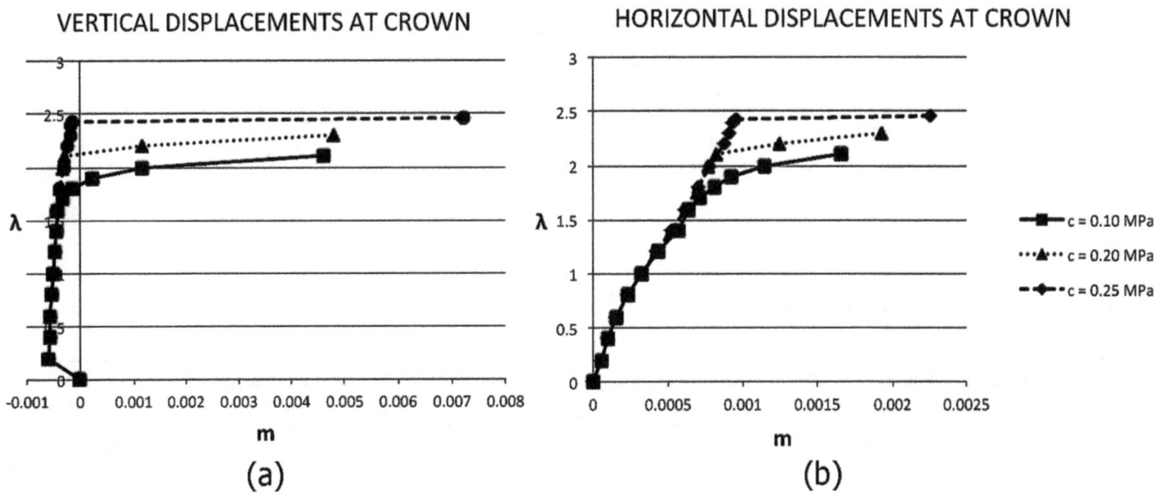

It is possible to notice that, increasing the value of cohesion, an increase of the collapse load multiplier is obtained, obviously in agreement with intuition. However also the collapse behavior changes: for low value of cohesion $c = 0.10$ MPa the collapse occurs with the typical 4 hinges mechanism, while with higher values of cohesion $c = 0.20$ or 0.25 MPa, the collapse occurs more swiftly because one of the joints cracks by slippage, as shown in Figure 16.

A second parametric analysis has been performed taking into account the backfill. The mechanical properties of the arch have been kept constant: $c = 0.10$ MPa and $\varphi = 37°$. The backfill has been considered deformable, as in the previous case. The same mechanical properties of the Prestwood Bridge backfill have been adopted. A parametric analysis has been performed varying the cohesion of the backfill: $c = 0.04$ MPa and $c = 0.08$ MPa have been adopted, while friction angle has been kept constant $\varphi = 50°$. The dimension of the smallest backfill element is $a = 0.18$ m. The mechanical properties of backfill and its internal joints are summarized in Tables 7 and 8.

Vertical and horizontal loads of backfill are distributed along the mesh, on the nodes belonging to each column of backfill over the arch. Boundary conditions applied are the same adopted for the abutments: they are fixed, while the external sides of backfill have horizontal restraints, in order to avoid contraction. The FEM/DEM model and the portions of backfill considered for self-weight are reported in Figure 17.

Results obtained confirm the capacity of FEM/DEM to take into account the effect of backfill to the global behavior. Graphs reported in Figure 18 show a comparison between the behavior of the single arch and of the arch with the backfill. In particular it is possible to notice that the presence of backfill delays the begin of the nonlinear behavior and increases the collapse load multiplier. The presence of

Figure 16. Different collapse mechanism of Venice trans-Lagoon Bridge arch, (a) cohesion c = 0.10 MPa, (b) cohesion c = 0.25 MPa

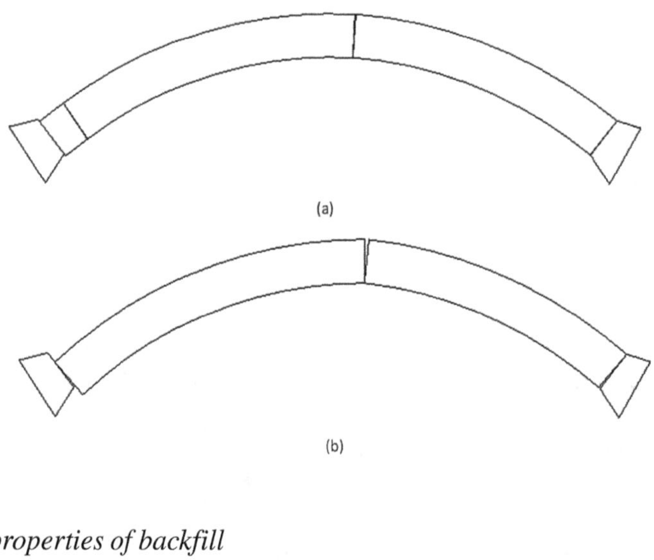

Table 7. Mechanical properties of backfill

Young Modulus E^F (MPa)	Poisson Ratio ν	Viscous Damping μ	Density ρ (kN/m^3)	Contact Penalty (MPa)	Tangential Penalty (MPa)
1.4E+03	0	6.02+02	20	1.4E+03	7E+02

Table 8. Mechanical properties of internal backfill joints

Cohesion c (MPa)	Friction φ (°)	Tensile Strength τ (MPa)	Fracture Energy I GI_c	Fracture Energy II GII_c
0.04	50	0.033	6.46E-07	4.40E-07
0.08		0.067	2.59E-06	1.81E-06

Figure 17. FEM/DEM model of the Venice trans-Lagoon Bridge (a) and loads applied (b)

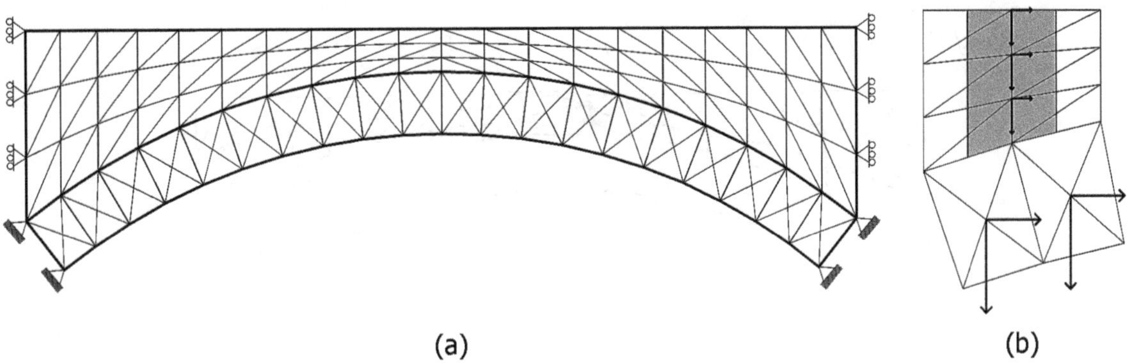

backfill reduces the horizontal displacements of the arch, as highlighted in the graph that plot horizontal displacements at crown. Moreover the stabilizing effect of backfill is clearly shown in the graph that plots vertical displacements at crown. In the case of the single arch, at the beginning the arch exhibits negative vertical displacements that became positive when the kinematic mechanism is activated. The presence of backfill reduces the initial negative vertical displacement of at least one order of magnitude and the kinematic mechanism is activated for higher values of load.

Figure 18. Vertical (a) and horizontal (b) displacements at crown comparison between single arch and arch + backfill

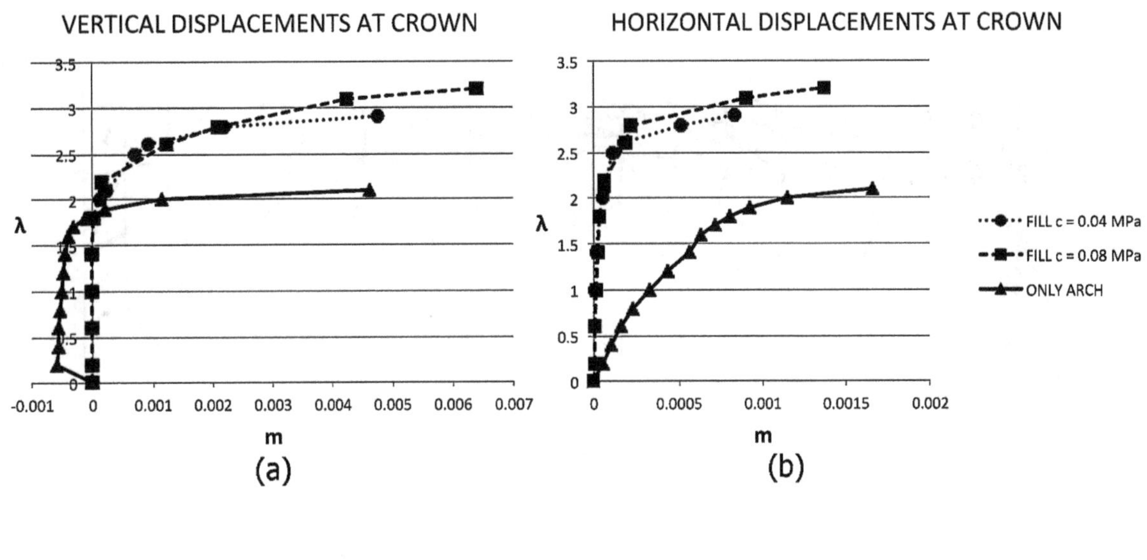

The backfill cracks before the arch, which collapses only when the stabilizing effect of the backfill ends. A plot of cracks distributions into backfill is reported in Figure 19.

Moreover, crack elements that activate in the model of both the single arch and the whole bridge (arch+backfill) are reported in Figure 20. The picture is obtained by means of an ad hoc MatLab script. Line types describe the type of failure: the mesh is in solid light line; crack elements broken represented by dashed line are subjected to de-cohesion, by dotted line to sliding and by dash-dot line to a mixed failure mode. It is possible to notice how the mechanism changes considerably due to the backfill presence.

Figure 19. Backfill cracking

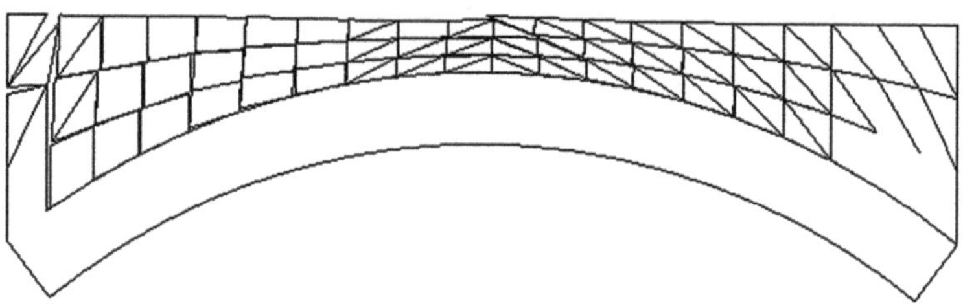

Figure 20. Crack elements broken, (a) single arch c = 0.10 MPa, (b) backfill c = 0.08 MPa (dashed line: de-cohesion; dotted line: sliding; dash-dot line: mixed mode)

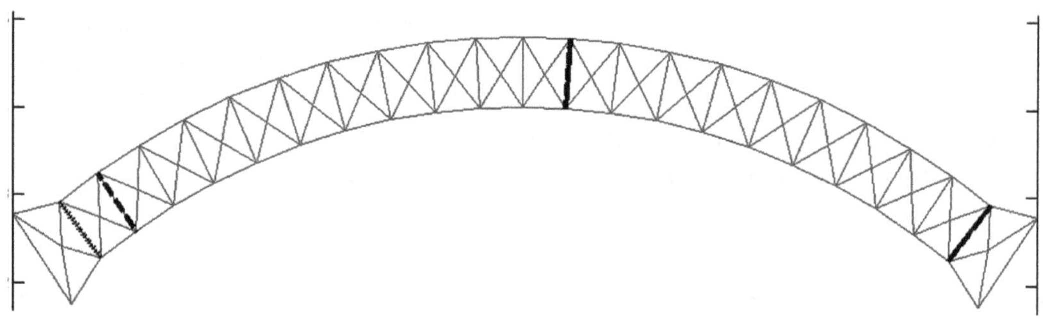

(a)

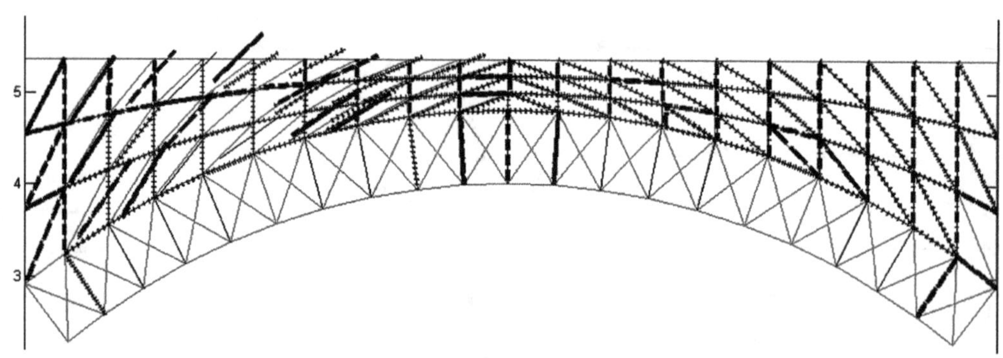

(b)

CONCLUSION

FEM/DEM modelling seems a procedure able to take into account the crucial phenomena characterizing the behaviour of a masonry arch, as the low and uncertain resistance to tensile stresses, the quite good compressive strength, the shear strength depending on the level of compression, on the base of a Coulomb's law of friction, a linear-elastic load/displacement relationship only up to very low levels of external loads, which early turns to strong nonlinearities.

The performed analyses regarding the role played by the infill, with reference to two famous case studies shows a clear stabilizing effect of the backfill, a decrease of the nonlinear behavior and an increase of the collapse load multiplier, as well as a reduction of the horizontal displacements of the arch.

FEM/DEM competes favorably with classic FEM discretizations, not only in the non-linear dynamic range but also when dealing with the pushover analysis of masonry arches interacting with the infill, because of the intrinsic nature of the deformations process, which exhibits clear plasticization zones where cracks propagate (and almost all the deformation occurs) and a moderate elastic deformation on blocks (which may be more or less important depending on the blocks typology considered).

REFERENCES

Alpa, G., & Monetto, I. (1994). Microstructural model for dry block masonry walls with in-plane loading. *Journal of the Mechanics and Physics of Solids*, *42*(7), 1159–1175. doi:10.1016/0022-5096(94)90065-5

Anthoine, A. (1995). Derivation of the in-plane elastic characteristics of masonry through homogenization theory. *International Journal of Solids and Structures*, *32*(2), 137–163. doi:10.1016/0020-7683(94)00140-R

Audenaert, A., Fanning, P., Sobczak, L., & Peremans, H. (2008). 2-D analysis of arch bridges using an elasto-plastic material model. *Engineering Structures*, *30*(3), 845–855. doi:10.1016/j.engstruct.2007.05.018

Azevedo, J., Sincraian, G., & Lemos, J. V. (2000). Seismic behavior of blocky masonry structures. *Earthquake Spectra*, *16*(2), 337–365. doi:10.1193/1.1586116

Baraldi, D., & Cecchi, A. (2014). Discrete element model for in-plane loaded viscoelastic masonry. *International Journal for Multiscale Computational Engineering*, *12*(2), 155–175. doi:10.1615/IntJ-MultCompEng.2014008118

Baraldi, D., Cecchi, A., & Tralli, A. (2015). Continuous and discrete models for masonry like material: A critical comparative study. *European Journal of Mechanics. A, Solids*, *50*, 39–58. doi:10.1016/j.euromechsol.2014.10.007

Baraldi, D., Reccia, E., Cazzani, A., & Cecchi, A. (2013). Comparative analysis of numerical discrete and finite element models: The case of in-plane loaded periodic brickwork. *Composites: Mechanics, Computations. Applications*, *4*(4), 319–344.

Baraldi, D., Reccia, E., & Cecchi, A. (2015). DEM & FEM/DEM Models For Laterally Loaded Masonry Walls. In *COMPDYN 2015, 5th ECCOMAS Thematic Conference on Computational Methods in Structural Dynamics and Earthquake Engineering.*

Barbosa, B. E. (1996). Discontinous structural analysis. In *Proceedings of the 11th world conference on earthquake engineering*.

Boothby, T. (1995). Collapse modes of masonry arch bridges. *Journal of the British Masonry Society*, *9*(2), 62-69.

Braga, F., & Dolce, M. (1982). A method for the analysis of antiseismic masonry multi-storey buildings. In *Sixth International Brick Masonry Conference*, Roma, Italy.

Brencich, A., & Morbiducci, R. (2007). Masonry Arches: Historical Rules and Modern Mechanics. *International Journal of Architectural Heritage*, *1*(2), 165–189. doi:10.1080/15583050701312926

Carpinteri, A. (1992). *Meccanica dei materiali e della frattura*. Bologna: Pitagora Editrice.

Cavicchi, A., & Gambarotta, L. (2005). Collapse analysis of masonry arch bridges taking into account arch-fill interaction. *Engineering Structures*, *27*(4), 605–615. doi:10.1016/j.engstruct.2004.12.002

Cavicchi, A., & Gambarotta, L. (2006). Two-dimensional finite element upper bound limit analysis of masonry bridges. *Computers & Structures*, *84*(31-32), 2316–2328. doi:10.1016/j.compstruc.2006.08.048

Cavicchi, A., & Gambarotta, L. (2007). Lower bound limit analysis of masonry bridges including arch-fill interaction. *Engineering Structures*, *29*(11), 3002–3014. doi:10.1016/j.engstruct.2007.01.028

Cecchi, A., & Milani, G. (2008). A kinematic FE limit analysis model for thick english bond masonry walls. *International Journal of Solids and Structures*, *45*(5), 1302–1331. doi:10.1016/j.ijsolstr.2007.09.019

Cecchi, A., & Sab, K. (2002). A multi-parameter homogenization study for modeling elastic masonry. *European Journal of Mechanics. A, Solids*, *21*(2), 249–268. doi:10.1016/S0997-7538(01)01195-0

Cecchi, A., & Sab, K. (2002). Out of plane model for heterogeneous periodic materials: The case of masonry. *European Journal of Mechanics. A, Solids*, *21*(5), 715–746. doi:10.1016/S0997-7538(02)01243-3

Cecchi, A., & Sab, K. (2004). A comparison between a 3D discrete model and two homogenized plate models for periodic elastic brickwork. *International Journal of Solids and Structures*, *41*(9-10), 2259–2276. doi:10.1016/j.ijsolstr.2003.12.020

Cundall, P. A. (1976) Explicit finite difference methods in geomechanics. In *Int. Conf. Numerical Methods in Geomechanics*, Blacksburg, VA.

Cundall, P. A. (1988). Formulation of a three-dimensional distinct element model-part I. A scheme to detect and represent contacts in a system composed of many polyhedral blocks. *International Journal of Rock Mechanics and Mining Sciences*, *25*(3), 107–116. doi:10.1016/0148-9062(88)92293-0

D'Asdia, P., & Viskovi, A. L. (1994). Analisi sismica degli edifici in muratura. *Ingegneria Sismica*, *1*, 32–42.

De Buhan, P., & De Felice, G. (1997). A homogenization approach to the ultimate strength of brick masonry. *Journal of the Mechanics and Physics of Solids*, *45*(7), 1085–1104. doi:10.1016/S0022-5096(97)00002-1

De Felice, G., & Giannini, R. (2001). Out-of-plane seismic resistance of masonry walls. *Journal of Earthquake Engineering*, *5*(2), 253–271. doi:10.1080/13632460109350394

Drosopoulos, G. A., Stavroulakis, G. E., & Massalas, C. V. (2006). Limit analysis of a single span masonry bridge with unilateral frictional contact interfaces. *Engineering Structures*, *28*(13), 1864–1873. doi:10.1016/j.engstruct.2006.03.016

Gambarotta, L., & Lagomarsino, S. (1997). Damage models for the seismic response of brick masonry shear walls. part I: The mortar joint model and its applications. *Earthquake Engineering & Structural Dynamics*, *26*(4), 423–439. doi:10.1002/(SICI)1096-9845(199704)26:4<423::AID-EQE650>3.0.CO;2-#

Gambarotta, L., & Lagomarsino, S. (1997). Damage models for the seismic response of brick masonry shear walls. part II: The continuum model and its applications. *Earthquake Engineering & Structural Dynamics*, *26*(4), 441–462. doi:10.1002/(SICI)1096-9845(199704)26:4<441::AID-EQE651>3.0.CO;2-0

Gilbert, M., Casapulla, C., & Ahmed, H. M. (2006). Limit analysis of masonry block structures with non-associative frictional joints using linear programming. *Computers & Structures*, *84*(13-14), 873–887. doi:10.1016/j.compstruc.2006.02.005

Gilbert, M., & Melbourne, C. (1994). Rigid-block analysis to masonry arches. *Structural Engineering*, *72*, 356–361.

Hart, R., Cundall, P. A., & Lemos, J. (1988). Formulation of a three-dimensional distinct element model-part II. mechanical calculations for motion and interaction of a system composed of many polyhedral blocks. *International Journal of Rock Mechanics and Mining Sciences*, *25*(3), 117–125. doi:10.1016/0148-9062(88)92294-2

Heyman, J. (1969). The safety of masonry arches. *International Journal of Mechanical Sciences*, *43*, 209–224.

Hughes, T. G., & Blackler, M. J. (1997). A review of the UK masonry arch assessment methods. *Proceedings - Institution of Civil Engineers*, *122*, 305–315.

Lemos, J. V. (2007). Discrete element modelling of masonry structures. *International Journal of Architectural Heritage*, *1*(2), 190–213. doi:10.1080/15583050601176868

Lofti, H. R., & Benson Shing, P. (1994). Interface model applied to fracture of masonry structures. *Journal of Structural Engineering*, *120*(1), 63–80. doi:10.1061/(ASCE)0733-9445(1994)120:1(63)

Lourenço, P. B., Milani, G., Tralli, A., & Zucchini, A. (2007). Analysis of masonry structures: Review of and recent trends in homogenization techniques. *Canadian Journal of Civil Engineering*, *34*(11), 1443–1457. doi:10.1139/L07-097

Lourenço, P. B., & Rots, J. G. (1997). Multisurface interface model for analysis of masonry structures. *Journal of Engineering Mechanics*, *123*(7), 660–668. doi:10.1061/(ASCE)0733-9399(1997)123:7(660)

Luciano, R., & Sacco, E. (1997). Homogenization technique and damage model for old masonry material. *International Journal of Solids and Structures*, *34*(24), 3191–3208. doi:10.1016/S0020-7683(96)00167-9

Magenes, G., & Calvi, G. M. (1997). In-plane seismic response of brick masonry walls. *Earthquake Engineering & Structural Dynamics*, *26*(11), 1091–1112. doi:10.1002/(SICI)1096-9845(199711)26:11<1091::AID-EQE693>3.0.CO;2-6

Mahabadi, O. K., Grasselli, G., & Munjiza, A. (2010). Y-GUI: A graphical user interface and preprocessor for the combined finite-discrete element code, Y2D, incorporating material inhomogeneity. *Computers & Geosciences*, *36*(2), 241–252. doi:10.1016/j.cageo.2009.05.010

Mahabadi, O. K., Lisjak, A., Munjiza, A., & Grasselli, G. (2012). Y-Geo: A new combined finite-discrete element numerical code for geomechanical applications. *Geomechanics*, *12*(6), 676–688. doi:10.1061/(ASCE)GM.1943-5622.0000216

Masiani, R., Rizzi, N., & Trovalusci, P. (1995). Masonry as structured continuum. *Meccanica*, *30*(6), 673–683. doi:10.1007/BF00986573

Masiani, R., & Trovalusci, P. (1996). Cosserat and cauchy materials as continuum models of brick masonry. *Meccanica*, *31*(4), 421–432. doi:10.1007/BF00429930

Massart, T. J., Peerlings, R. H. J., & Geers, M. G. D. (2007). An enhanced multi-scale approach for masonry wall computations with localization of damage. *International Journal for Numerical Methods in Engineering*, *69*(5), 1022–1059. doi:10.1002/nme.1799

Milani, G., Lourenço, P., & Tralli, A. (2006). Homogenization approach for the limit analysis of out-of-plane loaded masonry walls. *Journal of Structural Engineering*, *132*(10), 1650–1663. doi:10.1061/(ASCE)0733-9445(2006)132:10(1650)

Mirabella Roberti, G., & Calvetti, F. (1988). Distinct Element analysis of stone arches. In *Arch Bridges* (pp. 181–186). Rotterdam: Balkema.

Munjiza, A. (2004) *The finite/discrete element method.* John Wiley and Sons.

Munjiza, A., Owen, D. R. J., & Bicanic, N. (1995). Combined finite-discrete element method in transient dynamics of fracturing solids. *Engineering Computations*, *12*(2), 145–174. doi:10.1108/02644409510799532

Owen, D. R. J., Peric, D., Petrinic, N., Brookes, C. L., & James, P. J. (1998). Finite/Discrete element model for assessment and repair of masonry structures. In *Arch Bridges* (pp. 181–186). Rotterdam: Balkema.

Owen, D. R. J., Peric, D., Petrinic, N., Smokes, C. L., & James, P. J. (1998). *Finite/discrete element models for assessment and repair of masonry structures*. Paper presented at Second International Arch Bridge Conference, Venice, Italy.

Page, J. (1993). *Masonry Arch Bridge – TRL state of the art review*. London: HMSO.

Pippard, A. J. S. (1948). The approximate estimation of safe loads on masonry bridges. Civil engineer in war. *Institution of Civil Engineers*, *1*, 365.

Pippard, A. J. S., & Ashby, E. R. J. (1936). An experimental study of the voissour arch. *Journal of the Institution of Civil Engineers*, *10*, 383–403. doi:10.1680/ijoti.1939.14554

Reccia, E. (2013) *Conservation of masonry arch bridges. A procedure for modelling and strengthening*. (Unpublished PhD Dissertation). University of Nova Gorica.

Reccia, E., Cazzani, A., & Cecchi, A. (2012). FEM-DEM modeling for out-of-plane loaded masonry panels: A limit analysis approach. *Open Civil Engineering Journal*, *6*(1), 231-238.

Reccia, E., Cecchi, A., & Tralli, A. (2013). Homogenization of masonry vault bridges: Sensitivity to external stone arch. *Civil-Comp Proceedings*, 102.

Reccia, E., Milani, G., Cecchi, A., & Tralli, A. (2014). Full 3D homogenization approach to investigate the behavior of masonry arch bridges: The venice trans-lagoon railway bridge. *Construction & Building Materials*, *66*, 567–586. doi:10.1016/j.conbuildmat.2014.05.096

Sab, K. (2003). Yield design of thin periodic plates by a homogenization technique and an application to masonry walls. *Comptes Rendus. Mécanique*, *331*(9), 641–646. doi:10.1016/S1631-0721(03)00144-X

Salerno, G., & de Felice, G. (2009). Continuum modeling of periodic brickwork. *International Journal of Solids and Structures*, *46*(5), 1251–1267. doi:10.1016/j.ijsolstr.2008.10.034

Shi, G. H., & Goodman, R. E. (1988). Discontinuous deformation analysis – a new method for computing stress, strain and sliding of block systems. In *Key questions in rock mechanics* (pp. 381–393). Rotterdam: Balkema.

Shieh-Beygi, B., & Pietruszczak, S. (2008). Numerical analysis of structural masonry: Mesoscale approach. *Computers & Structures*, *86*(21-22), 1958–1973. doi:10.1016/j.compstruc.2008.05.007

Smoljanović, H., Živaljić, N., & Nikolić, Ž. (2013). A combined finite-discrete element analysis of dry stone masonry structures. *Engineering Structures*, *52*, 89–100. doi:10.1016/j.engstruct.2013.02.010

Tomaževic, M., & Turnšek, V. (1980). Lateral load distribution as a basis for the seismic resis- tance analysis of masonry buildings. In *International Research Conference on Earthquake Engineering*, Skopje.

Toth, A. R., Orban, Z., & Bagi, C. (2009). Discrete element analysis of a stone masonry arch. *Mechanics Research Communications*, *36*(4), 469–480. doi:10.1016/j.mechrescom.2009.01.001

Turnšek, V., & Cačovič, F. (1971). Some experimental results on the strength of brick masonry walls. In *Second International Brick Masonry Conference*, Stoke-on-Trent, UK.

Turnšek, V., & Sheppard, P. (1980). The shear and flexural resistance of masonry walls. In *International Research Conference on Earthquake Engineering*, Skopje.

UIC. (2005). *International Union of Railways: Improving Assessment, Optimization of Maintenance and Development of Database for Masonry Arch Bridges*. UIC.

KEY TERMS AND DEFINITIONS

Arch-Fill Interaction: The effects of arch-fill interaction can strongly affect the collapse behavior: the presence of backfill increases the load carrying capacity of masonry arch bridge.

Backfill: Also called backing or filling, it is the material, usually low quality fill, used to give support behind a structure. For a masonry arch bridge, backfill material is placed between the arch barrel and the road surface and retained laterally by the spandrel walls. It normally consists of granular material, e.g. gravel or building debris, which may have been excavated for the foundations or is waste from the construction.

Collapse Mechanism: A kinematic mechanism leading to the collapse of a voussoir arch, usually by the formation of four plastic hinges alternatively disposed ad intrados and extrados.

Masonry Arch: A curved structure for spanning an opening, designed to support a vertical load primarily by axial compression. Masonry arch is made of individual stone or brick voussoirs.

Masonry Material: A heterogeneous and anisotropic material exhibiting non-linear behavior, formed by an ordered set of interconnected blocks, joined together by means of dry or mortar joints.

Mohr-Coulomb Interface: A cohesive interface modelled on the base of Mohr-Coulomb criterion, therefore using two main parameters: cohesion and friction.

Nonlinear Static Analysis: Also known as "pushover" analysis, it consists of applying to the structures vertical and horizontal loads, monotonously increasing them until the collapse.

Prestwood Bridge: Located at Staffordshire (UK), it was a single span masonry arch bridge tested up to collapse at the beginning of 90's, and which is recognized as a standard benchmark.

Venice Trans-Lagoon Bridge: The bridge that connects Venice (Italy) to its mainland. The original masonry arch bridge was built in 1846 and represents an important piece of heritage in the European panorama of arch bridges.

Voussoir: Any of the wedge shaped units in an arch or vault, having side cuts converging at one of the arch centers.

Chapter 15
Discrete Finite Element Method for Analysis of Masonry Structures

Iraj H. P. Mamaghani
University of North Dakota, USA

ABSTRACT

Masonry structures are comprised of a finite number of distinct interacting rock blocks that have a length scale relatively comparable to the structure. Therefore, they are ideal candidates for modeling as discrete systems. This chapter covers the Discrete Finite Element Method (DFEM) developed by the author to model discontinuous media consisting of blocks of arbitrary shapes. The DFEM is based on the finite element method incorporating contact elements. The DFEM considers blocks as sub-domains and represents them as solid elements. Contact elements are used to model block interactions such as sliding or separation. In this chapter, through some illustrative examples, the applicability of the DFEM to static and dynamic analysis of masonry structures, including arch bridges, walls, slopes, and underground openings, is discussed. The DFEM provides an efficient tool for researchers and practical engineers in designing, analyzing, and studying the behavior of masonry structures under static and dynamic loadings.

INTRODUCTION

Masonry structures, such as arch bridges, are the primary engineering structures designed. Most historical bridges, temples, walls, walls with openings, building facades, arches, and towers, which are found all over the world, are masonry structures. A sample masonry arch bridge is shown in Figure 1. Because of modern civilization and land problems, there is a need to demolish some old masonry structures and replace them with modern structures. On the other hand, it is necessary to preserve some of these structures, which are historically valuable. Thus, analysis of these structures and, if required, repairing and reinforcing them against failure, is of paramount importance. Therefore, a well-defined numerical analysis method for these structures is needed.

DOI: 10.4018/978-1-5225-0231-9.ch015

Figure 1. Masonry Arch Bridge, (Ghari Koprisi), Tabriz, Iran

The discrete element method (DEM), originally developed by Cundall (1971), is an innovative numerical method for solving a wide spectrum of problems involving the interaction of rock, soil, and structures and is a widely recognized technology for modelling geomaterials. The DEM is applied to the simulation of problems characterized by severe discontinuities. The DEM assumes that the analyzed structure can be modelled as an assembly of rigid particles interacting among themselves. The overall behavior of the system is determined by cohesive and frictional contact laws (Cundall, 1971; Munjiza, 2004).

The DEM has been used for years in different industries (e.g., mining, civil, and nuclear waste disposal) to solve problems involving deformation, damage, fracturing, and stability of the fractured rock masses and masonry structures (among others: Baggio & Trovalusci, 1998, 1993; Cundall, 2011; Itasca Consulting Group, Inc., 2014; Giamundo *et al.*, 2014; Lermos, 2007; Sarhosis & Sheng, 2014; Toth *et al.*, 2009). Over the last 25 years, a number of different modeling techniques have been developed to simulate coupled hydro-mechanical problems with the DEM. These methods, used in different applications and for different modes of hydro-mechanical behavior and coupling, have been reviewed by Furtney *et al.* (2013). For example, the Universal Distinct Element Code (UDEC), developed by Itasca Consulting Group, Inc. (2014), is two-dimensional numerical software that simulates the quasi-static or dynamic response to loading of media containing multiple, intersecting joint structures. UDEC utilizes an explicit solution scheme that can model complex, non-linear behaviors. Models may contain a mix of rigid or deformable blocks. Deformable blocks are defined by a continuum mesh of finite-difference zones, with each zone behaving according to a prescribed stress-strain law. The relative motion of the discontinuities is also governed by force-displacement relations for movement in both the normal and shear directions. Joint models and properties can be assigned separately to individual discontinuities or sets thereof. The analysis of rock mass stimulation by fluid injection requires analytical tools, such as numerical models based on DEM, which can represent discontinuities explicitly (Damjanac *et al.*, 2015). A similar approach for simulation of fracturing and hydraulic fracturing of rocks is based on combined finite element method (FEM) and DEM. The formulation of the method and some example applications have been published by Rougier *et al.* (2011, 2012), Zhao *et al.* (2015), and Lisjak *et al.* (2015).

The discrete finite element method (DFEM), combining both the DEM and FEM, is a relatively new computational tool aimed at problems involving static and/or dynamic behavior of systems involving the deformability of one solid body or a large number of solid deformable bodies (Aydan *et al.*, 2011; Mamaghani, 2015a, 2006, 1993; Mamaghani *et al.*, 1994). Such systems include masonry structures, underground openings and structures, fragmentation using explosives (*e.g.*, rock blasting applications in which a more or less intact rock mass is transformed into a pile of solid rock fragments of different sizes which interact with each other), impacts, demolition (*e.g.*, collapsing buildings), blast loads, digging and loading processes, and powder technology. The topic is growing in importance and is at the forefront of current efforts in computational modeling of the failure of solids.

Oñate has presented recent developments in the integration of the Particle Finite Element Method (PFEM) and DEM for the analysis of coupled problems in mechanics involving particulate flows and their interactions with structures (Oñate, 2015; Oñate *et al.*, 2014, 2011). The particle discrete finite element method (PDFEM) uses a unified updated Lagrangian description to model the motion of material points in a domain containing a fluid and a variety of solids (such as particles of different sizes and rigid or deformable structures). In the PDFEM approach, a mesh connects the material points defining the discretized domain where the governing equations for each of the constituent materials are solved as in the standard FEM. Both a moving mesh and a fixed mesh are adopted in the PFEM procedure, which can be used for solving the equations of continuum mechanics for both fluids and solids using the FEM. Barbosa and Ghaboussi proposed a numerical technique for the analysis of multiple interacting deformable bodies undergoing large displacements and rotations. Each body is considered an individual discrete unit, which is idealized by a finite element model. Each element interacts with its surroundings through contact stresses, which are continually updated as the elements move and deform (Barbosa & Ghaboussi, 1990).

The analysis of unreinforced masonry structures is of particular interest to civil engineers. In recent years, several techniques have been developed to analyze rock masses consisting of distinct blocks. A comprehensive review of these techniques was presented by Kawamoto and Aydan (1999). The limiting equilibrium analysis by Hoek and Bray (1977) and Aydan, Shimizu, and Ichikawa (1989) as well as numerical analysis methods, including the FEM with joint or interface elements by, among others, Goodman, Taylor, and Brekke (1968), the DEM by Cundall (1971), and discontinuities deformation analysis (DDA) by Shi (1988), have all been proposed. In spite of all these techniques, it is difficult to say that a unique technique which guarantees satisfactory results has been developed. Although DEM and DDA can be used for the static and dynamic analysis of discontinuous media, the treatment of rate-dependent behavior of materials in these methods is inconsistent with actual behavior. For example, DEM introduces a forced damping to suppress oscillations, while DDA adopts very large time steps so that artificial damping occurs as a result of numerical integration.

The main objective of this chapter is to present the DFEM developed by the author to model discontinuous media consisting of blocks of arbitrary shapes. The author proposed the DFEM, which is based on the principles of the finite element method, for analysis of blocky systems under static and dynamic loadings (Aydan *et al.*, 2005; Mamaghani, 2015a, 2015b, 2008, 2006, 1993; Mamaghani and Aydan, 1999; Mamaghani *et al.*, 2014, 1999, 1994; Tokashiki *et al.*, 2001, 1997). The DFEM consists of a mechanical model to represent the deformable blocks and contact models that specify the interactions among them. In the DFEM, a visco-elastic constitutive law for linear behavior and a visco-elasto-plastic constitutive law for nonlinear behavior of blocks and contacts are used together with the updated Lagrangian scheme. The DFEM can handle large block motions within the framework of the finite element method. In this chapter, the modeling of discontinuities in masonry structures and DFEM formulation are first presented. Then, the applicability of the DFEM to static and dynamic analysis of unreinforced masonry structures is examined and discussed.

MODELING OF MASONRY STRUCTURES DISCONTINUITIES

A discontinuous medium consisting of blocks of arbitrary shapes is distinguished from a continuous medium by the existence of discontinuities at contacts between the discrete bodies that comprise the system. The actual geometry of contacts is never smooth and has asperities of varying amplitude and wave length (Aydan *et al.*, 1989). Relative sliding or separation in such localized zones presents a difficult problem in mechanical modeling and numerical analysis. The most suitable and mechanically sound approach in modeling discontinuities in masonry structures is band type modeling. In this approach, contacts between neighboring blocks are treated as bands with finite thicknesses. The thickness of the bands is related to the thickness of shear-bands observed in tests or in nature, and if data exists, the height of asperities along the plane (Aydan *et al.*, 1989). For an idealized contact of unit depth (in the out-of-plane direction) as shown in Figure 2, the average normal and shear stresses and plane strains are defined as follows:

$$\sigma_n = \frac{F_n}{A} \tag{1}$$

$$\varepsilon_n = \frac{\delta_n}{h} \tag{2}$$

$$\tau_s = \frac{F_s}{A} \tag{3}$$

$$\gamma_s = \frac{\delta_s}{h} \tag{4}$$

where $A = L \times 1 = L$, L and h are the area, length, and the thickness of the band; F_n and F_s are the normal and tangential forces; and δ_n and δ_s denote the normal and tangential deformations, respectively (Figure 2). Furthermore, it is also possible to define the average plain strain rates $\dot{\varepsilon}_n$ and $\dot{\gamma}_n$. As a result, this model also enables one to objectively define stress-strain rate dependent responses. The problem is, then, to select a constitutive model, such as an elastic, elasto-plastic, or elasto-visco-plastic constitutive law, which is appropriate for modeling the mechanical behavior of contacts between neighboring blocks (Mamaghani *et al.*, 1999).

Figure 2. Mechanical model of a contact as a band

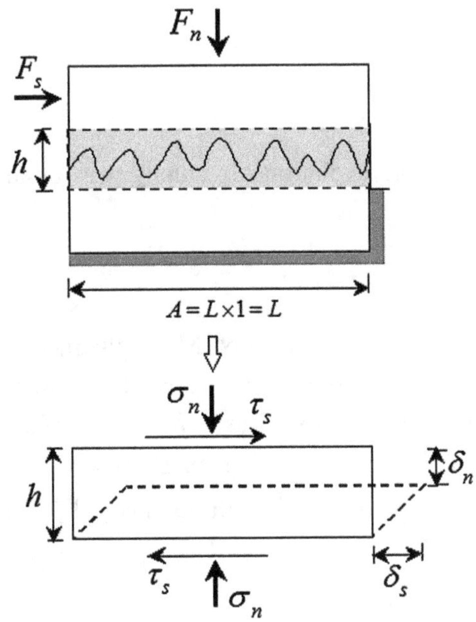

DISCRETE FINITE ELEMENT METHOD (DFEM)

The DFEM, which may be used to analyze and assess the stability of rock block systems such as masonry bridges and masonry underground openings, is based on the FEM. It consists of a mechanical model to represent the deformable blocks and contact models that specify the interactions among them. The deformation of blocks is assumed to be small unless they are allowed to rupture. Small displacement theory is applied to the deformable blocks, while blocks can take finite displacement. The large deformation of blocky systems is associated with the separation, translation, and rotation of blocks. Blocks are polygons with an arbitrary number of sides which are in contact with the neighboring blocks and are idealized as single or multiple finite elements. Block contacts are represented by contact elements.

Mechanical Modeling

The general equation of motion is given by

$$\nabla \cdot \sigma + \boldsymbol{b} = \rho \ddot{\boldsymbol{u}} \tag{5}$$

where σ , $\boldsymbol{b}$, ρ, and $\ddot{\boldsymbol{u}}$ are stress tensor, body force, mass density, and acceleration, respectively. The following presentation is restricted to the framework of the small-strain theory. The strain ε -displacement $\boldsymbol{u}$ relations are represented by

$$\varepsilon = \frac{1}{2}[\nabla \boldsymbol{u} + (\nabla \boldsymbol{u})^T] \tag{6}$$

The strain rate $\dot{\varepsilon}$ -velocity $\boldsymbol{v}$ relations are given by

$$\dot{\varepsilon} = \frac{1}{2}[\nabla \boldsymbol{v} + (\nabla \boldsymbol{v})^T] \tag{7}$$

where $\boldsymbol{v} = \dot{\boldsymbol{u}}$. The following constitutive relationship among stresses and strains and strain ratės holds:

$$\sigma = \boldsymbol{D}_e \varepsilon + \boldsymbol{D}_v \dot{\varepsilon} \tag{8}$$

where $\boldsymbol{D}_e$ and $\boldsymbol{D}_v$ are elasticity and viscosity tensors (Mamaghani, 1993), respectively. However, they can be replaced by elasto-plastic and visco-plastic tensors, if necessary. This type of constitutive law allows for the modelling of intact blocks as well as contacts, interfaces, or rock discontinuities. The boundary conditions are $\boldsymbol{u} = \hat{\boldsymbol{u}}$ on Γ_u and $\hat{\boldsymbol{t}} = \sigma \cdot \boldsymbol{n}$ on Γ_t, where $\hat{\boldsymbol{u}}$ is the displacement on boundary Γ_u and $\hat{\boldsymbol{t}}$ is the surface traction in the *n* direction on boundary Γ_t. The initial conditions are $\boldsymbol{u}_o$ and *Universidad Huelva, Spain* at $t = 0$.

Finite Element Modeling

In the following discussion, the finite element form of the equation of motion is derived. Taking a variation on $\delta \boldsymbol{u}$, the integral form of Eq. (5) can be written as

$$\int_\Omega (\nabla \cdot \sigma) \cdot \delta \boldsymbol{u} d\Omega + \int_\Omega \boldsymbol{b} \cdot \delta \boldsymbol{u} d\Omega = \int_\Omega \rho \ddot{\boldsymbol{u}} \cdot \delta \boldsymbol{u} d\Omega \tag{9}$$

With the use of the Gauss divergence theorem and the boundary conditions, the weak form of the governing equation takes the following form:

$$\int_{\Gamma_t} \hat{\boldsymbol{t}} \cdot \delta \boldsymbol{u} d\Gamma + \int_\Omega \boldsymbol{b} \cdot \delta \boldsymbol{u} d\Omega = \int_\Omega \sigma \cdot (\nabla \delta \boldsymbol{u}) d\Omega + \int_\Omega \rho \ddot{\boldsymbol{u}} \cdot \delta \boldsymbol{u} d\Omega \tag{10}$$

Eq. (10) is discretized in the space domain by assuming displacements are approximated by the following expression:

$$\boldsymbol{u} = \boldsymbol{N}\boldsymbol{U}(t) \tag{11}$$

where $\boldsymbol{N}$ is the shape function. Using the approximate form and the constitutive law, the following expressions, in condensed form, are obtained for a typical finite element (Mamaghani, 1993; Mamaghani *et al.*, 1999):

$$\boldsymbol{M}\ddot{\boldsymbol{U}} + \boldsymbol{C}\dot{\boldsymbol{U}} + \boldsymbol{K}\boldsymbol{U} = \boldsymbol{F} \tag{12}$$

where $\boldsymbol{F}$ is the force vector and M, C, and K, are the mass, damping, and stiffness matrices, respectively. They are defined as follows (Mamaghani, 1993):

$$\boldsymbol{M} = \int_{\Omega_e} \rho \boldsymbol{N}^T \boldsymbol{N} d\Omega \tag{13}$$

$$\boldsymbol{C} = \int_{\Omega_e} \boldsymbol{B}^T \boldsymbol{D}_v \boldsymbol{B} d\Omega \tag{14}$$

$$\boldsymbol{K} = \int_{\Omega_e} \boldsymbol{B}^T \boldsymbol{D}_e \boldsymbol{B} d\Omega \tag{15}$$

$$\boldsymbol{F} = \int_{\Omega_e} \boldsymbol{N}^T \boldsymbol{b} d\Omega + \int_{\Gamma_{te}} \bar{\boldsymbol{N}}^T \boldsymbol{t} d\Gamma \tag{16}$$

Modeling of Contacts

The contact element is used to model contacts of blocks in masonry structure discontinuities. Consider a two-noded element *lm* in two-dimensional space and take two coordinate systems *oxy* and *o'x'y'* as shown in Figure 3 (Mamaghani, 1993; Mamaghani *et al.*, 1999). Assuming that the strain component $\varepsilon_{y'y'}$ is negligible, the remaining strain components take the following forms:

Figure 3. Mamaghani-Aydan contact model

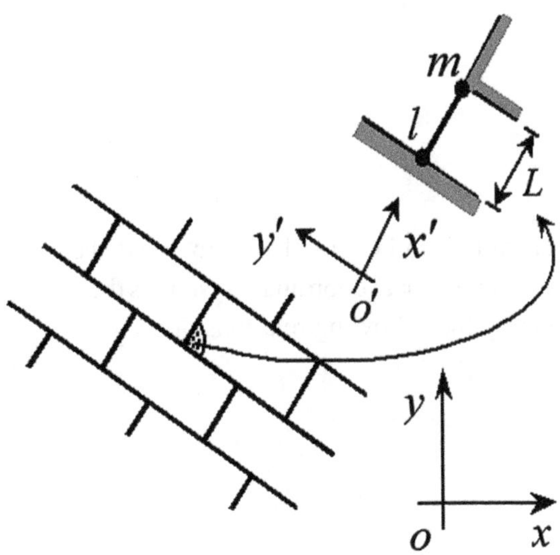

$$\varepsilon_{x'x'} = \frac{\partial u'}{\partial x'}, \qquad \gamma_{x'y'} = \frac{\partial v'}{\partial x'} \tag{17}$$

Let us assume that the shape functions are linear such that

$$N_l = 0.5(1-\xi), \quad N_m = 0.5(1+\xi) \tag{18}$$

where $\xi = (-2\,x' + x'_l + x'_m) / L$, and $L = x'_l - x'_m$. Then the relation between the strains and nodal displacements becomes

$$\begin{Bmatrix} \varepsilon_{x'x'} \\ \gamma_{x'y'} \end{Bmatrix} = \frac{1}{L}\begin{bmatrix} -1 & 0 & 1 & 0 \\ 0 & -1 & 0 & 1 \end{bmatrix}\begin{Bmatrix} U'_l \\ V'_l \\ U'_m \\ V'_m \end{Bmatrix} \tag{19}$$

Thus, the stiffness matrix of contact element in the local coordinate system is explicitly defined as:

$$\boldsymbol{K}' = \begin{bmatrix} k'_n & 0 & -k'_n & 0 \\ 0 & k'_s & 0 & -k'_s \\ -k'_n & 0 & k'_n & 0 \\ 0 & -k'_s & 0 & k'_s \end{bmatrix} \tag{20}$$

in which

$$k'_n = E_n \cdot \frac{A_c}{x'_m - x'_l} \tag{21}$$

$$k'_s = G_s \cdot \frac{A_c}{x'_m - x'_l} \tag{22}$$

Here, A_c is the contact area and E_n and G_s are the normal and shear elastic moduli of discontinuity, respectively. The stiffness matrix in the local coordinate system is then transformed to the stiffness matrix in the global coordinate system by the following relationship:

$$\boldsymbol{K} = \boldsymbol{T}^T \boldsymbol{K}' \boldsymbol{T} \tag{23}$$

where

$$\boldsymbol{T} = \begin{bmatrix} cos\theta & \sin\theta & 0 & 0 \\ -\sin\theta & \cos\theta & 0 & 0 \\ 0 & 0 & cos\theta & \sin\theta \\ 0 & 0 & -\sin\theta & \cos\theta \end{bmatrix} \tag{24}$$

$$\theta = tan^{-1}(\frac{y_m - y_l}{x_m - x_l}) \tag{25}$$

The viscosity (damping) matrix of contact elements in the local coordinate system can be also obtained in a similar manner, as given below (Mamaghani, 1993):

$$\boldsymbol{C}' = \begin{bmatrix} c_n' & 0 & -c_n' & 0 \\ 0 & c_s' & 0 & -c_s' \\ -c_n' & 0 & c_n' & 0 \\ 0 & -c_s' & 0 & c_s' \end{bmatrix} \tag{26}$$

$$c_n' = E_n^*.\frac{A_c}{x_m' - x_l'} \tag{27}$$

$$c_s' = G_s^*.\frac{A_c}{x_m' - x_l'} \tag{28}$$

where E_n* and G_s* are the normal and shear viscosity moduli of discontinuity, respectively. In the above equations, the values of the coefficients in the stiffness and viscosity matrices, as well as the value of θ, are affected by updating geometric changes of blocks and contacts. It is worth noting that on the basis of simplification of the finite element modeling of block contacts, using the small strain theory for modeling of the large deformation, a small error is always present in the computed strains of contacts. Nevertheless, such an error is negligible, as the geometry of the block system is incrementally updated, which allows one to take into account the effect of higher-order terms in the definition of the finite strain tensor.

NUMERICAL RESULTS AND DISCUSSIONS

In the numerical study, when the inertia term is considered, contacts and blocks are assumed to behave as elasto-visco-plastic materials or visco-elastic materials. On the other hand, if the inertia term is omitted, then the behavior of contacts and blocks is assumed to be elasto-plastic or elastic. Application of the DFEM to the numerical analysis of rock slopes (Aydan *et al.*, 1996; Mamaghani & Aydan, 2000) and stability of a single block on an incline and some masonry structures has been reported by the author and his co-workers (Mamaghani, 1993, 2006, 2008, 2015a, 2015b; Mamaghani *et al.*, 1994, 1999, 2014;

Tokashiki *et al.*, 2001). In this section, some typical numerical results for the static and pseudo-dynamic response of masonry structures obtained by the DFEM will be presented and discussed. For the analyses reported herein, the tensile strength of the contact element was assumed to be zero. The Mohr-Coulomb yield criterion was implemented in the present codes; nevertheless, one can easily implement any yield criterion which is appropriate for plastic or visco-plastic behavior. The contact area A_c was assumed to be half the area of the side of a block to which the contact element is attached. The thickness of the bands was taken as twice the weighted asperity height. Based on results reported by Aydan *et al.* (1989), the thickness of the bands was selected to be 10 mm. The secant stiffness method together with the updated Lagrangian scheme was employed to deal with non-linear behavior. The constant strain triangular element with two degrees of freedoms at each node, formed by properly joining the corners and contact nodes of an individual block, was adopted for finite element meshing of the blocks (Mamaghani, 1993). However, it must be noted that the method is not restricted to the use of such elements and one can easily implement finite elements with any number of chosen nodes.

The analysis is a pseudo-time stepping incremental procedure. First, the initial configuration of the structural system, boundary conditions, and material properties are specified. Then, iterations are carried out by forming the global stiffness matrix and solving the equilibrium equations of the system. Later, the strains and stresses of elements are computed. The no-tension condition and Mohr-Coulomb's yield criterion are checked, and the excess forces at contacts are applied to the updated configuration as the penalty load in the subsequent iteration until the norm of excess force vector converges to a very small value of convergence tolerance. The computation is terminated when a stable configuration is achieved or the global stiffness matrix becomes ill-conditioned as single or multiple blocks tend to move without any interaction with each other, corresponding to the failure of the system. The details of the numerical algorithm and computational procedure are given in previous work by the author (Mamaghani, 2015a, 2006, 1993; Mamaghani *et al.*, 1999, 1994).

STABILITY OF ONE BLOCK ON AN INCLINE

Static Stability

A very simple, yet meaningful, problem analyzed by DFEM is the stability of one block on an incline. The theoretical kinematic conditions for sliding and toppling of one block on an incline, under gravity, have been charted by Hoek and Bray (1977), hereafter referred to as the H-B chart. The H-B chart for the friction angle between the block and the incline $\varphi = 20^{\circ}$ is shown in Figure 4. In the H-B chart, four modes of behavior, namely (a) stability, (b) sliding without toppling, (c) sliding and toppling, and (d) toppling without sliding, are delineated by the boundaries I, II, III, and IV. DFEM is applied to study the stability of one block on an incline, and the results are compared with those predicted by the H-B chart.

In the numerical analysis, the assumed Lame's constants $\lambda = 56GPa$ and $\mu = 21GPa$ and unit weight ρ is $25kN / m^3$. The properties of contacts are assumed as follows: normal stiffness $E_n = 50GPa$ and shear stiffness $G_S = 0.5GPa$. For a methodical comparison, the slope angle α and the aspect ratio of the block .. (b = breadth of the block and d = height of the block, see Figure 4) were varied systematically, while the friction angle φ was fixed at 20°. Different symbols representing different modes of

Figure 4. Kinematic conditions of one block on an incline (DFEM versus theoretical results)

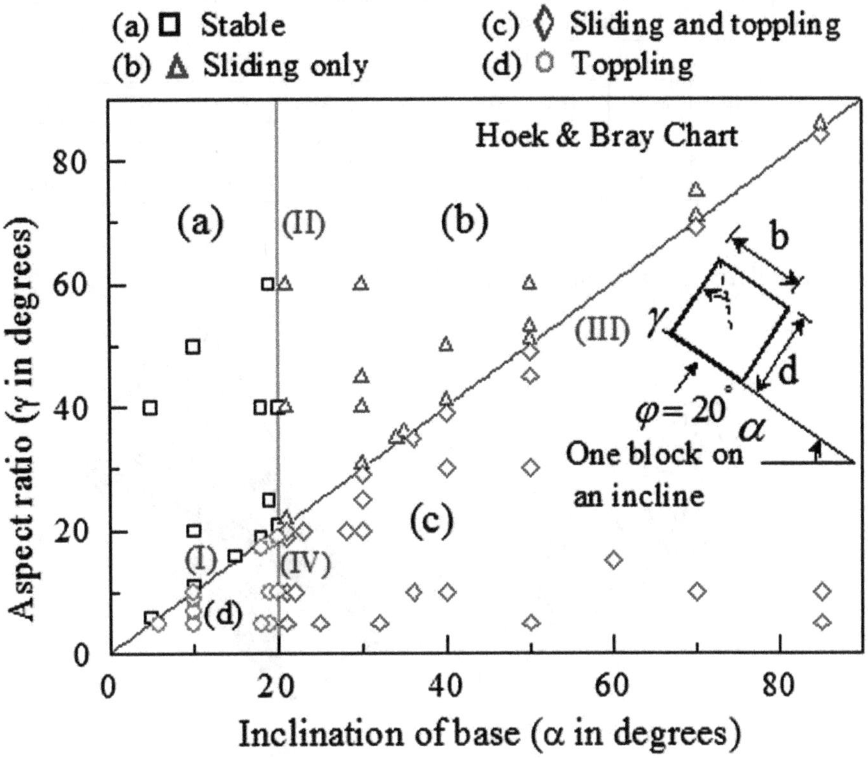

behavior obtained by the DFEM are plotted on the H-B chart as shown in Figure 4. As can be seen from these plots, the results obtained using DFEM are in complete agreement with the theoretical results. Since the theoretical solutions are validated by experiments (Hoek and Bray, 1977; Aydan *et al.* 1989), it can be concluded that the DFEM is a promising method for studying the mechanics of blocky media.

Dynamic Stability

The dynamic stability of square and rectangular blocks on a plane with an inclination of 30° was analyzed using DFEM. The rectangular block was assumed to have a height-to-breadth ratio, h / b, of ⅓. The assumed material properties of the intact rock blocks and mechanical properties of the contact elements used in numerical analyses are given in Table 1. In the table, λ, μ, λ^*, and μ^* indicate the elastic and viscous Lame's constants. ρ denotes the unit weight of the blocks. E_n, G_s, E_n^*, and G_s^* indicate the elastic and viscous normal and shear modulus of the contact. The friction angles for square and rectangular blocks are $\varphi = 25°$ and $\varphi = 35°$. The band width of contact elements, h, is 5 mm (Mamaghani, 1993). Figure 5 shows computed configurations of the $4m \times 4m$ square block and a $12m \times 4m$ rectangular block. The square block slides on the incline (time step $\Delta t = 0.04$ sec), while the rectangular block topples (time step $\Delta t = 0.01$ sec). These predictions are consistent with the kinematic conditions for the stability of a single block in previous studies (Mamaghani, 1993; Mamaghani *et al.*, 1999) as well as with the experimental results reported by Aydan *et al.* (1989). It should, however, be noted that the discretization of the domain, mechanical properties of blocks and contacts, and time steps may cause artificial oscillations and numerical instability.

Table 1. Properties of intact rock and contacts

One Block on an Incline	Intact Block					Contact					
	λ	μ	λ*	μ*	ρ	E_n	G_s	E_n*	G_s*	h	φ
	MPa	MPa	MPa	MPa	kN/m³	MPa	MPa	MPa.s	MPa.s	mm	(°)
Square Block	10	10	5	5	25	0.1	0.1	0.05	0.05	5	25
Rectangular Block	10	10	5	5	25	0.1	0.1	0.05	0.05	5	35

Figure 5. Dynamic stability of one block on an incline

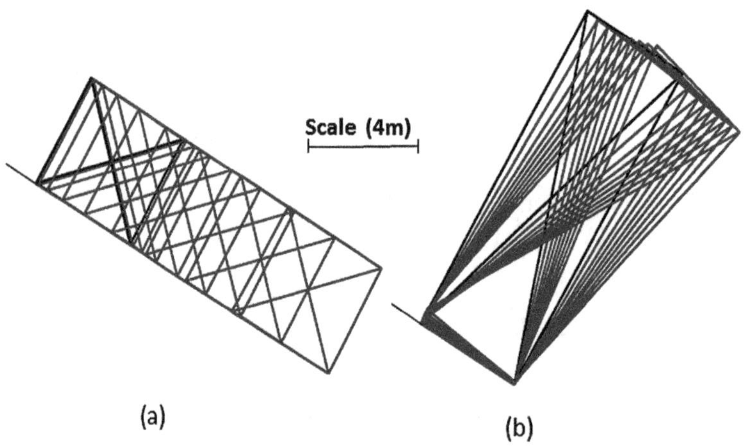

Static Analysis of Masonry Structures

This section is concerned with the application of the DFEM to the static analysis of two masonry structures – the arch and the pyramid. In the numerical analysis, the assumed material and mechanical properties of intact blocks and contacts for both arch and pyramid are listed in Table 2. In the table, λ and μ indicate the elastic Lame's constants. ρ denotes the unit weight of the blocks. $E_{n,}$ and $G_{s,}$ indicate the elastic normal and shear modulus of the contact. Lame's constants are set as $\lambda = 56$ GPa and $\mu = 21$ GPa and unit weight is set as $\rho = 25\ \text{kN/m}^3$. The contact properties were set as follows: normal stiffness $E_n = 50$ GPa and shear stiffness $G_S = 0.5$ GPa. The friction angle between blocks is assumed to be $\varphi = 25°$. The band width of contact elements is set to $h = 5$ mm (Mamaghani, 1993).

Static Analysis of a Masonry Arch

Figure 6a shows an arch structure analyzed using the DFEM under static loading. The dimension of the blocks perpendicular to the *xy* plane is $w = 1.0$ m (Figure 6a). The configuration of the arch in time step nos. 24, 27, and 30 is shown in Figures 6b, 6c, and 6d, respectively. The arch is stable under its own weight as shown in Figure 6b. It is still stable when the distributed uniform traction load per unit length over the arch is less than 1.47 kN/m. However, if the traction load reaches that level, the arch becomes

Table 2. Material and mechanical properties of intact rock and contacts for the analyzed masonry arch and pyramid

Structures	Intact Blocks			Contacts			
	λ	μ	ρ	E_n	G_s	h	φ
	GPa	GPa	kN/m³	GPa	GPa	mm	(°)
Arch	56	21	25	50	0.5	5	25
Pyramid	56	21	25	50	0.5	5	25

unstable. Figure 6c and 6d show the configuration of the arch at failure, which takes place at step no. 27 and no. 30 (Mamaghani, 2006; Mamaghani *et al.*, 1994, 1999).

Static Analysis of Masonry Pyramid

A two-dimensional pyramid was considered as shown in Figure 7. The dimension of the blocks perpendicular to the *xy* plane is taken as $w = 1.0$ m (Figure 7a). Two concentrated loads ($F_1 = 4.9$kN and $F_2 = 127.4$ kN) were applied together with the application of a gravity load, as shown in Figure 7a. As seen from the computed configurations in Figures 7b, 7c, and 7d, inter-block sliding occurs and the pyramid tends to become unstable after each computation step. The results in Figures 7b, 7c, and 7d indicate that the inter-block sliding becomes larger as the number of time steps increases. This observation is

Figure 6. Initial and deformed configurations of a masonry arch

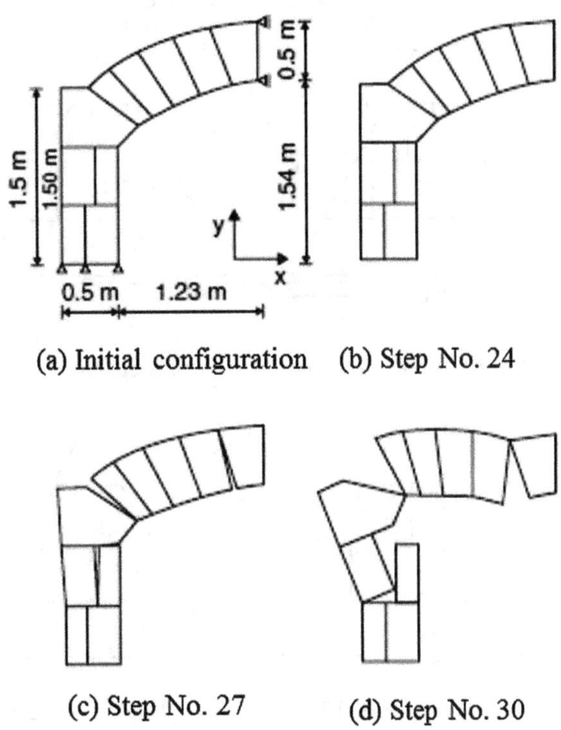

attributed to the fact that the contact material becomes softer as the sliding progresses (Mamaghani, 1993; Mamaghani *et al.*, 1999).

Pseudo-Dynamic Analysis of Masonry Structures

This section is concerned with the application of DFEM to the dynamic analysis of three masonry structures – the arch, the tower, and the wall (Mamaghani *et al.*, 1999). In the analysis, the foundation of the structures was subjected to two types of lateral acceleration waves – Acc. No. 1 with a large period:

$$Acc = 0.8te^{-0.5t}\sin(t)\times 981 \tag{29}$$

and Acc. No. 2 with a small period:

$$Acc = 0.8te^{-0.5t}\sin(3t)\times 981 \tag{30}$$

for which t = time and Acc = lateral acceleration in *gal*, as shown in Figure 8.

The assumed accelerations are used to check the response of analyzed masonry structures by DFEM under two different waveforms. The material and mechanical properties of blocks, foundations, and contacts for the analyzed masonry arch, tower, and wall are given in Table 3, where λ, μ, λ^*, and μ^* indicate the elastic and viscous Lame's constants. ρ denotes the unit weight of the rock mass. E_n, G_s, E_n^*, and G_s^* indicate the elastic and viscous normal and shear modulus of the contact. h and φ indicate band width of

Figure 7. Initial and deformed configurations of the analyzed pyramid

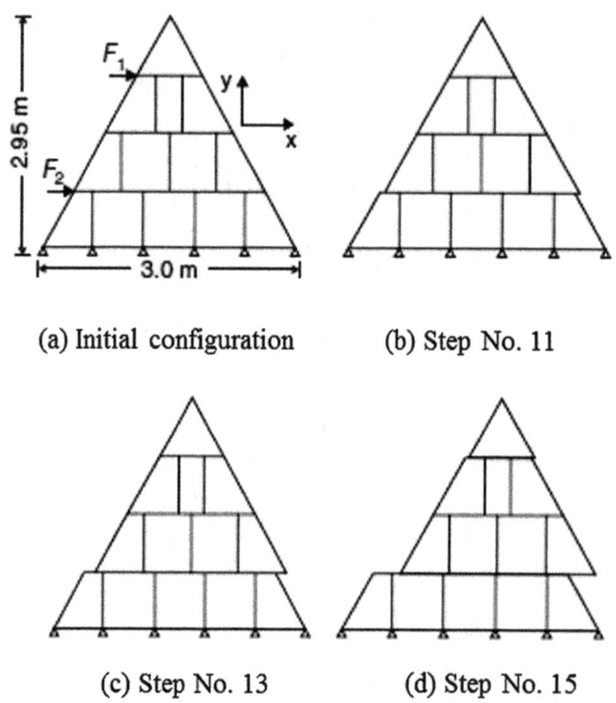

Figure 8. Imposed lateral acceleration waves on foundation

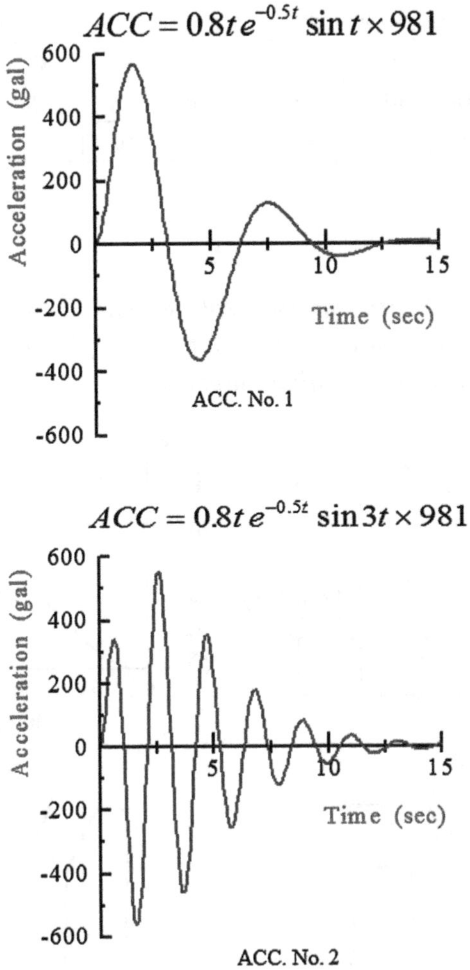

contact elements and friction angle, respectively. In all examples, the time step was chosen as 0.2 sec. In the following dynamic analysis of the three masonry structures, in the plots of the deformed configurations, the displacement in the deformed configurations is amplified by 50 times to make the deformed configuration (mode of failure) more visible from the initial configuration (Mamaghani *et al.*, 1999).

Pseudo-Dynamic Analysis of Masonry Arch

Figures 9a, c, and d show the initial and deformed configurations of a masonry arch at the 23rd time step (4.6 seconds) and 50th time step (10 seconds) subjected to Acc. No. 1 and Acc. No. 2, respectively. Figure 9c shows that the arch is sliding at the base at the 23rd time step under Acc. No. 1, and the crown blocks of the arch start to fall apart while the side columns are still stable. Figure 9c shows that, under Acc. No. 1 at the 50th time step, the arching action disappears and the crown blocks fall apart. The columns slide relative to the base, and they tend to topple in two opposite directions. The blocks tend to separate within the side columns (Figure 9c).

Table 3. Material properties of rock blocks and contacts for the analyzed masonry arch, tower, and wall

Structures	Properties of Blocks					Properties of Contacts					
	λ	μ	λ^*	μ^*	ρ	E_n	G_s	E_n^*	G_s^*	h	φ
	MPa	MPa	MPa	MPa	kN/m³	MPa	MPa	MPa.s	MPa.s	mm	(°)
Arch, Tower, Wall	30	30	30	30	25	50	25	5	2.5	5	35

Figure 9. Initial and deformed configurations and displacement response with time of the arch: (a) Initial configuration, (b) Displacement response with time at the top right corner, (c) Deformed configuration under acceleration No. 1, (d) Deformed configuration under acceleration No. 2

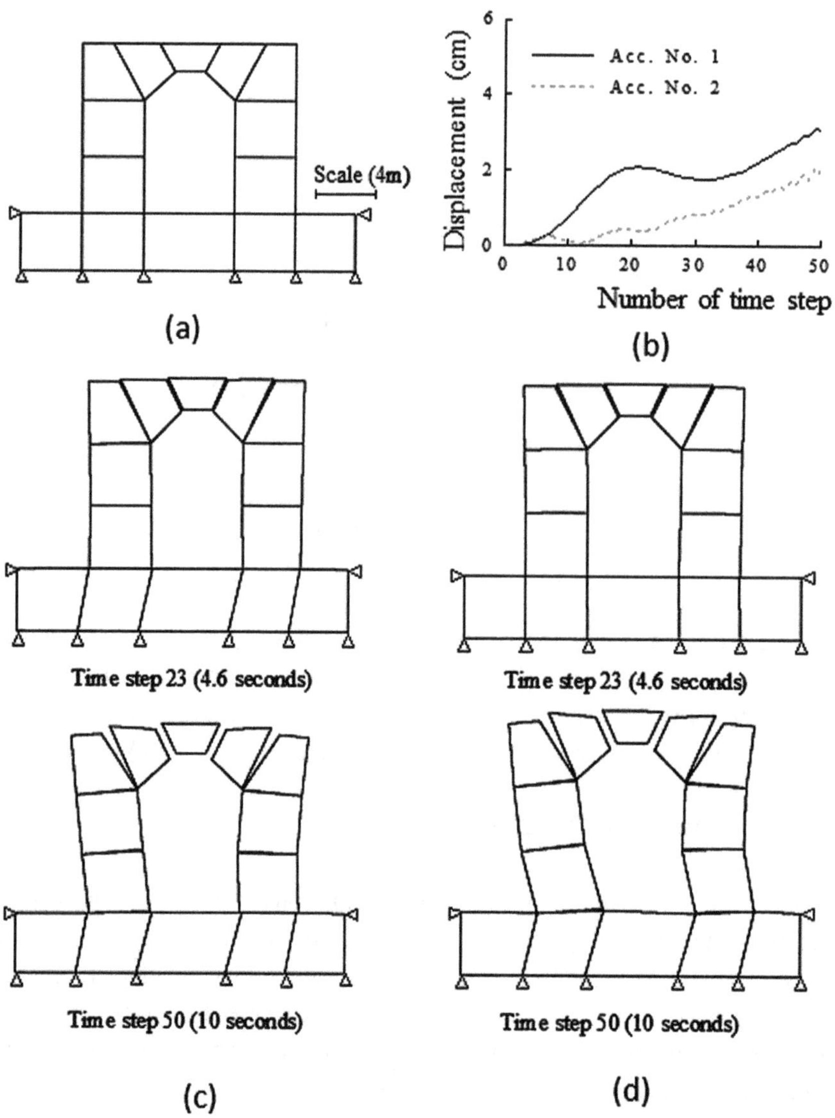

Figure 9 d shows that, under Acc. No. 2 at the 23rd time step, there is no sliding at the base of the arch, while the crown blocks are separated and tend to fall apart. At the 23rd time step, the side columns of the arch exhibit relatively stable behavior under Acc. No. 2 as compared with the Acc. No. 1 (Figures 9c and 9d). However, under Acc. No. 2 at the 50th time step (10 seconds), the side columns of the arch slide at the base, and the arching action disappears while the blocks start to fall apart. As expected, the toppling (failure) modes of the side columns of the arch differ depending on the nature of the imposed form of acceleration waves, as shown in Figures 9c and 9d for the 50th time step.

Figure 9b shows the displacement responses with time of a nodal point at the top most-right corner of the arch corresponding to Acc. No. 1 and Acc. No. 2. The results in Figure 9. b indicate that, as expected, the displacement of the side column of the arch with time is much severe under Acc. No. 1 as compared with Acc. No. 2, especially in the early stage of loading. Figures 9c and 9d show that, under both of the imposed acceleration waves, the reaction of the toppled columns forces the crown block to move upward. This is because of the geometrically symmetric configuration of the structure and outward inclination of the crown block contact interfaces at the center of symmetry (Figure 9a). As can be observed by examining the displacement response curves in Figure 9b, the real value of the displacement is very small as compared with the dimension of the crown block.

Pseudo-Dynamic Analysis of Masonry Tower

Figures 10a, 10b, and 10c show the initial and deformed configurations of a masonry tower at the 22nd time step (4.4 seconds) and 50th time step (10 seconds) corresponding to Acc. No. 1 and Acc. No. 2, respectively. Figure 10b shows that when the tower is subjected to Acc. No. 1, there is sliding at the base of the tower at the 22nd time step, and the uppermost blocks start to detach at the top of the tower. At the 50th time step, relative sliding and separation occurs along block contacts, and the two uppermost blocks tend to topple in opposite directions. Figure 10c shows that, under Acc. No. 2, there is no sliding of the tower at the base while the uppermost blocks of the tower are separated and tend to topple at the 22nd time step. At the 50th time step, there is a relative sliding at the base of the tower, and blocks are slid and detached along block contacts. Figure 11 shows the displacement response versus the number of time steps for a nodal point at the topmost right corner of the tower (monitoring node) corresponding with both imposed acceleration waves. As shown in Figure 11, the toppling of the topmost right block of the tower is more severe under Acc. No. 1 as compared with Acc. No. 2.

Comparison of the responses in Figures 10 and 11 also show that the failure mode of the tower depends on the nature of the imposed acceleration wave. The tower shows relatively stable behavior under Acc. No. 2 with a small period as compared with that of Acc. No. 1 with a large period at time step 22. However, the tower does not return to its original position and ceases to be stable at the end of shaking under both of the imposed forms of the acceleration waves (Figure 10). It is worth noting that, as shown in Figure 10, the response of the tower is quite similar to that expected in actual earthquakes.

Pseudo-Dynamic Analysis of Masonry Wall

Figures 12a, 12b, and 12c show the initial and deformed configurations of a 12m wide, 12m high masonry wall at the 22nd time step (4.4 seconds) and the 50th time step (10 seconds), respectively, subjected to Acc.

Figure 10. Initial and deformed configuration of the masonry tower: (a) Initial configuration, (b) Deformed configuration under acceleration No. 1, (c) Deformed configuration under acceleration No. 2

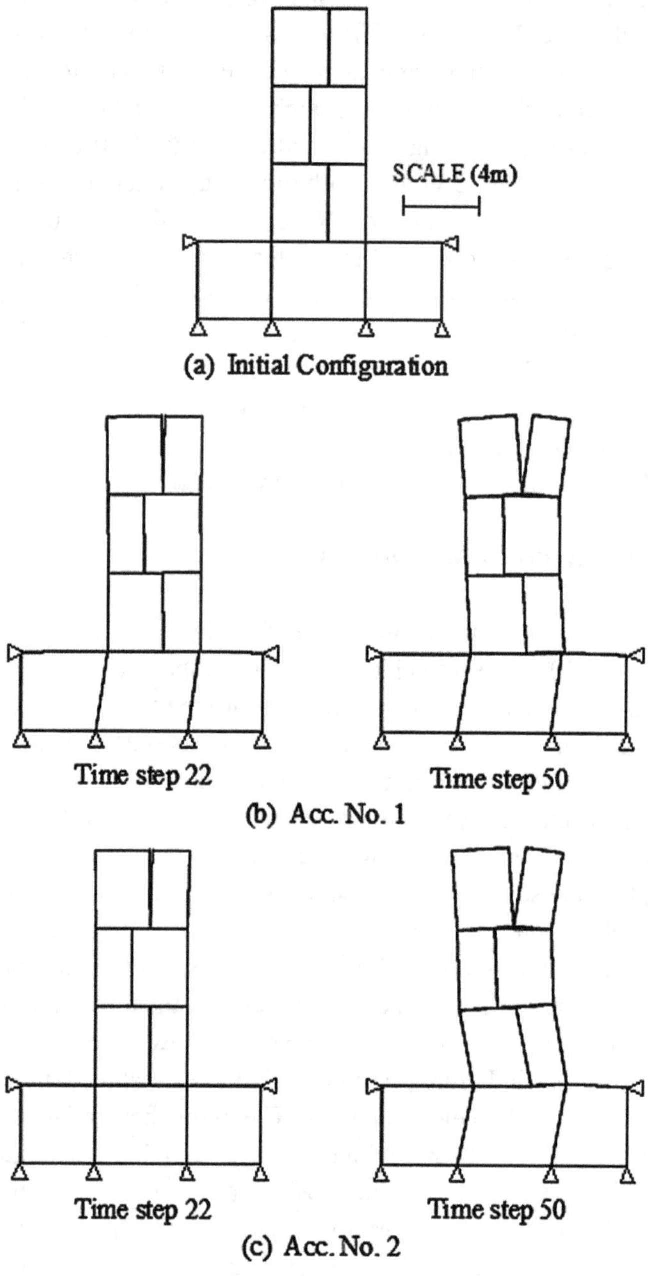

Figure 11. Displacement response with time at the top most-right corner of the tower

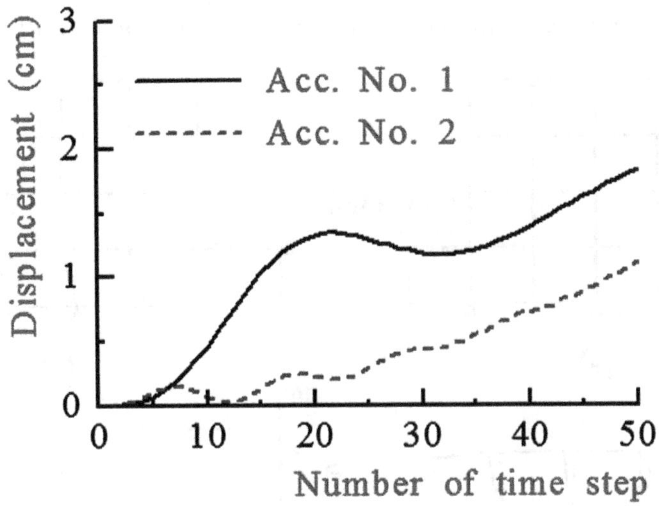

No. 1 (Mamaghani *et al.*, 1999). Figure 12d shows the displacement response versus the number of time steps for a nodal point at the topmost right corner of the wall (monitoring node) corresponding with both of the imposed acceleration waves. As shown in Figure 12b, there is a sliding at the base of the wall at the 22nd time step, and detaching of the blocks occurs starting from the top of the wall. At the 50th time step (10 seconds), there is a relative sliding at the base of the wall and separation and rotation of blocks occurs within the wall (Figures 12c). The blocks within the wall are separated along the vertical discontinuities and form columns which tend to topple in two opposite directions. The tendency of toppling is more severe for the outermost columns as compared with the inner columns, as shown in Figure 12c.

Although the examples discussed above involve very simple structures, they illustrate the fundamental features of the developed DFEM. It is also possible to analyze more complicated masonry structures by DFEM. However, more experimental information is required on the constitutive parameters of blocks and contacts before conducting the analysis of such structures.

CONCLUSION

This chapter was concerned with the analysis of masonry structures, which are comprised of a finite number of distinct, interacting blocks that have length scales relatively comparable with the structure of interest, using the discrete finite element method (DFEM) developed by the author. The DFEM is based on the principles of the finite element method incorporating contact elements. It considers blocks as sub-domains and represents them using solid elements. Contact elements are used to model the block interactions such as sliding or separation. The DFEM calculates displacements at the joints as well as deformation within the blocks, which can be used to follow the processes of the failure mechanism of masonry structures under static and pseudo-dynamic loadings. Through some illustrative examples, the applicability of the DFEM to static and pseudo-dynamic analysis of masonry structures was investigated and discussed. It has been shown that the DFEM is capable of simulating large displacement of masonry blocky systems. It was found that the DFEM is a promising method for studying the behavior of masonry

Figure 12. Initial and deformed configuration of the masonry wall under acceleration No. 1:(a) Initial configuration, (b) Deformed configuration at time step no. 22, (c) Deformed configuration at time step no. 50, (d) Displacement response with time at the top right corner of the wall

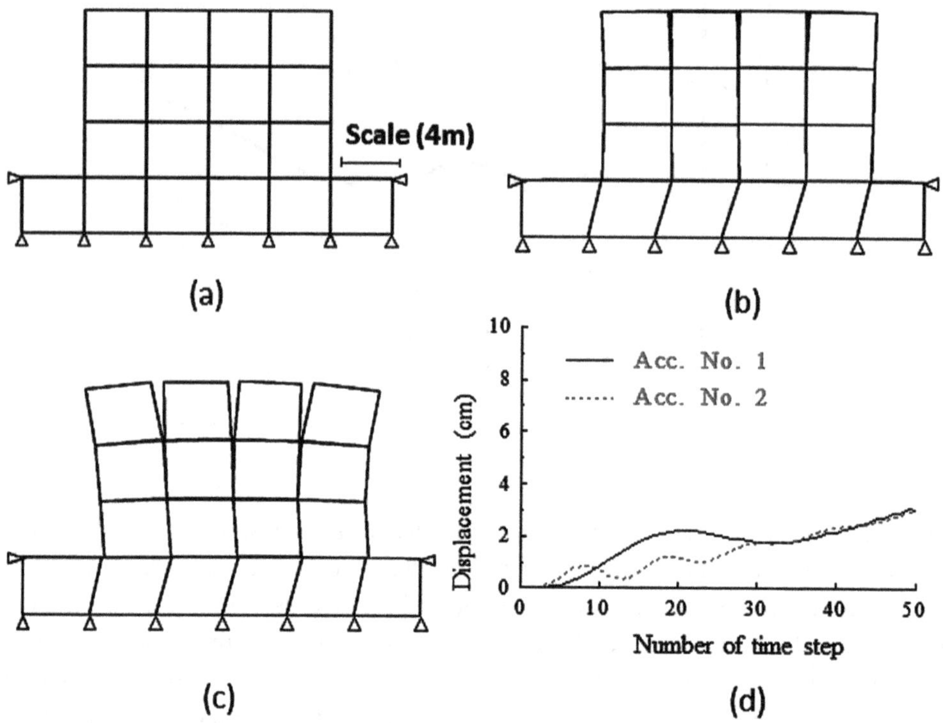

structures under static and dynamic loadings. However, the hyperbolic scheme of the DFEM is still in its formative phase for which both experiments on viscous characteristics of blocks and contacts as well as a numerically stable time-discretization scheme are necessary.

ACKNOWLEDGMENT

The author would like to sincerely thank Professor O. Aydan of Tokai University (Japan) for his collaboration in developing the DFEM and help in some of the computations. The author dedicates this chapter to his beloved mother, Mrs. Rakhshan Saberi Mamaghani, for her unconditional continuous support and encouragement.

REFERENCES

Aydan, Ö., Mamaghani, I. H. P., & Kawamoto, T. (1996). Applications of discrete finite element method (DFEM) to rock engineering structures. In*2nd North American Rock Mechanics Symposium Tools and Techniques in Rock Mechanics*.

Aydan, Ö., Ohta, Y., Daido, M., Kumsar, H., Geni¸, M., Tokashiki, N., & Amini, M. et al. (2011). Earthquakes as a rock dynamic problem and their effects on rock engineering structures. In *Advances in Rock Dynamics and Applications*. Taylor & Francis Publisher.

Aydan, Ö., Shimizu, Y., & Ichikawa, Y. (1989). The effective failure modes and stability of slopes in rock mass with two discontinuity sets. *Rock Mechanics and Rock Engineering*, *22*(3), 163–188. doi:10.1007/BF01470985

Aydan, O., Tokashiki, N., & Mamaghani, I. H. P. (2005). Modeling and Analysis of Rock structural systems by Discrete Finite Element Method. Journal of Geotechnical Engineering and Surveying, 106, 8-15. (in Japanese)

Baggio, C., & Trovalusci, P. (1993). Discrete models for jointed block masonry walls. In A. A. Hamid & H. G. Harris (Eds.), The Sixth North American Masonry Conference (Vol. 2, pp. 939–949). Lancaster, PA: Technomic Publishing Co. Retrieved from http://dsg.uniroma1.it/trovalusci/pubblicazioni_pdf/B2_Baggio-Trovalusci_NAMC09.pdf

Baggio, C., & Trovalusci, P. (1998). Limit Analysis for No-Tension and Frictional Three-Dimensional Discrete Systems. *Mechanics of Structures and Machines, 26*(3), 287-304. Retrieved from http://dsg.uniroma1.it/trovalusci/pubblicazioni_pdf/A4_Baggio-Trovalusci_MSM98.pdf

Barbosa, R. E., & Ghaboussi, J. (1990). Discrete finite element method for multiple deformable bodies. *Finite Elements in Analysis and Design, Elsevier Science Publishers, 7*(2), 145–158. doi:10.1016/0168-874X(90)90006-Z

Cundall, P. A. (1971). A computer model for simulating progressive large-scale movements in blocky rock systems.*Proc. Int. Symp. on Rock Fracture.*

Cundall, P. A. (2011). Lattice method for brittle, jointed rock. In *Continuum and distinct element numerical modeling in geomechanics, number*. Itasca International, Inc.

Damjanac, B., Detournay, C., & Cundall, P. A. (2015). *Application of particle and lattice codes to simulation of hydraulic fracturing*. Retrieved from http://www.itascacg.com/application-of-particle-and-lattice-codes-to-simulation-of-hydraulic-fracturing

Furtney, J., Zhang, F., & Han, Y. (2013). Review of methods and applications for incorporating fluid flow in the discrete element method. In P. Zhu, C. Detournay, R. Hart, & M. Nelson (Eds.), *Continuum and distinct element numerical modeling in geomechanics*. Minneapolis, MN: Itasca International Inc.

Giamundo, V., Sarhosis, V., Lignola, G. P., Sheng, Y., & Manfredi, G. (2014). Evaluation of different computational modelling strategies for modelling low strength masonry. *Engineering Structures*, *73*, 160–169. doi:10.1016/j.engstruct.2014.05.007

Goodman, R.E., Taylor, R. & Brekke, T.L. (1968). A model for the mechanics of jointed rock. *J. of Soil Mechs. and Found. Eng. Div.*, *94*(3), 637-659.

Hoek, E., & Bray, J. W. (1977). Rock slope engineering (2nd ed.). The Institution of Mining and Metallurgy.

Itasca Consulting Group, Inc. (2014). *UDEC Version 6., Distinict-Element Modeling of jointed and Blocky Material in 2D*. Retrieved from http://www.itascacg.com/software/udec

Kawamoto, T., & Aydan, Ö. (1999). A review of numerical analysis of tunnels in discontinuous rock masses. *International Journal for Numerical and Analytical Methods in Geomechanics*, *23*(13), 1377–1391. doi:10.1002/(SICI)1096-9853(199911)23:13<1377::AID-NAG932>3.0.CO;2-S

Lermos, J. V. (2007). Discrete Element Modeling of Masonry Structures. *International Journal of Architectural Heritage: Conservation, Analysis, and Restoration*, *1*(2), 190–213. doi:10.1080/15583050601176868

Lisjak, A., Kaifosh, P., Mahabadi, O. K., Tatone, B., & Grasselli, G. (2015). Fully 3D simulation of fluid-pressure-driven fracturing using a novel continuum-discontinuum approach: Preliminary results. In GeoConvention 2015: New Horizons.

Mamaghani, I. H. P. (1993). *Numerical analysis for stability of a system of rock blocks.* (Master Thesis). Department of Civil Engineering, Nagoya University, Nagoya, Japan.

Mamaghani, I. H. P. (2006). Analysis of Masonry Bridges: Discrete Finite Element Method. Transportation Research Record: Journal of Transportation Research Board, 1976, 13-19.

Mamaghani, I. H. P. (2008). Analysis of Underground Structures by Discrete Finite Element Method. Canadian Society for Civil Engineering.

Mamaghani, I. H. P. (2015a). *Discrete Finite Element Method Application for Analysis of Unreinforced Masonry Underground Structures. In Transportation Research Record: Journal of Transportation Research Board, No. 2522* (pp. 131–136). Washington, DC: Transportation Research Board of the National Academies. http://trrjournalonline.trb.org/doi/pdf/10.3141/2522-13

Mamaghani, I. H. P. (2015b). Analysis of Masonry Structures by Discrete Finite Element Method. *12th North American Masonry Conference*.

Mamaghani, I. H. P., & Aydan, O. (1999). Discrete Finite Element Method for Analysis of Masonry and Rock Structures. *International Journal of Science and Technology, Scientia Iranica, Tehran, Iran*, *6*(3 & 4), 157–164.

Mamaghani, I. H. P., & Aydan, Ö. (2000). Stability analysis of slopes by discrete finite element method. *GeoEng*, *2000*(November), 19–24.

Mamaghani, I. H. P., Aydan, Ö., & Akhoundi, F. (2014). Analysis of Masonry Bridges under Static and Dynamic Loading by Discrete Finite Element Method. *9th International Masonry Conference*. International Masonry Society, University of Minho.

Mamaghani, I. H. P., Aydan, Ö., & Kajikawa, Y. (1999). Analysis of masonry structures under static and dynamic loading by discrete finite element method. Journal of Structural Mechanics and Earthquake Engineering, 626, 1-12.

Mamaghani, I. H. P., Baba, S., Aydan, Ö., & Shimizu, Y. (1994). Discrete finite element method for blocky systems.*Proc. of the Eighth Int. Conf. on Computer Methods and Advances in Geomechanics (IACMAG)*.

Munjiza, A. A. (2004). The Combined Finite-Discrete Element Method. Wiley.

Oñate, E. (2015). A Particle-Discrete-Finite Element Method for Analysis of Particulate Flows and Their Interaction with Structures. *13th US National Congress on Computational Mechanics*.

Oñate, E., Celigueta, M. A., Idelsohn, S. R., Salazar, F., & Suárez, B. (2011). Possibilities of the particle finite element method for fluid–soil–structure interaction problems. *Computational Mechanics*, *48*(3), 307–318. doi:10.1007/s00466-011-0617-2

Oñate, E., Celigueta, M. A., Latorre, S., Casas, G., Rossi, R., & Rojek, J. (2014). Lagrangian analysis of multiscale particulate flows with the particle finite element method. *Computational Particle Mechanics*, *1*(1), 85–102. doi:10.1007/s40571-014-0012-9

Rougier, E., Knight, E. E., & Munjiza, A. (2012). Fluid driven rock deformation via the combined FEM/DEM methodology. In American rock mechanics association 46thUS rockmechanics/ geomechanics symposium.

Rougier, E., Knight, E. E., Munjiza, A., Broome, A. J. S. S. T., Swift, R. P., & Bradley, C. R. (2011). The combined finite-discrete element method applied to the study of rock fracturing behavior in 3D. In American Rock Mechanics Association 45 th US rock mechanics/geomechanics symposium.

Sarhosis V., Sheng Y. (2014). Identification of material parameters for low bond strength masonry. *Engineering Structures, 60*, 100-110. DOI: 10.1016/j.engstruct.2013.12.013

Shi, G. H. (1988). *Discontinuous deformation analysis: a new numerical model for the statics and dynamics of block system.* (PhD Thesis). Dept. of Civil Engineering, University of California, Berkeley, CA.

Tokashiki, N., Aydan, Ö., Mamaghani, I. H. P., & Kawamoto, T. (1997). The stability of a rock block on an incline by discrete finite element method (DFEM). In IACMAG '97.

Tokashiki, N., Aydan, O., Shimizu, Y., & Mamaghani, I. H. P. (2001). A stability analysis of masonry walls by discrete finite element method (DFEM).*Proc. of the 10th Int. Conf. on Computer Methods and Advances in Geomechanics (IACMAG'01).*

Toth, A. R. (2009). Orban, z., Bagi, k. (2009). Discrete element analysis of a masonry arch. *Mechanics Research Communications*, *36*(4), 469–480. doi:10.1016/j.mechrescom.2009.01.001

Zhao, Q., Lisjak, A., Mahabadi, O. K., & Grasselli, G. (2015). Numerical investigation of the influences of rock fabrics on hydraulic fracturing operations. In GeoConvention 2015: New Horizons.

KEY TERMS AND DEFINITIONS

Band Type Modeling: A model in which contacts between neighboring blocks in masonry structures are treated as bands with a finite thickness.

Contact Model: A model which specifies the interactions among blocks.

Contacts: Block interfaces or rock discontinuities.

Finite Displacement: The large deformation of blocky systems that is associated with the separation, translation, and rotation of blocks.

Kinematic Conditions: Sliding and/or toppling of a block on an incline.

Mechanical Model: A model which represents the deformable blocks.

Unreinforced Masonry Structures: Structures comprised of a finite number of distinct interacting rock blocks that have a length scale relatively comparable to the structure.

Chapter 16
A Semi-Discrete Approach for the Numerical Simulation of Freestanding Blocks

Fernando Peña
Instituto de Ingenieria, Universidad Nacional Autonoma de Mexico, Mexico

ABSTRACT

This chapter addresses the numerical modeling of freestanding rigid blocks by means of a semi-discrete approach. The pure rocking motion of single rigid bodies can be easily studied with the differential equation of motion, which can be solved by numerical integration or by linearization. However, when we deal with sliding and jumping motion of rigid bodies, the mathematical formulation becomes quite complex. In order to overcome this complexity, a Semi-Discrete Model (SMD) is proposed for the study of rocking motion of rigid bodies, in which the rigid body is considered as a mass element supported by springs and dashpots, in the spirit of deformable contacts between rigid blocks. The SMD can detect separation and sliding of the body; however, initial base contacts do not change, keeping a relative continuity between the body and its base. Extensive numerical simulations have been carried out in order to validate the proposed approach.

INTRODUCTION

The study of the dynamic behavior of rigid bodies is an important task in the seismic assessment of structures, since selected structures or structural elements can be modeled as rigid bodies. Typical examples, among others, are: constructions formed by large stone blocks (Lemos, 2007; Papaloizou & Komodromos, 2012; Park & Kim, 2013); simple masonry structures which often fail forming large macro-blocks under seismic loadings (Sorrentino et al., 2008; Mohammadi & Yasrebi, 2010; Costa et al., 2013; Lagomarsino, 2015); machines, furniture, equipment, nuclear reactors, statues and art objects (Di Egidio & Contento, 2009; Erdik et al., 2010; Konstantinidis & Makris, 2009; Sharif et al., 2009); and base isolated buildings (Hoseini & Alavi, 2014; Komodromos et al., 2007, Palmeri & Makris, 2008).

DOI: 10.4018/978-1-5225-0231-9.ch016

Initial studies are from the late 19th century (Milne, 1881; Perry, 1881); however, Housner's work (Housner, 1963) is considered as the first systematic study about the dynamics of rigid bodies. Housner has dealt only with pure rocking motion. After Housner, many authors have studied the dynamic motion of rigid bodies considering complex motion, as for example rocking plus sliding or bouncing (Anooshelpoor et al., 2004; Boroschek & Iruretagoyena, 2006; Hogan, 1990; Ishiyama, 1982; Plaut et al., 1996; Purvance et al., 2008; Taniguchi, 2002; Taniguchi & Miwa, 2007; Tso & Wong, 1989).

The pure rocking motion of single rigid bodies can be easily studied with the differential equations of motion proposed by Housner (1963) which can be solved by numerical integration or by linearization. However, when we deal with complex motion, the mathematical formulation becomes quite complex (Hogan, 1992; 1994; Nozaki et al., 2009; Spanos et al., 2001; Di Egidio & Contento, 2010; Voyagaki et al., 2014). In order to overcome this complexity, several authors have used other mathematical formulations, as the Discrete Element Method (Lemos, 2007; Peña et al., 2007; Komodromos et al., 2008) or have proposed novel analytical and numerical models (Andreus and Casini, 1999; Michaltsos and Raftoyiannis 2008; Nozaki et al., 2009; Prieto & Lourenço, 2005; Zulli et al., 2012). Thus, despite significant advances from past research, the study of complex motion of rigid bodies remains a challenging task.

In this chapter, a Semi-Discrete Model (SDM) is proposed for the study of freestanding blocks, in which the block is considered as a mass element supported by springs and dashpots, in the spirit of deformable contacts between rigid blocks (Andreaus & Casini, 1999).

The advantage of the proposed model with respect to a full Discrete Model is that the Semi-Discrete approach can also detect separation (jumping) and sliding of the body. However, initial contacts do not change, in order to simplify the computational effort, keeping a relative continuity between the body and its base. In fact, the computational codes based on discrete analysis techniques must include routines for identifying changes in the contacts during the analysis, without user intervention. In theory, the detection contacts requires simple geometric calculations but detection test may require a relatively long computer time (Azevedo et al., 2000). Thus, the model is able to reproduce the six motion states of the rigid bodies (Ishiyama, 1982): rest, slide, rotation, slide-rotation, translation-jump and rotation-jump.

The main limitation of the model is that cannot detect changes in the geometry (new base contacts). However, there are many practical applications where it is not necessary to consider changes in the initial geometry, since it means collapse of the body. Thus, the model is pointed to those applications.

DYNAMIC BEHAVIOR OF FREESTANDING BLOCKS

Freestanding blocks are objects particularly vulnerable to lateral seismic loading, since they have no tensile strength and stability is ensured if the resultant force from all actions falls inside the base of the block. The study based upon the assumption of continuum structures is not realistic and models based on rigid-block assemblies provide a suitable framework for the study of dynamic response under seismic actions. In this context, the problem is primarily concerned with Rocking Motion dynamics (Peña et al., 2008).

Rocking motion is defined as the oscillation of the rigid bodies present in a structure when the center of rotation instantly change from one reference point to another. This instantaneous change produces a loss of energy due to an impulsive force (Prieto et al., 2004) and it has been shown that it is not possible to study a structure that presents rocking motion as an oscillator of one degree of freedom (Makris & Konstantinidis, 2003).

The Housner's work (1963) is considered as the first systematic study about the dynamics of rigid bodies and is known as the classical theory. Housner (1963) obtained the equation for the period of the system, which depends on the amplitude of rocking, and the equation for the restitution coefficient. This author also proposed expressions to calculate the minimum acceleration required to overturn a single rigid body. The main hypotheses of Housner are:

- The block and its base are perfectly rigid.
- The surface of the base is horizontal.
- The block is symmetric with respect to the vertical central axis.
- There is no sliding of the rigid body.
- Only in-plane (2D) motion is considered.
- The impact of the block during rocking is not elastic, meaning that the block does not jump. Therefore, at least one contact point always exists between the rigid body and its base.
- Angular moment conservation exists (before and after impact).

With these hypotheses, there are three main parameters for the planar rocking motion problem. These quantities depend solely on the geometry of the specimens, as follows:

- The critical angle α defined as the angle at which the stone overturns due to static forces.
- The frequency p associated to the system due to the interaction between the block and its base.
- The coefficient of restitution μ defined as the angular velocity reduction ratio, between two consecutive impacts.

After Housner, several experimental and analytical studies have been made on the rigid bodies under different conditions. The following conclusions can be drawn from these studies (Peña et al., 2008):

- The three main parameters are different if they are obtained from laboratory tests.
- The friction between the base and the block is not infinite, thus it is possible to have slide behavior. According to Shenton (1996), sliding will occur when the following conditions are fulfilled $\eta_s < a_g$; $\eta_s < b/h$. Where η_s is the static coefficient of friction and a_g is the peak ground acceleration as fraction of the gravity acceleration.
- The motion of a rigid body is a combination of rocking, sliding and jumping. In fact, there are six motion states (Ishiyama, 1982): rest, slide, rotation, slide-rotation, translation-jump and rotation-jump.
- The response of a rigid body is very sensitive to the boundary conditions, the impact (coefficient of restitution) and the type of the base motion (harmonic, random, frequency contents, etc.).

Classical Formulation

Consider the rigid body of Figure 1, which has any geometrical shape and can rock on the rocking points RP1 and RP2, which are not necessarily placed at the same level. The differential equations of the rocking motion around rocking points RP1 and RP2, subjected to horizontal ground accelerations $a(t)$, are respectively (Houner, 1963):

Figure 1. Rigid body: a) rocking points and variables; b) rocking angle

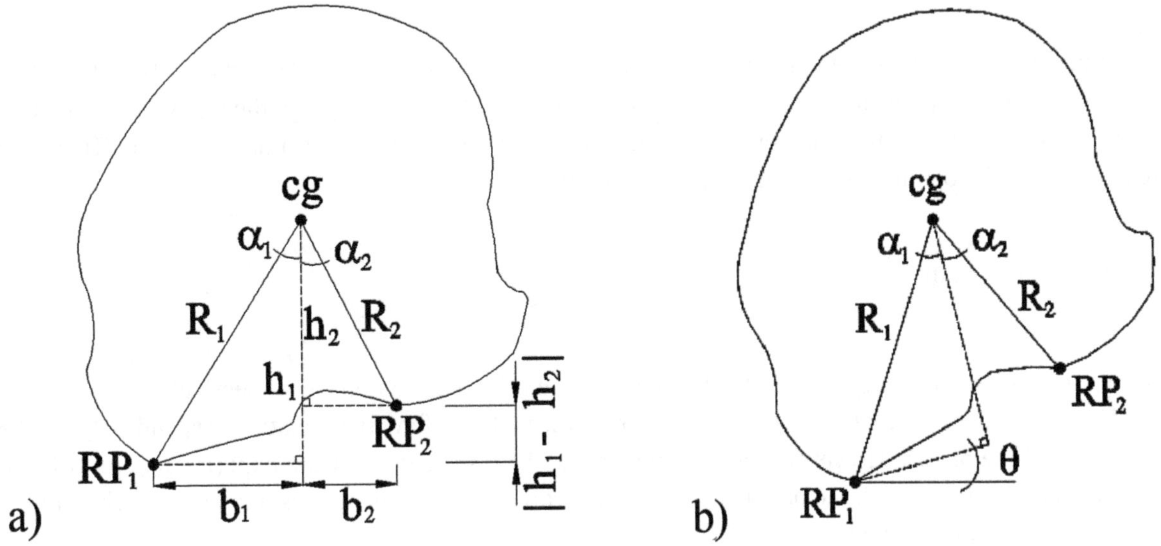

$$\ddot{\theta} + p_1^2 \sin(\alpha_1 - \theta) = p_1^2 \cos(\alpha_1 - \theta)\frac{a(t)}{g} \tag{1}$$

$$\ddot{\theta} - p_2^2 \sin(\alpha_2 + \theta) = p_2^2 \cos(\alpha_2 - \theta)\frac{a(t)}{g} \tag{2}$$

where θ and $\ddot{\theta}$ are the rocking angle and the rocking angular acceleration respectively, g is the acceleration of gravity.

The critical angle α_i is the angle formed by a perpendicular line from the level of the rocking point RP_i to the gravity center cg, called height h_i, and a line from the rocking point RP_i to the point where the height crosses the level of the rocking point, named base b_i, and can be calculated by means of (see Figure 1a):

$$\alpha_i = \tan^{-1}\left(\frac{b_i}{h_i}\right) \tag{3}$$

Parameter p_i is related to the frequency of the body and it is defined as:

$$p_i^2 = \frac{MgR_i}{I_i} \tag{4}$$

where M is the mass, I_i is the corresponding moment of inertia (defined with respect to RP_i) and R_i is the distance from the gravity center to the rocking point RP_i and is defined as:

$$R_i = \sqrt{b_i^2 + h_i^2} \tag{5}$$

In order to take into account the impact mechanism, it is appropriate to assume a coefficient of restitution μ_i that multiplies the angular velocity $\dot{\theta}$ when the body passes through the equilibrium position at θ=0. By considering that angular momentum conservation exists during impact, the coefficient of restitution can be obtained by means of (Housner, 1963):

$$\mu_i = \frac{\dot{\theta}^a}{\dot{\theta}^b} = 1 - \frac{3}{2}\sin^2(\alpha_i) \tag{6}$$

where $\dot{\theta}^a$ and $\dot{\theta}^b$ are the angular velocities just after and before the impact, respectively.

The three main parameters α, p and μ, which governs the differential equation, depend only on the geometry of the blocks. However these parameters can be obtained also from experimental test of free rocking motion. In general, the parameters obtained from the experimental test may not match with the theoretical ones due to (Peña et al., 2008):

- Detailed measurements in the real dimensions of the blocks. Small variations in the geometry give differences in the classical parameters.
- Real position of the gravity center. In the case of non-homogenous material or arbitrary shape of the block.
- Rotation surface. When the block is completely rigid, the rotation point is located at the corner of the base. But when the block is not completely rigid, the rotation point does not exist. In this case, the block will tend to rotate on a finite surface. The size of this surface will depend on the material stiffness. It will tend to zero (single point) when the stiffness of the material tends to infinite.
- Damage. Even minor damage at the base of the block changes the values of the classical parameters.
- Three dimensional effects.

Symmetry of the Rocking Motion

Considering for a while that the system of Figure 1 is undamped and $\alpha_1=\alpha_2$, $p_1=p_2$, then Equations 1 and 2 will be equal except for a sign that will depend on the sign of the rocking angle θ. Therefore, a symmetry in the system due to the rocking motion exists, by which the dynamics is invariant under specular reflections with respect to a vertical line passing through the rocking point. This symmetry is equivalent to an invariance with respect to the sign of rocking angle (Prieto and Lourenço, 2005).

In other words, this means that, for the same base motion, the rigid body will have the same response independently of the direction of the motion. However, this is only true if $\alpha_1=\alpha_2$ and $p_1=p_2$ (see Figure 2). On the one hand, the critical angle α will be equal for both sides if the following condition is fulfilled:

$$\frac{b_1}{h_1} = \frac{b_2}{h_2} \tag{7}$$

Figure 2. Dynamic symmetry: a) asymmetric rigid body with symmetric rocking motion; b) symmetric rocking motion

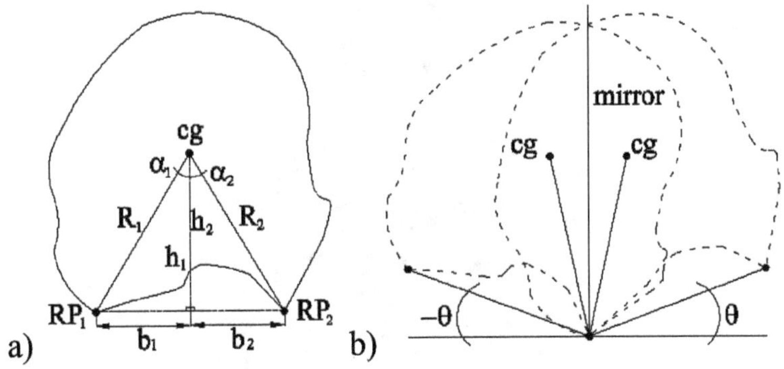

On the other hand, the parameter p_i depends on the mass of the body M, the moment of inertia I_i and the distance R_i (see Equation 4). The mass is independent of the position of the rocking point. Therefore, p_i will only depend on I_i and R_i. The moment of inertia *Ii* can be calculated by means of:

$$I_i = \int_V \rho r_i^2 dV \tag{8}$$

where ρ is the mass density of the material, r_i is the distance from the rocking point to a particle of mass and V is the volume of the body. If we consider a homogeneous material and a body with constant thickness z, Equation 8 becomes:

$$I_i = \rho z \int_A r_i^2 dA = \rho z \int_A \left(x^2 + y^2\right) dA = \rho z I_{pi} \tag{9}$$

where A is the area of the body and I_{pi} is the geometrical polar moment of inertia respect to an axis that passes through the rocking point RP_i. Considering a geometrical polar moment of inertia with respect to an axis that passes through the gravity center I_{pc}, Equation 9 becomes:

$$I_i = \rho z I_{pc} + MR_i^2 \tag{10}$$

Since I_{pc} is independent of the positions of the rocking points, the moment of inertia I_i depends only on R_i; thus parameter p_i is only dependent on the distance from the gravity center to the rocking point R_i. Therefore, to maintain the symmetry in the system, the other condition that must be fulfilled is ($p_1=p_2$):

$$R_1 = \sqrt{b_1^2 + h_1^2} = \sqrt{b_2^2 + h_2^2} = R_2 \tag{11}$$

The only possibility in which both conditions (Equations 7 and 11) will be fulfilled at the same time is when $h_1=h_2$ and $b_1=b_2$. Therefore, it is possible to conclude that the symmetry of a rigid body depends only on the symmetry of the rocking points with respect to the gravity center (same base b_i and same height h_i) and not on the geometrical shape of the body (Figure 3).

Figure 3. Asymmetric rigid body with symmetric rocking motion

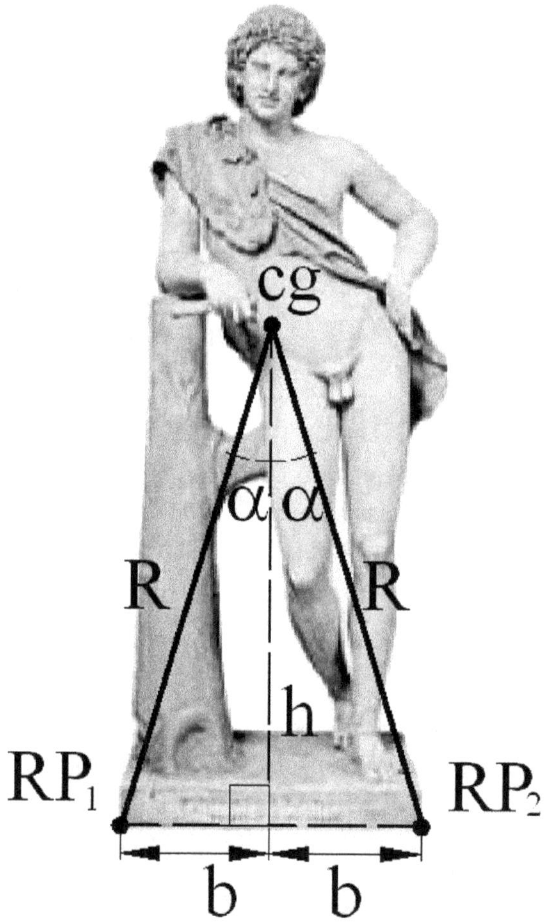

Causes of Asymmetry

Some authors have studied asymmetric rigid bodies (Anooshelpoor et al., 2004; Fielder et al., 1997; Plaut et al., 1996; Purvance et al., 2008; Di Egidio & Contento, 2010). It has been shown that asymmetry changes the dynamics of the rocking motion of rigid bodies. For example, according with Purvance et al. (2008), bodies with multiple rocking points are significantly more fragile than bodies with simple basal contact. This means that multiple rocking points are equivalent to a more slender block with simple basal contact condition.

It is possible to establish two types of asymmetry (Peña, 2015). The first one is the asymmetry of the rocking points with respect to the gravity center, here named base asymmetry. This type of asymmetry is due mainly to the geometry, material and damage of the body. The second type is the boundary asymmetry caused by the boundary conditions, especially due to more than two supports or rocking points.

Geometry is the most common cause of non-symmetry in rigid bodies. However, it must have in mind that geometrical asymmetry does not necessary produce asymmetry on the rocking motion. Consider, for example, the rigid body of Figure 3 made with a homogeneous material. This body will have symmetric rocking motion, since they have symmetry of the rocking points with respect to the gravity center, even if the geometrical shape is not symmetric.

A non-homogeneous material of the rigid body can be a cause of asymmetry. Since the gravity center does not coincide with the geometrical centroid. Damage and degradation of the material are another cause of asymmetry. Even small damage at the base changes the response of the rigid body. For example, Figure 4 shows a typical free rocking motion of two different blocks, with and without symmetry respectively. In the case of symmetry (Figure 4a), the amplitude of the rocking angle decrease constantly in each half cycle, since the motion lost energy due to the impact of the block. However, when the motion is asymmetric (Figure 4b), the amplitude of the rocking angle is different depending on the direction of rotation, yielding non-symmetric response and the maximum rocking angle in each cycle is always half degree greater in the damaged side, when compared with the undamaged side (Peña et al., 2008).

SEMI-DISCRETE MODEL

Base asymmetry can be easily modeled with Equations 1 and 2. For this procedure only parameters α_i, p_i and μ_i are necessary and they can be obtained from the geometry of the block (Equations 3 to 6). The differential equations of motion can be solved by numerical integration or by linearization. However, Equations 1 and 2 cannot be used when sliding or jumping are involved in the motion. As well as, when we deal with boundary asymmetry, the mathematical formulation becomes quite complex (Hogan, 1992; 1994; Spanos et al., 2001).

In order to overcome this complexity, a Semi-Discrete Model (SDM) is proposed for the study of rocking motion of slender rigid bodies, in which the rigid body is considered as a mass element supported by springs and dashpots, in the spirit of deformable contacts between rigid blocks (Andreaus & Casini, 1999; Palmeri & Markis, 2008).

The SDM is based on the Rigid Element Model (REM) proposed by Casolo and Peña (2007). However, the main difference between the models is that in REM the springs represent the mechanical characteristics of the material, and the strains and stresses of each spring represent the average strains and stresses

Figure 4. Typical free rocking response of a rigid body: a) symmetric rocking motion; b) asymmetric rocking motion

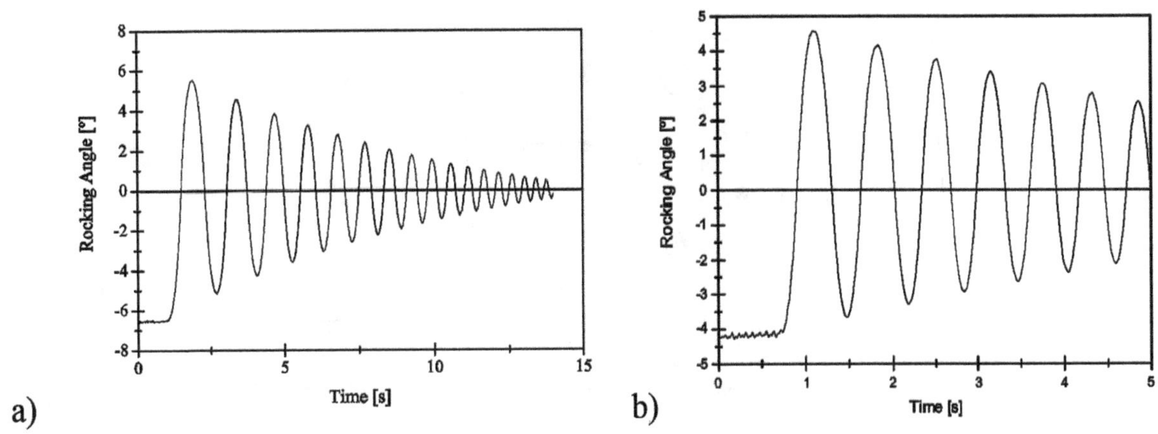

inside each element, according to a tributary volume. Nevertheless, the SDM considers that the springs represent the mechanical characteristics of the joint, while the element is fully rigid.

It is worth noting that the SDM proposed here is only for the study of slender rigid bodies. A slender rigid body can be understood as a body that has a slenderness angle α lesser than about 20° or a slenderness ration *h/b* greater than 2.75. The advantage to deal with a slender rigid body is that the sine of the angle may be approximated by the value of the angle in radians. Therefore, it is possible to model the rocking behavior considering only small rotations.

Formulation

The rigid body is considered as a two-dimensional plane solid body with *n* sides. The global Cartesian coordinate frame $\{O, x, y\}$ is placed in order to have the gravity acceleration *g* applied in the negative *y*-axis direction. A local reference frame $\{o, \xi, \eta\}$, with the ξ-axis initially parallel to the global *x*-axis, is fixed the barycenter of the element. The deformed configuration of the model is described by the variations of position of this local reference frame with respect to the global one. The translations *u*, *v* and the rotation angle ψ, associated with the element's barycenter, are collected into the Lagragian coordinates $\{u\}$. The loads, including inertial forces, are condensed into three resultants: the forces f_x and f_y applied to the element centroid and the couple Θ (Figure 5). They are assembled into the vector of external loads $\{f_e\}$ which is conjugated in virtual work with {u} as follows:

$$\{u\}^T = \{u, v, \psi\} \tag{12}$$

$$\{f_e\}^T = \{f_x, f_y, \Theta\} \tag{13}$$

Figure 5. Reference frame, forces and displacements in the SDEM

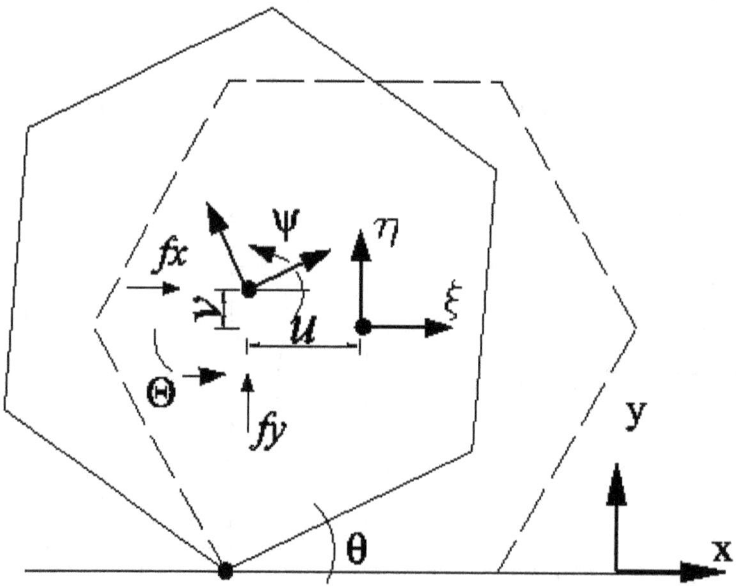

Figure 6. Example of a rigid body with three initial base contacts: a) springs and dampers; b) variables

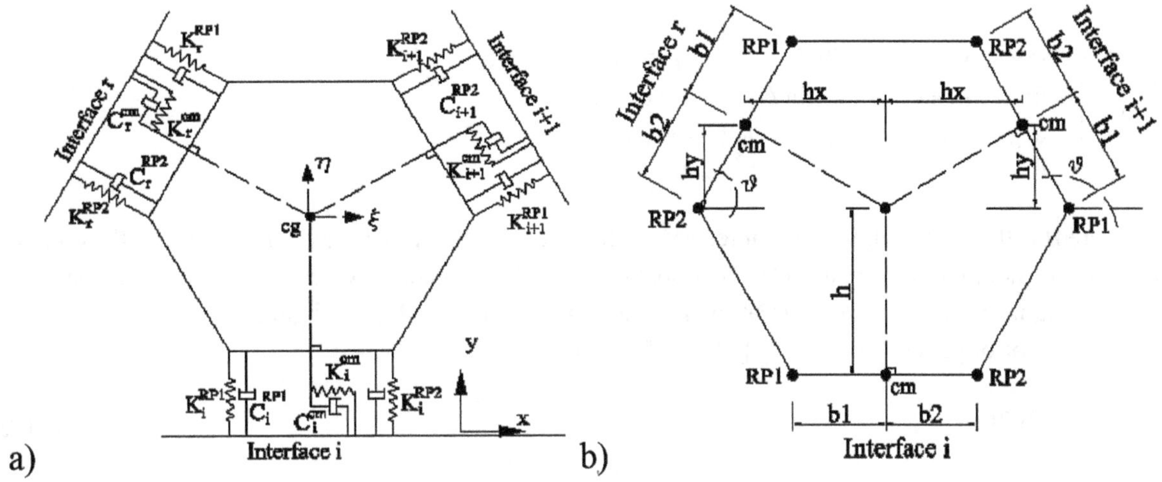

The r base contacts of the element are defined as a set of r interfaces. Each interface is defined by three linear springs and dampers (Figure 6). Two axial springs K_i^{RP} located at the corners of the interface RP_i and one shear spring K_i^{cm} located in the middle of the interface *cm* are considered. For each spring, one viscous damper C_i is defined. It is noted that the direction of the axial springs and dashpots are perpendicular to the interface, while the shear spring is parallel to it (see Figure 6a).

The dissipation of energy due to impact can be taken into account by using a viscous damping C_i, if it is regarded only as a stiffness K-dependent quantity by means of the Rayleigh formulation $C_i^{RP}=\beta_i^{RP}K_i^{RP}$, where β_i^{RP} is the stiffness proportional damping constant (Peña et al., 2007).

Thus, the equation of motion of the Lagrangian system is:

$$[M]\{\ddot{u}\}+[C]\{\dot{u}\}+\{f_s\}=\{f_e\} \tag{14}$$

where $[M]$ is the diagonal mass matrix defined with the mass of the element m and its polar moment of inertia I about the gravity center:

$$[M\} = Diag\{m, m, i\} \tag{15}$$

Under small displacement assumption, the deformation-displacement relationship can be expressed by considering a $3r$ x 3 matrix $[B]$ as follows:

$$\{\delta\} = [B]\{u\} \tag{16}$$

where the vector of deformation of the springs $\{\delta\}$ is defined as:

$$\{\delta\}^T = \left\{\delta_1^{RP1}, \delta_1^{cm}, \delta_1^{RP2}, \delta_2^{RP1}, \delta_2^{cm}, \delta_2^{RP2}, \ldots, \delta_r^{RP1}, \delta_r^{cm}, \delta_r^{RP2}\right\} \tag{17}$$

and matrix $[B]$ is the deformation-displacement matrix, which for each interface can be defined as:

$$\left[B\right]=\begin{bmatrix} \cos\vartheta & \sin\vartheta & (h_x \sin\vartheta - h_y \cos\vartheta - b_1) \\ -\sin\vartheta & \cos\vartheta & (h_x \sin\vartheta + h_y \cos\vartheta) \\ \cos\vartheta & \sin\vartheta & (h_x \sin\vartheta - h_y \cos\vartheta + b_2) \end{bmatrix} \tag{18}$$

Here ϑ is the initial angle of the interface with respect to the global x-axis; h_x and h_y are the components of the height h, whit respect to the global axes x and y, respectively; and b_i is the distance between the rocking point RP_i and the point at the middle of the interface cm (see Figure 6b).

The vector of generalized forces $\{f_s\}$ is defined as:

$$\{f_s\} = \{F_s\}^T[B] \tag{19}$$

and $\{F_s\}$ is the vector of the spring forces of each interface, which are obtained from the constitutive laws (see next subsection):

$$\left\{F_s\right\}^T = \left\{F_1^{RP1}, F_1^{cm}, F_1^{RP2}, F_2^{RP1}, F_2^{cm}, F_2^{RP2}, \ldots, F_r^{RP1}, F_r^{cm}, F_r^{RP2}\right\} \tag{20}$$

$[C]$ is the damping matrix defined as:

$$[C] = [B]^T[A][B] \tag{21}$$

where $[A]$ is the diagonal matrix of viscous dampers defined as:

$$\left[A\right] = Diag\left[\beta_1^{RP1}K_1^{RP1}, \beta_1^{cm}K_1^{cm}, \beta_1^{RP2}K_1^{RP2}, \ldots, \beta_r^{RP1}K_r^{RP1}, \beta_r^{cm}K_r^{cm}, \beta_r^{RP2}K_r^{RP2}\right] \tag{22}$$

Finally, Equation 14 is numerically integrated by means of the central difference method (Chopra, 2001) by using a constant time step Δt.

Mechanical Characteristics of the Interfaces

The axial spring is linear-elastic in compression, while zero tensile strength is assumed for tension. The shear spring is defined as linear-elastic-perfectly plastic and a Coulomb-type behavior is assumed (Figure 7). Thus, six parameters are needed to define the mechanical behavior of each interface: two axial stiffness K_i^{RP}, one shear stiffness K^{cm}, the friction angle ϕ, cohesion c and the stiffness proportional damping constant β_i^{RP}.

However, the shear stiffness can be considered equal to axial stiffness (Papantonopoulos et al., 2002; Peña et al., 2007), the cohesion c is considered null and friction angle ϕ equal to 30°, as typical of dry joints (Lourenço & Ramos, 2004). In this way, only two parameters are needed: the axial stiffness K_i^{RP} and the stiffness proportional damping constant β_i^{RP}.

Figure 7. Constitutive laws of springs: a) axial; b) shear

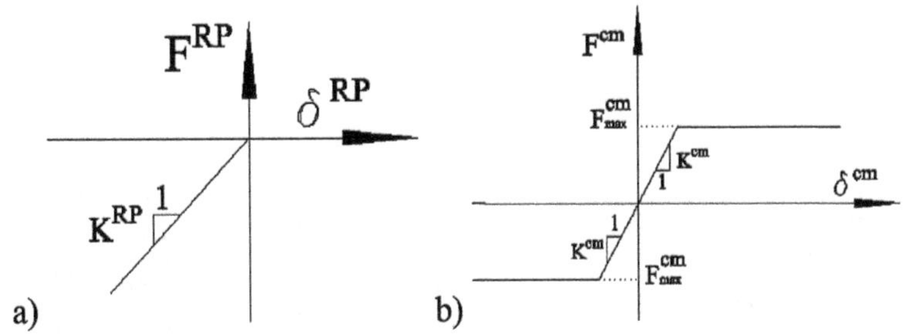

One limitation of the different models of rigid blocks supported on springs and dampers are the way, which the values of the mechanical characteristics of the model (mainly stiffness and damping) are defined; since the body and its base are fully rigid. Being a major limitation of these methods for engineering applications.

In this context, the mechanical properties of the proposed model can be assigned in two different ways. The first one is by given directly the value of the stiffness K_i^{RP} and the stiffness proportional damping constant β_i^{RP}. These values can be obtained by calibration with experimental or numerical tests, or by in situ measurements. On the other hand, the mechanical properties can be obtained from the geometry of the block, following the methodology proposed by Peña et al. (2007), by means of:

$$K_i^{RP} = Mp_i^2 = M\left(\frac{WR_i}{I_i}\right) \tag{23}$$

$$\sqrt{\beta_i^{RP}} = 0.0057\ln\left(\frac{K_i^{RP}R_i}{\mu_i^2}\right) - 0.0336 \tag{24}$$

VALIDATION

The SDM has been validated by comparing its response with the extensive experimental tests on the rocking response of rigid bodies performed at the shaking table of the National Laboratory of Civil Engineering (LNEC) of Portugal (Peña et al., 2007; 2008). The tests were carried out on four blue granite stones, subjected to different types of base motion. For details on the experimental testing program, please refer to Peña et al. (2008).

Pure Rocking Motion

Figure 8 shows a typical response under different types of base motion (free rocking, harmonic and random motion) of the rigid body of Figure 8a. Since this body is symmetric, the axial stiffness (K_i=6471 N/m) and the stiffness proportional damping constant (β_i=0.0212) are equal for both rocking points (Peña et al., 2007).

Figure 8. Typical response of a rigid body under different types of motion: a) block geometry; b) free rocking motion; c) harmonic motion (sine of 4hz and 4 mm of amplitude); d) random motion

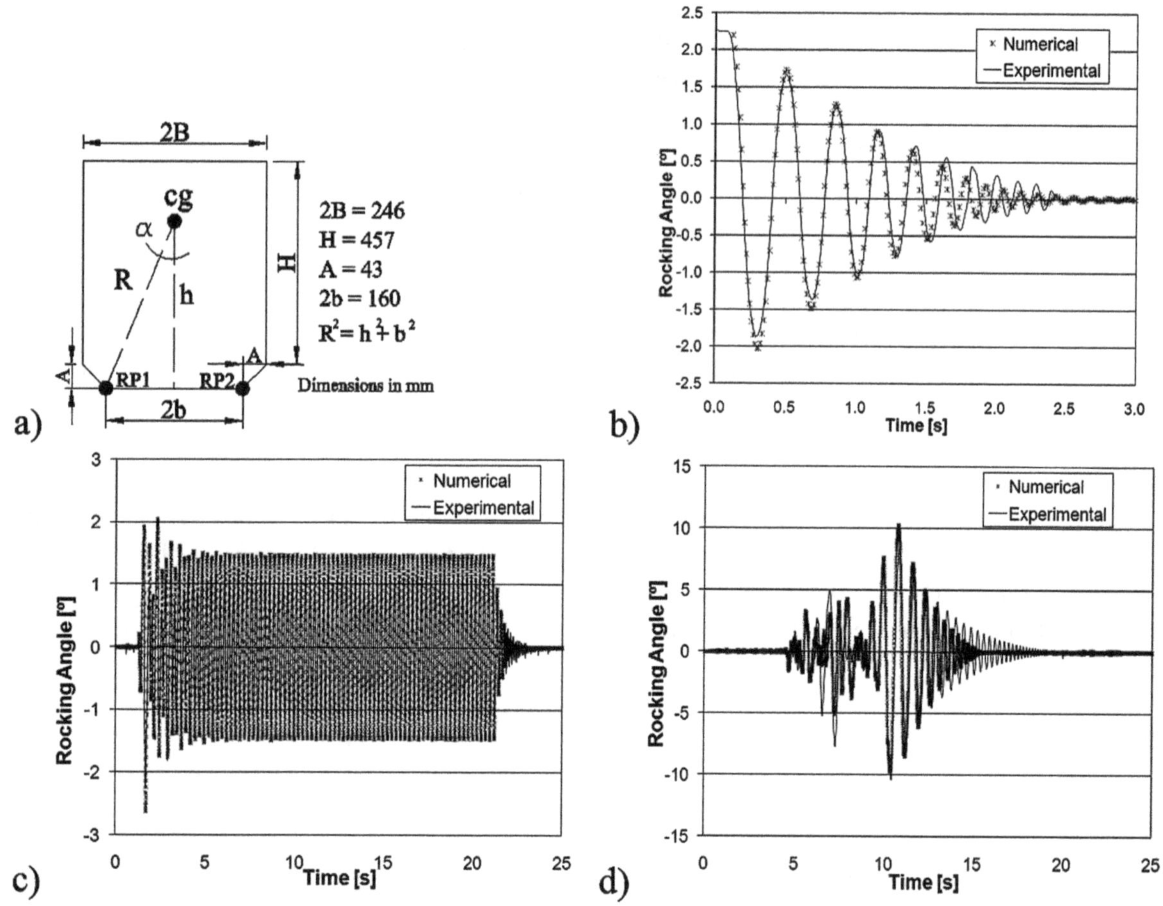

The proposed approach is able to fully reproduce the free rocking motion (Figure 8b). Figure 8c shows a typical response under constant sinusoidal base motion. The response of the model is almost the same as the experimental one. The three states of the harmonic motions (transient, stationary and free rocking) are well reproduced. On the other hand, Figure 8d shows a typical result for a random motion. The numerical model is in good agreement with the experimental test. These results validate the hypothesis of small rotations, since the model is able to reproduce satisfactorily the rocking behavior under different base loads.

The numerical model shows greater attenuation at the very end of the test (free rocking motion) than the experimental tests. This is due to the damping used in the SDM. For simplicity, the damping was considered viscous by means of the Raleigh formulation. This damping formulation considers an exponential decay of the peaks. However, the dissipation of energy is due to impact, which has a parabolic like decay of the peaks. That is the reason that the numerical model experiments a greater attenuation in the free rocking motion.

Overturning

As the proposed approach is a semi-discrete model, the initial contacts do not change during the analysis. Therefore, the overturning of the blocks cannot be detected automatically. Thus, it is necessary that the code can control the maximum rocking angle in order to warn about the possible overturning of the block. One possible control point is the static critical angle α, since it is the angle at which the block overturns due to static loads. However, some authors (Makris & Roussos, 1998; Winlker et al., 1995; Peña et al., 2008) have reported that under dynamic loads the blocks does not necessarily overturn when the static critical angle is reached. When this happens, there is a small lapse of time in which the block can return to a stable motion or can become unstable, depending on the ground motion. This stage is denoted as transition. If the block becomes stable again, the process of rocking continues. Otherwise, the block becomes unstable and the overturning process is irreversible (Figure 9). Thus, the dynamic critical angle that defines stability of the rocking block under dynamic loadings is always lager than the static critical angle (Peña et al., 2008).

Therefore, the code must take into account the transition stage. Thus, the control point that is taken into account is the dynamic critical angle. If the block reaches this value, then the code halts with a message of overturning. The dynamic critical angle is considered 20% greater than the static critical angle as the experimental test have shown (Peña et al., 2008). Figure 10 shows a typical response until collapse of the rigid body of Figure 8a due to random base motion.

Figure 9. Typical process of overturning
Peña et al. (2008)

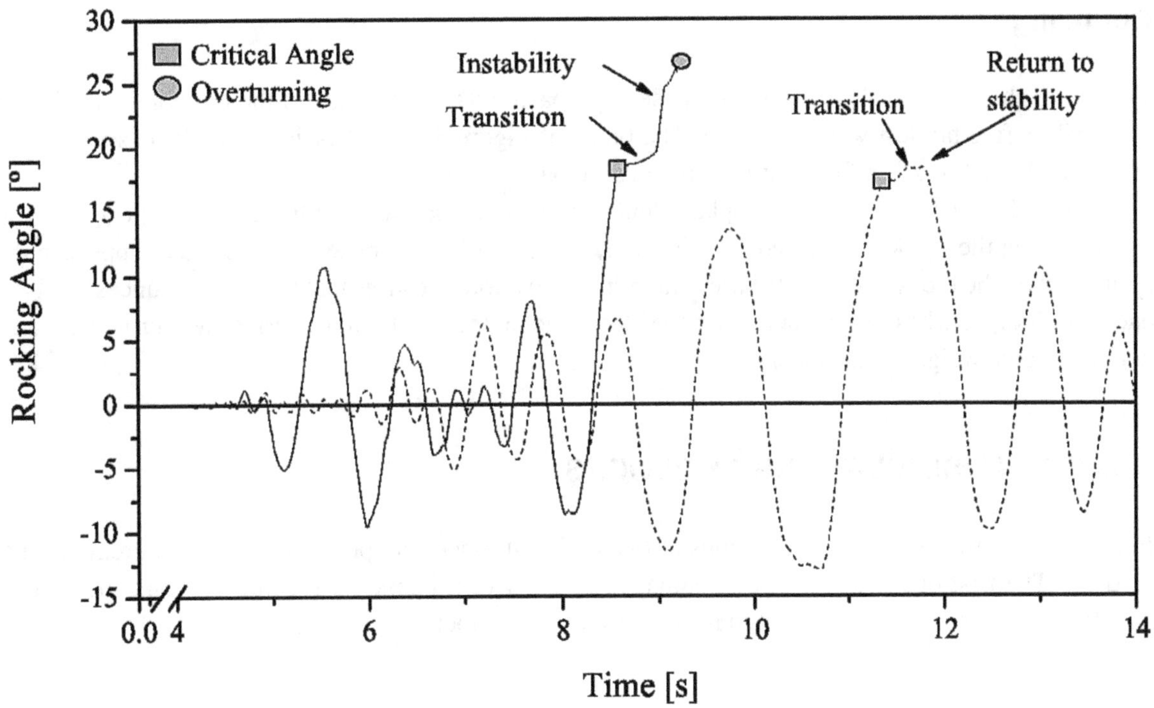

Figure 10. Typical collapse of the rigid body of Figure 8a due to random base motion

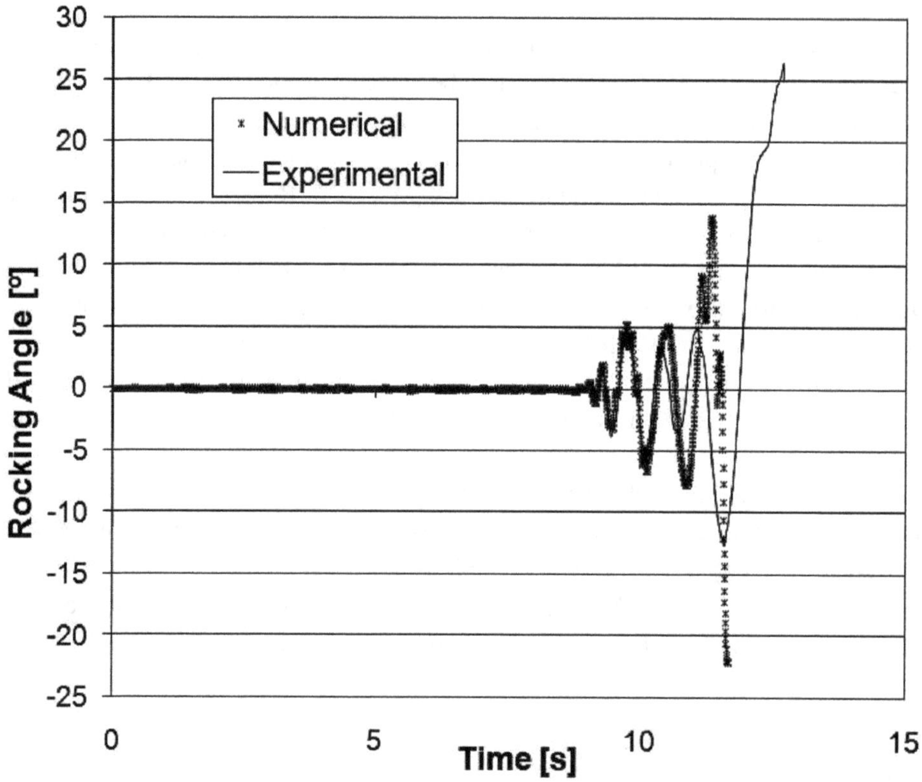

Bouncing

In order to show the capability of the proposed approach to take into account the possible bouncing of the block, a free fall test was performed. The block of Figure 8a was detached from its base 34.4 mm and was released in a free fall, as it is depicted schematically in Figure 11a.

Figure 11 b shows the vertical displacement of the block and the axial forces in the springs. It can see that when the block is detached from its base, the spring forces are zero. Axial forces appear in the springs when the block contacts its base. These impulsive forces causes that the block bounces until the energy is dissipated by the dampers. When the block reaches the rest, the axial springs remain compressed due to the own weight of the block.

NUMERICAL SIMULATIONS OF BLOCKS

In this section, the numerical simulations of two different blocks are presented under different considerations. The first one is a rectangular damaged block with asymmetric rocking motion. The second simulation corresponds to a one-sided rocking rectangular block.

Figure 11. Typical free fall test: a) schematic representation of the test at different times; b) vertical displacement and axial spring forces

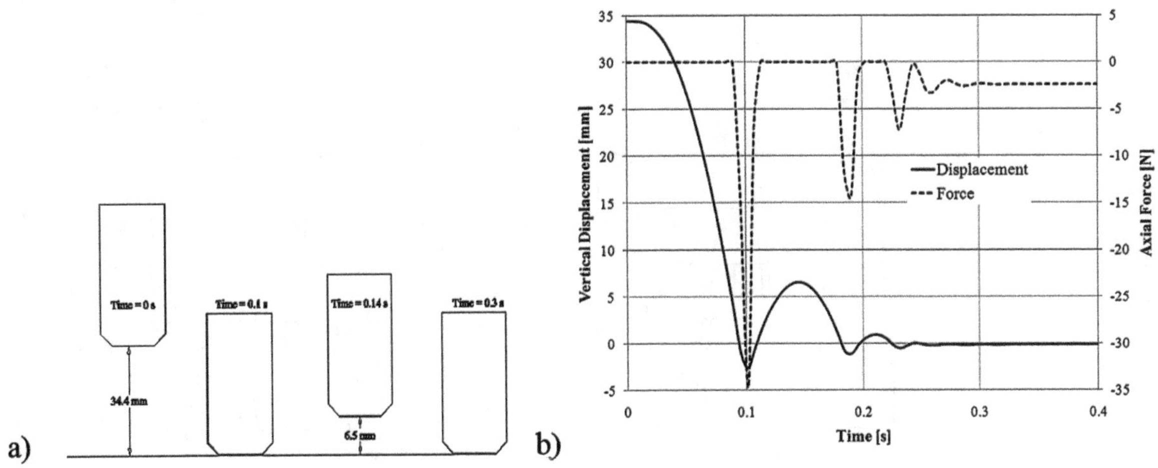

Damaged Block

The numerical simulation of a rectangular damaged block is presented. The rectangular undamaged block is depicted in Figure 12a. This block was initially part of the specimens tested at LNEC. However, this block was damaged due to incorrect handling before testing, as shown schematically in Figure 12b. Therefore, only free rocking test was performed on this block and it was not considered as part of the specimens tested (Peña et al., 2008).

Figure 13 shows the free rocking test of the block, as well as the numerical simulation of the damaged and undamaged blocks. The mechanical properties of both blocks used in the numerical model are shown in Figure 12. The properties of the damaged block were obtained by calibrating the model with the free rocking motion test, while the properties of the undamaged block were taken equal to the properties of the undamaged side of the damaged block (right side).

Figure 12. Geometry and properties of the rectangular block used in the numerical simulation: a) undamaged block; b) damaged block

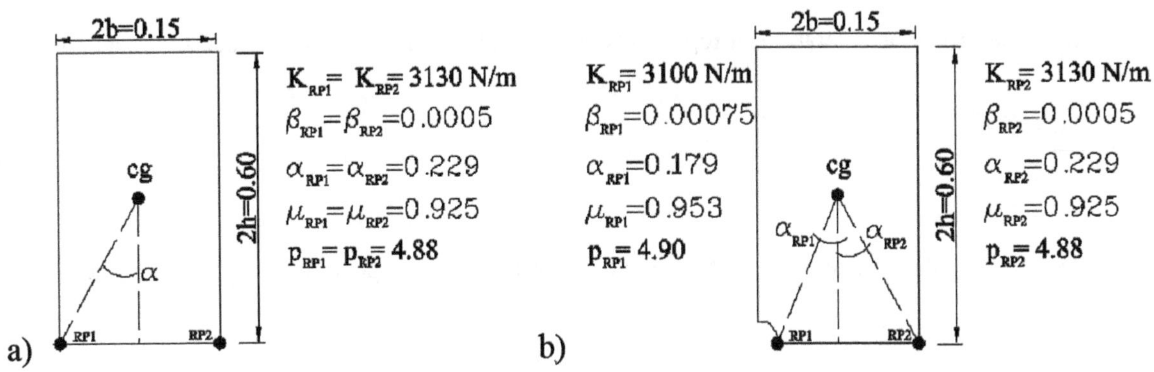

Figure 13. Free rocking motion of a rectangular damaged block

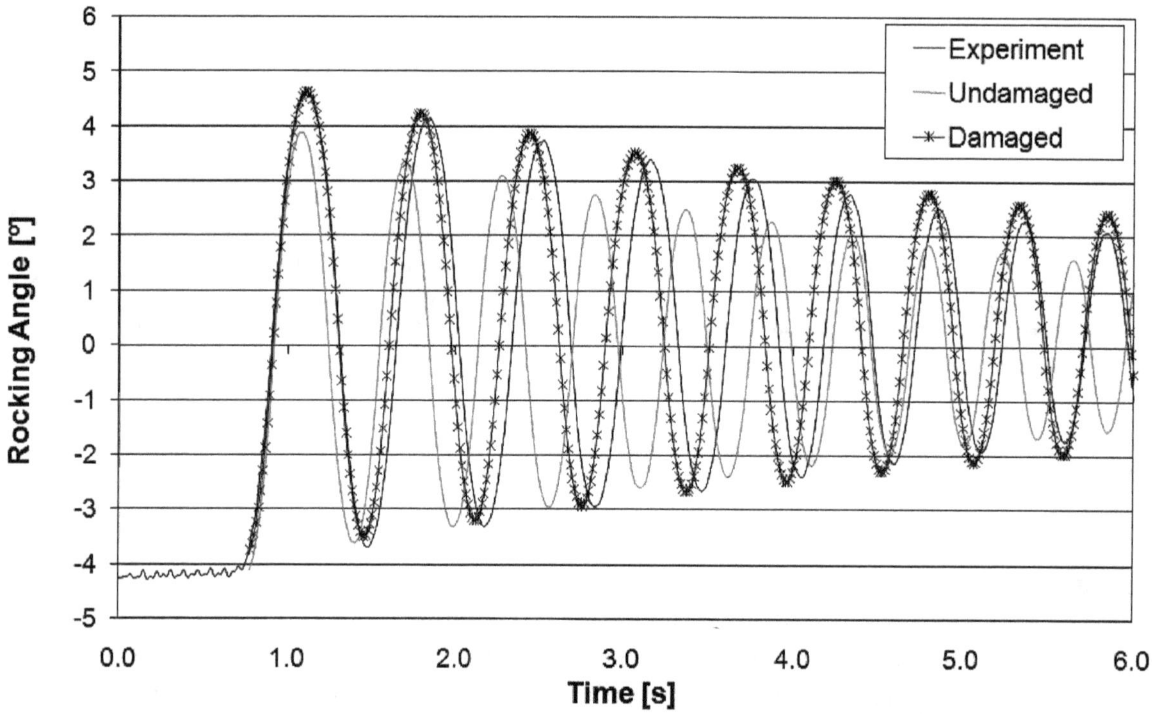

The undamaged block presents the typical free rocking motion, in which the amplitude of the rocking angle decreases constantly on both sides. In this example, the initial rocking angle is 4° and the rocking angle after the first impact is 3.9°. On the other hand, the numerical simulation of the damaged block fits correctly the experimental test. In this case, the rocking angle after the first impact is 4.75°, i.e., one degree greater than the initial rocking angle. This is a consequence of damage on one side, yielding a non-symmetric response.

Both blocks were submitted to a pulse type Ricker (Figure 14a). This type of pulse is mostly used in studies of seismic engineering as synthetic ground motion, since it can be associated to near faults motions (Arredondo and Reinoso, 2008). A Ricker pulse can be defined as:

$$a_g = A(2a - 1)e^{-a} \tag{25}$$

where a_g is the ground acceleration; A represents peak acceleration and a is obtained by means of:

$$a = \left|\frac{w(t - T_s)}{2}\right|^2 \tag{26}$$

where w is the angular frequency; T_s is the time of occurrence of the peak intensity; and t is the time.

Figure 14. Typical response of damaged and undamaged blocks under Ricker pulse: a) ground acceleration; b) rocking response (amplitude of 2.5 m/s²); c) rocking response (amplitude of 3.0 m/s²)

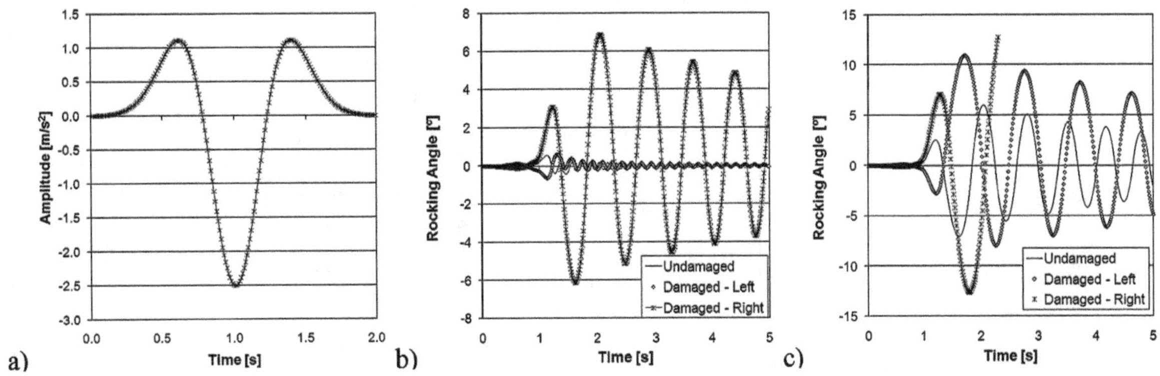

Figures 14b,c show the typical rocking response of both blocks under different amplitudes of the Ricker pulse (2.5 and 3.0 m/s², respectively). It is worth noting that for the damaged block the pulse was applied in both directions (right and left) in order to see the influence of the damaged side.

On the one hand, the undamaged block practically does not experience rocking motion, when the amplitude of the pulse is 2.5 m/s² (Figure 14b). The maximum rocking angle is equal to 0.6°. The damaged block has a similar response to the undamaged block when the pulse is initially applied in the direction of the damaged side (left), because the maximum acceleration of the pulse is applied to the undamaged side. This allows to the block to have a similar response as the undamaged block. In the other case (when the pulse is initially applied in the direction of the undamaged side), the maximum rocking angle is equal to 6.7° (on the damaged side). Therefore, for this particular case and load, the most unfavorable direction of the applied load is in the direction of the undamaged side (right direction).

On the other hand, when the amplitude of the pulse is 3.0 m/s² (Figure 14c), the maximum rocking angle of the undamaged block is equal to 7.0°. However, the damaged block overturns when the pulse is applied in the direction of the undamaged side (right). For this case, the dynamic critical angle is equal to 12.3° and 15.6° for damaged and undamaged sides, respectively. Therefore, even when the block reaches a rocking angle equal to 13° on the undamaged side, the block does not overturn and continues rocking. In the other case (when the load is applied in the direction of the damaged side), the maximum rocking angle is equal to 10.6° for the damaged side.

One-Sided Rocking Block

A masonry wall having a top restraint was modeled as a one-sided rocking rectangular block (Figure 15a). For this case, two initial base contacts were defined (see Figure 15b) and the mechanical properties of the interfaces were obtained from the geometry of the block by means of Equations 23 and 24. The first interface corresponds to the base of the block by the side defined by nodes 1-2 (K_i=22,873 N/m; β_i=0.000926). The second interface defines the top restraint and was modeled by the side defined by nodes 3-4 (K_i=22,873 N/m; β_i=0.00152). The wall was subjected to a real earthquake record (Figure 15c).

Figure 16 shows the rocking angle of the wall when the ground motion is initially applied at the left or at the right of the wall. Only positive angle are obtained since the wall is restrained in one direction. The

Figure 15. One-sided block: a) dimensions; b) model and interfaces; c) ground acceleration

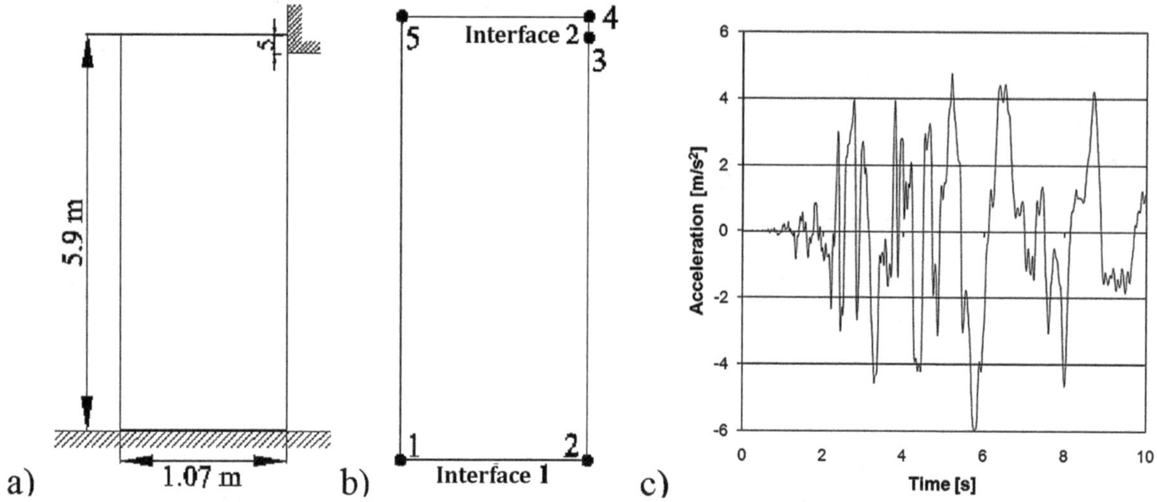

response of the wall will depend on the direction of the ground motion. In this way, the maximum rocking angle is 8° and 6°, when the acceleration is initially applied at the right and at the left, respectively. It is worth noting that small sliding was detected in the wall when the ground acceleration is initially applied at the right of the wall. The lateral displacement due to the sliding is of 20 mm and occurs at time 6.85 s.

In order to see the influence of the sliding in the rocking motion, a new analysis restraining the slide motion was performed. The comparison with and without sliding is presented in Figure 17. When the

Figure 16. Typical response of one-sided block under a real earthquake motion

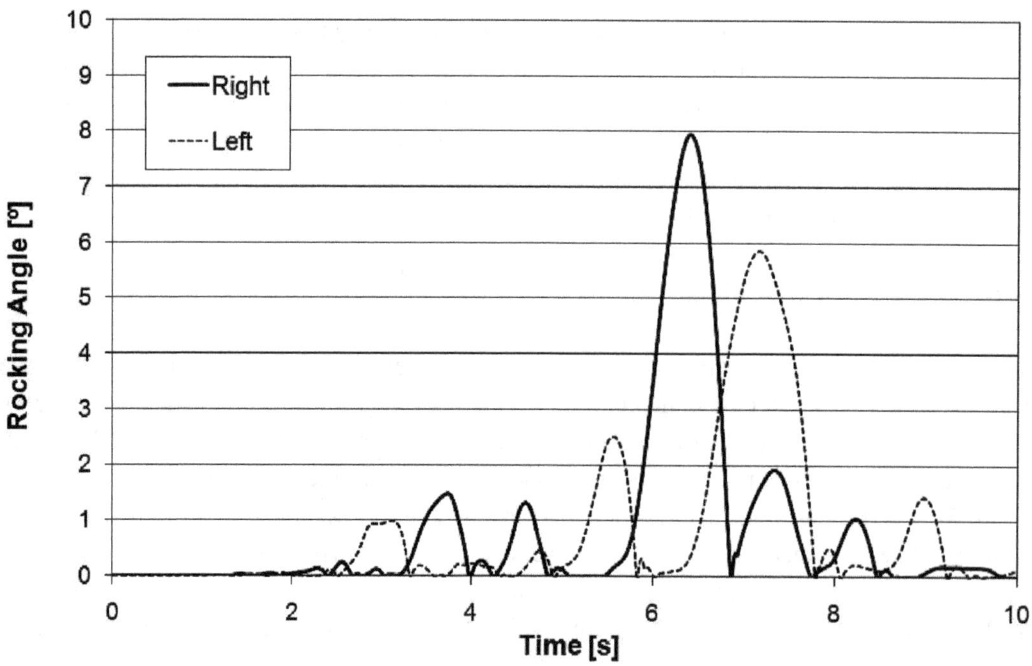

load is initially applied in the right direction, the rocking response is equal in both models until the slide is present at time 6.85 s. After this time, the rocking response is reduced considerably due to the sliding motion. When the load is applied in the left direction, the rocking motion is practically the same with and without sliding. There is a very small sliding motion at time 6 s, which can be neglected. Due to this slide, a difference in maximum rocking motion is around 5%. Therefore, for practical cases in this case the sliding motion can be neglected.

CONCLUSION

A Semi-Discrete Approach was proposed to analyze slender rigid blocks. This approach is based on a modification of the Rigid Element Method. The simple mathematical formulation of the proposed model allows that with one simple and well-known equation, the model can detect rocking, sliding and jumping of the block. Thus, it is not necessary complex mathematical formulation for each possible motion state and multiple rocking points.

The numerical results show that the hypothesis of small rotations is valid when slender bodies are studied, since the rocking angle is not greater than 20°. The model is able to reproduce satisfactorily the rocking behavior of rigid bodies. It is worth to note that the model proposed can take into account the possible sliding and bouncing of the blocks.

The dynamics of rocking motion of asymmetric rigid bodies was also described. It has been demonstrated that the symmetric rocking motion of a rigid body depends only on the symmetry of the rocking points with respect to the gravity center (same base and same height) and not on the geometrical shape of the body. Four causes that could produce asymmetric rocking motion have been detected: geometry, material, damage and boundary conditions.

Therefore, it is possible to establish two types of asymmetry. The first one is the asymmetry of the rocking points with respect to the gravity center, here named base asymmetry. This type of asymmetry

Figure 17. Response of one-sided block considering sliding: a) right direction; b) left direction

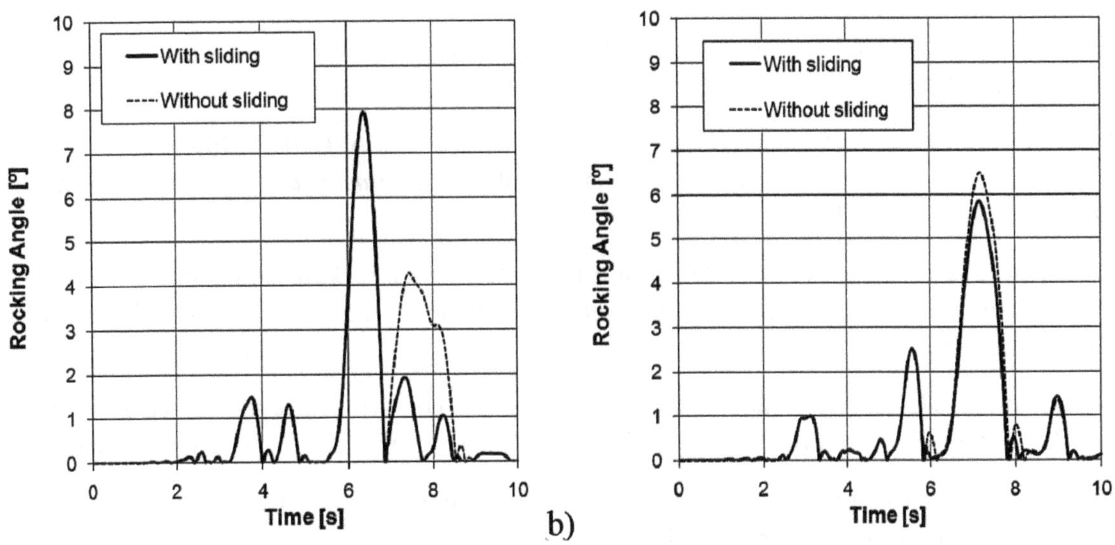

a) b)

is due mainly by the geometry, material and damage of the body. The second type is the boundary asymmetry produced by the boundary conditions, especially for more than two supports or rocking points.

In asymmetric rocking motion, the maximum response is a function of the type of asymmetry, as well as of the direction of the ground motion. Therefore, the direction of the acceleration should become a new variable in the study of asymmetric rocking motion.

The sliding motion modifies the rocking response of the block. Therefore, it is recommended to take into account in the study of the motion of rigid bodies. In this way, the numerical models must take into account the different motion states of the blocks.

REFERENCES

Andreaus, U., & Casini, P. (1999). On the rocking-uplift motion of a rigid block in a free and forced motion: Influence of sliding and bouncing. *Acta Mechanica*, *138*(3-4), 219–241. doi:10.1007/BF01291846

Anooshelpoor, A., Brune, J. N., & Zeng, Y. (2004). Methodology for obtaining constrains on ground motion from precariously balanced rocks. *Bulletin of the Seismological Society of America*, *94*(1), 285–303. doi:10.1785/0120020242

Arredondo, C., & Reinoso, E. (2008). Influence of frequency content and peak intensities in the rocking seismic response of rigid bodies. *Journal of Earthquake Engineering*, *12*(4), 517–533. doi:10.1080/13632460701672755

Azevedo, J., Sincrain, G., & Lemos, J. V. (2000). Seismic behavior of blocky masonry structures. *Earthquake Spectra*, *16*(2), 337–365. doi:10.1193/1.1586116

Boroschek, R., & Iruretagoyena, A. (2006). Controlled overturning of unanchored rigid bodies. *Earthquake Engineering & Structural Dynamics*, *35*(6), 695–711. doi:10.1002/eqe.554

Casolo, S., & Peña, F. (2007). Rigid element model for in-plane dynamics of masonry walls considering hysteretic behavior and damage. *Earthquake Engineering & Structural Dynamics*, *36*(8), 1029–1048. doi:10.1002/eqe.670

Chopra, A. K. (2001). *Dynamics of structures: Theory and applications to earthquake engineering* (2nd ed.). Prentice Hall.

Costa, A. A., Arêde, A., Campos Costa, A., Penna, A., & Costa, A. (2013). Out-of-plane behavior of a full scale stone masonry façade. Part 2: Shaking table tests. *Earthquake Engineering & Structural Dynamics*, *42*, 2097–2111.

Di Egidio, A., & Contento, A. (2009). Base isolation of slide-rocking non-symmetric rigid blocks under impulsive and seismic excitations. *Engineering Structures*, *31*(11), 2723–2734. doi:10.1016/j.engstruct.2009.06.021

Di Egidio, A., & Contento, A. (2010). Seismic response of a non-symmetric rigid block on a constrained oscillating base. *Engineering Structures*, *32*(10), 3028–3039. doi:10.1016/j.engstruct.2010.05.022

Erdik, M., Durukal, E., Erturk, N., & Sungay, B. (2010). Earthquake risk mitigation in Istanbul museums. *Natural Hazards*, *53*(1), 97–108. doi:10.1007/s11069-009-9411-2

Fielder, W. T., Virgin, L. N., & Plaut, R. H. (1997). Experiments and simulation of overturning of an asymmetric rocking block on an oscillating foundation. *European Journal of Mechanics. A, Solids, 16*, 905–923.

Hogan, S. J. (1990). The many steady states responses of a rigid block under harmonic forcing. *Earthquake Engineering & Structural Dynamics, 19*(7), 1057–1071. doi:10.1002/eqe.4290190709

Hogan, S. J. (1992). On the motion of a rigid block, tethered at one corner, under harmonic forcing. *Proccedings: Mathematical and Physical Science, 439*(1905), 35-45.

Hogan, S. J. (1994). Rigid block dynamics confined between side-walls. *Philosophical Transactions: Physical Science and Engineering, 37*(1683), 411-419.

Hoseini, M., & Alavi, S. S. (2014). A kind of repairable steel buildings for seismic regions based on building's rocking motion and energy dissipation at base level. *International Journal of Civil Engineering, 1*(3).

Housner, G. (1963). The behavior of inverted pendulum structures during earthquakes. *Bulletin of the Seismological Society of America, 53*(2), 403–417.

Ishiyama, Y. (1982). Motions of rigid bodies and criteria for overturning by earthquake excitations. *Earthquake Engineering & Structural Dynamics, 10*(5), 635–690. doi:10.1002/eqe.4290100502

Komodromos, P., Papaloizou, L., & Polycarpou, P. (2008). Simulation of the response of ancient columns under harmonic and earthquake excitations. *Engineering Structures, 30*(8), 2154–2164. doi:10.1016/j.engstruct.2007.11.004

Komodromos, P., Polycarpou, P., Papaloizou, L., & Phocas, M. (2007). Response of seismically isolated buildings considering poundings. *Earthquake Engineering & Structural Dynamics, 36*(12), 1605–1622. doi:10.1002/eqe.692

Konstantinidis, D., & Makris, N. (2009). Experimental and analytical studies on the response of freestanding laboratory equipment to earthquake shaking. *Engineering Structures, 38*, 827–848.

Lagomarsino, S. (2015). Seismic assessment of rocking masonry structures. *Bulletin of Earthquake Engineering, 13*(1), 97–128. doi:10.1007/s10518-014-9609-x

Lemos, J. V. (2007). Discrete element modeling of masonry structures. *Journal of Architectural Heritage, 1*(2), 190–213. doi:10.1080/15583050601176868

Lourenço, P. B., & Ramos, L. (2004). Characterization of cyclic behavior of dry masonry joints. *Journal of Structural Engineering, 130*(5), 779–786. doi:10.1061/(ASCE)0733-9445(2004)130:5(779)

Makris, N., & Konstantinidis, D. (2003). The rocking spectrum and the limitations of practical design methods. *Earthquake Engineering & Structural Dynamics, 32*(2), 265–289. doi:10.1002/eqe.223

Makris, N., & Roussos, Y. (1998). *Rocking response and overturning of equipment under horizontal pulse-type motions. PEER Report 1998/05*. Berkeley, CA: University of California.

Michaltsos, G. T., & Raftoyiannis, G. I. (2008). A mathematical model for the rocking, overturning and shifting problems in bridges. *Engineering Structures*, *30*(12), 3587–3594. doi:10.1016/j.engstruct.2008.06.007

Milne, J. (1881). Experiments on observational seismology. *Transaction of the Seismological Society of Japan*, *3*, 12–64.

Mohammadi, G. M., & Yasrebi, F. (2010). Out of plane behaviour of walls, using rigid block concepts. *Structural Engineering & Mechanics*, *34*(3), 335–350. doi:10.12989/sem.2010.34.3.335

Nozaki, K., Terumichi, Y., Nishimura, K., & Sogabe, K. (2009). Study of rocking motion of rigid body with slide contact. *Journal of Mechanical Science and Technology*, *23*(4), 1001–1007. doi:10.1007/s12206-009-0329-0

Palmieri, A., & Makris, N. (2008). Response analysis of rigid structures rocking on viscoelastic foundation. *Earthquake Engineering & Structural Dynamics*, *37*(7), 1039–1063. doi:10.1002/eqe.800

Papaloizou, L., & Komodromos, P. (2012). Investigating the seismic response of ancient multi-drum colonnades with two rows of columns using an object-oriented designed software. *Advances in Engineering Software*, *44*(1), 136–149. doi:10.1016/j.advengsoft.2011.05.030

Papantonopoulos, C., Psycharis, N., Papastamatiou, D. Y., Lemos, J. V., & Mouzakis, H. P. (2002). Numerical prediction of the earthquake response of classical columns using the distinct element method. *Earthquake Engineering & Structural Dynamics*, *31*(9), 1699–1717. doi:10.1002/eqe.185

Park, H. J., & Kim, D. S. (2013). Centrifuge modelling for evaluation of seismic behaviour of stone masonry structure. *Soil Dynamics and Earthquake Engineering*, *53*, 187–195. doi:10.1016/j.soildyn.2013.06.010

Peña, F. (2015). Modelo simplificado para el estudio del balanceo asimétrico de cuerpos rígidos esbeltos. *Revista Internacional de Métodos Numéricos para Cálculo y Diseño en Ingeniería*, *31*(1), 1–7. doi:10.1016/j.rimni.2013.10.005

Peña, F., Lourenço, P. B., & Campos-Costa, A. (2008). Experimental dynamic behaviour of free standing multi-block structures under seismic loadings. *Journal of Earthquake Engineering*, *12*(6), 953–979. doi:10.1080/13632460801890513

Peña, F., Prieto, F., Lourenço, P. B., Campos-Costa, A., & Lemos, J. V. (2007). On the dynamics of rocking motion of single rigid-block structures. *Earthquake Engineering & Structural Dynamics*, *36*(15), 2383–2399. doi:10.1002/eqe.739

Perry, J. (1881). Note on the rocking of a column. *Transaction of the Seismological Society of Japan*, *3*, 103–106.

Plaut, R. H., Fielder, W. T., & Virgin, L. N. (1996). Fractal behaviour of an asymmetric rigid block overturning due to harmonic motion of a tilted foundation. *Chaos, Solitons, and Fractals*, *7*(2), 177–196. doi:10.1016/0960-0779(95)00059-3

Prieto, F., & Lourenço, P. B. (2005). On the rocking behaviour of rigid objects. *Meccanica*, *40*(2), 121–133. doi:10.1007/s11012-005-5875-7

Prieto, F., Lourenço, P. B., & Oliveira, C. S. (2004). Impulsive Dirac-delta forces in the rocking motion. *Earthquake Engineering & Structural Dynamics*, *33*(7), 839–857. doi:10.1002/eqe.381

Purvance, M. D., Anooshehpoor, A., & Brune, N. (2008). Freestanding block overturning fragilities: Numerical simulation and experimental validation. *Earthquake Engineering & Structural Dynamics*, *37*(5), 791–808. doi:10.1002/eqe.789

Sharif, V., Ghafory Ashtiany, M., Eshghi, S., & Soroushian, A. (2009). Dynamic assessment of constrained rigid equipments. *Journal of Applied Sciences*, *9*(13), 2362–2371. doi:10.3923/jas.2009.2362.2371

Shenton, H. III. (1996). Criteria for initiation of slide, rock and slide-rock rigid-body blocks. *Journal of Engineering Mechanics*, *122*(7), 690–693. doi:10.1061/(ASCE)0733-9399(1996)122:7(690)

Sorrentino, L., Kunnath, S., Monti, G., & Scalora, G. (2008). Seismically induced one-sided rocking response of unreinforced masonry façades. *Engineering Structures*, *30*(8), 2140–2153. doi:10.1016/j.engstruct.2007.02.021

Spanos, P. D., Roussis, P. C., & Politis, N. P. A. (2001). Dynamic analysis of stacked rigid blocks. *Soil Dynamics and Earthquake Engineering*, *21*(7), 559–578. doi:10.1016/S0267-7261(01)00038-0

Taniguchi, T. (2002). Non-linear response analyses of rectangular rigid bodies subjected to horizontal and vertical ground motion. *Earthquake Engineering & Structural Dynamics*, *31*(8), 1481–1500. doi:10.1002/eqe.170

Taniguchi, T., & Miwa, T. (2007). A simple procedure to approximate slip displacement of freestanding rigid body subjected to earthquake motions. *Earthquake Engineering & Structural Dynamics*, *36*(4), 481–501. doi:10.1002/eqe.639

Tso, W. K., & Wong, C. M. (1989). Steady state rocking response of rigid blocks to earthquake. Part 1: Analysis. Part 2: Experiment. *Earthquake Engineering & Structural Dynamics*, *18*, 89–120. doi:10.1002/eqe.4290180109

Voyagaki, E., Psycharis, I. N., & Mylonakis, G. (2014). Complex response of a rocking block to a full-cycle pulse. *Journal of Engineering Mechanics*, *140*(6), 04014024. doi:10.1061/(ASCE)EM.1943-7889.0000712

Winkler, T., Meguro, K., & Yamazaki, F. (1995). Response of rigid body assemblies to dynamic excitation. *Earthquake Engineering & Structural Dynamics*, *24*(10), 1389–1408. doi:10.1002/eqe.4290241008

Zulli, D., Contento, A., & Di Egidio, A. (2012). 3D model of rigid block with a rectangular base subject to pulse-type excitation. *International Journal of Non-linear Mechanics*, *47*(6), 679–687. doi:10.1016/j.ijnonlinmec.2011.11.004

Chapter 17
Application of Discrete Finite Element Method for Analysis of Unreinforced Masonry Structures

Iraj H. P. Mamaghani
University of North Dakota, USA

ABSTRACT

In this chapter, through some illustrative examples, the applicability of the Discrete Finite Element Method (DFEM) to analysis of unreinforced masonry structures such as rock pillars, open rock slopes, underground openings, tunnels, fault propagations, and fault-structure interactions is examined and discussed. In the numerical study, the behavior of contacts and blocks is assumed to be elasto-plastic or elastic. The Mohr-Coulomb yield criterion, representing material behavior of contacts, is implemented in the developed codes for DFEM used in the analysis. The secant stiffness method with the updated Lagrangian scheme is employed to deal with non-linear behavior. The constant strain triangular element with two degrees of freedoms at each node, formed by properly joining the corners and contact nodes of an individual block, is adopted for finite element meshing of the blocks. The DFEM provides an efficient and promising tool for designing, analyzing, and studying the behavior of unreinforced masonry structures.

INTRODUCTION

Masonry structures, consisting of interacting distinct blocks, have been constructed since the earliest days of civilization, are still commonly built in many countries all over the world, and constitute a significant percentage of current structures. Many of these structures are located in seismically active regions and were built before the establishment of any design code requirements for earthquake-resistant construction. Compared with modern structures built of materials with well-understood constitutive laws, the mechanics of masonry structures are still not clearly understood in spite of their long use. The failures and damage reported in recent earthquakes attest to the need for efficient strengthening procedures and therefore an efficient analytical method for analysis of masonry structures.

DOI: 10.4018/978-1-5225-0231-9.ch017

The analysis of unreinforced masonry and rock engineering structures excavated in discontinuous rock masses has been receiving particular interest among civil engineers, rock mechanics, and rock engineers. Since rock masses consist of distinct blocks due to geological discontinuities, several techniques have been developed to analyze masses consisting of distinct blocks. A literature review shows that during the last three decades, the limiting equilibrium analysis (Aydan *et al.*, 1989; Hoek & Bray, 1977) and some numerical analysis methods such as the finite element method (FEM) (Aydan *et al.*, 1990; Barbosa & Ghaboussi, 1990; Ghaboussi *et al.*, 1973; Goodman *et al.*, 1968; Kawamoto & Aydan, 1999; Oda *et al.*, 1993), distinct element method (DEM) (Cundall, 1971), and discontinuities deformation analysis (DDA) (Shi, 1988) have been developed for the analysis of problems involving discontinuities in rock mechanics. The DEM has been used for years in different industries (*e.g.*, mining, civil engineering, and nuclear waste disposal) for the solution of problems involving deformation, damage, fracturing, and stability of fractured rock masses and masonry structures (among others: Baggio & Trovalusci, 1993, 1998; Cundall, 2011; Itasca Consulting Group, Inc., 2014; Lermos, 2007; Mamaghani, 1994, 2006). Recently, a number of modeling techniques have been developed to simulate coupled hydro-mechanical problems with the DEM; these methods have been reviewed by Furtney *et al.* (2013). For example, the Universal Distinct Element Code (*UDEC*), developed by Itasca Consulting Group, Inc. (2014), utilizes an explicit solution scheme that can model the complex, non-linear behavior of media containing multiple intersecting joint structures. Joint models and properties can be assigned separately to individual discontinuities or sets thereof. The analysis of rock mass stimulation by fluid injection requires analytical tools, such as numerical models based on DEM, which can represent discontinuities explicitly (Damjanac *et al.*, 2015). A similar approach for simulation of fracturing and hydraulic fracturing of rocks is based on the combined finite element method (FEM) and DEM. The formulation of the method and some example applications are found in Rougier *et al.* (2011, 2012), Zhao *et al.* (2015), and Lisjak *et al.* (2015). In spite of all these techniques, it is difficult to say that a unique technique that guarantees satisfactory results has been developed. Although DEM and DDA can be used for the static and dynamic analysis of discontinuous media, the treatment of rate-dependent behavior of materials in these methods is not realistic. For example, DEM introduces a forced damping to suppress oscillations, while DDA adopts very large time steps so that artificial damping occurs as a result of numerical integration.

In the previous chapter, the discrete finite element method (DFEM) developed by the author for analysis of blocky systems under static and dynamic loading (Mamaghani 1993, 2006, 20015a; Mamaghani *et al.*, 1994, 1999) was presented. The DFEM consists of a mechanical model which represents the deformable blocks and contact models that specify the interactions among them. In the DFEM, a visco-elastic constitutive law for linear behavior and a visco-elasto-plastic constitutive law for nonlinear behavior of blocks and contacts are used together with an updated Lagrangian scheme. The DFEM can handle large block motions within the framework of the finite element method. In this chapter, the applicability of the DFEM to the analysis of unreinforced masonry structures such as open rock slopes, underground openings, tunnels, and fault-structure interactions will be presented and discussed.

NUMERIC ANALYSIS

In the numeric analysis, when the inertia term is considered, contacts and blocks are assumed to behave as elasto-visco-plastic materials or visco-elastic materials. On the other hand, if the inertia term is omitted, then the behavior of contacts and blocks is assumed to be elasto-plastic or elastic. Application of the DFEM to the stability analysis of a single block on an incline and numerical analysis of some masonry structures has been previously reported by the author and his co-workers (Aydan *et al.*, 1996; Mamaghani, 1993, 2006, 2015a, 2015b; Mamaghani & Aydan, 1999, 2000; Mamaghani et al., 1994, 1999, 2014; Tokashiki *et al.*, 1997, 2001). In this section, some typical numerical results for static response of unreinforced masonry open rock slopes, underground openings, tunnels, fault propagation, and fault-structure interactions obtained by the DFEM will be presented and discussed (Aydan *et al.*, 2011; Aydan *et al.*, 2005; Mamaghani, 2008; Mamaghani & Aydan, 1999). In the analyses reported herein, the tensile strength of the contact element was assumed to be zero. The Mohr-Coulomb yield criterion was implemented in the present codes; nevertheless, one can easily implement any yield criterion which is appropriate for plastic or visco-plastic behavior. The contact area A_c was assumed to be half of the area of the side of a block to which the contact element is attached. The thickness of the bands was taken as twice the weighted asperity height. Taking into account the results reported by Aydan *et al.* (1989), the thickness of the bands was set at 10 mm. The secant stiffness method together with the updated Lagrangian scheme was employed to handle non-linear behavior.

The constant strain triangular element with two degrees of freedoms at each node, formed by properly joining the corners and contact nodes of an individual block, was adopted for finite element meshing of the blocks (Mamaghani, 1993). However, it must be noted that the method is not restricted to the use of such elements and one can implement finite elements with any number of chosen nodes.

The DFEM flow chart for static analysis is shown in Figure 1 (Mamaghani, 1993). The analysis is a pseudo-time stepping incremental procedure. First, the initial configuration of the structural system, boundary conditions, and material properties are specified. Then, iterations are carried out by forming the global stiffness matrix and solving the equilibrium equations of the system. Next, the strains and stresses of the elements are computed. The no-tension condition and Mohr-Coulomb's yield criterion are checked, and the excess forces at contacts are applied to the updated configuration as the penalty load in the subsequent iteration until the norm of the excess force vector converges to a very small value of convergence tolerance. The computation is terminated when a stable configuration is achieved or the global stiffness matrix becomes ill-conditioned as single or multiple blocks tend to move without any interaction with each other, corresponding to the failure of the system. The details of the numerical algorithm and computational procedure are given in previous work by Mamaghani (1993).

In the previous chapter, the application of the DFEM to the static analysis of two unreinforced masonry structures, the arch and the pyramid (Mamaghani *et al.*, 1994), and to the dynamic analysis of three unreinforced masonry structures, the arch, the tower, and the wall, was presented and discussed (Mamaghani *et al.*, 1999). The following presents the application of the DFEM to the static analysis of unreinforced masonry structures such as rock pillars, open rock slopes, underground openings, tunnels, fault propagations, and fault-structure interactions (Aydan *et al.*, 1996; Mamaghani, 2015a, 2008; Mamaghani & Aydan, 2000).

Stability of a Rock Pillar

A numerical simulation of the stability of a rock pillar having a discontinuity plane inclined at an angle of 45° was performed. The material properties used in the numerical analyses are given in Table 1. Figure 2 shows the configuration of the pillar at each iteration step. As expected, sliding occurs along the inclined discontinuity, and the displacement of the sliding block increases as the iteration number, which may be regarded as fictitious time, increases.

Figure 1. Flow chart of DFEM for static analysis

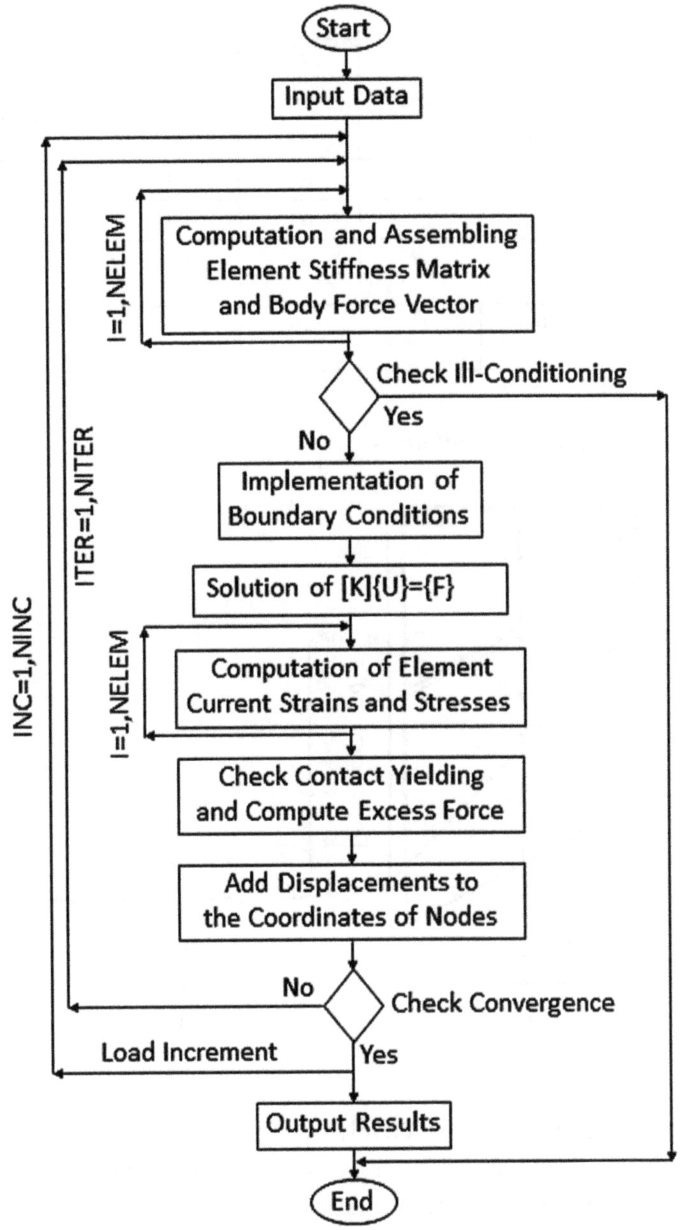

Open Rock Slopes

The first example regarding rock slopes is related to the numerical simulation of the sliding failure of a rock slope under the force of gravity, as shown in Figure 3. It is assumed that the upper blocks are bounded by the slope surface, vertical discontinuities, and a discontinuity inclined at an angle of 31°. The material properties used in the numerical analyses are given in Table 1, where λ and μ indicate the elastic Lame's constants. ρ denotes the unit weight of the rock mass. E_n and G_s indicate the elastic

Table 1. Material properties

Structure	Intact Blocks			Contacts				
	λ	μ	ρ	E_n	G_s	c	σ_t	φ
	MPa	MPa	kN/m³	MPa	MPa	MPa	MPa	(°)
Rock Pillar	10	10	25	0.3	0.3	0	0	30
Open Slope-I	10	10	25	0.3	0.3	0	0	20
Open Slop-II	10	10	25	0.1	0.1	0	0	10
Underground Opening-I	50	50	25	0.9	0.1	0	0	30
Underground Opening-II	10	10	25	10	0.3	0	0	15

Figure 2. Sliding failure mode of a rock pillar

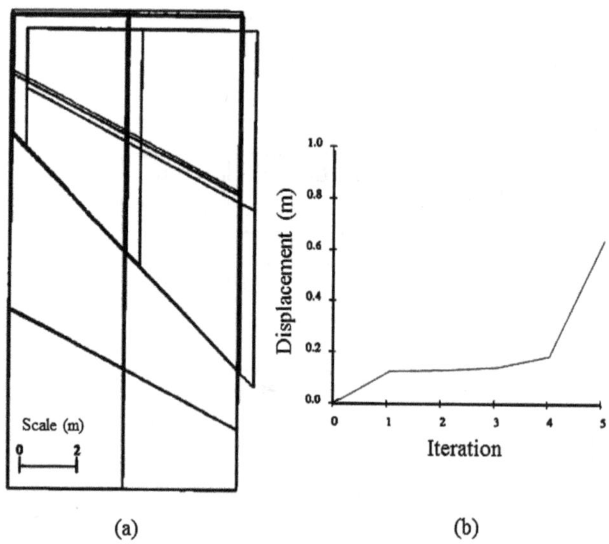

normal and shear moduli of the contact, respectively. σ_t and φ indicate contact tensile stress and friction angle. Figure 3 shows the configuration of the slope at various iteration steps. As expected, separation at the vertical discontinuity and sliding along the inclined discontinuity occurs. The displacement of the sliding block increases as the iteration number increases (Figure 3b).

The second example on open rock slopes is related to numerical simulation of the toppling failure of rock slopes. The material properties used in the numerical analysis are given in Table 1. It was assumed that a basal plane inclined at angle of 12° dipping into slope and discontinuities dipping into the mountainside with an inclination of 70° existed within the slope. Figure 4 shows the configuration of the slope at each iteration step and the vertical displacement of the slope crest. As expected, each rock column tends to rotate clock-wise, and the slope fails in the toppling mode. The toppling of blocks increases as the iteration number increases (Figure 4b).

Figure 3. Simulated sliding failure mode of an open rock slope-I

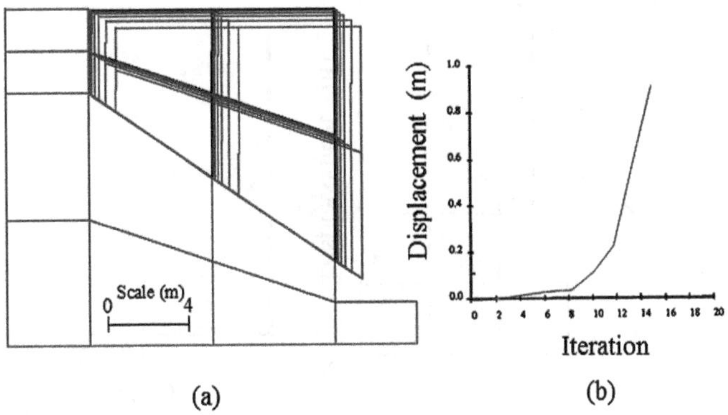

Figure 4. Simulated toppling failure mode of an open rock slope-II

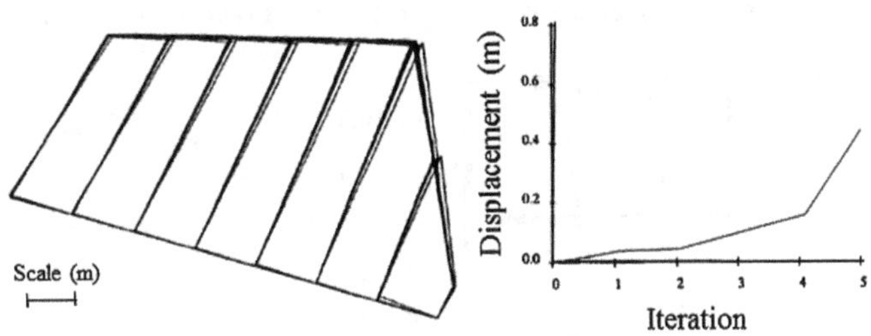

Analysis of Underground Openings

The first example of an underground opening is concerned with the stability of a shallow underground opening having two thoroughgoing discontinuity planes emanating from the side walls. The material properties used in numerical analyses are given in Table 1 (underground opening-I). Figure 5 shows the configuration of the underground opening at each iteration step and the downward vertical displacement of the opening crown and ground surface. During the initial stage, downward sliding of the crown blocks takes place on the discontinuity planes near the opening and propagates upward, and the blocks above the crown slide into the opening like a rigid block. The downward sliding of blocks increases as the iteration number increases (Figure 5b).

The second example of an underground opening is concerned with the stability of tunnels adjacent to steep slopes. It is assumed that two discontinuity planes exist; one is inclined at an angle of 80° emanating from the left shoulder upward, and the other one emanates from the right bottom corner and is inclined at an angle of 28° and dips towards the slope (Figure 6a). The material properties used in numerical analyses are given in Table 1 (underground opening-II). Figure 6a shows the configuration of the underground opening at each iteration step, and Figure 6b shows the vertical displacement of the opening crown at each iteration step. The deformed configuration of the unstable region tends to fail in a rotational and sliding mode as shown in Figure 6a.

Figure 5. Simulated sliding failure mode of an underground opening-I

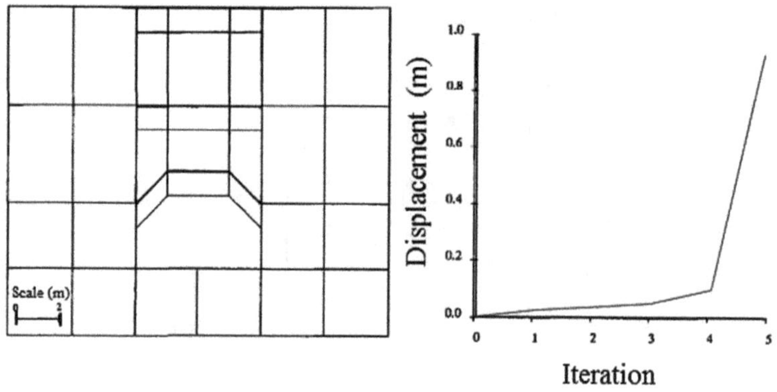

(a) (b)

Figure 6. Simulated toppling-sliding failure mode of an underground opening-II

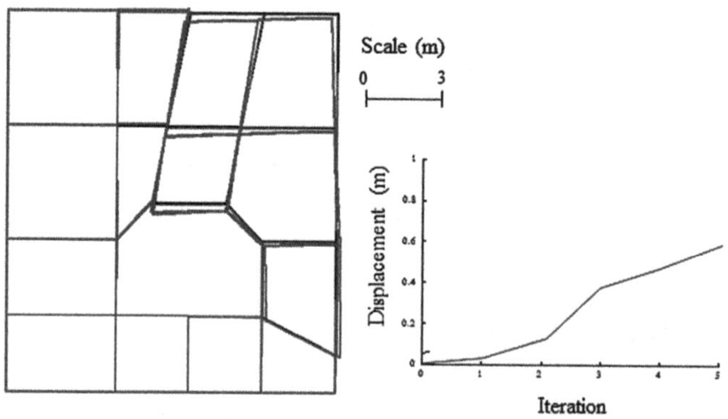

(a) (b)

Stability Analysis of an Underground Tunnel

It is well known that underground structures, such as tunnels and powerhouses, are generally resistant to earthquake-induced motions. However, they may be damaged when permanent ground movements occur in or along the underground structures. Permanent ground deformation-induced damage is generally caused either by faulting or slope movements. Most tunnels have unreinforced concrete linings; since the lining is brittle, permanent ground movements may induce the rupture of the linings, and falling debris may have tremendous consequences on vehicles passing through. Figure 7a shows the configuration of such a tunnel analyzed using DFEM. The characteristics of the materials for seven solid layers and four discontinuity (fault) planes in the tunnel surroundings, shown in Figure 7a, are listed in Table 2, where *h* indicates the band width of contact elements (Mamaghani, 2015a; Tokashiki *et al.*, 1997). The presence of cracks and discontinuities has an essential role in the behavior of rock mass under applied forces. The deformed configuration of the tunnel under gravity load (self-weight) is shown in Figure 7b. The physical behavior of such a system involves the opening of discontinuities and sliding along them. The analysis of this tunnel shows that the unsupported tunnel is not stable due to the existence of steep joints.

Figure 7. Stability analysis of an underground tunnel

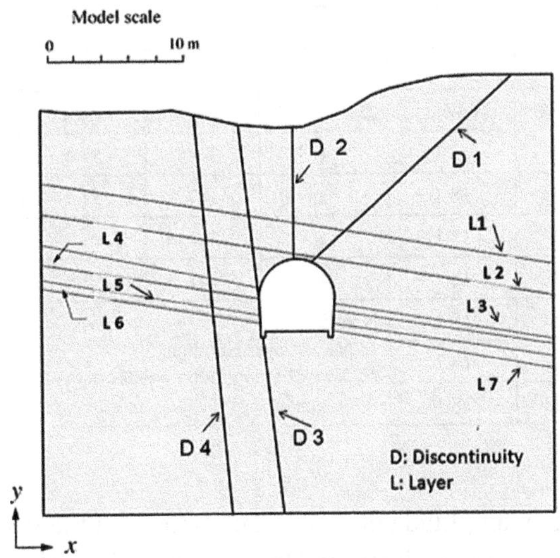

(a) Analyzed tunnel discontinuities

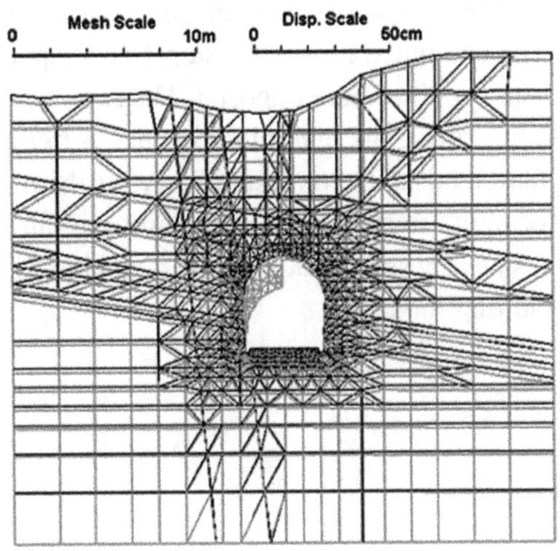

(b) Deformed configuration

Analysis of Fault Propagation

A fault is geologically defined as a discontinuity in a geological medium along which a relative displacement has taken place. Faults are broadly classified into three groups: normal faults, thrust faults, and strike-slip faults (Figure 8). Figures 9a, 10a, and 11a show the geometry and dimensions of the meshes and imposed boundary conditions for three groups of normal, thrust, and strike-slip faults used in analysis. Table 3 gives the material properties used in the DFEM analysis. The thickness of the fault plane was set at 10 mm based on previous work. In the analysis, the fault plane was modeled using contact elements. The fault plane behaves elastically when the normal and shear stresses are below its yield strength. However, if yielding takes place, its behavior is elastic to perfectly plastic. In normal and thrust

Table 2. Material properties for analyzed tunnel

Intact Blocks				Contacts					
	λ	μ	ρ		E_n	G_s	h	σ_t	φ
L. No.	MPa	MPa	kN/m³	D. No.	MPa	MPa	cm	MPa	(°)
1	490	0.282	18.1	1	147	57.4	0.6	0.01	35.0
2	980	0.282	18.1	2	147	57.4	0.6	0.01	0.0
3	343	0.282	18.1	3	147	57.4	0.6	0.01	0.0
4	245	0.282	18.1	4	147	57.4	0.6	0.01	35.0
5	441	0.282	18.1	Note:					
6	343	0.282	18.1	*L. No.* = Layer Numbers					
7	147	0.282	18.1	*D. No.* = Discontinuity Numbers					

faults, displacements having an amplitude of 10 cm are imposed at the boundary nodes as indicated in Figure 9a and 10a in both the x and y directions. In the strike-slip faulting simulation, the prescribed displacements are imposed only in the y direction at selected points, as shown in Figure 11c. Figures 9b and c, 10b and c, and 11b and c show the deformed configurations of the analyzed faults at computation steps 1 and 15, respectively. The computed results for step 1 correspond to the initial elastic responses after the prescribed displacement conditions are imposed. The results at computation step 15 correspond to the fault propagation if yielding along the fault plane occurs. After a certain number of computation steps, the deformation of the fault tends to become stationary for the prescribed displacement boundary conditions. The propagation of faults starts at the bottom and migrates towards the ground surface, as expected. These three examples clearly show that DFEM is capable of simulating the fault propagation processes for three different faulting modes.

Figure 8. Analyzed fault types

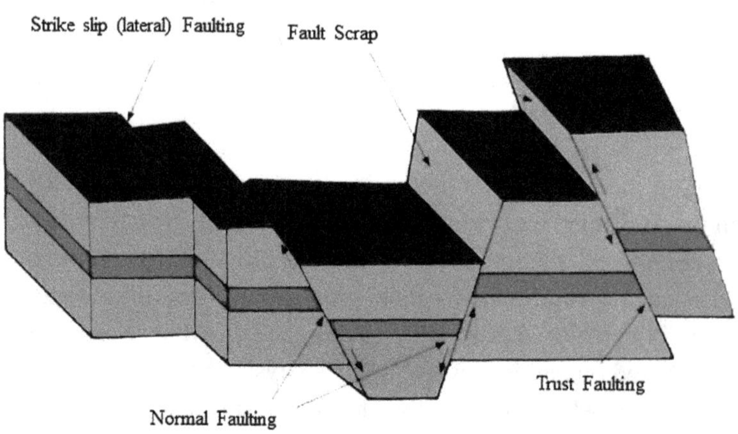

Figure 9. Initial and deformed configurations at steps 1and 15 for normal fault simulation

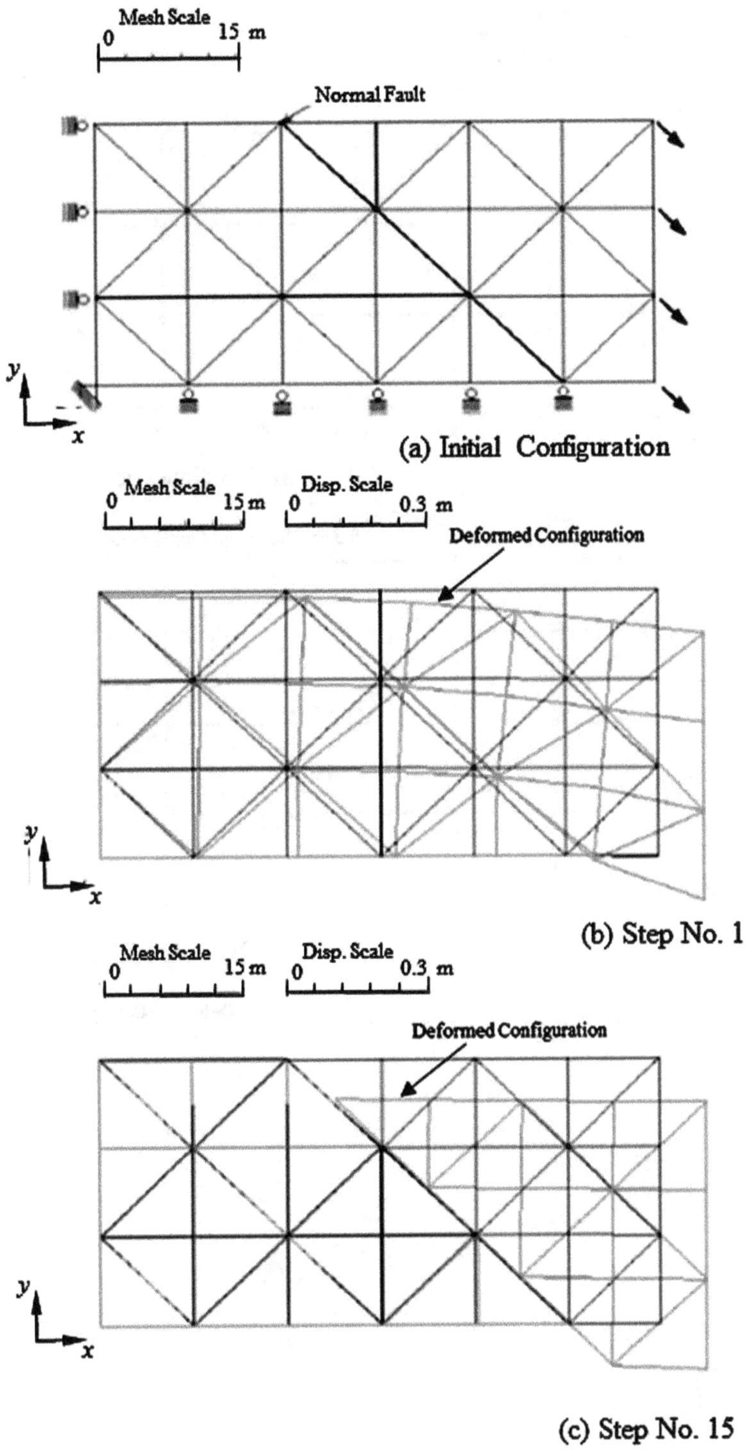

Figure 10. Initial and deformed configurations at steps 1and 15 for thrust fault simulation

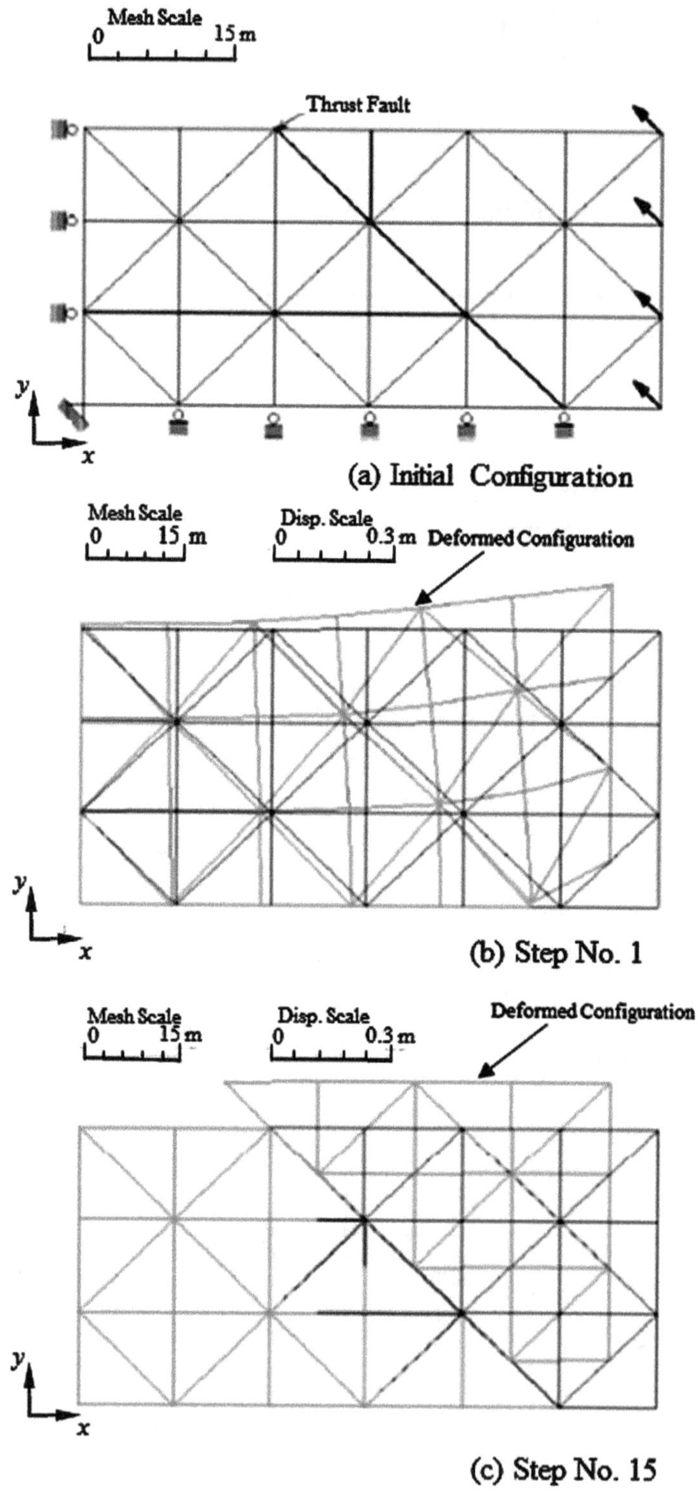

Figure 11. Initial and deformed configurations at steps 1and 15 for strike-slip fault simulation

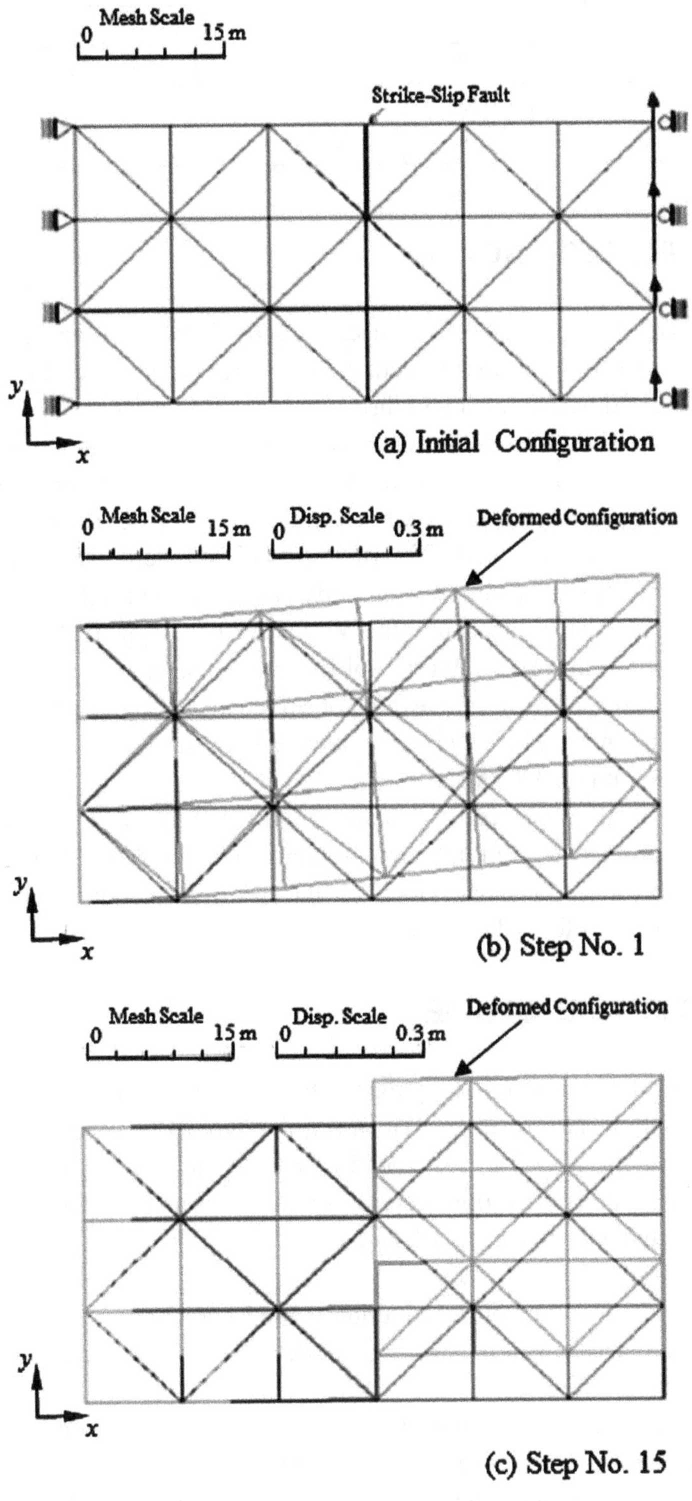

Table 3. Material properties for analyzed faults

Material	λ	μ	ρ	c	σ_t	φ
	MPa	MPa	kN/m^3	MPa	MPa	(°)
Solid	2000	2000	26	-	-	-
Fault	50	50	-	0.0	0.0	40

FAULT-STRUCTURE INTERACTION

The most important aspect of earthquake engineering is the interaction between structures and fault breaks. To study this interaction, a truss structure straddling the projected fault trace on the ground surface was considered, and normal fault and thrust fault conditions were imposed through prescribed displacement at selected points as in the previous computations. The material properties for the solid and fault are given in Table 3. The elastic modulus and the cross-section area of a typical truss were set at 90 GPa and 0.1 m^2, and their behavior was assumed to be elastic. Figures 12a and 13a show the finite element meshes and boundary conditions used in analysis. Figure 12b and c and Figure 13b and c show the deformed configurations at computation steps 1 and 15 for normal fault and thrust fault modes, respectively. In both cases, the truss structure tilts. While the thrust-type fault causes the contraction of trusses, the normal fault results in the extension of trusses and separation of the supporting members fixed to the ground. These responses resemble those observed in recent earthquakes (Aydan *et al.*, 2011). Although trusses were assumed to be behaving elastically in these simulations, it is easy to implement elasto-plastic behavior within the DFEM.

The above computational results clearly demonstrate that the DFEM is capable of simulating both the fault propagation and fault-structure interaction, although a simple version of this method is used. Nevertheless, it is possible to use the other versions of the DFEM depending upon the available information on the characteristics of solid and fault planes.

RECOMMENDED FUTURE STUDIES

The DFEM is a promising method for studying the behavior of unreinforced masonry underground openings under static and dynamic loadings. However, the hyperbolic scheme of the DFEM is still in its formative phase, and both experiments on the viscous characteristics of blocks and contacts as well as a numerically stable time-discretization scheme are needed. The main issues which require further consideration, for both dynamic and static behavior, are:

- Appropriate element modelling of the continuum and discrete regions with a view to incorporating the deformation mechanisms necessary to model stress, strain, and frictional contact/adhesive conditions;
- The inclusion of elasto-plastic/viscoplastic behavior in both the solid and contact elements;
- Algorithms for treatment of the progressive element fracturing;
- Detection procedures for monitoring contact between discrete elements and continuous regions; and
- Representation of frictional/adhesive contact conditions for contacting elements.

Figure 12. Initial and deformed configurations at steps 1and 15 for normal fault-structure interaction simulation

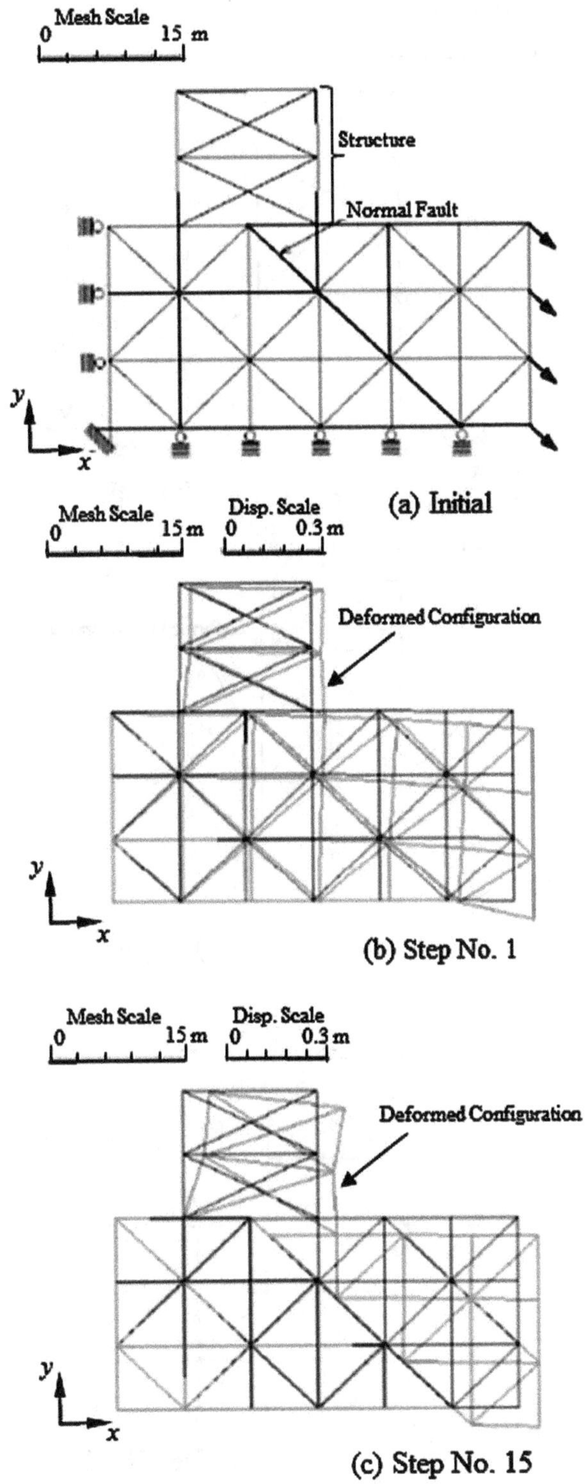

Figure 13. Initial and deformed configurations at steps 1and 15 for thrust fault-structure interaction simulation

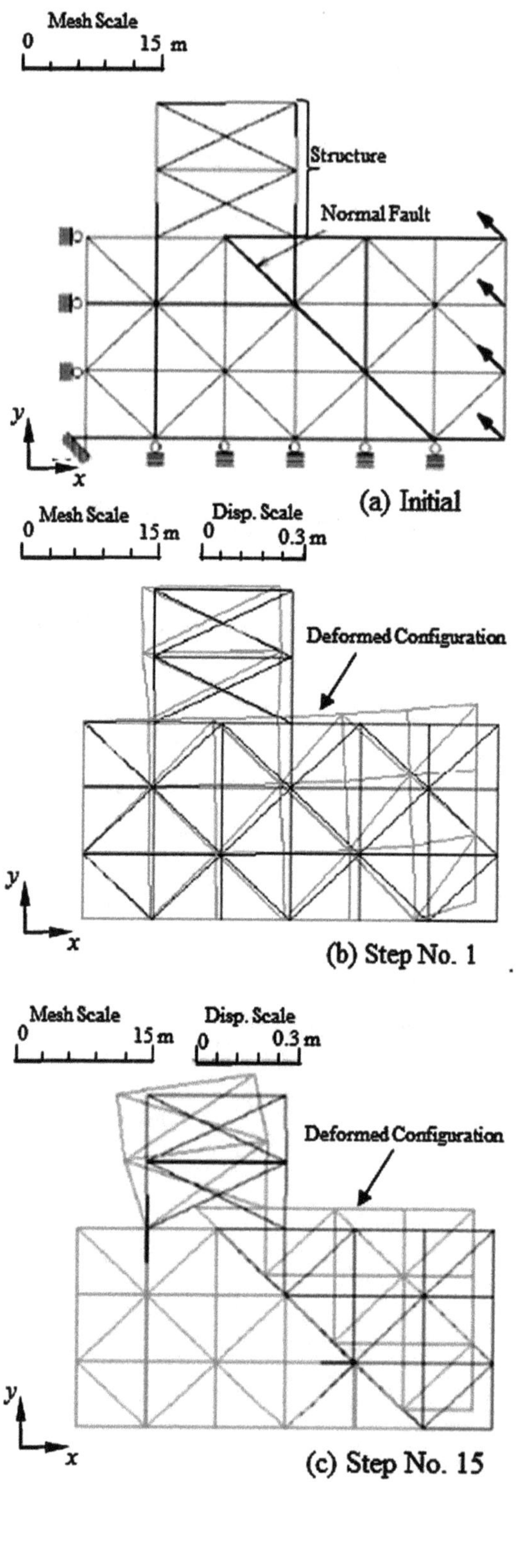

CONCLUSION

This chapter was concerned with the application of the DFEM to the stability analysis of unreinforced masonry structures, which are comprised of a finite number of distinct, interacting blocks that have length scales relatively comparable with the structure of interest. The DFEM, recently developed by the author, is based on the principles of the finite element method incorporating contact elements. It considers blocks as sub-domains and represents them by solid elements. Contact elements are used to model block interactions, such as sliding or separation. The DFEM calculates displacements at the joints as well as deformations within the blocks, which can be used to follow the processes of the failure mechanism of masonry structures under static and/or dynamic loadings. Through some illustrative examples, the applicability of the DFEM to the stability analysis of unreinforced masonry structures was investigated and discussed. It has been shown that the DFEM is capable of simulating large displacements of blocky systems, such as open rock slopes, underground openings, tunnels, fault propagation, and fault-structure interaction. It was found that the DFEM is a promising method for studying the stability of unreinforced masonry structures. However, the hyperbolic scheme of the DFEM is still in its formative phase, and both experiments on viscous characteristics of blocks and contacts as well as a numerically stable time-discretization scheme are necessary.

ACKNOWLEDGMENT

The author would like to sincerely thank Professor O. Aydan of Tokai University (Japan) for his collaboration in developing the DFEM and help in some of the computations. The author dedicates this chapter to his wife *Mrs. Sara Karbalay Gholami* and his son *Sahand Mamaghani* for their patience, understanding, continuous support, and encouragement.

REFERENCES

Aydan, O., Ichikawa, Y., & Kawamoto, T. (1990). Numerical modelling of discontinuities and interfaces in rock mass. *Proc. 4th Computational Mechanics Symp.*

Aydan, Ö., Mamaghani, I. H. P., & Kawamoto, T. (1996). Applications of discrete finite element method (DFEM) to rock engineering structures. *2nd North American Rock Mechanics Symposium Tools and Techniques in Rock Mechanics.*

Aydan, Ö., Ohta, Y., Daido, M., Kumsar, H., Geni¸, M., Tokashiki, N., & Amini, M. et al. (2011). *Earthquakes as a rock dynamic problem and their effects on rock engineering structures.*In Y. Zhou & J. Zhao (Eds.), *Advances in Rock Dynamics and Applications* (pp. 345–421). Taylor & Francis.

Aydan, Ö., Shimizu, Y., & Ichikawa, Y. (1989). The effective failure modes and stability of slopes in rock mass with two discontinuity sets. *Rock Mechanics and Rock Engineering*, *22*(3), 163–188. doi:10.1007/BF01470985

Aydan, O., Tokashiki, N., & Mamaghani, I. H. P. (2005) Modeling and Analysis of Rock structural systems by Discrete Finite Element Method. Journal of Geotechnical Engineering and Surveying, 106, 8-15. (in Japanese)

Baggio, C., & Trovalusci, P. (1993). Discrete models for jointed block masonry walls. In A. A. Hamid & H. G. Harris (Eds.), The Sixth North American Masonry Conference (Vol. 2, pp. 939–949). Lancaster, PA: Technomic Publishing Co. Retrieved from http://dsg.uniroma1.it/trovalusci/pubblicazioni_pdf/B2_Baggio-Trovalusci_NAMC09.pdf

Baggio, C., & Trovalusci, P. (1998). Limit Analysis for No-Tension and Frictional Three-Dimensional Discrete Systems. *Mechanics of Structures and Machines, 26*(3), 287-304. Retrieved fromhttp://dsg.uniroma1.it/trovalusci/pubblicazioni_pdf/A4_Baggio-Trovalusci_MSM98.pdf

Barbosa, R. E., & Ghaboussi, J. (1990). Discrete finite element method for multiple deformable bodies. *Finite Elements in Analysis and Design, Elsevier Science Publishers, 7*(2), 145–158. doi:10.1016/0168-874X(90)90006-Z

Cundall, P. A. (1971). A computer model for simulating progressive large-scale movements in blocky rock systems.*Proc. Int. Symp. on Rock Fracture*.

Cundall, P. A. (2011). Lattice method for brittle, jointed rock. In *Continuum and distinct element numerical modeling in geomechanics, number*. Itasca International, Inc.

Damjanac, B., Detournay, C., & Cundall, P. A. (2015). *Application of particle and lattice codes to simulation of hydraulic fracturing*. Retrieved from http://www.itascacg.com/application-of-particle-and-lattice-codes-to-simulation-of-hydraulic-fracturing

Furtney, J., Zhang, F., & Han, Y. (2013). Review of methods and applications for incorporating fluid flow in the discrete element method. In P. Zhu, C. Detournay, R. Hart, & M. Nelson (Eds.), *Continuum and distinct element numerical modeling in geomechanics*. Minneapolis, MN: Itasca International Inc.

Ghaboussi, J., Wilson, E. L., & Isenberg, J. (1973). Finite element for rock joints and interfaces. *J. Soil Mech. and Found. Engng. Div. ASCE, SM10*(99), 833–848.

Giamundo, V., Sarhosis, V., Lignola, G. P., Sheng, Y., & Manfredi, G. (2014). Evaluation of different computational modelling strategies for modelling low strength masonry. *Engineering Structures, 73*, 160–169. doi:10.1016/j.engstruct.2014.05.007

Goodman, R.E., Taylor, R. & Brekke, T.L. 1968. A model for the mechanics of jointed rock. *J. of Soil Mechs. and Found. Eng. Div., 94*(3), 637-659.

Hoek, E., & Bray, J. W. (1977). Rock slope engineering (2nd ed.). The Institution of Mining and Metallurgy.

Itasca Consulting Group, Inc. (2014). *UDEC Version 6., Distinict-Element Modeling of jointed and Blocky Material in 2D*. Retrieved from http://www.itascacg.com/software/udec

Kawamoto, T., & Aydan, Ö. (1999). A review of numerical analysis of tunnels in discontinuous rock masses. *International Journal for Numerical and Analytical Methods in Geomechanics, 23*(13), 1377–1391. doi:10.1002/(SICI)1096-9853(199911)23:13<1377::AID-NAG932>3.0.CO;2-S

Lermos, J. V. (2007). Discrete Element Modeling of Masonry Structures. *International Journal of Architectural Heritage: Conservation, Analysis, and Restoration, 1*(2), 190–213. doi:10.1080/15583050601176868

Lisjak, A., Kaifosh, P., Mahabadi, O. K., Tatone, B., & Grasselli, G. (2015). Fully 3D simulation of fluid-pressure-driven fracturing using a novel continuum-discontinuum approach: Preliminary results. In GeoConvention 2015: New Horizons.

Mamaghani, I. H. P. (1993). *Numerical analysis for stability of a system of rock blocks.* (Master Thesis). Department of Civil Engineering, Nagoya University, Nagoya, Japan.

Mamaghani, I. H. P. (2006). Analysis of Masonry Bridges: Discrete Finite Element Method. Transportation Research Record: Journal of Transportation Research Board, 1976, 13-19.

Mamaghani, I. H. P. (2008). Analysis of Underground Structures by Discrete Finite Element Method. Canadian Society for Civil Engineering.

Mamaghani, I. H. P. (2015a). Discrete Finite Element Method Application for Analysis of Unreinforced Masonry Underground Structures. In Transportation Research Record: Journal of Transportation Research Board, No. 2522 (pp. 131–136). Washington, DC: Transportation Research Board of the National Academies. Retrieved from http://trrjournalonline.trb.org/doi/pdf/10.3141/2522-13

Mamaghani, I. H. P. (2015b). Analysis of Masonry Structures by Discrete Finite Element Method. *12th North American Masonry Conference*, Denver, CO.

Mamaghani, I. H. P., & Aydan, O. (1999). Discrete Finite Element Method for Analysis of Masonry and Rock Structures. *International Journal of Science and Technology, Scientia Iranica, Tehran, Iran, 6*(3 & 4), 157–164.

Mamaghani, I. H. P., & Aydan, Ö. (2000). Stability analysis of slopes by discrete finite element method. *GeoEng, 2000*(November), 19–24.

Mamaghani, I. H. P., Aydan, Ö., & Akhoundi, F. (2014). Analysis of Masonry Bridges under Static and Dynamic Loading by Discrete Finite Element Method. *9th International Masonry Conference*. International Masonry Society, University of Minho.

Mamaghani, I. H. P., Aydan, Ö., & Kajikawa, Y. (1999). Analysis of masonry structures under static and dynamic loading by discrete finite element method. Journal of Structural Mechanics and Earthquake Engineering, 626, 1-12.

Mamaghani, I. H. P., Baba, S., Aydan, Ö., & Shimizu, Y. (1994). Discrete finite element method for blocky systems.*Proc. of the Eighth Int. Conf. on Computer Methods and Advances in Geomechanics (IACMAG).*

Oda, M., Yamabe, T., Ishizuka, Y., Kumasaka, H., Tada, H., & Kimura, K. (1993). Elastic stress and strain in jointed rock masses by means of crack tensor analysis. *Rock Mechanics and Rock Engineering, 26*(2), 89–112. doi:10.1007/BF01023618

Rougier, E., Knight, E. E., & Munjiza, A. (2012). Fluid driven rock deformation via the combined FEM/DEM methodology. In American rock mechanics association 46thUS rockmechanics/geomechanics symposium.

Rougier, E., Knight, E. E., Munjiza, A., Broome, A. J. S. S. T., Swift, R. P., & Bradley, C. R. (2011). The combined finite-discrete element method applied to the study of rock fracturing behavior in 3D. In American Rock Mechanics Association 45 th US rock mechanics/geomechanics symposium.

Sarhosis V., Sheng Y. (2014). Identification of material parameters for low bond strength masonry. *Engineering Structures, 60*, 100-110. DOI: 10.1016/j.engstruct.2013.12.013

Shi, G. H. (1988). *Discontinuous deformation analysis: a new numerical model for the statics and dynamics of block system.* (PhD Thesis). Dept. of Civil Engineering, University of California, Berkeley, CA.

Tokashiki, N., Aydan, Ö., Mamaghani, I. H. P., & Kawamoto, T. (1997a). The stability of a rock block on an incline by discrete finite element method (DFEM). IACMAG '97.

Tokashiki, N., Aydan, Ö., Shimizu, Y., & Kawamoto, T. (1997b): The assessment of the stability of a very old tunnel by discrete finite element method (DFEM).*Proceedings of the Sixth International Symposium on Numerical Methods in Geomechanics, NUMOG VI.*

Tokashiki, N., Aydan, O., Shimizu, Y., & Mamaghani, I. H. P. (2001). A stability analysis of masonry walls by discrete finite element method (DFEM).*Proc. of the 10th Int. Conf. on Computer Methods and Advances in Geomechanics (IACMAG'01).*

Toth, A. R. (2009). Orban, z., Bagi, k. (2009). Discrete element analysis of a masonry arch. *Mechanics Research Communications, 36*(4), 469–480. doi:10.1016/j.mechrescom.2009.01.001

Zhao, Q., Lisjak, A., Mahabadi, O. K., & Grasselli, G. (2015). Numerical investigation of the influences of rock fabrics on hydraulic fracturing operations. In GeoConvention 2015: New Horizons.

KEY TERMS AND DEFINITIONS

Discontinuity: Contact surfaces of interacting rock blocks in rock mass or blocky rock systems.

Failure Mode: Type of failure, such as sliding or toppling.

Fault: A discontinuity in geological medium along which a relative displacement has taken place.

Fault-Structure Interaction: Forces imposed on structures due to fault propagation.

Rock Slopes: Geological discontinuities causing rock mass consists of distinct interacting blocks to slip or topple.

Sliding: Linear displacement of rock blocks over discontinuities.

Static Analysis: Analysis under static loads that are applied slowly so that no inertia is induced.

Toppling: Rotational displacement of rock blocks.

Underground Opening: Opening caused by excavation below the ground surface.

Unreinforced: Discontinuous media consisting of interacting blocks of arbitrary shapes without any reinforcement.

Compilation of References

Abdou, L., Ami, S. R., Meftah, F., & Mebarki, A. (2006). Experimental investigations of the joint-mortar behaviour. *Mechanics Research Communications*, *33*(3), 370–384. doi:10.1016/j.mechrescom.2005.02.026

Acary, V., & Jean, M. (1998). Numerical simulation of monuments by the contact dynamics method. In *Procs. Monument-98 Workshop on seismic performance of monuments.*

Acary, V., & Jean, M. (2000). Numerical modeling of three-dimensional divided structures by the Non Smooth Contact Dynamics method: Application to masonry buildings In *Procs. Fifth International Conference on Computational Structures Technology.*

Achyutha, H., Jagadish, R., Rao, P. S., & Shakeebur, R. S. (1986). Finite element simulation of the elastic behavior of infilled frames with openings. *Computers & Structures*, *23*(5), 685–696. doi:10.1016/0045-7949(86)90077-5

Adami, C. E., & Vintzileou, E. (2008). Investigations of the bond mechanism between stones or bricks and grouts. *Materials and Structures*, *41*(2), 255–267. doi:10.1617/s11527-007-9235-z

Ahmed, H. M., & Gilbert, M. 2003. The computational efficiency of two rigid block analysis formulations for application to masonry structures. In 9th *International Conference on Civil and Structural Engineering Computing*. Egmond-aan-Zee, Holland: Civil-Comp Press.

Albuerne, A., & Huerta, S. (2010). Coulomb's theory of arches in Spain ca. 1800: the manuscript of Joaquín Monasterio. In *Proceedings of the 6th international conference on Arch Bridges* (pp. 354-362). Fuzhou, China: Academic Press.

Alexakis, H. (2013). *Limit State Analysis and Earthquake Resistance of Masonry Arches*. (Doctoral dissertation). Department of Civil Engineering, University of Patras, Greece.

Alexakis, H., & Makris, N. (2014b). Limit equilibrium analysis of masonry arches. *Archive of Applied Mechanics*. doi: 10.1007/s00419-014-0963-6

Alexakis, H., & Makris, N. (2013). Structural stability and bearing capacity analysis of the tunnel-entrance to the stadium of ancient Nemea. *International Journal of Architectural Heritage*, *7*(6), 673–692. doi:10.1080/15583058.2012.662262

Alexakis, H., & Makris, N. (2013b). Minimum thickness of elliptical masonry arches. *Acta Mechanica*, *224*(12), 2977–2991. doi:10.1007/s00707-013-0906-2

Alexakis, H., & Makris, N. (2014a). Limit equilibrium analysis and the minimum thickness of circular masonry arches to withstand lateral inertial loading. *Archive of Applied Mechanics*, *84*(5), 757–772. doi:10.1007/s00419-014-0831-4

Alexandris, A., Protopapa, E., & Psycharis, I. (2004). Collapse mechanisms of masonry buildings derived by the distinct element method. In *Proc. 13th World Conf. Earthquake Engineering.*

Alexandris, A., Protopapa, E., & Psycharis, I. 2004. Collapse mechanisms of masonry buildings derived by the distinct element method. In *13th World Conference on Earthquake Engineering*.

Alexandris, A., Psycharis, I., & Protopapa, E. (2014). The collapse of the ancient temple of Zeus at Olympia revisited. In *2nd European Conference on Earthquake Engineering and Seismology*.

Allen, R. H., Oppenheim, I. J., Parker, A. P., & Bielak, J. (1986). On the dynamic response of rigid body assemblies. *Earthquake Engineering & Structural Dynamics*, *14*(6), 861–876. doi:10.1002/eqe.4290140604

Alpa, G., & Monetto, I. (1994). Microstructural model for dry block masonry walls with in-plane loading. *Journal of the Mechanics and Physics of Solids*, *42*(7), 1159–1175. doi:10.1016/0022-5096(94)90065-5

Ambraseys, N., & Psycharis, I. N. (2011). Earthquake stability of columns and statues. *Journal of Earthquake Engineering*, *15*(5), 685–710. doi:10.1080/13632469.2010.541549

Ambraseys, N., & Psycharis, I. N. (2012). Assessment of the long-term seismicity of Athens from two classical columns. *Bulletin of Earthquake Engineering*, *10*(6), 1635–1666. doi:10.1007/s10518-012-9388-1

Anderson, C., & Held, L. C. (1986). The effect of sand grading on mortar properties and tensile bond strength of brickwork specimens. *Proceedings of the Brick Masonry Society*, *1*, 1–6.

Andreaus, U., & Casini, P. (1999). On the rocking-uplift motion of a rigid block in a free and forced motion: Influence of sliding and bouncing. *Acta Mechanica*, *138*(3-4), 219–241. doi:10.1007/BF01291846

Andreu, A., Gil, L., & Roca, P. (2007). Computational analysis of masonry structures with a funicular model. *Journal of Engineering Mechanics*, *133*(4), 473–480. doi:10.1061/(ASCE)0733-9399(2007)133:4(473)

Andreu, A., Gil, L., & Roca, P. (2010). Analysis of masonry structures by funicular networks. *Engineering and Computational Mechanics*, *163*(3), 147–154.

Anooshelpoor, A., Brune, J. N., & Zeng, Y. (2004). Methodology for obtaining constrains on ground motion from precariously balanced rocks. *Bulletin of the Seismological Society of America*, *94*(1), 285–303. doi:10.1785/0120020242

Anthoine, A. (1995). Derivation of the in-plane elastic characteristics of masonry through homogenization theory. *International Journal of Solids and Structures*, *32*(2), 137–163. doi:10.1016/0020-7683(94)00140-R

Arredondo, C., & Reinoso, E. (2008). Influence of frequency content and peak intensities in the rocking seismic response of rigid bodies. *Journal of Earthquake Engineering*, *12*(4), 517–533. doi:10.1080/13632460701672755

Asteris, P. G. (2003). Lateral stiffness of brick masonry infilled plane frames. *Journal of Structural Engineering*, *129*(8), 1071–1079. doi:10.1061/(ASCE)0733-9445(2003)129:8(1071)

Asteris, P. G., Cotsovos, D. M., Chrysostomou, C. Z., Mohebkhah, A., & Al-Chaar, G. K. (2013). Mathematical micromodeling of infilled frames: State of the art. *Engineering Structures*, *56*, 1905–1921. doi:10.1016/j.engstruct.2013.08.010

Asteris, P. G., Sarhosis, V., Mohebkhah, A., Plevris, V., Papaloizou, P., Komodromos, P., & Lemos, J. V. (2015). *Numerical Modeling of Historic Masonry Structures. In Seismic Assessment and Rehabilitation of Historic Structures*. IGI Global. doi:10.4018/978-1-4666-8286-3

Atkinson, G. W., & Silva, W. (2000). Stochastic modeling of California ground motions. *Bulletin of the Seismological Society of America*, *90*(2), 255–274. doi:10.1785/0119990064

Atkinson, R. H., Amadei, B. P., Saeb, S., & Sture, S. (1989). Response of masonry bed joint in direct shear. *Journal of Structural Engineering*, *115*(9), 2276–2296. doi:10.1061/(ASCE)0733-9445(1989)115:9(2276)

Audenaert, A., Fanning, P., Sobczak, L., & Peremans, H. (2008). 2-D analysis of arch bridges using an elasto-plastic material model. *Engineering Structures, 30*(3), 845–855. doi:10.1016/j.engstruct.2007.05.018

Aydan, O., Tokashiki, N., & Mamaghani, I. H. P. (2005) Modeling and Analysis of Rock structural systems by Discrete Finite Element Method. Journal of Geotechnical Engineering and Surveying, 106, 8-15. (in Japanese)

Aydan, O., Tokashiki, N., & Mamaghani, I. H. P. (2005). Modeling and Analysis of Rock structural systems by Discrete Finite Element Method. Journal of Geotechnical Engineering and Surveying, 106, 8-15. (in Japanese)

Aydan, O., Ichikawa, Y., & Kawamoto, T. (1990). Numerical modelling of discontinuities and interfaces in rock mass. *Proc. 4th Computational Mechanics Symp.*

Aydan, Ö., Mamaghani, I. H. P., & Kawamoto, T. (1996). Applications of discrete finite element method (DFEM) to rock engineering structures. *2nd North American Rock Mechanics Symposium Tools and Techniques in Rock Mechanics.*

Aydan, Ö., Mamaghani, I. H. P., & Kawamoto, T. (1996). Applications of discrete finite element method (DFEM) to rock engineering structures. In*2nd North American Rock Mechanics Symposium Tools and Techniques in Rock Mechanics.*

Aydan, Ö., Ohta, Y., Daido, M., Kumsar, H., Geni¸, M., Tokashiki, N., & Amini, M. et al. (2011). Earthquakes as a rock dynamic problem and their effects on rock engineering structures. In *Advances in Rock Dynamics and Applications.* Taylor & Francis Publisher.

Aydan, Ö., Ohta, Y., Daido, M., Kumsar, H., Geni¸, M., Tokashiki, N., & Amini, M. et al. (2011). *Earthquakes as a rock dynamic problem and their effects on rock engineering structures.*In Y. Zhou & J. Zhao (Eds.), *Advances in Rock Dynamics and Applications* (pp. 345–421). Taylor & Francis.

Aydan, Ö., Shimizu, Y., & Ichikawa, Y. (1989). The effective failure modes and stability of slopes in rock mass with two discontinuity sets. *Rock Mechanics and Rock Engineering, 22*(3), 163–188. doi:10.1007/BF01470985

Azevedo, J., & Drei, A. (1995). *Characterisation of Typical Historical Constructions from a Structural Point of View: Application to the Seismic Behaviour assessment of the Águas Livres aqueduct in Lisbon.* Report IC-IST AI 8/95. Instituto Superior Técnico, Lisboa.

Azevedo, J. J., Sincraian, G. E., & Lemos, J. V. (2000). Seismic behavior of blocky masonry structures. *Earthquake Spectra, 16*(2), 337–365. doi:10.1193/1.1586116

Azevedo, N. M., Lemos, J. V., & Almeida, J. R. (2015). Discrete Element Particle Modelling of Stone Masonry. In Sarhosis et al. (Eds.), *Computational Modeling of Masonry Structures Using the Discrete Element Method.* IGI Global.

Baggio, C., & Trovalusci, P. (1993). Discrete models for jointed block masonry walls. In A. A. Hamid & H. G. Harris (Eds.), The Sixth North American Masonry Conference (Vol. 2, pp. 939–949). Lancaster, PA: Technomic Publishing Co. Retrieved from http://dsg.uniroma1.it/trovalusci/pubblicazioni_pdf/B2_Baggio-Trovalusci_NAMC09.pdf

Baggio, C., & Trovalusci, P. (1995). Stone assemblies under in-plane actions comparison between non-linear discrete approaches. In J. Middleton & G.N. Pande (Eds.), Computer Methods in Structural Masonry – 3. Books and Journals International Ltd.

Baggio, C., & Trovalusci, P. (1995). Stone assemblies under in-plane actions. Comparison between nonlinear discrete approaches, Computer methods in structural masonry. In *Proceedings of the 3rd International Symposium* (pp. 184-193). Swansea, UK: Academic Press.

Baggio, C., & Trovalusci, P. (1998). Limit Analysis for No-Tension and Frictional Three-Dimensional Discrete Systems. *Mechanics of Structures and Machines, 26*(3), 287-304. Retrieved from http://dsg.uniroma1.it/trovalusci/pubblicazioni_pdf/A4_Baggio-Trovalusci_MSM98.pdf

Baggio, C., & Trovalusci, P. (1998). Limit Analysis for No-Tension and Frictional Three-Dimensional Discrete Systems. *Mechanics of Structures and Machines, 26*(3), 287-304. Retrieved fromhttp://dsg.uniroma1.it/trovalusci/pubblicazioni_pdf/A4_Baggio-Trovalusci_MSM98.pdf

Baggio, C., & Trovalusci, P. 1993. Discrete models for jointed block masonry walls. In *The Sixth North American Masonry Conference*. The Masonry Society TMS.

Baggio, C., De Felice, G., Sguerri, L., & Callipo, C. (2002). The resistance to earthquakes of structures made by dry stone blocks: The case of the Coliseum hypogeum. In *Proc. 12th European Conference on Earthquake Engineering*. Elsevier.

Baggio, C., & Trovalusci, P. (1998). Limit analysis for no-tension and frictional three-dimensional discrete systems. *Mechanics of Structures and Machines, 26*(3), 287–304. doi:10.1080/08905459708945496

Bagi, K. (2014). When Heyman's Safe Theorem of rigid block systems fails: Non-Heymanian collapse modes of masonry structures. *International Journal of Solids and Structures, 51*(14), 2696–2705. doi:10.1016/j.ijsolstr.2014.03.041

Bagneris, M., Dubois, F., & Martin, A. (2013). Numerical analysis of historical masonry structures for stone degradation diagnosis: An application to the Roman Amphitheater of Nîmes. In *2013 Digital Heritage International Congress (DigitalHeritage)*. doi:10.1109/DigitalHeritage.2013.6743792

Bakun-Mazor, D., Hatzor, Y. H., & Glaser, D. (2012). Dynamic sliding of tetrahedral wedge: The role of interface friction. *International Journal for Numerical and Analytical Methods in Geomechanics, 36*(3), 327–343. doi:10.1002/nag.1009

Bandis, S. C., Lumsden, A. C., & Barton, N. R. (1983). Fundamentals of rock joint deformation. *International Journal of Rock Mechanics and Mining Sciences & Geomechanics Abstracts, 20*(6), 249–268. doi:10.1016/0148-9062(83)90595-8

Baraldi, D., Reccia, E., & Cecchi, A. (2015). DEM & FEM/DEM Models For Laterally Loaded Masonry Walls. In *COMPDYN 2015, 5th ECCOMAS Thematic Conference on Computational Methods in Structural Dynamics and Earthquake Engineering.*

Baraldi, D., & Cecchi, A. (2014). Discrete element model for in-plane loaded viscoelastic masonry. *International Journal for Multiscale Computational Engineering, 12*(2), 155–175. doi:10.1615/IntJMultCompEng.2014008118

Baraldi, D., Cecchi, A., & Tralli, A. (2015). Continuous and discrete models for masonry like material: A critical comparative study. *European Journal of Mechanics. A, Solids, 50*, 39–58. doi:10.1016/j.euromechsol.2014.10.007

Baraldi, D., Reccia, E., Cazzani, A., & Cecchi, A. (2013). Comparative analysis of numerical discrete and finite element models: the case of in-plane loaded periodic brickwork. *Composites: Mechanics, Computations, Applications: International Journal (Toronto, Ont.), 4*(4), 319–344.

Baraldi, D., Reccia, E., Cazzani, A., & Cecchi, A. (2013). Comparative analysis of numerical discrete and finite element models: The case of in-plane loaded periodic brickwork. *Composites: Mechanics, Computations. Applications, 4*(4), 319–344.

Baratta, A., Zuccaro, G., & Binetti, A. (2004). Strength capacity of a no tension portal arch-frame under combined seismic and ash loads. *Journal of Volcanology and Geothermal Research, 133*(1-4), 369–376. doi:10.1016/S0377-0273(03)00408-6

Barbosa, B. E. (1996). Discontinous structural analysis. In *Proceedings of the 11th world conference on earthquake engineering.*

Barbosa, R. E., & Ghaboussi, J. (1990). Discrete finite element method for multiple deformable bodies. *Finite Elements in Analysis and Design, Elsevier Science Publishers, 7*(2), 145–158. doi:10.1016/0168-874X(90)90006-Z

Bear, J. (1988). *Dynamics of fluids in porous media*. New York: Dover Publications.

Beattie, G. (2003). *Joint fracture in Reinforced and Unreinforced Masonry under Quasi-static and Dynamic Loading*. (Ph.D thesis). University of Liverpool, Liverpool, UK.

Begg, D. W., & Fishwick, R. J. (1995). Numerical analysis of rigid block structures including sliding. In: J. Middleton & G.N. Pande (eds.), *Proceedings of the 3rd International Symposium in Computational Methods Structural Masonry*. Books and Journals International.

Bélidor, B. F. (1729). *La science des ingénieurs dans la conduite des travaux de fortification et d'architecture civile*. Paris: C-A Jombert.

Belytschko, T., Liu, W. K., & Moran, B. (2000). *Nonlinear finite elements for continua and structures*. John Wiley & Sons Ltd.

Beolchini, G.C., Cifani, G., Pucci, G.G., et al. (2002) *Repertorio dei meccanismi di danno, delle tecniche di intervento e dei relativi costi negli edifici in muratura*. Academic Press.

Berton, S., & Bolander, J. (2006). Crack band model of fracture in irregular lattices. *Computer Methods in Applied Mechanics and Engineering*, *195*(52), 7172–7181. doi:10.1016/j.cma.2005.04.020

Bettzieche, V., Deutsch, R., & Heitfuss, C. (2004). *100 Years of experience in ageing of masonry dams and life-time-based rehabilitation*. Paper presented at the 2nd International Conference on Lifetime Oriented Design Concept, Bochum.

Bicanic, N., & Stirling, C. (2001). DDA analysis of the Couplet/Heyman minimum thickness arch problem. In N. Bicanic (Ed.), Procs. ICADD-4.

Bićanić, N., Ponniah, D., & Robinson, J. (2001). Computational modelling of masonry, brickwork and blockwork structures. In J. W. Bull (Ed.), *Discontinuous Deformation Analysis of Masonry Bridges* (pp. 177–196). Edinburgh, UK: Saxe-Coburg Publications.

Bićanić, N., Stirling, C., & Pearce, C. J. (2003). Discontinuous modelling of masonry bridges. *Computational Mechanics*, *31*(12), 60–68. doi:10.1007/s00466-002-0393-0

Binda, L., Cardani, G., Saisi, A., & Valluzzi, M.R. (2006). *Vulnerability analysis of the historical buildings in seismic area by a multilevel approach*. Academic Press.

Block, P. (2009). *Thrust Network Analysis: Exploring Three-dimensional Equilibrium*. (Doctoral dissertation). Department of Architecture, Massachusetts Institute of Technology, Cambridge, MA.

Block, P., & Ochsendorf, J. (2007). Thrust network analysis: a new methodology for three-dimensional equilibrium. *Journal of the International Association for Shell and Spatial Structures*, *48*(3).

Block, P., DeJong, M., & Ochsendorf, J. (2006). As Hangs the Flexible Line: Equilibrium of Masonry Arches. *Nexus Network Journal*, *8*(2), 13–24. doi:10.1007/s00004-006-0015-9

Bobet, A., Fakhimi, A., Johnson, S., Morris, J., Tonon, F., & Yeung, M. R. (2009). Numerical models in discontinuous media, review of advances for rock mechanics applications. *Journal of Geotechnical and Geoenvironmental Engineering*, *135*(11), 1547–1561. doi:10.1061/(ASCE)GT.1943-5606.0000133

Boore, D. M. (2003). Simulation of Ground Motion Using the Stochastic Method. *Pure and Applied Geophysics*, *160*(3), 653–676. doi:10.1007/PL00012553

Boothby, T. (1995). Collapse modes of masonry arch bridges. *Journal of the British Masonry Society*, *9*(2), 62-69.

Boothby, T. E. (1994). Stability of masonry piers and arches including sliding. *Journal of Engineering Mechanics*, *120*(2), 304–319. doi:10.1061/(ASCE)0733-9399(1994)120:2(304)

Boroschek, R., & Iruretagoyena, A. (2006). Controlled overturning of unanchored rigid bodies. *Earthquake Engineering & Structural Dynamics*, *35*(6), 695–711. doi:10.1002/eqe.554

Braga, F., & Dolce, M. (1982). A method for the analysis of antiseismic masonry multi-storey buildings. In *Sixth International Brick Masonry Conference*, Roma, Italy.

Brencich, A., & Morbiducci, R. (2007). Masonry Arches: Historical Rules and Modern Mechanics. *International Journal of Architectural Heritage*, *1*(2), 165–189. doi:10.1080/15583050701312926

Bretas, E. M., Batista, A. L., & Lemos, J. V. (2015). *Stability analysis of an old masonry gravity dam*. Paper presented at the XXX Seminário Nacional de Grandes Barragens, Foz do Iguaçu.

Bretas, E. M., Léger, P., Lemos, J. V., & Lourenço, P. B. (2010). *Analysis of a gravity dam considering the application of passive reinforcement*. Paper presented at the II International Congress on Dam Maintenance and Rehabilitation, Zaragoza, Spain.

Bretas, E. M., Lemos, J. V., & Lourenço, P. B. (2010). *Masonry Gravity Dams - A numerical application for stability analysis*. Paper presented at the 8th International Masonry Conference, Dresden, Germany.

Bretas, E. M., Lemos, J. V., Léger, P., & Lourenço, P. B. (2011). *Structural analysis of the rehabilitation works of Bhandardara dam*. Paper presented at the XXVIII Seminário Nacional de Grandes Barragens, Rio de Janeiro.

Bretas, E. M., Lemos, J. V., & Lourenço, P. B. (2012). Masonry dams: Analysis of the historical profiles of Sazilly, Delocre, and Rankine. *International Journal of Architectural Heritage*, *6*(1), 19–45. doi:10.1080/15583058.2010.501399

Bretas, E. M., Lemos, J. V., & Lourenço, P. B. (2013). Hydromechanical analysis of masonry gravity dams and their foundations. *Rock Mechanics and Rock Engineering*, *46*(2), 327–339. doi:10.1007/s00603-012-0305-3

Bretas, E. M., Lemos, J. V., & Lourenço, P. B. (2014). A DEM based tool for the safety analysis of masonry gravity dams. *Engineering Structures*, *59*(Feb), 248–260. doi:10.1016/j.engstruct.2013.10.044

Bretas, E. M., Lemos, J. V., & Lourenço, P. B. (2015). Seismic analysis of masonry gravity dams using the Discrete Element Method: Implementation and application. *Journal of Earthquake Engineering*.

Brocken, H. J. P., & Pel, L. (1995). Moisture transport over the brick/mortar interface. In: National Research Council Canada (NRCC). *International Symposium on Moisture Problems in Building Walls*.

BS 5628. (2005). *Code of practice for the use of masonry. Materials and components, design and workmanship*.

Cakti, E., Saygili, O., Lemos, J. V., & Oliveira, C. S. (2014). A parametric study on earthquake behavior of masonry minarets. *Tenth U.S. National Conference on Earthquake Engineering, Frontiers of Earthquake Engineering*.

Cakti, E., Oliveira, C. S., Lemos, J. V., Saygili, O., Görk, S., & Zengin, E. (2015). Ongoing Research on Earthquake Behavior of Historical Minarets in Istanbul. In I. N. Psycharis et al. (Eds.), *Seismic Assessment, Behavior and Retrofit of Heritage Buildings and Monuments, Computational Methods in Applied Sciences, 37*. Springer.

Camenen, J.-F., Ceh, N., Jelenic, G., Koziara, T., & Bicanic, N. (2015). Dynamic Sensitivity of a Multi-block Stack Subjected to Horizontal Harmonic Excitation.*22ème Congrès Français de Mécanique*, Lyon, France.

Carpinteri, A. (1992). *Meccanica dei materiali e della frattura*. Bologna: Pitagora Editrice.

Casapulla, C., & D'Ayala, D. (2001). Lower bound approach to the limit analysis of 3D vaulted block masonry structures. In *Proceedings of the 5th International Symposium on Computer Methods in Structural Masonry* (pp. 28-36). Rome: STRUMAS V.

Casolo, S., & Milani, G. (2010). A simplified homogenization-discrete element model for the non-linear static analysis of masonry walls out-of-plane loaded. *Engineering Structures*, *32*(8), 2352–2366. doi:10.1016/j.engstruct.2010.04.010

Casolo, S., & Milani, G. (2013). Simplified out-of-plane modeling of three-leaf masonry walls accounting for the material texture. *Construction & Building Materials*, *40*, 330–351. doi:10.1016/j.conbuildmat.2012.09.090

Casolo, S., Milani, G., Uva, G., & Alessandri, C. (2013). Comparative seismic vulnerability analysis on ten masonry towers in the coastal Po Valley in Italy. *Engineering Structures*, *49*, 465–490. doi:10.1016/j.engstruct.2012.11.033

Casolo, S., & Peña, F. (2007). Rigid element model for in-plane dynamics of masonry walls considering hysteretic behavior and damage. *Earthquake Engineering & Structural Dynamics*, *36*(8), 1029–1048. doi:10.1002/eqe.670

Cavicchi, A., & Gambarotta, L. (2005). Collapse analysis of masonry arch bridges taking into account arch-fill interaction. *Engineering Structures*, *27*(4), 605–615. doi:10.1016/j.engstruct.2004.12.002

Cavicchi, A., & Gambarotta, L. (2006). Two-dimensional finite element upper bound limit analysis of masonry bridges. *Computers & Structures*, *84*(31-32), 2316–2328. doi:10.1016/j.compstruc.2006.08.048

Cavicchi, A., & Gambarotta, L. (2007). Lower bound limit analysis of masonry bridges including arch-fill interaction. *Engineering Structures*, *29*(11), 3002–3014. doi:10.1016/j.engstruct.2007.01.028

Cecchi, A., & Milani, G. (2008). A kinematic FE limit analysis model for thick english bond masonry walls. *International Journal of Solids and Structures*, *45*(5), 1302–1331. doi:10.1016/j.ijsolstr.2007.09.019

Cecchi, A., & Sab, K. (2002). A multi-parameter homogenization study for modeling elastic masonry. *European Journal of Mechanics. A, Solids*, *21*(2), 249–268. doi:10.1016/S0997-7538(01)01195-0

Cecchi, A., & Sab, K. (2002). Out of plane model for heterogeneous periodic materials: The case of masonry. *European Journal of Mechanics. A, Solids*, *21*(5), 715–746. doi:10.1016/S0997-7538(02)01243-3

Cecchi, A., & Sab, K. (2004). A comparison between a 3D discrete model and two homogenized plate models for periodic elastic brickwork. *International Journal of Solids and Structures*, *41*(9-10), 2259–2276. doi:10.1016/j.ijsolstr.2003.12.020

Ceh, N., Pellegrino, A., Camenen, J.-F., Bicanic, N., Petrinic, N., & Tuhtan, M. (2015b). *Overturning of multiple-block stack - dynamic sensitivity parameters and scaling effect.* 8th International Congress of Croatian Society of Mechanics, Opatija, Croatia.

Ceh, N., Pellegrino, A., Camenen, J.-F., Petrinic, N., Jelenic, G., Koziara, T., & Bicanic, N. (2015a). Dynamic sensitivity of multi-block stacks subjected to pulse base excitation – Experimental evidence and non-smooth contact dynamics simulations. In Procs. COMPDYN 2015.

Chakrabarty, J. (2006). *Theory of plasticity*. London, UK: Butterworth Heinemann.

Chang, K., & Meegoda, J. (1997). Micromechanical Simulations of Hot Mix Asphalt. *Journal of Engineering Mechanics*, *123*(5), 495–503. doi:10.1061/(ASCE)0733-9399(1997)123:5(495)

Chen, W. F. (1982). *Plasticity in reinforced concrete*. New York: McGraw-Hill.

Chen, X., & Liu, Y. (2015). Numerical study of in-plane behaviour and strength of concrete masonry infills with openings. *Engineering Structures*, *82*, 226–235. doi:10.1016/j.engstruct.2014.10.042

Chetouane, B., Dubois, F., Vinches, M., & Bohatier, C. (2005). NSCD discrete element method for modelling masonry structures. *International Journal for Numerical Methods in Engineering*, *64*(1), 65–94. doi:10.1002/nme.1358

Chopra, A. K. (2001). *Dynamics of structures: Theory and applications to earthquake engineering* (2nd ed.). Prentice Hall.

Churilov, S., & Dumova-Jovanoska, E. (2008). Calibration of a numerical model for masonry with applications to experimental results. Journal of Architecture, Civil Engineering and Environment, 3, 41-48.

Claxton, M., Hart, R. A., McCombie, P. F., & Walker, P. J. (2005). Rigid block distinct-element modeling of dry-stone retaining walls in plane strain. *Journal of Geotechnical and Geoenvironmental Engineering*, *131*(3), 381–389. doi:10.1061/(ASCE)1090-0241(2005)131:3(381)

Clemente, P. (1998). Introduction to dynamics of stone arches. *Earthquake Engineering & Structural Dynamics*, *27*(5), 513–522. doi:10.1002/(SICI)1096-9845(199805)27:5<513::AID-EQE740>3.0.CO;2-O

Clemente, P., Occhiuzzi, A., & Raithel, A. (1995). Limit behavior of stone arch bridges. *Journal of Structural Engineering*, *121*(7), 1045–1050. doi:10.1061/(ASCE)0733-9445(1995)121:7(1045)

Cocchetti, G., Colasante, G., & Rizzi, E. (2012). On the Analysis of Minimum Thickness in Circular Masonry Arches. *Applied Mechanics Reviews*, *64*(5).

Comité Français des Barrages et Réservoirs (CFBR). (2013). *Guidelines for the justification of the stability of gravity dams*. French Dams and Reservoirs Committee.

Como, M. (2012). Statics of bodies made of a compressionally rigid no tension material. In M. Frémond & F. Maceri (Eds.), *Mechanics, Models and Methods in Civil Engineering, LNACM* (Vol. 61, pp. 61–78). Springer Berlin Heidelberg. doi:10.1007/978-3-642-24638-8_2

Cook, D. A., & Pegam, G. (1993). Analysis of crack damage in three Georgian facades in the city of Bath, England, with a simple approach to the repair problem. In Structural repair and Maintanence of historical building III. Computational Mechanics Publications.

Costa, A. A., Arêde, A., Campos Costa, A., Penna, A., & Costa, A. (2013). Out-of-plane behavior of a full scale stone masonry façade. Part 2: Shaking table tests. *Earthquake Engineering & Structural Dynamics*, *42*, 2097–2111.

Coulomb, C. A. (1773). Essai sur une application des regles des maximis et minimis à quelques problémes de statique relativs a l'arquitecture. In Mémoires de mathematique et de physique présentés à l'académie royal des sciences per divers savants et lus dans ses assemblées (vol. 7, pp. 343-382).

Couplet, P. (1729, 1730). De la poussée des voûtes. In Histoire de l'Académie Royale des Sciences (pp. 79-117 & pp. 117-141). Paris.

Culmann, K. (1866). *Die graphische Statik*. Zurich: Verlag von Meyer und Zeller.

Cundall, P. A. (1971). A computer model for simulating progressive large scale movements in blocky rock systems. In *Proc. Int. Symposium on Rock Fracture (ISRM)*, Nancy (vol. 1, paper II-8).

Cundall, P. A. (1971). *A computer model for simulating progressive large scale movements in blocky rock systems*. Paper presented at the International symposium, international society of rock mechanics, Nancy, France.

Cundall, P. A. (1971). A computer model for simulating progressive large-scale movements in blocky rock system. In *Proceedings of the Symposium of the International Society for Rock Mechanics* (*vol. 1*, paper no II-8). Nancy, France: Academic Press.

Cundall, P. A. (1971). A computer model for simulating progressive large scale movements in blocky rock systems. I: *Proceedings of the Symposium of the International Society of Rock Mechanics*.

Cundall, P. A. (1971). A computer model for simulating progressive large scale movements in blocky rock systems. In *Proc. of the Symposium of the International Society of Rock Mechanics*.

Cundall, P. A. (1971). A computer model for simulating progressive large scale movements in blocky rock systems. In *Proc. Symp. Rock Fracture (ISRM).*

Cundall, P. A. (1971). *A Computer Model for Simulating Progressive Large Scale Movements in Blocky Rock Systems.* In *Proceedings of the Symposium of the International Society for Rock Mechanics.*

Cundall, P. A. (1971). A computer model for simulating progressive large-scale movements in blocky rock systems. *Proc. Int. Symp. on Rock Fracture.*

Cundall, P. A. (1976) Explicit finite difference methods in geomechanics. In *Int. Conf. Numerical Methods in Geomechanics*, Blacksburg, VA.

Cundall, P. A. (1987). Distinct element models for rock and soil structure. In E. T. Brown (Ed.), *Analytical and computational methods in engineering rock mechanics* (pp. 129–163). London: Allen & Unwin.

Cundall, P. A. (1987). Distinct element models of rock and soil structure. In E. T. Brown (Ed.), *Analytical and Computational Methods in Engineering Rock Mechanics*. London: Allen and Unwin.

Cundall, P. A. (1988). Formulation of a three-dimensional distinct element model - Part I: A scheme to detect and represent contacts in a system composed of many polyhedral blocks. *International Journal of Rock Mechanics and Mining Sciences*, *25*(3), 107–116. doi:10.1016/0148-9062(88)92293-0

Cundall, P. A. (2011). Lattice method for brittle, jointed rock. In *Continuum and distinct element numerical modeling in geomechanics, number.* Itasca International, Inc.

Cundall, P. A., & Hart, R. D. (1989). Numerical modelling of discontinua. In *Proceedings of the 1st US Conference on Discrete Element Methods*. CSM Press.

Cundall, P. A., & Strack, O. D. L. (1979). A discrete numerical model for granular assemblies. *Geotechnique*, *29*(1), 47–65. doi:10.1680/geot.1979.29.1.47

D'Asdia, P., & Viskovi, A. L. (1994). Analisi sismica degli edifici in muratura. *Ingegneria Sismica*, *1*, 32–42.

Damjanac, B., Detournay, C., & Cundall, P. A. (2015). *Application of particle and lattice codes to simulation of hydraulic fracturing*. Retrieved from http://www.itascacg.com/application-of-particle-and-lattice-codes-to-simulation-of-hydraulic-fracturing

Dasiou, M.-E., Mouzakis, H. P., Psycharis, I. N., Papantonopoulos, C., & Vayas, I. (2009a). Experimental investigation of the seismic response of parts of ancient temples. In *Prohitech Conference*.

Dasiou, M.-E., Psycharis, I. N., & Vayas, I. (2009b). Verification of numerical models used for the analysis of ancient temples. In *Prohitech Conference*.

Daudon, D., Lanier, J., & Jean, M. (1997). A micromechanical comparison between experimental results and numerical simulation of a biaxial test on 2D granular material. In Behringer et al. (Eds.), *Powders and Grains 97* (pp. 219–222). Balkema.

Dawe, J. L., & Seah, C. K. (1989). Behavior of masonry infilled steel frames. *Canadian Journal of Civil Engineering*, *16*(6), 865–876. doi:10.1139/l89-129

Dawe, J. L., Seah, C. K., & Liu, Y. (2001). A computer model for predicting infilled frame behavior. *Canadian Journal of Civil Engineering*, *28*(1), 133–148. doi:10.1139/l00-083

De Buhan, P., & De Felice, G. (1997). A homogenization approach to the ultimate strength of brick masonry. *Journal of the Mechanics and Physics of Solids*, *45*(7), 1085–1104. doi:10.1016/S0022-5096(97)00002-1

De Felice, G. (2011). Out-of-plane seismic capacity of masonry depending on wall section morphology. *International Journal of Architectural Heritage*, *5*(4-5), 466–482. doi:10.1080/15583058.2010.530339

De Felice, G., & Giannini, R. (2001). Out-of-plane seismic resistance of masonry walls. *Journal of Earthquake Engineering*, *5*(2), 253–271. doi:10.1080/13632460109350394

De Lorenzis, L., DeJong, M., & Ochsendorf, J. (2007). Failure of masonry arches under impulse base motion. *Earthquake Engineering & Structural Dynamics*, *36*(14), 2119–2236. doi:10.1002/eqe.719

De Lorenzis, L., Dimitri, R., & Ochsendorf, J. (2012a). Structural study of masonry buttresses: The trapezoidal form. *Proceedings of the Institution of Civil Engineers. Structures and Buildings*, *165*(9), 483–498. doi:10.1680/stbu.10.00042

De Lorenzis, L., Dimitri, R., & Ochsendorf, J. (2012b). Structural study of masonry buttresses: The stepped form. *Proceedings of the Institution of Civil Engineers. Structures and Buildings*, *165*(9), 499–521. doi:10.1680/stbu.10.00043

De Lorenzi-Venneri, G., Lee, R. C., Luscher, D. J., Bronkhorst, C. A., Rougier, E., Knight, E. E., . . . Salmon, M. W. (2014). *Proceedings of the workshop on the structural cracking of the cupola of Santa Maria del Fiore*. International Report.

De Luca, A., Giordano, A., & Mele, E. (2004). A simplified procedure for assessing the seismic capacity of masonry arches. *Engineering Structures*, *26*(13), 1915–1929. doi:10.1016/j.engstruct.2004.07.003

DeJong, M. (2009). *Seismic Assessment Strategies for Masonry Structures*. (Doctoral dissertation). Department of Architecture, Massachusetts Institute of Technology, Cambridge, MA.

DeJong, M. I. (2009). *Seismic assessment strategies for masonry structures*. (Doctoral Dissertation). Massachusetts Institute of Technology, Cambridge, UK.

DeJong, M. J., & Ochsendorf, J. (2006). *Analysis of vaulted masonry structures subjected to horizontal ground motion*. Paper presented at the 5th International conference on structural analysis of historical constructions, New Delhi, India.

DeJong, M. J., De Lorenzis, L., & Ochsendorf, J. (2008). Rocking stability of masonry arches in seismic regions. *Earthquake Spectra*, *24*(4), 847–865. doi:10.1193/1.2985763

DeJong, M. J., & Vibert, C. (2012). Seismic response of stone masonry spires: Computational and experimental modeling. *Engineering Structures*, *40*, 566–574. doi:10.1016/j.engstruct.2012.03.001

Delbecq, J. M. (1982a). Analyse de la stabilité des voûtes en maçonnerie par le calcul à la rupture. *Journal de Mécanique théorique et appliquée*, *1*, 91-121.

Delbecq, J. M. (1982b). *Les Ponts en Maçonnerie – Constitution et Stabilité*. Paris: Ministère des Transports, SETRA.

Delbecq, J. M. (1981). Analyse de la stabilité des voûtes en maçonnerie de Charles Augustin Coulomb à nos jours. *Annales des Ponts et Chaussées*, *19*, 36–43.

Dhanasekar, M., & Page, A. W. (1986). Influence of brick masonry nfill properties on the behaviour of infilled frames. Proc, Inst Civil Engrs. *Lond*, *81*(Part 2), 593–605.

Di Egidio, A., & Contento, A. (2009). Base isolation of slide-rocking non-symmetric rigid blocks under impulsive and seismic excitations. *Engineering Structures*, *31*(11), 2723–2734. doi:10.1016/j.engstruct.2009.06.021

Di Egidio, A., & Contento, A. (2010). Seismic response of a non-symmetric rigid block on a constrained oscillating base. *Engineering Structures*, *32*(10), 3028–3039. doi:10.1016/j.engstruct.2010.05.022

Dialer, C. (1990). *Fracture and deformation behaviour of stress, biaxial tests on scale model masonry*. (MSc dissertation). Technical University of Munich, Germany.

Dialer, C. (1992). A distinct element approach for the deformation behaviour of shear stressed masonry panels. In *6th Canadian Masonry Symposium*. University of Saskatchewan.

Dialer, C. (2002). Typical masonry failures and repairs, a German Engineer's view. *Progress in Structural Engineering and Materials*, *4*(3), 332–339. doi:10.1002/pse.125

Dimitri, R. (2009). *Stability of Masonry Structures under Static and Dynamic.* (Doctoral Dissertation). Università del Salento, Lecce, Italy.

Dimitri, R., De Lorenzis, L., & Zavarise, G. (2011). Numerical study on the dynamic behavior of masonry columns and arches on buttresses with the discrete element method. *Engineering Structures*, *33*(12), 3172–3188. doi:10.1016/j.engstruct.2011.08.018

Dimitri, R., & Tornabene, F. (2015). A parametric investigation of the seismic capacity for masonry arches and portals of different shapes. *Engineering Failure Analysis*, *52*, 1–34. doi:10.1016/j.engfailanal.2015.02.021

Dolezalova, M. (2004). Numerical analysis of an old masonry dam using UDEC. In Konietzky (Ed.), Numerical Modeling of Discrete Materials (pp. 269-277). London: Taylor & Francis Group.

Doolin, D. M., & Sitar, N. (2002). Displacement accuracy of discontinuous deformation analysis method applied to sliding block. *Journal of Engineering Mechanics*, *128*(11), 1158–1168. doi:10.1061/(ASCE)0733-9399(2002)128:11(1158)

Drei, A., & Fontana, A. (2001). Load carrying capacity of multiple-leaf masonry arches. In *Proc. 3rd Historical Construction Conference.*

Drei, A., & Oliveira, C. S. (2001). The seismic behaviour of the 'Aqueduto da Amoreira' in Elvas using distinct element modeling. In P. B. Lourenço & P. Roca (Eds.), *Historical Constructions* (pp. 903–911). Guimarães.

Drosopoulos, G. A., Stavroulakis, G. E., & Massalas, C. V. (2006). Limit analysis of a single span masonry bridge with unilateral frictional contact interfaces. *Engineering Structures*, *28*(13), 1864–1873. doi:10.1016/j.engstruct.2006.03.016

Dubois, F., & Mozul, R. (2013). LMGC90. In *CSMA 2013,11e Colloque National en Calcul des Structures.*

Dubois, F., & Jean, M. (2006). The non smooth contact dynamic method: recent LMGC90 software developments and application. In P. Wriggers & U. Nackenhorst (Eds.), *Analysis and Simulation of Contact Problems* (Vol. 27, pp. 375–378). Lecture Notes in Applied and Computational Mechanics. doi:10.1007/3-540-31761-9_44

Durand-Claye, A. (1867). Note sur la vérification de la stabilité des voûtes en maçonnerie et sur l'emploi des courbes de pression. *Annales des Ponts et Chaussées*, *13*, 63–93.

Ebeling, R. M., Nuss, L. K., & Tracy, F. T. (2000). *Evaluation and comparison of stability analysis and uplift criteria for concrete gravity dams by three federal agencies.* Washington, DC: Academic Press.

El-Dakhakhni, W. W. (2002). *Experimental and analytical seismic evaluation of concrete masonry-infilled steel frames retrofitted using GFRP laminates.* (Ph.D. thesis). Drexel University, Philadelphia, PA.

El-Dakhakhni, W. W., Elgaaly, M., & Hamid, A. A. (2003). Three-strut model for concrete masonry-infilled steel frames. *Journal of Structural Engineering*, *129*(2), 177–185. doi:10.1061/(ASCE)0733-9445(2003)129:2(177)

Erdik, M., Durukal, E., Erturk, N., & Sungay, B. (2010). Earthquake risk mitigation in Istanbul museums. *Natural Hazards*, *53*(1), 97–108. doi:10.1007/s11069-009-9411-2

Erguler, Z. A., & Ulusay, R. (2009). Water-induced variations in mechanical properties of clay-bearing rocks. *Int J Rock Mech Min*, *46*(2), 355–370. doi:10.1016/j.ijrmms.2008.07.002

Fanning, P. J., Boothby, T. E., & Roberts, B. J. (2001). Longitudinal and transverse effects in masonry arch assessment. *Construction & Building Materials*, *15*(1), 51–60. doi:10.1016/S0950-0618(00)00069-6

Ferris, M. C., & Tin-Loi, F. (2002). Limit analysis of frictional block assemblies as a mathematical program with complementary constrains. *International Journal of Mechanical Sciences*, *43*(1), 209–224. doi:10.1016/S0020-7403(99)00111-3

Fielder, W. T., Virgin, L. N., & Plaut, R. H. (1997). Experiments and simulation of overturning of an asymmetric rocking block on an oscillating foundation. *European Journal of Mechanics. A, Solids*, *16*, 905–923.

Fjær, E. (2009). Static and dynamic moduli of a weak sandstone. *Geophysics*, *74*(2), WA102–WA112. doi:10.1190/1.3052113

Flanagan, R. D., & Bennett, R. M. (1999). In-plane behavior of structural clay tile infilled frames. *Journal of Structural Engineering*, *125*(6), 590–599. doi:10.1061/(ASCE)0733-9445(1999)125:6(590)

Foce, F., & Aita, D. (2003). The masonry arch between «limit» and «elastic» analysis. A critical re-examination of Durand-Claye's method. In S. Huerta (Ed.), *Proceedings of the 1st International Congress on Construction History* (*vol. 2*, pp. 895-905). Madrid: Instituto Juan de Herrera.

Foce, F. (2005). On the safety of the masonry arch. Different formulations from the history of structural mechanics. In S. Huerta (Ed.), *Essays in the history of theory of structures* (pp. 117–142). Madrid: Instituto Juan de Herrera.

Foce, F. (2007). Milankovitch's Theorie der Druckkurven: Good mechanics for masonry architecture. *Nexus Network Journal*, *9*(2), 185–210. doi:10.1007/s00004-007-0039-9

Franzoni, E., Gentilini, C., Graziani, G., & Bandini, S. (2014). Towards the assessment of the shear behaviour of masonry in on-site conditions: A study on dry and salt/ water conditioned brick masonry triplets. *Construction & Building Materials*, *65*, 405–416. doi:10.1016/j.conbuildmat.2014.05.002

Furtney, J., Zhang, F., & Han, Y. (2013). Review of methods and applications for incorporating fluid flow in the discrete element method. In P. Zhu, C. Detournay, R. Hart, & M. Nelson (Eds.), *Continuum and distinct element numerical modeling in geomechanics*. Minneapolis, MN: Itasca International Inc.

Gambarotta, L., & Lagomarsino, S. (1997). Damage models for the seismic response of brick masonry shear walls. part II: The continuum model and its applications. *Earthquake Engineering & Structural Dynamics*, *26*(4), 441–462. doi:10.1002/(SICI)1096-9845(199704)26:4<441::AID-EQE651>3.0.CO;2-0

Gambarotta, L., & Lagomarsino, S. (1997a). Damage models for the seismic response of brick masonry shear walls. Part I: The mortar joint model and its applications. *Journal of Earthquake Engineering and Structural Dynamics*, *26*(4), 423–439. doi:10.1002/(SICI)1096-9845(199704)26:4<423::AID-EQE650>3.0.CO;2-#

Gao, F., & Stead, D. (2014). The application of a modified Voronoi logic to brittle fracture modelling at the laboratory and field scale. *International Journal of Rock Mechanics and Mining Sciences*, *68*, 1–14. doi:10.1016/j.ijrmms.2014.02.003

Garrity, S. W. (1995). Testing of small scale masonry arch bridges with surface reinforcement. In *Proceedings of the 6th International Conference on Structural Faults and Repair*. Engineering Technics Press.

Garrity, S. W. (2004). *Independent testing of the Bersche-Rolt system for strengthening a brickwork lintel containing a soldier course. Technical Report*. Bradford, UK: University of Bradford, School of Engineering, Design and Technology.

Garrity, S. W., Ashour, A. F., & Chen, Y. (2010). An experimental investigation of retro-reinforced clay brick arches. In *Proceedings of the 8th International Masonry Conference*.

Garza-Cruz, T. V., & Pierce, M. (2014). A 3DEC Model for Heavily Veined Massive Rock Masses. In *Proceedings of the 48th US Rock Mechanics/Geomechanics Symposium*. American Rock Mechanics Associated.

Gentilini, C., Franzoni, E., Bandini, S., & Nobile, L. (2012). Effect of salt crystallisation on the shear behaviour of masonry walls: An experimental study. *Construction & Building Materials*, *37*, 181–189. doi:10.1016/j.conbuildmat.2012.07.086

Ghaboussi, J., & Barbosa, R. (1990). Three-dimensional discrete element method for granular materials. *International Journal for Numerical and Analytical Methods in Geomechanics*, *14*(7), 451–472. doi:10.1002/nag.1610140702

Ghaboussi, J., Wilson, E. L., & Isenberg, J. (1973). Finite element for rock joints and interfaces. *J. Soil Mech. and Found. Engng. Div. ASCE*, *SM10*(99), 833–848.

Ghosh, A. K., & Made, A. M. (2002). Finite element analysis of infilled frames. *Journal of Structural Engineering*, *128*(7), 881–889. doi:10.1061/(ASCE)0733-9445(2002)128:7(881)

Giamundo, V., Sarhosis, V., Lignola, G. P., & Cosenza, E. (2014). Discrete element modelling of the archaeological colonnade in Pompeii. In *9th International Masonry Conference*.

Giamundo, V., Sarhosis, V., Lignola, G. P., Sheng, Y., & Manfredi, G. (2014). Evaluation of different computational strategies for modelling low strength masonry. *Engineering Structures*, *73*, 160–169. doi:10.1016/j.engstruct.2014.05.007

Gilbert, M. (2007). Limit analysis applied to masonry arch bridges, state of the art and recent developments. In *ARCH'07 - 5th International Conference on Arch Bridges*. University of Minho.

Gilbert, M. (2007). Limit analysis applied to masonry arch bridges: state-of-the-art and recent developments. In P.B. Lourenço, D.B. Oliveira & A. Portela (Eds.), *Proceedings of the 5th International Conference on Arch Bridges* (pp.13-28). Funchal, Madeira: Multicomp, Lda Publishers.

Gilbert, M., Casapulla, C., & Ahmed, H. M. (2006). Limit analysis of masonry block structures with non-associative frictional joint using linear programming. *Computers & Structures*, *84*(1), 873–887. doi:10.1016/j.compstruc.2006.02.005

Gilbert, M., & Melbourne, C. (1994). Rigid block analysis of masonry structures. *The Structural Engineer*, *72*(21), 356–361.

Gilbert, M., & Melbourne, C. (1994). Rigid-block analysis of masonry structures. *Structural Engineer*, *72*(21), 356–361.

Gilbert, M., & Melbourne, C. (1994). Rigid-block analysis to masonry arches. *Structural Engineering*, *72*, 356–361.

Giordano, A., Mele, E., & Luca, A. (2002). Modelling of historical masonry structures. Comparison of different approaches through a case study. *Engineering Structures*, *24*(8), 1057–1069. doi:10.1016/S0141-0296(02)00033-0

Gonen, H., Dogan, M., Karacasu, M., Ozbasaran, H., & Gokdemir, H. (2013). Structural failures in refrofit historical murat masonry arch bridge. *Engineering Failure Analysis*, *35*, 334–342. doi:10.1016/j.engfailanal.2013.02.024

Goodman, R.E., Taylor, R. & Brekke, T.L. (1968). A model for the mechanics of jointed rock. *J. of Soil Mechs. and Found. Eng. Div.*, *94*(3), 637-659.

Goodman, R.E., Taylor, R. & Brekke, T.L. 1968. A model for the mechanics of jointed rock. *J. of Soil Mechs. and Found. Eng. Div.*, *94*(3), 637-659.

Goodman, R. E., Taylor, R. L., & Brekke, T. L. (1968). A model for the mechanics of jointed rock. *Journal of the Soil Mechanics and Foundations Division*, *94*(3), 637–659.

Goodwin, J.F., & West, H.W.H. (1982). *A review of the literature on brick/mortar bond.* The British Ceramic Research Association. Technical Note No. 308.

Grandet, J., Javelza, R., Perrin, B., & Thenoz, B. (1972). Role of ettrignite in the binding of the mechanical type between clay and Portal cement paste. *Revue Terre Cuite*, *48*, 21–28.

Gregory, D. (1697). Catenaria. *Philosophical Transactions of the Royal Society*, *19*, 637–652.

Halabian, A. M., Mirshahzadeh, L., & Hashemol-Hosseini, H. (2014). Non-linear behavior of unreinforced masonry walls with different Iranian traditional brick-work settings. *Engineering Failure Analysis*, *44*, 46–65. doi:10.1016/j.engfailanal.2014.04.018

Hall, J. F. (2006). Problems encountered from the use (or misuse) of Rayleigh damping. *Earthquake Engineering & Structural Dynamics*, *35*(5), 525–545. doi:10.1002/eqe.541

Hamdi, P., Stead, D., & Elmo, D. (2014). Damage characterization during laboratory strength testing: A 3D-finite-discrete element approach. *Computers and Geotechnics*, *60*, 33–46. doi:10.1016/j.compgeo.2014.03.011

Hart, R. D., Cundall, P. A., & Lemos, J. V. (1988). Formulation of a three-dimensional distinct element model - Part II: Mechanical calculations. *International Journal of Rock Mechanics and Mining Sciences*, *25*(3), 117–125. doi:10.1016/0148-9062(88)92294-2

Hatzor, Y. H., Arzi, A. A., & Tsesarsky, M. (2002). Realistic dynamic analysis of jointed rock slopes using DDA. In Y. Hatzor (Ed.), *Procs. ICADD-5* (pp. 47–56). Abingdon, UK: Balkema.

Hendry, A. W. (1998). *Structural masonry* (2nd ed.). London, UK: Palgrave Macmillan. doi:10.1007/978-1-349-14827-1

Hendry, A. W., Sinha, B. P., & Davis, S. R. (2004). *Design of masonry structures* (3rd ed.). London, UK: Taylor and Francis.

Hentz, S., Daudeville, L., & Donze, V. (2004). Identification and Validation of a discrete element model for concrete. *Journal of Engineering Mechanics*, *6*(6), 709–719. doi:10.1061/(ASCE)0733-9399(2004)130:6(709)

Heyman, J. (2009). La Coupe des Pierres. In K.E. Kurrer, W. Lorenz, & V. Wetzk (Eds.), *Proceedings of the 3rd International Congress on Construction History* (*vol. 2*, pp. 807-812). Berlin: Neunplus1.

Heyman, J. 1998. Structural Analysis: A historical approach. University Press. doi:10.1017/CBO9780511529580

Heyman, J. (1966). The Stone Skeleton. *International Journal of Solids and Structures*, *2*(2), 249–279. doi:10.1016/0020-7683(66)90018-7

Heyman, J. (1969). The safety of masonry arches. *International Journal of Mechanical Sciences*, *43*, 209–224.

Heyman, J. (1969). The safety of masonry arches. *International Journal of Mechanical Sciences*, *11*(4), 363–385. doi:10.1016/0020-7403(69)90070-8

Heyman, J. (1972). *Coulomb's Memoir on Statics: An Essay in the History of Civil Engineering*. Cambridge, UK: Cambridge University Press.

Heyman, J. (1982). *The Masonry Arch*. Chichester, UK: Ellis Horwood.

Heyman, J. (1982). *The masonry arch*. West Sussex, UK: Ellis Horwood-Wiley.

Heyman, J. (1992). Leaning towers. *Meccanica*, *27*(3), 153–159. doi:10.1007/BF00430041

Hoek, E., & Bray, J. W. (1977). Rock slope engineering (2nd ed.). The Institution of Mining and Metallurgy.

Hogan, S. J. (1992). On the motion of a rigid block, tethered at one corner, under harmonic forcing. *Proccedings: Mathematical and Physical Science, 439*(1905), 35-45.

Hogan, S. J. (1994). Rigid block dynamics confined between side-walls. *Philosophical Transactions: Physical Science and Engineering, 37*(1683), 411-419.

Hogan, S. J. (1990). The many steady states responses of a rigid block under harmonic forcing. *Earthquake Engineering & Structural Dynamics*, *19*(7), 1057–1071. doi:10.1002/eqe.4290190709

Holmes, D., Williams, J. R., & Tilke, P. (2010). Events Based Algorithm for Distributing Concurrent Tasks on Multi-Core Architectures. *Computer Physics Communications*, *181*(2), 341–354. doi:10.1016/j.cpc.2009.10.009

Hooke, R. (1675). *A description of helioscopes, and some other instruments*. London.

Hoseini, M., & Alavi, S. S. (2014). A kind of repairable steel buildings for seismic regions based on building's rocking motion and energy dissipation at base level. *International Journal of Civil Engineering*, *1*(3).

Housner, G. W. (1963). The Behavior of Inverted Pendulum Structures During Earthquakes. *Bulletin of the Seismological Society of America*, *53*(2), 403–417.

Housner, G. W. (1963). The behavior of inverted pendulum structures during earthquakes. *Bulletin of the Seismological Society of America*, *53*, 403–417.

Huerta, S., & Foce, F. (2003). Vault theory in Spain between XVIIIth and XIXth century: Monasterio's unpublished manuscript Nueva teorica sobre el empuje de bóvedas. In S. Huerta (Ed.), *Proceedings of the 1st International Congress on Construction History* (vol. 2, pp. 1155-1166). Madrid: Instituto Juan de Herrera.

Huerta, S. (2004). *Arcos, Bóvedas y Cúpulas – Geometria y Equilibrio en el Cálculo Tradicional de Estructuras de Fábrica*. Spain: Instituto Juan de Herrera, Escuela Técnica Superior de Arquitectura de Madrid.

Hughes, T. G., & Blackler, M. J. (1997). A review of the UK masonry arch assessment methods. *Proceedings - Institution of Civil Engineers*, *122*, 305–315.

I.N.C.M. (1983). *Regulamento de segurança e acções para estruturas de edifícios e pontes*. Lisbon.

International Commission on Large Dams (ICOLD). (1994). *Ageing of dams and appurtenant works: review and recommendations (Bulletin 93)*. Paris: International Commission on Large Dams.

International Commission on Large Dams (ICOLD). (2000). *Rehabilitation of dams and appurtenant works: State of the art and case histories (Bulletin 119)*. Paris: International Commission on Large Dams.

International Commission on Large Dams (ICOLD). (2013). *Historical review on ancient dams (Bulletin 143)*. Paris: International Commission on Large Dams.

International Commission on Large Dams (ICOLD). (n.d.). *Dam Safety Management: operational phase of the dam life cycle (Bulletin 154)*. France: International Commission on Large Dams.

Isfeld, A., & Shrive, N. (2015). Discrete Element Modeling of Stone Masonry Walls With Varying Core Conditions: Prince of Wales Fort Case Study. *International Journal of Architectural Heritage*, *9*(5), 564–580. doi:10.1080/15583058.2013.819135

Ishiyama, Y. (1982). Motions of rigid bodies and criteria for overturning by earthquake excitations. *Earthquake Engineering & Structural Dynamics*, *10*(5), 635–690. doi:10.1002/eqe.4290100502

Itasca (2014). *PFC (Particle Flow Code) - Version 5.0*. Minneapolis, MN: Itasca.

Itasca Consulting Group, Inc. (1998). Three-Dimensional Distinct Element Code, version 2.0 – User's manual. Minneapolis, MN: Author.

Itasca Consulting Group, Inc. (2000). Universal Distinct Element Code– User's manual. Minneapolis, MN: Author.

Itasca Consulting Group, Inc. (2004). UDEC- Universal Distinct Element Code, version 4.0- User's manual. Minneapolis, MN: Author.

Itasca Consulting Group, Inc. (2004). UDEC: Universal Distinct Element Code. Version 4.0. Minneapolis: Itasca.

Itasca Consulting Group, Inc. (2014). *UDEC Version 6., Distinict-Element Modeling of jointed and Blocky Material in 2D*. Retrieved from http://www.itascacg.com/software/udec

Itasca Consulting Group, Inc. (n.d.). *3DEC – Three-dimensional distinct element code*. Minneapolis, MN: Itasca

Itasca Consulting Group. (2004). UDEC Version 4.0 User's Guide. Minneapolis, MN: Author.

Itasca. (2004a). *UDEC - Universal Distinct Element Code Manual. Theory and Background*. Itasca Consulting Group.

Itasca. (2004b). *UDEC -Universal Distinct Element Code Manual. User's manual*. Itasca Consulting Group.

Itasca. (2013). *3DEC - Three-dimensional Distinct Element Code*. Itasca Consulting Group.

Itasca. (2014). *UDEC - Universal Distinct Element Code*. Itasca Consulting Group.

Jaky, J. (1944). The coefficient of earth pressure at rest. *Journal of Society of Hungarian Architects and Engineers*, *78*(22), 355–358.

James, J. A. (1973). Investigation of the effect of workmanship and curing conditions on the strength of brickwork. In L. Fortig, & K. Gobel (eds.), *Proceedings of the 3rd International Brick Masonry Conference*.

Jean, M., & Moreau, J. J. (1992). Unilaterality and dry friction in the dynamics of rigid body collections. In *Procs. Contact Mechanics International Symposium*. Presses Polytechniques et Universitaires Romandes.

Jean, M. (1999). The non-smooth contact dynamics method. *Computer Methods in Applied Mechanics and Engineering*, *177*(3-4), 235–257. doi:10.1016/S0045-7825(98)00383-1

Jiang, H., Wang, L., Li, L., & Guo, Z. (2014). Safety evolution of an ancient maonry seawall structure with modified DDA method. *Computers and Geotechnics*, *55*, 277–289. doi:10.1016/j.compgeo.2013.09.012

Jiang, W., & Zheng, H. (2015). An efficient remedy for the false volume expansion of DDA when simulating large rotation. *Computers and Geotechnics*, *70*, 18–23. doi:10.1016/j.compgeo.2015.07.008

Jimenez-Gonzalez, I., & Scherer, G. W. (2004). Effect of swelling inhibitors on the swelling and stress relaxation of clay bearing stones. *Environ Geol*, *46*(3-4), 364–377. doi:10.1007/s00254-004-1038-8

Jing, L. (2003). A review of techniques, advances and outstanding issues in numerical modelling for rock mechanics and rock engineering. *International Journal of Rock Mechanics and Mining Sciences*, *40*(3), 283–353. doi:10.1016/S1365-1609(03)00013-3

Jukes, P., & Riddington, J. R. (1997). A review of masonry joint shear strength test methods. *Masonry International*, *11*(2), 37–43.

Jung, E. (1988). The binding between mortar and brick. In J.W. Courcy (Ed.), *Proceedings of the 8th International Brick and Block Masonry Conference (IBMAC)*. Trinity University.

Kamai, R., & Hatzor, Y. H. (2005). Dynamic back analysis of structural failures in archeological sites to obtain paleoseismic parameters using DDA. In M. MacLaughlin, N. Sitar (Eds.), Procs. ICADD-7.

Kamai, R., & Hatzor, Y. H. (2008). Numerical analysis of block stone displacements in ancient masonry structures: A new method to estimate historic ground motions. *International Journal for Numerical and Analytical Methods in Geomechanics*, *32*(11), 1321–1340. doi:10.1002/nag.671

Kawamoto, T., & Aydan, Ö. (1999). A review of numerical analysis of tunnels in discontinuous rock masses. *International Journal for Numerical and Analytical Methods in Geomechanics*, *23*(13), 1377–1391. doi:10.1002/(SICI)1096-9853(199911)23:13<1377::AID-NAG932>3.0.CO;2-S

Keller, C.F., Jr., Salmon, M.W., Zubelewicz, A., Farrar, C.R., Luscher, D.J., Rougier, E., & De Lorenzi-Venneri, G. (2012). *Characterization and computer modeling for Brunelleschi's Dome of Santa Maria del Fiore in Florence.* International Report.

Khan, M. S. (2010). *Investigation of Discontinuous Deformation Analysis for Application in Jointed Rock Masses.* (PhD Thesis). Department of Civil Engineering, University of Toronto.

Khan, M. S., Riahi, A., & Curran, J. H. (2010). Effects of time-step size on the efficiency of Discontinuous Deformation Analysis. In *Procs.6th Asian Rock Mechanics Symposium.*

King, G. J. W., & Pandey, P. C. (1978). The analysis of infilled frames using finite elements. *Proceedings - Institution of Civil Engineers, 65*(Part 2), 749–760. doi:10.1680/iicep.1978.2707

Kjaer, E. (2010). Bond strength: What is influencing the bond strength? In what way the bond strength influencing the masonry properties. In *Proceedings of the 8th International Masonry Conference.*

Komodromos, P., Papaloizou, L., & Polycarpou, P. (2008). Simulation of the response of ancient columns under harmonic and earthquake excitations. *Engineering Structures, 30*(8), 2154–2164. doi:10.1016/j.engstruct.2007.11.004

Komodromos, P., Polycarpou, P., Papaloizou, L., & Phocas, M. (2007). Response of seismically isolated buildings considering poundings. *Earthquake Engineering & Structural Dynamics, 36*(12), 1605–1622. doi:10.1002/eqe.692

Konstantinidis, D., & Makris, N. (2005). Seismic response analysis of multidrum classical columns. *Earthquake Engineering & Structural Dynamics, 34*(10), 1243–1270. doi:10.1002/eqe.478

Konstantinidis, D., & Makris, N. (2009). Experimental and analytical studies on the response of freestanding laboratory equipment to earthquake shaking. *Engineering Structures, 38*, 827–848.

Kooharian, A. (1952). Limit Analysis of Voussoir (Segmental) and Concrete Arches. *Journal Proceedings of American Concrete Institute, 49*(12), 317–328.

Koutromanos, I. (2011). *Numerical analysis of masonry-infilled reinforced concrete frames subjected to seismic loads and experimental evaluation of retrofit techniques.* (Ph.D. thesis). University of California, San Diego, CA.

Koutromanos, I., Stavridis, A., Shing, P. B., & Willam, K. (2011). Numerical modeling of masonry-infilled RC frames subjected to seismic loads. *ComputStruct., 89*, 1026–1037.

Koziara, T. & Bićanić, N. (2008). Semismooth Newton method for frictional contact between pseudo-rigid bodies. *Computer Methods in Applied Mechanics and Engineering, 197*(33-40), 2763–2777.

Kuhn, M. R., & Bagi, K. (2009). Specimen size effect in discrete element simulations of granular assemblies. *Journal of Engineering Mechanics, 135*(6), 485–492. doi:10.1061/(ASCE)0733-9399(2009)135:6(485)

Kuo, J. S. H. (1982). *Fluid-structure interactions: added mass computations for incompressible fluid (UCB/EERC-82/09).* Berkeley, CA: Earthquake Engineering Research Centre, University of California.

La Hire, P. (1712). Sur la construction des voûtes dans les édifices. In Mémoires de l'Académie Royale des Sciences (pp. 69-77). Paris: Academic Press.

La Hire, P. (1695). *Traité de mécanique, où l'on explique tout ce qui est nécessaire dans la pratique des Arts, et les proprietés des corps pesants lesquelles ont eu plus grand usage dans la Physique.* Paris: Imprimerie Royale.

Lagomarsino, S. (1998). Seismic Damage Survey of the Churches in Umbria.*Proc. of the Workshop on Seismic Performance of Monuments.*

Lagomarsino, S. (2006). On the vulnerability assessment of monumental buildings. *Bulletin of Earthquake Engineering*, *4*(4), 445–463. doi:10.1007/s10518-006-9025-y

Lagomarsino, S. (2015). Seismic assessment of rocking masonry structures. *Bulletin of Earthquake Engineering*, *13*(1), 97–128. doi:10.1007/s10518-014-9609-x

Lagrange, J. L. (1788). *Mécanique analytique*. Paris: Chez La Veuve Desaint, Libraire.

Lamé, M. G., & Clapeyron, E. (1823). Mémoire sur la stabilité des voûtes. *Annales des mines*, *8*, 789-836.

Lancioni, G., Lenci, S., Piattoni, Q., & Quagliarini, E. (2013). Dynamics and failure mechanisms of ancient masonry churches. *Engineering Structures*, *56*, 1527–1546. doi:10.1016/j.engstruct.2013.07.027

Lang, A. F., & Benzoni, G. (2014). Modeling of nonlinear behavior of confined masonry using discrete elements. In *Proceedings of the 10th National Conf. on Earthquake Engineering*. Frontier of Earthquake Engineering.

Lauenstein, R., & Bastine, P. (1913). *Die graphische Statik: Elementares Lehrbuch fur den Schul- und Selbstunterricht sowie zum Gebrauch in der Praxis*. Leipzig: Alfred Kruner.

Law, H. K., & Lam, I. P. (2003). Evaluation of seismic performance for tunnel retrofit project. *Journal for Geotechnical and Geoenviromental Engineering*, *129*(7), 575–589. doi:10.1061/(ASCE)1090-0241(2003)129:7(575)

Lawrence, S. J., & Cao, H. T. (1987). An experimental study of the interface between brick and mortar. In *Proceedings of the 4th North American Masonry Conference*. University of California.

Lawrence, S. J., & Cao, H. T. (1988). Microstructure of the interface between brick and mortar. In: J.W. Courcy (Ed.), *Proceedings of the 8th International Brick and Block Masonry Conference (IBMAC)*. Trinity University.

Lawrence, S. J., Sugo, H. O., & Page, A. W. (2008). Masonry bond strength and the effects of supplementary cementitious materials. *Australian Journal of Structural Engineering*, *8*(2), 101–115.

Lei, Z., Rougier, E., Knight, E. E., & Munjiza, A. (2014). A framework for grand scale parallelization of the combined finite discrete element method in 2D. *Computational Particle Mechanics*, *1*(3), 307–319. doi:10.1007/s40571-014-0026-3

Lemaitre, J., & Chaboche, J. L. (1990). *Mechanics of solid materials*. New York: Cambridge University Press. doi:10.1017/CBO9781139167970

Lemos, J. V. (1995). Assessment of the ultimate load of a masonry arch using discrete elements. In J. Middleton & G.N. Pande (Eds.), Computer Methods in Structural Masonry. Books & Journals International.

Lemos, J. V. (1997). Discrete element modeling of the seismic behavior of stone masonry arches. Computer Methods in Structural Masonry- 4.

Lemos, J. V. (2001). Modelling the behaviour of a stone masonry arch structure under cyclic loads. In T.G. Hughes & G.N. Pande (Eds.), Computer Methods in Structural Masonry - 5. Computers & Geotechnics Ltd.

Lemos, J. V. (2007). *Numerical issues in the representation of masonry structural dynamics with discrete elements*. Paper presented at the ECCOMAS Thematic conference on computational methods in structural dynamics and earthquake engineering, Crete, Greece.

Lemos, J. V. (2007b). Numerical issues in the representation of masonry structural dynamics with discrete elements. In M. Papadrakakis, D.C. Charmpis, N.D. Lagaros, Y. Tsompanakis (Eds.), Compdyn2007. Rethymno.

Lemos, J. V., & Cundall, P. A. (1999). Earthquake analysis of concrete gravity dams on jointed rock foundations. In V. M. Sharma, K. R. Saxena, & R. D. Woods (Eds.), Distinct element modelling in geomechanics (pp. 117-143). New Delhi: Oxford & IBH Publishing CO. PVT. LTD.

Lemos, J. V., Azevedo, F. S., Oliveira, C. S., & Sincraian, G. (1998). Three dimensional analysis of a block masonry pillar using discrete elements. In *Proc. Monument 98,Workshop on Seismic Performance of Monuments*.

Lemos, J. V., Costa, A. C., & Bretas, E. M. (2011). Assessment of the seismic capacity of stone masonry walls with block models. In M. Papadrakakis, M. Fragiadakis & N.D. Lagaros (Eds.), Computational Methods in Earthquake Engineering. Springer. doi:10.1007/978-94-007-0053-6_10

Lemos, J. V. (1995). Assessment of the ultimate load of a masonry arch using discrete elements. In J. Middleton & G. N. Pande (Eds.), *Computer Methods in Structural Masonry – 3* (pp. 294–302). Swansea, UK: Books and Journals International.

Lemos, J. V. (1995). Assessment of the ultimate load of a masonry arch using discrete elements. In J. Middleton & G. N. Pande (Eds.), *Computer Methods in Structural Masonry* (pp. 294–302). Swansea: Books & Journals International.

Lemos, J. V. (1997). *Discrete element modelling of the seismic behaviour of stone masonry arches.*In G. N. Pande, J. Middleton, & B. Kralj (Eds.), *Computer Methods in Structural Masonry - 4* (pp. 220–227). London: E&FN Spon.

Lemos, J. V. (1998). Discrete element modelling of the seismic behaviour of stone masonry arches. In J. Middleton, G. N. Pande, & B. Kralj (Eds.), *Computer Methods in Structural Masonry - 4* (pp. 220–227). London: E&FN Spon.

Lemos, J. V. (1998). Discrete element modelling of the seismic behaviour of stone masonry arches. In G. N. Pande, J. Middleton, & B. Kralj (Eds.), *Computer Methods in Structural Masonry — 4* (pp. 220–227). London, UK: E & FN Spon.

Lemos, J. V. (2001). Modelling the behaviour of a stone masonry arch structure under cyclic loads. In T. G. Hughes & G. N. Pande (Eds.), *Computer Methods in Structural Masonry - 5* (pp. 101–108). Swansea, UK: Computers & Geotechnics Ltd.

Lemos, J. V. (2004). Modelling stone masonry dynamics with 3DEC. In H. Konietzky (Ed.), *Numerical modelling of discrete materials in geotechnical engineering, civil engineering and earth sciences* (pp. 7–13). London, UK: Taylor and Francis Group.

Lemos, J. V. (2007). Discrete element modelling of masonry structures. *International Journal of Architectural Heritage*, *1*(2), 190–213. doi:10.1080/15583050601176868

Lemos, J. V., Oliveira, C. S., & Navarro, M. (2015). 3D nonlinear behavior of an obelisk subjected to the Lorca May 11, 2011 strong motion record. *Engineering Failure Analysis*, *58*, 212–228. doi:10.1016/j.engfailanal.2015.09.001

Lemos, J. V., Sincraian, G. E., & Oliveira, C. S. (2000). Modelling the seismic behaviour and structural restoration of a damaged masonry tower. In *Proc. 5th Int. Congress on Restoration of Architectural Heritage*. CICOP.

Lengyel, G., & Bagi, K. (2015). Numerical analysis of the mechanical role of the ribs in groin vaults. *Computers & Structures*, *158*, 42–60. doi:10.1016/j.compstruc.2015.05.032

Liauw, T. C., & Kwan, K. H. (1985). Unified plastic analysis for infilled frames. *Journal of Structural Engineering*, *111*(7), 1427–1448. doi:10.1061/(ASCE)0733-9445(1985)111:7(1427)

Lilliu, J., & Van Mier, M. (2003). 3D lattice type fracture model for concrete. *Engineering Fracture Mechanics*, *70*(7-8), 841–927. doi:10.1016/S0013-7944(02)00158-3

Lisjak, A., Kaifosh, P., Mahabadi, O. K., Tatone, B., & Grasselli, G. (2015). Fully 3D simulation of fluid-pressure-driven fracturing using a novel continuum-discontinuum approach: Preliminary results. In GeoConvention 2015: New Horizons.

Lisjak, A., & Grasselli, G. (2014). A review of discrete modeling techniques for fracturing processes in discontinuous rock masses. *Journal of Rock Mechanics and Geotechnical Engineering*, *6*(4), 301–314. doi:10.1016/j.jrmge.2013.12.007

Liu, L. P., Tang, D. X., & Zhai, X. M. (2006). Failure criteria for grouted concrete block masonry under biaxial compression. *Advances in Structural Engineering*, *9*(2), 229–239. doi:10.1260/136943306776987001

Liu, Y. J., & Buchanan, R. B. (2004). Free vibration of stepped cantilever Mindlin plates. *Journal of Sound and Vibration*, *271*(3-5), 1083–1092. doi:10.1016/S0022-460X(03)00777-6

Liu, Y., & Manesh, P. (2013). Concrete masonry infilled steel frames subjected to combined in-plane lateral and axial loading – An experimental study. *Engineering Structures*, *52*, 331–339. doi:10.1016/j.engstruct.2013.02.038

Livesley, R. K. (1978). Limit analysis of structures formed from rigid blocks. *International Journal for Numerical Methods in Engineering*, *12*(12), 1853–1871. doi:10.1002/nme.1620121207

Lofti, H. R., & Shing, P. B. (1994). Interface model applied to fracture of masonry structures. *Journal of Structural Engineering*, *120*(1), 63–80. doi:10.1061/(ASCE)0733-9445(1994)120:1(63)

Lourenço, P. B. (1996). *Computational strategies for masonry structures.* (Ph.D thesis). Delft University of Technology, Delft, The Netherlands.

Lourenco, P. B. (1996). *Computational strategies for masonry structures.* (Ph.D. thesis). Delft University of Technology, Delft, The Netherlands.

Lourenço, P. B. (1996). *Computational strategies for masonry structures. (PhD).* Delft University of Technology.

Lourenço, P. B. (1998a). Experimental and numerical issues in the modelling of the mechanical behaviour of masonry. In P. Roca, J. L. Gonzalez, E. Onate, & P. B. Lourenço (Eds.), *Structural Analysis of Historical Constructions II* (pp. 57–92). Barcelona, Spain: CIMNE.

Lourenço, P. B. (1998b). Sensitivity analysis of masonry structures. In *Proceedings of the 8th Canadian Masonry Symposium.*

Lourenço, P. B. (2001). *Analysis of historical constructions: From thrust-lines to advanced simulations.* Guimaraes, Portugal: Historical Constructions.

Lourenço, P. B. (2002). Computations on historic masonry structures. *Progress in Structural Engineering and Materials*, *4*(3), 301–319. doi:10.1002/pse.120

Lourenço, P. B., Barros, J. O., & Oliveira, J. T. (2004). Shear testing of stack bonded masonry. *Construction & Building Materials*, *18*(2), 125–132. doi:10.1016/j.conbuildmat.2003.08.018

Lourenço, P. B., Milani, G., Tralli, A., & Zucchini, A. (2007). Analysis of masonry structures: Review of and recent trends in homogenization techniques. *Canadian Journal of Civil Engineering*, *34*(11), 1443–1457. doi:10.1139/L07-097

Lourenco, P. B., Oliveira, D. V., Roca, P., & Orduna, A. (2005). Dry joint masonry walls subjected to in-plane combined loading. *Journal of Structural Engineering*, *131*(11), 1665–1673. doi:10.1061/(ASCE)0733-9445(2005)131:11(1665)

Lourenço, P. B., & Pina-Henriques, J. (2006). Validation of numerical and continuum numerical methods for estimating the compressive strength of masonry. *Computers & Structures*, *84*(29), 1977–1989. doi:10.1016/j.compstruc.2006.08.009

Lourenço, P. B., & Ramos, L. (2004). Characterization of cyclic behavior of dry masonry joints. *Journal of Structural Engineering*, *130*(5), 779–786. doi:10.1061/(ASCE)0733-9445(2004)130:5(779)

Lourenço, P. B., & Rots, J. G. (1997). A multi-surface interface model for the analysis of masonry structures. *Journal of Engineering Mechanics, 123*(7), 660–668. doi:10.1061/(ASCE)0733-9399(1997)123:7(660)

Luciano, R., & Sacco, E. (1997). Homogenization technique and damage model for old masonry material. *International Journal of Solids and Structures, 34*(24), 3191–3208. doi:10.1016/S0020-7683(96)00167-9

Lukas, T., Schiava D'Albano, G. G., & Munjiza, A. (2014). Space decomposition based parallelization solutions for the combined finite-discrete element method in 2D. *Journal of Rock Mechanics and Geotechnical Engineering, 6*(6), 607–615. doi:10.1016/j.jrmge.2014.10.001

Lysmer, J., & Kuhlemeyer, R. L. (1969). Finite dynamic model for infinite media. *Journal of the Engineering Mechacnics Division*, 859-877.

MacLaughlin, M. M. (1997). *Discontinuous deformation analysis of the kinematics of landslides.* (PhD Dissertation). Dept. of Civil and Envr. Eng., University of California Berkeley.

MacLaughlin, M. M., & Doolin, D. M. (2006). Review of validation of the discontinuous deformation analysis (DDA) method. *International Journal for Numerical and Analytical Methods in Geomechanics, 30*(4), 271–305. doi:10.1002/nag.427

Madan, A., Reinhorn, A. M., Mander, J. B., & Valles, R. E. (1997). Modeling of masonry infill panels for structural analysiss. *Journal of Structural Engineering, 123*(10), 1295–1302. doi:10.1061/(ASCE)0733-9445(1997)123:10(1295)

Magenes, G., & Calvi, G. M. (1997). In-plane seismic response of brick masonry walls. *Earthquake Engineering & Structural Dynamics, 26*(11), 1091–1112. doi:10.1002/(SICI)1096-9845(199711)26:11<1091::AID-EQE693>3.0.CO;2-6

Mahabadi, O. K., Grasselli, G., & Munjiza, A. (2010). A graphical user interface and pre-processor for the combined finite-discrete element code, Y2D, incorporating material heterogeneity. *Computers & Geosciences, 36*(2), 241–252. doi:10.1016/j.cageo.2009.05.010

Mahabadi, O. K., Lisjak, A., Munjiza, A., & Grasselli, G. (2012). Y-Geo: A new combined finite-discrete element numerical code for geomechanical applications. *Geomechanics, 12*(6), 676–688. doi:10.1061/(ASCE)GM.1943-5622.0000216

Makris, N., & Alexakis, H. (2015). Limit equilibrium analysis of masonry buttresses and towers under lateral and gravity loads. *Archive of Applied Mechanics*. doi: 10.1007/s00419-015-1027-2

Makris, N., & Psychogios, T. (2004). Static and dynamic analysis of columns and part of the entablature of the NE corner of the temple of Zeus at Nemea. Report to American School of Classical Studies, Athens.

Makris, N., & Alexakis, H. (2012). *From Hooke's "Hanging Chain" and Milankovitch's "Druckkurven" to a variational formulation: The Adventure of the Thrust-line of Masonry Arches. Report series in EEAM 2012-02*. University of Patras.

Makris, N., & Alexakis, H. (2013). The effect of stereotomy on the shape of the thrust-line and the minimum thickness of semicircular masonry arches. *Archive of Applied Mechanics, 83*(10), 1511–1533. doi:10.1007/s00419-013-0763-4

Makris, N., & Konstantinidis, D. (2003). The rocking spectrum and the limitations of practical design methods. *Earthquake Engineering & Structural Dynamics, 32*(2), 265–289. doi:10.1002/eqe.223

Makris, N., & Roussos, Y. (1998). *Rocking response and overturning of equipment under horizontal pulse-type motions. PEER Report 1998/05*. Berkeley, CA: University of California.

Makris, N., & Roussos, Y. (2000). Rocking response of rigid blocks under near-source ground motions. *Geotechnique, 50*(3), 243–262. doi:10.1680/geot.2000.50.3.243

Makris, N., & Zhang, J. (2001). Rocking Response of Anchored Blocks under Pulse-Type Motions. *Journal of Engineering Mechanics*, *127*(5), 484–493. doi:10.1061/(ASCE)0733-9399(2001)127:5(484)

Ma, M. A., Pan, A. D. E., Luan, M., & Gebara, J. M. (1996). Seismic analysis of stone arch bridges using discontinuous deformation analysis. In *Proceedings of the 11th World Conference on Earthquake Engineering*.

Mamaghani, I. H. P. (1993). *Numerical analysis for stability of a system of rock blocks*. (Master Thesis). Department of Civil Engineering, Nagoya University, Nagoya, Japan.

Mamaghani, I. H. P. (2006). Analysis of Masonry Bridges: Discrete Finite Element Method. Transportation Research Record: Journal of Transportation Research Board, 1976, 13-19.

Mamaghani, I. H. P. (2008). Analysis of Underground Structures by Discrete Finite Element Method. Canadian Society for Civil Engineering.

Mamaghani, I. H. P. (2015a). Discrete Finite Element Method Application for Analysis of Unreinforced Masonry Underground Structures. In Transportation Research Record: Journal of Transportation Research Board, No. 2522 (pp. 131–136). Washington, DC: Transportation Research Board of the National Academies. Retrieved from http://trrjournalonline.trb.org/doi/pdf/10.3141/2522-13

Mamaghani, I. H. P., Aydan, Ö., & Kajikawa, Y. (1999). Analysis of masonry structures under static and dynamic loading by discrete finite element method. Journal of Structural Mechanics and Earthquake Engineering, 626, 1-12.

Mamaghani, I. H. P., Aydan, Ö., & Kajikawa, Y. (1999). Analysis of masonry structures under static and dynamic loading by discrete finite element method. Journal of Structural Mechanics Earthquake Engineering, 626(16), 75-86.

Mamaghani, I. H. P. (1999). Analysis of masonry structures under static and dynamic loading by discrete finite element method. *Structural Engineering / Earthquake Engineering*, *16*(2), 75–86.

Mamaghani, I. H. P. (2015a). *Discrete Finite Element Method Application for Analysis of Unreinforced Masonry Underground Structures. In Transportation Research Record: Journal of Transportation Research Board, No. 2522* (pp. 131–136). Washington, DC: Transportation Research Board of the National Academies. http://trrjournalonline.trb.org/doi/pdf/10.3141/2522-13

Mamaghani, I. H. P. (2015b). Analysis of Masonry Structures by Discrete Finite Element Method. *12th North American Masonry Conference*.

Mamaghani, I. H. P., & Aydan, O. (1999). Discrete Finite Element Method for Analysis of Masonry and Rock Structures. *International Journal of Science and Technology, Scientia Iranica, Tehran, Iran*, *6*(3 & 4), 157–164.

Mamaghani, I. H. P., & Aydan, Ö. (2000). Stability analysis of slopes by discrete finite element method. *GeoEng*, *2000*(November), 19–24.

Mamaghani, I. H. P., Aydan, Ö., & Akhoundi, F. (2014). Analysis of Masonry Bridges under Static and Dynamic Loading by Discrete Finite Element Method. *9th International Masonry Conference*. International Masonry Society, University of Minho.

Mamaghani, I. H. P., Baba, S., Aydan, Ö., & Shimizu, Y. (1994). Discrete finite element method for blocky systems. *Proc. of the Eighth Int. Conf. on Computer Methods and Advances in Geomechanics (IACMAG)*.

Mann, W., & Muller, H. (1982). Failure of shear-stressed masonry – An enlarged theory, tests and application to shear walls. In H.W.H West (Ed.), *Proceedings of the British Ceramic Society.*

Markulak, D., Radic´, I., & Sigmund, V. (2013). Cyclic testing of single bay steel frames with various types of masonry infill. *Engineering Structures*, *51*, 167–277. doi:10.1016/j.engstruct.2013.01.026

Marquis, E. L., & Borchelt, G. (1986). Bond strength comparison of laboratory and field tests. In *Proceedings of the 4th Canadian Masonry Symposium*.

Martìnez-Martìnez, J., Benavente, D., & Garcìa-del-Cura, M. A. (2011). Comparison of the static and dynamic elastic modulus in carbonate rocks. *Bulletin of Engineering Geology and the Environment*, 1–6.

Mascheroni, L. (1785). *Nuove ricerche sull'equilibrio delle volte*. Bergamo.

Masiani, R., Rizzi, N., & Trovalusci, P. (1995). Masonry as structured continuum. *Meccanica*, *30*(6), 673–683. doi:10.1007/BF00986573

Masiani, R., & Trovalusci, P. (1996). Cosserat and cauchy materials as continuum models of brick masonry. *Meccanica*, *31*(4), 421–432. doi:10.1007/BF00429930

Massart, T. J., Peerlings, R. H. J., & Geers, M. G. D. (2007). An enhanced multi-scale approach for masonry wall computations with localization of damage. *International Journal for Numerical Methods in Engineering*, *69*(5), 1022–1059. doi:10.1002/nme.1799

Matsuda, Y., & Iwase, Y. (2002). Numerical simulation of rock fracture using three-dimensional extended discrete element method. *Earth, Planets, and Space*, *54*(4), 367–378. doi:10.1186/BF03352426

Mavroeidis, G. P., & Papageorgiou, A. S. (2003). A Mathematical Representation of Near-Fault Ground Motions. *Bulletin of the Seismological Society of America*, *93*(3), 1099–1131. doi:10.1785/0120020100

Mazor, D. B. (2011). *Modelling dynamic rock mass deformattion with the numerical DDA method.* (PhD Dissertation). Ben-Gurion University of the Negev.

McInerney, J., & DeJong, M. J. (2014). Discrete element modelling of groin vault displacement capacity. *International Journal of Architectural Heritage*. doi:10.1080/15583058.2014.923953

McKibbins, L., Melbourne, C., Sawar, N., & Galliard, C. S. (2006). *Masonry arch bridges, condition appraisal and remedial treatment*. Report C656. CIRIA, London, UK.

Meguro, K., & Hakuno, M. (1989). Fracture analysis of concrete structures by the modified distinct element method. *Structural Engineering/Earthquake Engineering. Japanese Society for Civil Engineers*, *6*(2), 283–294.

Mehrabi, A. B., & Shing, P. B. (1997). Finite element modeling of masonry-infilled RC frames. *Journal of Structural Engineering*, *123*(5), 604–613. doi:10.1061/(ASCE)0733-9445(1997)123:5(604)

Melbourne, C., & Tomor, A. K. (2005). *Effect of weak/deteriorated masonry on the performance of arch bridges (Arch tests R-S). Test Report*. Salford, UK: University of Salford.

Meyer, P., Ochsendorf, J., Germaine, J., & Kausel, E. (2007). The impact of high-frequency/low-energy seismic waves on unreinforced masonry. *Earthquake Spectra*, *23*(1), 77–94. doi:10.1193/1.2431211

Michaltsos, G. T., & Raftoyiannis, G. I. (2008). A mathematical model for the rocking, overturning and shifting problems in bridges. *Engineering Structures*, *30*(12), 3587–3594. doi:10.1016/j.engstruct.2008.06.007

Milani, E., Milani, G., & Tralli, A. (2008). Limit analysis of masonry vaults by means of curved shell Finite Elements and homogenization. *International Journal of Solids and Structures*, *45*(20), 5258–5288. doi:10.1016/j.ijsolstr.2008.05.019

Milani, G. (2008). 3D upper bound limit analysis of multi-leaf masonry walls. *International Journal of Mechanical Sciences*, *50*(4), 817–836. doi:10.1016/j.ijmecsci.2007.11.003

Milani, G. (2011). Simple homogenization model for the non-linear analysis of in-plane loaded masonry walls. *Computers & Structures*, *89*(17-18), 1586–1601. doi:10.1016/j.compstruc.2011.05.004

Milani, G., & Lourenço, P. B. (2012). 3D non-linear behavior of masonry arch bridges. *Computers & Structures*, *110-111*, 133–150. doi:10.1016/j.compstruc.2012.07.008

Milani, G., Lourenço, P., & Tralli, A. (2006). Homogenization approach for the limit analysis of out-of-plane loaded masonry walls. *Journal of Structural Engineering*, *132*(10), 1650–1663. doi:10.1061/(ASCE)0733-9445(2006)132:10(1650)

Milankovitch, M. (1904). *Beitrag zur Theorie der Druckkurven*. Dissertation zur Erlangung der Doktorwürde. Vienna: K.K. technische Hochschule.

Milankovitch, M. (1907). Theorie der Druckkurven. *Zeitschrift für Mathematik und Physik*, *55*, 1–27.

Milankovitch, M. (1910). Zur Statik der massive Widerlager. *Zeitschrift für Mathematik und Physik*, *58*, 120–128.

Millard, A. (2003). CASTEM 2000, User Guide, Report CEA 93/007. Saclay, France: Academic Press..

Miller, S. G. (2001). *Excavations at Nemea II: The early Hellenistic Stadium*. Berkeley, CA: University of California Press.

Milne, J. (1881). Experiments on observational seismology. *Transaction of the Seismological Society of Japan*, *3*, 12–64.

Minghini, F., Milani, G., & Tralli, A. (2014). Seismic risk assessment of a 50m high masonry chimney using advanced analysis techniques. *Engineering Structures*, *69*, 255–270. doi:10.1016/j.engstruct.2014.03.028

Mirabella Roberti, G., & Calvetti, F. (1988). Distinct Element analysis of stone arches. In *Arch Bridges* (pp. 181–186). Rotterdam: Balkema.

Moghaddam, H. A. (2004). Lateral load behavior of masonry infilled steel frames with repair and retrofit. *Journal of Structural Engineering*, *130*(1), 56–63. doi:10.1061/(ASCE)0733-9445(2004)130:1(56)

Moghaddam, H. A., & Dowling, P. J. (1987). *The state-of-the-art in infilled frames. ESEE Res. Rep. No. 87-2, Imperial Coll. of Sci. and Technol.* London: Civ. Eng. Dept.

Moghaddam, H. A., Mohammadi, M. G. H., & Ghaemian, M. (2006). Experimental and analytical investigation into crack strength determination of infilled steel frames. *Journal of Constructional Steel Research*, *62*(12), 1341–1352. doi:10.1016/j.jcsr.2006.01.002

Mohammadi, G. M., & Yasrebi, F. (2010). Out of plane behaviour of walls, using rigid block concepts. *Structural Engineering & Mechanics*, *34*(3), 335–350. doi:10.12989/sem.2010.34.3.335

Mohebkhah, A. (2007). *A nonlinear-seismic model for brick masonry-infill panels with openings in steel frames*. (Ph.D. Dissertation). TarbiatModares University, Tehran, Iran.

Mohebkhah, A. and Tasnimi, A.A. (2012). Distinct element modeling of masonry-infilled steel frames with openings. *The Open Construction and Building Technology Journal*, *6*(Suppl 1-M2), 42-49.

Mohebkhah, A., & Sarv-Cheraghi, A. A. (2015). Nonlinear analysis of unreinforced masonry buildings using distinct element method[in Persian]. *Modares Civil Engineering Journal,*15(3), 85-91, *Tehran, Iran.*

Mohebkhah, A., & Tasnimi, A. A. (2007). Seismic behavior of brick masonry walls recommended by IS-2800: experimental and numerical approaches. In *Proceedings of the 5thConf. on Seismology and Earthquake Engineering*. International Institute of Earthquake Engineering and Seismology.

Mohebkhah, A., Tasnimi, A. A., & Moghadam, H. A. (2008). Non-linear analysis of masonry in-filled frames with opening using discrete element method. *Journal of Constructional Steel Research, 64*(21), 1463–1472. doi:10.1016/j.jcsr.2008.01.016

Monteiro Azevedo, N. (2003). A rigid particle discrete element model for the fracture analysis of plain and reinforced concrete. PhD Thesis, Heriot-Watt University, UK.

Monteiro Azevedo, N., Candeias, M., & Gouveia, F. (2015). A Rigid Particle Model for Rock Fracture Following the Voronoi Tessellation of the Grain Structure: Formulation and Validation. *Rock Mechanics and Rock Engineering, 48*(2), 535–557. doi:10.1007/s00603-014-0601-1

Monteiro Azevedo, N., & Lemos, J. V. (2005). A generalized rigid particle contact model for fracture analysis. *International Journal for Numerical and Analytical Methods in Geomechanics, 29*(3), 269–285. doi:10.1002/nag.414

Monteiro Azevedo, N., & Lemos, J. V. (2013). A 3D generalized rigid particle contact model for rock fracture. *Engineering Computations, 30*(2), 277–300. doi:10.1108/02644401311304890

Moradabadi, E., & Laefer, D. F. (2014). Numerical modelling options for cracked masonry buildings. *9th International Masonry Conference*, Guimarães, Portugal.

Moreau, J. J. (2006). Facing the plurality of solutions in nonsmooth mechanics. In C.C. Baniotopoulos (Ed.), *Nonsmooth/Nonconvex Mechanics with Applications in Engineering, II. NNMAE 2006 (Proc. International Conference in Memoriam of P. D. Panagiotopoulos).*

Moreau, J. J. (1988). Unilateral contact and dry friction in finite freedom dynamics. In J. J. Moreau & P. D. Panagiotopoulos (Eds.), *CISM Courses and Lectures* (Vol. 302). Vienna: Springer. doi:10.1007/978-3-7091-2624-0_1

Mosalam, K. M., White, R. N., & Gergely, P. (1997). *Computational strategies for frames with infill walls: discrete and smeared crack analysis and seismic fragility.* Technical Report NCEER-97-0021. National Center for Earthquake Engineering Research, State Univ. of N.Y. at Buffalo.

Mosalam, K. M., White, R. N., & Gergely, P. (1997). Static response of infilled frames using quasi-static experimentation. *Journal of Structural Engineering, 123*(11), 1462–1469. doi:10.1061/(ASCE)0733-9445(1997)123:11(1462)

Moseley, H. (1843). *The Mechanical Principles of Engineering and Architecture.* London, UK: Longman, Brown, Green and Longmans.

Moseley, H. (1843). *The mechanical principles of engineering and architecture.* London: Longman.

Mouzakis, H., Psycharis, I. N., Papastamatiou, D. Y., Carydis, P. G., Papantonopulos, C., & Zambas, C. (2002). Experimental investigation of the earthquake response of a model of a marble classical column. *Earthquake Engineering & Structural Dynamics, 31*(9), 1681–1698. doi:10.1002/eqe.184

Mullett, P. J., Briggs, M., & Minton, K. (2006). Archtec-Strengthening and preventing masonry arch bridges in Cumbria. In *Proceedings of the 11th International Conference on Structural Faults and Repair.* Edinburgh Press.

Munjiza, A. (2004) *The finite/discrete element method.* John Wiley and Sons.

Munjiza, A. A. (2004). The Combined Finite-Discrete Element Method. Wiley.

Munjiza, A. (2004). *The combined finite-discrete element method.* John Wiley. doi:10.1002/0470020180

Munjiza, A., & Andrews, K. R. F. (1998). NBS contact detection algorithm for bodies of similar size. *International Journal for Numerical Methods in Engineering, 43*(1), 131–149.

Munjiza, A., & Andrews, K. R. F. (2000). Penalty function method for combined finite-discrete element systems comprising large number of separate bodies. *International Journal for Numerical Methods in Engineering*, *49*(11), 1377–1396. doi:10.1002/1097-0207(20001220)49:11<1377:AID-NME6>3.0.CO;2-B

Munjiza, A., Andrews, K. R. F., & White, J. K. (1999). Combined single and smeared crack model in combined finite-discrete element method. *International Journal for Numerical Methods in Engineering*, *44*(1), 41–57. doi:10.1002/(SICI)1097-0207(19990110)44:1<41::AID-NME487>3.0.CO;2-A

Munjiza, A., & John, N. W. M. (2002). Mesh size sensitivity of the combined FEM/DEM fracture and fragmentation algorithms. *Engineering Fracture Mechanics*, *69*(2), 281–295. doi:10.1016/S0013-7944(01)00090-X

Munjiza, A., Knight, E. E., & Rougier, E. (2012). *Computational mechanics of discontinua*. Chichester, UK: John Wiley & Sons.

Munjiza, A., Knight, E. E., & Rougier, E. (2015). *Large strain finite element method: A practical course*. Chichester, UK: John Wiley & Sons.

Munjiza, A., Lei, Z., Divic, V., & Peros, B. (2013). Fracture and fragmentation of thin shells using the combined finite-discrete element method. *International Journal for Numerical Methods in Engineering*, *95*(6), 478–498. doi:10.1002/nme.4511

Munjiza, A., Owen, D. R. J., & Bićanić, N. (1995). A combined finite/discrete element method in transient dynamics of fracturing solids. *Engineering Computations*, *12*(2), 145–174. doi:10.1108/02644409510799532

Munjiza, A., Rougier, E., & John, N. W. M. (2006). MR linear contact detection algorithm. *International Journal for Numerical Methods in Engineering*, *66*(1), 46–71. doi:10.1002/nme.1538

Nagai, K., Sato, Y., & Ueda, T. (2005). Mesoscopic simulation of failure of mortar and concrete by 3D RBSM. *Journal of Advanced Concrete Technology*, *3*(3), 385–402. doi:10.3151/jact.3.385

Navier, C. L. M. H. (1826). *Résumé des Leçons données à l'École des Ponts et Chaussées sur l'application de la mécanique à l'établissement des constructions et des machines*. Paris: Dibot.

Nayeri, S. A. (2012). *Seismic Assessment of the Roman Temple in Évora, Portugal*. (MS thesis). University of Minho, Guimarães, Portugal.

Nikolić, Z., Smoljanović, H., & Živaljić, N. (2014). *Seismic analysis of dry stone masonry structures based on combined finite-discrete element method*. Paper presented at EURODYN, Porto, Portugal.

Nozaki, K., Terumichi, Y., Nishimura, K., & Sogabe, K. (2009). Study of rocking motion of rigid body with slide contact. *Journal of Mechanical Science and Technology*, *23*(4), 1001–1007. doi:10.1007/s12206-009-0329-0

O'Connor, K. M., & Dowding, C. M. (1992). Distinct Element Modeling and Analysis of Mining-induced Subsidence. *Rock Mechanics and Rock Engineering*, *25*(1), 1–24. doi:10.1007/BF01041873

O'Dwyer, D. (1999). Funicular analysis of masonry vaults. *Computers & Structures*, *73*(1-5), 187–197. doi:10.1016/S0045-7949(98)00279-X

Ochsendorf, J. (2002). *Collapse of masonry structures*. (Doctoral dissertation). Department of Engineering, University of Cambridge, Cambridge, UK.

Ochsendorf, J., Huerta, S., & Hernando, J. I. (2004). Collapse of masonry buttresses. *Journal of Architectural Engineering*, *10*(3), 88–97. doi:10.1061/(ASCE)1076-0431(2004)10:3(88)

Oda, M., Yamabe, T., Ishizuka, Y., Kumasaka, H., Tada, H., & Kimura, K. (1993). Elastic stress and strain in jointed rock masses by means of crack tensor analysis. *Rock Mechanics and Rock Engineering*, *26*(2), 89–112. doi:10.1007/BF01023618

Oetomo, J. J., Vincens, E., Dedecker, F., & Morel, J.-C. (2014). Discrete Element Method to assess the 2D failure of dry stone retaining walls, In *9th International Masonry Conference*, Guimarães, Portugal.

Oliveira, C. S., Martins, A., & Lopes, M. S. (1995). Seismic studies for the 'Águas Livres' Aqueduct in Lisbon. In *Proc. 10th European Conf. on Earthquake Engineering*.

Oliveira, D. V. (2003). *Experimental and numerical analysis of blocky masonry structures under cyclic loading*. (PhD thesis). University of Minho, Portugal.

Oliveira, D. V. C. (2003). *Experimental and numerical analysis of blocky masonry structures under cyclic loading*. (Ph.D thesis). University of Minho, Minho, Portugal.

Oliveira, C. S., Lemos, J. V., & Sincraian, G. E. (2002). Modelling large displacements of structures damaged by earthquake motions. *European Earthquake Engineering*, *3*, 56–71.

Oliveira, C. S., Portugal, A., Lopes, M. S., & Martins, A. (1991). *Comportamento sísmico do Aqueduto das Águas Livres, Report 202/91 NDA*. Lisboa, Portugal: Laboratório National de Engenharia Civil.

Oñate, E. (2015). A Particle-Discrete-Finite Element Method for Analysis of Particulate Flows and Their Interaction with Structures. *13th US National Congress on Computational Mechanics*.

Oñate, E., Celigueta, M. A., Idelsohn, S. R., Salazar, F., & Suárez, B. (2011). Possibilities of the particle finite element method for fluid–soil–structure interaction problems. *Computational Mechanics*, *48*(3), 307–318. doi:10.1007/s00466-011-0617-2

Oñate, E., Celigueta, M. A., Latorre, S., Casas, G., Rossi, R., & Rojek, J. (2014). Lagrangian analysis of multiscale particulate flows with the particle finite element method. *Computational Particle Mechanics*, *1*(1), 85–102. doi:10.1007/s40571-014-0012-9

Oppenheim, I. J. (1992). The masonry arch as a four-link mechanism under base motion. *Earthquake Engineering & Structural Dynamics*, *21*(11), 1005–1017. doi:10.1002/eqe.4290211105

Orduna, A. (2006). Three-dimensional limit analysis of ancient masonry buildings with rigid block models. *WIT Transactions on the Built Environment*, *85*, 673–682.

Orduña, A., & Lourenço, P. B. (2003). Cap model for limit analysis and strengthening of masonry structures. *Journal of Structural Engineering*, *129*(10), 1367–1375. doi:10.1061/(ASCE)0733-9445(2003)129:10(1367)

Orduña, A., & Lourenço, P. B. (2005). Three-dimensional limit analysis of rigid block assemblages. Part I: Torsion failure on frictional interfaces and limit analysis formulation. *International Journal of Solids and Structures*, *42*(18-19), 5140–5160. doi:10.1016/j.ijsolstr.2005.02.010

Orduña, A., & Lourenço, P. B. (2005). Three-dimensional limit analysis of rigid blocks assemblages. Part II: Load-path following solution procedure and validation. *International Journal of Solids and Structures*, *42*(18-19), 5161–5180. doi:10.1016/j.ijsolstr.2005.02.011

Ostergaard, J. (2010). *The influence of the grain size distribution on the mortar and masonry properties. Test Report*. Denmark: Danish Technological Institute.

Owen, D. R. J., Feng, Y. T., Han, K., & Peric, D. (2000). *Dynamic domain decomposition and load balancing in parallel simulation of finite/discrete elements.* Paper presented at European Congress on Computational Methods in Applied Sciences and Engineering, Barcelona, Spain.

Owen, D. R. J., Peric, D., Petrinic, N., Brookes, C. L., & James, P. J. (1998). *Finite/discrete element models for assessment and repair of masonry structures.* Paper presented at Second International Arch Bridge Conference, Venice, Italy.

Owen, D. R. J., Peric, D., Petrinic, N., Smokes, C. L., & James, P. J. (1998). *Finite/discrete element models for assessment and repair of masonry structures.* Paper presented at Second International Arch Bridge Conference, Venice, Italy.

Owen, D. R. J., & Feng, Y. T. (2001). Parallelised finite/discrete element simulation of multi-fracturing solids and discrete systems. *Engineering Computations, 18*(3-4), 557–576. doi:10.1108/02644400110387154

Owen, D. R. J., Peric, D., & Petrinic, N. (1998). Finite/discrete element models for assessment and repair of masonry structures. In A. Sinopoli (Ed.), *Proceedings of the 2nd International Conference on Arch Bridges: History, Analysis, Assessment, Maintenance and Repair.*

Owen, D. R. J., Peric, D., Petrinic, N., Brookes, C. L., & James, P. J. (1998). Finite/Discrete element model for assessment and repair of masonry structures. In *Arch Bridges* (pp. 181–186). Rotterdam: Balkema.

Page, A. W. (1978). Finite element model for masonry. *Journal of the Structural Division, 104*(8), 1267–1285.

Page, J. (1993). *Masonry Arch Bridge – TRL state of the art review.* London: HMSO.

Pagnoni, T. (1994). Seismic analysis of masonry and block structures with the discrete element method. In G. Duma (Ed.), *Proceedings of the 10th European Conference on Earthquake Engineering* (vol. 3, pp. 1669-1674). Rotterdam: Balkema.

Pagnoni, T., & Vanzi, I. (1995). Experimental and numerical study of the seismic response of block structures. In J. Middleton & G. N. Pande (Eds.), *Computer Methods in Structural Masonry* (pp. 213–222). Swansea, UK: Books & Journals International.

Palmieri, A., & Makris, N. (2008). Response analysis of rigid structures rocking on viscoelastic foundation. *Earthquake Engineering & Structural Dynamics, 37*(7), 1039–1063. doi:10.1002/eqe.800

Papa, E. (2001). Damage and failure modes. In J. W. Bull (Ed.), *Computational modelling of masonry, brickwork and blockwork structures* (pp. 1–26). Stirling, UK: Saxe-Coburg Publications. doi:10.4203/csets.6.1

Papaloizou, L., & Komodromos, P. (2009). Planar investigation of the seismic response of ancient columns and colonnades with epistyles using a custom-made software. *Soil Dynamics and Earthquake Engineering, 29*(11-12), 1437–1454. doi:10.1016/j.soildyn.2009.06.001

Papaloizou, L., & Komodromos, P. (2012). Investigating the seismic response of ancient multi-drum colonnades with two rows of columns using an object-oriented designed software. *Advances in Engineering Software, 44*(1), 136–149. doi:10.1016/j.advengsoft.2011.05.030

Papantonopoulos, C., Psycharis, I. N., Papastamatiou, D. Y., Lemos, J. V., & Mouzakis, H. P. (2002). Numerical prediction of the earthquake response of classical columns using the distinct element method. *Earthquake Engineering & Structural Dynamics, 31*(9), 1699–1717. doi:10.1002/eqe.185

Papastamatiou, D., & Psycharis, I. (1993). Seismic response of classical monuments – A numerical perspective developed at the Temple of Apollo Bassae, Greece. *Terra Nova, 5*(1), 591–601. doi:10.1111/j.1365-3121.1993.tb00309.x

Park, H. J., & Kim, D. S. (2013). Centrifuge modelling for evaluation of seismic behaviour of stone masonry structure. *Soil Dynamics and Earthquake Engineering, 53*, 187–195. doi:10.1016/j.soildyn.2013.06.010

Paulay, T., & Priestley, M. J. N. (1992). *Seismic design of reinforced concrete and masonry buildings*. New York, NY: John Wiley & Sons, Inc. doi:10.1002/9780470172841

Pegon, P., Pinto, A. V., & Geradin, M. (2001). Numerical modelling of stone-block monumental structures. *Computers & Structures*, *79*(22), 2165–2181. doi:10.1016/S0045-7949(01)00070-0

Peña, F. (2015). Modelo simplificado para el estudio del balanceo asimétrico de cuerpos rígidos esbeltos. *Revista Internacional de Métodos Numéricos para Cálculo y Diseño en Ingeniería*, *31*(1), 1–7. doi:10.1016/j.rimni.2013.10.005

Peña, F., Lourenço, P. B., & Campos-Costa, A. (2008). Experimental dynamic behaviour of free standing multi-block structures under seismic loadings. *Journal of Earthquake Engineering*, *12*(6), 953–979. doi:10.1080/13632460801890513

Peña, F., Prieto, F., Lourenço, P. B., Costa, A. C., & Lemos, J. V. (2007). On the dynamics of rocking motion of single rigid-block structures. *Earthquake Engineering & Structural Dynamics*, *36*(15), 2383–2399. doi:10.1002/eqe.739

Pérez-Aparicio, J. L., Bravo, R., & Ortiz, P. (2013). Refined element discontinuous numerical analysis of dry-contact masonry arches. *Engineering Structures*, *48*, 578–587. doi:10.1016/j.engstruct.2012.09.027

Perkins, E., & Williams, J. (2002). Generalized spatial binning of bodies of different sizes. *Discrete Element Methods*, *52-55*, 52–55. doi:10.1061/40647(259)10

Perry, J. (1881). Note on the rocking of a column. *Transaction of the Seismological Society of Japan*, *3*, 103–106.

Persy, N. (1825). *Cours sur la stabilité des constructions, à l'usage des élèves de l'école royale de l'Artillerie et du Génie*. Metz: Lithographie de l'École Royale de l'Artillerie et du Génie.

Pina-Henriques, J., & Lourenço, P. B. (2006). Masonry compression: A numerical investigation at the meso-level. *Engineering Computations*, *23*(4), 382–407. doi:10.1108/02644400610661163

Pippard, A. J. S. (1948). The approximate estimation of safe loads on masonry bridges. Civil engineer in war. *Institution of Civil Engineers*, *1*, 365.

Pippard, A. J. S., & Ashby, E. R. J. (1936). An experimental study of the voissour arch. *Journal of the Institution of Civil Engineers*, *10*, 383–403. doi:10.1680/ijoti.1939.14554

Plaut, R. H., Fielder, W. T., & Virgin, L. N. (1996). Fractal behaviour of an asymmetric rigid block overturning due to harmonic motion of a tilted foundation. *Chaos, Solitons, and Fractals*, *7*(2), 177–196. doi:10.1016/0960-0779(95)00059-3

Polyakov, S. V. (1956). Masonry in framed buildings. In *Translated by G. L. Cairns in 1963*. Boston Spa, UK: National Lending Library for Science and Technology.

Pöschel, T., & Schwager, T. (2005). *Computational Granular Dynamics: Models and Algorithms*. Berlin: Springer.

Potyondy, D., & Cundall, P. A. (2004). A bonded-particle model for rock. *International Journal of Rock Mechanics and Mining Sciences*, *41*(8), 1329–1364. doi:10.1016/j.ijrmms.2004.09.011

Potyondy, D., Cundall, P., & Lee, C. (1996). Modelling rock using bonded assemblies of circular particles. In: M Aubertin et al. (eds). *Proceedings of the 2nd North American Rock Mechanics Symposium*, 1937-1944, Rotterdam, Balkema.

Powrie, W., Harkness, R. M., Zhang, X., & Bush, D. I. (2002). Deformation and failure of drystone retaning walls. *Geotechnique*, *52*(6), 435–446. doi:10.1680/geot.2002.52.6.435

Prieto, F., & Lourenço, P. B. (2005). On the rocking behaviour of rigid objects. *Meccanica*, *40*(2), 121–133. doi:10.1007/s11012-005-5875-7

Prieto, F., Lourenço, P. B., & Oliveira, C. S. (2004). Impulsive Dirac-delta forces in the rocking motion. *Earthquake Engineering & Structural Dynamics*, *33*(7), 839–857. doi:10.1002/eqe.381

Psycharis, I., Drougas, A., & Daisou, M. (2011). Seismic Behaviour of the Walls of the Parthenon: A Numerical Study. In Computational Methods in Earthquake Engineering. Computational Methods in Applied Sciences. doi:10.1007/978-94-007-0053-6_12

Psycharis, I. N. (1990). Dynamic behaviour of rocking two-block assemblies. *Earthquake Engineering & Structural Dynamics*, *19*(4), 555–575. doi:10.1002/eqe.4290190407

Psycharis, I. N., Fragiadakis, M., & Stefanou, I. (2013). Seismic reliability assessment of classical columns subjected to near-fault ground motions. *Earthquake Engineering & Structural Dynamics*, *42*, 2061–2079.

Psycharis, I. N., & Jennings, P. C. (1985). Upthrow of objects due to horizontal impulse excitation. *Bulletin of the Seismological Society of America*, *75*(2), 543–561.

Psycharis, I. N., Lemos, J. V., Papanastasiou, D. Y., Zambas, C., & Papantonopoulos, C. (2003). Numerical study of the seismic behaviour of a part of the Parthenon Pronaos. *Earthquake Engineering & Structural Dynamics*, *32*(13), 2063–2084. doi:10.1002/eqe.315

Psycharis, I. N., Papastamatiou, D. Y., & Alexandris, A. P. (2000). Parametric investigation of the stability of classical columns under harmonic and earthquake excitations. *Earthquake Engineering & Structural Dynamics*, *29*(8), 1093–1109. doi:10.1002/1096-9845(200008)29:8<1093::AID-EQE953>3.0.CO;2-S

Psycharis, I. N., Papastamatiou, D. Y., Cundall, P. A., & Lorig, L. (1993) Numerical and analytical modeling of Greek classical columns.*Proc. 3rd Pan American Congress of Applied Mechanics - PACAM III.*

Purvance, M. D., Anooshehpoor, A., & Brune, N. (2008). Freestanding block overturning fragilities: Numerical simulation and experimental validation. *Earthquake Engineering & Structural Dynamics*, *37*(5), 791–808. doi:10.1002/eqe.789

Radjai, F., & Richefeu, V. (2009). Contact dynamics as a nonsmooth discrete element method. *Mechanics of Materials*, *41*(6), 715–728. doi:10.1016/j.mechmat.2009.01.028

Radjai, F., Wolf, D. E., Jean, M., & Moreau, J. J. (1998). Bimodal character of stress transmission in granular packings. *Physical Review Letters*, *80*(1), 61–64. doi:10.1103/PhysRevLett.80.61

Rafiee, A., & Vinches, M. (2013). Mechanical behaviour of a stone masonry bridge assessed using an implicit discrete element method. *Engineering Structures*, *48*, 739–749. doi:10.1016/j.engstruct.2012.11.035

Rafiee, A., Vinches, M., & Bohatier, C. (2008a). Application of the NSCD method to analyse the dynamic behaviour of stone arched structures. *International Journal of Solids and Structures*, *45*(25-26), 6269–6283. doi:10.1016/j.ijsolstr.2008.07.034

Rafiee, A., Vinches, M., & Bohatier, C. (2008b). Modelling and analysis of the Nimes arena and the Arles aqueduct subjected to a seismic loading, using the Non-Smooth Contact Dynamics method. *Engineering Structures*, *30*(12), 3457–3467. doi:10.1016/j.engstruct.2008.05.018

Rafiq, M. Y., Bugmann, G., & Easterbrook, D. J. (1998). Artificial neural networks for modelling some of the activities of the computational stage of the design process. In K.C.P. Wang, T. Adams, M.L. Maher, & A. Songer (Eds.), *Proceedings of the International Computing Congress.*

Raman, P. G. (2004). Structural masonry and architectural expression. *Construction & Building Materials*, *18*(2), 133–139. doi:10.1016/j.conbuildmat.2003.08.016

Rankine, W. J. M. (1858). *A Manual of Applied Mechanics*. London, UK: Charles Griffin.

Reccia, E. (2013) *Conservation of masonry arch bridges. A procedure for modelling and strengthening*. (Unpublished PhD Dissertation). University of Nova Gorica.

Reccia, E., Cazzani, A., & Cecchi, A. (2012). FEM-DEM modeling for out-of-plane loaded masonry panels: A limit analysis approach. *Open Civil Engineering Journal*, *6*(1), 231-238.

Reccia, E., Cecchi, A., & Tralli, A. (2013). Homogenization of masonry vault bridges: Sensitivity to external stone arch. *Civil-Comp Proceedings*, 102.

Reccia, E., Milani, G., Cecchi, A., & Tralli, A. (2014). Full 3D homogenization approach to investigate the behavior of masonry arch bridges: The venice trans-lagoon railway bridge. *Construction & Building Materials*, *66*, 567–586. doi:10.1016/j.conbuildmat.2014.05.096

Reda, T. M. M., & Shrive, N. G. (2001). The use of pozzolans to improve bond strength. In *Proceedings of the 9th Canadian Symposium*. University of Rots.

Riddington, J. R., & Stafford-Smith, B. (1977). Analysis of infilled frames subjected to racking with design recommendations. *The Structural Engineer*, *55*(6), 263–268.

Rizzi, E., Rusconi, F., & Cocchetti, G. (2014). Analytical and numerical DDA analysis on the collapse mode of circular masonry arches. *Engineering Structures*, *60*, 241–257. doi:10.1016/j.engstruct.2013.12.023

Roberti, G. M., & Spina, O. (2001). Numerical analysis of the Sardinian 'Nuraghe. In T. G. Hughes & G. N. Pande (Eds.), *Computer Methods in Structural Masonry - 5* (pp. 190–197). Swansea, UK: Computers & Geotechnics Ltd.

Roca, P., Cervera, M., Gariup, G., & Pela, L. (2010). Structural analysis of masonry historical constructions. Classical and advanced approaches. *Archives of Computational Methods in Engineering*, *17*(3), 299–325. doi:10.1007/s11831-010-9046-1

Rokugo, K., Iwasa, M., Susuki, T., & Koyanagi, W. (1989). Testing methods to determine tensile strain softening curve and fracture energy of concrete. In H. Mihasi et al. (Eds.), *Fracture toughness and fracture energy: Test methods for concrete and rock* (pp. 153–163). Rotterdam: Balkema.

Rots, J. G. (1991). Numerical simulation of cracking in masonry. *HERON*, *36*(2), 49–63.

Rots, J. G. (1997). *Structural masonry: An experimental/numerical basis for practical design rules*. A.A. Balkema Publishers.

Rots, J. G. (1997). *Structural masonry: an experimental/numerical basis for practical design rules. Center for Civil Engineering Research and Codes (CUR)*. Balkema.

Rougier, E., Knight, E. E., & Munjiza, A. (2012). Fluid driven rock deformation via the combined FEM/DEM methodology. In American rock mechanics association 46thUS rockmechanics/ geomechanics symposium.

Rougier, E., Knight, E. E., & Munjiza, A. (2012). Fluid driven rock deformation via the combined FEM/DEM methodology. In American rock mechanics association 46thUS rockmechanics/geomechanics symposium.

Rougier, E., Knight, E. E., Munjiza, A., Broome, A. J. S. S. T., Swift, R. P., & Bradley, C. R. (2011). The combined finite-discrete element method applied to the study of rock fracturing behavior in 3D. In American Rock Mechanics Association 45 th US rock mechanics/geomechanics symposium.

Rougier, E., Knight, E. E., Broome, S. T., Sussman, A. J., & Munjiza, A. (2014). Validation of a three-dimensional Finite-Discrete Element Method using experimental results of the Split Hopkinson Pressure Bar test. *International Journal of Rock Mechanics and Mining Sciences*, *70*, 101–108.

Rougier, E., Munjiza, A., & John, N. W. M. (2004). Numerical comparison of some explicit time integration schemes used in DEM, FEM/DEM and molecular dynamics. *International Journal for Numerical Methods in Engineering*, *61*(6), 856–879. doi:10.1002/nme.1092

Rouxinol, G. A. F., Providência, P., & Lemos, J. V. (2007). Bridgemill bridge bearing capacity assessment by a discrete element method. In P. B. Lourenço, D. B. Oliveira & A. Portela (Eds.), *Arch'07 - Proc. 5th International Conference on Arch Bridges* (pp. 669-676). Madeira, Portugal: Multicomp Lda Publishers.

Rupakhety, E., Sigurdsson, S. U., Papageorgiou, A. S., & Sigbjörnsson, R. (2011). Quantification of ground-motion parameters and response spectra in the near-fault region. *Bulletin of Earthquake Engineering*, *9*(4), 893–930. doi:10.1007/s10518-011-9255-5

Sab, K. (2003). Yield design of thin periodic plates by a homogenization technique and an application to masonry walls. *Comptes Rendus. Mécanique*, *331*(9), 641–646. doi:10.1016/S1631-0721(03)00144-X

Salerno, G., & de Felice, G. (2009). Continuum modeling of periodic brickwork. *International Journal of Solids and Structures*, *46*(5), 1251–1267. doi:10.1016/j.ijsolstr.2008.10.034

Saneinejad, A., & Hobbs, B. (1995). Inelastic design of infield frames. *Journal of Structural Engineering*, *121*(4), 634–650. doi:10.1061/(ASCE)0733-9445(1995)121:4(634)

Sarhosis, V. (2012). *Computational modelling of low bond strength masonry*. (PhD thesis). University of Leeds, UK.

Sarhosis, V., Garrity, S. W., & Sheng, Y. (2008). *Distinct element modelling of masonry wall panels with openings*. 9th International Conference on Computational Structures Technology, Athens, Greece.

Sarhosis, V., Lignola, G. P., & Asteris, P. (2014). Seismic Vulnerability of Ancient Colonnade: Numerical analysis of the two storey colonnade of the Forum in Pompeii. In Seismic Assessment and Rehabilitation of Historic Structures. IGI Global.

Sarhosis, V., Lignola, G. P., & Asteris, P. (2015). Seismic Vulnerability of Ancient Colonnade: Numerical analysis of the two storey colonnade of the Forum in Pompeii. In P. G. Asteris & V. Plevris (Eds.), Seismic Assessment and Rehabilitation of Historic Structures. IGI Global.

Sarhosis, V., Lignola, G. P., & Asteris, P. (2016). Seismic Vulnerability of Ancient Colonnade: Numerical analysis of the two storey colonnade of the Forum in Pompeii. In Asteris & Repapis (Eds.), Seismic Assessment and Rehabilitation of Historic Structures. IGI Global.

Sarhosis, V., Garrity, S. W., & Sheng, Y. (2008). Distinct element modelling of masonry wall panels with openings.*9th International Conference on Computational Structures Technology, CST 2008.*

Sarhosis, V., Garrity, S. W., & Sheng, Y. (2015). Influence of the brick-mortar interface on the mechanical response of low bond strength masonry lintels. *Engineering Structures*, *88*, 1–11. doi:10.1016/j.engstruct.2014.12.014

Sarhosis, V., Oliveira, D. V., Lemos, J. V., & Lourenço, P. B. (2014b). The effect of the angle of skew on the mechanical behaviour of arches. *Journal of Mechanics Research Communications*, *61*, 53–49. doi:10.1016/j.mechrescom.2014.07.008

Sarhosis, V., Oliveira, D. V., & Lourenço, P. B. (2015). On the mechanical behaviour of masonry structures. In V. Sarhosis, J. V. Lemos, K. Bagi, & G. Milani (Eds.), *Computational Modeling of Masonry Structures Using the Discrete Element Method*. IGI Global.

Sarhosis, V., & Sheng, Y. (2014). Identification of material parameters for low bond strength masonry. *Engineering Structures*, *60*, 100–110. doi:10.1016/j.engstruct.2013.12.013

Sarhosis, V., Tsavdaridis, K., & Giannopoulos, G. (2014). Discrete Element Modelling of masonry in-filled steel frames with multiple window openings subjected to lateral load variations. *Open Construction and Building Technology Journal*, *8*(1), 93–103. doi:10.2174/1874836801408010093

Sassoni, E., Franzoni, E., Pigino, B., Scherer, G. W., & Naidu, S. (2013). Consolidation of calcareous and siliceous sandstones by hydroxyapatite: Comparison with a TEOS-based consolidan. *Journal of Cultural Heritage*, *14*(1), 103–108. doi:10.1016/j.culher.2012.11.029

Sauvé, R. G., & Metzger, D. R. (1995). Advances in dynamic relaxation techniques for nonlinear finite element analysis. *Journal of Pressure Vessel Technology*, *117*(2), 170–176. doi:10.1115/1.2842106

Scheldt, T., Lu, M., & Myrvang, A. (2002). Numerical analysis of Gjovick cavern. In Y. Hatzor (Ed.), *Procs. ICADD-5* (pp. 125–132). Abingdon, UK: Balkema.

Schiappa de Azevedo, F., Sousa, M. R., Reis, M. O. B., Casaca, J. M. M., Rodrigues, J. D., Veiga, M. R., & Silva, A. S. (1996). *Estudo dos efeitos das vibrações no Aqueduto das Águas Livres, produzidas pelas construção e tráfego do eixo Norte-Sul. Relatório 207/96 - NDA*. Lisboa: Laboratório Nacional de Engenharia Civil.

Schiava D'Albano, G. G. (2014). *Computational and algorithmic solutions for large scale combined finite-discrete elements simulations.* (Unpublished doctoral dissertation). Queen Mary, University of London.

Schiava D'Albano, G. G., Munjiza, A., & Lukas, T. (2013). *Novel MS (MunjizaSchiava) contact detection algorithm for multi-core architectures.* Paper presented at Particles, Stuttgart, Germany.

Schlangen, E & Garboczi, E. (1997). Fracture simulations of concrete using lattice models:Computational aspects. *Engineering Fracture Mechanics*, 57(2/3), 319:332.

Schlegel, R., & Rautenstrauch, K. (2004a). Failure analyses of masonry shear walls. In H. Konietzky (Ed.), *Numerical modelling of discrete materials in geotechnical engineering, civil engineering and earth sciences* (pp. 15–20). UK: Taylor and Francis Group London.

Schlegel, R., & Rautenstrauch, K. (2004b). Comparative computations of masonry arch bridges using continuum and discontinuum mechanics. In H. Konietzky (Ed.), *Numerical modelling of discrete materials Numerical modelling of discrete materials in geotechnical engineering, civil engineering and earth sciences* (pp. 3–6). London, UK: Taylor and Francis Group.

Schneider, S. P., Zagers, B. R., & Abrams, D. P. (1998). Lateral strength of steel frames with masonry infills having large openings. *Journal of Structural Engineering*, *124*(8), 896–904. doi:10.1061/(ASCE)0733-9445(1998)124:8(896)

Schnitter, N. J. (1994). *A history of Dams – The useful pyramids.* Rotterdam: A.A. Balkema.

Schubert, I. P. (1988). The influence of mortar on the strength of masonry. In J.W. Courcy (Ed.), *Proceedings of the 8th International Brick and Block Conference*. Elsevier.

Shames, I. H., & Dym, C. L. (1985). *Energy and Finite Elements Methods in Structural Mechanics.* New York: Hemisphere Publishing Corporation.

Sharif, V., Ghafory Ashtiany, M., Eshghi, S., & Soroushian, A. (2009). Dynamic assessment of constrained rigid equipments. *Journal of Applied Sciences*, *9*(13), 2362–2371. doi:10.3923/jas.2009.2362.2371

Sharma, V. M., Saxena, K. R., & Woods, R. D. (1999). *Distinct Element Modeling in Geomechanics*. Rotterdam: A. A. Balkema.

Shenton, H. III. (1996). Criteria for initiation of slide, rock and slide-rock rigid-body blocks. *Journal of Engineering Mechanics*, *122*(7), 690–693. doi:10.1061/(ASCE)0733-9399(1996)122:7(690)

Shi, G. H. (1988). *Discontinuous deformation analysis: a new numerical model for the statics and dynamics of block system.* (PhD Thesis). Dept. of Civil Engineering, University of California, Berkeley, CA.

Shi, G. H. (1988). *Discontinuous deformation analysis: A new numerical model for the statics and dynamics of block systems.* (Unpublished doctoral dissertation). University of California, Berkeley, CA.

Shi, G.-H. (1988). *Discontinuous deformation analysis – A new model for the statics and dynamics of block systems.* (PhD thesis). University of California Berkeley.

Shi, G.-H. (2001). Three dimensional discontinuous deformation analysis. In N. Bicanic (Ed.), Procs. ICADD-4.

Shi, G.-H., & Goodman, R. E. (1988). Discontinuous deformation analysis – A new method for computing stress, strain and sliding of block systems. In Cundall, Sterling & Starfield (Eds.), Key Questions in Rock Mechanics. Balkema.

Shieh-Beygi, B., & Pietruszczak, S. (2008). Numerical analysis of structural masonry: Mesoscale approach. *Computers & Structures*, *86*(21-22), 1958–1973. doi:10.1016/j.compstruc.2008.05.007

Shi, G. H., & Goodman, R. E. (1988). Discontinuous deformation analysis – a new method for computing stress, strain and sliding of block systems. In P.A. Cundall, R. Sterling, & A. Starfield (Eds.), *Proceedings of the 29th U.S. Symposium: Key Questions in Rock Mechanics.*

Shi, G. H., & Goodman, R. E. (1988). Discontinuous deformation analysis – a new method for computing stress, strain and sliding of block systems. In *Key questions in rock mechanics* (pp. 381–393). Rotterdam: Balkema.

Shi, G. H., & Goodman, R. E. (1988). Discontinuous deformation analysis – a new method for computing stress, strain and sliding of block systems. In: P.A. Cundall, R. Sterling, & A. Starfield (eds.), *Proceedings of the 29th U.S. Symposium: Key Questions in Rock Mechanics*. Balkema.

Shi, G.-H. (1992). Discontinuous deformation analysis: A new numerical model for the statics and dynamics of deformable block structures. *Engineering Computations*, *9*(4), 157–168. doi:10.1108/eb023855

Simon, J., & Bagi, K. (2014). Discrete element analysis of the minimum thickness of oval masonry domes. *International Journal of Architectural Heritage*. doi:10.1080/15583058.2014.996921

Sincraian, G. E. (2001). *Seismic behaviour of block masonry structures – A discrete element method approach.* (PhD Thesis). Instituto Superior Técnico, Lisbon, Portugal.

Sincraian, G. E., & Azevedo, J. J. (1997). Earthquake analysis of structural topologies with masonry infill panels using the distinct element method. In Proc. 3rd Encontro Sobre Sismologia e Engenharia Sismica.

Sincraian, G. E., & Azevedo, J. J. (1998). Numerical simulation of the seismic behaviour of stone masonry structures using the discrete element method. In Bisch, Labbé & Pecker (Eds.), *Proc. 11th European Conf. on Earthquake Eng.* Balkema.

Sincraian, G. E., Oliveira, C. S., & Lemos, J. V. (1998). Assessment of the seismic behaviour of a stone masonry aqueduct using the discrete element method. In Bisch, Labbé & Pecker (Eds.), *Proc. 11th European Conf. on Earthquake Eng*. Balkema.

Sincraian, G. E., & Azevedo, J. J. (1998). Numerical simulation of the seismic behavior of stone and brick masonry structures using the discrete element method. In *Proc. 11th European Conference on Earthquake Engineering*. Balkema.

Sincraian, G. E., & Azevedo, J. J. (1998). Numerical simulation of the seismic behavior of stone and brick masonry structures.*Proc. of the 11th European Conference on Earthquake Engineering*. Balkema.

Sincraian, G. E., & Lemos, J. V. (1998). *A discrete element program based on a rigid block formulation. Report 40/98-NDE/NEE*. Lisbon: Laboratorio National de Engenharia Civil.

Sincraian, G. E., & Lemos, J. V. (1999). Seismic analysis of a stone masonry acqueduct using discrete elements. In *Proc. 8th Canadian Conference on Earthquake Engineering*.

Sincraian, G. E., & Lemos, J. V. (1999). Seismic analysis of a stone masonry aqueduct using discrete elements. In *Proc. 8th Canadian Conference on Earthquake Engineering*.

Sincraian, G. E., & Lemos, J. V. (1999). Seismic analysis of a stone masonry aqueduct using Discrete Elements. In *Proceedings of the 8th Canadian Conference on Earthquake Engineering*.

Sincraian, G. E., Lemos, J. V., & Oliveira, C. S. (1998). Assessment of the seismic behavior of a stone masonry acqueduct using the discrete element method. In *Proc. 11th European Conference on Earthquake Engineering*. Balkema.

Sincraian, G. E., Lemos, J. V., & Oliveira, C. S. (1998). Assessment of the seismic behavior of stone masonry aqueduct using the discrete element method.*Proc. 11th European Conference on Earthquake Engineering*. Balkema.

Sinopoli, A. (2003). The role of geometry in the theories on vaulted structures by Lorenzo Mascheroni. In S. Huerta (Ed.), *Proceedings of the 1st International Congress on Construction History* (pp. 1865-1873). Madrid: Instituto Juan de Herrera.

Sinopoli, A. (1991). Dynamic Analysis of a Stone Column Excited by a Sine Wave Ground Motion. *Applied Mechanics Reviews*, *44*(11S), 246–256. doi:10.1115/1.3121361

Sinopoli, A., Corradi, M., & Foce, F. (1997). Modern formulation for preelastic theories on masonry arches. *Journal of Engineering Mechanics*, *123*(3), 204–213. doi:10.1061/(ASCE)0733-9399(1997)123:3(204)

Smith, N. (1971). *A history of dams*. London: Peter Davies.

Smoljanović, H., Živaljić, N., & Nikolić, Ž. (2013). A combined finite-discrete element analysis of dry stone masonry structures. *Engineering Structures*, *52*, 89–100. doi:10.1016/j.engstruct.2013.02.010

Smoljanović, H., Živaljić, N., & Nikolić, Ž. (2013). Nonlinear analysis of engineering structures by combined finite-discrete element method. *Gradevinar*, *65*, 331–334.

Sorrentino, L., Kunnath, S., Monti, G., & Scalora, G. (2008). Seismically induced one-sided rocking response of unreinforced masonry façades. *Engineering Structures*, *30*(8), 2140–2153. doi:10.1016/j.engstruct.2007.02.021

Spanos, P. D., Roussis, P. C., & Politis, N. P. A. (2001). Dynamic analysis of stacked rigid blocks. *Soil Dynamics and Earthquake Engineering*, *21*(7), 559–578. doi:10.1016/S0267-7261(01)00038-0

Stafford-Smith, B. (1962). Lateral stiffness of infilled frames. *Journal of the Structural Division*, *88*(6), 183–199.

Stafford-Smith, B. (1966). Behavior of square infilled frames. *Journal of the Structural Division*, *92*(1), 381–403.

Stavridis, A. (2009). *Analytical and experimental study of seismic performance of reinforced concrete frames infilled with masonry walls*. (PhD thesis). Department of Structural Engineering, University of California, San Diego, CA.

Stavridis, A., & Shing, P. B. (2010). Finite-element modeling of nonlinear behavior of masonry-infilled RC frames. *Journal of Structural Engineering*, *136*(3), 285–296. doi:10.1061/(ASCE)ST.1943-541X.116

Stefanou, I., Psycharis, I., & Georgopoulos, I. O. (2011). Dynamic response of reinforced masonry columns in classical monuments. *Journal of Construction and Building Materials*, *25*(1), 4325–4337. doi:10.1016/j.conbuildmat.2010.12.042

Stefanou, I., Sab, K., & Heck, J.-V. (2015). Three dimensional homogenization of masonry structures with building blocks of finite strength: A closed form strength domain. *International Journal of Solids and Structures*, *54*, 258–270. doi:10.1016/j.ijsolstr.2014.10.007

Stefanou, I., Vardoulakis, I., & Mavraganis, A. (2011b). Dynamic motion of a conical frustum over a rough horizontal plane. *International Journal of Non-linear Mechanics*, *46*(1), 114–124. doi:10.1016/j.ijnonlinmec.2010.07.008

Sugo, H. O., Page, A. W., & Lawewnce, S. J. (2001). The development of mortar/unit bond. In *Proceedings of the* 9th-*Canadian masonry symposium*. University of New Brunswick.

Taniguchi, T. (2002). Non-linear response analyses of rectangular rigid bodies subjected to horizontal and vertical ground motion. *Earthquake Engineering & Structural Dynamics*, *31*(8), 1481–1500. doi:10.1002/eqe.170

Taniguchi, T., & Miwa, T. (2007). A simple procedure to approximate slip displacement of freestanding rigid body subjected to earthquake motions. *Earthquake Engineering & Structural Dynamics*, *36*(4), 481–501. doi:10.1002/eqe.639

Tasnimi, A. A., & Farzin, M. (2006). Inelastic behavior of RC columns under cyclic loads based on cohesion and internal friction angle of concrete. *Modarres Technical and Engineering Journal*, *23*, 29–40.

Tasnimi, A. A., & Mohebkhah, A. (2011). Investigation on the behavior of brick-infilled steel frames with openings, experimental and analytical approaches. *Engineering Structures*, *33*(3), 968–980. doi:10.1016/j.engstruct.2010.12.018

Thavalingam, A., Bicanic, N., Robinson, J. I., & Ponniah, D. A. (2001). Computational framework for discontinuous modelling of masonry arch bridges. *Computers & Structures*, *79*(19), 1821–1830. doi:10.1016/S0045-7949(01)00102-X

Timoshenko, S. P. (1953). *History of Strength of Materials*. New York: McGraw-Hill Book Company, Inc.

Tokashiki, N., Aydan, Ö., Mamaghani, I. H. P., & Kawamoto, T. (1997). The stability of a rock block on an incline by discrete finite element method (DFEM). In IACMAG '97.

Tokashiki, N., Aydan, Ö., Mamaghani, I. H. P., & Kawamoto, T. (1997a). The stability of a rock block on an incline by discrete finite element method (DFEM). IACMAG '97.

Tokashiki, N., Aydan, Ö., Shimizu, Y., & Kawamoto, T. (1997b): The assessment of the stability of a very old tunnel by discrete finite element method (DFEM).*Proceedings of the Sixth International Symposium on Numerical Methods in Geomechanics, NUMOG VI*.

Tokashiki, N., Aydan, O., Shimizu, Y., & Mamaghani, I. H. P. (2001). A stability analysis of masonry walls by discrete finite element method (DFEM).*Proc. of the 10th Int. Conf. on Computer Methods and Advances in Geomechanics (IACMAG'01)*.

Tomaževic, M., & Turnšek, V. (1980). Lateral load distribution as a basis for the seismic resis- tance analysis of masonry buildings. In *International Research Conference on Earthquake Engineering*, Skopje.

Tóth, A. R., & Bagi, K. (2011).Analysis of a Lunar Base Structure Using the Discrete-Element Method.*Journal of Aerospace Engineering*, *24*(3), 397–401.

Toth, A. R., Orban, Z., & Bagi, K. (2009). Discrete element analysis of a masonry arch. *Mechanics Research Communications*, *36*(4), 469–480. doi:10.1016/j.mechrescom.2009.01.001

Toumbakari, E., & Psycharis, I. N. (2010). Parametric investigation of the seismic response of a column of the Aphrodite Temple in Amathus. In *14th European Conference on Earthquake Engineering (ECEE)*. Ohrid: FYROM.

Tran, V. H., Vincens, E., Morel, J. C., Dedecker, F., & Le, H. H. (2014). 2D-DEM modelling of the formwork removal of a rubble stone masonry bridge. *Engineering Structures*, *75*, 448–456. doi:10.1016/j.engstruct.2014.05.048

Tsesarsky, M., Hatzor, Y. H., & Sitar, N. (2005). Dynamic displacement of a block on an inclined plane: Analytical, experimental and DDA results. *Rock Mechanics and Rock Engineering*, *38*(2), 153–167. doi:10.1007/s00603-004-0043-2

Tso, W. K., & Wong, C. M. (1989). Steady state rocking response of rigid blocks to earthquake. Part 1: Analysis. Part 2: Experiment. *Earthquake Engineering & Structural Dynamics*, *18*, 89–120. doi:10.1002/eqe.4290180109

Turnšek, V., & Čačovič, F. (1971). Some experimental results on the strength of brick masonry walls. In *Second International Brick Masonry Conference*, Stoke-on-Trent, UK.

Turnšek, V., & Sheppard, P. (1980). The shear and flexural resistance of masonry walls. In *International Research Conference on Earthquake Engineering*, Skopje.

UIC. (2005). *International Union of Railways: Improving Assessment, Optimization of Maintenance and Development of Database for Masonry Arch Bridges*. UIC.

Underwood, P. (1983). Dynamic relaxation. In T. Belytschko & T. Hughes (Eds.), *Computation Methods for Transient Analysis* (pp. 246–265). New York: North-Holland.

Underwood, P. (1983). Dynamic relaxation. In T. Belytschko & T. J. R. Hughes (Eds.), *Computational Methods for Transient Analysis* (pp. 1–65). North Holland.

Unger, T., Wolf, D. E., & Kertész, J. (2004). *Force indeterminacy in the jammed state of hard disks*. cond-mat/0403089

Unger, T., & Kertész, J. (2003). The contact dynamics method for granular media. In *Modeling of Complex Systems* (pp. 116–138). Melville, NY: American Institute of Physics. doi:10.1063/1.1571300

US Bureau of Reclamation (USBR). (1976). *Design of gravity dams*. Denver, CO: Bureau of Reclamation, US Department of the Interior.

Valluzi, M. R., da Porto, F., & Modena, C. (2001). Behaviour of multi-leaf stone masonry walls strengthened by different intervention techniques. In P. B. Lourenço & P. Roca (Eds.), *Historical Constructions* (pp. 1023–1032). Guimarães, Portugal.

Van der Pluijm, R. (1992). Material properties of masonry and its components under tension and shear. In *Proceedings of the 6th Canadian Masonry Symposium*. University of Saskatchewan.

Van der Pluijm, R. (1999). *Out-of-plane bending of masonry behaviour and strength*. (Ph.D thesis). Eindhoven University of Technology, The Netherlands.

Van der Pluijm, R. (1993). Shear behaviour of bed joints. In: A.A. Hamid, & H.G. Harris (Eds.), *Proceedings of the 6th North American Masonry Conference*. Drexel University.

Vanderbei, T. J. (2001). *Linear Programming: Foundations and extensions* (2nd ed.). London, UK: Springer. doi:10.1007/978-1-4757-5662-3

Vasarhelyi, B. (2003). Some observations regarding the strength and deformability of sandstones in dry and saturated conditions. *Bulletin of Engineering Geology and the Environment*, *62*(3), 245–249. doi:10.1007/s10064-002-0186-x

Vásárhelyi, B., & Ván, P. (2006). Influence of water content on the strength of rock. *Engineering Geology*, *84*(1), 70–74. doi:10.1016/j.enggeo.2005.11.011

Vasconcelos, G. F. M. (2205). Experimental investigations on the mechanics of stone masonry: Characterization of granites and behaviour of ancient masonry shear walls. PhD Thesis, Universidade do Minho, Portugal.

Vermeltfoort, A. T., Martens, D. E. W., & van Zijl, G. P. A. G. (2007). Brick mortar interface effects on masonry under compression. *Canadian Journal of Civil Engineering*, *34*(1), 1475–1485. doi:10.1139/L07-067

Verstrynge, E., Adriaens, R., Elsen, J., & Van Balen, K. (2014). Multi-scale analysis on the influence of moisture on the mechanical behavior of ferruginous sandstone. *Construction & Building Materials*, *54*, 78–90. doi:10.1016/j.conbuildmat.2013.12.024

Viola, E., Panzacchi, L., & Tornabene, F. (2005). General analysis and application to redundant arches under static loading. *Construction & Building Materials*, *21*(5), 1129–1143. doi:10.1016/j.conbuildmat.2005.10.001

Voyagaki, E., Psycharis, I. N., & Mylonakis, G. (2014). Complex response of a rocking block to a full-cycle pulse. *Journal of Engineering Mechanics*, *140*(6), 04014024. doi:10.1061/(ASCE)EM.1943-7889.0000712

Voyiadjis, G. Z., & Kattan, P. I. (2005). *Damage mechanics*. London, UK: Taylor and Francis. doi:10.1201/9781420027839

Walker, P., McCombie, P., & Claxton, M. (2006). Plane strain numerical model for drystone retaining walls. *Proceedings of the ICE - Geotechnical Engineering*, *160*(2), 97–103. doi:10.1680/geng.2007.160.2.97

Wang, F., Feng, Y. T., & Owen, D. R. J. (2004). Parallelization for finite-discrete element analysis in a distributed-memory environment. *International Journal of Computational Engineering Science*, *5*(1), 1–23. doi:10.1142/S146587630400223X

Wang, F., Feng, Y., Owen, D., Jing, Z., & Yang, L. (2004). Parallel analysis of combined finite/discrete element systems on PC cluster. *Acta Mechanica Sinica*, *20*(5), 534–540. doi:10.1007/BF02484276

Wang, L., Jiang, H., Yang, Z., Xu, Y., & Zhu, X. (2013). Development of discontinuous deformation analysis with displacement-dependent interface shear strength. *Computers and Geotechnics*, *47*, 91–101. doi:10.1016/j.compgeo.2012.06.006

Wang, Z. M., Kwan, A. K. H., & Chan, H. C. (1999). Mesoscopic study of concrete I: Generation of random structure and finite element mesh. *Computers & Structures*, *70*(5), 533–544. doi:10.1016/S0045-7949(98)00177-1

Wegmann, E. (1888). *The desing and construction of dams*. New York: John Wiley & Sons.

Westergaard, H. M. (1933). Water pressure on dams during earthquakes. *Transactions of the American Society of Civil Engineers*, *98*, 418–433.

West, H. W. H., Hodgkinson, H. R., Haseltine, B. A., & De Vekey, R. C. (1986). Research results on brickwork and aggregate blockwork since 1977. *The Structural Engineer*, *64*(11), 255–267.

Winkler, T., Meguro, K., & Yamazaki, F. (1995). Response of rigid body assemblies to dynamic excitation. *Earthquake Engineering & Structural Dynamics*, *24*(10), 1389–1408. doi:10.1002/eqe.4290241008

Wittke, W., Wittke, M., & Kiehl, J. R. (2012). Interaction of a masonry dam and the rock foundation. *Geotechnical and Geological Engineering*, *30*(3), 581–601. doi:10.1007/s10706-012-9493-6

Wolfe, W. S. (1921). *Graphical Analysis: a text book on Graphic Statics*. New York: McGraw-Hill Book Company, Inc.

Xiang, J., Munjiza, A., & Latham, J.-P. (2009). Finite strain, finite rotation quadratic tetrahedral element for the combined finite-discrete element method. *International Journal for Numerical Methods in Engineering*, *79*(8), 946–978. doi:10.1002/nme.2599

Xu, D., Kaliviotis, E., Munjiza, A., Avital, E., Ji, C., & Williams, J. (2013). Large scale simulation of red blood cell aggregation in shear flows. *Journal of Biomechanics*, *46*, 1810–1817. PMID:23809770

Yagoda-Biran, G., & Hatzor, Y. H. (2010). Constraining paleo PGA values by numerical analysis of overturned columns. *Earthquake Engineering & Structural Dynamics*, *39*, 463–472.

Yagoda-Biran, G., & Hatzor, Y. H. (2016). Benchmarking the numerical Discontinuous Deformation Analysis mehod. *Computers and Geotechnics*, *71*, 30–36. doi:10.1016/j.compgeo.2015.08.003

Yim, C. S., Chopra, A. K., & Penzien, J. (1980). Rocking response of rigid blocks to earthquakes. *Earthquake Engineering & Structural Dynamics*, *8*(6), 565–587. doi:10.1002/eqe.4290080606

Young, M. P., Schultz, A. E., & Lemos, J. V. (2015). Seismic analysis of the Panhellenic Sanctuary of Nemea, Greece. In *Proc. 12th North American Masonry Conference*.

Zang, M., Gao, W., & Lei, Z. (2011). A contact algorithm for 3D discrete and finite element contact problems based on penalty function method. *Computational Mechanics*, *48*(5), 541–550. doi:10.1007/s00466-011-0606-5

Zanotti, S. (2015). *Seismic Analysis of the Church of Kuño Tambo (Peru)*. (MS thesis). University of Minho, Guimarães, Portugal.

Zhang, Y., Zhou, G. C., Xiong, Y., & Rafiq, M. Y. (2010). Techniques for predicting the cracking pattern of masonry wallet using artificial neural networks and cellular automata. *Journal of Computing in Civil Engineering*, *24*(2), 161–172. doi:10.1061/(ASCE)CP.1943-5487.0000021

Zhao, Q., Lisjak, A., Mahabadi, O. K., & Grasselli, G. (2015). Numerical investigation of the influences of rock fabrics on hydraulic fracturing operations. In GeoConvention 2015: New Horizons.

Zhao, Sh., Salami, M. R., & Rahman, M. Sh. (2000). Three dimensional spherical DDA model for granular media. In J.L. Tassoulas (Ed.), *Procs.14th Engineering Mechanics Conf.*

Zhou, G. C. (2002). *Application of stiffness/strength corrector and cellular automata in predicting response of laterally loaded masonry panel*. (Ph.D thesis). University of Plymouth, Plymouth, UK.

Zhou, G. C., & Rafiq, M. Y. (2006). A CA & ANN techniques of predicting failure load and failure pattern of laterally loaded masonry panel. In *Proceedings of the 11th Joint International Conference on Computing and Design Making in Civil and Building Engineering*.

Zhu, W., Zhang, Q., & Jing, L. (1999). Stability analysis of the ship-lock slopes of the Three Gorges project by three-dimensional FEM and DEM techniques. In B. Amadei (Ed.), Procs. ICADD-3. American Rock Mechanics Association.

Zhuge, Y. (2002). Micro-modelling of masonry shear panels with distinct element approach. In L. Chowdhury & Fragomeni (Eds.), Advances in Mechanics of Structures and Materials. Swets & Zeitinger.

Zhuge, Y., Jin, F., & Hunt, S. (2003). The prediction of damage to masonry houses caused by foundation movements. *Advances in Structural Engineering*, *7*(1), 81–93. doi:10.1260/136943304322985783

Zucchini, A., & Lourenço, P. B. (2006). Mechanics of masonry in compression: Results from a homogenisation approach. *Computers & Structures*, *85*(3), 193–204.

Zulli, D., Contento, A., & Di Egidio, A. (2012). 3D model of rigid block with a rectangular base subject to pulse-type excitation. *International Journal of Non-linear Mechanics*, *47*(6), 679–687. doi:10.1016/j.ijnonlinmec.2011.11.004

About the Contributors

Vasilis Sarhosis completed his Undergraduate and Master's degree in Civil and Structural Engineering at the University of Leeds in 2005. He then spent time working in the industry as a graduate Civil Engineer in two of the largest consult engineering companies in UK, MWH Global and Gronmij. In 2007, he returned to Leeds to study for a PhD in Computational Modelling of Materials and Structures where he developed a tool that can be used to improve understanding of the in-service and near collapse behaviour of low bond strength masonry and compare and evaluate alternative methods of rehabilitation, repair or strengthening of masonry structures. He is currently a Lecturer in Civil Engineering at Newcastle University.

Katalin Bagi, Doctor of Science, Dr. Habil, is a professor of Structural Mechanics at TU Budapest. She has been dealing with discrete element modelling since the late 1980ies. Until 2006 she studied granular micromechanics as a researcher of the Hungarian Academy of Sciences. In 2007 she went to a teaching position at TU Budapest, and since then her main field of research is the mechanical analysis of masonry vaults. She is married since 1993, they have two daughters.

José V. Lemos is a Civil Engineer from the University of Porto, Portugal. PhD from the University of Minnesota, Minneapolis, MN, USA. Presently, he is Principal Research Officer at LNEC, Portugal.

Gabriele Milani has been Associate Professor at POLIMI since 2014, previously Assistant Professor (2009-2013) and Post Doctoral Researcher at ETHZ CH (2008). His research focuses on masonry modeling and safety assessment of historic masonry in seismic area. He works on FEM limit analysis, rubber vulcanization (with Pirelli) and seismic isolation. He has been awarded with a Telford Premium (2012) by ICE and a Bathe award in 2014 as the best young researcher on FEs in 2014. He has co-authored 104 ISI papers, he is the 2nd author in Scopus under the keyword "masonry". He is EIC of a Journal dedicated to masonry (International Journal of Masonry Research and Innovation) and co-editor of a Scopus journal generalist for civil engineering. He will chair the International Masonry Conference (2018).

* * *

Haris Alexakis is currently adjunct Lecturer at the University of Patras, Greece. He has one and a half years post-doctoral research experience in experimental (shake table tests) and numerical study for the stability and dynamic response of articulated structures. He holds a bachelor's and master's degree in Civil Engineering, a master's degree in Seismic Design of Structures and a PhD, all from the same institution. His PhD thesis clarifies fundamental concepts of Limit (Equilibrium) Analysis with the introduction of an innovative and rigorous energy method strategy and a parallel critical use of the majority of the available modern methods of masonry discontinuous modelling. He has 6 peer-reviewed journal publications and several conference papers. He has more than 10 years practical experience as a chartered civil engineer. He has been leading a 12-engineers-group that undertook the project of restoration and reinforcement of several masonry structures and monuments seriously damaged from earthquakes.

João Rocha de Almeida is an Associate Professor at the Civil Engineering Department of the New University of Lisbon.

Antonella Cecchi is Full Professor of "Strength of Materials" at University IUAV of Venice from 1st October 2011. Director of department of Architecture Construction Conservation DACC of Università IUAV di Venezia from 24 July 2015. Associate Professor of "Strength of Materials" at University IUAV of Venice from 1st October 2006 to 30 September 2011. In 1997 (IX Cycle), the PhD degree had been obtained about "History of structural Mechanics and building techniques", University of Rome "La Sapienza". The knowledge has been increased, during the academic career, attending some specialization courses at CISM – International Centre for Mechanical Sciences Author of about 70 publications in international journals and proceedings in national and international conferences.

Rossana Dimitri is an Assistant Professor at the School of Engineering, Department of Innovation Engineering, University of Salento, Lecce, Italy. She obtained a PhD in Materials and Structural Engineering in 2009 and a PhD in Industrial and Mechanical Engineering in 2013 at the University of Salento. She is a researcher in Fracture Mechanics, Debonding Problems, Contact Mechanics, Isogeometric Methods, and Structural Analysis of Masonry Structures with various journal papers and one book. She collaborates as reviewer with different international journals.

Michalis Fragiadakisis Assistant Professor at the School of Civil Engineering of the National Technical University of Athens (NTUA), Greece. He received his PhD from NTUA and holds MSc degrees form NTUA (2002) and Imperial College, London (2001). He has also served as adjunct lecturer at the Department of Civil Engineering of the University of Thessaly and as a postdoctoral researcher at the University of Cyprus and the University of Pavia. His teaching includes structural analysis and earthquake engineering courses at both under and post-graduate level. His research activity is focused on earthquake engineering and on the seismic risk assessment of structures and lifelines. His interests also extend to computational methods in structural analysis and on the application of novel soft-computing methodologies for structural analysis and design. Dr Fragiadakis has been involved in various national and international research projects and has also organised a number of mini-symposia and special sessions in international conferences. He has authored and co-authored more than 17 book chapters and 33 journal papers, while his work has been included in the proceedings of several international conferences.

Nicos Makris received his Ph.D (1992) and his Master of Science (1990) from the State University of New York at Buffalo, USA; while he holds a Diploma in Civil Engineering from the National Technical University, Athens, Greece (1988). He has more than twenty-five (25) years of research, academic and professional experience in the areas of structural-earthquake engineering and applied mechanics-dynamics. Ha has served as: Assistant Professor at the University of Notre Dame (1992-1996) and at the University of California, Berkeley (1996-1998); Associate Professor at the University of California, Berkeley (1998-2002) and Professor at the University of California, Berkeley (2002-2005), at the University of Patras (2003-present) and at the University of Central Florida (2014-present). He has published more than 100 papers in archival journals, while he has supervised ten (10) Ph.D thesis and more than forty (40) MSc and 5th year Diploma thesis. He has served as Associate Editor for the Journal of Engineering Mechanics, ASCE, and the Chair of the Dynamics Committee on the same Journal. He is a member of Academia Europaea "The Academy of Europe", a member of the Congress Committee of IUTAM; while, he has been honored with several international prizes and awards including the Walter L. Huber Civil Engineering Research Prize from the American Society of Civil Engineering (ASCE), the T. K. Hsieh Award from the Institution of Civil Engineers, U.K., the Shah Family Innovation Prize from the Earthquake Engineering Research Institute (EERI), USA and the CAREER Award from the National Science Foundation, USA.

Iraj H. P. Mamaghani is an Associate Professor of Civil Engineering at University of North Dakota. He received his B.Sc. in Civil Engineering from Istanbul Technical University with Honors in 1989. He continued his Master and PhD studies at University of Nagoya, Japan, where he obtained his Master and Doctor of Engineering degrees in Civil Engineering. Dr. Mamaghani has published several papers in professional journals and in conference proceedings. Dr. Mamaghani works in the area of civil engineering, with emphasis on structural mechanics and structural engineering. He focuses on modeling and analysis of masonry structures, cyclic elastoplastic material modeling, structural stability, seismic design, advanced finite element analysis and ductility evaluation of steel and composite (concrete-filled steel tubular) structures.

Amin Mohebkhah received his PhD degree, in 2007, from Tarbiat Modares University, Tehran, Iran, in the field of seismic design of masonry-infilled steel frames, and is currently Assistant Professor of Structural Engineering at Malayer University, Malayer, Iran. His research interests include seismic design of masonry structures, ductile design of steel structures and buckling of thin-walled structures.

Fernando Peña is a Professor at the Instituto de Ingeniería, Universidad Nacional Autónoma de México (Institute of Engineering of the National Autonomous University of Mexico) . He is a Civil Engineer from the Universidad Nacional Autónoma de México. He holds a PhD in Earthquake Engineering from the Politecnico di Milano, Italy. As well as made a Post-doctoral stay of two years at the University of Minho, campus Guimarães in Portugal. He has done several research stays at different centers, among which are: the National Laboratory of Civil Engineering (LNEC) in Lisbon, Portugal, the Polytechnic University of Catalonia, Spain and the Politecnico di Milano, Italy. His main research line is the Mathematical and Computational Modelling of Systems, which the main systems of study are the historical structures and tunnels. His areas of interest are: computational mechanics, dynamics and non-linear analysis. He is a lifetime member of the Association of Civil Engineers of Mexico and member of the European Mechanics Society; as well as Mexican Societies of Structural Engineering and Earthquake Engineering.

Ioannis Psycharis is full Professor in the School of Civil Engineering of the National Technical University of Athens, Greece (NTUA). He holds a Civil Engineering diploma from NTUA and a M.Sc. and a Ph.D. degree in Earthquake Engineering from the California Institute of Technology (CALTECH),

USA. His main areas of specialization and research interests include: seismic response of modern structures, of historical buildings and of classical monuments, analytically and experimentally; repairing and strengthening of structures; seismic isolation; innovative methods in seismic design; soil-structure interaction; and engineering seismology. He is the author or co-author of more than 120 papers in refereed journals and international and national conferences and has participated in about 30 research projects funded by the European Commission or national funds, in one half of which he was the Principal Investigator. He is a Founding Member and Executive Board Member of the Hellenic Society of Earthquake Engineering and, presently, serves as the National Delegate of Greece in the International Association for Earthquake Engineering. Currently, he is the vice President of the Hellenic Earthquake Planning and Protection Organization (OASP). He has been the designer or reviewer of the design of more than 150 bridges and many buildings and tunnels, and has participated in the design of interventions for the restoration of several monumental structures.

Emanuele Reccia (Quartu S.E. 24/06/1981, Cagliari, Italy) Eng., M.Phil., Ph.D., is research fellow at the DACC - Department of Architecture Construction Conservation - of the University IUAV of Venice. His research interests regard the mechanics of historical buildings, with focus on the adoption of continuous models (obtained by means of homogenization procedures) and discrete models (DEM and in particular the FEM/DEM method) for the study of masonry structures. In particular, attention has been paid to masonry panels, masonry arch bridges, masonry domes and bell-towers. In addition to theoretical research, he deals with experimental tests in lab and in situ.

Gabriela Sincraian, formerly at Instituto Superior Tecnico and LNEC in Lisbon, Portugal, is currently affiliated with Vancouver Coastal Health, Canada.

Ioannis Stefanou, is Researcher at Navier Laboratory and Assistant Professor at the Ecole des Ponts ParisTech (ENPC). He holds a Diploma in Civil Engineering from the National Technical University of Athens (NTUA) and a Master Degree in Applied Mechanics from the School of Applied Mathematical and Physical Sciences (NTUA). He did his Thesis on upscaling procedures in Micromorphic Continua with the late Professor Ioannis Vardoulakis. His research activities involve homogenization of heterogeneous structures in the frame of higher order continuum theories, such as Cosserat continuum, bifurcation theory and strain localization, (multisurface) plasticity theory, finite and discrete element methods (development of higher order FE and applications). The applications of his research span from deep below the earth surface, where interesting questions rise regarding earthquake nucleation, to the dynamic behavior of modern and historic structures. He teaches the Finite Element program Abaqus, Mechanics of structures and Soil Dynamics at undergraduate and postgraduate level at the ENPC.

Giorgio Zavarise is a Full Professor of Structural Mechanics at the University of Salento from 2006, chairman of several International Conferences on Computational Mechanics, referee for several international journals, author of more than 180 publications, among them 52 journal contributions.

Index

E

F

G

H

I

J

K

L

M

N

O

P

R

S

T

U

V

www.ingramcontent.com/pod-product-compliance
Lightning Source LLC
LaVergne TN
LVHW060350300926
849162LV00006B/1238